OCEAN YEARBOOK 1

**OCEAN YEARBOOK 1
BOARD OF EDITORS**

Francis Auburn
  University of Western Australia
Frank Barnaby
  Stockholm International Peace Research Institute
Thomas S. Busha
  Inter-Governmental Maritime Consultative Organization
Sydney Holt
  Food and Agriculture Organization
Arvid Pardo
  University of Southern California
Lord Ritchie-Calder
  Center for the Study of Democratic Institutions
Mario Ruivo
  Ministry for Fisheries, Republic of Portugal

Assistant to the Editors: Daniel Dzurek

**INTERNATIONAL OCEAN INSTITUTE
BOARD OF TRUSTEES**

H. S. Amerasinghe *(Chairman)*
  President, United Nations Conference on the Law of the Sea
Zakaria Ben Mustapha
  Ministere de l'Agriculture, Republic of Tunisia
Elisabeth Mann Borgese
  Center for the Study of Democratic Institutions
Edwin J. Borg Costanzi *(Treasurer)*
  Rector, Old University of Malta
Gunnar Myrdal
  University of Stockholm
Aurelio Peccei
  President, The Club of Rome
Jan Pronk
  RIO Foundation; M.P., The Netherlands
Roger Revelle
  University of California, San Diego
Mario Ruivo
  Ministry for Fisheries, Republic of Portugal
Hernan Santa Cruz
  Food and Agriculture Organization
Anton Vratusa
  President, Socialist Republic of Slovenia, Yugoslavia
Layachi Yaker
  Deputy President, National Popular Assembly, Republic of Algeria

**INTERNATIONAL OCEAN INSTITUTE
PLANNING COUNCIL**

Elisabeth Mann Borgese *(Chairman)*
Silviu Brucan
  University of Bucharest
Maxwell Bruce, Q.C.
  (Canada)
Thomas S. Busha
Jorge Castaneda
  Head, Delegation of Mexico to UNCLOS
Edwin J. Borg Costanzi
Peter Dohrn
  Mediterranean Association for Marine Biology and Oceanology (Italy)
R. J. Dupuy
  University of Nice
Reynaldo Galindo Pohl
  UNFICYP, Cyprus
J. King Gordon
  International Development Research Center (Canada)
G. L. Kesteven
  Director, International Ocean Institute; Food and Agriculture Organization
Anatoly L. Kolodkin
  Research Institute of Maritime Transport (USSR)
Frank LaQue
  (U.S.A.)
Arvid Pardo
Jacques Piccard
  Fondation pour l'Etude et la Protection de la Mer et des Lacs (Switzerland)
Christopher Pinto
  Ambassador to the Federal Republic of Germany, Sri Lanka
Sir Egerton Richardson
  Ministry of the Public Service, Jamaica
Lord Ritchie-Calder
Peter Serracino-Inglott
  University of Malta
Jun Ui
  Tokyo University
V. K. S. Varadan
  Director-General, Geological Survey of India
Joseph Warioba
  Ministry for Foreign Affairs, United Republic of Tanzania
Alexander Yankov
  Ambassador to the United Nations, Bulgaria

OCEAN YEARBOOK 1

Sponsored by the
International Ocean Institute

Edited by
Elisabeth Mann Borgese and
Norton Ginsburg

University of Chicago Press

Chicago and London

The University of Chicago Press, Chicago 60637
The University of Chicago Press, Ltd., London

© 1978 by the University of Chicago
All rights reserved. Published 1978
Printed in the United States of America

International Standard Book Number: 0-226-06602-9

COPYING BEYOND FAIR USE. The code on the first page of an article in this volume indicates the copyright owner's consent that copies of the article may be made beyond those permitted by Sections 107 or 108 of the U.S. Copyright Law provided that copies are made only for personal or internal use, or for the personal or internal use of specific clients and provided that the copier pay the stated per-copy fee through the Copyright Clearance Center Inc. To request permission for other kinds of copying, such as copying for general distribution, for advertising or promotional purposes, for creating new collective works, or for resale, kindly write to the publisher. If no code appears on the first page of an article, permission to reprint may be obtained only from the author.

Contents

Editor's Preface   xv

Issues and Prospects

Man and the Oceans   *Elisabeth Mann Borgese*   1
The Evolving Law of the Sea: A Critique of the Informal Composite Negotiating Test (1977)   *Arvid Pardo*   9

Living Resources

Marine Fisheries   *Sidney Holt*   38
Progress of Aquaculture   *T. V. R. Pillay*   84
The Cradle of Sea Fisheries   *Rudolf Kreuzer*   102

Nonliving Resources

Offshore Oil and Gas   *Edward Symonds*   114
Oil and Gas Exploration and Exploitation in the North Sea   *Peter R. Odell*   139
Other Ocean Resources   *R. H. Charlier*   160

Transportation and Communication

The Organization of Shipping   *R. A. Ramsay*   211
A Safe Voyage to a New World   *Thomas Busha and James Dawson*   217
International Maritime Satellite-Communication System: History and Principles Governing Its Functioning   *M. E. Volosov, A. L. Kolodkin, and Y. M. Kolosov*   240

Marine Science and Technology

Perspectives on the Sciences of the Sea   *Lord Ritchie-Calder*   271

Environment

The Oceans: Health and Prognosis   *Peter S. Thacher and Nikki Meith-Avcin*   293
Radioactive Waste Disposal in the Oceans   *Robert A. Frosch, Charles D. Hollister, and David A. Deese*   340

Coastal Management

Coastal Area Development and Management and Marine and Coastal Technology   350

Military Activities

Strategic Submarines and Antisubmarine Warfare *Frank Barnaby* 376
The ASW Problem: ASW Detection and Weapons Systems 380
The Seabed Treaty *Jozef Goldblat* 386
Naval Forces *Ronald Huisken* 412
Military Impact on Ocean Ecology *Arthur H. Westing* 436

Regional Developments

The Southern Ocean *G. L. Kesteven* 467
Legal Implications of Petroleum Resources of the Antarctic Continental Shelf *F. M. Auburn* 500

Appendices

A. Reports from Organizations 517
 Engineering Committee on Oceanic Resources (ECOR): An International Nongovernmental Society 518
 Maritime Activities of the ILO 520
 Excerpts from the Annual Report of the Inter-Governmental Maritime Consultative Organization 1976/1977 531
 Report of the Intergovernmental Oceanographic Commission on Its Activities 545
 The Scientific Committee on Ocean Research (SCOR) 563
 United Nations, Ocean Economics and Technology Office: Activities in the Field of Marine Science and Its Applications 569
 UNESCO: Activities of the Division of Marine Sciences during 1976 573
 UN, Environmental Programme: Activities for the Protection and Development of the Mediterranean Region 584
 UN Institute for Training and Research (UNITAR): Activities Related to Ocean Problems 598
 WHO: Recent Activities in the Field of Marine Pollution and Related Subjects 600
B. Selected Documents and Proceedings 605
 United Nations Third Conference on the Law of the Sea: Informal Composite Negotiating Text 606
 International Conventions and Other Agreements for Control of Marine Pollution 692
 Joint Oceanographic Assembly 697
 Conference of Plenipotentiaries of the Coastal States of the Mediterranean Region for the Protection of the Mediterranean Sea 702
 Convention on the International Maritime Satellite Organization (INMARSAT) 734

Convention concerning Minimum Standards in Merchant Ships   756
SCAR/SCOR Group on the Living Resources of the Southern Ocean
    (SCOR Working Group 54)   762
Convention on Limitation of Liability for Maritime Claims, 1976   773
C. Directory of Institutions   785
D. Tables, Living Resources   801
E. Tables, Nonliving Resources   813
F. Tables, Transportation and Communications   821
G. Tables, Military Activities   835

Contributors   873

Index   879

List of Tables

Issues and Prospects

Arvid Pardo
Breadth of Territorial Seas and Fishing Jurisdictions in Excess of 12 Nautical Miles   35
Summary of the Breadths of Territorial Seas   37
Summary of Fishing Jurisdiction Claims   37

Living Resources

Sidney Holt
World Marine Catches   41
Change in World Catches of Fish and Shellfish, 1974 to 1975   42
Catches of Selected Fish Species   44
Percentage of Marine Fish and Shellfish Catch Taken from Major Fishing Areas and Zones   45
Species Groups in Catches   49
Prices and Values, 1971 Marine Catch   54
Estimated Value of Catches at 1971 Prices   55
Disposition of Catches Used for Direct Human Consumption   56
Percentage of Catch Reduced to Meal and Oil or Used for Miscellaneous Purposes Other than Direct Human Consumption   56
Average Annual Rates of Growth of Sea-Fish Catches for Direct Consumption and for Reduction to Oil and Meal   57
Utilization of Fish Catches, by Type   57
Food Supply in Live-Fish Weight Equivalent   59
Increases in Food Production per Capita, 1961–74   60
Equivalent Protein Yield, Expressed as a Percentage of the Protein Contained in the Landed Fish: Live Weight   66
World Nominal Marine Catches, by Continent   68
Ratio of Marine to Total Aquatic Catches   69
World Marine Catches, by Economic Groupings   70
Marine Fisheries Production per Capita, by Economic Groupings   72
Marine Production by Economic Groups and by Disposition   73
International Trade in Commodities Derived from Living Aquatic Resources   74
Food Originating from Marine Resources in 1970, by Economic Groupings   75
The Place of Fish in Diet: 1964–66 Average   77
Consequences of Interclass Trade in Marine Fish Products, 1970   78
Location of Marine Catches in 1972, by Economic Groupings   79

T. V. R. Pillay
Estimated World Production through Aquaculture in 1975   86

x   List of Tables

Rudolf Kreuzer
Fish Prices of the Time of the Third Dynasty of UR   107

Appendix D
1D. World Nominal Marine Catch, by Continent   802
2D. World Nominal Marine Catch, by Major Fishing Area   803
3D. World Nominal Fish Catch, Disposition   805
4D. World Nominal Marine Catch, by Country   806
5D. Trade of Fishery Commodities, by Major Exporting and Importing Countries   810

Nonliving Resources

Edward Symonds
Oil and Natural Gas Liquids Estimated Ultimate Recoverable Reserves in Offshore Fields   116
Minimum Economic Field Size for Offshore Development   117
Offshore Oil Production   119
Offshore Gas Production   120
Worldwide Offshore Drilling Activity   123
OCS Acreage Leased, 1954–1976   127
Company Shares of Crude and Condensate, Federal Gulf of Mexico   132
Estimated Savings Resulting from an East Coast U.S. Monobuoy Superport   134

Peter R. Odell
North Sea Oil and Gas Province: Offshore Discoveries to June 1977   142
Natural Gas Resources of the Southern North Sea Basin, excluding Associated Gas   143
Western Europe: Its Currently "Proven" and Possible Natural Gas Resources and an Estimate of Their Development by the Early 1980s   145
North Sea Oil Reserves, December 31, 1976   147
North Sea Oil Reserves: Estimates and Forecasts Compared   148
Forties, Montrose, and Piper Fields: Contrasts in Benefits for the U.K Arising from Company and Country Optimal Developments of the Fields   158

R. H. Charlier
Estimated Value of By-Products from a Sea Thermal Power Plant   175
Recoverable Minerals from the Marine Environment   183
World Annual Production of Minerals from Oceans and Beaches   184
Classification of Marine Mineral Resources   188
Average Analyses of Polymetallic Nodules   191
Reserves of Metals in Polymetallic Nodules of the Pacific Ocean   192
Polymetallic Nodules: Commercially Attractive Constituents Projections for Future Demand   193
1973 Projected Market Value of Metals from Polymetallic Nodules in 1985 and 2000   195

List of Tables xi

Typical Ongoing and Possible Future Marine Mining of Unconsolidated Surficial Deposits   200
Probable Production from Nodules, Estimated World Demand, and Estimated Net Import Requirement of Industrial Countries   202
Estimated Value of Mineral Production   202
Sulfur, Outer Continental Shelf of the United States, 1960–75   204
Salt, Outer Continental Shelf of the United States, 1960–75   205
Magnesium Production from Seawater   207

Appendix E
1E. World Production of Crude Oil, Total and Offshore   814
2E. Offshore Crude Oil Production, by Region and Country   815
3E. World Production of Natural Gas, Total, and Offshore   817
4E. Offshore Natural Gas Production, by Region and Country   818

Transportation and Communication

Appendix F
1F. World Shipping Tonnage, by Type of Vessel   822
2F. Estimated Average Size of Selected Types of Vessels: Existing World Fleets and Vessels on Order   823
3F. World Shipping Tonnage, by Groups of Countries   824
4F. World Merchant Fleets, by Region and Country   826
5F. World Shipbuilding: Merchant Vessels under Construction, by Country   831
6F. Vessels Lost, 1976   833

Environment

Peter S. Thacher and Nikki Meith-Avcin
Estimated Inputs of Petroleum Hydrocarbons Entering the Ocean Annually, circa 1969–71   296
Tar Densities in the World Oceans   298
Petroleum Hydrocarbon Levels in Marine Macroorganisms   299
Production of PCBs in Certain Countries, 1971   302
DDT Residues and PCBs in a Variety of Plankton Samples   303
Fluxes of Heavy Metals into Sediments of California Coastal Basins   308
Fluxes of Materials to Southern California Coastal Region   309
Characteristics of Diseases in Man   310
Total Litter Estimates   317
Number of Research Institutions in Each Mediterranean Country Participating in the Mediterranean Pollution Monitoring and Research Program's Seven Pilot Projects   322
Main Human Activities and the Main Pollutants Associated with Them in East Asia   327
Major Marine Pollutants in East Asia in Approximate Order of Priority by Groups and the Countries Having Special Concern with Them or Being Affected by Them   328

xii   List of Tables

Coastal Management

Marine and Coastal Technology Programme   374

Military Activities

Jozef Goldblat
Countries Which Have Signed, Ratified, Acceded, or Succeeded to the Convention on the Territorial Sea and the Contiguous Zone as of December 31, 1975   406
Countries Which Have Signed, Ratified, or Acceded to the Seabed Treaty as of December 31, 1976   408

Ronald Huisken
World Naval Stock, 1950–76   417
World Stock of Fighting Ships: Estimated Growth Rates in Value of Stock   418
Distribution of the World Stock of Fighting Ships   418
Percentage of U.S. and Soviet Naval Stock in Strategic Submarines   420
Structure of U.S. Naval Stock, excluding Strategic Submarines   421
Structure of Soviet Naval Stock, excluding Strategic Submarines   422
Estimated Value of Naval Stock: Selected NATO Countries   424
Naval Stock, Miscellaneous   425
Estimated Value of Naval Stock: Third World Countries   426
Attack Aircraft Carriers   427
Nuclear-powered Ballistic-Missile Submarines   429
Nuclear-powered Attack Submarines   430
Conventionally Powered Patrol Submarines   431
Major Surface Warships   431
Missile-armed Patrol Boats   433

Arthur H. Westing
Navies of the World   439
Lethal Blast Zones Resulting from Underwater Explosions   446
Oceanic and Atmospheric Nuclear Explosions   450
Nuclear-powered Ships of the World   452
Characteristics of Selected Fission Products   455
Quantities and Decay Rates of Selected Nuclear-Bomb Fission Products   456
Quantities and Decay Rates of Selected Submarine-Reactor Fission Products after a Month at Sea   457

Appendix G

1G. World Stock of Fighting Ships, Estimated Value   836
2G. World Stock of Aircraft Carriers, by Country   837
3G. World Stock of Strategic Submarines, by Groups of Countries   838
4G. World Stock of Patrol Submarines, by Groups of Countries   839

List of Tables xiii

- 5G. World Stock of Coastal Submarines, by Groups of Countries   840
- 6G. World Stock of Major Surface Warships, by Groups of Countries   840
- 7G. World Stock of Patrol Boats, Torpedo Boats, and Gunboats, by Groups of Countries   842
- 8G. Warship Construction Under Way or Firmly Planned, 1976, by Type and Country   844
- 9G. U.S. and Soviet Strategic Missile-launching Submarines and Missiles   851
- 10G. U.S. and Soviet Submarines, Submarine-launched Ballistic Missiles, Intercontinental Ballistic Missiles, and Strategic Bombers, 1976   852
- 11G. USSR Antisubmarine-equipped Surface Ships, 1976   853
- 12G. U.S. Antisubmarine Fixed-Wing Aircraft, 1976   854
- 13G. Antisubmarine-equipped Helicopters, 1976   856
- 14G. U.S. Mobile Antisubmarine Warfare Sensors, 1976   858
- 15G. Fixed Acoustic Arrays and Sonobuoys   861
- 16G. Strategic Nuclear Submarines, by Country and Class   864
- 17G. Nuclear-powered Attack Submarines, by Country and Class   867

Regional Developments

G. L. Kesteven
Conversion Factors for Chlorophyll and for Carbon Value   476
Whale Catch   481

Preface

Before announcing that the International Ocean Institute takes pride in presenting the first volume of the *Ocean Yearbook*, we owe the reader an introduction to the International Ocean Institute itself.

The International Ocean Institute is an offspring of an initiative of the Center for the Study of Democratic Institutions in Santa Barbara, California. Under the leadership of the later Robert Hutchins, the Center embarked, as early as 1967, on a research project on a new order for the oceans, based on the concept—first proposed by Ambassador Arvid Pardo of Malta—that the oceans and their resources are the common heritage of mankind. Within the general program of the Center, this was one of the studies on World Order. It turned out to be the most challenging, the most continuous, and the most action-oriented of these studies.

In 1968 the Center published a first monograph, *The Ocean Regime*, on the subject. This was one of the earliest blueprints for an ocean-space constitution, proposing to regulate, harmonize, and manage the uses and resources of ocean space for the benefit of all mankind. It was in the wake of this publication that the government of Malta, through Ambassador Pardo, invited the Center to organize a major international Conference in Malta to probe deeper into the complex and interdependent issues of ocean management. In the meantime, the United Nations had, on the initiative of Malta, instituted the United Nations Seabed Committee (Committee on the Peaceful Uses of the Seabed and the Ocean Floor beyond the Limits of National Jurisdiction). Its scope, however, was restricted to the *seabed,* excluding, for the time being, the other dimensions of ocean space.

After 2 years of preparatory work, including a series of exploratory conferences and the compiling of seven volumes of research papers, analyses, and projections, a major conference was held in Malta in 1970 under the title *Pacem in Maribus,* "Peace on the Oceans." The title was derived from *Pacem in Terris,* "Peace on Earth," the title of the great Encyclical of Pope John XXIII, which played such an important role in promoting détente and to which the Center had dedicated a number of major international convocations.

Pacem in Maribus was attended by about 300 diplomats, legal experts, marine scientists, industrialists, students, and junior civil servants. It is not too much to claim that it was a pioneering event. This, certainly, was the feeling of the participants at the time when they, spontaneously, insisted that this effort should be continued and institutionalized, and for this purpose established a Continuing Committee. With the help of the United Nations Development Programme and the government and the University of Malta, this Continuing Committee founded the International Ocean Institute at the University of Malta in 1972. The Continuing Committee was transformed into the Planning Council of the IOI to which a board of distinguished trustees was added. The Board of Trustees and the Planning Council are the governing bodies of IOI, an independent, international, and interdisciplinary institution at the University of Malta. The purpose of the Institute is to study in depth the issues underlying the establishment of a new international order in the oceans, to propose new approaches and solutions, and to widen the scope of the dialogue on ocean affairs.

The IOI has continued its annual Pacem in Maribus convocations, of which eight have been held to date—II–V in Malta, VI in Okinawa, VII in Algiers, and VIII in Mexico. It undertook a series of research projects on regional development and organization, especially in the Mediterranean and in the Caribbean, on Disarmament in the Oceans, and on Energy and the Oceans. It followed closely, and analyzed meticulously, the work of the Seabed Committee and of the U.N. Conference on the Law of the Sea and published a series of studies on the New International Economic Order and the Law of the Sea.

At an early date the IOI paid special attention not only to the ecological but also to the economic aspects and to the implications of the uses of ocean space and resources. These studies were initiated by a monograph by Bertrand de Jouvenel on the *Economic Potential of the Oceans* (1971) and continued with a study, directed by John Eatwell of Cambridge University, on an "Ocean Development Tax," based on the simple notion that wealth might be redistributed for the benefit of the poorer nations by the levy of a very modest tax or contribution (1 percent) on all uses of the oceans (extraction of living and nonliving resources and energy, navigation, communication, waste disposal, other uses). As we tried to design modalities for the collection of such a "tax," it became increasingly clear that there simply existed no adequate data base on many of the economic uses of the oceans, much less on their totality and interaction. Such data, on which a marine economic policy must be based, were nowhere available. Pacem in Maribus III (1972), therefore, concluded with the recommendation

> that Pacem in Maribus, or an affiliated organization, should encourage, in collaboration with the U.N. statistical office, the establishment of a statistical team which will produce a *Yearbook of Ocean Statistics*. Discussions of oceanographic problems tend to fail at the crucial step of providing data to justify or illustrate their arguments. This is as true of the ocean development tax discussion as any others. It must be recognized that no real progress with any scheme or other international oceanic projects will be made until such a yearbook exists. Until accurate information on oceanographic affairs is available, no state will be willing to commit itself to vague projects advanced with good intentions.

The voices of Malta and of the IOI were rather lonely at that time. Not enough people were aware of, or interested in, the marine revolution that was in the offing, carrying the industrial revolution from the land out into the sea. Not enough people were aware of, or interested in, the impact of the oceans on the building of a new international economic order—an expression that was yet to be coined by the Sixth Special Session of the General Assembly. The Seabed Committee had just barely recognized the interdependence of all parts of the oceans and the interaction of all uses and enlarged the scope of its endeavor to cover the oceans as a whole and to call for a conference on the law of the sea rather than a seabed conference.

All this has changed. The Law of the Sea Conference finds itself steeped in issues of the economics of mining, of fishing, of navigation, of technology and technology-transfer. Without clarity about such issues, without concepts and data readily available, it is not possible to frame a meaningful new law of the sea; no new order for the seas and oceans can be built. The new law of the sea has to straddle law, economics, and science in unprecedented ways.

The people who have to make this new law need the facts set in an orderly and logical framework: not specialized facts in specialized books that are hard to come by and go through but an overview of concepts and data in their interaction. The people who will have to ratify the new Convention—hopefully the people of all countries—will need the same kinds of information and understandings in order to vote meaningfully. Policymakers, students, concerned citizens, planners, conservationists, industrialists, fishermen, all will need this information and understandings. The *Ocean Yearbook* is intended as a contribution to the satisfaction of this newly felt need.

Each subsequent *Ocean Yearbook* will provide an authoritative, impartial, and up-to-date review of major ocean issues. Each volume will contain an overview of the principal events of the year under review—significant incidents, legislative acts, negotiations, treaties, scientific discoveries, etc. This will be followed by a series of special articles on ocean resources (living, nonliving, energy, ocean/land interface); ocean surface and subsurface communications (navigation, pipelines and cables, ocean/air interface); ocean services including air/sea rescue, satellite communications, pollution surveillance, hovercraft; marine sciences; military activities, statistical data and technological developments; treaties and other international acts; publications, and notes. Each volume will contain a special feature article on *regional development* which is an essential part and, in some instances, a forerunner of global development and organization. Thus volume 2 will deal with regional development and organization in the western Pacific.

That the time has come for the *Ocean Yearbook* is confirmed by the gratifyingly positive response, the readiness to cooperate, of so many institutions and grant-giving organizations. In particular, we wish to thank FAO, IOC, IMCO, UNEP, UNCTAD, WHO, UNESCO (Department of Marine Sciences), UNITAR, SPIRI, SCOR, SCAR, and ECOR for their active cooperation; and the General Service Foundation, the government of the Netherlands (Minister for Development Cooperation), and the Gulbenkhian Foundation of Portugal, for their financial contributions which have made this publication possible. Last, but by no means least, we wish to thank the staff of the University of Chicago Press for admirable assistance and cooperation; the personnel of the *Yearbook* Project at the University of Chicago for invaluable research contributions: Lynn Chu, Daniel Dzurek, and Albert Modiano; and Jean Muller and Peter Tagger at the Center for the Study of Democratic Institutions for logistic support.

THE EDITORS

Issues and Prospects

Man and the Oceans

Elisabeth Mann Borgese
International Ocean Institute, Malta

One of the showcases of the Anthropological Museum in Mexico presents a reconstruction of Paleolithic life: a hunting scene. Tiny cavemen can be seen swarming over a vast plain, attacking a mammoth. They are attacking him with the technology of their times, with darts and spears that look like pins on the hairy giant; some of the men are working together wielding heavy staves to stab in his chest; other men, wounded by the irate mammoth, are lying bleeding on the swampy ground, with yet others tending their wounds. But there are so many of them, and more are coming from all sides with darts and spears and staves and stones, and one must assume that, in the end, the mammoth would fall. They would make a feast of him and leave his rotting mass of flesh to the winged dinosaurs, and his bones to history.

It seems an arduous task for the tiny men to have tackled so enormous a quarry. But they did it. Had they attempted to tame him and put him to work, their task would have been even more arduous—too complex for their tools and brains.

The world ocean is like that irate mammoth, ringed by tiny men with their darts and staves. But while already drawing blood, they change their minds: perhaps those tending the wounded hunters get the idea. They decide to save and harness the mammoth and to design rules of behavior for the hunters turned husbandmen and their relations to the wild and mighty beast.

That is the stage we are at now. The task is almost overwhelming. The changes required in the concepts and behavior of people everywhere are profound, and it will take long, perhaps forever, to establish this new relationship between humankind and nature. But this is the path on which we have embarked; and the new order we are attempting to impose on the oceans, and on ourselves in the oceans, is a milestone on this path.

This is volume 1 of the *Ocean Yearbook,* and we have barely scratched the surface of the issues requiring comment. Yet the facts, acts, and figures put together here are awesome. The role the oceans play in the economy of every nation and in the global household is enormous and bound to increase further as advancing technologies push the industrial revolution deeper and deeper into the oceans. Besides assembling economic and ecological data related to the exploration and exploitation of the oceans, the *Ocean Yearbook* attempts to analyze trends and to present them in their interaction.

If one were to summarize the content of this volume in a few paragraphs, one might reach the following conclusions:

© 1978 by The University of Chicago. 0-226-06602-9/77/1978-1000$00.81

1. The world fishing industry is undergoing a process of transformation. Overkill technologies, lack of sound management principles and legal infrastructure, pollution and other uses and abuses of the ocean space—drilling and dredging, coastal development, and the destruction of many breeding and fishing areas—have led to the collapse of some of the major commercial fisheries (the anchoveta, the herring; the North Atlantic tuna, etc.), reduced the yearly rate of increase in world catches, and set the limits of the sustainable annual yield at a much lower level than the more optimistic forecasts of a decade ago. Rising fuel prices and legal restrictions resulting from the new law of the sea may accelerate the downward trend and the displacement of the industry from distant-water fishing to the development of fisheries in coastal and inshore waters. This development, in turn, may be accelerated by the progress of *aquaculture* (including *mariculture*) technologies. Sea ranching, cage cultures, the stocking of natural and artificial ponds, enclosures, raceways and tanks in salt water, brackish water, or fresh water are producing an increasing proportion of finfish, mollusks, crustaceans, and seaweeds. Some experts predict that at least half of the world fisheries will be affected by aquaculture over the next two to three decades; that is, that there will be human intervention at least once, perhaps more times, in the life cycle of captured species and that capture will become merely a phase of culture. The scientific, economic, and legal problems to be solved remain imposing. Food production from the seas—like all other productive systems—will have to come to grips with the twofold revolution of our time: the technological revolution and the revolution in international relations that is transforming the relations between the industrialized states and the preindustrial nations claiming a fairer share of technologies and resources (see "Marine Fisheries" by Sidney Holt and "Progress of Aquaculture" by T. V. R. Pillay).

2. The production of offshore oil and gas is rapidly expanding, and the proportion of offshore oil and gas within the world's total production is dramatically rising and may reach 50 percent in the next couple of decades. The industry is diversifying geographically, and as new regions—the North Sea, the China Seas, and Latin American and African offshore waters—are being discovered and developed, the balance of the world's energy supply is shifting. The political implications of these developments may be far-reaching. There is a high conflict potential in the absence of clear and unambiguous international rules for the determination of boundaries, as well as in the relationship between States, on the one hand (whether developed or developing), and, on the other, the private multinational companies whose interests often determine approaches to resource exploration and exploitation that are not optimal for the States who may own the resource but depend on the companies for their exploitation (see "Offshore Oil and Gas" by Edward Symonds and "Oil and Gas Exploration and Exploitation in the North Sea" by Peter Odell).

3. The oceans contain vast "unconventional" energy resources (tides, currents, waves, thermogradients, salinity gradients) the exploitation of which is

still largely in an experimental stage but may make a substantial contribution to the world energy household over the next few decades (see "Other Ocean Resources" by Roger Charlier).

4. The mineral industries are poised for a significant breakthrough in the mining of deep-sea nodules, with major economic and, even more important, political implications. For, according to the Declaration of Principles adopted in 1970 by the United Nations General Assembly, the deep-sea minerals are the common heritage of mankind, are to be exploited for the benefit of mankind as a whole, and are giving rise to a new type of international institution for their management. The creation of this institution is one of the main tasks—and the most difficult one—now before the Third United Nations Conference on the Law of the Sea.

5. The technological revolution and the concurrent revolution in international relations, referred to above, are affecting the shipping industry and navigation in two major ways:

a) The increase in size and speed of ships, together with the increased toxicity of fuels and cargoes, require new safety standards and "rules of the road." These have to be determined in the context of the new Law of the Sea, expanding national jurisdiction along with the inevitable expansion of the functions and responsibilities of the international organization charged with such matters, that is, the Inter-Governmental Maritime Consultative Organization (IMCO). The *Yearbook* reports some major progress along these lines ("A Safe Voyage to a New World" by Thomas Busha and James Dawson).

b) Structural changes in the relations between industrialized and preindustrialized nations with regard to the determination of freight rates, the participation in liner conferences and other international agreements, and the ominously growing role of "flags of convenience" are in the making and are urgently needed considering the importance of ocean shipping to the economy of many developing countries (see "The Organization of Shipping" by R. A. Ramsay).

6. The interaction between oceans and atmosphere has a number of components, some of which are mentioned in this volume of the *Yearbook*, inasmuch as they have touched off major developments during the last few years. Others will be more fully dealt with in volume 2.

a) The world climate is largely determined by the ocean/atmosphere interaction. Meetings and programs that have been held to advance knowledge in this crucially important sector will be described in later issues of this *Yearbook*. Activities in this sector are necessarily interdisciplinary, requiring the cooperation of meteorologists and oceanographers, physicists and computer experts, among others, at the intergovernmental as well as the nongovernmental level.

b) The marine satellites—that is, satellites used for the specific purpose of communication between ships and meteorological stations and other warning and support systems—symbolize another aspect of the ocean/atmosphere interaction. The adoption and ratification of the INMARSAT Convention, completed at the initiative and under the auspices of IMCO, is a milestone on the road toward rational management of ocean/atmosphere related services, with at least an opening for the participation of developing countries in these technologically highly sophisticated activities (see "International Maritime Satellite-Communication System" by Anatoly Kolodkin and colleagues and the INMARSAT Convention in this volume).

c) It is interesting to note that the actual and impending changes in sea traffic regulations may be influenced by the far better developed system of air traffic controls (see "A Safe Voyage to a New World" by Thomas Busha and James Dawson).

d) The ocean/atmosphere interface plays an important role in the military uses of both oceans and atmosphere, as borne out by recent developments in the construction of aircraft carriers, on the one hand, and, on the other, in antisubmarine warfare involving satellites, planes, ships, buoys and submarines in one coordinated and computerized system (see "Strategic Submarines and Antisubmarine Warfare" and "The AWS Problem" by Frank Barnaby and others of the staff of SIPRI).

7. The military uses of the oceans are increasing. An underwater arms race between the superpowers is well under way, and the expansion of national jurisdiction over a 200-mile economic zone and its resources puts the burden of increased naval power for surveillance and enforcement purposes on the shoulders of developing coastal States. Their share of warships and weaponry is in fact rapidly increasing (see "Naval Forces" by Ronald Huisken). Measures to advance disarmament and arms control in the oceans have been as frustrated as they have been on land. The Treaty Prohibiting the Emplacement of Atomic Weapons and Other Weapons of Mass Destruction on the Seabed is inadequate to stem the arms race. No link has been established, furthermore, between this treaty and the emerging Law of the Sea Convention which reserves not only the seabed but also the high seas for peaceful purposes, without, however, providing adequate definitions or operational guidelines to this end.

8. Inextricable are the relations between the military uses of the oceans and scientific research, since a great deal of this research is carried out under military auspices or may have military (besides commercial) implications. This obviously complicates the problem of international scientific cooperation as well as the question of the freedom of scientific research. It is remarkable, under the circumstances, that international cooperation in marine scientific research is as highly developed and as active as it is. The number of bilateral, regional, or global projects, seminars, meetings, and expeditions is staggering and still increasing (see reports by ECOR, IOC, SCOR, UNESCO, UN [Ocean Economics and Technology], UNEP, UNITAR, WHO). If one should single out only one event for the comprehensiveness of its scope, it would be the Joint Oceanographic Assembly, held in 1976 under the auspices of SCOR, IOC, and UNESCO (see text among Selected Documents and Proceedings in Appendices).

A breakthrough on the subject of freedom of scientific research was made during the sixth session of the Law of the Sea Conference. A new article (Art. 248), included in the text released at the end of that session, provides that "A coastal State which is a member of a regional or global organization or has a bilateral agreement with such an organization, and in whose exclusive economic zone or on whose continental shelf the organization wants to carry out a marine scientific research project, shall be deemed to have authorized the project to be carried out, upon notification to the duly authorized officials of the coastal State by the organization, if that State approved the project when the decision was made by the organization for the undertaking of the project or is willing to participate in it." This article, for the first time, distinguishes international research from national research and breaks the dilemma between coastal State control (likely to hamper, in some cases, research by technologically developed countries) and freedom of research (not acceptable to technologically underdeveloped coastal States which are unable to control any military or commercial implications of research carried out by developed nations in areas under their jurisdiction) by providing a third alternative: international control, based on cooperation and participation in decision making on research projects.

Another promising event is the establishment, in 1976, after a number of years of cooperation among IOC, WMO, FAO, WHO, the International Atomic Energy Agency, the International Council for the Exploration of the Sea, IMCO, and UNEP, of a Marine Environment Data Information (MEDI) referral system, reflecting the interdisciplinary nature of marine scientific research and the interdependence of all uses of the oceans.

9. Still in the context of areas under coastal State jurisdiction, the development of coastal management principles and guidelines has assumed major importance in recent years (see "Coastal Area Development and Management"). In this immensely complex, delicate, and rapidly developing environment, new forms of governance are taking shape, integrating local, regional, and national authorities in their interaction with international regimes and articulating cooperation among science, civil engineering, landscape architecture, urbanistics, tourism, agriculture, fisheries, mining, harbor construction, navigation, protection of the environment, etc., both at the governmental and nongovernmental level. The United Nations published, in 1976, a *Manual on Coastal Development* with detailed guidelines for this whole range of activities and their interactions. Quite conceivably, some of the approaches and solutions to coastal management problems could eventually be carried over to the management of international ocean space, where many of the problems, and their interactions, are analogous.

This applies to regional organization in general. Quite particularly, it may apply to the problem of the transfer of technology. Here the program of the Role of Marine Coastal Technology (MSCTEC) is a pioneering event. As in the case of the MEDI program, a central referral service is being created at the UN Secretariat, involving the whole network of UN agencies and organizations as well as other competent organizations. One of the assignments is the compilation of an inventory of marine and coastal technologies, which is urgently needed as a basis for any discussion on technology transfers.

10. A great many activities have taken place in the past couple of years with regard to the protection of the marine environment, mostly under the auspices of the United Nations Environment Program (UNEP), to cope with the most acute problems of the pollution of the oceans (see "The Oceans: Health and Prognosis" by Peter S. Thacher and Nikki Meith-Avcin). The most important of these is the adoption of the Convention of Barcelona (see text of Convention in Appendices) in 1976 by 15 out of 18 Mediterranean States plus the EEC. The Convention sets up a general framework for cooperation among governments and the development of additional legal protocols, institutional arrangements, and procedures for the settlement of disputes. By August 1977, it had been signed by Cyprus, Egypt, France, Greece, Israel, Italy, Lebanon, Libya, Malta, Monaco, Morocco, Spain, Tunisia, Turkey, Yugoslavia, and the European Economic Commission (EEC). A regional oil-combating center has been established in Malta to serve as the central information service and to advise those concerned of the occurrence of oil spills, the availability of clean-up assistance, the dangers of environmental damage, and similar matters. The center, under the auspices of UNEP and IMCO, was inaugurated in December 1976. In February 1977, UNEP convened an intergovernmental meeting in Athens to discuss principles for a Draft Protocol for the Protection of the Mediterranean from Land-based Sources of Pollution—this being the principal source of pollution and the most difficult to cope with since the required action implies rather profound rethinking of the concept of national sovereignty and eventual modification of national production systems. The adoption of the Barcelona Convention is being followed up by a host of activities in regional economic and social planning.

Other important regional developments are in course with regard to the Southern Ocean (see articles on the Southern Ocean by G. L. Kesteven and F. M. Auburn), East Asian waters, the Caribbean, the Persian Gulf, the Gulf of Guinea, the Red Sea, the Baltic, the North Sea, and the South Pacific. Policies of environmental conservation interact with all the uses of ocean space and resources (as well as with land uses). Regional organization interacts with global organization which it must articulate and which it sometimes precedes.

11. Nongovernmental organizations also have dealt with the management of ocean space and resources and the law of the sea. Two of the major annual events, in this context, are the Conference of the Law of the Sea Institute of the University of Rhode Island (which has just been transferred to the University of Hawaii) and Pacem in Maribus, organized by the International Ocean Institute, Malta.

The Eleventh Annual Conference of the Law of the Sea Institute took place in November 1977 in Hawaii. The overall theme was "The Regionalization of the Law of the Sea," with particular emphasis on the regional fisheries and other resources management and potentials within the gigantic Pacific Basin, around which live nearly half the world's population.

Pacem in Maribus VII took place in Algiers in conjunction with a conference on the New International Economic Order; its main emphasis was to stress the interdependence between the making of the new law of the sea and the making of a new global economic order.

Pacem in Maribus VIII took place in Mexico in December 1977 at the Center for Economic and Social Studies of the Third World. Its overall theme was the impact of the new law of the sea on developing countries.

12. The number of intergovernmental organizations and programs dealing with the oceans has increased at such a rate that it would now require a substantial dictionary of acronyms to list them all. On the whole, their activities are still handicapped by the sectoral organization of the United Nations system as a whole, as well as by the lack of operational capacity, limiting their functions to those of coordinating national activities (in which only highly developed States can actively engage), to the exclusion of genuine international operations in which developing countries could participate more fully. There is, however, in many of the organizations involved, an unambiguous trend toward restructuring: toward enlarging their membership, democratizing their decision-making processes, and widening their scope. This trend responds to the generally recognized need for a restructuring of the UN system as a whole. It is accelerated by developments at the Third United Nations Conference on the Law of the Sea, which confers new responsibilities on the ocean-oriented UN specialized agencies (COFI/FAO, IOC, IMCO, UNEP), explicitly in matters relating to dispute settlement, implicitly in a number of others: management of living resources, scientific research, transfer of technology, navigation, and the protection of the marine environment.

Along with the trend toward intra-agency restructuring, there is a noticeable trend toward increased inter-agency cooperation, no longer restricted to the intersecretariat level but involving common efforts in common projects and the provision of common services. Nor is this kind of expanded, interdisciplinary cooperation limited to the agencies. It comprises regional economic commissions and other regional or functional institutions (regional banks, economic communities), all of which are beginning to play a role in the making of the new order for the oceans. The difficulty at this stage is that all these developments are as yet not properly or purposively coordinated; they are not integrated with the work of the UN Law of the Sea Conference; there is, so far, no great design for a new international economic order in the oceans to respond to the exigencies of both the technological revolution and the revolution in international relations.

13. Slowly and gropingly, however, the Third United Nations Conference on the Law of the Sea is moving in this direction. Its progress, since the first working session in Caracas in 1974, can be measured by comparing the papers released at the end of that session ("Trends," the Informal Single Negotiating

Text Released after the Third Session in Geneva (1975); the Revised Single Negotiating Text (New York, 1976); and the Informal Composite Negotiating Text (New York, 1977). The Composite Text, with its 303 Articles in 16 parts plus seven annexes, is the most complex text ever contemplated by an international community which, in turn, is more numerous and heterogeneous than it has ever been before. The Composite Text, which is reproduced in this volume, is in fact a nascent world constitution, a document radically different from and far more advanced than the United Nations Charter. Its provisions move from national ocean space—which is greatly expanded and consists of internal waters, the territorial sea, the contiguous zone, the exclusive economic zone, archipelagic waters, and the legal continental shelf—to international ocean space—which consists of the high seas and the international seabed area. Needless to say, the problems of delimitation and jurisdiction are as numerous as the provisions. The Text moves from noninstitutional provisions—the codification and updating of the traditional law of the sea—to institutional provisions—the establishment of an entirely new type of international organization, an International Seabed Authority, which would manage the common heritage of mankind, and a dispute-settlement mechanism which would tie together the whole system of organizations dealing with the uses of ocean space and resources.

The Text, in one way or another, deals with all peaceful uses of the oceans: mining, fishing, navigation, communications, and scientific research. In a wider sense it faces problems of food, minerals and metals, energy, trade and communication, science policy, the transfer of technology, multinational corporations, the arms race, arms reduction and control, environmental conservation, and economic development. It also confronts the problem of accommodating in one system socialist and market-economy countries, and developing and industrialized States. In other words, it confronts the whole range of basic issues facing the international community in the broadest sense in its attempt to create a new world order. It is in this sense that the new Law of the Sea is not only an essential part of, but conceivably a model and forerunner of, a new world order.

Ten brief years have passed since Malta's call for a new order for the oceans, based on the principle of the common heritage of mankind. One should marvel that, by 1977, with the Composite Text, the international community has moved as far as it has in the direction Malta indicated in 1967!

Issues and Prospects

The Evolving Law of the Sea: A Critique of the Informal Composite Negotiating Text (1977)

Arvid Pardo
University of Southern California

Over the last couple of decades the traditional legal order in the oceans, based on freedom beyond a narrow belt of sea adjacent to the coast under coastal State sovereignty, has been shaken by multiplying claims of coastal States to expanded jurisdiction in the marine environment. These claims have resulted both from the impact of technological advance and from the emergence of many new States more concerned to control activities and to establish exclusive rights to resources in marine areas near their coasts than in the maintenance of the integrity of the High Seas.

In 1970 the United Nations General Assembly decided to convene in 1973 a general conference on the law of the sea in order to revise existing law in a manner acceptable to the international community and to establish an entirely new legal regime for the seabed beyond national jurisdiction, based on the principle that this part of the marine environment and its resources are a "common heritage of mankind."[1] The conference, attended by representatives of nearly 150 sovereign States, held its first organizational session in New York in December 1973. Five working sessions have subsequently been held.[2]

Progress at the conference has been hampered by a variety of factors, including procedural complications, an inappropriate conference organization, and the many difficulties inherent in a complex political negotiation

---

1. This concept was proposed by the government of Malta in 1967 and endorsed by the United Nations General Assembly in 1970. In 1971 the government of Malta suggested that a common-heritage regime replace the traditional regime of the High Seas in all ocean space beyond national jurisdiction. The basic elements of the common-heritage concept, as proposed by the government of Malta, were as follows: (*a*) the area under the common-heritage regime may not be appropriated (it may be used but not owned); (*b*) all rights to resources in the common-heritage area are vested in mankind as a whole acting through an international organization; (*c*) the common-heritage area and its resources are managed through an international organization in which all States have the right to participate; (*d*) benefits, both financial and deriving from participation in management and exchange and transfer of technologies, are shared; (*e*) the common-heritage area may be used only for peaceful purposes; (*f*) the common-heritage area must be transmitted environmentally unimpaired to future generations.

2. One session in 1974 (Caracas), one in 1975 (Geneva), two in 1976 (New York), and one in 1977 (New York).

© 1978 by The University of Chicago. 0-226-06602-9/77/1978-1001$02.35

between a great number of regional, geographical, and other groups with varied interests. The main obstacles to progress, however, have been of a substantive nature: the vastness, importance, and complexity of the subject matter; the conflicting interests of States; and the difficulty of elaborating an effective and efficient international regime for the seabed beyond national jurisdiction (international seabed area, hereafter referred to as the Area) which can equitably balance the interests of developed and developing countries.[3]

A further important factor has retarded and complicated the work of the conference. Ocean space[4] is acquiring ever-increasing military and economic importance.[5] Hence, in dealing with major issues, the conference must face, directly or indirectly, the entire range of basic problems confronting the international community today—military security and arms control, food, raw materials, energy, and economic development—which are discussed in other international fora. Many believe that these problems can be solved only through the creation of a more equitable and cooperative world order, and developing countries have proposed the concept of a New International Economic Order. Whatever the merits of this concept, many representatives are aware that the decisions taken by the conference may decisively influence the manner in which the international community will consider the basic problems which have been mentioned above. The new Law of the Sea being elaborated by the conference could become a model for New International Economic Order or a serious obstacle to its achievement by recognizing national control over the preponderance of marine resources and failing to establish viable regimes in marine areas beyond national jurisdiction. The conference has struggled with this dilemma, and its response is reflected in the documents issued at the end of each session.[6] The most recent of these, the Informal Composite Negotiating

---

3. The terms "developed countries" and "developing countries," although frequently used even in official documents, are imprecise. Often the term "developing countries" is used to indicate States which are members of the Group of 77 established at the United Nations some years ago.

4. The term ocean space is used to indicate the surface of the sea, the water column, the seabed, and its subsoil.

5. Until recently ocean space was used essentially for navigation and fishing. Over the century 1850–1950, intensive exploration was initiated and some new uses were developed, such as the laying of submarine cables, submerged navigation, extraction of minerals from seawater, and offshore hydrocarbon exploitation in shallow water. The postwar scientific and technological explosion is now making all parts of ocean space accessible, exploitable, and usable for an increasing variety of purposes. Ocean space contains more than 95 percent of the world's water, probably more hydrocarbons, and greater quantities of a wide range of minerals than are found on land. It contains vast living resources and is an immense potential source of unconventional energy. Ocean space is also an essential medium for international trade. Ocean space, finally, has acquired vital importance in the nuclear age for the maintenance of the strategic balance of power and for national security purposes.

6. The documents are the Main Trends paper in 1974, the Single Negotiating Text in 1975, and the Revised Single Negotiating Text in 1976.

Text (hereafter referred to as the Text), was issued after the Sixth session in the Summer of 1977. It is reproduced in full as an Appendix to this volume.

The Text, with its nearly 200 pages, 303 articles in 16 Parts, and seven annexes, is perhaps the most comprehensive and complex document drafted by an international conference. The document is in the form of a draft treaty which is intended to constitute a legal framework for man's activities in the marine environment. As such, it may be regarded as a draft constitution for ocean space.

The Text reflects closely in its various Parts the uncertainties and contrasting trends at the conference—innovative and conservative, nationalist and internationalist. Those portions of the Text (Parts II–X) which deal with marine areas under national jurisdiction, the rights and duties of States therein, and the regime of the High Seas reflect a strong feeling of nationalism and considerable legal traditionalism by reproducing with little substantive change concepts and provisions contained in the 1958 Geneva Conventions on the Law of the Sea and by proposing international recognition of the results of the widespread enclosure movement in the marine environment of the past 30 years. The major innovation here is a shift in the balance of the present law of the sea from freedom over the greater part of ocean space to control by coastal States over vast areas of previously High Seas. Many uses of the marine environment—fisheries, navigation, offshore installations and structures, cables and pipelines, marine scientific research, and control of marine pollution—are dealt with in these Parts of the Text largely for the purpose of clarifying the rights of the coastal State. Part XI of the Text (with its two annexes), on the other hand, contains the highly innovative proposals to create an international regime, based on the principle of common heritage of mankind, and to establish an international organization (International Seabed Authority, hereafter referred to as the Authority) to implement the international regime. Parts XII and XIII develop in great detail the rights and (in more general terms) the duties of the coastal State with respect to protection of the marine environment and marine scientific research in marine areas under national jurisdiction; in addition, they enunciate a number of innovative general principles, but effective implementation of these is lacking. Part XIV is intended to encourage international cooperation in the development and transfer of marine technology; no effective mechanisms, however, are suggested for this purpose. Finally, Part XV of the Text, with its four annexes, contains a detailed outline of a system for the binding settlement of most disputes arising from the interpretation or application of the proposed convention, a system which, if implemented, would represent a very significant development of present international law.

Although the Informal Composite Negotiating Text is "informal," it is nevertheless intended to be a basis for final negotiations at the next session of the conference in 1978, and many of its provisions are supported by the great majority of States. While other provisions remain controversial, particularly those dealing with the future International Seabed Authority, it is certain that the Text will exert a profound influence on the law of the sea and on the practice of States, even if no agreement is eventually reached at the conference. Both because of its status and its comprehensive contents, therefore, the Text is a document which deserves to be summarized and briefly analyzed.

Most of the Text (Parts II–VI, Part VIII, and much of Parts XII and XIII) is concerned with problems of allocation of marine areas to coastal States and with establishing the right of these States to regulate activities and to exploit marine resources on an exclusive basis in marine areas under their jurisdiction.

Marine areas under coastal State sovereignty or jurisdiction have increased from the four[7] recognized under the 1958 Geneva Conventions on the Law of the Sea to six by the addition of "archipelagic waters"[8] and of an "exclusive economic zone."[9] Coastal State control in ocean space will now extend at least to 200 nautical miles from the coast. In some cases, this control will extend much

---

7. The 1958 Geneva Conventions on the Law of the Sea recognized the following marine areas to be under coastal State sovereignty or jurisdiction: (1) *Internal waters* are on the landward side of baselines drawn by the coastal State. Coastal States have full sovereignty over their internal waters. (2) The *territorial sea* is a belt of sea, adjacent to the coast or seaward of straight baselines drawn by the coastal State, over which the coastal State exercises sovereignty subject to the obligation to permit "innocent passage" of foreign vessels. The breadth of the territorial sea was not defined, but in 1958 the majority of coastal States claimed a 3-nautical-mile territorial sea, and it was generally recognized that the breadth of the territorial sea should not exceed 12 nautical miles measured from the appropriate baselines. The Informal Composite Text proposes a breadth of 12 nautical miles. The breadths of the territorial seas as of December 1, 1977, are shown in table A2, along with coastal State claims to jurisdiction over fishing beyond 12 miles from their coasts or other appropriate baselines. (3) The *contiguous zone* is a zone adjacent to the territorial sea in which the coastal State may exercise the control necessary to prevent and punish infringement of its customs, fiscal, immigration, or sanitary regulations within the territorial sea. The breadth of the contiguous zone was 12 nautical miles measured from the appropriate baselines. The Text now proposes a breadth of 24 nautical miles. (4) The *continental shelf* is ambiguously defined in the 1958 Convention on the Continental Shelf in terms of adjacency to the coast, depth (200 meters), and exploitability of resources over which the coastal State exercises sovereign rights for the purpose of exploration and exploitation of natural resources (including living organisms belonging to so-called sedentary species).

8. Archipelagic waters are defined as the waters enclosed by straight baselines drawn by an archipelagic State joining the outermost points of the outermost islands and drying reefs of the archipelago. The length of the baselines must not exceed 100 nautical miles (3 percent, however, may be up to 125 miles long) and must include the main islands of the archipelago and an area in which the ratio of the area of water to that of land is between one to one and nine to one. The archipelagic State exercises sovereignty over its archipelagic waters (including the airspace above and the seabed) subject to the right of innocent passage and of archipelagic sea-lanes passage, i.e., the right of foreign vessels and aircraft safely, continuously, and expeditiously to transit through sea-lanes or air lanes designated by the archipelagic State.

9. The exclusive economic zone is an area beyond and adjacent to the territorial sea in which the coastal State exercises (*a*) sovereign rights for the purpose of exploration,

further with respect to the seabed.[10] Islands, whatever their size (apart from "rocks which cannot sustain human habitation or economic life of their own"),[11] are given the same extensive jurisdiction as other land territory. The Text avoids a clear definition of the limits of national jurisdiction by setting no limit to the length of straight baselines drawn by coastal States and by defining the legal continental shelf in terms of natural prolongation of land territory.[12]

Within the third or more of ocean space in which they will exercise immediate control, coastal States have extensive rights but few duties: the latter are usually related to an obligation to respect the innocent passage [13] of foreign vessels in the territorial sea and not to hinder overflight and the transit of foreign vessels in other areas under national jurisdiction and in straits used for international navigation.[14] There are indeed provisions in the Text purporting to create other obligations, both general and specific, for coastal States in marine areas under their jurisdiction, particularly in the exclusive economic

---

exploitation, conservation, and management of natural resources, whether living or nonliving, and with regard to other activities of economic value. It also has *(b)* jurisdiction with regard to (i) establishment and use of artificial islands, installations, and structures; (ii) marine scientific research; and (iii) preservation of the marine environment. Finally, it exercises *(c)* other rights and duties provided for in the proposed convention. All States in the exclusive economic zone enjoy the freedoms of navigation, overflight, and of the laying of submarine pipelines and cables and other internationally lawful uses of the sea, related to the proposed convention. In exercising these rights, States must have regard to the rights and duties of the coastal State and must comply with the latter's laws and regulations. The breadth of the exclusive economic zone is 200 nautical miles measured from the baselines from which the breadth of the territorial sea is measured.

10. The Text redefined the legal continental shelf in terms of the natural prolongation of a coastal State's land territory "to the outer edge of the continental margin or to a distance of 200 nautical miles from the baselines from which the breadth of the territorial sea is measured where the outer edge of the continental margin does not extend to that distance." Determination of the outer edge of the continental margin is left to the discretion of the coastal State.

11. These rocks, regardless of size, are, however, recognized to have a territorial sea and a contiguous zone.

12. Natural prolongation of land territory, particularly when its determination is left to the discretion of the coastal State, is a highly uncertain criterion for the definition of national jurisdictional limits because the outer edge of the continental margin is often difficult to determine with precision.

13. Innocent passage is defined as passage not prejudicial to the peace, good order, or security of the coastal State. The Text enumerates a number of activities considered prejudicial to the peace, good order, or security of the coastal State.

14. Straits used for international navigation must connect one area of the High Seas or an exclusive economic zone with another area of the High Seas or an exclusive economic zone. In archipelagic waters, the Text creates a right of archipelagic sea-lanes passage. In the exclusive economic zone, coastal States must respect, subject to some qualifications, "freedoms of navigation and overflight and of the laying of submarine pipelines and cables, and other internationally lawful uses of the sea related to these freedoms."

zone, but the general obligations are couched in vague language, and the specific obligations are often either heavily qualified or negated by other articles.[15]

Existing rules are maintained for the delimitation of the territorial sea between States which are adjacent or opposite each other; delimitation of the exclusive economic zone and of the legal continental shelf, however, is effected "by agreement between the States concerned in accordance with equitable principles, employing where appropriate the median or equidistance line and taking account of all the relevant circumstances."

Only the briefest of references is required to Parts IX and X of the Text. Part IX urges States bordering enclosed or semienclosed seas to cooperate with each other in the exercise of their rights and duties under the proposed convention, particularly with regard to scientific research, preservation of the marine environment, and the management and exploitation of the living resources of the sea. Part X, devoted to the right of access of landlocked States to and from the sea, does not significantly enlarge the rights already recognized to these States under the 1958 Geneva Convention on the High Sea, or under the 1965 Convention on the Transit Trade of Landlocked States.

According to traditional law of the sea, all parts of the sea not included in the territorial sea or internal waters of a State were part of the High Seas. It is now proposed to establish two radically different regimes in marine areas beyond national sovereignty or jurisdiction by maintaining, on the one hand, a somewhat modified regime of the High Seas "for all parts of the sea that are not included in the exclusive economic zone, in the territorial sea or in the internal waters of a State, or in the archipelagic waters of an archipelagic State" and by creating, on the other hand, a new regime for the seabed based on the principle that the seabed and its resources "beyond the limits of national jurisdiction" are a common heritage of mankind.

The traditional regime of the High Seas remains essentially unchanged in the reduced area to which it now applies. The High Seas are open to all States; the traditional freedoms[16] are maintained, and to these are added the freedom of scientific research and the freedom to construct artificial islands and other installations permitted under international law. Nevertheless, some interesting developments of traditional law are proposed: the exercise of the freedoms of construction of artificial islands, of the laying of submarine pipelines and cables, and of scientific research are made subject to the rights of the coastal State in the legal continental shelf; the freedom of fishing is made subject to the

---

15. For instance, failure of coastal States to discharge their fishery-conservation and management obligations cannot be called into question.

16. The reference is to the freedoms specifically mentioned in the 1958 Geneva Convention on the High Seas, i.e., the freedom of navigation, the freedom of overflight, the freedom of fishing, and the freedom to lay submarine pipelines and cables.

"primary responsibility" of the coastal State for the management of anadromous and catadromous living resources of the sea[17] and to a duty of cooperation in the conservation and management of living resources. The duty of States to exercise their jurisdiction and control over ships flying their flag has been made stricter. It is proposed that States cooperate in the suppression of illicit traffic in narcotic drugs by ships on the High Seas, and unauthorized broadcasting from the High Seas is, in many respects, assimilated to piracy.[18]

As distinguished from the High Seas regime, the common-heritage regime proposed for the seabed beyond national jurisdiction (international seabed area) is innovative in concept. The main characteristics of the regime proposed are *(a)* nonappropriability of the Area and of its resources; *(b)* use of the Area exclusively for peaceful purposes; *(c)* all rights in the resources of the Area "are vested in mankind as a whole"; *(d)* international cooperation in marine scientific research, protection, and conservation of the marine environment and transfer of technology relating to mineral resource exploration and exploitation; *(e)* promotion of the effective participation of developing countries in mineral resource exploration and exploitation; and *(f)* creation of a system of administration for the mineral resources of the Area implemented through an International Seabed Authority. This Authority must (i) undertake scientific research in the Area, (ii) establish a system for the equitable sharing of benefits from the Area, and (iii) undertake mineral resource exploration and exploration in the Area either directly or in association with State Parties or public and private entities in accordance with a number of objectives, including "the protection of developing countries from any adverse effect on their economies or earnings resulting from a reduction in the price of an affected mineral[19] or in the volume of that mineral exported," through participation in commodity conferences, restrictions on mineral production from the Area, and compensation to affected developing countries.

The Authority is "the organization through which State Parties organize and control activities[20] in the Area." Principal organs of the Authority are an Assembly, a Council, a Secretariat, and an Enterprise. In addition, the Text

---

17. According to the Text, these stocks may be harvested only in waters under the jurisdiction of the coastal State of origin or, in the case of catadromous stocks, of the State where they "spend the greater part of their life cycle."

18. Any person engaged in unauthorized broadcasting may be prosecuted before a court of *(a)* the flag State of the vessel, *(b)* the place of registry of the installation, *(c)* the State of which the person is a national, *(d)* any place where the transmissions can be received, or *(e)* any State where authorized radio communication is suffering interference. Any of these States may arrest the ship and any person on board and seize the broadcasting apparatus.

19. The Text refers here to minerals contained in manganese nodules.

20. The term "activities" is narrowly defined in the Text "as all activities of exploration for, and exploitation of, the resources of the Area." The term "resources" in turn has the narrow meaning of "mineral resource in situ."

provides for three subsidiary organs of the Council—an Economic Planning Commission, a Technical Commission, and a Rules and Regulations Commission—and for such other subsidiary organs as may be necessary.

The Assembly is "the supreme organ of the Authority and as such shall have the power to establish the general policies ... to be pursued by the Authority" on any matters within its competence. Meeting in regular session every year, it consists of all members of the Authority, each of whom has one vote. Decisions on questions of substance are taken by a two-thirds majority of the members present and voting. When a matter of substance "comes up for voting for the first time, the President may, and shall, if requested by at least one fifth of the members of the Assembly, defer the question of taking a vote ... for a period not exceeding five calendar days."

The Council is the executive organ of the Authority: it has the power to establish the specific policies to be pursued by the Authority in conformity with the provisions of the proposed convention and with the general policies established by the Assembly. It consists of 36 members[21] elected by the Assembly for 4 years, and it meets not less than three times a year. Decisions on questions of substance are taken by a three-fourths majority of members present and voting.

The Economic Planning Commission, composed of 18 experts appointed by the Council taking into account equitable geographical distribution and the need to ensure a fair balance between experts from countries which export and which import minerals derived from the Area, reviews "the trends of, and factors affecting supply, demand and prices of raw materials which may be obtained from the Area" and makes recommendations[22] to the Council on the basis of an affirmative vote of two-thirds of its members present and voting.

The Technical Commission, composed of 15 members appointed by the

---

21. The Text proposes that members of the Council be elected as follows: (i) four members from countries which have made the greatest contributions to the exploration and exploitation of the resources of the Area (including at least one State from the Eastern European region); (ii) four members from countries which are major importers of the categories of minerals derived from the Area (including at least one State from the Eastern European region); (iii) four members from countries which are major exporters of the categories of minerals derived from the Area, including at least two developing countries; (iv) six members from developing countries representing special interests (landlocked States, States with large populations, States which are major importers of the categories of minerals to be derived from the Area, and least-developed countries); (v) 18 members "elected according to the principle of ensuring an equitable geographical distribution of seats in the Council as a whole," provided that each region has at least one member. The regions are Asia, Africa, Eastern Europe (Socialist), Latin America, and Western Europe and Others.

22. The Economic Planning Commission may advise the Council on the exercise of the latter's powers to take measures for the protection of developing countries from adverse effects on their economies or earnings resulting from a reduction in the price or volume of an affected mineral. In addition, the Commission must investigate any situation likely to lead to such a situation and make recommendations to the Council on measures which the Authority should take.

Council, is principally concerned with *(a)* making recommendations to the Council on measures needed to carry out the Authority's functions with respect to scientific research and the transfer of technology; *(b)* preparing assessments of environmental implications of activities in the Area; and *(c)* supervising all operations with respect to activities in the Area, including directing and supervising a staff of inspectors, auditing accounts, and reviewing plans of work for activities in the Area.

The Rules and Regulations Commission, composed of 15 members expert in legal matters related to ocean mining, formulates and submits to the Council rules, regulations, and procedures relating to prospecting, exploration, and exploitation in the Area.[23] The Secretariat, comprising a Secretary General and staff, is modeled generally after the United Nations Secretariat.[24] The Enterprise is the organ of the Authority which is empowered to undertake mineral resource exploration and exploitation in the Area. It is subject to the general policies laid down by the Assembly and to the directives and control of the Council. It has such legal capacity as is necessary for the performance of its functions. The funds and assets of the Enterprise are separate from those of the Authority and are not subject to seizure. Property and revenues of the Enterprise are immune from taxation. States are encouraged to provide "special incentives, rights, privileges and immunities" to the Enterprise in their territories. It is governed by a Governing Board of 15 "qualified, competent and experienced members," elected by the Assembly for 4 years, taking into account special interests and equitable geographical distribution. Each member of the Board has one vote, and decisions are taken by a majority of the votes cast.

The legal representative and chief operating officer of the Enterprise is the Director-General, who is elected by the Assembly of the Authority for a term not exceeding 5 years upon recommendation of the Council. The Director-General appoints the staff and conducts the ordinary business of the Enterprise under the direction of the Governing Board.

The Text contains a number of provisions concerning the finances of the Authority, its legal status, and the immunities and privileges to which assets of the Authority and certain persons connected with the Authority have right.[25]

---

23. The subjects of these regulations are enumerated at considerable length in Annex II of the Text.

24. The staff must have no financial interest in mineral exploration and exploitation activities in the Area and are obligated not to reveal industrial secrets or proprietary data "coming to their knowledge by reason of their official duties."

25. The Authority has international legal personality and such legal capacity as may be necessary for the fulfillment of its purpose. Property and assets of the Authority are immune from legal process, search, and seizure and are exempt from taxation or custom duties. Members of any organ of the Authority, the Secretary General, and staff of the Authority are accorded immunity from legal process with respect to acts performed in the exercise of their functions and the same immunities and facilities as are accorded by States Parties to representatives and officials of comparable rank of other States Parties.

Provision is also made for the binding settlement of disputes arising from the interpretation or application of Part XI of the Convention,[26] either through arbitration or through judicial proceedings before the Seabed Disputes Chamber of the proposed Law of the Sea Tribunal.

Part XII of the Text proposes to establish a comprehensive system for the prevention and control of marine pollution to replace articles 24 and 25 of the Geneva Convention on the High Seas and to supplement multilateral conventions in force. The system proposed is based on the explicit obligation of States to protect and preserve the marine environment and to prevent and control its pollution from any source, subject to their "sovereign right to exploit their natural resources pursuant to their environmental policies." States are urged to cooperate on a global and regional basis and through "competent international organizations" *(a)* in elaborating rules, standards, and recommended practices for the protection of the marine environment; *(b)* in promoting contingency plans for responding to pollution incidents; *(c)* in promoting scientific research and the exchange of information about pollution in the marine environment; *(d)* in formulating scientific criteria for the elaboration of rules, standards, and recommended practices for the prevention of marine pollution; *(e)* in monitoring the marine environment for pollution; and *(f)* in promoting scientific, technical, and other programs of assistance to developing countries for the prevention of marine pollution.

The Text prescribes that States must establish national laws and regulations for the prevention and control of marine pollution from land-based sources, from seabed activities, and from dumping and urges them to develop, through competent international organizations or diplomatic conferences,

---

26. The Seabed Disputes Chamber also has jurisdiction *(a)* in disputes between a State Party and the Authority concerning an allegation that an act of an organ of the Authority is in violation of Part XI of the convention or that the organ lacks jurisdiction or has misused its power; *(b)* in similar disputes between a national of a State Party and the Authority with regard to decisions or measures directed specifically to that person; *(c)* in disputes between the Authority and a State Party (or one of its nationals) relating to the interpretation or application of a contract concerning activities in the Area or concerning alleged violations by a State Party of the provisions of Part XI of the convention; *(d)* in disputes between a State Party and a national of another State Party, or between nationals of such State, regarding the interpretation of any contract between them or in respect of their activities in the Area; *(e)* in disputes where a State Party has been suspended from the exercise of the rights and privileges of membership of the Assembly of the Authority because of allegations that it had "grossly and persistently" violated the provisions of Part XI or contractual arrangements entered into pursuant to this Part of the Convention; *(f)* in disputes where a State Party, or an entity sponsored by a State Party, alleges that an official of the Authority has violated the responsibilities stated in the Text (Article 167). However, the question whether rules, regulations, and procedures adopted by the Assembly or Council of the Authority are in conformity with the provisions of the proposed convention is not subject to the jurisdiction of the Seabed Disputes Chamber. Similarly, the Chamber has no jurisdiction with regard to the exercise by any organ of the Authority of its discretionary powers.

global and regional rules, standards, and recommended practices to prevent and control marine pollution. Enforcement is by the States concerned or in cooperation with the Authority in the international seabed area.

The Text pays particular attention to the prevention and control of vessel-source pollution. All States must both cooperate in developing international rules and standards for the prevention and control of vessel-source pollution and enact national laws and regulations for this purpose that have at least the same effect as generally accepted international rules and standards. In this connection, the Text enumerates in detail the rights of coastal States (and port States) in establishing and enforcing national laws and regulations for the prevention and control of vessel-source pollution[27] and attempts to balance these rights against the traditional jurisdiction of the State whose flag the vessel flies. Warships and government-owned or operated vessels employed on government noncommercial service are explicitly exempted from the provisions of the convention relating to marine pollution.

The first articles of Part XIII of the Text affirm the right of all States and competent international organizations to conduct marine scientific research subject to the rights and duties of other States and their obligation to promote its development. The conduct of marine scientific research must conform to four general principles: it must be conducted "exclusively for peaceful purposes," "with appropriate scientific methods and means" compatible with the proposed convention; it must not "unjustifiably interfere with other legitimate uses of the sea" and it must comply "with all relevant regulations" established in conformity with the proposed convention. The Text contains a number of provisions obligating States to cooperate in the promotion and creation of

---

27. In the territorial sea, coastal States in the exercise of their sovereignty may establish laws and regulations for the control of marine pollution which do not hamper the innocent passage of foreign vessels. In the exclusive economic zone, such laws and regulations must comform to generally accepted international rules and standards established through a competent international organization or general diplomatic conference, but in ice-covered areas, coastal States may freely establish pollution-control regulations having "due regard to navigation and the protection of the marine environment." The coastal State may also, in special clearly defined areas of its exclusive economic zone where "special mandatory methods for the prevention of pollution from vessels is required," establish with the consent of the competent international organization such laws and regulations for the control of vessel-source pollution as that organization has made applicable to special areas. Enforcement of laws and regulations with respect to vessel-source pollution is (a) by the flag State for vessels flying its flag, (b) by the port State with respect to violations of international standards and rules in international waters, and (c) by the coastal State. The coastal State may inspect and arrest a vessel in its territorial sea for violation of national laws and regulations; in the exclusive economic zone, the coastal State may inspect a vessel suspected of having caused significant pollution in the territorial sea or exclusive economic zone in violation of national laws and regulations or applicable international standards and rules. The coastal State may arrest the vessel if it is suspected of a "flagrant or gross violation" causing major damage or threat of major damage to the environment.

favorable conditions for[28] marine scientific research and in the exchange and transfer of scientific knowledge; at the same time, however, the Text stresses in great detail the right of coastal States in the exercise of their sovereignty or jurisdiction "to regulate, authorize and conduct" marine scientific research in marine areas under their control[29] and their discretionary right to require the cessation of foreign research activities in these areas. Part XIII is completed by articles on the legal status of scientific research installations and on responsibility and liability for damage caused by marine scientific research activities.

Part XIV of the Text is mainly of a hortatory nature: States are urged actively to promote within their capabilities "the development and transfer of marine science and marine technology on fair and reasonable terms and conditions" and with "proper regard for all legitimate interests." To this end States are urged to promote a number of objectives and to cooperate in a number of activities.[30] The International Seabed Authority is obligated to make its technical documentation available to all States upon request and to engage nationals of developing countries as members of its managerial, research, and technical staff for training purposes. Finally, the Text purports to obligate States to establish regional marine scientific and technological centers, the functions of which are specified in considerable detail.

28. Cooperation is also to be exercised by States in an effort "to integrate the efforts of scientists in studying the essence of and interrelations between phenomena and processes occurring in the marine environment." The meaning of this sentence is not clear.
29. In the territorial sea, marine scientific research activities may "be conducted only with the express consent of and under conditions set forth by the Coastal State." Marine research activities in the exclusive economic zone or legal continental shelf require the consent of the coastal State, which must normally be given for research projects conducted by other States or competent international organizations when such projects are "exclusively for peaceful purposes" and "in order to increase scientific knowledge of the marine environment for the benefit of mankind as a whole," and, in addition, are not of direct significance to the exploration or exploitation of natural resources, do not involve drilling, the use of explosives, or the construction or operation of artificial islands or other structures. The entity conducting the research must supply the coastal State with detailed information on its research project before obtaining permission to conduct the research, and it must comply with a number of burdensome conditions during the research. If, on the other hand, the project is submitted by a competent international institution, the consent of the coastal State is assumed to have been given if that State voted in favor of the project when it was adopted by the international organization (Art. 248). This is at least a beginning in the right direction, encouraging the internationalization of marine research as a way out of the dilemma between freedom of inquiry and coastal State control.
30. The objectives are the acquisition and dissemination of marine technological knowledge, the development of "appropriate" marine technology, and "the development of the necessary technological infrastructure to facilitate the transfer of marine technology." The activities in which States are urged to cooperate range from technical cooperation to conferences and seminars.

Negotiations on the question of dispute settlement were initiated informally at the Caracas session of the Law of the Sea Conference. A draft text on dispute settlement attached to the Informal Single Negotiating Text was issued by the president of the conference in 1975. This text was twice revised at the New York sessions of the Conference in 1976 and 1977. The latest revision forms Part XV of the Informal Composite Negotiating Text. While the approaches of States have differed in many important respects, it appears that there is now general recognition that the future Law of the Sea Convention should contain an effective system for the settlement of disputes.

The present text takes into account the different views of States by flexibly combining comprehensive and functional, compulsory and noncompulsory approaches to dispute settlement. The system proposed is based on the assumption that when a dispute arises it is advisable to encourage amicable settlement through nonjudicial procedures by the method preferred by the parties to the dispute rather than to prescribe rigid procedures. Accordingly, the stress throughout the text is on flexibility.

After reaffirming the obligation of States under the United Nations Charter to settle disputes between them by peaceful means, the Text prescribes that its provisions do not apply if a State has accepted an obligation to settle the dispute by resort to a final and binding procedure under a general, regional, or special agreement or if the Parties to the dispute have agreed to seek a settlement by peaceful means of their choice.[31]

In the event that no settlement is reached through the procedures mentioned, the dispute must be submitted to judicial settlement at the request of any of the parties to the dispute, but the parties are offered a choice of tribunal—the Law of the Sea Tribunal; the International Court of Justice; or an arbitral tribunal or a special arbitral tribunal for disputes concerning fisheries, protection of the marine environment, marine scientific research, and navigation. If the parties to a dispute have not accepted the same dispute-settlement procedure, the dispute may be submitted only to arbitration "unless the parties otherwise agree."

Provision is made for scientific or technical experts to assist the Court or tribunal at the request of any party to the dispute. The Court or tribunal is empowered to prescribe provisional measures and, at the request of the flag State, the prompt release of vessels unduly detained by a coastal State. The law applicable is the proposed convention and rules of international law not incompatible with it. The decisions of the Court or tribunal are final and binding.

Whereas in principle the dispute-settlement system proposed is comprehensive, the exercise by a coastal State of sovereign rights or jurisdiction is subject to binding judicial procedures only in a limited number of specified

---

31. In this connection, the Text makes specific mention of conciliation procedures.

cases,[32] and the Court or tribunal is specifically prohibited from substituting its discretion for that of the coastal State in matters concerning (a) the granting or withholding of consent for, or ordering the cessation of, marine scientific research in the exclusive economic zone or legal continental shelf, and (b) the conservation or utilization of living resources in the exclusive economic zone, including determination of the allowable catch and fishery regulations. In addition, when expressing its consent to be bound by the proposed convention, a State may declare that it does not accept "one or more" of the proposed procedures for dispute settlement with respect to "one or more" of the following: (a) disputes regarding sea-boundary delimitations between adjacent or opposite States, provided that the State concerned declares it accepts some other procedure entailing binding decision and that such procedure or decision excludes determination of any claim to sovereignty over land territory; (b) disputes concerning military activities by government vessels or law-enforcement activities in the exercise of sovereign rights or jurisdiction; (c) disputes in respect of which the Security Council of the United Nations is exercising the functions assigned to it by the Charter.

The structure of the dispute-settlement system proposed in the Text comprises a Law of the Sea Tribunal, establishment of ad hoc arbitral tribunals, special arbitration procedures, and a conciliation procedure. The Tribunal is composed of 21 independent judges elected for a term of 9 years by a two-thirds vote at a conference of States Parties convened periodically by the Secretary-General of the United Nations. The representation of the principal legal systems of the world and equitable geographical representation must be assured in the Tribunal as a whole. Provision is made for the mandatory establishment of an 11-member Seabed Disputes Chamber (the members of which are elected by the assembly of the Authority) and for optional creation by the Tribunal of special chambers composed of three or four members to deal with particular categories of disputes. Decisions of the Tribunal and of its chambers are final and binding between the parties to the dispute.[33] The

---

32. The cases are (a) when it is alleged that a coastal State has acted in contravention of the provisions of the convention with regard to the freedoms of the High Seas in the exclusive economic zone; (b) where it is alleged that a coastal State has acted in contravention of specified international rules and standards for the protection ... of the marine environment applicable to the coastal State and which have been established by the convention or by a competent international organization or diplomatic conference; (c) when it is alleged that a State when exercising the freedoms of the High Seas in the exclusive economic zone has acted in contravention of the provisions of the convention or of the laws and regulations established by the coastal State. In addition, in all cases involving the exercise by a coastal State of its sovereign rights or jurisdiction, the complainant State must establish a prima facie case that the claim is well founded before the Tribunal calls upon the other party to respond.

33. Decisions of the Seabed Disputes Chamber are "enforceable in the territories of States Parties in the same manner as judgments or orders of the highest court of the State Party where enforcement is sought."

awards of the arbitral tribunals[34] proposed in the Text are also final and without appeal, unless the parties to the dispute have agreed, to an appellate procedure; on the other hand, the recommendations or conclusions of the proposed Conciliation Commission are not binding upon the parties to a dispute.[35]

There can be no doubt that preparation of the Informal Composite Negotiating Text is in itself a major achievement. For the first time in history a heterogeneous international community has attempted to establish a comprehensive conventional framework for man's activities in the marine environment, instead of leaving the development of law to the practice of States and to the traditional process of claim and counterclaim supplemented by treaties dealing with a limited subject matter.

The Text reflects the contemporary need for management and regulation in wide areas of the marine environment, which is the inevitable consequence of technological advance and of our intensified and diversifying uses of ocean space. The proposed creation of an international organization with the specific task of administering the mineral resources of more than half the seabed of the world ocean "for the benefit of mankind as a whole" is a precedent of far-reaching importance.[36] The dispute-settlement system represents a re-

34. The general and special arbitration procedures proposed in the Text are of interest. In the general arbitration procedure, a list of arbitrators is maintained by the Secretary General of the United Nations; each State Party is entitled to nominate four arbitrators for inclusion in the list. The arbitral tribunal consists of five members. Each party to a dispute appoints one member, preferably selected from the list of arbitrators, and the remaining three members, preferably selected from the list of arbitrators, are appointed by agreement of the parties or, in case of disagreement, by the President of the Law of the Sea Tribunal. In the special arbitration procedure, separate lists of experts are maintained in the field of fisheries, protection of the marine environment, marine scientific research, and navigation, including vessel-source pollution. The lists are prepared and maintained by FAO (fisheries), by UNEP (protection of the marine environment), by IOC (scientific research), and IMCO (navigation). Each State Party is entitled to nominate two experts to each list. The special arbitral tribunal consists of five members, to which each party to the dispute may appoint two members, preferably chosen from the appropriate list; the remaining member, preferably chosen from the appropriate list, is appointed by agreement between the parties or, in case of disagreement, by the Secretary General of the United Nations. An interesting provision of the Text provides that the parties to a dispute may at any time agree to request a special arbitral tribunal "to carry out an inquiry and establish the facts giving rise to a dispute on the interpretation or application" of the convention. The facts so established are conclusive as between the parties.

35. Conciliation procedure: A list of conciliators, to which each State Party may appoint four persons, is maintained by the Secretary General of the United Nations. The Conciliation Commission consists of five members; each party to the dispute appoints two conciliators, and the four conciliators appoint the fifth member by agreement. In case of disagreement, the fifth member is appointed by the Secretary General of the United Nations.

36. It could, for instance, in future provide a model for the exploitation of the resources of the moon.

markable advance in international law. General principles are formulated with respect to control of marine pollution, which may have long-term beneficial effects on cooperation between States in this field. Finally, there are numberless provisions, some technical and others substantive, which are a constructive development of present law of the sea.[37]

Inevitably a document such as the Text, which must often deal with highly controversial political matters involving important interests of States, cannot be expected to be without shortcomings, and these are certainly not lacking. We can ignore the numerous exhortations and the great number of provisions prescribing vague duties for States; these may be considered superfluous or undesirable in a treaty, but they do not affect the substance of the agreement proposed in the Text. The latter may be evaluated from different sectoral and national points of view, but if stability of expectations in our varied activities in the marine environment, if the promotion of efficient development of ocean space which will benefit all States, and if a law of the sea which does not hamper progress toward a New International Economic Order are among the goals of the future convention, the response to three questions appears particularly important. First, does the Text attempt to further solutions to global problems, such as arms control, food, mineral resources, and energy? Second, are viable legal regimes proposed for marine areas beyond national jurisdiction? Third, does the Text as a whole promote the avoidance of conflict in ocean space?

The Text avoids as far as possible any mention of uses of ocean space which might be relevant to arms control. The phrase "exclusively for peaceful purposes" recurs with a certain frequency,[38] but the meaning of this phrase, if it has a specific meaning, is nowhere explained. Warships retain their immunities, and indirect provision is made for the submerged passage of submarines through international straits.[39] Apart from these few references, the Text is remarkably silent on the legal questions connected with military uses of ocean space.[40] It must, therefore, be concluded that the Text is unlikely either to promote or to hinder arms control in the marine environment.

The situation, however, is quite different with regard to economic uses of

---

37. The more detailed obligations of flag States to ensure safety at sea should be particularly noted together with the effort to encourage States to exercise effective control over ships flying their flag (Article 94 [6]).

38. For instance, "marine scientific research must be conducted exclusively for peaceful purposes," the international seabed area must be used "exclusively for peaceful purposes," etc.

39. The reference here is to the new legal concept of transit passage.

40. Many such questions are unresolved. The legality of the following, among many other, uses of the marine environment, is controversial: (a) exclusion of foreign shipping from vast areas of the High Seas, because these have been unilaterally reserved for missile-testing purposes; (b) use by a State of the legal continental shelf of another

the marine environment. Here the Text reflects the highly acquisitive aspirations of many coastal States. At least one-third of ocean space,[41] by far the most valuable part in terms of economic uses and accessible resources, is placed under national jurisdiction. This means that all exploitable offshore hydrocarbons; all commercially exploitable minerals in unconsolidated sediments, from sand and gravel to tin; most phosphorite nodule deposits and a significant proportion of exploitable manganese nodule deposits of the deep seabed; over 90 percent of commercially exploited living resources of the sea; and nearly all marine plants and nearly all potentially exploitable sources of unconventional energy are recognized as the exclusive property of coastal States if the proposals contained in the Text are retained in a future treaty. The value of these resources must be estimated at many trillions of dollars. It has been pointed out that the magnitude of this appropriation, in terms both of area and of resources, is unprecedented in history. This appropriation not only seriously prejudices the interests of landlocked and geographically disadvantaged States which lose the right of access to resources vital for their future, but it is also inequitable as between coastal States themselves. Ten such States obtain more than half the area which the Text proposes to place under national control, and six of these are considered to be wealthy.[42]

Nor is this all; since the Text does not clearly define the limits of national jurisdiction in ocean space, coastal States fronting on the open ocean can legally continue to extend their control over the marine environment as their marine capability increases and their national interests dictate.

Within the third or more of ocean space in which they will exercise control immediately, coastal States have authority ranging from full sovereignty to sovereign rights over natural resources. The Text in all its parts exhibits overwhelming concern to safeguard in every possible manner the rights of the coastal State and considerable reluctance to mention any specific coastal State duties (other than duties directly related to overflight[43] and navigation) in the

---

State without the latter's consent for the emplacement of antisubmarine-warfare devices; (c) use by a State of the exclusive economic zone of another State, without the latter's consent, for naval exercises, etc.

41. It is recalled that, in addition, the Text in some cases permits the legal continental shelf to extend beyond the exclusive economic zone, and thus more than one-third of the seabed will pass under national control.

42. The proposals in the Text favor a group of wealthy coastal States not only for the reasons stated, but also because adequate scientific capability, appropriate technology, and considerable financial resources are required effectively to develop offshore resources, particularly mineral resources. Poor countries are unlikely to be able independently to develop the mineral resources in their exclusive economic zone.

43. It is noted that the duties of coastal States in matters not directly connected with navigation are usually stated in vague general terms which are often qualified by other provisions in the Text.

vast marine areas under national control. Indeed, the concern to safeguard the interests of the coastal State sometimes extends to the marine area, which remains beyond national jurisdiction.[44]

Particularly unfortunate is the lack of standards or criteria for the conservation, utilization, and management of living resources in rapidly expanding internal waters,[45] in the territorial sea, on the legal continental shelf, or in archipelagic waters and the fact that failure of the coastal State to discharge the obligations in the Text relating to the conservation and management of living resources in the exclusive economic zone cannot be called into question before an international tribunal. Thus coastal States remain free to utilize living resources found in areas under their national jurisdiction in a manner which may cause serious prejudice to the interests of neighboring States.

In proposing a New International Economic Order, the developing countries sought a solution of global economic problems related to resources through cooperation between States, through participation of all States in international decision making, and through implementation of the concept of equity at the international level. These concepts, it was believed, should replace the economic domination by a few States over the majority of the international community and lead in due course to a world order based on cooperation rather than competition between States.

It is clear that the provisions of the Text do not promote the practical realization of the objectives of a New International Economic Order. The vast expansion of national control in the marine environment and the appropriation by coastal States of the preponderance of marine resources favor only the interests of oceanic island and archipelagic States or of States with long coastlines fronting on the open oceans. The emphasis throughout the Text on the sovereign rights of the coastal State exercised in a discretionary manner in all matters relating to the economic uses of ocean space certainly does not promote cooperation between States. Furthermore, fragmentation of ocean space between more than 100 different sovereignties is likely significantly to obstruct vital transnational uses of the marine environment and make difficult rational management of fisheries and effective control of marine pollution, particularly in areas where many States have relatively short coastlines. Thus it may be concluded that adoption by the conference of the provisions of the Text concerning the limits of national jurisdiction in ocean space and the rights of coastal States within these limits will hamper management of living resources

---

44. For instance, the coastal State is recognized primary interest in, and responsibility for, anadromous and catadromous stocks of fish, even in marine areas outside its national jurisdiction.

45. It is recalled that internal waters are waters on the landward side of baselines drawn by the Coastal State. In recent years, there has been a tendency on the part of some coastal States to draw straight baselines enclosing large areas of the sea.

and the conduct of some other important marine activities.[46] It will also have a negative impact on the prospects for achieving a New International Economic Order[47] and on the possibility of reaching equitable solutions to global resource-related problems.

There is no necessary contradiction between international recognition of an expanded coastal State jurisdiction in the marine environment (and in particular of the concept of a 200-nautical-mile exclusive economic zone) and the objectives of a New International Economic Order. Extension of coastal State jurisdiction is unavoidable for security and economic reasons, and the concept of an economic zone, adjacent to the territorial sea, under coastal State control conveniently consolidates into a single regime a variety of coastal State claims advanced over the past 30 years. But extension of coastal State authority must, at least, be balanced by (1) establishing clear and definitive limits to national jurisdiction in ocean space. This implies, inter alia, setting strict criteria for the drawing of straight baselines, abolishing the concept of the legal continental shelf extending beyond the exclusive economic zone, and making reasonable rules with regard to the extent of marine jurisdiction which may be claimed by islands. It must also be balanced by (2) the development of an expanded concept of State responsibility. Coastal State authority in marine areas within national jurisdiction can no longer be "sovereign" in the traditional sense: it must be limited by clear obligations, enforceable through binding international judicial procedures, in matters relating to utilization and management of living resources, scientific research, pollution control, and others. (3) There must be established beyond national jurisdiction an international regime for ocean space through which compensation can be offered to those States that have been adversely affected by the expansion of coastal State jurisdiction.

Unfortunately, such balancing provisions are lacking in the Text, and little thought appears to have been given to the viability of the legal regimes proposed for marine areas beyond national jursidiction. As has already been noted, these areas are now explicitly divided into two parts: a slightly modified traditional regime of the High Seas is maintained for the surface of the sea and

---

46. Marine scientific research, which is the prerequisite of rational resource management, will, for instance, be made more difficult and expensive by subjecting its conduct to coastal State consent in marine areas under national jurisdiction. The impact of the provisions in the Text on the conduct of commercial navigation should be noted. Despite attempts to ensure reasonable freedom of navigation, it is probable that international seaborne trade will be hampered, and it is certain that it will become more expensive.

47. Adoption of the proposals in the Text will also have far-reaching political consequences; inter alia, the economic control over resources by a few States in a favorable geographical position will be confirmed, the solidarity of the so-called Group of 77 may be shaken, and world tensions could be aggravated.

water column, while the seabed and its resources are declared to be subject to a special regime as a common heritage of mankind.

There is serious question whether even an amended regime of the High Seas, such as that proposed in the Text, is likely to be viable in contemporary circumstances, except in increasingly remote areas of the oceans, in view of the need to apply agreed criteria for the utilization of ever larger areas of the marine environment. In particular, it seems doubtful that freedom of fishing can be long maintained without serious impairment of the living resources of the High Seas.[48] Certainly, the jurisdictional vacuum existing with respect to the waters beyond national jurisdiction permits further expansion of coastal State jurisdiction as technology advances and as the interests of coastal States in the marine environment expand. These reasons alone would justify serious consideration being given to establishing for the waters of the ocean beyond national jurisdiction a regime based on the principle of common heritage of mankind.[49] It is unfortunate that such consideration has not been possible at the conference owing to the opposition of some influential States.

It has already been observed that the common-heritage regime proposed for the seabed beyond national jurisdiction is highly innovative in concept and could establish a new legal order which could serve as a model for the management of other areas beyond national jurisdiction. The Text contains many useful proposals in this connection, but the new regime proposed is not established on solid foundations, largely owing to lack of consensus on the purpose of the proposed international regime for the seabed. For some, the basic purpose of the new regime is the creation of a new equitable legal order for the Area and its resources in the management of which all countries would participate and from the management of which all countries would benefit; for others, the basic objective is to create a system of international administration of the mineral resources of the Area which can be used to restrict production; still others desire a system of administration, ostensibly international, through which developing countries would obtain some financial benefits, but which would not significantly restrict the free enterprise of States or corporations. A

---

48. The Text proposes a general obligation to the effect that States should cooperate with each other in the management and conservation of the living resources of the High Seas and that, to this end, they should, as appropriate, establish regional or subregional fisheries organizations. The record of existing fishery organizations, however, has been uneven, and displacement of long-distance fishery vessels from the exclusive economic zones of coastal States is likely to increase pressure on High Seas fisheries beyond optimum sustainable yields. The imbalance between the capacity to harvest and the yield of stocks will make effective cooperation between the States concerned very difficult.

49. A common-heritage regime would also offer a number of positive benefits. It could, for instance, provide a framework for the effective management of living resources beyond national jurisdiction and for some compensation to landlocked countries and to States with long-distance fishing fleets, whose interests have been damaged by the establishment of the exclusive economic zone.

combination of the latter two approaches has prevailed at the conference and is reflected in the Text. Accordingly, the competence of the Authority is limited to dealing with the exploration and exploitation of mineral resources in situ,[50] particularly manganese nodules. An Authority established on such a narrow basis cannot, however, implement effectively the principle of common heritage, nor can it impede further extensions of the legal continental shelf at the expense of the international seabed area, which would diminish the economic significance of the international area and hence undermine the stated purposes of the Authority.

It is impossible to analyze briefly the large number of questions which arise in connection with the manner in which the international regime for seabed minerals is implemented in the Text. Comment on three important questions, however, is necessary. These questions are *(a)* the limits of the international seabed area, *(b)* access to the mineral resources of the Area, and *(c)* methods of exploitation and production controls.

There has been little discussion at the conference about the limits of the Area. The Text merely states that State Parties to the convention must notify the Authority of the limits of their national jurisdiction over the seabed defined by coordinates of longitude and latitude and that the Authority must register and publish such notifications. The Authority is obligated to accept without question the decisions of States; it is not even permitted to remind States of their obligations if the required notifications are delayed. States, furthermore, remain free to redefine as often as they wish the limits of their legal continental shelves within the limits of the flexible criteria contained in Part VI of the Text. Nor, finally, does the Text contain any provisions for the delimitation of the international seabed area with respect to areas having an uncertain legal status, such as the Antarctic, where notifications are likely to be delayed.

In view of the lack of any clear provision on limits, it is obvious that the international seabed area may be subject to unforeseeable change which may diminish its already reduced significance. It is surprising that the inadequacy of the present Text on the matter of limits has not aroused comment at the conference. One would have expected that the Authority, at least, would have been authorized to remind coastal States of their obligations and to establish provisional boundaries to the Area in the event that one or more coastal States omitted to notify the limits of their legal continental shelf within a reasonable period of time after the entry into force of the convention.

The question of access to the mineral resources of the Area and related methods of exploitation has been the subject of considerable controversy at the

---

50. Thus the Authority may take measures to ensure protection of human life or the preservation of the marine environment only when these matters are directly related to exploration and exploitation of seabed mineral resources. Provision is made only for the transfer of technology directly related to mineral resource exploration and exploitation, etc. Living resources in the Area are not subject to management by the Authority.

conference. Developed countries stress the requirement of substantial freedom of access to the international seabed area for mining entities and limited, precisely defined, control by the future Authority over these entities. The majority of developed countries also does not favor production controls on the mining of mineral resources in the Area.

Developing countries, on the other hand, favor exploitation of seabed minerals directly by the Authority through the Enterprise. At the very least, in their opinion, there should be strict and direct control by the Authority over such operations as are not conducted directly by it, and they do not favor limiting too precisely the discretionary powers of the Authority. Also, developing countries want to establish special schemes to compensate those developing countries which are adversely affected by mineral production in the international seabed area. This is schematically the position at the conference.

The sharp divergence of views has resulted in the suggestion, accepted in the Text, to create a so-called parallel system of exploitation. Exploitation of mineral resources is to be undertaken directly by the Authority through the Enterprise; provision, however, is also made for exploitation by entities other than the Enterprise with the consent of the Authority and under the latter's strict supervision and subject to a number of obligations.[51] Complicated production controls would be applicable to manganese nodules extracted from the Area.

The system proposed in the Text is simply not viable. In this connection, it is recalled, first, that ample mineable manganese-nodule deposits are found within national jurisdiction because of the provisions on the limits of national jurisdiction contained in other Parts of the Text. Consequently, manganese-nodule production in the Area will have to compete with production under national jurisdiction. The major effect of production controls applicable only to manganese nodules mined in the international seabed area would be merely to undermine the viability of the Authority. If production controls are desired, they should be applied to all manganese-nodule production, whether within or outside national jurisdiction. Second, the world mineral market is highly volatile. This means that nodule exploitation in the international area must not be subject to the heavy bureaucratic controls provided for in the Text and must be

---

51. Some of the major obligations are that (a) the entity comply with the rules and regulations adopted by the Authority and the decisions of its organs, (b) undertake to negotiate an agreement making available to the Enterprise under license the technology to be used in the exploitation of the minerals of the Area on fair and reasonable terms, and (c) accept control by the Authority of all stages of operations. In addition, "the proposed contract area shall be sufficiently large and of sufficient value to allow the Authority to determine that one-half shall be reserved solely for the conduct of activities by the Authority through the Enterprise or in association with developing countries. Upon such determination by the Authority, the contractor shall indicate the coordinates dividing the area into two halves of equal estimated commercial value, and the Authority shall designate the half which is to be reserved."

managed by persons with considerable business experience: the diplomats and civil servants who will govern the Enterprise are unlikely to possess the required experience. Third, manganese-nodule harvesting and processing is a promising but also a costly and still marginal technology. This means that at least first-generation production should be encouraged, and not discouraged, by the Authority, as the Text proposes, since if there is no production in the Area the Authority will have no purpose and will only constitute a financial burden on the international community.

More consonant with the concept of common heritage and with the realities of manganese-nodule mining would be a system which *(a)* guaranteed access to the mineral resources of the seabed to all qualified applicants under general conditions designed to avoid control by one State or group of States of an excessive number of mineable manganese-nodule sites in the international seabed area, and *(b)* established a system of joint ventures on an equal basis between the Authority and entities, whether public or private, wishing to exploit seabed mineral deposits in the international area. Half the directors in each joint venture would be appointed by the Council of the Authority, with due regard to equitable geographical distribution. The Authority would contribute to each joint venture the manganese nodules in situ, the guarantee to exclusive rights to the deposit, and a relatively modest amount of cash, while the entity, whether public or private, in partnership with the Authority would contribute capital and technology. As an equal partner, the Authority would have access to technology and could insist on the introduction of training programs for nationals of developing countries. Profits and losses of each joint venture would be equally divided. Such a system would imply abolition of the bureaucratic Enterprise contemplated in the Text, elimination of production controls, and a considerable streamlining of the present heavy structure of the Authority.

A third complex of matters on which the conference would appear to have reached unfortunate conclusions concerns the problem of avoidance of conflicts and disputes in ocean space. Conflicts and disputes between States originating from ocean-related activities are still relatively uncommon. Nevertheless, they will tend to increase as activities of States in the marine environment diversify and expand. Accordingly, it is important to examine whether the Text as a whole is likely to promote avoidance of disputes and, when disputes arise, whether adequate provision is made for their settlement.

Provisions in the Text settle some questions, such as that of the limits of the territorial sea, which have caused dispute for many years. The Text also provides for a comprehensive dispute-settlement system, which previously did not exist, as an integral part of the future convention. Nevertheless, a convention based on the present Text could be more conducive to multiplying than to diminishing disputes between States, both because the Text raises more questions than it settles and because of the vagueness or ambiguity of several crucial provisions designed either to reconcile opposing points of view or to suggest obligations the precise content of which it will be extremely difficult to determine.[52] At the same time, compulsory and binding dispute-settlement procedures with regard to disputes arising in the one-third of ocean space now subject to coastal State jurisdiction are in effect limited to rights specifically reserved to third States (essentially navigational uses of the sea); this excludes from binding third-party dispute settlement several categories of disputes which are likely to be frequent.[53] Another important category of potential disputes specifically excluded from the dispute-settlement system are those concerning the exercise of discretionary powers by organs of the Authority.

Even more important, perhaps, the Text fails to provide a mechanism for review of the Convention as a whole.[54] This is a serious deficiency which may give rise not merely to dispute but also to serious conflict. Many provisions in the Text are already obsolete;[55] others are vague or ambiguous. Technology is advancing rapidly; uses of the sea are multiplying. Whatever conventional law may be elaborated now can, in a sense, be only experimental. The scope and functions of the International Seabed Authority will certainly need to be

---

52. Article 74 on the delimitation of the exclusive economic zone between adjacent or opposite States is an example of a crucial article which is excessively vague. The fact that different principles are employed to measure the extent of the exclusive economic zone and of the legal continental shelf (the breadth of the former is measured on the basis of distance from baselines, that of the latter on the basis of natural prolongation of the coastal State's land territory) could cause serious difficulties since the legal continental shelf of one State could extend under the exclusive economic zone of another. Numerous articles throughout the Text, but particularly in Parts XII–XIV, purport to create legal obligations the precise content of which it is virtually impossible to ascertain.

53. Lawrence Hargrove, in *Law of the Sea: Conference Outcomes and Problems of Implementation* (Proceedings of the Law of the Sea Institute, Tenth Annual Conference, 1976) (Cambridge, Mass.: Ballinger, 1977), has stated that the general approach of the conference has been to regard the coastal ocean as a possession of the coastal State with certain specific easements in favor of other States. There is no enforceable obligation on the part of the coastal State to manage the maritime area under its control in a manner that is not prejudicial to the interests of neighboring States or to those of the international community as a whole. This lack of enforceable obligations could give rise to serious disputes not covered by the proposed dispute-settlement system when, for instance, a coastal State depletes fish stocks shared with neighboring States or fails to preserve spawning areas within its jurisdiction.

54. Provision, however, is made for convening a conference to review Part XI of the Text (international seabed area) 20 years after the entry into force of the proposed convention. This is quite inadequate since nobody can know either the specific, practical requirements of manganese-nodule mining on a commercial basis or whether the elaborate international organization proposed in the Text is capable of functioning in a reasonable manner.

55. For instance, safety zones around artificial islands, installations, and structures in the marine environment are limited to a radius of 500 meters. This provision has been taken over from the 1958 Convention on the Continental Shelf without considering that modern supertankers cannot change course within such a short distance.

reviewed at an early date. Controversies will multiply if the law, as embodied in the Convention, is not subject to discussion and review in an organized forum to adapt it to the evolving realities in ocean space.

The need for some continuing mechanism has been noted by some representatives at the conference, but views on its nature and functions vary. The Text would be considerably improved could the proposed periodic (triennial) conference of States Parties to elect the judges of the future Law of the Sea Tribunal be also empowered to discuss problems related to the interpretation and application of the convention which may require clarification in a political forum and to adopt, in defined circumstances, amendments for ratification by governments.[56] In conclusion, the modern problems of ocean space cannot be resolved on the basis of the present text—fragmented but virtually unfettered sovereignty in marine areas under national jurisdiction (subject to some assurances of unhampered navigation) coupled with an unrealistic High Seas regime and a seabed mineral resource regime which is not viable.

The Informal Composite Negotiating Text is a political document, negotiated with political considerations in mind. This reflects the realities of the contemporary world and is also the reason for its basic deficiencies.

The law of the sea negotiations cannot be successfully concluded merely through political accommodation of the perceived national interests of significant States and through the search, where these interests are in opposition, for a vague formula which will postpone conflict until after ratification of the convention. Political accommodation is certainly essential; nothing can be achieved without accommodation of national interests. But present negotiations on the law of the sea cannot be compared to negotiations between States on ordinary matters of international intercourse. The U.N. Conference on the Law of the Sea is engaged in the creation of a viable legal framework for man's activities in the marine environment. Accommodation of national interests avails little if the result is not viable. A convention negotiated on this basis alone will bedevil international relations for years to come. In addition to the national interest, a viable convention requires that appropriate consideration be given to the realities of a marine environment exploited with increasing intensity. This means taking into due account the need for effective management of ocean resources, for harmonization of ocean uses, for effective preservation of the marine environment,[57] for an expanded responsibility of States for their

---

56. In this connection it is interesting to note that some countries at the Conference—for instance, Portugal—have suggested that this mechanism could be used to create an institutional framework for ocean uses other than seabed mining beyond national jurisdiction.

57. The Text takes a narrow view of the preservation of the marine environment by equating this term with prevention and control of marine pollution.

acts, and a host of other considerations.[58] Equally important is a sense of international equity. The Text proposes that the greater part of the world's unexploited, or comparatively lightly exploited, resources should pass under the exclusive control of a few States without significant compensation to the remainder of the international community. This is excessive. There can never be perfect justice, but there must be elementary equity in a future convention if it is to survive. If equity is too blatantly ignored, the future treaty will be a permanent source of international friction. A global law of the sea is required so as to provide security of expectations for increasingly varied activities in the marine environment. A treaty which is not viable will not provide such security.

The effort to implement the principles of equity and of common heritage of mankind in the seas was the major impulse in the decision of the United Nations General Assembly to convene the present Law of the Sea Conference. This effort gave a focus to the early stages of the negotiations. If the effort is abandoned and if the principles are forgotten, the conference cannot achieve enduring results, even if agreement on a treaty is reached. The present challenge requires looking beyond the immediate and the obvious so as to ensure the necessary protection of national interests within a framework which is recognized as fair; which takes into account technological, economic, and environmental factors; and which flexibly reconciles a global order with regional needs. It is not yet too late for the conference to reach a result which will mark the beginning of an era of peaceful development of ocean space.

58. In particular, a means must be found to preserve vital transnational uses of the sea such as marine scientific research and commercial navigation. It is simplistic to believe that the only alternative to coastal State control is total freedom. These uses of the sea can be internationally regulated in a manner that will not impair their constructive purposes.

APPENDIX

TABLE A1.—BREADTH OF TERRITORIAL SEAS AND FISHING JURISDICTIONS IN EXCESS OF 12 NAUTICAL MILES
(As of December 1, 1977)

Country	Territorial Sea (Nautical Miles)	Fishing Limits (Nautical Miles)
Albania	15	15
Angola	20	200
Argentina	200*	200
Bahamas	3	200
Bangladesh	12	200
Benin	200	200
Brazil	200	200
Burma	12	200
Cameroon	50	50
Canada	12	200
Cape Verde	100	100
Chile	3	200
Comoro Islands	12	200
Congo	30	30
Costa Rica	12	200
Cuba	12	200
Denmark	3	200†
Dominican Republic	6	200
Ecuador	200	200
El Salvador	200	200
France	12	200‡
Gabon	100	150
Gambia	50	50
Germany, Federal Republic	3	200
Ghana	200	200
Guatemala	12	200
Guinea	200	200
Guinea-Bissau	150	150
Guyana	12	200
Haiti	12	200
Iceland	4	200
India	12	200
Iran	12	50
Ireland	3	200§
Japan	12	200
Korea, North	12	200
Liberia	200	200
Madagascar	50	150
Maldives	A	A
Malta	6	20
Mauritania	30	36
Mauritius	12	200
Mexico	12	200
Morocco	12	70

TABLE A1. Continued

Country	Territorial Sea (Nautical Miles)	Fishing Limits (Nautical Miles)
Mozambique	12	200
Nicaragua	3	200
Nigeria	30	30
Norway	4	200
Oman	12	200
Pakistan	12	200
Panama	200	200
Peru	200	200
Philippines	A	A
Portugal	12	200
Senegal	150	200
Seychelles	3	200
Sierra Leone	200	200
Somalia	200	200
South Africa	6	200
Sri Lanka	12	200
Tanzania	50	50
Togo	30	200
Tonga	A	A
U.S.S.R.	12	200
United Kingdom	3	200¶
United States	3	200#
Uruguay	200*	200
Vietnam	12	200

Source.—U.S. Department of State, Office of the Geographer.
Note.—A = modified archipelago claims.
*Overflight and navigation permitted beyond 12 nautical miles.
†Includes Greenland and Faroe Islands.
‡Includes St. Pierre and Miquelon and French Guiana.
§Ireland, in addition, has claimed a 50-nautical-mile exclusive fishing zone.
¶Includes British Virgin Islands and Bermuda.
#Includes Puerto Rico, U.S. Virgin Islands, American Samoa, Guam, Johnston Atoll, Wake Island, Jarvis Island, Howland, and Baker Island.

TABLE A2.—SUMMARY OF THE BREADTHS OF TERRITORIAL SEAS
(As of December 1, 1977)

Breadth (Nautical Miles)	States (N)	Coastal States* (%)
3	25	19.4
4	4	3.1
6	8	6.2
10	1	.8
12	60	46.5
15	1	.8
20	1	.8
30	4	3.1
50	4	3.1
100	2	1.6
150	2	1.6
200	13	10.1
Archipelago	3	2.3
No legislation	1	.8
Total	129	100.2

SOURCE.—U.S. Department of State, Office of the Geographer.
*At this time there are 29 independent landlocked states.

TABLE A3.—SUMMARY OF FISHING-JURISDICTION CLAIMS
(As of December 1, 1977)

Breadth (Nautical Miles)	States (N)	Coastal States (%)
3	8	6.2
6	3	2.3
12	50	38.8
15	1	.8
20	1	.8
30	2	1.6
36	1	.8
50	4	3.1
70	1	.8
100	1	.8
150	3	2.3
200	51	39.5
Archipelago	3	2.3
Total	129	100.1

SOURCE.—U.S. Department of State, Office of the Geographer.

Living Resources

Marine Fisheries[1]

Sidney Holt
Food and Agriculture Organization

It is difficult, in mid-1977, to be optimistic about the state of world sea fisheries. As this review is written, a team of Peruvian scientists assisted by foreign experts from the FAO are investigating the second collapse in 5 years of the fishery for anchoveta *(Engraulis ringens)*, once by far the world's biggest fishery. On the other side of the world, the Commission of the European Economic Committee (EEC) has recommended the complete closure for the time being of the herring *(Clupea harengus)* fishery in the North Sea which, with the cod *(Gadus morhua)*, has been for centuries the main source of food from the Atlantic. Catches of anchoveta and herring have been regulated for some years—the first by the nation concerned, the second internationally. In both cases the causes for collapse, as far as can be judged, are a combination of a fluctuation in the ocean environment, which could not be predicted, and the failure of a natural population which had been much reduced and perhaps otherwise stressed, by intense fishing, to respond adequately to the environmental stresses.

Back in the Pacific area, the International Whaling Commission (IWC) has just held its annual meeting, this year in Canberra. Its decisions will continue the trend of recent years in reducing the production of meat and oil from whales. The smaller catch quotas reflect some declines in the numbers of whales left in the sea but mainly arise from new scientific assessments which reveal that earlier views of the states of whale populations were overoptimistic. In this case there is cause for hope, in that the IWC has adopted a "new management policy," under which restorative action is taken when any whale population is judged to have been reduced by whaling to less than about half its original size. But apart from the continuing difficulty in making that judgment, environmental factors again appear as complications, in that the productivity of one species is found to depend, via the food chain, on the abundance of other species of whales as well as of seals and other competitors. Furthermore, in this period of rather rigorous restraints on the whaling operations of its member states, the Commission is worried about escapes from those restraints through transference of vessels by sale or other arrangement to the flags of nonmember states.

---

1. The views expressed here are those of the author and do not necessarily reflect the policy or position of the FAO.

© 1978 by the University of Chicago. 0-226-06602-9/77/1978-1002$03.47

Meanwhile, the UN Conference on the Law of the Sea (UNCLOS) drags on, paying remarkably little attention to the nature of problems which threaten to undermine in the coming years the contribution of marine resources to world food supplies. Both hopes and fears are expressed concerning the possible effects of the unilateral declarations by states of 200-mile exclusive economic zones (EEZ) on the redistribution of economic benefits from living marine resources. There is still talk of latent resources in some areas such as the Indian Ocean, which should be accessible to developing countries; and there is some excitement about the possible huge potential of krill *(Euphausia superba)* in the Southern Ocean, which, however, is accessible at present only to a few of the industrialized nations. Against this optimism must be set persistent questions about our ability to ensure that benefits from such resources accrue to those now in most need of the products from them and that they remain available in good shape for future generations of humanity who may need them even more than we do. At this time there is an increasing realization that "in good shape" means not only not overfished[2] but also not adversely and perhaps permanently affected by pollution, seabed and reef dredging and mining, and the like. There is still very great uncertainty about the long-term and often indirect effects of these man-caused environmental disturbances on fish resources, but enough instances of adverse consequences are accumulating to reinforce uneasiness among ocean scientists.

It is not practicable at this time to make a quantitative appraisal of the world fisheries in 1976; the compilation even of basic catch statistics for that year will not be finished before this chapter of the first *Ocean Yearbook* is in press. In any case it is perhaps more useful to view 1976 as the last year in which the greater part of fisheries throughout the world were conducted under a legal regime characterized by the principle of freedom of fishing on the high seas, and to establish 1976 as a baseline against which to compare future developments under another regime, involving new international arrangements and perhaps new technologies. For this purpose data for 1975 will serve as the most recent statistical information, and I shall be concerned mainly with the evolution of trends in the 1 or 2 preceding decades.

Marine fisheries are economically important primarily as sources of high-quality protein food as well as edible oils, for human consumption either directly or indirectly, through livestock feeds. Locally, of course, fisheries may be valuable as sources of precious shells, sponges, pharmaceuticals, and, if we include hunting of marine mammals within the term "fisheries," pelts and ivory. In this overview I concentrate on the nutritional use of living resources.

---

2. In this article, taking into account the synergy of fishing pressure and deleterious environmental changes, whether natural or man caused, "overfishing" means conducting fishing operations in such a way as to prejudice present or possible future values of marine living resources.

PRODUCTION

The general increasing trend of marine production over the past 2 decades is shown in table 1. The figures are based on official statistics published annually by the FAO but include estimates of the weights of whale catches. Over a quarter of a century the rate of growth exceeded by a comfortable margin the rate of increase in the human population; the performance of fisheries development, in gross-product terms, was better than that of most other forms of food production (last row of table 1). If, however, we break the 25 years down into 5-year periods, it is evident that, while the growth rate was maintained at a high level for 20 years—with perhaps a slight tendency to diminish—a drastic change occurred at the beginning of the present decade: stagnation, even a slight decline in production. The first sign of this might be the decrease from 1968 to 1969. This is followed by recovery for 2 years to the highest levels ever attained and then another drop of nearly 10 percent from 1971 to 1972, from which production has not yet risen.

Until 1963 whales steadily contributed just over 2 million MT to the annual catch, although because of the increase in other species their percentage of contribution declined from 10 to 5 percent. Seventy to 90 percent of the whales caught were species of baleen whale, originally for their oil and then increasingly also for their meat; the rest were sperm whales which were and still are used as a source of industrial inedible oils and waxes, though recently their meat has been increasingly consumed in Japan. After 1963 the catches of baleen whale declined rapidly and continuously as a consequence of overfishing, and a small increase in sperm whale catches for a few years did not prevent an overall decline. This decline in catches has been reinforced for the past 3 years by the application of stringent quotas by the International Whaling Commission under its new management policy in a last-minute effort to hold populations at levels which can give an economically worthwhile sustainable yield.

If the consequences of the dismal history of whaling are put aside, we can examine the sequence of catches of true fishes and shellfish (mollusks and crustaceans). This is shown in the second column of the table. The figures given include small quantities of seaweeds (algae), contributing about 2 percent to the total catch, and assorted food items such as sea urchins, sea cucumbers, and other invertebrate animals. It will be seen that performance looks a little better without the whales, but year-to-year fluctuations are somewhat more marked, and the last 5 years show a tendency toward decline.

A phenomenon during the 25-year period which, because of its magnitude, must distort our interpretation of these statistics is the rise and fall of the anchoveta fishery of Peru and, to a much smaller extent, Chile. After small catches for many years, mainly for local consumption, an extremely rapid growth started in 1957. This growth continued practically unchecked until the peak year 1970, when the catch reached 13 million MT (over one-fifth of the world catch), after which a crash occurred which was catastrophic for the

TABLE 1.—WORLD MARINE CATCHES

A. ANNUAL CATCH IN MILLIONS MT

	Total	Total, excluding Whales	Fish and Shellfish, Excluding Anchoveta	Annual % Increase
1938	21.9	18.8	18.8	...
1948	19.8	17.7	17.7	...
1950	20.7	18.6	18.6	...
1955	27.7	25.5	25.4	4.1
1956	29.3	27.0	26.9	5.9
1957	29.7	27.5	26.7	−.7
1958	30.3	28.0	27.2	1.9
1959	33.8	31.5	29.5	8.5
1960	36.2	33.9	30.4	3.1
1961	39.4	37.0	31.7	4.3
1962	42.3	40.1	33.0	4.1
1963	43.1	41.2	34.0	3.0
1964	47.2	45.4	35.6	4.7
1965	47.3	45.6	37.9	6.5
1966	50.7	49.4	39.8	5.0
1967	53.9	52.7	42.2	6.0
1968	56.9	55.9	44.6	5.7
1969	54.4	54.4	44.7	.2
1970	61.9	60.9	47.8	6.9
1971	61.9	61.0	49.8	4.2
1972	56.9	56.2	51.4	3.2
1973	57.4	56.7	54.7	6.4
1974	60.9	60.4	56.4	3.1
1975	59.6	59.3	55.8	−1.1

B. AVERAGE ANNUAL % INCREASE

	Total	Total, excluding Whales	Annual % Increase
1950–55	5.7	6.2	6.2
1955–60	5.8	5.9	3.7
1960–65	5.5	6.1	4.5
1965–70	5.5	6.0	4.7
1970–75	−.8	−.5	3.1
1950–75	4.3	4.7	4.5

economy of Peru and changed the nature and pattern of world trade in animal-feed supplements.

No other single species of fish possesses or has possessed as much significance in the total picture as the anchoveta, and it is rewarding to examine the trend in food catches, *excluding* the anchoveta; this is done in the third column of table 1. We find that the trend of diminishing growth rate is weakened but still present. The 1970–75 average (3.1 percent) still exceeds the human population growth rate (2 percent) but from 1974 to 1975 we see for the first time a decrease in catch.

The reason for this most recent decline is not to be found in the collapse of a particular fishery. While some countries reported increases from 1974 to 1975, many more reported decreases. In fact the overall decrease might well have exceeded 1.1 percent; many of the countries in the "developing market economies" category and two in the "developed market economy" category had not reported their 1975 catches by the time of publication of the FAO *Yearbook of Fishery Statistics* for that year, so that estimated figures were used, usually assuming no change from the previous year. But it is in these economic categories that the decreases occurred, as table 2 shows. On the basis of these figures, the actual decline may better be estimated as 1.8 percent, excluding anchoveta, or 2.5 per cent including anchoveta.[3]

TABLE 2.—CHANGE IN WORLD CATCHES OF
FISH AND SHELLFISH, 1974 TO 1975

	Market Economies			Centrally Planned Economies	
	Developed	Developing		Developed	Developing
		All	Excluding Peru		
% of 1974 catch	37	32	(28)*	15	(16)*
% change, 1974 to 1975:					
All countries, including estimates	−2.5	−4.2	−1.3	+7.7	0
Reporting countries only	−2.7	−5.9	−2.6	+7.7	...†

*Estimate.
†Data unavailable.

3. According to an FAO Secretariat paper, COFI/77/6 to the 1977 Session of the FAO Committee on Fisheries, the preliminary 1976 catch figures (at mid-February 1977) indicate an increase of 3.6 percent over the 1975 catch, including the catch from inland waters. The latter has been increasing at a variable rate, average 2.3 percent annually during 1970–75; it increased by 3.6 percent from 1974 to 1975. Inland catches now compose 15 percent of the total. These figures therefore suggest that the marine increase may have been 3.6 percent or a little more. Much of the increase was in 800,000 MT by Peru, mainly of anchoveta. If this is excluded, the global increase was 2.6 percent, which is among the lowest recorded in the 25-year period I consider here. There is thus no cause for optimism, even though a new catch record appears to have been established. A change in the pattern of expansion may be starting; whereas the total catch of the group of developed-market economies increased by 3.4 percent and that of the developing-market economies (excluding Peru) by 3.9 percent, catches of the developed centrally planned economies (the USSR and eastern Europe) apparently increased by only 0.9 percent.

In order to judge whether the changes over the past 5 years, and particularly from 1974 to 1975, reflect an approach toward an upper limit of practicable catch determined by a finite resource base or merely changes in the number, power, and deployment of fishing vessels, or both, it would be necessary to have measures of the effort expended in fishing each year. Unfortunately, such measures do not exist.

The recent marine-fish catches are composed of species which have given fairly steady yields from year to year; species whose yield has declined, nearly always because the resource has become less productive, usually through overfishing; and species whose catches are rapidly increasing as a result of the opening of a market for them, the introduction of new or modified fishing gears which take them more efficiently, or the application of new processing methods—or commonly a combination of these factors. Examples of these, species which contribute or have contributed substantially to total catches, are given in table 3. The species tabulated together accounted for 28 percent of the 1975 fish catch, excluding anchoveta. The same species accounted for 27 percent of the 1965 catch, but in very different proportions. Stagnancy of total fish catches seems to result from the fact that the opening up of new fisheries now does barely more than compensate for the decline of some older ones.

It will be seen that, apart from the three species of tunas, all of which have a worldwide distribution in tropical and subtropical waters, the examples given are from the Northern Hemisphere, and all but two are from the North Atlantic. The temperate waters generally have large stocks of relatively few species, while the variety of species in warmer waters is much greater. But apart from this, the production from northern waters continues to dominate the fishery situation, as table 4 shows. The geographic breakdown used is that adopted by the FAO for statistical purposes; in certain areas—notably in the North Atlantic and the Mediterranean—the breakdown corresponds to the areas of authority of regional international fisheries commissions and councils (see fig. 1).

Some changes in the distribution of fish catches over the 25-year period are noteworthy, although adjustments of some important statistical boundaries make comparisons difficult. While the relative contributions from the Indian Ocean has hardly changed, there has been an inversion of the relative importance of the Atlantic and Pacific Oceans. This latter change, which is still apparent even when anchoveta is excluded, reflects changes mainly in the Northern Hemisphere; increased catches in the Southeast Atlantic are not sufficient to make up for relative declines in the North Atlantic, particularly in the northwestern area. The contribution from the South Pacific as a whole (excluding anchoveta) has not changed, reductions on the western side being balanced by increases on the eastern side.

The statistical areas differ greatly in size, of course. In the last column of table 4 are shown, for the sake of comparison, the 1975 catches from each area in tons per 1,000 km^2 of sea surface. These values should not, however, be taken as indicating either the natural productivity of the resources or the

TABLE 3.—CATCHES OF SELECTED FISH SPECIES (Millions MT)

FAO Spp. Group	1938	1950	1955	1960	1964	1965	1969	1970	1971	1972	1973	1974	1975
32 Cod (*Gadus morhua*), North Atlantic	2.08	2.32	2.88	2.74	...	2.77	3.66	3.14	2.85	2.74	2.54	2.81	2.42
32 Polar cod (*Boreogadus saida*), Northeast Atlantic	...	...	...	...	...	...	.13	.24	.35	.17	.08	.13	.06
32 Alaska pollack (*Thelagra chalcogramma*), North Pacific	.19	.15	.27	.48	...	1.05	2.55	3.06	3.59	3.21	4.62	4.90	5.03
32 Haddock (*Melanogrammus aeglefinus*), Atlantic:													
Northwest	...	...	...	...	.14	.20	.07	.05	.05	.03	.03	.02	.03
Northeast	...	...	...	...	.50	.50	.89	.87	.46	.52	.60	.56	.50
23 Capelin (*Mallotus villosus*), North Atlantic	.02	.03	.08	.14	...	.28	.85	1.51	1.58	1.95	2.05	1.91	2.24
33 Sand eels (*Ammodytes spp.*), Northeast Atlantic	...	...	...	.11	...	.15	.11	.19	.40	.36	.31	.53	.44
35 Herring (*Clupea harengus*), North Atlantic	1.75	2.08	2.88	2.64	...	3.87	3.33	2.32	2.14	1.92	1.98	1.57	1.52
35 Sardine (*Sardinops caerulea*), East Central Pacific	.51	.25	.05	.04	...	.02	.03	.04	.05	.05	.06	.08	.12
35 Sprat (*Sprattus sprattus*), Northeast Atlantic	.16	.16	.27	.27	...	.31	.20	.23	.30	.33	.51	.64	.98
37 Mackerel (*Scomber scombrus*), North Atlantic	...	...	...	...	.18	.36	...	.66	.74	.78	1.02	.98	1.09
36 Yellowfin tuna (*Thunnus albacares*), all oceans	.05	.10	.14	.26	...	.25	.34	.33	.31	.39	.38	.39	.39
36 Skipjack (*Katsuwonus pelamis*), all oceans	.14	.14	.18	.20	...	.25	.32	.39	.43	.42	.54	.63	.50
23 Salmons (*Onchorhynchus spp.*), North Pacific*	.77	.42	.48	.40	...	.42	.38	.42	.44	.34	.40	.34	.40

NOTE.—Values for 1950, 1955, 1960, and 1965 are 3-year averages; e.g., 1950 = average for 1949, 1950, and 1951.
*Including river catches.

Marine Fisheries 45

TABLE 4.—PERCENTAGE OF MARINE FISH AND SHELLFISH CATCH TAKEN FROM MAJOR FISHING AREAS AND ZONES

	Including Anchoveta							Excluding Anchoveta							Catch/1,000 km² MT 1975
	1975	1974	1970	1965	1960	1955	1950	1975	1974	1970	1965	1960	1955	1950	
North Atlantic	26.9	26.3	24.4	30.2	31.4			28.5	28.1	31.1	35.0	34.9	41.5	43.5	723
Northwest (21)	6.4	6.7	6.9	8.6	9.1			6.8	7.2	8.8	10.3	10.1	11.2	12.9	734
Northeast (27)	20.5	19.6	17.5	20.6	22.3			21.7	20.9	22.3	24.7	24.8	30.3	30.6	719
Central Atlantic	8.5	8.7	7.2	5.6	5.8			9.1	9.3	9.1	6.6	6.5	5.2	6.0	178
West central (31)	2.6	2.5	2.3	2.8	2.3			2.8	2.6	2.9	3.3	2.6	3.6	3.8	107
East central (31)	5.9	6.2	4.9	2.8	3.5			6.3	6.7	6.2	3.3	3.9	1.6	2.2	252
South Atlantic	5.8	6.3	5.8	5.6	4.1			6.1	6.8	7.4	6.7	4.6	4.4	3.3	89
Southwest (41)	1.4	1.5	1.8	1.1	.9			1.5	1.6	2.3	1.3	1.0	.8	1.1	42
Southeast (47)	4.4	4.8	4.0	4.5	3.2			4.6	5.2	5.1	5.4	3.6	3.6	2.2	139
Atlantic Ocean	43.5	43.6	39.2	42.5	43.4			46.2	46.7	49.9	50.9	48.4	56.5		235
North Pacific	32.5	31.4	25.6	11.8	12.0			34.5	33.6	32.6	14.1	13.4	12.8	13.4	700
Northwest (61)	28.7	27.6	21.3					30.5	29.5	27.1					850
Northeast (67)	3.8	3.8	4.3					4.0	4.1	5.5					298
Central Pacific	10.6	10.2	8.3	23.4	27.3			11.2	10.9	10.6	28.0	30.4	25.6	21.5	68
West central (71)	8.4	8.4	6.8	22.3	25.8			8.9	9.0	8.7	26.7	28.8	24.0	17.7	146
(61 + 67 + 71)	(40.9)	(39.8)	(32.4)	(34.1)	(37.8)			(43.4)	(42.6)	(41.3)	(40.8)	(42.2)	(36.8)	(31.1)	
East central (77)	2.2	1.8	1.5	1.1	1.5			2.3	1.9	1.9	1.3	1.6	1.6	3.8	22
South Pacific	8.3	9.5	22.8	18.0	2.1			2.6	3.2	1.7	1.8	2.3	2.4	2.2	99 (29)*
Southwest (81)	.5	.7	.3	.6	.9			.5	.8	.3	.8	1.0	1.2	1.1	9
Southeast (87)	7.8	8.8	22.5	17.4	1.2			2.1	2.4	1.4	1.0	1.3	1.2	1.1	280 (70)*
Pacific Ocean	51.3	51.2	56.7	53.2	41.4			48.3	47.7	44.9	43.9	46.1	40.8	37.1	170 (150)*
Mediterranean, Black Sea (37)	2.2	2.3	1.8	2.1	2.1			2.4	2.4	2.3	2.6	2.3	2.8	3.8	444

46 Living Resources

TABLE 4. Continued

	Including Anchoveta						Excluding Anchoveta							Catch/1,000 km² MT
	1975	1974	1970	1965	1960	1975	1974	1970	1965	1960	1955	1950	1975
Temperate and tropical Indian Ocean	5.2	5.2	4.0	4.3	5.0	5.5	5.6	5.0	5.1	5.6	5.2	6.5	50
West (51)	3.4	3.5	2.7	2.6	2.9	3.6	3.8	3.4	3.1	3.3	2.8	3.8	63
East (57)	1.8	1.7	1.3	1.7	2.1	1.9	1.8	1.6	2.1	2.3	2.4	2.7	35
Indian Ocean, Antarctic (58)	.002	...	...	...	...	.002	...	...	...	...	...	...	.1
World	100	100	100	100	100	100	100	100	100	100	100	100	164† (155)*
Temperate Northern Hemisphere‡	61.6	60.0	50.0	(42.0)§	(43.4)§	65.4	64.1	63.7	(49.1)§	(49.0)§	(54.3)§	(56.9)§	695
Tropics and subtropics, Southern Hemisphere	38.4	40.0	50.0	(58.0)§	(56.6)§	34.6	35.9	36.3	(50.9)§	(51.0)§	(45.7)§	(43.1)§	81† (69)*

NOTE.—Between 1965 and 1970 several statistical boundaries of fishing areas were changed, principally (a) boundary of 61 with 71 moved southward 20°, (b) areas 61 and 67 separated, (c) boundary of 77 with 81 moved southward 15°, (d) boundary of 57 with 81 moved eastward 20°, (e) boundary of 71 and 77 moved eastward 5°.
*Excluding anchoveta.
†Excluding polar seas except the Indian Ocean and Antarctic.
‡Includes the Mediterranean and Black Seas.
§Parentheses are used to distinguish data for earlier years during which the southern boundary of the western Pacific statistical area was 20° further south than it is now.

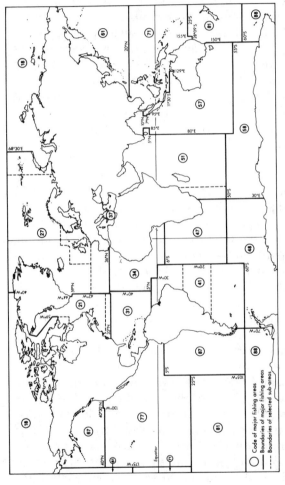

FIG. 1.—Major fishing areas (estimated surface areas) for statistical purposes, with proportions of total marine area (361,060,000 km²) and area of local ocean and adjacent seas, respectively, in parentheses. *Atlantic Ocean and adjacent seas* (109,708,000 km²; 30.4 percent of total marine area): *18*. Arctic Sea = 7,336,000 km² (2.0 percent of total marine area, 6.7 percent of local area); *21*. Atlantic, Northwest = 5,207,000 km² (1.4 percent, 4.7 percent); *27*. Atlantic, Northeast = 16,877,000 km² (4.7 percent, 15.4 percent); *31*. Atlantic, West Central = 14,681,000 km² (4.1 percent, 13.4 percent); *34*. Atlantic, East Central = 13,979,000 km² (3.9 percent, 12.7 percent); *37*. Mediterranean and Black Seas = 2,980,000 km² (0.8 percent, 2.7 percent); *41*. Atlantic, Southwest = 20,224,000 km² (5.6 percent, 18.4 percent); *47*. Atlantic, Southeast = 18,594,000 km² (5.2 percent, 17.0 percent); *48*. Atlantic, Antarctic = 9,830,000 km² (2.7 percent, 9.0 percent). *Indian Ocean and adjacent seas* (72,307,000 km²; 20.0 percent of total marine area): *51*. Indian Ocean, West = 32,096,000 km² (8.9 percent, 41.0 percent); *57*. Indian Ocean, East = 29,681,000 km² (8.2 percent, 41.0 percent); *58*. Indian Ocean, Antarctic = 10,530,000 km² (2.9 percent, 14.6 percent). *Pacific Ocean and adjacent seas* (179,045,000 km²; 49.6 percent of total marine area): *61*. Pacific, Northwest = 20,006,000 km² (5.5 percent, 11.2 percent); *67*. Pacific Northeast = 7,503,000 km² (2.1 percent, 4.2 percent); *71*. Pacific, West Central = 34,000,000 km² (9.4 percent, 19.0 percent); *77*. Pacific, East Central = 57,467,000 km² (15.9 percent, 32.1 percent); *81*. Pacific, Southwest = 33,212,000 km² (9.2 percent, 18.5 percent); *87*. Pacific, Southeast = 16,471,000 km² (4.6 percent, 9.2 percent); *88*. Pacific, Antarctic = 10,386,000 km² (2.9 percent, 5.8 percent).

degree of present utilization of them. Fishing is not carried on throughout any of the areas, and particularly in the tropics and Southern Hemisphere there are vast areas hardly if ever visited by distant-water fishing vessels. Comparison of productivity calls for calculation of catches per kilometer of coast or, for fish other than oceanic species, per square kilometer of continental shelf, but such discussion is beyond the scope of this article.

In a period during which the annual catch has exactly tripled and considerable changes have occurred in the technology of taking and processing it, one might expect changes in the proportions of the various types of organisms being caught. Table 5 shows that there have in fact been such changes, but the trends have been gentle ones. The predominance of fishes over shellfish has slightly strengthened. Among the fishes the gadoids (Class 32) and the clupeoids (Class 35) continue to be by far the most important groups. If anchoveta is included, the clupeoids nearly always have been more important by weight; but if anchoveta is excluded (as it is throughout table 5 except in the first column), we see that over the period the two groups have exchanged ranks, the clupeoids being in relative decline. The catches of pleuronectid flat fishes (Class 31) have slightly declined relatively, and those of the redfishes, etc. (33), have increased. Catches of tunas have also increased slightly relatively, but the jacks, etc. (34), and the mackerels, etc. (37), have both nearly doubled in importance. The catches of cartilaginous fishes (elasmobranchs, Class 38) have relatively declined, as might be expected of animals with rather low total biomass. Last, the contribution to the total catch from the group including salmons and shads, which live partly in the ocean and partly in inland waters (Classes 23–25), has substantially increased, notwithstanding the pressures of other uses by man on the fresh and brackish waters on which these mostly diadromous fishes depend for survival. However, this increase can be attributed mainly to the growth of the North Atlantic fishery for capelin *(Mallotus villosus)*, which is included in the same class (see table 3).

Among the shellfish, the crustaceans have remained relatively constant, but the mollusks have diminished in importance. There are, however, different trends within each of these major classes. Within the first, the shrimps and prawns (45) have increased relatively, and the lobsters and crabs (42–44) have decreased. Within the second, both bivalves (Classes 53–56) and the squids (57) have decreased. Since a large part of the bivalve harvest is from cultivated stocks, it seems that despite its promise mariculture of mollusks has not yet been able to increase as fast as the capture of fishes and shellfish from wild stocks, even though these latter are increasingly subject to overfishing.[4] Finally, the seaweeds have maintained a fairly constant contribution of just under 2 percent.

---

4. Interpretation is complicated, on one hand, by some declines in oyster and mussel production attributed to pollution and disease and, on the other hand, by an uneven development of mariculture. Thus the culture of oysters and mussels developed long ago but is still gaining slowly over harvest of natural stocks; 70 percent of the oyster harvest and 80 percent of the mussel harvest are now from cultured stocks, and these

Marine Fisheries   49

TABLE 5.—SPECIES GROUPS IN CATCHES (%)

FAO Taxonomic Classification		Total including Anchoveta 1975*	Total excluding Anchoveta								
			1975	1974	1973†	1971‡	1970	1965	1960	1955	1950
23–25	Diadromous fishes§	1.7	1.8 (75)	1.7 (68)	1.6 (62)	1.8 (65)	2.0 (69)	1.4 (66)	1.5 (66)	2.2 (66)	1.9 (66)
23¶	Capelin	3.8	4.0	3.4	3.8	3.2	3.2	1.7	.3	.2	.1
31	Flounders, halibuts, soles, etc.	1.9	2.1	2.1	2.4	2.8	2.7	2.5	4.0	2.5	2.6
32	Cods, hakes, haddocks, etc.	20.0	21.2	22.4	22.5	21.4	21.9	17.8	16.4	19.0	19.3
33	Redfishes, basses, congers, etc.	8.4	8.9	8.5	8.0	8.1	8.0	8.5	7.7	7.4	7.5
34	Jacks, mullets, sauries, etc.	5.9	6.3	6.4	6.9	6.4	5.0	5.7	5.7	5.9	3.9
35	Herrings, sardines, anchovies, etc.	23.2	18.4	18.0	18.1	17.2	17.8	24.2	21.9	24.7	27.6
36	Tunas, bonitos, billfishes, etc.	3.2	3.4	3.7	3.7	3.5	3.1	3.6	3.5	3.2	3.0
37	Mackerels, snoeks, cutlass fishes, etc.	6.0	6.4	6.4	6.4	6.6	6.6	4.0	3.6	3.2	3.6
38	Sharks, rays, chimaeras, etc.	1.0	1.0	1.0	1.1	1.0	1.1	1.1	1.2	1.2	1.7
	Fishes#	88.6	87.8	88.1	88.4	88.1	88.1	87.7	86.9	87.2	86.1
42–44	Large crustaceans**	1.0	1.0	1.0	1.0	1.1	1.2	1.2	1.0	1.2	1.3
45–46	Shrimps, prawns, krill, etc.	2.1	2.3	2.3	2.4	2.1	2.0	1.8	1.7	1.9	2.0
	Crustaceans#	3.2	3.4	3.4	3.5	3.4	3.4	3.1	3.0	3.3	3.5
52	Gastropod mollusks††	.1	.1	.1	.1	.1	.1	.1	.1	.1	.1
53–56	Bivalve mollusks	4.0	4.2	3.7	4.0	4.4	3.9	4.7	5.2	4.4	4.4
57	Squids, cuttlefishes, octopuses, etc.	1.9	2.1	1.9	2.0	1.8	2.0	2.2	2.1	2.0	2.8
	Mollusks#	6.3	6.7	6.0	6.4	6.6	6.8	7.5	8.1	7.7	8.7
	Other invertebrate animals‡‡	.1	.11	.1	.1	.1	.1	.1	.1	.1	.1
	"Shellfish"§§	9.6	10.2	9.5	10.0	10.1	10.3	10.7	11.2	11.1	12.2
	Algae¶¶	1.8	1.9	2.2	2.0	1.8	1.7	1.5	1.8	1.7	1.7
	(Various "miscellaneous")##	(13.9)	(14.8)	(15.1)	(15.8)	(15.2)	(17.5)	(17.5)	(21.6)	(21.5)	(19.4)

*The figures in this column, in which anchoveta catches are included throughout, are comparable with those for the year 1971 given in Holt and Vanderbilt (1974), table 11, last column.
†Data for 1973 are included here to establish baselines in relation to table 11.
‡Data for 1971 are included here to permit comparison with the analyses given by Holt and Vanderbilt (1974).
§Includes sea catches of salmons, trouts, smelts (except capelin), shads, milkfishes, and other diadromous species, i.e., species which migrate between the seas and fresh or brackish waters. The figures in this row are less reliable than those in other rows because of uncertainties as to the distribution of catches between sea and inland waters. In parentheses are given percentages of catches thought to have been taken at sea.
¶Part.
#Includes miscellaneous catches not assigned to categories above.
**Crabs, sea spiders, spiny rock lobsters, squat lobsters, *Nephrops*, etc.
††Abalones, winkles, conches, etc.
‡‡Mainly sea urchins, sea cucumbers, and other echinoderms; also sea squirts and sponges, etc.
§§Crustaceans, mollusks, and other invertebrates.
¶¶Brown, red, and green seaweeds.
##Marine mammals (whales, dolphins and porpoises, seals, and sea lions) are excluded.

RESOURCE POTENTIAL

During the period under study, many individual scientists have published predictions of the upper limits of catches which the marine resources could sustain. More often than not these predictions have turned out to be excessively pessimistic. This contrasts with the preceding 25 years, during which many observers maintained, like Thomas Huxley in the late nineteenth century, that the seas were practically inexhaustible; those who then perceived and were convinced by the evidence of incipient overfishing were perhaps in a minority.

The study of the limits of marine-fishery resources to which the most specialists and organizations have contributed and which used the greatest range of methods for cross-checking as well as the greatest volume of data was that conducted by the FAO in connection with the Indicative World Plan for Agricultural Development (IWP). The results were first published in 1970, but the work was carried out mainly during 1966–68. No other comprehensive study has been made in the decade since, but from time to time partial reviews and revisions of the situation have been presented by the FAO Secretariat to the annual sessions of the FAO's intergovernmental Committee on Fisheries (COFI).

In 1966 the marine catch, excluding whales but including anchoveta, was just under 50 million MT, and that included nearly 10 million MT of anchoveta. It was concluded that the potential of the shoaling pelagic (surface and midwater) fish and the medium-sized demersal (bottom-dwelling) fish together was just over 100 million MT. The squids over the continental shelf were thought to have a potential several (perhaps seven) times that of the catches then current—about 1 million MT—and to this could be added increased catches of crustaceans and harvests of bivalve mollusks. It was therefore predicted that the total production of these more familiar organisms could be increased to—and sustained at—100–150 million MT annually. It was realized at the time that such an increase would only be possible if management arrangements were improved. It was also assumed—although data were and are still not available to quantify this—that a doubling, say, of the catch would require more, perhaps much more, than a doubling of the fishing effort to take it. It was suggested that no more than 80 percent of the potential in an area may be harvestable in practice. The IWP study led to the conclusion that the further increases in catches of "conventional" types of fishes and shellfish would come mainly from the relatively lightly exploited areas such as the Indian Ocean (see fig. 2).

---

two subgroups account for 60 percent of the total bivalve harvest. The cultivated proportion of the clam harvest increased from 6 to 12 percent during 1960–70 and is now probably nearly 20 percent. Scallop culture arrived later still—only 2 percent in 1960, 5 percent in 1970, but 20 percent in 1973. See A. Purchell Maschke, *Production, Trade and Consumption of Gastropods and Bivalve Molluscs*, FAO Fisheries Circular no. 345 (Rome: FAO, 1976).

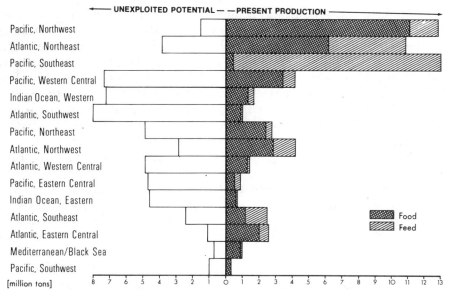

FIG. 2.—Adapted from Robinson and Crispoldi, "Trends in World Fisheries," *Oceanus* 18 (Winter 1975): 26. Production and potential of the world's oceans, 1970. Some 60 percent of the world's marine-fish catch came from three areas: the Northeast Atlantic (in close proximity to industrialized northwest European countries), the Northwest Pacific (heavily exploited by the USSR and Japan), and the Southwest Pacific (dominated by the anchoveta fishery; most exploited area in terms of harvested potential). Areas of greatest unexploited potential are in tropical or southern temperate latitudes, but none has the productive capacity of the currently heavily exploited areas, according to present knowledge. Chart is based on physical quantities, but the Mediterranean Sea and West Indian Ocean are much more important in terms of value; however, the Northwest Pacific is still in first place.

At that time the great potential for harvest of krill in the Southern Ocean and of lantern fishes (Myctophids) in warm oceanic waters was recognized, and it was considered that these "unconventional" resources, with oceanic squids from all areas, would yield at least 100 million MT and possibly much more. The volume of the oceanic squids is estimated variously as from eight to 60 times the potential of the shelf squids, on the basis of assumed consumption rates by sperm whales.

The baseline for the IWP was 1965; 1975 and 1985 were the first and second target dates. It is therefore appropriate at this time, when we are in possession of data for 1975, to see how the earlier predictions have fared. A continued rate of increase in catch of 5.5 percent annually, the overall rate then prevailing, would have led to a 1975 total of 78 million MT; the actual catch of 59 million MT fell far short of this. If one had been able to foresee trouble in the anchoveta fishery, one might have predicted that 46 million MT, excluding anchoveta, would increase at 6 percent annually to 68 million MT and have come closer to reality but still have erred on the optimistic side.

In the IWP an attempt was made to estimate how production increases would affect the utilization of the resources. Trends in the pattern of fishing effort suggest that the IWP considerably overestimated the extent to which fishing effort would be diverted to or generated in the less heavily exploited areas. Thus, of the increase of 20 million MT in the world catch of marine fish and crustaceans in the period 1962–75, nearly half has come from the Northwest Pacific and the Northeast Atlantic, while lightly fished areas such as the Southwest and West Central Atlantic, the East Central Pacific, and the Indian Ocean have remained so.

With the benefit of hindsight, the FAO Secretariat reported to the COFI in April 1977 that

> the main sources of growth now remaining and available to present technology lie in the relatively lightly fished areas, which have a theoretical unexploited potential of approximately 50 million tonnes; allowing for fluctuations caused by environmental factors and the impossibility of exploiting stocks at one hundred per cent of potential, the world catch is probably capable of increase by some 30/35 million tonnes under optimum conditions of exploitation. Bearing in mind the factors now inhibiting the harvesting of this potential, and especially the types of fish concerned and the marketing problems involved, it seems that in the absence of any major development the annual rate of increase in the world catch is on average unlikely to exceed one or two per cent over the next decade.[5]

VALUES OF FISHERIES

The value of marine fisheries to a country, a group of countries, or the world as a whole may be judged in a number of ways. Principally two of these are considered here: the first-sale monetary value and the nutritional value to people. Various secondary measures have been used from time to time: for example, the capacity to give employment and the availability of seagoing craft and experienced sailors to be called upon in times of war. This latter value is, I imagine, now obsolescent.

As to employment, it has been estimated that in 1970 the full-time equivalent of about 5 million people were engaged in primary fishing operations, that is, 0.13 percent of the total world population. Many more than that are fishermen, at least part time. (For purposes of estimation, three "part-time" fishermen are taken as equivalent to one full-time.) Most of the catch is "distributed," in the sense of being consumed elsewhere than in the place at which it is landed, and three-quarters of it is processed in some way. This means that probably two

---

5. FAO, "Prospects and Requirements for the Better Utilization of Fishery Resources" (COFI/77/7).

or three times as many more people than fishermen are engaged in secondary fisheries industries, including wholesale marketing.

The world average of 0.13 percent obscures, moreover, a wide range of different levels of engagement in fisheries among countries. In the Philippines, for example, 5 percent of the employed population are fishermen, and in that country sea fish provide as much as 50 percent of the animal protein and 15 percent of the total protein in the national diet. On the other hand, a half million fishermen in Japan, 1 percent of the employed population, supply 60 percent of the animal protein and 20 percent of the total protein in the Japanese diet.

The groups of developed countries are remarkably similar to each other: in the United States and Canada together, the USSR and eastern Europe, and EEC countries fishermen form about 0.1 percent of the total population; the annual catch per fisherman is 20–30 MT, and they supply about 3 percent of the dietary protein.

Somewhat surprisingly, in view of the great differences in degree of industrialization, there are not a particularly large number of fishermen in the developing countries as a whole: 49 percent of the people and 56 percent of the fishermen live in developing countries with market economies; within this group there does, however, tend to be a concentration in Asia and the Far East, which has 30 percent of the world's population but 40 percent of its fishermen.

Marine catches, excluding whales, were valued at about US$11 billion in 1971 and US$15 billion in 1974, about 0.44 percent of the world GNP. Detailed data on values of fish landings were compiled, though incompletely, from about 1950, but the FAO ceased collecting them in 1973. A review of existing historical data is beyond the scope of this article, but an idea can be gained of the relative monetary values of total catches by considering the changes over time of their composition by species groups and of the forms in which the different types are landed. The relative prices of the different species groups have not in most cases changed much in recent years (although the price of fish in the same group in the same year may vary greatly from one region to another). An exception is the category of salmons, etc., because of the increasing percentage of capelin, which is all reduced to fish meal. Table 6 shows the 1971 prices by species groups and the percentage of contribution by each group to the total value. In the last column is shown the breakdown by weight, for comparison.

It is evident that the shellfish, especially the crustaceans, are far more important in value than catches by weight would suggest. There are also considerable differences among the groups of fishes. Thus, while the gadoids retain their preeminence, the clupeoids become less significant than any other major group except salmonids, etc., and the sharks; the pleuronectids, jacks, and tunas all appear relatively much more important. Among the groups of salmonids, smelts, and shads, etc., the capelin becomes very low in relative value; but of course the rest of the species such as salmons and milkfish, which are valuable foodfish, double with respect to their contribution by weight.

TABLE 6.—PRICES AND VALUES, 1971 MARINE CATCH*
(Total Value $11 Billion)

	Price $/MT	% Contribution By Value	% Contribution By Weight
Salmonids (excluding capelin) and other diadromous fishes	340	3.6	1.5
Capelin	32	.6	2.9
Flounders, etc.	399	5.0	2.2
Cods, etc.	182	17.7	17.5
Redfishes, etc.	286	9.9	6.2
Jacks, etc.	282	7.6	4.8
Herrings, etc.	34 (61)	5.9 (4.6)	31.8 (13.4)
Tunas, etc.	462	6.8	2.7
Mackerels, etc.	202	5.9	5.3
Sharks, etc.	200	.9	.8
Miscellaneous fishes	181	13.9	13.8
Fishes	117 (194)	77.8 (76.5)	89.0 (70.6)
Large crustaceans	875	4.0	.9
Shrimps, etc.	876	8.1	1.6
Miscellaneous crustaceans	716	.5	.1
Crustaceans	850	12.7	2.7
Gastropods	280	.1	.1
Bivalves	156	3.0	3.1
Cephalopods	465	3.7	1.5
Miscellaneous mollusks	348	1.2	.6
Mollusks	209	7.8	5.2
Shellfish	428	20.5	7.9
Other invertebrates	216	.1	.1
Algae	124	1.1	1.5
All	141 (217)	100 (98.6)	100 (81.6)
All "miscellaneous" categories	192	16.9	12.4

*Numbers in parentheses represent price or percentage excluding anchoveta.

The prices listed for 1971 can be applied to catches in other years to calculate an average value per ton in 1971 dollars, which would take some account of the trends of changes in species composition. The ratios of these averages to the 1971 average provide a rough index of quality of catch (excluding the anchoveta catch) as measured by prices paid. It turns out that over the entire period 1950–75 this index does not in any year fall outside the range of 0.98–1.02. This seems surprising in view of the fact that the percentage of the catch being reduced to fish meal has been increasing. However, any trend in average price of catch would reflect a combination of two factors: among fish destined for direct human consumption, the relative change of those kinds having lower or higher value, and the relative increase in catches of kinds of fish which are destined for reduction to meal and oil rather than for direct consumption. Evidently declines in a low-priced group such as the clupeods, insofar as they are consumed directly, would tend to raise the average, as would the combined increase of jacks, etc., and shrimps and prawns. Such changes have been sufficent to counterbalance the low values of the increasing quantities of other categories being reduced to meal and oil.

If anchoveta catches are included in the quality index, the picture changes somewhat, as table 7 shows. From this it appears that, although the catch by weight decreased by an average of 0.5 percent a year from 1970 to 1975, in that period its "value" increased at nearly 3 percent annually, as a result of the declining anchoveta catches (which reached their peak in 1970) almost being replaced by equivalent catches of relatively more valuable species.

This type of analysis cannot be pursued much further without looking in more detail at changes within the taxonomic groups or considering categories other than taxonomic ones by which catches can be grouped.

During the period under consideration, the ways have changed in which fish for direct consumption are disposed of, as table 8 shows.

While about one-half of the fish is still consumed fresh, the proportion has diminished, but this has been more than compensated for by the increase in freezing. Of the older methods of preservation, curing has declined, but canning continues to increase in relative importance. In absolute terms, the quantity cured has hardly changed over 25 years, but the other three categories have increased, though at different rates.

Starting in the early 1950s a rapid increase began in the proportion of the total catch destined for reduction to fish meal and oil. The first column of table 9 suggests that after a peak in 1970—which was in fact the highest percentage recorded—there might have begun a slow decline. However, this is due entirely to the subsequent collapse of the anchoveta fishery, as the second column of the table shows. In fact the proportion reduced continues to increase, apparently toward an upper limit not higher than 35 percent under existing conditions of markets and processing technology.

Over the period 1950–75, the catch of fish for processing has increased at

TABLE 7.—ESTIMATED VALUE OF CATCHES AT 1971 PRICES

Year	Quality Index	Catch Weight Relative to 1971	Catch Value Relative to 1971
1950	1.21	.30	.36
1955	1.21	.42	.51
1960	1.08	.56	.60
1965	.97	.75	.73
1970	.96	1.00	.96
1971	1.00	1.00	1.00
1973	1.20	.93	1.12
1974	1.13	.99	1.12
1975	1.14	.97	1.11

Note.—All quantities include anchoveta.

TABLE 8.—DISPOSITION OF CATCHES USED
FOR DIRECT HUMAN CONSUMPTION (%)

Year	Not Processed*	Frozen	Cured†	Canned
1938	61	0	31	8
1948	57	6	29	8
1950	52	6	30	10
1955	51	8	29	11
1960	53	11	24	12
1965	49	17	22	12
1970	45	22	19	14
1975	42	26	17	15

*Fresh or chilled.
†Dried, salted, or smoked.

double the rate of growth in catches for direct consumption (table 10). Since 1 MT of fish reduced to meal and fed through livestock results in less than 1 MT of food for human consumption, this means that the contribution of sea fish to world food supplies has not increased as fast as the overall catch increase of 4.5 percent (table 1) would suggest. I return to this matter below, but before doing so it is worth examining more closely the ways in which ecologically and taxonomically different types of organisms are disposed of. Ninety percent of

TABLE 9.—PERCENTAGE OF CATCH REDUCED TO MEAL
AND OIL OR USED FOR MISCELLANEOUS PURPOSES
OTHER THAN DIRECT HUMAN CONSUMPTION*

Year	All Catch	Catch excluding Anchoveta
1938	14.4	14.4
1948	14.1	14.1
1950	16.1	16.1
1955	18.0	17.7
1960	25.4	16.8
1965	36.0	23.0
1970	43.5	28.0
1971	41.8	28.7
1972	36.3	30.4
1973	32.8	30.3
1974	34.8	30.1
1975	35.4	31.4

NOTE.—The quantity recorded as used for "miscellaneous purposes," e.g., fertilizer, has stayed constant throughout the period at about 1 million MT. Thus as the total amount of the catch used for purposes other than direct human consumption has increased from 3 million MT in 1950 to 21 million MT in 1975, the proportion used for miscellaneous purposes has fallen to less than 5 percent. This means that the proportion of the total catch reduced to fish meal has increased faster, to the present level, from an initial much lower level than the figures above indicate.
*Excluding whales.

TABLE 10.—AVERAGE ANNUAL RATES OF GROWTH
OF SEA-FISH CATCHES FOR DIRECT CONSUMPTION
AND FOR REDUCTION TO OIL AND MEAL

		Reduction	
Period	Direct Consumption	Including Anchoveta	Excluding Anchoveta
1950–55	6.0	8.9	8.4
1955–60	3.9	13.3	2.5
1960–65	3.0	13.6	12.0
1965–70	3.3	10.2	9.3
1970–75	2.2	−4.5	5.5
1950–75	3.7	8.1	7.3

the cephalopods, all the crustaceans, and all the bivalve and gastropod mollusks are directly consumed, but the recorded total weights of the latter group, which include shells, obviously give no measure of the edible quantities taken. The fishes may be split, roughly equally, into three groups: the small pelagic fishes from lower trophic levels, used very largely for reduction; the demersal fishes from higher trophic levels, which are mostly consumed directly; and the larger pelagic fishes also from higher trophic levels, which are practically all consumed directly. Table 11 shows that, after a decade during which just over 40 percent of small pelagic fish catches were being reduced, there seems to have been some improvement (1969–71), but now the percentage so utilized has begun to climb to higher levels than attained before. While much less demersal fish is reduced, the percentage has increased fairly steadily for 15 years, from 4 per cent to nearly 10 percent. However, the demersal catch itself has risen faster than this, both absolutely and relative to the other ecological groups.

Fish oils are refined and enter human diets directly in various ways. The meals, which are more voluminous and valuable, enter livestock feeds and reach the human diet as eggs, as the flesh of domestic fowl and pigs, and, to a

TABLE 11.—UTILIZATION OF FISH CATCHES, BY TYPE*

	% of All Fish			% of Fish Reduced	
Year	Small Pelagic	Demersal	Large Pelagic	Small Pelagic	Demersal
1960	31.7	31.4	36.9	38.5	4.4
1965	37.3	36.1	26.6	42.0	6.1
1970	32.3	40.7	27.0	34.4	7.5
1973	33.0	39.1	27.9	52.0	9.7

NOTE.—1975 data not available.
*Excluding anchoveta.

much lesser extent, as cultivated freshwater fishes. The separation of uses is not complete, especially in recent years: while the flesh of demersal fish, much of which is mechanically filleted from the bone and frozen, goes directly to human consumption, residues are reduced to meal. On the other hand, an unmeasured but not insignificant amount of fish recorded as canned goes to feed pets rather than people. In addition, a significant amount of the actual catch is in some fishing operations discarded at sea as "trash fish" but in others may be retained and processed in some way. These quantities are not taken into account in the statistical tables presented here, and this should be borne in mind when the tables are interpreted.

In order to determine the trends in the supply to mankind of food from marine sources, a conversion must be made back from fish meal to live weight of livestock or to protein equivalents. Various ratios have been used for the conversion of live-weight equivalents, ranging from $11:1$[6] and $10:1$[7] to $3:1$[8] and apparently $2:1$.[9] There are, moreover, big differences in protein yield between, for example, frozen fish and canned fish, "white" and oily fish, utilization efficiency in developing and developed countries, and feeding meal to chickens and pigs or to freshwater fishes. One could attempt to allow for these variables in compiling total figures, but the present state of data does not seem to justify such refinement. A conversion ratio of $5:1$ is used here, which may be somewhat generous with respect to the average efficiency of utilization. The quantities given in table 12 are in equivalents of whole fish. Trends of annual rates of increases are to be compared with the gross rates given at the bottom of table 1 and with an approximately 2 percent annual rate of growth in human population over the period.

In the 15-year period from 1952 to 1967, for which comparable statistics are available, the total sea-fish catch increased by 146 percent, while the human population increased by 34 percent. In that period the marine contribution to nutrition—direct and indirect—increased by 74 percent. Corresponding figures for dairy products and for beef, buffalo, sheep, and goat meat range from 44 to 61 percent. Among the recorded meats, only pork increased more than fish—by 93 percent—and part of that is attributable to the fish-meal contribution to pig feeds.

More recently the fish contribution has certainly slowed down. From tables 12 and 13 it may be concluded that the increase in marine production is now

---

6. S. J. Holt, "Marine Fisheries and World Food Supplies," in *The Man-Food Equation: Proceedings of a Symposium Held at the Royal Institution,* ed. F. Steele and A. Bourne (London: Academic Press, 1975).

7. S. J. Holt and C. Vanderbilt, "Materials for an Appraisal of the Dependence of Nations on the Living Resources of the Sea," mimeographed (Malta: International Ocean Institute, 1974). This is a detailed analysis of 1971 and 1970 statistics on catches, values, employment in sea fisheries, etc.

8. *Expanding the Utilization of Marine Fishery Resources for Human Consumption,* FAO Fisheries Reports, no. 175 (Rome: FAO, 1975).

9. Maschke (n. 4 above).

TABLE 12.—FOOD SUPPLY
IN LIVE-FISH WEIGHT EQUIVALENT

A. MILLIONS MT*

Year	Including Anchoveta	Excluding Anchoveta
1950	16.2	16.2
1955	21.8	21.8
1960	27.0	26.3
1965	32.5	30.9
1970	39.7	37.1
1975	42.5	41.8

B. % ANNUAL RATE OF INCREASE

1950–55	6.1	6.1
1955–60	4.4	3.8
1960–65	3.8	3.3
1965–70	4.1	3.7
1970–75	1.4	2.4
1950–75	3.93	3.86

*Calculated as total weight produced for direct human consumption plus one-fifth weight of fish reduced to meal.

barely keeping pace with population increase in terms of its direct and indirect contribution to human diet. To put this conclusion in perspective, on the basis of the years 1964–66, we note that the global contribution of marine catches, direct and indirect, was about 13 percent of the animal protein and 4 percent of the total protein consumed.[10] Table 13 also relates food of marine origin to the total supply of terrestrial food between 1961 and 1974. The relative contribution of marine products varies very greatly, however, among areas of the world, among different economic groups of countries, among coastal and landlocked states, and, within countries, among town dwellers and the rural and coastal populations and among income groups.

The trend of relative increase in demersal fish catches shown in table 11

---

10. In terms of dry protein weight, marine sources yielded 9 million MT in 1975, which may be compared with 95 million MT from cereals, 30 million MT from legumes, 6 million MT from other vegetables, and 33 million MT from livestock. Only one-third of fish production is fed to livestock, and about one-third of vegetable proteins is so disposed; the utilization ratio in this case, however, is 11:1. Just before the collapse of the anchoveta fishery, fish provided 11 percent of the recorded supply of animal-feed protein; after the collapse it fell to 6 percent or less. In its "Mid-Term Review and Appraisal" of the Second UN Development Decade, the FAO reported that the direct contribution of fish, including freshwater fish, accounted for 4.3 percent of world protein supplies in 1960 and 5.6 percent (14 percent of animal-protein supplies) in 1971. In some areas, however, including the Congo, Indonesia, Japan, the Philippines, and Vietnam, total fish consumption exceeded that of meats.

TABLE 13.—INCREASES IN FOOD PRODUCTION PER CAPITA, 1961–74: AVERAGE ANNUAL RATES (%)*

	Terrestrial Food‡		Aquatic Food Fish§		Food of Marine Origin, 1970–74¶	
	1961–74	1970–74	1961–74	1970–74	Including Anchoveta	Excluding Anchoveta
Market economies	...†	−.3	...	...	...	...
Developed	1.3	1.6	...	...	...	...
Developing	...†	−1.0	...	...	...	...
Centrally planned economies	1.3	1.0	...	...	...	...
Developed	2.5	2.2	...	...	...	...
Developing	.8	.5	...	...	...	...
Developed countries	1.8	2.0	.9	.2	.6	.6
Developing countries	.3	−.7	2.6	1.8	−1.0	2.7
World	.7	.2	2.1	1.3	.5	2.1

NOTE.—Conversion of these production figures to consumption indices calls for analysis of trade patterns. The patterns of change are markedly different for agricultural products and fisheries products. Thus the volume of agricultural exports grew at an average annual rate of 3.3 percent from 1961 to 1974, but much faster from the developed countries (4.4 percent) than from the developing countries (1.9 percent). For fisheries products the corresponding world figure was 4.0 percent, but the volume from developed countries increased by only 3.3 percent annually, while that from developing countries increased by 6.0 percent. Most of the imports, of whatever origin, were into developed countries. The growth rate of exports from developing countries varied greatly, however, from region to region. Thus from Latin America there was actually a decrease (−1.2 percent) due mainly to the anchoveta failure, while from Africa, the Near East, and the Far East the increases reached 5.1, 3.8, and 10.5 percent annually, respectively. This last high figure, reflecting a nearly 10-fold growth over the period, was achieved mainly by increases in relatively high-priced crustacean and fresh-fish products. The largest proportion of the value of trade is now, as it was in the 1960s, in the shipments of fresh, frozen, and canned products between developed countries; developing countries have retained their dominance, however, in exports of crustaceans, mainly fresh and frozen shrimp.

*World fishery production (total aquatic, excluding China) increased at an average rate of 4.3 percent annually over the period 1961–64 and at 2.5 percent from 1970 to 1974. The corresponding rates for agriculture (crops plus livestock) were 2.6 and 2.1 percent, respectively, and for forestry 1.8 and 1.0 percent. The 1970–74 per capita production rates were .5 percent for fisheries and .2 percent for agriculture. However, the rates of growth in food production from terrestrial and aquatic sources were very different, as the table shows.
†Not available.
‡Excludes inedible products such as tobacco, fibers, rubber, and inedible oil seeds.
§Marine plus freshwater, excluding marine catches for reduction.
¶Calculated as marine food-fish catches plus one-fifth of the catches for reduction.

appears to contradict the commonly expressed opinion that, as mankind has exploited the living resources of the sea more intensively, more effort has been directed to the organisms at lower trophic levels. It is true that in recent years important fisheries have developed, for example, for the capelin *(Mallotus villosus)*—an important food of cod, finwhales and seals—and for sand eels *(Ammodytes spp.)*. However, these have been more than counterbalanced by declines, and in some cases virtual collapses, of fisheries for some small pelagic species which previously had been very important. Thus the story of the anchoveta, which itself followed an earlier eclipse of the sardine fishery of California, has been repeated with respect to other clupeoids, including the herring.

Some of the species fisheries which have undergone massive changes over the years are shown in table 3. These examples are drawn mainly from industrialized fisheries of the Northern Hemisphere. Developing countries are not, however, immune to effects of drastic man-caused changes in stocks. Thailand provides a good example. A trawl fishery for demersal species began in the Gulf of Thailand in the 1960s, in part as a result of effective, technical assistance from the Federal Republic of Germany. As catches increased, catches per unit effort fell. Now for 3 years the total marine catch by Thailand has been dropping despite continued and possibly increasing fishing effort.

It would be useful if this analysis of gross value of production could be followed by conclusions regarding the trends in net value. Unfortunately, just as there are no comprehensive data on fishing effort to relate to yields by weight, there are no data from which may be estimated the relative changes in monetary costs of fishing or in energy expenditure in fishing on a global basis. It is certain, however, that, despite some increase in efficiency resulting from technical improvements, the costs of each increment in production and in derived food supply are increasing. A number of case histories support this statement. The relative small growth in tuna production, for example, has been secured by a much larger increase in deployment of a fishing effort, measured in terms of number of vessels, their sizes, their capital costs, and their fuel expenditures. The Mexican shrimp fishery in the Gulf of California provides another example. In the first decade of the period of review, an annual catch of 20,000 MT was obtained by an average of about 400 small vessels; in the second decade 800 larger and much more expensive (both to construct and to operate) vessels brought in the same level of total catch; and since 1973, when the number of large vessels exceeded 800, the total catch has fallen. The possible ways out of this situation are discussed in the following sections, but first, reference should be made to the few recent studies of energy expenditures in fishing. These have followed the lines of pioneer work in the last decade on the energetics of food production by agriculture, but they have not yet attained that degree of detail. A considerable but unmeasured portion of the marine-fish catch is taken by methods which use little energy input other than human labor. Where fishing is industrialized, however, the consumption of other natural resources, especially of fossil fuels, can be rather high. Thus in the North Sea fisheries, 8,000 kcal are expended in harvesting each kilogram of fish; that is, 20 kcal of fossil energy for every 1 kcal of gutted fish. This 20:1 ratio is very much higher than for vegetable-protein production, which ranges from below 1:1 to 10:1. It is higher also than for range beef (10:1) and for lamb (16:1), is of the same order as for dairy production, but is much less than for feedlot beef or pork.

DIRECTIONS FOR GLOBAL IMPROVEMENT

Vast increases in total investments in vessels and equipment, and thereby fishing effort, is evidently not in the future going to be the way to continue growth in the fisheries contribution to food supplies or to maintain even present levels of net yield from "conventional" types of living marine resources. Large-scale production from "unconventional" resources may prove feasible; indeed a number of industrialized countries have been studying for several years, on a pilot scale, the exploitation of the krill population of the Southern Ocean. The costs of this are still high, because of the distance fishing vessels from those countries must travel and because the dense concentration of krill are somewhat elusive and transitory. The FAO reported in March 1977 that,

> with the present fish meal prices and the considerably improved catching efficiency achieved by some expeditions during the 1975/76 season, commercial production of krill meal, probably combined with smaller quantities of other higher priced products for human consumption (e.g., krill paste, deep frozen whole krill, minced meat, protein concentrate) may well eventually become economic. So far only small quantities of 5,000–15,000 tonnes/year are being utilized, still largely on an experimental or semi-commercial scale, using conventional long-distance factory trawlers and research vessels with fishing and processing capacity. Apart from krill meal for animal feeding, the best known products are krill paste (USSR) and frozen whole krill (Japan), but development work for other products is being intensified and accelerated. Krill meal for feed should not present any marketing problems but other, mostly new, products for human consumption will require acceptability tests and sales promotion efforts. The hostile conditions of the Southern Ocean and the seasonality will require the development of suitable bulk harvesting, processing, storage and transport technology and logistics for which, however, considerable vessel and fishing capacities may become available through enforced diversion from traditional grounds under a new international fisheries regime.[11]

11. FAO, "Review of the State of Exploitation of the World Fish Resources: Living Resources of the Southern Ocean" (COFI/77/5, suppl. 2).

Other unconventional resources which have been cited, such as the mesopelagic lantern fishes and oceanic squids, though great in quantity are, as far as is known, even more thinly spread and technically less accessible. There may be a considerable potential for catch of small swimming crabs, but little is yet known about these, and the considerations of efficiency above probably apply. In all cases it may turn out to be beneficial to exploit the general small and widely dispersed animals, if they can be processed efficiently into a form for direct consumption, but any benefit is unlikely in my opinion if they are merely caught for conversion to flesh through animal husbandry and aquaculture.

Another factor in the appraisal of these possibilities in the long run, as well as consideration of the future of the conventional fisheries, is the growing evidence of substantial interactions among the different species which are subject to exploitation. Thus the abundance of krill is a limiting factor in the growth and reproduction of baleen whales as well as of other krill feeders such as crab-eater seals and penguins. When blue and fin whales were depleted in the Southern Ocean, sei whales, possibly minke whales, and seals and birds increased in numbers and biological productivity. The North Pacific fur seal seems to have responded adversely to the impact of fishing on its important dietary item, the Alaska pollack. Changes in North Sea fish stocks suggest that interaction among species may now be significant; in particular high reproductive success of some demersal species has coincided with reductions in some of the pelagic species, though there is no evidence of a causal relationship. The cod in the North Atlantic as a whole does not seem to have been drastically affected yet by the growth of the fishery for one of its important food items, the capelin. The cod stocks yielded catches of 3–4 million MT annually through the 1960s, but catches have been falling gradually since; the capelin catch grew from practically nothing in 1961 to 2 million MT by 1973 and has remained at that level since. It is possible that the cod is now beginning to suffer from the effects of very high catches in the late 1960s, peaking at 4 million MT in 1968, combined with stress on its food supply. In other cases too, the cause of gradual decline or collapse may be a combination of overfishing and an adverse natural or man-made environmental change at a time when the stock is small and less stable because of a reduction in the number of year classes present.

The observation of interspecific effects, especially between predators and their prey, poses new questions regarding management objectives. How are the values of fur-seal skins and pollack catches to be balanced, or the high monetary value of a relatively small catch of predatory animal which is esteemed as human food balanced against a larger catch of its smaller prey, for reduction to meal? Resolutions of such questions may become a major political issue in the next decade.

Apart from the special case of the large whales, the earlier examples of overfishing—of some demersal flatfishes and some gadoids—involved change in age and size structure of the populations rather than in their reproductive performance. Theoretically this could be corrected by increasing the size at which fish were first likely to become captured (by protecting "nursery grounds" and increasing sizes of meshes and hooks) and restraining growth in fishing effort or even reducing it. Regional fishery-management organizations in the North Atlantic tended to concentrate on the former, because effort limitation was judged to pose enormous political problems. In other areas, particularly with respect to halibut in the North Pacific, indirect fishing-effort regulation was attempted through the establishment of catch quotas. In both cases such "management" in fact was followed by and might well have stimulated higher rather than lesser concentrations of fishing power.

Later cases of collapse of stocks of shorter-lived pelagic fishes have also involved reductions in reproduction and survival of young fish to commercially useful size. Mathematical models based on earlier experience thus tend to give an overoptimistic impression of the resilience of these resources; if used to predict or determine catch levels, they may even encourage the deployment of excessive fishing effort. When combined with statutory management objectives embodied in existing treaties for the regulation of fisheries which define a "maximum sustainable yield" (MSY) and even suggest that it is "wasteful" not to take that yield, the consequences have been serious.

Thus it can be said that inefficient use of the conventional living resources results from a combination of three factors: weakness of our ability to make predictions on the basis of present scientific understanding of the dynamics of marine populations and ecosystems; poorly defined management objectives, especially in international agreements (this includes the 1958 Convention on Conservation of Fisheries Resources of the High Seas, most of the conventions establishing regional or specialized fishery commissions, and ambiguous phrases in the UNCLOS negotiating texts); and the inability of states to reach agreement promptly, and to implement and enforce such agreement, on the restraints to be placed on the primary inputs to fishing.

Some of the interactions in modern fisheries arise less from biological relationships than from the unselective nature of the fishing techniques. The most important of these is the destruction of young fish during so-called "industrial" fishing for fish-meal production. In some cases such industrial fisheries compete with fisheries for food fish by taking huge numbers of the young of the very species on which the latter depend; the herring is a conspicuous example. Calculations can be made in principle of the best age to begin catching a species, so that its capacity for reproduction is not impaired and that for growth is exploited, provided that it is known at what age the species matures sexually and how quickly indivduals increase in weight relative to the rate at which they die naturally. Nevertheless, there are complications in this, and the relations between these industrial fisheries and the food fisheries, often pursued by enterprises from different countries, is a matter of persistent controversy and conflict.

In other cases industrial fisheries are based on small species rather than young fish, but there may be incidental catches of the young of more valuable food fishes. If the incidental catches are big enough, as they sometimes are, they can affect the yield capacity of the food-fish stocks. There is some biological basis for this interaction, since the different species—small adults of one group and the young of the others—may often be competing for the same food supply. However, there is usually some zonal or geographic separation between them, and this can be used, by setting limits to the maximum permissible incidental catch, to mitigate the conflict between the two types of fishing.

The factors described above may all be regarded as leading to forms of waste of natural resources—either of the resources (including in this case human resources) used to generate excessive inputs or of the biological productivity of the natural systems. Another form of waste, which is certainly now among the most important, is the discard at sea of unwanted incidental catches. These are always of less valued species: less valued at least in the context of the particular fishery concerned. For example, the tens or hundreds of thousands of dolphins drowned in tuna purse seines in the eastern tropical Pacific and thrown overboard represent, at least in the short run, a prodigious waste of meat which would be fit for human consumption. In the long run such incidental catches may be reducing the productivity of the dolphins as well as their usefulness to the very fishermen who kill them. They are discarded because they are, to the U.S. fishermen mainly concerned, much less valuable than the hold space they would occupy if retained on the boat, which would be better occupied by more tuna. Other fishermen, supplying other markets, might behave differently. The bitter controversy now raging over this question is, however, between different groups—the tuna-boat operators on one hand and public groups concerned with environmental and resource conservation on the other—who place entirely different values on the dolphins, dead or alive.

While dolphins may be depleted by seining for tunas, the existence of marine turtles is made even more precarious than it is in any case as a result of overharvesting, egg collecting, and hindrances to reproduction on beaches, by greatly increased incidental capture in trawls and suspended nets in tropical and subtropical waters. In this instance, however, the incidental catch is usually consumed, even welcomed as a bonus, by the fishermen.

In terms of quantity, the biggest waste by incidental catches is of a wide variety of fishes taken during trawling for shrimp. As with the dolphins, the deciding factor is that hold space is valuable because shrimp are valuable. It has been estimated that for every ton of shrimp caught, 3 MT of fish may be discarded at sea, a total waste of 3–4 million MT annually, amounting to 5 percent of the present world landed-fish catch.[12] To this can be added discarded by-catches from other kinds of fisheries, for example, 15 percent of

---

12. Maschke.

the 7.5 million MT of pleuronectid and gadoid landings, bringing the total to 4–6 million MT. These by-catches are largely composed of fishes with relatively low fat content which, if retained, would be acceptable for use either for human consumption or to produce high-grade livestock feeds. Several approaches to making them utilizable have been suggested, including accompanying the trawler fleet by large collector vessels and separation of the by-catch on board and preserving it by salting, freezing, or making minced-fish products such as sausage or fish protein concentrate (FPC) for enrichment of human diets. However, there are some problems in tropical areas, where most shrimp are caught, in that the unsorted fish can be a cause of deadly ciguatera poisoning.

Yet another source of waste is in the processing of fish destined for direct consumption. The edible part of the fish is, it seems, more fully recovered by mechanized than by traditional manual methods, and in modern fisheries the residues can be more fully utilized by conversion to fish meal. The degree of recovery under various conditions is illustrated by the tentative figures in table 14.

TABLE 14.—EQUIVALENT PROTEIN YIELD, EXPRESSED AS A PERCENTAGE OF THE PROTEIN CONTAINED IN THE LANDED FISH: LIVE WEIGHT*

Product	Developing Countries		Developed Countries	
	"White" Fish	Oily Fish	"White" Fish	Oily Fish
Fresh†	63	54	72	62
Dried	50	44	72	62
Salted and dried	42	36	59	53
Smoked and dried	54	46	72	62
Canned	...	...	58	58
Skinned and frozen	...	...	70	60
FPC:‡				
Type A	70	79–91	...	...
Type A (functionalized)	...	...	77§	88
Type B—small fish	97	97	...	...
Type B—large fish, headed and gutted	77	72	...	...
Fish meal:				
Via chicken or pigs	...	...	31	31
Via fish culture	...	...	45	45

Source.—*Expanding the Utilization of Marine Fishery Resources for Human Consumption,* FAO Fisheries Reports, no. 175 (Rome: FAO, 1975).

*The protein content of fish constitutes 17–20 percent of the live weight; its nutritional quality is similar to but somewhat higher than that in meat and milk and less than that in eggs. The amino acid pattern is about the same in all species, and the lysine content is high. The energy supply from lean ("white") species such as gadoids and pleuronectids is low compared with that from oily species such as clupeoids and mackerels. The fat has a high content of polyunsaturated fatty acids. Oily species are a source of vitamins A and D, which are found only in the livers of lean species. Livers and roes (gonads) are good sources of B vitamins. Sea fish are also good sources of minerals (calcium, phosphorous, iron), and trace elements—especially in bones and therefore in canned small fish and FPC—and also of iodine. Protein quality of cooked sea fish expressed as net protein utilization (NPU)—the percentage of ingested protein which is retained in the human body—is 65–80 percent, as compared with meat = 67–79 percent; chicken = 71 percent; hen eggs = 94 percent; cow's milk = 76–82 percent; cheese = 70 percent; and FPC = 70–75 percent.

†Including processing of off cuts and offal.

‡In FPC type B production, losses only occur in icing the raw material; in type A production, further losses occur during fat extraction.

§Assumed fish headed and gutted.

Finally, we arrive at the question of upgrading the present use of much of the catches as animal feeds. Research has been carried out with varying success over the past 15 years to develop purified meals for direct human consumption. The main idea has been to use these FPCs as ingredients in carbohydrate foods such as breads and in soups, especially in regions where protein deficiency is chronic, in emergencies, and in feeding of special groups such as children and the sick. Statistical data are not sufficiently refined for quantitative assessment of trends in use of specific processes and products. The same must be said regarding the preparation of hydrolyzed and fermented products and of frozen minced fish, especially from lean species. Many such developments are being pursued in a number of countries.

Although losses from inadequate preservation of products are much higher for "traditional" processing methods than for "modern" ones, the latter are not free of cost, and it remains to be demonstrated that there is usually a net gain from introducing them.

Last, it must be said that the wastage at the point of consumption is likely to be higher among the more affluent consumers of industrial products than among the consumers of traditional products in developing countries. That is a question going beyond the scope of discussion of marine fisheries, but it conveniently leads to the next section of this review.

THE DISTRIBUTION OF NUTRITIONAL BENEFITS FROM LIVING MARINE RESOURCES

The distribution of landings of marine catches, according to the conventional continental classification used by the United Nations, is shown in table 15, which gives the marine catch in each continent as a percentage of the world marine catch, excluding Peru. In addition to the prevailing dominance of Asia and Europe, clear continuing trends are evident. Africa, South America, Asia, and the USSR all increased in relative importance throughout the period, while North America and, to a lesser extent, Europe declined. For orientation and to demonstrate how the inclusion of Peru would obscure the other trends, the last column shows the rise and fall of Peru's percentage share of the total marine catch. The table also shows the marine-catch percentage of the total fish and shellfish harvest from the oceans and inland waters.

This series shows a slight steady increase in the relative importance of freshwater-fish catches. The overall trend results from a combination of two factors. The first is that the areas which increased in importance are those in which inland fisheries are more significant. The second factor is that the relative importance of freshwater fisheries shows significant trends in certain areas, as table 16 illustrates.

Because freshwater-fish catches are almost all consumed directly, these factors will be seen to be important when we consider the contribution of fish in general to human nutrition and the relative importance of the marine catches

TABLE 15.—WORLD NOMINAL MARINE CATCHES, BY CONTINENT*

Year	Catch as Percentage of Total, excluding Peru							Marine Total (%)†	Peru Marine (% of Total)
	Africa	America, North	America, South	Asia	Europe	Oceania	USSR		
1950	4.3	20.0	1.7	34.5	32.9	.5	6.1	87.9	.5
1955	5.3	15.4	2.0	39.0	31.1	.4	6.8	86.9	.8
1960	5.6	13.2	2.2	49.9	25.8	.4	8.0	83.8	8.9
1965	6.1	11.4	3.5	39.2	28.1	.4	11.3	83.3	14.4
1970	6.4	9.6	4.4	40.8	24.7	.4	13.4	83.7	18.0
1971‡	5.9	9.7	5.0	42.1	24.0	.5	12.9	83.4	15.0
1975	6.1	9.4	4.8	47.1	24.9	.5	14.4	82.7	6.5
	% Rate of Annual Increase								
1950–55	10.7	.9	10.1	8.8	5.0	2.1	8.5	...	...
1955–60	5.2	.8	6.0	7.1	.4	5.4	7.6	...	...
1960–65	6.4	1.6	14.8	1.7	6.2	2.9	11.8	...	...
1965–70	5.4	1.9	9.6	5.5	2.1	5.8	8.4	...	...
1970–75	.1	−.4	2.5	3.7	.9	3.2	2.3	...	...
1950–75	5.5	.9	8.5	5.3	2.9	3.9	7.7	...	...

*Excludes whales. All columns except last exclude Peru, most of whose catch since 1957 has been anchoveta.
†Marine catch percentage of total aquatic catches (marine plus inland), both excluding Peru.
‡1971 is included to facilitate comparison with analyses of supply and consumption given in Holt and Vanderbilt 1974.

TABLE 16.—RATIO OF MARINE
TO TOTAL AQUATIC CATCHES (%)

A. 1960–75 AVERAGE

Area	Average 1960–75	Trend of Inland Fisheries
Africa	70.9	Increasing
America, North	97.2	Declining
America, South*	87.1	No trend
Asia	75.2	No trend
Europe	97.8	No trend
Oceania	98.7	No trend
USSR	85.7	Declining

B. YEAR BY YEAR

Year	Africa	America, South	USSR
1955	73	83	68
1960	74	77	80
1965	73	85	84
1970	69	93	88
1975	67	94	90

*Excluding Peru.

in this regard. However, it must also be borne in mind that fishery statistics have improved in several ways over the period, and certain catches which were perhaps not at one time recorded now enter into the statistics. All the calculations of growth rates may be influenced by this—causing them to be overestimated—but to different degrees according to the area in question and always to an unknown extent. Catches which are consumed locally or by the fisherman himself are often not recorded. This means not only that catches in developing countries, where much subsistence fishing continues, tend to be more underestimated than catches in industrialized countries but also that inland fish catches might be more underestimated than marine catches.

The continental classification groups extremely diverse countries within the same category. It is therefore necessary to examine the statistical breakdown by economic groupings, again following the United Nations definitions of these. This is shown in table 17. In this case the nature of the data base, as well as changes in classification during the period, necessitate a rather different format and consideration of a shorter time span for establishing trends. Details of subdivisions of the economic classes are given in Holt and Vanderbilt (1974), where the base date is 1971 and trends over the 5 years 1964–71 are

TABLE 17.—WORLD MARINE CATCHES, BY ECONOMIC GROUPINGS*

	1964†	1970†	1971	1974	1975	1964–70†	1971–75	1974–75
Market economies	80.1	78.0	77.0	75.2	73.6	3.9	2.0	−3.2
Developed‡	56.9	53.3	53.6	48.8	48.0	3.0	.3	−2.7
Developing§	23.2	24.7	23.4	26.4	25.6	5.9	5.5	−4.1
Centrally planned economies	19.9	22.0	(23.0)	(24.8)	(26.4)	5.6	(6.8)	(5.4)
Developed	12.6	15.6	15.1	17.4	18.7	7.2	8.8	6.6
Developing	7.3	6.4	(7.9)	(7.4)	(7.7)	2.6	(2.6)	(2.6)
Developed countries	69.5	68.9	68.7	66.1	66.7	3.8	2.4	−.2
Developing countries	30.5	31.1	31.3	33.9	33.3	5.2	4.8	−2.6
World	100.0	100.0	100.0	100.0	100.0	4.3	3.1	−1.1

NOTE.—Figures in parentheses are estimates.
*As % of total, excluding anchoveta.
†Including whales; the years 1971–75 exclude whales.
‡Including Yugoslavia.
§Including Cuba.

considered.[13] Average growth rates for 1964–71 can be taken as close to those for 1965–70 and for 1971–75 as close to those for 1970–75. Table 17 must be interpreted with care also because recent data are not available for the centrally planned developing countries of Asia (People's Republic of China, etc.), and it has been assumed that the rate of increase in that economic class in 1971–75 was the same as in 1964–71, that is, 2.6 percent annually. This assumption also affects the calculated growth rates for the centrally planned countries as a whole and the developing countries as a whole, to whose 1971 catches the Asian centrally planned economies are thought to have contributed 30 percent and 20 percent respectively.

First we note that the developed countries catch about twice as much as the developing countries and that the countries with market economies catch three or four times as much as the centrally planned countries. Some shifts have been occurring, however, with an overall shift in favor of the developing countries. Most of this gain took place in the first part of the period; in the second part corresponding gains were made by the USSR and the eastern European group. The developed market economies have declined steadily from more to less then half of the total. However, the rates of growth of both developed and developing market economies dropped considerably in the 1971–75 period, while the already high growth rate of the USSR plus eastern Europe moved even higher. Growth rates of developed and developing countries as a whole have both fallen, but the latter group have stayed ahead, provided that of China did not decline substantially.

It is more meaningful to examine the catches of economic classes and their increase rates in relation to their population sizes and population growth rates, respectively; this is done in table 18. In 1975 only 29 percent of the population lived in the developed countries; their production per capita was five times that of the developing countries, and it has been increasing. This increase, however, is now entirely due to increases per capita in the countries with centrally planned economies, the developed market economies having declined somewhat.

These catch levels do not provide any indication of the contribution by the living resources of the sea to the nutrition of people in the various economic groups, partly because there are differences in the proportion of catch taken for direct consumption and for fish meal, but mainly because of the pattern of international trade in fish products. The FAO has published data for total aquatic catches for the years 1962, 1966, 1970, and 1975[14] by economic class and type of utilization, that is, for food or feed. These data have been adjusted to estimate the same breakdown for marine catches only; the results are given in table 19; anchoveta is excluded, as usual. Unfortunately, the figures for the

13. Holt and Vanderbilt (n. 6 above).
14. These years represent averages of 3 years; i.e., 1962 = average of 1961, 1962, and 1963, etc.

TABLE 18.—MARINE FISHERIES PRODUCTION PER CAPITA, BY ECONOMIC GROUPINGS*

	Human Population		Catch per Capita (kg)		
	% of 1971 Total†	Rate of Growth (%)‡	1964	1971	1975
Market economies	69.0	2.0	12.4	15.3	15.4
Developed	20.1	1.0	28.4	35.1	34.5
Developing	48.9	2.4	5.2	7.2	7.5
Centrally planned economies	31.0	1.6	7.9	9.8	11.4
Developed	9.6	1.0	15.8	21.2	27.5
Developing	21.4	1.9	4.2	4.7	(4.5)
Developed countries	29.7	1.0	24.3	30.6	32.2
Developing countries	70.3	2.3	4.9	6.4	(6.6)
World	100.0	1.9	11.0	13.5	14.1

*Peru excluded.
†Taken as 3,690 million minus Peru (14 million) = 3,676 million.
‡Average rate, 1963–71, according to UN demographic review, 1974. For the period 1970–75, UN estimates published at that time gave a value of 2.36 percent as the annual rate of population increase in developing countries and .88 percent in developed countries. These values led to an estimate of the 1975 total population (excluding Peru) at 3,975 million, of which 28.4 percent were in developed countries. However, according to the Worldwatch Institute, the overall growth rate had dropped by 1975 to 1.64 percent and the actual 1975 population was 3,920 million (3,905 million, excluding Peru).

centrally planned countries are lumped and are in any case based on very incomplete data.

The proportions of the catches destined for production of feed increased in all economic groupings through the period; thus

Developed market economies = 27 (1961–63) to 40 percent (1975).
Developing market economies = 10 (1961–63) to 22 percent (1975).
Centrally planned economies = 6 (1965–67) to 14 percent (1975).

If anchoveta is included, the proportions in the class of developing market economies become close in recent years to those in the developed market economies.

Evidently, in the market economies, a much higher proportion of the catch (apart from anchoveta) than in the developing countries is for reduction. In the developed countries production for direct consumption has changed little over the 13-year period, and the rate of increase is below the rate of population increase. On the other hand, both types of catch have increased in the developing countries, and it is significant that the rate of increase in production for direct consumption has overtaken the rate of increase for reduction in the last 5 years. This does not mean, however, that the diet of the people of the developing countries is getting substantially better as far as fish is concerned, because just as in earlier years a large-volume export trade in meal was built up, more recently exports of preserved fish, especially frozen fish, have been increasing particularly to the developed countries.

TABLE 19.—MARINE PRODUCTION BY
ECONOMIC GROUPINGS AND BY DISPOSITION

Year	Market Economies				Centrally Planned Economies	
	Developed		Developing			
	Food	Feed	Food	Feed	Food	Feed
	Millions MT					
1962	13.8	5.0	7.1	.8	...	...
1966	15.4	7.2	7.9	1.9	8.2	.5
1970	15.4	9.5	9.8	2.8	9.4	1.2
1975	15.6	10.5	12.8	3.6	11.5	1.9
	% Annual Increase					
1970–75	.3	2.0	5.5	5.2	4.1	9.6
1962–75	.9	5.9	4.6	12.3	...	...

Note.—"Food": direct human consumption; "feed": for reduction to meal.

Nearly one-third of world fish production crosses international frontiers in product form. Available trade statistics do not facilitate separation of freshwater and marine products. Thus, since it appears likely that a higher fraction of products of marine than of freshwater origin are exported from the country in or by which they are caught, the percentages given in table 20 may be underestimates of the relative volume of trade in marine products. The figures given are based on the catches of 150–60 reporting countries or territories which themselves account for about 90 percent of the total aquatic catches. The first column of table 20 shows that the relative total volume of trade doubled between 1950 and 1965 but has declined since then. As in other statistical series, however, the inclusion of anchoveta in the catch data and of meal from anchoveta in the product data confuses the general picture. In this case not only is the overall importance of the trade flow somewhat exaggerated, but the time sequence and pattern of change is also distorted, as can be seen from the second column of table 20. First there was a period of growth from 20 to 25 percent from 1948 to the early 1950s, then another shift up to 30 percent in the early 1970s.

To see how trade might affect diets in the various economic groupings, a synthesis of international trade flows and of utilization patterns is required. Such a synthesis has been made for the year 1970 only; in a subsequent issue of *Ocean Yearbook* it is hoped to show the development of trade and utilization patterns over time. Table 21 summarizes the derived data for total supplies and consumption and for these quantities per capita. In this context the term "supplies" means marine catch (production) plus imports, less exports, all in

TABLE 20.—INTERNATIONAL TRADE
IN COMMODITIES DERIVED FROM
LIVING AQUATIC RESOURCES*

Year†	Including Anchoveta	Excluding Anchoveta‡
1938	23	23
1948	20	20
1949	20	20
1950	22	22
1951	23	23
1952	24	24
1953	25	25
1954	25	25
1955	26	26
1960	32	24
1965	41	26
1970	39	24
1971	39	27
1972	37	29
1973	35	31
1974	32	30
1975	(34)§	(30)§

*% of total catches. Exported product weights are converted to live-weight equivalents; whale products excluded. Catches of anchoveta were small until the late 1950s and did not affect trade figures until that time.

†The figures for years 1949–74 are averages of triplets; e.g., 1949 is the average of 1948, 1949, and 1950. This grouping smooths the series which fluctuates partly because products traded in a particular year do not all originate from catches taken the same year.

‡Anchoveta meal excluded from product totals and anchoveta from catch totals.
§Estimates.

live-weight equivalents; "consumption" is calculated as the supply of food plus one-fifth of the supply of feed.[15]

Table 21 shows the net results of the flow of trade mainly between the developed and developing market economies, which leads to supplies in the former exceeding their production. There is a substantial amount of movement of fish both as food and as meal within the two classes; trade by the developed centrally planned economies with the other classes is roughly balanced. The increased supplies in the developed market economies naturally increases their consumption, but not proportionately, because fish meal predominates in the net flow. The column headed "Per Capita Consumption" (col. 8) shows the final result to be a consumption rate of protein derived directly or indirectly from marine resources (excluding anchoveta) in the developed market economies about four times that in the developing countries. The last two columns show per capita the supplements of aquatic origin that were also available in 1970, that is, from inland catches consumed domestically and directly and from fish meal derived from anchoveta caught by Peru and Chile

15. The method of computation is explained in Holt and Vanderbilt.

TABLE 21.—FOOD ORIGINATING FROM MARINE RESOURCES IN 1970, BY ECONOMIC GROUPINGS*

	Supplies (Millions MT)			% of Total			Per Capita (kg)			
	Food	Feed	All	Supply	Con-sumption	Pro-duction	Supply	Con-sumption	+In-land	+Anchoveta
Market economies	26.1	10.5	36.6	78	75	78	15	11	1.5	.8
Developed	16.5	10.0	26.5	56	49	52	33	25	.6	2.1
Developing	9.6	.5	10.1	22	26	26	6	6	1.8	.2
Centrally planned economies	(8.9)	(2.0)	(10.9)	(22)	(25)	(22)	(10)	(9)	4.6	0
Developed	6.0	1.8	7.8	15	17	15	22	19	2.8	0
Developing	(2.9)	(.2)	(3.1)	(7)	(8)	(7)	(4)	(4)	5.4	0
Developed countries	22.5	9.3	31.8	72	66	67	30	23	1.3	1.4
Developing countries	12.5	3.2	15.7	28	34	33	5	5	2.9	.2
World	35.0	12.5	47.5†	100	100‡	100§	13	10	2.4	.3

*Excluding Peru and anchoveta, except as indicated in last column. Figures in parentheses are estimates.
†This total differs from the 1970 catch of 47.8 million MT because of changes in fishmeal inventories and lags in trade generally, and because of the weighing procedures used in allowing for missing information for some countries. In fact it is largely accidental that the two figures are so close.
‡Total consumption was 37.5 million MT live-weight equivalent.
§47.8 million MT, as in table 1, third column.

in 1969 or 1970. The two kinds of supplements together added about the same amount to each class of market economies, but the inland supplement was important in the centrally planned economies.

Published figures relating roughly to the middle of the review period—1964–66—allow quantification of the role of fish in diets in the economic classes of countries. These are summarized in table 22, which refers to nutritional levels indexed as average daily intake of calories and proteins. The range of levels as well as of the balance of vegetable protein, fish, and other animal protein is evident, but of course the country groupings obscure the great differences between one country and another within an economic grouping. The fish-protein intakes quoted are for direct consumption only, and they include consumption of freshwater fishes. Information in table 21 could in principle be used to estimate what fraction of the "animal, not fish" intake category given in table 22 was derived from marine-fish catches, via fish meal and livestock. It is in practice hazardous to do this, because the two series of data refer to different years, in which the quantities of fish meal produced differed. Making rough allowance for the fact that world fish-meal production per capita, both including and excluding anchoveta, was 40–50 percent higher in 1970–71 than in 1964–66, it may be concluded that in the developed countries with market economies not more than 2 percent of the intake of nonfish animal protein in 1970 was derived from marine sources. This calculation is, however, sensitive to the factor used for conversion of live-fish weights to livestock production. If the factor is in fact nearer to 2:1 than to 5:1, the estimate of the indirect marine contribution to protein intake would rise to 5 percent.

Last, by putting together data from tables 19 and 21, it is possible to see directly the apparent nutritional results of the interclass trade. From table 19 can be calculated the consumption patterns that would result if there had been no interclass exchanges; the results are shown in table 23. It is assumed for the purposes of table 23 that the fish meal produced in the developing countries, instead of being largely exported, could have been fully utilized in those countries. This is of course quite unrealistic, but the comparison shows that local utilization of fish meal would go only a certain way toward achieving a different nutritional balance; more attention would need to be given either to increasing food-fish production by developing countries or reversing the net flow of trade in that commodity. Naturally the actual dietary trend depends on the use to which the monetary gain is put from the present trade by the developing countries, the prices secured, and the distribution of the revenue within the developing countries.

The current claims of EEZs are already causing changes, directly and indirectly, in the patterns of production, but it is too soon to attempt to evaluate these. Such evaluation will in any case be extremely complex, even after the new pattern has emerged over the next few years, because of the conclusion of bilateral agreements, establishment of joint ventures, arrangements for transshipments of fish and fish products, and so on. There has been no compilation of statistics on the amounts of fish caught at distances less or greater than 200

TABLE 22.—THE PLACE OF FISH IN DIET: 1964–66 AVERAGE

	Fish Consumption (kg/Year per Capita)*	Protein				Calories			
		Consumption (g/Day)	Animal Origin (%)	% Fish of Total	% Fish of Animal	Consumption (cal/Day)	Animal Origin (%)	% Fish of Total	% Fish of Animal
Market economies	11.1	63.9	35	5.0	14	2,350	17	1.1	6
Developed	21.5	86.5	57	6.4	11	2,960	32	1.3	4
Developing	6.9	54.6	21	4.0	20	2,100	9	.6	7
Centrally planned economies	10.6	67.9	26	3.7	14	2,390	15	.7	5
Developed	16.5	90.5	40	3.1	8	3,132	23	.6	3
Developing	7.8	57.8	15	4.2	27	2,040	9	.7	8
Developed countries ...	19.8	87.8	51	5.2	10	3,019	29	1.1	4
Developing countries ..	7.2	55.6	19	4.1	22	2,083	9	.6	7
World	11.0	65.3	32	4.4	14	2,370	17	.8	5

*Marine and freshwater combined.

TABLE 23.—CONSEQUENCES OF INTERCLASS TRADE
IN MARINE FISH PRODUCTS, 1970

Economic Class	Excluding Anchoveta		Including Anchoveta	
	Trade	No Trade	Trade	No Trade
Market economies:				
Developed	25 (26)	23 (24)	27 (28)	23 (24)
Developing	6 (8)	10 (12)	6 (8)	11 (13)
Centrally planned economies	9 (14)	9 (14)	9 (14)	9 (14)

NOTE.—Figures in parentheses represent totals for all aquatic products.

miles from coasts and by whom—in particular in the present context whether they were taken by vessels from the same economic-class country or not. In 1970–75 the situation was roughly as follows: (1) Between 85 and 95 percent of the total catch was taken within 200 miles of shore; that fraction taken offshore was taken almost entirely by a few developed countries. (2) About 50 percent of catches within 200 miles of shore were taken in areas off countries other than the flag states of the fishing vessels. (3) A very small proportion of the total catches were taken off the coasts of developed countries by vessels flying flags of developing countries. (4) Of the catches by vessels from developed countries which were taken off the coasts of other countries, over half were taken off the coasts of other developed countries. Table 24 gives details available for the year 1972; in that year, which happens to be the year of the first anchoveta collapse, it does not make much difference whether or not Peru is exluded.

The longer-term pattern will also depend greatly on the interpretations that may eventually be given to provisions in the various "negotiating texts" of the current Conference on the Law of the Sea—for determination of "total allowable catches" (TAC) by coastal states within their EEZs, the levels at which these are set with reference to estimates of maximum sustainable yields, and hence the practical meaning of the concept of the "surpluses" that may by agreement be taken by other states. A formulation of management objectives more in accordance with modern biological and social concepts of renewable resource use, as well as with the requirements of the emerging legal regime, is now being attempted by the scientific community.

Nearly all current unilateral EEZ claims are a source of dispute between neighbors. While the justification commonly given is that they will facilitate conservation, they are in fact more directed to securing a redistribution among nations of fish catches and/or economic benefits from fishing. They reflect skepticism as to the likelihood of the UNCLOS reaching agreement that would effect an equitable redistribution and a recognition that regional fisheries bodies have not in general been successful in placing effective restraints on fishing effort. This is particularly true where they have been established in

TABLE 24.—LOCATION OF MARINE CATCHES IN 1972, BY ECONOMIC GROUPINGS

Economic Category	% of 1972 Catch Taken	% of Catch Taken by Location			
		Off Own Coast	Off Other Coasts	Off Coast of Developed Countries	Off Coast of Developing Countries
Developed countries ..	61	58.3	41.7	31.0	10.8
Developing countries:					
Including Peru	39	95.1	4.9	2.7	2.3
Excluding Peru	32	93.6	6.2	3.4	2.9

areas adjacent to developing countries but which have been exploited in recent years by Northern Hemisphere fishing powers.

There are continued attempts to distinguish the twin issues of setting overall "development" and "conservation" targets and agreeing on shares of yields or other benefits among interested nations, and to resolve them through different international arrangements. Experience shows, however, that these twins are in fact Siamese, with a single blood stream which may contain nutrients or poisons. It has in specific cases (such as the regulation of whaling) proven necessary to negotiate agreements simultaneously on overall quotas and on national allocations. The status quo has not provided a very stable basis for such negotiations. Even if EEZ claims are not directly related to the needs of peoples for food, they might at least provide such a base, if boundary disputes can be resolved. However, there are a number of problems to which the UNCLOS has not devoted sufficient attention, polarized as it has been on questions concerning seabed minerals and the rights of states in coastal zones. These problems were recently raised in COFI with reference to the Northeast Atlantic. In that area most stocks of fish are shared by at least two and sometimes by a dozen nations. Nearly all the fish live wholly or mainly within 200 miles of some coasts. When an overall TAC has been agreed (on a basis yet to be determined), the question arises of how it is to be subdivided, with respect to both fishing nation and coastal zone. If "historical" data are to be used, over what time periods are past national catches to be averaged? Further, fishing has been concentrated more in some parts of the area of stock distribution than in others for various practical reasons, so a particular stock might have been exploited until now mainly in one or a few zones. The distribution of fishing has not been unchanging, but neither has been the distribution of fish. Environment factors have caused both year-to-year variations and medium- and long-term trends of change in geographic distribution. Fishing has itself changed the distribution of the fish, and to varying degrees, depending on the type and intensity of the fishing. Different life stages of a species may occupy parts of the total area, and it then becomes necessary to weigh the contribution to stock "productivity" in one EEZ that happens to be a spawning or nursery area against that in another EEZ where adult fish feed. No guidelines for such decisions have yet been devised, and there are very few areas where enough is known about the biology of the fish to permit reasonable discussion.

A beginning has recently been made, after several years of relative stagnation, in elaborating a useful economic theory of marine fisheries management. This stagnation can be attributed in part to the fact that open and serious economic discussion has been effectively banned from international fishery bodies. There attention has been directed to regulating output while ignoring input and avoiding questions of economic efficiency. Recently the implications of taking account of the rate at which future values are discounted in cost-benefit analysis, as it might be applied to renewable resources, have been analyzed and suggestions made, on the basis of new theory, as to how yield allocations in biologically and technologically interdependent fisheries might be made rationally.[16]

In the UNCLOS texts, an amount of fish which may be allocated to noncoastal states when the coastal state has decided how much it wishes to take from an area is defined as a "surplus." This usage derives from the concept that a resource not "fully utilized" is a resource wasted. The concept can no longer be regarded as a biologically valid one, and its use generally serves the special interests of the more powerful fishing nations. It is remarkable that those who have maintained that it is wasteful not to take a maximum sustainable yield appear not to notice that it is wasteful to expend fuel oil, and other natural resources in limited supply, by striving for catches at MSY level when one could often secure, say, 80 percent of that level with half the fishing effort and hence with considerably greater total net profits.

The economic and ecological desirability of permitting catches only up to levels somewhat less than calculated MSY is reinforced by the great uncertainties in calculations. Such considerations have led to attempts in recent years to reformulate management objectives. The following principles have been proposed:

1. Ecosystems should be maintained in such a state that both their consumptive and nonconsumptive values[17] to humanity can be realized on a continuing basis.
2. Options for different use by present and succeeding human generations should be ensured.
3. Risk of irreversible changes or long-term adverse effects of exploitation should be minimized.
4. Decisions should include a safety factor to allow for limitations of knowledge and inevitable imperfections of management institutions.
5. Measures to conserve one resource should not be wasteful of another.
6. Survey or monitoring, analysis, and assessment should precede planned use and accompany actual use of a resource, and the results should be made available promptly for critical public review.

It will not be easy to put such principles into operational form, but in some agencies, for example, safety factors are now being incorporated in management decisions. A change of approach which could give more time to learn how to manage fisheries on ecological principles concerns principle 5. By giving as much consideration to restraints on growth of fishing power and effort as has been given in the past to setting catch quotas, application of most of the other principles would probably be assured.

16. See C. Clark, *Mathematical Bioeconomics* (New York: Wiley, 1976); and L. G. Anderson, "Optimum Economic Yield of an Internationally Utilized Common Property Resource," *Fishery Bulletin USA* 73 (1975): 51–65.

17. Nonconsumptive in the sense of, e.g., aesthetic or ecological values in the case of whales.

Principle 2, concerning options for future use of resources, has been given virtually no attention with respect to its practical application. Some new kinds of institutional arrangements to this end may be called for; it has been suggested that international resource-management institutions should, in making decisions regarding types or levels of exploitation and allocations of present benefits, necessarily consult and heed a new official entity established as a guardian of the interests of future generations of mankind. The entity would be empowered and enabled to collect data and evidence and authorized to intervene in the deliberations of the managing institutions. As far as marine resources, living and nonliving, are concerned, the guardian might be given a special responsibility with respect to the areas beyond national jurisdiction.

The idea that the offshore living resources might be regarded as a common heritage of mankind just as much as minerals of the deep seabed has been advanced tentatively in some quarters, but this is far from being accepted, and its institutional consequences have barely been discussed.

The UNCLOS is now beginning to recognize the need for continuing, flexible, and comprehensive instruments for ocean management. Not only is there a great deal to be done to improve the management of fishing, if continuing benefits are to be expected from the living resources of the sea, but a move toward integrated management of different uses of ocean space is now required. There are growing ecological and physical interactions between fisheries and mariculture; oil, metallic ore, and other hard-mineral extraction; energy extraction; navigation; communications; waste disposal; recreational industry; and other ocean uses.[18]

ADDITIONAL REFERENCES[19]

Expanding the Utilization of Marine Fishery Resources for Human Consumption. FAO Fisheries Reports, no. 175. Rome: FAO, 1975.

FAO Department of Fisheries. *Atlas of the Living Resources of the Seas.* Rome: FAO, 1972. Published in a trilingual edition (English, French, Spanish).

Goldberg, E., and Holt, S. "Whither Oceans and Seas?" Paper prepared for the

---

18. The institutional implication of these interactions have been analyzed in Arvid Pardo and E. M. Borgese, *The New International Order and the Law of the Sea*, I.O.I. Occasional Papers, no. 4 (revision) (Malta: International Ocean Institute, 1976). This 220-page review analyzes the institutional arrangements at the global level, for management of the uses of ocean space, including fishing. A more popular illustrated account is given in E. M. Borgese, *Drama of the Oceans* (New York: Abrams, 1976). Results of the Pardo and Borgese study are summarized in J. Tinbergen, *Reshaping the International Order: A Report to the Club of Rome* (New York: Dutton, 1976).

19. A list of major data sources used for this review and references to prior partial analyses prepared under the sponsorship of the International Ocean Institute, Malta.

Second International Conference on Environmental Futures, Reykjavik, June 1977.

Gulland, J. A., ed. *The Fish Resources of the Ocean.* West Byfleet, Surrey: Fishing News, Books, by arrangement with FAO, 1971. Reports the FAO-conducted World Appraisal of Fisheries Resources; previously issued in 1970 in the FAO Fisheries Reports.

Holt, S. J. "The Food Resources of the Ocean." *Scientific American* 221 (1969): 178–97. An overview of the state of world fisheries in the late 1960s.

Holt, S. J., and Vanderbilt, C. "Mediterranean Fisheries." *Options méditerranéenes* 35 (1976): 30–39.

Kreuzer, R., ed. *Fishery Products.* West Byfleet, Surrey: Fishing News, Books, by arrangement with FAO, 1974. Report and proceedings of a technical conference containing a wealth of information about recent developments in fish utilization. The general reader is recommended to read at least Kreuzer's introductory paper, "Fish and Its Place in Culture," in the section on "The Influence of Tradition and Change."

"Mediterranean Fisheries." *Options méditerranéenes* 19 (1973): 80–87. Reprinted in *The Mediterranean Marine Environment and the Development of the Region,* edited by N. Ginsburg, S. J. Holt, and W. Murdoch. Malta: International Ocean Institute, 1974. Revised version printed in *Tides of Change: Peace, Pollution, and Potential of the Ocean,* edited by E. M. Borgese and D. Kreiger. New York: Mason/Charter, 1975. Updated and expanded anlysis of this data appears in Holt and Vanderbilt 1976.

The State of Food and Agriculture. Rome: FAO, published annually. Each volume includes a section on fisheries, and occasional volumes contain review articles on particular aspects of world fisheries and marine affairs. The 1975 volume (published 1976) contains, in addition to the usual "World Review," a "Mid-Term Review and Appraisal of the Second United Nations Development Decade" (DD2), which began in 1971. The review covers both production and trade in the 4 years 1971–74, but trends are also shown from 1961, the beginning year of DD1. Data given permit comparisons between fisheries developments and agriculture and forestry.

Yearbooks of Fisheries Statistics. Rome: FAO, published annually. Publication occurs about 1 year after the end of the calendar year to which statitistics refer. Appears in two volumes, *Catches and Landings* (even numbers) and *Production and Trade* (odd numbers). Catches of large whales, by weight, are given by S. J. Holt in FAO Document ACMR/MM/SC/7. Rome: FAO, 1976.

Progress of Aquaculture

T. V. R. Pillay
Food and Agriculture Organization

INTRODUCTION

Aquaculture, or the culture of aquatic animals and plants in fresh, brackish, and marine waters, is an ancient occupation in some parts of the world, whereas in others it is a relatively new means of food production. In the last decade there has been worldwide recognition of the need to adopt on a large scale methods of husbandry to meet the world's future requirements for many aquatic products. In a number of industrially advanced countries substantial support for scientific research in this field became available, and many marine and freshwater biological institutions utilized this opportunity to initiate significant investigations, particularly into problems related to marine aquaculture. The volume of literature on the subject has shown a marked increase, and national and international aquaculture meetings have become frequent. An increasing number of private companies, especially in Western countries, have invested in aquaculture enterprises, although a number of them are still only in the research and development phase. This general upsurge of interest in aquaculture has made some of the developing countries realize that they have been neglecting a traditional means of food production that has a great potential. Its role in integrated rural development, generation of employment, and saving of foreign exchange is being increasingly appreciated. The impending changes of the Law of the Sea have given an urgency to the need for developing alternative or additional sources of fishery production and means of employment for surplus fishermen, even in some of the industrially advanced distant-water fishing countries. As a result of this, aquaculture has come to be accepted as a high priority in the national economic plans of at least some countries, and consequently some notable expansion of the industry has occurred.

AQUACULTURE PRODUCTION

Although there has been a worldwide expansion of aquaculture activities, only a few countries have established suitable systems for the collection and

---

This essay is partly based on a paper entitled "State of Aquaculture, 1976," presented at the FAO Technical Conference on Aquaculture held in Kyoto, May 26–June 2, 1976.

© 1978 by The University of Chicago. 0-226-06602-9/77/1978-1003$01.51

compilation of aquaculture statistics. The highly dispersed location of production units, the lack of specialized enumerators for data collection, and the unwillingness of producers to provide detailed statistics have all contributed to this situation. However, the task of estimating production has become slightly easier in most countries, as some of the basic information is now available for making estimates, in contrast to the situation in 1966–67, when the production of fish through culture could be calculated only on a rough percentage basis. It was then estimated to be about 7 percent of the world catch of freshwater fish, or 1 million tons annually.[1] In 1970, a partial estimate was made on the basis of information collected from 36 countries: the total fish production through aquaculture amounted to 2.6 million tons.[2] A more comprehensive but again rough estimate was made in 1973, which showed that the production of fish was about 3.7 million tons and, together with the production of crustaceans, mollusks, and seaweeds, amounted to about 5 million tons.[3]

The estimates of production in 1975, based on data provided by different countries, are given in table 1, which shows that the world production through aquaculture has now risen to over 6 million tons. Of this, 66 percent consists of freshwater, brackish-water, and marine fish; about 16.2 percent, mollusks; 17.5 percent, seaweeds; and 0.3 percent, crustaceans.

Owing to the incompleteness of previous and indeed even the current data, one hesitates to make comparisons. Certainly better estimates are now available for some countries, and the coverage has also increased. However, by also using associated information, one can determine where production is increasing or declining. Although the percentage of increase varies considerably, in almost all countries there has been an increase in production during the last decade and also since the 1973 estimate. As is only to be expected, percentages of increase are the highest in countries in which aquaculture has been newly introduced or a new system adopted. The major increases have occurred in countries in which sufficient importance has been given to this industry in national economic-development plans and the essential investment and support services have been at least partially provided. For example, the production of farmed fish rose almost five times in a period of 7 yr in Japan.[4] In the Philippines it increased 32 percent in a period of about 3 yr. In Poland it has

---

1. President's Science Advisory Committee, *The World Food Problem—a Report of the President's Science Advisory Committee*, vol. 2, *Report of the Panel on the World Food Supply*, (Washington, D.C.: Government Printing Office, 1967).

2. T. V. R. Pillay, "Problems and Priorities in Aquaculture Development," in *Progress in Fishery and Food Science: University of Washington College of Fisheries Fiftieth Anniversary Celebration Symposium*, University of Washington Publications in Fishery, no. 5 (Seattle: University of Washington Press, 1972), pp. 203–8.

3. T. V. R. Pillay, "The Role of Aquaculture in Fishery Development and Management," *Journal of the Fish Reserve Board of Canada* 30, no. 12, pt. 2 (1973): 2202–17.

4. Japan Fisheries Association, *Fish Farming in Japan* (Tokyo: Japan Fisheries Association, 1975).

TABLE 1.—ESTIMATED WORLD PRODUCTION
THROUGH AQUACULTURE IN 1975

Country	Tons	Country	Tons
Finfish:*		Kenya	400
China (excluding		Nepal	400
Taiwan Province	2,200,000	Venezuela	332
Taiwan Province of China	81,236	Switzerland	300
India	490,000	Ireland	207
USSR	210,000	South Korea	169
Japan	147,291	Netherlands	129
Indonesia	139,840	Ecuador	90
Philippines	124,000	Central African Empire	43
Thailand	80,000	Cyprus	40
Bangladesh	76,485	Ghana	40
Nigeria	75,000	Zambia	29
Poland	38,400	Paraguay	23
South Vietnam	30,000	Ivory Coast	10
Yugoslavia	27,000	Puerto Rico	9
Rumania	25,000	Shrimps and prawns:*	
Hungary	23,515	India	4,000
USA	22,333	Indonesia	4,000
Italy	20,500	Thailand	3,300
Madagascar	17,392	Japan	2,779
East Germany	16,000	Ecuador	900
France	15,000	Taiwan	549
Czechoslovakia	12,222	Singapore	105
Israel	12,169	South Korea	30
Denmark	12,120	Oysters:*	
Brazil	12,000	Japan	229,899
West Germany	8,900	USA	129,060
Sri Lanka	7,659	France	71,448
Egypt	7,000	South Korea	56,008
Mexico	7,000	Mexico	45,000
Malaysia	6,559	Thailand	23,000
Zaire	5,000	Taiwan	13,359
Cuba	4,500	Australia	9,200
Hong Kong	4,019	Canada	5,080
Norway	3,500	United Kingdom	3,000
Austria	2,500	Spain	2,289
United Kingdom	2,000	Netherlands	1,500
Finland	1,940	Chile	870
Belgium	1,800	Philippines	782
Tanzania	1,500	New Zealand	700
Burma	1,500	Senegal	191
El Salvador	1,208	Mussels:*	
Canada	1,103	Spain	160,000
Greece	900	Netherlands	100,000
Chile	800	Italy	30,000
Uganda	700	France	17,000
Singapore	680	West Germany	14,000

TABLE 1. *Continued*

Country	Tons	Country	Tons
South Korea	5,578	Cockles/other mollusks:*	
Chile	1,260	Malaysia†	28,000
Yugoslavia	287	Taiwan	1,243
Philippines	182	South Korea	733
New Zealand	150	Philippines	11
Tunisia	60	Seaweeds:*	
Clams:*		Japan	502,651
South Korea	24,920	China	300,000
Taiwan	13,898	South Korea	244,795
Philippines	33	Taiwan	7,347
Scallops:*			
Japan	62,600	World total aquafood	6,102,289

*Total tons: finfish, 3,980,492; shrimps and prawns, 15,663; oysters, 591,386; mussels, 328,517; clams, 38,851; scallops, 62,600; cockles/other mollusks, 29,987; seaweeds, 1,054,793.
†Cockles.

increased about four times, and in Rumania, over two times in the same period. Increase in seaweed production in South Korea has been over 14 times in the last 5 yr, and in Japan, over two times in that period. Except in a small number of localized cases, there have been no major problems of domestic marketing of aquaculture products because of sustained demand. Stricter quality controls imposed by importing countries have, however, affected somewhat the export of products such as oysters and trout, but this has led to greater attention to the environmental conditions under which culture operations are carried out and the sanitary quality of products.

It is fairly well known that, despite the loud pronouncements of interest in aquafarming, only a few countries have so far implemented a well-balanced development program with adequate financial and technical support. If this is considered, the increase in production now reported is certainly encouraging and justifies guarded optimism about the achievement of further increases.

Some significant changes have occurred in the general concept of aquaculture. Traditional practices being largely governed by local conditions and needs, farmers seldom felt the need to intensify operations. Low-density culture with minimum inputs and low production per unit area has often been more economical than intensive farming, which involves the raising of dense populations and heavy inputs. For various reasons, this is changing fast, and many countries are now turning to intensive and semi-intensive systems. For example, in Israel fish ponds producing less than 2.5 tons/ha are no longer profitable.[5] Even in some of the eastern European countries that have fish

5. S. Sarig, "Fisheries and Fish Culture in Israel in 1973," *Bamidgeh* 3, no. 26 (1974): 57–83.

farms with ponds too large for intensive culture, higher overall productivity and profits are achieved through a combination of duck raising and fish farming.

Aquaculturists now try to exercise greater control over the environment and stocks in their farms, even in what are called "trap ponds." They seek to produce seed by artificial propagation and resort in most cases to at least supplemental feeding. Even in culture systems in which natural food such as algae or plankton is raised through fertilization, improved systems of management are adopted to intensify production.

Several projections of aquaculture-production potentials have been made; one often quoted is a five- to 10-fold increase in 2–3 decades. As part of the preparatory activities of the FAO/UNDP Aquaculture Development and Coordination Programme, representatives of some 34 developing countries have been assisted through regional workshops in Asia, Africa, and Latin America to prepare 10-yr aquaculture-development plans, setting out targets of production based on existing and proven systems of culture.[6] These targets, together with the production increases from other countries, are expected to contribute to a doubling of world production through aquaculture in 10 yr. Based on present world production, this would amount to nearly 12 million tons by the end of 1985, and, if this pace of increase is maintained, it may be reasonable to expect at least a fivefold increase by the end of this century.

In considering present and future production through aquaculture, it may be interesting to examine some of the factors that have affected the industry in recent years. At least three of these have served to focus renewed attention on aquaculture in the last few years. One is the increased cost of fishing due to steep increases in fuel cost; second, the fear of decrease in fishery production by countries that depend on fishing in foreign waters as a result of new laws of the sea; and third, the need in some countries to relocate and find alternative or additional employment for large numbers of excess fishermen or underemployed farmers. The continued high demand in developed countries for high-valued species like shrimps and prawns has also served to promote interest in aquaculture in countries that wish to increase foreign-exchange earnings. On the other hand, steep increases in the cost of some of the essential inputs, particularly feed, have hit the intensive culture systems, as for catfish (*Ictalurus* spp.) and rainbow trout (*Salmo gairdneri*). This was largely due to an increase in the cost of fish meal, a major ingredient in most fish feeds. A similar

---

6. Food and Agriculture Organization/United Nations Development Programme, *Aquaculture Planning in Africa,* Report of the First Regional Workshop on Aquaculture Planning in Africa, Accra, July 2–17, 1975, ADCP/REP/75/1 (Rome: FAO, 1975); *Aquaculture Planning in Asia,* Report of the Regional Workshop on Aquaculture Planning in Asia, Bangkok, October 1–17, 1975, ADCP/REP/76/2 (Rome: FAO, 1976); and *Planificación de la acuicultura en América Latina,* Informe de la Reunión Consultiva Regional de Planificación sobre Acuicultura en América Latina, Caracas, noviembre 24–diciembre 10, 1975, ADCP/REP/76/3 (Rome: FAO, 1976).

increase in price and indeed even of availability has occurred also for a widely used larval food, the brine shrimp *(Artemia)*. The situation regarding feed prices has eased somewhat recently, although it still remains a major element in the cost of production. This has led to a search for cheap substitutes, and some progress has already been reported in this direction. The shortage of fertilizers in developing countries and the allocation of all production for agricultural purposes has also been an adverse factor.

Even though the prevailing environmental concern has resulted in increased interest in protecting the aquatic environment and has in many cases helped aquaculture, in some countries it has had the reverse effect. Aquaculture has been classed as a polluter, and waste-discharge regulations meant for terrestrial animal production have been made applicable also to aquaculture. This has greatly hampered the expansion of the industry.

Water pollution has affected aquaculture production in coastal waters in some countries; for example, oyster production has remained more or less at a stagnant level since 1965 in Japan, and in the United States there has been marked decrease in oyster production due to the closure of polluted oyster beds. Large-scale mortality of oysters, from known and unknown causes in important culture areas like the French coast, has resulted in a serious decline of production.

The absence of a legal framework under which aquaculture enterprises could be established and operated has also stood in the way of the development of the industry, particularly in industrially advanced countries. Some of the problems faced by entrepreneurs in getting the necessary permits from different, only peripherally interested agencies and the need to conform to regulations not relevant to aquaculture have been revealed in recent discussions on the subject.[7]

Although there has been considerable interest in the private sector in investing in aquaculture, it is largely confined to the culture of exportable products, and in many cases the technologies for such types of culture are still under development. Only a small percentage of the present aquaculture production relates to such high-valued exportable products, and therefore investment support for aquaculture as a whole remains at a very low level. These factors, together with a general lack of appreciation of the economic viability of aquaculture, have stood in the way of a speedier development of the industry. The inevitable failure of some badly conceived enterprises has also given some credibility to arguments of detractors of aquaculture.

---

7. T. Loftas, "New Fish Farming Lobby," *New Scientist* 65, no. 928 (1974): 878–79; and T. V. R. Pillay, "Planning Aquaculture Development—an Introductory Guide," mimeographed (Rome: FAO, 1975) (partly based on the discussions at the FAO/NORAD Round Table on "The Strategy for Development of Aquaculture as an Industry," Svanøy, Norway, July 1–10, 1974).

TECHNOLOGICAL ADVANCES

During the last decade, a number of technical advances have been made and some new technologies developed in the field of aquaculture. Some of the research done during this period has been extremely valuable in understanding the scientific bases of certain traditional practices and has made possible the improvement and modernization of old technologies. Cage and enclosure/pen culture are outstanding examples of this. Though such forms of culture have existed for many years, particularly in Asia and the Far East, they were not very well suited to application elsewhere. Modifications or changes in the design, building materials, installation, and operation, together with the preparation of suitable feeds, including floating pellets, have made these into technologies capable of application in other regions. Cage culture of catfish, salmon *(Salmo salar* and *Oncorhynchus* spp.), trout, and yellowtail *(Seriola* spp.) and enclosure/pen culture of salmon and milkfish *(Chanos chanos)* have achieved the level of commercial-scale operations in some countries, particularly in the United States, Norway, West Germany, the Philippines, and Japan. Several types of cage have been designed for use under different hydrological conditions. The marked increase in the production of trout and salmon in Norway is due to the introducion of cage and enclosure culture. Cage culture accounts for a very high proportion of the present-day production of yellowtail in Japan. In a period of about 5 yr, over 5,000 ha of fish pens have been established in Laguna de Bay in the Philippines, producing 7,500–10,000 tons of milkfish annually.

Similarly, polyculture of fish is an ancient technology in Asia, especially China and India.[8] Experimental work in recent years has given a fuller understanding of the significance and value of the system. This is becoming an accepted practice in the culture of Chinese and Indian carps *(Ctenopharyngodon idella, Hypophthalmichthys molitrix, Aristichthys nobilis* and *Catla catla, Labeo rohita, Cirrhina mrigala),* especially because of the increasing need for the adoption of low-energy systems. In fact a system of combined culture of these groups of carps along with the common carp *(Cyprinus carpio)* has been developed in India, yielding up to 8,500 kg/ha with only modest supplemental feeding. Polyculture of common carp and tilapia *(Tilapia* spp.) has served to increase average production in Israel and provide an additional crop, which is of special importance from the point of view of marketing in that country because of the ceilings imposed on the production of common carp.

Catfish farming in the United States is a good example of the development of new technologies to keep up with rapid expansion of production requirements. Little was known about catfish farming until 1960. Research and

---

8. T. V. R. Pillay, ed., *Proceedings of the FAO World Symposium on Warm-Water Pond Fish Culture, Rome, Italy, 18–25 May 1966,* FAO Fisheries Reports, no. 44 (Rome: FAO, 1967), vol. 1.

promotional efforts were effective in developing new techniques of production and creating a demand for cultured catfish. Consequently a considerable increase in production took place in the late 1960s and early 1970s. From 1966 to 1975 the area under catfish culture increased by more than 10-fold; so did the production (50,223 tons valued at U.S. $40 million). The many advances in catfish-culture techniques, including breeding and fry production, feeding, pond management, disease control, and harvesting, have served to develop a well-advanced technology that has the potential for transfer to other areas.

Pond culture of tilapia has existed in tropical countries, particularly those in Africa, for many years. However, the problem of overpopulation of ponds and consequent stunting due to frequent breeding of this fish made tilapia culture generally unattractive. Two broad systems of culture have evolved in recent years which make culture of tilapia viable and capable of being practiced on a commercial scale. One consists of the use of selected species such as *Tilapia nilotica,* which grow fast and attain a fairly large size, and feeding them with suitable pelleted or other feeds so that a majority of the stock will attain marketable size before large-scale breeding occurs, that is, within 3–5 months.[9] The second system is by the use of hybrids, which are all or mainly males, and this reduces breeding and overpopulation of ponds. Although neither of these systems is entirely satisfactory in overcoming the problem of prolific breeding, they do serve to raise up to 5,000 kg/ha of marketable fish a year and provide a fair return on investment. Further work on hybridization may succeed in consistently producing all-male hybrids, but it has been shown that, if tilapia were cultured in suspended cages, breeding of the stock could be eliminated.

Although traditional systems of extensive shrimp and prawn farming have existed in Asia for many years, it was only in the 1960s that intensive culture based on hatchery-produced larvae and juveniles was developed in Japan. Because of the high demand for shrimps in world markets and the inability of natural fisheries to meet the demand, an almost worldwide interest in their culture has developed in the last 10 yr. Initial attempts were made to transfer the Japanese technology, but this has not been generally successful. Significant advances have, however, been made in a number of countries, notably the United States, South Korea, Taiwan Province of China, the Philippines, Indonesia, France (and Tahiti), and the United Kingdom, in establishing successful methods of artificial propagation and hatchery production of juveniles of a number of *Penaeus* species. Development of techniques for the maturation of penaeid shrimps such as *P. merguiensis* and *P. monodon* has added

---

9. Food and Agriculture Organization/United Nations Development Programme, *Perfectionnement et recherche en pisciculture—Cameroun, Gabon, République Centreafricaine, Congo,* Rapport préparé par le Centre Technique Forestier Tropical, FI:DP/RAF/66/054 (Rome: FAO, 1973).

to the efficiency of seed production.[10] However, large-scale intensive monoculture of shrimps to marketable size on an economically viable basis has not yet developed outside of Japan, with the probable exception of Taiwan Province of China. Improvements and intensification of traditional practices in some of the Southeast Asian countries, particularly Indonesia, have led to better yields in polyculture with brackishwater fishes, mainly the milkfish.[11]

The culture of the freshwater prawn *Macrobrachium rosenbergii* existed in Asia on a very small scale, but interest in commercial fishing of this species developed as a result of the very high demand for shrimps and prawns in the world market and the methods devised in 1961–69 for the propagation of this species, which has the advantage of a shorter life history than panaeid shrimps.[12] Many improvements have been brought about in the technique of mass raising of their larvae.[13] There are now a large number of institutions, agencies, and private companies throughout the world engaged in research and development activities related to *Macrobrachium*. In the United States alone there were at least 25 of them in 1974.[14] While the hatchery technology and procedures are considered adequate, high-density commercial farming techniques have yet to be perfected. Most commercial enterprises use earth ponds for production purposes, and, where physical and economic conditions are favorable, some operations have been profitable. Further technological advances are needed, however, to bring prawn farming to the level of large-scale commercial ventures.

The "hanging method" of oyster culture (in which oysters are suspended from rafts, long lines, or racks) is not really a new system, but it has been adopted during the last decade on a much wider scale in many countries and has undergone modification and adaptation. This system permits high

---

10. Food and Agriculture Organization, *Informe del Simposio sobre Acuicultura en América Latina, Montevideo, Uruguay, 26 noviembre–2 diciembre de 1974*, FAO Fisheries Reports, no. 159 (Rome: FAO, 1975), pp. 17–61; and K. H. Alikunhi et al., "Preliminary Observations on Induction of Maturity and Spawning in *Penaeus monodon* Fabricus and *Penaeus merguiensis* de Man by Eye Stalk Extirpation," *Bulletin of the Shrimp Culture Research Centre, Jepara, Indonesia* 1, no. 1 (1975): 1–11.

11. P. G. Padlan, B. S. Ranoemihardjo, and E. Hanami, "Improved Methods of Milkfish Culture. I. Increasing Production in Shallow, Undrainable Brackishwater Ponds," *Bulletin of the Shrimp Culture Research Centre, Jepara, Indonesia* 1, no. 1 (1975): 33–39.

12. S. W. Ling, *Methods of Rearing and Culturing Macrobrachium rosenbergii (de Man)*, FAO Fisheries Reports, no. 57 (Rome: FAO, 1969), 3:607–19.

13. T. Fujimura and H. Okamoto, "Notes on Progress Made in Developing a Mass Culturing Technique for *Macrobrachium rosenbergii* in Hawaii," in *Coastal Aquaculture in the Indo-Pacific Region*, ed. T. V. R. Pillay (West Byfleet, Surrey: Fishing News [Books], 1972).

14. H. L. Goodwin and J. A. Hanson, "The Aquaculture of Freshwater Prawns: *Macrobrachium* Species," in *Summary of the Proceedings of the Workshop on the Culture of Freshwater Prawns, St. Petersburg, Florida, 1974* (Waimanalo, Hawaii: Oceanic Institute, 1975).

production rates per unit area and reduces losses from predation. The use of nylon net bags for holding oysters and mussels for raising is becoming common in some European countries.

A system of aquaculture that has received world attention in recent years is the culture of eels *(Anguilla* spp.*)* in stagnant or flowing-water ponds or in net enclosures. Although the technique of eel culture originated many years ago, it is only in the last decade that it has received wide attention, largely as a result of the expansion of Japanese eel culture and a rapid increase in demand for the product. The culture technology has undergone considerable improvement, particularly in the production of satisfactory feeds, which accounts to a large extent for the expansion of the industry in Japan. However, the culture is still based on elvers caught from the wild, and a shortage of elvers in Japan has resulted in a worldwide search for new sources and the establishment of a sizeable export trade for elvers in many countries, particularly in Europe.

Use of domestic and farm wastes for fish culture is an age-old practice in Asia, especially in China, Malaysia, and Indonesia, but only recently has this practice attracted wider attention as a means of recycling wastes to protect the environment and at the same time contribute to food production. This has led, on one hand, to wider use of human and animal wastes in aquaculture and, on the other, to critical studies on the benefits and risks involved. Many developing countries are adopting fish farming in association with duck, pig, or cattle raising so as to utilize the wastes for fertilizing fish ponds. The area of sewage-fed fishponds in India has expanded to over 12,000 ha. Duck-*cum*-fish farming, which is an efficient means of recycling duck droppings, has become widespread in eastern Europe (Czechoslovakia, Hungary, Poland, and Rumania) and has now been introduced in the Central African Empire and Nepal. Experimental studies on the use of domestic sewage have been undertaken in some countries, and the results of these studies are in many respects encouraging. It has been shown conclusively that productivity in aquaculture of fish or shellfish can be significantly increased by the controlled use of treated sewage for increased production of plankton.

Experiments in the culture of shellfish fed on algae raised in sewage effluents in Woods Hole, Massachusetts, have provided the basis for a "pilot-plant" multiple-production system.[15] Simpler systems suited for rural communities, in which animal and domestic wastes are conditioned for use in aquaculture and also generate by-products such as methane for use as fuel and algae as animal feed and fertilizer, have been successfully used in some of the South Pacific islands.[16]

15. J. H. Ryther, "Preliminary Results with a Pilot-Plant Waste Recycling Marine Aquaculture System" (paper presented at the International Conference on the Renovation and Recycling of Wastewater through Aquatic and Terrestrial Systems, Bellagio, Italy, 1975).

16. G. L. Chan, "The Use of Pollutants for Aquaculture—Conditioning of Wastes for Aquaculture," *Proceedings of the Indo-Pacific Fisheries Council* 15, no. 2 (1974): 84–91.

Another type of waste recycling in aquaculture developed in recent years and already used on a production scale in temperate and subtropical climate is the use of waste heat. Extension of growing period, better feed conversions, acclimatization of organisms that cannot withstand lower temperature, and earlier attainment of marketable size or maturity have been possible through this practice.

A new technology that has developed at a rapid pace during the last decade is the reconditioning and reuse of water for aquaculture. The availability of water will become a limiting factor for aquaculture development in an increasing number of countries in the future, and the cost of pre- or posttreatment may sometimes be prohibitive. These factors and the restrictions that are imposed on discharges from aquaculture installations have given an added significance to this technology insofar as aquaculture is concerned. As a result, several systems of water reconditioning for reuse have been developed. Not all of them are equally economical, but the use of reconditioned water has become a fairly common practice in hatcheries, especially in the United States.

During recent years "sea ranching," "aquarange farming," or "artificial recruitment," as it is variously described, has become an accepted technology in aquaculture. Some years ago attempts to improve fish populations with hatchery-raised fish were generally considered unsuccessful. This concept has undergone considerable change with improvements in the methods adopted, particularly feeding, release of adequate numbers of hatchery-raised animals after they have grown to a size at which they can fend for themselves, and recapture of a satisfactory percentage of released fish to make the operation viable. Anadromous fish are obviously best suited for such aquarange farming, and it is estimated that over 2 billion juvenile anadromous fishes are artificially produced annually and released into the fresh and marine waters of the world, mainly from government hatcheries. Privately owned sea ranching has already made a beginning. According to recent studies in North America, for every dollar spent for the hatchery raising and release of coho salmon *(Oncorhynchus kisutch)* smolts, the return has been $7.00; and for chinook salmon *(O. tshawytscha)*, $3.50. Large-scale release of hatchery-raised *Penaeus japonicus* in the Inland Sea of Japan is believed to have resulted in substantial improvement in the local shrimp fishery with a cost/benefit ratio between 2:5 and 2:20. Appropriate and adequate stocking operations have led to the improvement of fishery resources of many inland reservoirs and natural lakes.

With regard to the techniques per se, two major problem areas have received special attention during recent years: the controlled reproduction of cultivated animals and the formulation and manufacture of artificial feeds. The Chinese carps, Indian carps, and the grey mullet *(Mugil* spp.), which do not generally breed in confined waters of culture installations, have been induced to breed by the administration of pituitary hormones. Induced breeding by pituitary injections or adjustment of photoperiod has become a recognized practice in fish culture, and a number of cultivated and cultivable species have been bred experimentally. Methods of mass raising of fish larvae have also been developed, and these are now employed by culturists in the case of Chinese and Indian carps. One of the early advances in shrimp-culture technology was artificial propagation based on gravid females collected from the wild. Collection of adequate number of breeders at the required time is often expensive and difficult. Consequently the successful maturation of shrimps in captivity in the laboratory by eyestalk ablation is a breakthrough of considerable significance, but much more remains to be done to perfect this technique for large-scale application.[17] Similarly, the controlled reproduction of oysters and hatchery production of oyster seed is a development of considerable significance in oyster farming, particularly because the supply of seed oysters from natural reproduction is decreasing, due to environmental degradation, and imported seed has become quite expensive. These circumstances contribute to making hatchery production of seed economically viable in countries like the United States. Controlled reproduction has also helped in genetic selection of strains for special qualities such as resistance to diseases.

With the introduction or expansion of aquaculture in wider geographical areas, international exchange of cultivated species has become widespread. This has given rise to considerable controversy about the introduction of nonindigenous species, some considering it unavoidable for rapid expansion of aquaculture, others terrified by the possible adverse effect of introductions on local fauna and flora. While introductions into certain countries or areas have been totally banned, indiscriminate introductions continue in others. Though appropriate and adequate guidelines are still lacking, scientists and many aquaculturists accept the need for extreme care and critical study of all relevant environmental and behavioral information before deciding on such introductions. Recognition of the dangers of transmission of communicable diseases through shipments of live organisms has led to consideration of an international convention to control the spread of communicable fish diseases,[18] and a draft convention is now being considered by interested countries.

ORGANIZATION OF AQUACULTURE

A good proportion of current fish production through aquaculture comes from China and the Socialist countries of Europe. In these countries fish culture is undertaken on state farms, in communes, or through cooperatives, and because of its role in communal welfare the industry seems to receive

17. Food and Agriculture Organization, *FAO Aquaculture Bulletin* 6, nos. 2–3 (1974): 4, 15; and Alikunhi et al.
18. W. A. Dill, ed., *Report of the Symposium on the Major Communicable Fish Diseases in Europe and Their Control, Amsterdam, April 20–22, 1972*, European Inland Fisheries Advisory Commission Technical Papers, no. 17 (Rome: FAO, 1972); and Food and Agriculture Organization, *Government Consultation on an International Convention for the Control of the Spread of Major Communicable Fish Diseases, Aviemore, Scotland, 30 April–1 May 1974*," FAO Fisheries Reports, no. 149 (Rome: FAO, 1974).

special attention. In other countries, particularly those industrially advanced, aquaculture production is largely undertaken by the private sector. Individual farmers dominate the scene, but many small and large companies have in recent years become interested and involved in research and development activities or commercial production. An incomplete survey in 1975 revealed the existence of some 833 companies in 26 countries, the majority of which are in North America, Japan, and western Europe. In developing countries most production is still in the hands of small-scale operators or subsistence-level farmers, although there are also some instances of involvement by large commercial firms. The farmers are in most cases dependent on government agencies for support services, including technical and financial assistance. Only in some exceptional cases has the government been sufficiently responsive to the needs of the aquaculture industry. The anomalous legal status of aquaculture—not recognized as an agricultural, animal husbandry, or truly fishery activity in the legal sense, to be eligible for governmental support and other incentives—the aquaculturists face formidable problems in establishing or operating their enterprises. Nevertheless, restrictive legal provisions designed for and relevant only to other industries are readily applied to aquaculture, like, for example, the application to aquaculture of waste-disposal and disease-control regulations meant for animal husbandry. Although in some countries such as the United States and the United Kingdom aquafarmers are organizing and attempting to influence government policies, in most others they do not have the political or social clout to force governmental action necessary to solve the problems of the industry.

In government organizations aquaculture forms part of the fishery sector. The tertiary phase of aquaculture industry has close similarity to that of the fishing industry, but since the production phase is more allied to agriculture and animal production, there is a large body of opinion favoring better linkage with these forms of food production in order to benefit from allied experience and from the many incentives offered by governments for their promotion. With increasing public interest and modest increases in investments, a distinctly negative attitude and rivalry appear to be developing between aquaculture and conventional-fisheries interests. This is an unfortunate and unnecessary situation, as at least in the foreseeable future there is little likelihood of aquaculture supplanting conventional fishing. On the other hand, the world demand for fishery products is increasing steadily, and it is generally accepted that it will not be possible to meet the demand by conventional fishing alone. Changes that may be brought about in the marketing and price structure as a result of enhanced production through aquaculture has to be accepted as a healthy development; and, through proper integration of production and marketing, stability can be ensured in both sectors.

Research and extension are two major support services required from governments in most countries. There has undoubtedly been considerable interest in aquaculture research in many developing and developed countries.

The number of experimental stations in developing countries has certainly increased, though very few of them have adequate research personnel, equipment, and necessary facilities. In many developed countries, research on aquaculture problems has expanded as a result of financial support available from funding agencies. It has generally been recognized that close coordination of research, which is highly diffused at present, is essential to obtain maximum benefit from the investment and effort expended. The multidisciplinary nature of aquaculture science and the need for teamwork to undertake systems-oriented research to improve or develop new aquaculture systems is gradually being understood.

One of the weakest areas of aquaculture activities at present is extension services, which creates serious problems in the implementation of development programs. This is closely connected to the shortage of adequately trained and experienced field personnel with the ability and knowledge to assist aquafarmers. The large majority of aquaculture personnel employed in government organizations and in senior positions even in the industry are those who developed their knowledge of the subject through specialized research and then tried it in the field, often learning by trial and error. This is no doubt a costly and slow means of acquiring expertise for large-scale development programs. Recognizing this, some countries have established extension-training centers. But there is an urgent need to upgrade the facilities for well-balanced theoretical and practical training in these centers to meet the requirements for extension personnel. Regional symposia and workshops organized by the FAO during the last 10 yr have emphasized the need for accelerated cooperative efforts in establishing multidisciplinary research programs and training core personnel, including extension staff.[19]

Information exchange has a vital role in a developing science like aquaculture, particularly since the industry has to depend to a large extent on transfer of technology for its expansion. During the last decade, a number of periodicals, including *Aquaculture, FAO Aquaculture Bulletin, Commercial Fish Farmer and Aquaculture News,* and *Fish Farming International,* have come into being. While these and a number of recent books represent significant progress, they fulfill only a small part of the information needs. The National Aquaculture Information System (NAIS), sponsored by the U.S. Sea Grant Program, has started a computerized file of information on the subject, and this is a good beginning in information collection, storage, and dissemination.

---

19. General Fisheries Council for the Mediterranean, *Report of the Symposium on Brackishwater Aquaculture,"* Report of the General Fisheries Council for the Mediterranean, vol. 11 (Rome: FAO, 1972), pp. 37–55; Indo-Pacific Fisheries Council, *Proceedings, 14th Session, Bangkok, Thailand, 18–27 November 1970,* in *Proceedings of the Indo-Pacific Fisheries Council,* sec. 1 (Bangkok: IPFC Secretariat, FAO, 1971); T. V. R. Pillay, ed., *Coastal Aquaculture in the Indo-Pacific Region* (West Byfleet, Surrey: Fishing News [Books]: 1973); and publications listed in n. 6 above.

REGIONAL AND INTERREGIONAL COOPERATION

In a field such as aquaculture, in which future developments would involve transfer of technology on a large scale, regional and interregional cooperation assume special significance. During the last 10 yr the regional fishery bodies of the FAO, especially the Indo-Pacific Fisheries Council (IPFC), the General Fisheries Council for the Mediterranean (GFCM), and the European Inland Fisheries Advisory Commission (EIFAC), have been actively engaged in promoting regional cooperation in aquaculture development, with particular reference to the assessment of available sites, evaluation of the economics of different types of operation, effect of environmental pollution, and control of communicable fish diseases. Regional symposia organized by these three bodies as well as by the newly formed Committee on Inland Fisheries of Africa (CIFA) led to the identification of high-priority problems for research in the respective regions. Efforts were made to organize programs of research through voluntary cooperation of national research institutions, viz., IPFC, GFCM, and EIFAC Cooperative Programmes of Research on Aquaculture. As was only to be expected, the progress of investigations through such voluntary efforts without adequate financial or technical backing has been extremely slow. The establishment of a subregional institution for aquaculture research, the Aquaculture Department of Southeast Asian Fisheries Development Center (SEAFDEC), in the Philippines in 1973 is an important attempt to bring about cooperation in aquaculture research and training.

A working group on aquaculture appointed by the Technical Advisory Committee (TAC) of the Consulative Group on International Agricultural Research (CGIAR), which met in Spoleto, Italy, February 4–8, 1973, made a detailed review of research needs in aquaculture in developing countries.[20] Later a TAC subcommittee on aquaculture prepared specific proposals for the establishment of coordinated regional networks of research centers in Asia, Africa, and Latin America to undertake systems-oriented interdisciplinary research to solve the problems faced in the large-scale application of selected technologies.[21] The FAO/UNDP Aquaculture Development and Coordination Programme organized a series of planning workshops in 1975, which considered *inter alia* the need for regional and interregional cooperation to implement national aquaculture-development plans.[22] In view of the investments involved in terms of scientific manpower, equipment, and facilities, the

---

20. Technical Advisory Committee, *Report of the TAC Working Group on Aquaculture, Technical Advisory Committee of the Consultative Group on International Aquaculture Research* (Rome: FAO, 1973).

21. Technical Advisory Committee, *Report of the TAC Sub-Committee on Aquaculture, Technical Advisory Committee of the Consultative Group on International Agricultural Research* (Rome: FAO, 1974).

22. Food and Agriculture Organization/United Nations Development Programme, *Aquaculture Planning in Africa; Aquaculture Planning in Asia;* and *Planificación de la Acuicultura en América Latina.*

workshop recommended that long-term research in this field and training of senior aquaculture personnel should be organized on a regional basis. This proposal was strongly endorsed by the FAO Technical Conference on Aquaculture held in Kyoto, Japan, May–June 1976,[23] and the FAO/UNDP Aquaculture Development and Coordination Programme has subsequently initiated action for the establishment of regional research and training centers in Africa and Latin America.

The emergence of the World Mariculture Society (established in 1970) as a truly international professional association and the decision to organize an International Aquaculture Federation through affiliation of national and regional societies are significant recent developments that could contribute to worldwide cooperation by the scientific community and industry in this important field.

OUTLOOK FOR THE FUTURE

As indicated earlier, available production estimates show steady increases in many countries. Areas under aquaculture are expanding, and improvements in technology are making it possible to intensify production and obtain higher yields. The fact that annual yields range from a few hundred kilograms per hectare to over 20 tons/ha shows what improved technology and provision of essential inputs could achieve in terms of increased production. New systems of culture to be developed through research and experimentation will also contribute to increased production. Even using existing technology, it is expected that a doubling of world production can be achieved in the next 10 yr. The area now under aquaculture is estimated to be on the order of 3–4 million ha. A 10-fold expansion of this area is considered feasible if the necessary investment becomes available. Improvement of techniques has already shown the possibility of increasing average production at least two to three times per unit area or unit volume of water in a relatively short period of time. One can therefore be reasonably optimistic about global increases in production, even though the rate of increase in individual countries or through individual systems of culture may vary considerably. Such an expansion of aquaculture is bound to increase the availability to people of acceptable animal-protein foods. Whether it will be equally available and within the reach of all segments of the population will largely depend on national policies rather than aquaculture technology.

There are species and systems of culture to produce aquafoods at prices that the "common man"—if that species can be identified—can afford. There are also high-valued species and systems of culture to meet the needs of the

23. Food and Agriculture Organization, *Report of the FAO Technical Conference on Aquaculture, Kyoto, Japan, 26 May–2 June 1976*, FAO Fisheries Reports, no. 188 (Rome: FAO, 1976).

luxury market and to export to earn foreign exchange. Aquaculture could become a major element in integrated rural development and serve to generate employment for a good number of unemployed and underemployed people in the rural areas of developing countries, thus helping to arrest the drift of populations to urban areas and mitigate the problems faced in the cities due to this drift. Wider application of aquaculture techniques for artificial recruitment and transplantation could help to build up new fishery resources or enhance existing stocks, giving rise to what is referred to by some as "farmed fish fisheries." Aquaculture can contribute to the development of sport fisheries and baitfish production for commercial or sport fishing. Mention should also be made of the possibility for the introduction or expansion of pearl culture in many countries. Greater use of inland and coastal waters for aquaculture would also inevitably involve effective measures for protecting the aquatic environment. It could provide efficient means of recycling agricultural and domestic wastes and thus help waste disposal and environmental protection in general. Besides all these benefits, it is well worth recognizing that aquaculture denotes a step in man's evolution from hunter to herdsman and husbandman. Aquaculture therefore deserves to be considered from a wider angle than that of only the increased production of animal protein or the creation of economically viable enterprises.

The planned doubling of production in the next 10 yr or the five- to 10-fold increase in 3 decades will need accelerated transfer of technology, massive financial investments, suitable legislation, intensive research, manpower training, and development of institutions and other essential infrastructures. If these are left to evolve through the inner pressures of an emerging industry, much valuable time will be lost, and it will take many more years for the industry to fulfill its potential. Recognizing this, the FAO Technical Conference on Aquaculture adopted the following "Declaration on Aquaculture" to serve as a policy instrument reflecting the determination of governments and the world community to elevate aquaculture to an appropriate level in national and international priorities:

Kyoto Declaration on Aquaculture

The FAO Technical Conference on Aquaculture, assembled in Kyoto, Japan, on 2 June 1976, after a week-long review of present status, problems, opportunities and potential for the culture of fish, crustaceans, molluscs and seaweeds, declares:

(1) That aquaculture has made encouraging progress in the past decade, producing significant quantities of food, income and employment; that realistic estimates place future yields of food at twice the present level in ten years, and five times the present level in 30 years if adequate support is provided.

(2) That aquaculture, imaginatively planned and intelligently applied, provides a means of revitalizing rural life and of supplying products of high nutritional value, and that aquaculture, in its various forms, can be

practiced in most countries, coastal and landlocked, developed and developing.

(3) That aquaculture has a unique potential contribution to make to the enhancement and maintenance of wild aquatic stocks and thereby to the improvement of capture fisheries, both commercial and recreational.

(4) That aquaculture forms an efficient means of recycling and upgrading low-grade food materials and waste products into high-grade protein-rich food.

(5) That aquaculture can, in many circumstances, be combined with agriculture and animal husbandry with mutual advantage, and contribute substantially to integrated rural development.

(6) That aquaculture provides intellectual challenge to skilled professionals of many disciplines, and a rewarding activity for farmers and other workers at many levels of skill and education.

(7) That aquaculture provides now, and will continue to provide, options for sound investment money, materials, labour and skills.

(8) That aquaculture merits the fullest support and attention by national authorities for integration into comprehensive renewable resource, energy, land and water use policies and programmes, and for ensuring that the natural resources on which it is based are enhanced and not impaired.

(9) That aquaculture could benefit greatly from support and assistance from international agencies, which should include the transfer of technology activily planned and executed, with research carried out in centres representative of various regions concerned.[24]

---

24. Ibid., pp. 43–47.

Living Resources

The Cradle of Sea Fisheries[1]
Rudolf Kreuzer
Food and Agriculture Organization

The scribe in the central office of the temple of Baba puts the stylus down. The landings delivered by the sea fishermen during the last month, recorded on small clay tablets, have been summed up and engraved on a larger tablet. The scribe allows the still-wet surface to dry in the hot sun and then brings the tablet to the office of the manager. Eniggal, the *nubanda* (inspector), technical director, and highest temple official, goes over the figures:

10 tar-fish	(*Pterois miles*? [Scorpaenidae or Thriacanthidae])
180 kin-fish	(*Sillago sihama*)
180 ki-KAxSAR-fish	(*Pampus argenteus*, or silver pomfret)
840 SE+SUHUR-fish	(*Polydactylus tetradactylus*, or Threadfin)
2,500 fresh gir-fish	(*Caranx sexfasciatus*)
39 Ka-lub-fish	(unidentified)
25 gis PI-fish	(unidentified)

The long list continues: 2,545 different kinds of fish delivered by Lugalsaklaltuk, 720 by Galatur, and 170, together with 134 mussels, by

---

Thanks are expressed to the administration of the British Museum, London, the Bibliotheca Apostolica Vaticana, Rome, and the Pontificio Instituto Biblico, Rome, for assistance in compiling research materials.

1. EDITORS' NOTE.—Elsewhere in this volume Sidney Holt gives a résumé, mainly in statistical terms, of the recent history of sea fisheries and an analysis of their contribution to human food supplies over the past quarter-century. It especially concerns events of the last 5 years. But fishery statistics did not begin in the twentieth century A.D., or even in the fourteenth century, when the kings of England received numerous petitions for the suppression of that devilish new instrument, the trawl, which was said to destroy huge quantities of young fish and animals of the seabed—initiating a controversy which continues to this day. Here, the editors of the *Ocean Yearbook* have decided to include the following short piece by Holt's colleague in Rome, Dr. Rudolf Kreuzer. By profession Kreuzer is a food technologist, but the breadth of his interest in the sea, and especially in fisheries as demonstrated by this article, exemplifies the multidisciplinary approach to marine affairs which the *Ocean Yearbook* was established to promote. Rome is the seat not only of FAO, the first established organization in the United Nations system, but also of one of the world's oldest "international organizations"—the Roman Catholic church. Kreuzer's off-duty hours in the Vatican library revealed this ancient story of the first documented sea fishery. It may add perspective to current attempts to bring more order into the dealings of humanity with the ocean.

© 1978 by The University of Chicago. 0-226-06602-9/77/1978-1004$01.09

Lugalmegalgal.[2] These fish are recorded as "fish for the offering table." The list closes with the name of the month, the name of Eniggal, who is responsible for correct recording, and the names of the ruling city governor and his wife. "No outstanding debts?" inquires Eniggal. "No, *nubanda*." "Then the fishermen can have their barley rations the next time they come to the temple."

The scribe goes back to his office. He puts the small tablets with the notes on the individual deliveries into a clay jar, labels and seals it, and sends the jar to the archives. He hands the large tablet with the monthly summary over to the warehouse so that the barley rations for the fishermen can be prepared. This tablet, then, goes to the archives too.

Some days later, Nesang and Lugalsaklaltuk, the foremen of two groups of sea fishermen, aided by some of their people, carry a load of fish to the temple of Baba, the city goddess of Lagash and the deity who "wills man's destiny." The fishermen had sailed in sultry heat about 20 kilometers up river from the coast of the Persian Gulf to the town of Lagash. They place large baskets full of fish in front of the spacious storehouse in the precincts of the temple. Eniggal appears with a scribe. He peers closely at the fish, and the scribe engraves on a small clay tablet their number, whether fresh or processed, and their names, as well as the quality assessed by Eniggal. Eniggal, promoted from scribe of the temple to its highest official, was sharp and had managed to serve under two city governors. No fish of bad quality could be hidden from his critical eyes.

Satisfied, he turns to the fishermen and orders them to claim their barley rations from the granary. This is their reward for harvesting the goddess' "property": about 60 *sila* (50 liters) of barley for a full fisherman and 30 for an assistant. Wool rations for this quarter of the year are also due: 2 *mina* (1 kilogram) for a fisherman and 1 for an assistant. The return to the fishing village is happy, and prayers of thanks are rendered to Nanse, the divine protectress of the fishermen. It is the year 2336 B.C.[3] in Lagash, in southern Sumaria.

Four thousand years later the clay tablets of the archives of the temple of Baba are dug up by grave robbers, and the world of the Sumerian peasants and fishermen who worked for the temple is restored to life. The tablets cover a period of 19 years and report mainly on the economy of the temple during the reign of two city governors, Lugalanda and Urukagina.

Lagash is thought to have consisted of four towns, each having a large estate centered on a temple or a group of temples. Baba's temple was the center of the town Lagash, and it controlled an estate of about 11,000 acres of arable land. This was managed by the wife of the city governor. A hierarchy of priests and officials administered the temple and estate, and a labor force of about 1,200 men and women worked there. Among them were 100 fishermen, 90

---

2. J. Bauer, "Altsumerische Wirtschaftstexte aus Lagash," *Studia POHL*, vol. 9 (Roma: Instituto Biblico, 1972).

3. Dating is according to Bauer (ibid.).

herdsmen, 80 regular soldiers, 25 brewers, 25 scribes, 21 bakers and cooks, 20 craftsmen, and 250–300 slaves.

The fishermen were free citizens, not slaves. They settled in fishing villages or in separate quarters outside the town, where they formed communities of their own. They were, according to a religious belief, bound to work for the temple as "servants of the gods." Ea, god of the waters, was believed to have taken clay from the sea, from which he created the king to care for the temples. Then he created ordinary man to serve the gods. Temple land was the land of the gods, and the produce of fields and waters was Ea's property. Barley was at Urukagina's time the traditional standard for measuring the value of goods. Silver was not yet in general use, although taxes, debts, and rents were calculated in silver which was weighed out for each commercial transaction. Coins were unknown. The amount of barley allotted to the individual depended, among other factors, on his age and status and on the type of work he performed. The temple supplied boats and fishing gear. It was not until 300 years later that the system of barley rations was replaced by wages.

Fisheries owed their eminent position in ancient Sumer to the importance fish had as food for men and in the cults of the gods. Legend has it that in the sea Ea, the god of waters, wisdom, and magic, created Adapa as the ideal man and taught him to navigate with the aid of the wind (see fig. 1). Adapa became priest of the god and established daily offerings of fish in Ea's temple at Eridu. To secure regular supplies of fish for the altar Adapa applied his knowledge of the sea and of sailing: "He [Adapa] enters the sailingboat and a wind blew and

FIG. 1.—Ea the god of waters with streams of water flowing from his shoulders and fish swimming in them. Detail of an Akkadian cylinder seal, about 2360–2150 B.C.

his ship rode the waves, with the gimusu [punting pole] he steers his ship over the far stretching sea," proclaims an almost 4,000-year-old Babylonian hymn.[4] It praises Adapa as: "The blameless, the clean of hands, the ointment priest, the observer of rights.... With his clean hands he arranges the offering table, without him the table cannot be cleared.... He steers the ship, he does the prescribed fishing for Eridu."[5]

The legend of Adapa demonstrates the religious dimensions of fishing in the cult of Ea at Eridu. Eridu was the most ancient and sacred town of Sumeria, where the first settlement on the plain between the Euphrates and the Tigris appeared perhaps as early as 5000 B.C. The legend relates the development of the sea fishery to this town, located by a lagoon off the Persian Gulf, a few days' journey from Lagash. Eridu may be regarded as the cradle of sea fishing. In fact, archaeologists have discovered a heap of fish bones on an offering table in the prehistoric remains of Ea's great temple at Eridu. Moreover, the world's oldest model of a sailboat was discovered at Eridu in a prehistoric grave.

Sea fishing had, over the centuries, made great headway along the coast of the Persian Gulf. By 2300 B.C. the numerous temples at Lagash employed about 600 fishermen, including those of the temple of Baba. Fishermen, together with sailors, were the second largest population group in the area of Lagash, following the peasants and outnumbering the herdsmen.

The grave robbers who uncovered the archives of the temple of Baba apparently did not steal the entire archives, since the tablets recovered do not account for the total catch of the fishermen or the yearly fish consumption of the temple. Nevertheless, the tablets convey an idea of the huge quantities of fish delivered. They indicate that in the month called Distribution of Wool 7,056 fish were brought to the temple by Nesang. In another month fishermen serving under Nesang, Lugalsaklaltuk, Galatur, and Lugalmegalgal delivered 7,209 sea fish of good quality. Later, 5,235 fresh and processed fish were credited to Nesang and Lugalsaklaltuk with the remark that these were the first deliveries. In the month of the feast of Eating of Malt of Ningirsu the group of Nesang and Lugalsaklaltuk supplied 11,510 pieces of fish, as well as 220 turtles and 24 *sila* (20.4 liters) of turtle oil. There is little information on fish prices. According to one tablet the merchant Ninulil bought 1,800 gir-fish of the best quality and paid 6 *sila* (about 5.1 liters) of barley. This corresponds to about 41.6 grams of silver. *Sila* and *gur* were the main units for measuring capacities. One *sila* was equal to about 0.85 liter, and 1 *gur* corresponded to 144 *sila*. Units for measuring weights were *mina* (mana) and *shekel* (gin). One *mina*, comprising 60 *shekels*, corresponded to about 500 grams.[6] On another occasion 10 trap nets

---

4. A Salonen, *Die Fischerei im alten Mesopotamien* (Helsinki: Suomalainen Tiedeakatemia, 1970).
5. S. N. Kramer, *The Sumerians* (Chicago: University of Chicago Press, 1963).
6. Ibid.

of marsh crabs were sold for 2½ *gur* (about 360 liters) of barley, worth about 20.83 grams of silver.

One *gur* (122.4 liters) of barley was worth 1 *shekel* of silver (about 8.33 grams). One *gur* of dates or 2 *mina* (about 1 kg) of wool had the same value. Five woolly sheepskins cost 2 *shekels* of silver and a cow 6½. Using the price for gin-fish (15 pieces for 6 *sila* of barley), the value of the 7,065 pieces of fish—4,620 of them were gir-fish—can be calculated. It corresponds to about 2,826 *sila* (about 2,402 liters) of barley. Assuming that Nesang's group consisted of 12 men, the production of each of these sea fishermen during the Distribution of Wool month was 588 fish, worth about 295 *sila* (about 250 liters of barley). One fisherman received about 50 liters of barley for 1 month's work.

The same calculation may be applied to the deliveries of sea fish (11,510 pieces) in the month of the Eating of Malt of Ningirsu. In this case the monthly production per fisherman serving under Nesang and Lugalsaklaltuk was 479 fish with a value corresponding to 214 liters of barley. These figures are impressive. They do not, however, reflect the average monthly production per fisherman bound to the temple of Baba.

Economic tablets from the time of the Third Dynasty of Ur (2133–2006 B.C.) give the value of fish in *shekel* (*gin*) and *se*, one *shekel* being equal to 180 *se*. Table 1 shows comparable fish prices. Notably, scaled and gutted fish fetched about double the price of whole ungutted fish. It also seems that by this time greater quantities of fish were offered in the open markets.

The Sumerians used 324 names for fish which denoted about 90 different species: 36 of these can be identified with fair certainty.[7] The temple fishermen of Lagash supplied in greatest numbers four-finger threadfin (*Polydactylus tretradactylus*) and horse mackerel (*Caranx sexfasciatus*). Nesang and Lugalsaklaltuk are credited with the supply of tens of thousands of these fish. Silver pomfret (*Pampus argenteus*) and gestug-ha (unidentified) were also abundant. Coastal and brackish water fishermen caught large numbers of *Sillago sihama*, and the giant carp (*Barbus esocinus*), the marsh carp (*B. sharpeyi*), and *Hilsa ilisha* were the most frequently harvested freshwater fish.

On shore, before being transported to the temple, the larger fish were scaled, gutted, washed, and split. Washing was considered so important that different terms were applied to the fish depending on the way they were washed: best quality, "immediately following" or second quality, and bad. Fish were processed in nine different ways. Besides drying, salting, and smoking they were preserved in oil, soused in brine, or fermented. The chief of the kitchen himself supervised preservation in oil. Six of the 30 storehouses in the temple precincts were equipped for preserving fish and storing fish products.

There can be no doubt that the large quantities of fish appearing in the records could neither be caught nor carefully handled or transported unless the fishermen felt responsible and were well organized. Cooperation and

---

7. Salonen.

TABLE 1.—FISH PRICES OF THE TIME OF THE THIRD DYNASTY OF UR

Document	Pieces of Fish	Quality of Fish	Shekel (gin)	Silver Equivalent (grams)	Silver, Grams to 100 pieces
TCL 6056	720	Scaled and gutted	2.00	16.66	2.3
TCL 6052	1,380	Scaled and gutted	3.50	29.16	1.8
TCL 6046	900	Scaled and gutted	2.00	16.66	1.8
TCL 6056	2,190	Whole fish	2⅓ + 18 se	20.27	1.0
TCL 6052	1,980	Whole fish	2 + 6 se	17.40	.8
TCL 6046	7,740	Whole fish	11.00	91.66	1.1
TCL 6056	3,300	2d quality	4.00	33.33	1.0
TCL 6052	12,450	2d quality	16 + 20 se	134.25	1.0
TCL 6046	27,600	2d quality	25 + 9 se	208.74	.7
TCL 6046	16,200	2d quality	20⅔ + 15 se	172.91	1.0

SOURCE.—A. Salonen, *Die Fischerei im alten Mesopotamien* (Helsinki: Suomalainen Tiedeakatemia, 1970).

discipline were ensured through guildlike organizations. Salonen mentions that Eniggal was the leader of the fishermen's guild at the temple of Baba. Under him served the fishermen foremen who were responsible for the landing, the quality of the catch, and its transport. The fishermen worked in groups of up to 12 under one foreman. During wartime, or when carrying out communal works such as the cleaning of canals, the fishermen were under the orders of the Chief of the Fishermen whose rank was higher than that of the *nubanda*.

Sea fishermen were obviously regarded as specialists. Freshwater fishermen and sea fishermen formed separate groups and in general used different gear, each group fishing specified waters (see fig. 2). There were fishermen who fished the canals which traversed the fields, others who fished in those crossing the date plantations, and still others who harvested the Euphrates. Sea fishermen were divided into coastal and brackish-water fishermen, specialists who used the large hand trawl net or the large cast net, such as Nesang and Lugalsaklaltuk, and turtle fishermen. There were spear fishermen in both groups.

The city governor imposed taxes on freshwater fisheries, and his overseers were stationed near the canals to collect them. Moreover, an important duty of these overseers was to promote fish culture, and simple fish culture continued for a long time. Even 200 years later, the city governor of Lagash considered the building and stocking of ponds one of his foremost duties.

Periodic festivals, days of rejoicing for all, filled the year of the Sumerians, and fishermen played an important role in this ritual life of Sumeria. Colorful processions proceeded from temple to temple, and distant sanctuaries were visited by boat. Mythological scenes were performed in ritual. Songs and music

Fig. 2.—Detail of Assyrian Palace relief showing Babylonian fisherman fishing with a line. (Photo from the British Museum.)

accompanied the religious ceremonies. All those who contributed to the feast in one way or another shared the meal which followed. The fishermen had their own divine patroness, Nanse, called the Queen of the Fishermen. She was the daughter of Ea and sister of Ningirsu, the highest deity of the state Lagash. Nanse's domain was social justice; she was the "one who searches the heart of the people, who knows the orphans, knows the widow, knows the oppression of man over man, is the orphan's mother, Nanse who cares for the widow, who seeks out justice for the poorest."[8]

Fish were regularly offered to her. Boat processions proceeded from Girsu to her temple at Siraran, near the coast, on the occasion of her great annual festivals, Eating of Malt of Nanse and Eating of Barley of Nanse. These were, it can be assumed, the main festivals for the fishermen, and many of them may

---

8. H. Ringgren, *Religions of the Ancient Near East* (London: S. P. C. K., 1926).

have had special functions in the processions. Galatur, the fishermen's foreman, and sometimes called follower of Nanse, received in the month of Harabimua of Nanse a cloak or other garment and perhaps a special position at the great temple feasts of Nanse.

Large quantities of fish were required for the celebration of Ningirsu's feast, Eating of Malt of Ningirsu and Eating of Barley of Ningirsu. Barley, wheat flour, beer, and fish were presented to the gods. Ningirsu and his wife Baba were, on the first day of the feast, offered two bundles of fish each, and one bundle each on the second day. Furthermore, 20 other gods and sanctuaries received one bundle of fish during the festival, and it was eaten by worshippers at the meals which followed the ceremonies. The temple of Baba certainly supplied these fish. The 11,510 fish which Nesang and Lugalsaklaltuk delivered in the month of the Eating of Malt of Ningirsu, as well as 220 turtles and turtle oil, were used for one of these feasts.

Of course, the harvests were not uniformly abundant. Any deficit in fish supplies was put on the fishermen's debt account. These debts were sometimes quite serious: "65 *shekel* of silver [about 541.65 grams] for fish which were paid for [in barley] but had not been delivered did Subur the nubanda [predecessor of Eniggal] charge to the fishermen's debt account," says one text, and another states: "15 [*shekel* silver for] Nesang, 10 [*shekel* silver for] Subur, 10 [*shekel* silver for] Lugalsaklaltuk, 5 [*shekel* silver for] Nammahne—sea fishermen they are—because they did not deliver the fish pursuant to the conditions Subur, the inspector, has charged to their debt account, Galatur, the follower, participated as commissioner."[9]

The temple fishermen were always able to pay back their debts, even if hard labor was involved. Unpaid debts at Lugalanda's time were rigorously collected, and the debtor and his family could be turned over as slaves to the creditor. The temple fishermen were obliged to perform military services if required, and in return they got a piece of temple land on loan. By selling the crops of these plots of land they could avoid confiscation of property or slavery. The settlement of debts was changed during Urukagina's time, and one tablet discloses that Eniggal had, instead of silver, put 300 dried fish, 60 turtles, and 1,740 fresh fish on Nesang's debt account. While at the beginning of Urukagina's time economic tablets were numerous, by the sixth year of his reign very few were kept, indicating the dominance of other affairs over fiscal bookkeeping.

Urukagina had succeeded Lugalanda in a time of strained social tensions. The "oppression of man over man" had, under the cloak of bureaucracy, spread considerably and had seriously affected the weak and helpless. Urukagina, who wished to see an end of social injustice in Lagash, issued a decree at the beginning of his reign in an attempt to take away privileges from the upper classes. Exemplifying this, one provision stated: "If a villein has a fish

9. Bauer.

pond, no-one of the gentry class may take away this fish."[10] The decree forbade the overseers of fishermen and shepherds to increase their income by taking directly from the produce of their subordinates. They were expected to live from their own income. These reforms brought Urukagina in conflict with people of rank and weakened his political position. And no state in the thriving delta region between the twin rivers could afford political weakness over an extended period. Also, Urukagina could not maintain external peace. Lugalzagesi, king of the neighboring state Umma, took advantage of the internal difficulties in the state of Lagash and prepared for war. A century-old quarrel about a strip of border land, which had changed hands several times and had been conquered by Lagash about 30 years previously, served as a pretext. Finally, Umma succeeded in occupying Lagash, except for a narrow strip of land along the coast and the heavily fortified town of Girsu. Because of this war all subsequent traces of Nesang and his fellow fishermen have been lost.

Lagash, however, rose from the ashes of war. Its commerce flourished, and fishermen again supplied the temples with fish. For centuries, in the southern part of Sumeria sea fish together with barley, remained the most important staple food. Such was the case also in Ur during the Third Dynasty. Tablets reported that when Ur, during its last struggle for survival, was cut off from its supply bases, the price of fish and barley increased 50–60 times.

The destruction of Ur marked the end of the Sumerian ruling classes, the end of Sumeria's political independence, and the end of Sumerian as a spoken language. Non-Sumerian peoples and their habits gained dominance, and they influenced food consumption. The Amorites from the west constituted the Babylonian Empire. But soon after the great king Hamurabi, the Cassites came from the Zagros Mountains and dominated the plain for several hundred years (about 1600–1100 B.C.). During this period fish taboos were introduced, and these became widespread after the rule of the Cassites. After about 1500 B.C. fish-eating appears to have lost its significance in most parts of Babylonia; and for most parts of the Assyrian Empire (about 740–620 B.C.) only myths and symbols recalled what fish had once meant in the Sumerian civilization.

In a wall relief which had once decorated the palace of the Assyrian king Sennacherib (now in the British Museum in London), the life of fishermen and marsh dwellers is depicted as it appeared by about 700 B.C. In this relief, the people live in primitive reed huts and on reed platforms which float on the water. Fish and a few domestic animals provide the bare necessities of life. But the life of these fishermen, surely, cannot be compared with the life of temple fishermen in the time of Nesang and Galatur 1,600 years earlier.

Though in some parts of the Assyrian kingdom fish had disappeared from the table, in others it continued to be of economic importance, as is evident from documents from the Neo-Babylonian period (600–450 B.C.). These documents refer mostly to legal problems. A priest who was responsible for making offerings complains in one document about difficulties encountered by

10. H. W. F. Saggs, *The Greatness That Was Babylon* (New York: Praeger, 1969).

the fishermen who fished for the temple of the Lady of Uruk (Ishtar, goddess of love, fertility, and war; the cult of Nanse was incorporated in her cult, and she was identified by the Romans with Venus). At that time fishermen fished 5 days per month only for the temple of the Lady of Uruk. Other documents from this time deal with the lease of fish ponds to private persons, the revenues of temples from fisheries, and poaching, a crime which was severely punished. In later times no documents exist which reveal fishing or the eating of fish in this region. Nevertheless, it can be assumed that in the rural areas of the south of Babylonia fish continued to be eaten, and fish is still an important food there today.

The Sumerians believed that luck as well as misfortune were natural consequences of the benevolence or wrath of the gods. Their benevolence kept evil demons in check; their wrath gave them free play. Fish on the offering altar pleased Ea, the mighty divine magician and exorcist, whose assistance was invoked against the demons of sickness. Amulets were believed to offer additional protection against evil, and fish emblems on such amulets were thought to bring luck. In some countries fish amulets are still counted among the symbols of luck.

The Zodiac sign Pisces appeared first in Babylonian times. It was, like the signs Capricorn and Aquarius, attributed to the southern region of the sky, the "watery" region reigned by Ea. Creatures symbolizing this region were depicted with fish tails, indicating their watery origin. Thus Capricorn appeared as goat-fish, a being half goat, half fish, and Aquarius appeared as fish-man. The Greeks, and later the Romans, who were familiar with the Sumerian and Babylonian fish-beings, adorned their minor sea gods—the Oceanides and Tritons, and daughters of sea gods, the Nereides—with fish tails. The Nereides became the fabled mermaids of the seafarers in the Middle Ages. The prophet Jeremiah, seemingly inspired by the grotesque water-born creatures of Ea, saw the ocean filled with numerous mixed beings with fish tails. To him these misshapen animals and men were monstrous creatures of Hell, and he describes them as a threat to the unfaithful on the Last Day. Fantastic sea monsters were depicted with delight by European artists of the Middle Ages and can still be seen high up in the Swiss Alps in the church of Zillis, a village located at the end of the Via Mala. This twelfth-century mural depicts biblical scenes, surrounded by the blue band of the ocean which is filled with sea monsters. Among them appears a mixed being—half goat, half fish (see fig. 3).

An ancient civilization reflects in many ways upon our time. Fish occupied a prominent place in Sumerian culture as a divine gift to man. The Sumerians organized their fishing around their religious beliefs and so created the world's first developed sea fishery. The fish harvested served simultaneously as man's food and, as Ea's sacred animal, a symbol of divinity. This Sumerian concept in fact encouraged the spectacular development of a strikingly sophisticated fishing industry and differs from the concept of other ancient populations, such as the Egyptians, who regarded those fish species sacred to a deity as taboo for man.

FIG. 3.—Goat or ram with fish tail, one of the monsters of the ocean depicted on the ceiling of the 12th century church in Zillis. (Photo from the Bibliotheca Apostolica Vaticana, Rome.)

ADDITIONAL REFERENCES

Civil, M. "The Home of the Fish." *Iraq*, vol. 23 (1961).
Deimel, P. A. *Sumerische Tempelwirtschaft zur Zeit Urukaginas und seiner Vorgänger*. Rome: Pontifico Instituto Biblico, 1931.
Ebeling, V. E., and Meissner, B., eds. *Reallexicon der Assyrologie*. Berlin: de Gruyter, 1957.
Falkenstein, A. *Die Inschriften Gudeas von Lagash*. Rome: Pontificum Institutum Biblicum, 1966.
Gadd, C. J. *Ideas of Divine Rule in the Ancient East*. London: Oxford University Press, for the British Academy, 1948.
Goff, B. L. *Symbols of Prehistoric Mesopotamia*. New Haven, Conn.: Yale University Press, 1963.
Gray, John. *Near Eastern Mythology*. London: Feltham, Hamlyn, 1969.
Hackman, G. G. *Sumerian and Akkadian Administrative Texts from Predynastic Times to the End of the Akkadian Dynasty*. New Haven, Conn.: Yale University Press, 1958.
Hawkes, Jacquetta. *The First Great Civilizations: Life in Mesopotamia, the Indus Valley, and Egypt*. New York: Knopf, 1973.

Hunger, J. "Babylonische Tieromina." *Vorderasiatisch-Aegyptische Gesellschaft Mitteilungen*, vol. 14 (1909).
Jastrow, M. *Die Religion Babyloniens und Assyriens*. Giessen: 1905.
King, Leonard William. *A History of Sumer and Akkad*. New York: Greenwood, 1968.
Lampl, P. *Cities and Planning in the Ancient Near East*. New York: Braziller, 1968.
Langdon, Stephen H. *Sumerian Epic of Paradise, the Flood and the Fall of Man*. Philadelphia: University Museum, 1915.
Legrain, L. *The Culture of the Babylonians from their Seals in the Collection of the Museum*. Philadelphia: University Museum, 1925.
Mellaart, Janes. *The Neolithic of the Near East*. London: Thames & Hudson, 1975.
Muller-Karpe, Hermann. *Geschichte der Steinzeit*. Munich: Beck, 1974.
Radcliffe, W. *Fishing from the Earliest Times*. London: Murray, 1926.
Reiner, E. "The Ethnological Myth of the Seven Sages." *Orientalia*, vol. 30 (1961).
Van Buren, E. D. "Fish Offerings in Ancient Mesopotamia." *Iraq*, vol. 10 (1948).

Nonliving Resources

Offshore Oil and Gas

Edward Symonds
Energy Economics and Finance

As the world's population multiplies and mankind's hunger for energy intensifies, the main hope for satisfying the new need must be sought in the oceans that cover 70 percent of the surface of the globe. From its earliest beginnings, the petroleum industry understandably avoided activity in the oceans, where the difficulties are so much greater and the costs so much higher. Fortunately, recent technological advances have allowed both the search and the recovery of petroleum to extend far out of sight of land and into much deeper waters than were formerly accessible. Moreover, the latest estimates are that the reserves of oil and gas underlying the oceans, although still comparatively little known, are large enough to make a dramatic difference to the world's supply outlook for many decades to come.

With the dramatic increase in petroleum prices during 1973–74 and the steady progress of technology, the major short-range uncertainty overhanging the future of offshore hydrocarbon development arises from the legal and political obstacles that still lie in the way.

The Geneva Convention of 1958 stands as a milestone in the successful solution of some of the major international problems concerning the boundaries between the various national sovereignties adjacent to such important petroliferous regions as the North Sea, the South China Sea, and the Indian Ocean, but the convention left unsolved the problems of jurisdiction over the more distant areas. The long-drawn-out negotiations attempting to draw up a Law of the Sea affecting the deep oceans have yet to reach an agreed outline of a satisfactory international agreement. In the meantime, however, they have been valuable in focusing political attention on the importance of the oceans' resources and on the need for an early agreement on their ownership and exploitation. This, in turn, has led to a refinement of the data available on the petroleum and other resources lying on or under the ocean bed.

RESERVE FRONTIERS PUSHED BACK

One of the most comprehensive and authoritative reviews of the seabed resources of the globe was published by the United Nations in 1973 in connection with the Law of the Sea Conference under the title, "Economic Significance in Terms of Sea Bed Mineral Resources of the Various Limits Proposed for

© 1978 by The University of Chicago. 0-226-06602-9/77/1978-1005$01.93

National Jurisdiction." Table 1, indicating the resources of oil and natural gas liquids of the major areas of the world, is taken from this review.

The impressive figures in table 1 indicate a total petroleum liquid endowment, excluding the Communist areas, of almost 115 billion bbl. Yet the search that has taken place since this tabulation was completed already indicates that the available resources likely to be found in offshore fields are in fact much greater than is here shown. The left-hand column of the table is limited to fields within 40 miles of the coast and in less than 200 m (656 feet) of water. As can be seen from the table, additional reserves are believed to exist in areas up to 200 miles from the coast.

Moreover, it has been estimated that, in the much deeper water lying at greater distances from shore, even larger reserves are present. According to Dr. J. D. Moody, a well-known reserve expert, recent developments suggest that the potential of the as-yet-unexplored deep oceans could conceivably double present estimated world resources. Reasons for this dramatic increment include promising studies of the geology of certain deep-ocean areas, such as those extending from the Voring Plateau off Norway to the waters north of Iceland. Moreover, the officially sponsored United States Deep-sea Drilling Project has revealed the presence of hydrocarbons in very deep water and organic-rich sediments at depths formerly assumed to be barren. In addition, turbidites, a typical form of deep-ocean sediment, have been found to contain substances resembling crude oil.

Nor does the prospect for major new discoveries in various parts of the world necessarily depend on success in overcoming the problems of exploration and production in extremely deep water. A recent study by the National Academy of Sciences finds that the ancient deltas of the world's largest rivers offer decidedly promising prospects of oil and gas production, particularly in South America, India, and the Far East. The study draws attention to the fact that all the continental shelves off Eastern Asia (except Indonesia), Southern Asia, Eastern Africa, Northwestern Africa, and various areas off South America and Antarctica offer high potential but have in the past attracted little exploratory effort. Inhospitable climatic conditions have been partly responsible for the neglect of these areas, which have, in addition, often been politically unfriendly to exploration activities. Because of the economic and technological advances of recent years, a surge of interest in these relatively unexplored areas is expected.

Notable progress is now being made in overcoming two of the major obstacles to further offshore development. In the past, estimates of field size have not made full allowance for unusually favorable circumstances that often exist offshore. Recent researches have indicated that organic-rich runoff from the continental shelf is likely to pile up behind reefs and other tectonic structures underwater, resulting in much larger producing zones, possibly up to 10 times as large as those regarded as normal onshore. The critical significance of size, once the economic threshold has been crossed, is indicated in table 2.

Assuming a 10 percent return on investment, the table indicates that a

TABLE 1.—OIL AND NATURAL GAS LIQUIDS
ESTIMATED ULTIMATE RECOVERABLE RESERVES
IN OFFSHORE FIELDS
(in Millions of Barrels, as of May 1972)

Area	0–40 Nautical Miles (<200 m Water Depth)	40–200 Nautical Miles (<200 m Water Depth)
North America:		
Alaska (Cook Inlet)	750	...
California	5,400	...
Louisiana	12,000	...
South America:		
Trinidad	725	...
Peru	30	...
Brazil	370	...
Europe:		
Norway	...	1,800
United Kingdom	...	3,000
Denmark	...	385
Netherlands	...	5
Spain	65	...
Middle East:		
Iran	2,000	7,700
Neutral Zone	...	10,000
Saudi Arabia	40,000	...
Abu Dhabi	1,000	15,000
Dubai	...	3,000
Africa:		
Egypt	1,800	...
Tunisia	130	...
Nigeria	3,400	...
Gabon	280	...
Congo Brazzaville	730	...
Angola	1,085	...
Far East:		
Japan	25	...
Malaysia	200	...
Brunei	800	...
Indonesia	1,240	...
Australasia:		
Australia	710	1,060
New Zealand	170	...
Total	72,910	41,950

SOURCE.—"Economic Significance in Terms of Sea-Bed Mineral Resources of the Various Limits Proposed for National Jurisdiction," *Report of the Secretary General* A/AC. 138/87 of June 4, 1973 (New York: United Nations, 1973).

TABLE 2.—MINIMUM ECONOMIC FIELD SIZE FOR OFFSHORE DEVELOPMENT (in Recoverable Reserves)

	Gas (Billions of Ft^3)			Oil (Millions of Bbl)		
Rate of Return	10%	15%	25%	10%	15%	25%
Atlantic	180	290	660	17	26	70
Gulf of Mexico	120	185	400	11	17	47
Pacific	220	300	770	18	30	74
Gulf of Alaska	660	1,100	5,400	60	97	425
Lower Cook Inlet	370	560	1,550	37	58	150
Bering Sea	600	930	4,400	49	80	260
Beaufort Sea	850	1,600	6,400	80	135	560

SOURCE.—Arthur D. Little, Inc.
NOTE.—Wellhead price of gas = $1.50/thousands of ft^3; wellhead price of oil = $14.40/bbl.

reserve discovery of 11 million bbl would be required to make a field economic in the Gulf of Mexico, compared with a field of 1 million or even less that might be economic onshore. In the much more arduous conditions of the Gulf of Alaska, a 60 million bbl field would be the minimum, and in the Beaufort Sea an 80 million bbl minimum is indicated.

In practice, a rate of return much higher than 10 percent is normally required by the producer. A 25 percent rate would not be considered excessive, in view of the scale of expenditures and the degree of risk. At that rate, the minimum economic field becomes 47 million bbl in the Gulf of Mexico, 425 million bbl in the Gulf of Alaska, and over ½ billion bbl in the Beaufort Sea.

In the case of gas, at a 10 percent rate of return, a minimum of 120 billion ft^3 is indicated in the Gulf of Mexico, compared with a 660 billion minimum in the Gulf of Alaska and an 850 billion minimum in the Beaufort Sea. For an assumed 25 percent rate of return, these minimum field sizes increase by progressively larger steps—to 400 billion in the Gulf of Mexico, 5,400 billion in the Gulf of Alaska, and 6,400 billion in the Beaufort Sea. The significance of these figures is that, despite the high economic threshold, the expectation of commercial discoveries is intense because of the more favorable prospects for giant fields in offshore areas.

The second frontier of offshore development is water depth. While the figures in table 1 are limited by the traditional outer boundary of 200 m (656 feet), drilling and production at far greater depths are now considered feasible.

As an example, much of the most productive area off Southern California is contained in the Santa Barbara Channel. Activity there was halted following the severe but short-lived coastal pollution caused by leaks in January 1969 from the Union Oil group drilling venture. After legal delays stretching over 7 years, the Department of the Interior completed its environmental impact statement, and activity was resumed—both close to shore and in areas 5 miles or more out to sea (where the ocean bed slopes off rapidly). A production platform has since been installed in 860 feet of water, and oil is expected to start flowing from this depth in mid-1978—to be followed, about 1 year later, by gas from the same area. A striking feature of this Exxon enterprise is that, because of objections raised by the state of California concerning the construction of onshore facilities, storage, treatment, and shipment, facilities may be installed far out to sea, at the production platform.

Although this venture in the Santa Barbara Channel represents the deepest offshore production program in the history of the petroleum industry, it does not represent the limit of today's technological capacity. Several wells have already been drilled in even deeper water. Off the west coast of Thailand, wells have been completed in depths of up to 3,460 feet and as far as 100 miles out to sea.

PROGRESS OF PRODUCTION

It was estimated in 1973 that offshore fields accounted for 26 percent of total world reserve. Allowing for the subsequent dramatic rise in wellhead prices and using the conservative totals of table 1, the value of offshore oil resources in 1976 can be estimated at $1,610 billion. Making a similar estimate for offshore gas, the wellhead value must exceed $335 billion. This total of $1,945 billion makes petroleum by far the most valuable resource at present under exploitation in the world's oceans. Moreover, the calculation certainly underestimates the significance of offshore oil and gas. Whereas proved reserves of hydrocarbons are usually developed only to the extent of market needs over the forseeable future—some 10 years at most—the ultimate availability of resources depends upon steadily improving estimates of undiscovered volumes. For obvious economic and technical reasons, these are more promising in the largely unexploited offshore zones than in the increasingly well-known onshore basins.

Notwithstanding the high values and favorable omens for offshore production, the record of the first half of the present decade was not exciting. As can be seen from table 3, the decade opened with worldwide offshore production estimated at 8,232 thousand bbl daily. By 1975, this total had risen a mere 0.4 percent, to 8,264 thousand bbl. But this increase of only 32 thousand over the half decade conceals the peak that had meanwhile been reached—10,067 thousand bbl daily in 1973. Undoubtedly, the decline in petroleum demand following the Arab oil embargo of 1973-74 accounted for the retreat of offshore as well as total oil production. Moreover, accurate data on offshore progress are not always available, since the practice is to record production for each field as a single total, thus obliterating the distinction between onshore and offshore production for fields that lie beneath the coastline.

The national totals in table 3 underline the important place already oc-

TABLE 3.—OFFSHORE OIL PRODUCTION

Country	Thousands of Bbl/Day		
	1975	1973	1971
Venezuela*	1,737.10	2,700.00	2,490.00
Saudi Arabia*	1,385.81	1,990.40	1,210.10
United States	909.59	1,697.46	1,692.00
Iran	481.19	452.41	444.00
Abu Dhabi	462.71	454.43	342.00
Nigeria	431.33	518.44	361.00
Australia	412.52	348.20	262.00
Neutral Zone	315.07	394.04	380.00
Brunei/Malaysia	141.22 (B) 84.49 (M)	264.16	138.00
Dubai	249.32	221.49	126.50
Indonesia	246.45	174.22	32.00
USSR	230.00	236.00	250.00
Norway	189.57	32.30	22.00
Gabon	179.88	57.20	27.20
Trinidad/Tobago	174.04	110.97	123.00
Egypt	165.00	130.75	121.00
Angola/Cabinda	141.20	144.23	131.00
United Kingdom	83.00		
Mexico	45.00	20.80	37.00
Tunisia	43.00	0	0
Sharjah	38.36		
Congo	37.25	34.80	.50
Spain	32.88	20.00	
Peru	28.86	31.35	22.00
Brazil	18.95	17.21	8.20
Italy	10.41	12.74	10.90
Denmark	3.30	2.68	
Japan	.86	1.00	2.00
Totals	8,264.36	10,067.28	8,232.30

SOURCE.—*Offshore* (June 20, 1976).
*Many fields in these communities lie both on and offshore. Since production is reported as a single total, it is impossible to separate the exact amount flowing from offshore.

cupied by offshore production in the region containing by far the largest fraction of the world's crude oil reserves, namely, the Middle East. Although it is noticeable that two of the world's largest producers—Iraq and Kuwait—are missing from the list, the Middle Eastern countries here tabulated accounted in 1975 for some 37 percent of the world's offshore production. Venezuela, where offshore activity centers in the relatively shallow waters of Lake Maracaibo, still heads the list, with production of 1,737 thousand bbl daily, followed by Saudi Arabia's 1,386 thousand bbl daily. The United States, which pioneered much of today's technology in the Gulf of Mexico, is shown in third place, with 910 thousand bbl daily of offshore production. Indonesia, Norway, Gabon, and the United Kingdom are shown as countries whose production from offshore wells has now reached sizeable volumes, but which started the decade with very low rates of flow.

At first sight, table 4 suggests a much stronger performance for offshore gas production. After holding steady at around 12 billion ft^3 daily in the early years of the decade, offshore production shows a 24 percent leap to 14,923 million ft^3 daily in 1975. On closer inspection, it is found that the jump in 1975 is accounted for by a change in the reporting procedures that concealed a further year of production at approximately unchanged levels. Traditionally, much of the gas produced in conjunction with oil in the major exporting countries has been flared and as such is not recorded in the production figures. Total flaring in the Middle East has recently exceeded 5 billion ft^3 daily. On most definitions, only such gas as is utilized—for reinjection or for sale—is included in the national totals. In 1975 for the first time, however, the Saudi Arabian government decided to include nonutilized gas in its offshore production total, bringing an increase of some 3 billion ft^3 daily in the reported figure. For most of the early years of this decade, marketed production from offshore areas in Saudi Arabia has been running at some 700 million ft^3 daily.

Another adverse trend that lies concealed within the apparently favorable 1975 totals is the drop to 5.5 billion ft^3 daily in the United States, representing a cut to approximately half the production rate achieved at the beginning of this

TABLE 4.—OFFSHORE GAS PRODUCTION

Country	Millions of Ft^3/Day		
	1975	1973	1971
United States	5,506.85	7,130.81	10,046.58
Saudi Arabia	±3,825.39*	721.74	587.12
United Kingdom	3,600.00	3,000.00	1,794.25
USSR	764.00	670.00	NA
Nigeria	314.00	NA	NA
Australia	203.00	173.60	98.10
The Netherlands	186.00	NA	NA
Indonesia	158.80	108.00	0
Trinidad/Tobago	123.00	16.79	10.69
Egypt	120.00	NA	NA
Peru	77.00	60.79	62.75
Brazil	25.00	NA	NA
Norway	16.50	NA	NA
Ghana	3.70	NA	NA
Totals	14,923.24	11,881.73	12,599.49

Source:—*Offshore* (June 20, 1976).

*Almost all the natural gas produced in Saudi Arabia is associated with oil production, and little of the gas is marketed. The jump in this figure is due to reporting changes by the government.

decade. Even so, the United States, with its prolific Gulf of Mexico fields, remains by a substantial margin the world leader in offshore gas production.

Next in line, allowing for the adjustment to the Saudi Arabian figures, is the United Kingdom. That nation has witnessed a dramatic increase in its gas production, to 3.6 billion ft^3 daily, since the discovery of the rich fields in the southern sector of the North Sea. Other countries that have made substantial progress during this decade are Nigeria, Australia, the Netherlands, Indonesia, Trinidad, and Egypt. It is noticeable that the list of substantial offshore gas producers is lengthening. It will doubtless continue to do so, as the market for this premium fuel expands and as the justification for investment in pipelines and terminals increases, in line with the higher prices now commanded by gas.

At *Ocean Yearbook* press time, full data for 1976 were not available. Preliminary estimates indicate that total world crude production outside the Sino-Soviet area increased by 8 percent over the 1975 total. With new offshore producers and mature producing areas such as the United States likely to show a trend that remains firmly upward, while the onshore trend will often be downward, the probability is that 1976 offshore production increased at an even higher rate. Assuming that offshore oil production advanced by 10 percent in 1976, the total will have approximately recovered the 1974 level of 9.3 million bbl daily. For natural gas, an increase of at least 10 percent can also be assumed. This would bring the worldwide total for 1976 close to 16.5 billion ft^3 daily.

LATEST DEVELOPMENTS

Salient events during 1976 indicate the continuing spread of offshore operations to new producing areas. In Central and South America, Petrobras, the state-owned oil company of Brazil, announced discoveries off the northeast coast and the Amazon delta. In Venezuela, activity in the south lake and elsewhere in Lake Maracaibo indicated that the potential of this prolific area is far from exhausted. In confirmation of this assessment, a major oil discovery, with an estimated production capacity of 5,300 bbl, was reported by Lagoven. In the Mediterranean, Campsa, the government-owned company, and Shell announced discoveries off the coast of Spain, one of which tested at 7,100 bbl. Off Casablanca in Morocco, Standard of California announced another discovery, as did Buttes Gas and Oil off Tunisia. Oceanic exploration of Greece discovered oil 4 miles northeast of production already established in the Prinos offshore field, where a well in the north Aegean Sea has tested at 3,700 bbl daily.

In the Middle East, several new discoveries were reported. These included finds in the Persian Gulf and in the Gulf of Suez. Iran's Ardeshir offshore field began production at the rate of 30,000 bbl daily from 10 wells and is expected ultimately to reach 300,000 bbl daily.

An important development in Asia was the continued development of the Bombay High offshore field, where production in this formerly oil-poor nation will rise to 80,000 bbl daily by the end of 1977. Another find of possible importance is located close to the Bombay High field, off the state of Maharashtra. In Indonesia, production continued to hold the level of about 260,000 bbl daily from offshore fields. A discovery by a Union Oil consortium in the Gulf of Thailand was reported. Encouraging shows were also reported by a Cities Service group drilling north of Palawan Island in the Philippines.

In the Communist countries, production continued from relatively shallow depths in old-established producing areas in the Caspian Sea. In China, production from shallow water in the Po Hai was expanded, allowing a rapid growth of total onshore and offshore production to a reported 1.7 million bbl daily.

The North Sea, featured in greater detail elsewhere in the *Yearbook*, remains the most active of the world's newly established offshore producing areas. With reserves now officially estimated at 22.5 billion bbl, of which approximately three-fourths are in the British sector and one-fourth in the Norwegian sector, exploration and production continue their rapid advance. Three very large gravity platforms have been installed, and some 120 wells are producing. Production in the sector allocated to Norway, whose Ekofisk field was the first oil discovery to be made, approached 200,000 bbl daily in 1975 but has since been overtaken by production in the British sector. The gas will mainly flow to Germany and Britain rather than Norway (although part of the production is in the Norwegian sector) and it can be seen from the figures in table 4 for Norway that progress in developing this resource has so far been modest.

Discoveries in the North Sea reported in 1976 included those by Continental, Occidental, Burmah, Phillips, Texaco, the Shell-Exxon group, and others. Of the 14 fields so far declared commercial in British waters, three—Montrose, Brent, and Beryl—were brought into production during the year. They are expected to flow at a rate of approximately 340,000 bbl daily by the end of this decade. Total North Sea production at that time (apart from small volumes in the Danish, German, and Dutch sectors) is expected to total 3.5 million bbl daily, of which nearly 1.5 million bbl daily will come from three major fields in the Norwegian sector and the remainder from the British fields.

In addition to those offshore areas mentioned above, particular interest attaches to the Beaufort Sea and Canadian Arctic areas, which will benefit from proximity to the world's largest market, once transportation problems have been overcome. Three oil fields and five gas fields have already been proved in the Arctic Islands. Plans for construction of the Polar Pipeline are being developed as a way of linking these important gas discoveries to the mainland. According to some experts, the petroleum potential of the deepwater off Labrador is even greater than in the Arctic.

Another index of the growth and focus of offshore activity can be obtained by an analysis of the deployment of deep-water rigs. Table 5 indicates that early in 1977 there were 299 active rigs, an advance of 5 percent over the preceding year. The main center of activity is shown as the Gulf of Mexico, with 86 rigs active off Louisiana and Texas. The next most active area is the North Sea. Traditional producing areas in South America, such as Lake Maracaibo, together with new areas on the west coast and in the extreme south, account for 34 rigs. In the Middle East, with the developments described above, 37 rigs were employed offshore early in 1977. Indonesia and the Burma-Thailand arc provided work for 25 rigs. In Africa, with activity centered along the west coast, 19 rigs are engaged. New activity in the Bay of Biscay is indicated by the rigs drilling in the Atlantic. Recent Mexican discoveries, although very substantial onshore, have not greatly affected deep-water activity. The new interest in the waters south and east of Ireland is reflected in the presence of two rigs in the Celtic Sea.

WHOSE RESOURCES?

One reason why the activity outlined in table 5 is so unevenly spread around the globe is that the jurisdictional status of the oceans remains most uncertain. Although the Geneva Convention solved some of the most pressing problems,

TABLE 5.—WORLDWIDE OFFSHORE DRILLING ACTIVITY
(Active Rigs, February 1977)

Area	Working	Idle	En Route
Africa	19	6	1
Atlantic	4	1	0
Australia	2	1	1
Canada and Great Lakes	6	6	0
Caribbean	6	0	0
Celtic Sea	2	1	0
Japan	3	0	0
Louisiana	56	5	2
Mediterranean	9	3	0
Mexico	3	0	0
Middle East	37	6	1
North Sea	47	12	0
Pacific	10	6	2
South America	34	0	1
Texas	30	2	0
U.S. East Coast	0	2	0
USSR	6	0	0
Total	299	72	9

SOURCE.—*Offshore* (February 1977).

the need for a fuller accord was recognized by the General Assembly of the United Nations. Consequently, it set up an ad hoc committee in 1967 to elaborate a new regime for the floor of the ocean and the mining of its mineral resources "for the common benefit of mankind." Problems concerning the ownership of the resources of the deep sea have been kept under constant review since that time. Since the 1974 Caracas meeting, a full-scale international conference on the subject has been under way, though with necessary adjournments for further study. The latest session of this conference was held in New York in the summer of 1977.

The areas of agreement so far identified are that an international authority shall be set up, with an enterprise as its operating and financial arm, in order to exploit the resources of the ocean deeps. This is in itself a striking advance, representing the first time that an international organization of global scope will own and be responsible for developing natural resources. The principles on which the authority will operate include the necessary legal safeguards, arrangements for settlement of disputes, special consideration for the less-developed countries, and guaranteed access of all nations and their nationals without discrimination to any resources taken from the seabed.

The primary unresolved issue, however, is that the proposed accord would provide that exploration of the resources would be conducted for all mankind and not for the exclusive gain of those possessing the technological wherewithal. The implication is that those with the ability to exploit the resources would be deprived of some of the advantages of their activities, thus throwing doubt upon the feasibility of financing them. A further problem is the need to protect the interests of the land-based producers of similar resources. The nations representing such producers have, however, shown cognizance of the need to allow exploitation to begin. Throughout the conference, in fact, issues have been settled with a remarkable degree of international understanding, proceeding by consensus rather than by the casting of votes on matters of substance.

Although an encouraging degree of progress is being made in these negotiations, the outcome has to some extent been overtaken by events. Because of the urgent need to protect both the fishing rights and the mineral resources of the seabed, the coastal states have increasingly been resorting to unilateral action. So long as the boundaries of the ocean deeps remain indeterminate, no treaty can regulate their exploitation. Hence, as a precondition of further progress on the Law of the Sea, the national boundaries of the coastal countries had first to be determined. In anticipation of actions by other countries, many have unilaterally extended their jurisdictions from the traditional 3 miles to 12 and their fishing and possibly other economic rights to new limits as much as 200 miles offshore. In doing so, they have asserted exclusive rights over a much larger area than was formerly considered acceptable.

In the years ahead, the impact of these actions on hydrocarbon resources is likely to become increasingly clear. In the ocean deeps, manganese nodules and other hard minerals are, for the time being, likely to be the most important resource. Yet in the shallow waters that mainly cover the area up to 200 miles offshore, petroleum development is likely to continue to take pride of place. The interest surrounding these relatively nearby areas is enhanced by the fact that jurisdiction over many of them is still in dispute. This is particularly true of the continental margin surrounding isolated islands, many of which had previously occupied only a minor place in national considerations.

The proposed Law of the Sea would disregard the rights of desert islands; their surrounding shelves would be considered part of the continental shelf of the adjoining mainland. But this proposal has not eliminated bitter disputes over islands lying in waters believed to offer a rich hydrocarbon potential.

In the Aegean, the presence of Greek islands close to the Turkish coast has caused a flare-up of hostility between the two countries, with the International Court of Justice in The Hague calling for written submissions by the two parties in dispute during the summer of 1977. The potentially rich Barents Sea, to the north of the Soviet Union, is in dispute because of the uncertain position of Spitsbergen—to which island both Norway and the Soviet Union lay partial claim. Farther west, Rockall, an uninhabited outcrop 300 miles west of Scotland, is claimed as part of the continental shelf by both Britain and Denmark.

In the South Atlantic, the Falkland Islands have long been under British jurisdiction and have maintained a small permanent population. But they are surrounded by the widest stretch of continental shelf off South America and are claimed by Argentina. Various islands in the South China Sea, particularly the Nansha Archipelago, are the subject of claims by the Philippines, Malaysia, Taiwan, and Vietnam. Malta, which clearly does not come within the desert-island ruling proposed for the United Nations Law of the Sea, is claiming jurisdiction over waters where the exploration rights have already been granted to Exxon by Libya. Disputes exist elsewhere in the eastern Mediterranean and the Middle East. Bitterest of all, the oil-rich dividing line down the Gulf of Suez remains in violent contention between Israel and Egypt.

Most of the established producing areas, where islands do not confuse the boundary-drawing problem, have by now reached reasonably amicable demarcation arrangements. In the Persian Gulf, the small islands of Bubiyan and Warba, off Kuwait, have been occupied by Iraqi troops, but no oil activity has been reported. In the rest of the gulf, the demarcation lines are more clearly drawn. The same is true of the North Sea, except for disagreements involving parts of the area north of the sixty-second parallel. To the west of the British Isles, petroleum exploration has been held up by disagreements among Britain, France, and Ireland over the dividing line through the Celtic Sea and the area to the south, known as the Western Approaches. The International Court of Justice has taken this dispute, as well as that concerning the median line for the English Channel, under advisement. During the summer of 1977, Britain and France reached complete agreement on the major points outstanding between them.

Even in the oldest offshore producing areas, however, certain boundary lines remain to be finalized. An agreement has been reached between the United States and Mexico regarding the boundaries in the Gulf of Mexico and the Pacific Ocean. But the lines between Canada and the United States, affecting both the Georges Bank off Nova Scotia and the Pacific waters off the Juan de Fuca Strait, remain under discussion.

According to one estimate, the extension to 200 miles of the area claimed by the 120 nations possessing coastlines would preempt one-third of the total area of international waters. Another estimate suggests that the offshore petroleum resources of the globe, including both proven and probable reserves, amount to 2.3 trillion bbl, making them larger than total presently known onshore reserves. So long as the various disputes over ownership remain unresolved, the exploitation of these massive resources will be severely hampered.

UNITED STATES POTENTIAL

Because the evaluation and exploitation of offshore oil and gas resources in most of the world are still at such an early stage, it is instructive to review experience and prospects in the areas most actively explored and to analyze the problems impeding further development. For this purpose, the United States can be taken as indicative of the offshore successes and problems that will be encountered in other parts of the world.

As to the scale of the reserves remaining to be discovered, intensive review by the U.S. Geological Survey and other groups leaves no doubt about the scope for further offshore development. Undiscovered recoverable crude oil resources are estimated to lie in the range of 10–49 billion bbl. This accounts for at least one-fifth and possibly nearly one-half of the total crude oil resources remaining to be discovered. The lower figure indicates that over three times as much offshore oil remains to be discovered as the 3.2 billion bbl so far identified. If the upper range of the estimate proves correct, offshore oil yet to be discovered will provide more than 15 times the offshore discoveries already made.

In the case of gas, undiscovered offshore resources are estimated in the range of 42 to 181 trillion ft^3. These resources account for at least one-eighth and possibly nearly one-fourth of the nation's total undiscovered gas. Even the lower estimate, if borne out by exploration, would more than double the present total of known offshore gas reserves. The higher estimate, if realized, would represent a five-fold increase in offshore gas reserves and a 76 percent addition to total discoveries already made, both onshore and offshore.

Progress in developing these vast oil and gas resources would undoubtedly have been faster had not legal problems dogged the footsteps of those willing to take the extraordinarily high risks of offshore development. Early initiatives were complicated by disagreements among the states and the federal government. After the end of the Second World War, the activity of the states in leasing their offshore waters was suspended on legal grounds and could not be resumed until the passage of the Submerged Lands Act of 1953. This act gave jurisdiction over a 3-mile zone (later amended to 9 miles for Florida and Texas) to be available for leasing by each state, with the more distant waters—the outer continental shelf—remaining under federal control.

Even after the passage of this legislation, there was no surge in leasing activity. As can be seen from table 6, the area leased each year, after an early surge, had fallen by 1971 to a mere 37,000 acres. In the wake of the Arab oil embargo, the administration announced a program to step up the rate of leasing to 10 million acres annually. By the end of the following year, this target had been abandoned, the new emphasis being to maximize production and encourage frontier development rather than to stress the total area leased.

At the same time, as can be judged from the Santa Barbara case mentioned above, environmental obstacles became much more severe than they had previously been. Especially long delays were imposed for the waters off Alaska, where the Secretary of the Interior has concluded that no more leasing may prove feasible for several years to come. Off the Atlantic Coast, the other major unexplored area remaining to the United States, a U.S. District Court decision, which came up for hearings in mid-1977, found that the environmental impact statement offered by the Department of the Interior in justification for the leasing program had been improperly prepared. Regardless of the outcome of this action, the delay to exploration will be costly to the companies, which have together spent over $1 billion on lease bonuses, and to the consumer, whose fuel supplies will inevitably deteriorate as other resources approach exhaustion.

TABLE 6.—OCS ACREAGE LEASED, 1954–1976

Year	Acres Leased	
	Annual	Cumulative
1954–1968	6,513,621	...
1969	114,282	6,627,903
1970	596,040	7,223,943
1971	37,222	7,261,165
1972	826,195	8,087,360
1973	1,032,570	9,119,930
1974	1,762,158	10,882,088
1975	1,679,877	12,561,965
1976	1,274,809*	13,836,774*

Sources.—Personal communication to John C. Whitaker from officials in Bureau of Land Management, Department of the Interior.
*Estimated on the basis of the official schedule.

With the rate of offshore leasing still running at only approximately 1¼ million acres annually, as shown in table 6, national concern has begun to focus on the need to release more of the resources locked up offshore. Yet the legislative proposals that have recently been under consideration in the Congress could well have an entirely opposite effect.

A number of changes in leasing procedures have indeed been made with the purpose of enhancing petroleum industry enthusiasm. In 1974, in order to attract smaller bidders, leases were issued on the basis of royalty bidding rather than bonus bidding, the effect being that payments to the federal government did not become due until production (and hence royalties) began. In 1975, some of the tracts off southern California were offered with a 33⅓ percent royalty rather than the traditional 16⅔ percent rate, the object being to reduce the front-end burden represented by the bonus. The formation of joint bidding groups containing more than one major company was barred in order to encourage small companies to participate. Regulations were issued requiring companies to file with the Geological Survey, and thereby make broadly available any significant geological data that they had acquired during their exploration programs.

A determined effort was also made to overcome the hostility to offshore development expressed by several of the coastal states. The Coastal Zone Management Act Amendments of 1976 set up a $1.2 billion fund to soften the impact of offshore development by providing $400 million over an 8-year period for the impacted states. The remaining $800 million is to be held in reserve for a 10-year period for use in assisting states in planning local energy facilities and in responding to new social and economic loads imposed by offshore development.

The Carter administration's energy program provided that oil and gas from the outer continental shelf should be accorded liberal treatment as new production, with the price allowed to move up by stages to the world level. Yet this relaxation was probably insufficient to compensate for other impediments, existing and prospective, to offshore development. In particular, some of the proposed outer continental shelf legislation would complicate the picture by bringing the federal government into offshore drilling ventures. The purpose would be to provide added environmental protection and to allow the federal government to carry part of the cost of this important type of exploration. Critics believe, however, that this proposal, along with the new, two-stage concept of splitting exploration and development leasing, would merely confuse the outlook and engender still further delays.

In any case, latest available data indicate that offshore development will be held back, even in the absence of the national and international legal and jurisdictional problems already outlined, by the extremely high level of costs. In 1975, the latest year for which full data are available, the average cost of drilling and equipping an offshore oil well was $892,000, more than six times the $139,000 average for all United States oil wells. For gas, the offshore average was $1,247,000, nearly five times the $262,000 average for all gas wells. In the following year, the cost of offshore drilling appears to have risen by some 20 percent and to have still further increased its lead over the average for onshore wells.

Moreover, these average costs conceal striking differences among the various regions. The cost of drilling off Alaska is several times greater than under the much more favorable conditions in the Gulf of Mexico. This disadvantage is only partially offset by the fact that jurisdictional problems have provided less of a barrier in Alaska, where the only boundary problem concerns that with the Soviet Union, on which the established accord appears to be standing the test of time. In the Southwest, on the other hand, the problems of the Submerged Lands Act and the ensuing delays have not yet been fully resolved.

The interstate difficulties in the Southwest are highlighted by the case of Mississippi, which has appealed for a change in a Gulf of Mexico boundary formula proposed by the National Oceanographic and Atmospheric Administration. According to the governor, after the extension of the national boundaries by the addition of an economic zone stretching 200 miles across the outer continental shelf, the proposed formula would virtually exclude Mississippi from offshore petroleum rights, thus costing the state millions of dollars in federal grants and petroleum royalties. Mississippi can be expected to contest this issue with corresponding vigor, leading to costly deferrals of offshore development in that part of the Gulf of Mexico.

LOSING THE INTERNATIONAL RACE

The high cost and the niceties of U.S. environmental and legal requirements can be judged by setting the national record within a framework of international comparisons. It is now a full 30 years since the first well was drilled out of sight of land from a mobile platform off the coast of Louisiana. United States initiative and technology were thus responsible for ushering in the new era of massive, worldwide offshore development. Yet it is clear from the data provided above that the progress achieved in subsequent years in the United States has been less than breathtaking.

A striking contrast is provided by experience in the North Sea, where no history of established oil and gas production existed and exploration did not begin until the British sector was partially opened in 1964. Within that same decade, large discoveries of gas had been made, allowing neighboring countries to benefit from a new, low-cost source of fuel whose percentage of their energy markets moved up rapidly from almost zero toward the United States figure of about 30 percent. In the Norwegian sector, oil was not discovered until 29 dry holes had been drilled and most of the groups receiving exploration permits had abandoned the search. Nevertheless, discovery of

Ekofisk and other fields has already established the North Sea as a major oil province, with reserves more than half those of the entire United States. The contrast between achievements in the United States and elsewhere is silhouetted in figure 1. While the area of the continental shelf is so much smaller in the North Sea than off the United States coast, at the end of a much shorter period of activity the tracts leased in the North Sea are approximately twice the size of those leased off the United States. The contrasts are even more striking in the cases of Canada, Australia, and Southeast Asia. Only off Africa, with its far more complex political structure and much larger total area, does the unleased acreage exceed that of the United States.

It is true that in figure 1 the contrast between leased and unleased areas is magnified by the use of a 3,000 m cut-off point, rather than the 200 m limit usually adopted. Even so, the observer is tempted to conclude that the United States, far from seeking to maximize its offshore production, has been deliberately holding back while other nations forge ahead.

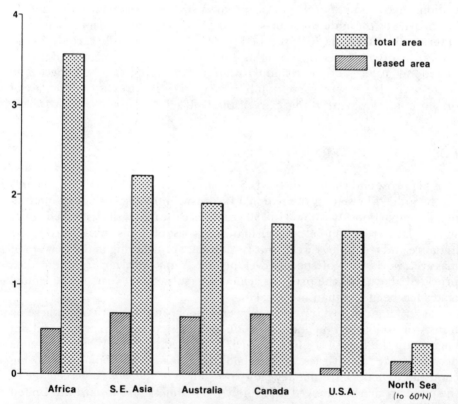

FIG. 1.—Leased area compared to total shelf area within a 3,000 m water-depth contour (in millions of mi^2). Source: Y. Bonillas, "New Reserves for U.S. Keyed to OCS," *Oil and Gas Journal* (October 14, 1974), p. 42.

THE POLLUTION THREAT

In the present era of gathering congestion and increasingly synthetic modes of living, the governments of the world have become more conscious than ever before of the dangers of environmental pollution. In the past, because of their vast size and (from the human viewpoint) their emptiness, the oceans have been regarded as the last stronghold of the unpolluted state of nature. In recent years, largely because of the growth of oil traffic and energy consumption, this pristine state has been increasingly threatened. It is frequently, but wrongly, believed that offshore petroleum production has made a major contribution to the threat.

In fact, offshore operations, in the sense in which the words are normally used in the technical literature, contribute only a minor amount of the pollution of oceans and beaches over which public opinion is justifiably aroused. A study published by the National Academy of Sciences in 1975 estimates that the worldwide total of petroleum pollution reaching the oceans amounts to some 42 million bbl a year. Although this is a large volume of waste, it is insignificant in relation to the world's total oil production, which in 1976 reached 58 million bbl daily. Moreover, the responsibility for the wastage lies almost entirely outside the sphere of oil and gas production. The academy estimates that the largest source of pollution is ship and tanker traffic, with 35 percent; followed by river and urban run-offs, with 31 percent; pollution from coastal refineries and industrial and municipal sources, with 13 percent; and a further 10 percent from each atmospheric fallout and seeps. This leaves barely 1 percent accounted for by offshore petroleum production.

While national governments bear the ultimate responsibility for any marine pollution that occurs, oil companies have in practice (with the benefit of a cooperative insurance program that they have set up) borne the burden of damage such as that caused at Santa Barbara and in other spills. Negotiations are now proceeding to draw up more comprehensive compensation arrangements, with worldwide application. A proposal is also under study for an international convention on civil liabilities for oil pollution damages. Another proposal would establish an international fund to compensate for oil pollution damages. Together with an improved accident record, the risks from offshore operations are thus being progressively reduced.

The costliness of the offshore installations that are now being put in place gives another indication of the lengths to which companies are going in their efforts to design rugged, accident-free equipment. A production platform designed for the Murchison field in the British sector of the North Sea will operate in more than 500 feet of water and will cost over $70 million. Equally massive structures are being installed in other parts of the world.

In the Gulf of Mexico, however, a lighter design of production platform, for water depths down to 2,000 feet, is being tried out by a group of 12 companies. To be secured to the seabed by guy lines attached to anchors, this new design could show the way to a stronger but less massive platform design. As another solution to the same problem, successful tests are being undertaken with various designs for submerged production systems. These would allow drilling and production in water depths below 1,000 feet, while eliminating the need for platforms above the surface.

Improvements are also being made in the efficiency and environmental compatibility of offshore drilling procedures. New methods are being introduced to compact and in other ways ease the handling of spent mud that has to be disposed of as drilling proceeds. Off the Canadian Arctic Islands, it has been found that artificial thickening of the ice allows wells to be drilled with conventional land rigs in water more than 1,000 feet deep and as much as 20 miles offshore.

The argument is sometimes offered that, because of the high costs and complex technologies required for offshore petroleum development, smaller companies are in danger of being squeezed out of the business. But this argument can be turned around. Without the financial and technical resources of the major companies, it is questionable whether development could go forward at anything approaching the speed that has already been achieved in some of the world's offshore areas — and that appears even more necessary to meet the needs of the future.

The dominant part played by the major companies in this high-stake activity can be judged from table 7. Taking the outer continental shelf of Louisiana in 1974, it can be seen that more than 64 percent of the crude oil and condensate production was accounted for by the eight largest oil companies. Other large

TABLE 7.—COMPANY SHARES OF CRUDE AND CONDENSATE, FEDERAL GULF OF MEXICO,
(Louisiana Outer Continental Shelf, 1974)

Companies	Thousands of Bbl	%
Shell	60,807	18.8
Exxon	44,828	13.9
SoCal	36,467	11.3
Gulf	21,610	6.7
ARCO	16,380	5.1
Texaco	11,701	3.6
Mobil	9,026	2.8
Standard (Indiana)	7,801	2.4
Top eight	208,620	64.5
Lesser majors	65,190	20.3
Others	49,217	15.2
Total	323,027	100.00

SOURCE. —94th Cong., 1st Sess., Senate Subcommittee on Antitrust and Monopoly, *Hearings on S.2387 and Related Bills*, Pt. 1, pp. 51-70 (statement of Walter S. Measday), September 23, 1975.

companies accounted for 20 percent of the production, leaving all the rest of the oil and related companies contributing only 15 percent of the total.

NEW USES FOR THE OCEAN

It is now believed by most industry and government experts that a global tightening of oil and gas supplies is inevitable within the foreseeable future—and possible within the next 10 years. For this reason, increased use of less-flexible energy resources (such as coal) will be necessary, with profound impacts on ocean traffic, ports, and so on. More significant still will be the effort to move the world's economy as rapidly as possible onto a basis of renewable energy sources. Most of the effort will be directed toward increased uses of solar energy developed through the intermediary of the oceans.*

Other proposed energy-related uses of the oceans arise from the fact that they offer space and safety no longer available onshore. The case of liquefied natural gas (LNG) illustrates the growing importance of such uses. Because it calls for reduction of the temperature of natural gas to -259°F, giving rise to dangers of escape and ignition of a highly volatile liquid, gas liquefaction places special burdens on transportation methods. Problems concerning the shipment and reception of LNG have aroused much public concern, resulting in highly vocal opposition to the siting of various LNG terminals and to an undertaking in the Carter energy program that such terminals would in future be located far from the main centers of population. As an extreme example of the desire to remove LNG terminals from any area where they might provide a new source of danger or add to the problems of industrial and traffic congestion, California legislators have proposed that the next LNG terminal required to supply gas to that state be located 20 miles out to sea.

An even more ambitious energy-related proposal for the use of ocean space is for the siting of nuclear plants. The search for sites remote from population centers and natural attractions and for designs that can be standardized and therefore readily approved has led to the Westinghouse proposal for floating nuclear plants to be located several miles out to sea. Further moves along these lines are expected under the administration's new drive to intensify the development of light-water reactors.

Another proposal would site methanol plants out to sea. These plants would be fed by onshore or offshore gas at present being flared because of the lack of a nearby market, with the methanol then shipped in conventional tankers to consuming centers in other parts of the world. A number of possible sites have been studied, especially in the Persian Gulf. The Japanese market

---

*EDITORS' NOTE.—For a more detailed discussion, see "Ocean Resources" elsewhere in this volume.

would be an attractive destination for the output of such plants, and Continental Oil and Mitsui have recently begun negotiations to build one.

The newest use of the oceans for petroleum transportation calls for the construction of major ports far out to sea. This offers the advantage of water deep enough to carry the size of tanker now being built and at the same time satisfies the objections of local groups opposing further inshore congestion and more severe navigational hazards. The construction of such an offshore port downstream from Rotterdam, where an artifical island is being used, provided an early example of this type of project. In the United States, construction will begin in 1978 on a deep-water superport to be built 18 miles off the coast by an oil-company consortium known as Louisiana Offshore Oil Port, Inc.

The purpose of this and other proposed ports, including one off Texas, is to allow supertankers to transport the massive volumes of crude oil now being imported by the major consuming nations. The ports have to provide water depths and berthing capacity to accommodate ships with a deadweight rating that may be as high as half a million tons.

Table 8 provides a calculation of the savings that could be achieved by the installation of such ports. In the example chosen, even using 1974 values and depending on the rate of throughput, the savings may rise to more than 15¢ per barrel received. For a supertanker with a typical cargo of 1 million barrels, this indicates a saving of $150,000 on each delivery.

TABLE 8.—ESTIMATED SAVINGS RESULTING FROM AN EAST COAST U.S. MONOBUOY SUPERPORT

	Cents per Bbl.	
Throughput*	Worst Case†	Best Case‡
0.600	−4.0	3.3
0.800	−1.6	5.7
1.000	−0.2	7.2
1.135	0.4	7.8
1.200	1.0	8.4
1.572	3.2	10.5
2.000	5.3	12.7
2.500	6.6	14.0
3.200	7.4	14.8
5.106	8.1	15.5
6.600	9.1	16.5

Source.—Department of the Interior, Final Environmental Impact Statement, "Deepwater Ports," mimeographed, vol. 2 (April 1974), Table B7, p. B-44.
*In millions of bbl per day.
†Tankers serving U.S. superports required to have double bottoms, while tankers serving foreign ports not so required.
‡For the most part, tankers serving both U.S. and foreign ports required to have double bottoms.

THE OUTLOOK

In the immediate future, the prospect around the globe is for increased offshore oil and gas activity. The slowdown that occurred in 1976, particularly in the North Sea and Southeast Asia, has given place to renewed and vigorous growth. The number of offshore rigs idle stopped growing, and the total of 315 active early in 1977 represented an all-time record. In addition to a high, continued rate of activity in traditional areas—the Middle East and the Gulf of Mexico—the prospect is for increased drilling and production in a wide range of new areas.

This broadening interest is likely to be reflected in further growth of offshore production in areas not previously known as large oil consumers (like North America) or as large oil producers (like members of the Organization of Petroleum Exporting Countries—OPEC). Taken as a group, offshore production in the non-oil-exporting developing countries in 1977 is estimated at 1½ million bbl daily this year. This represents a production pace that has more than doubled over the last 2 years, bringing the current contribution of offshore wells to almost one-third of the total estimated production of that group of countries.

For the longer term, the outlook is less certain. Because of the scale of investment required in modern drilling equipment, production platforms, and other offshore installations, a steady rate of expansion will be hard to maintain. As occurred recently, overbuilding of the new types of rig, service equipment, and other facilities will no doubt lead to occasional surpluses and idle capacity.

These cyclical uncertainties will continue to be aggravated by the political forces that come into play as new offshore tracts are offered for exploration. This type of unknown is highlighted by figures 2 and 3, which indicate the large differences in production expected from the U.S. outer continental shelf, depending on whether rapid or slow leasing policies are followed. It will be seen that, under the best circumstances, crude oil and condensate production (fig. 2) could rise to nearly 3½ million bbl daily by 1990, compared with a bare 1 million bbl daily in the later 1970s. For natural gas (fig. 3), the best prospect would be for as much as 4½ trillion ft^3 of annual production, compared with a volume that is expected to dwindle to some 3 trillion ft^3 annually by the end of the present decade.

Political uncertainties and year-to-year fluctuations should not obscure the long-term prospect. The offshore industry has made immense strides since its birth in the late nineteenth century, when the first offshore oil wells were cautiously drilled using wooden piers built out to sea in southern California or into the Caspian Sea and Lake Maracaibo. Strong markets for the oil and gas flowing from today's offshore wells are assured by the seemingly unending growth of demand and increase in prices.

This does not mean that every known reserve of offshore hydrocarbon,

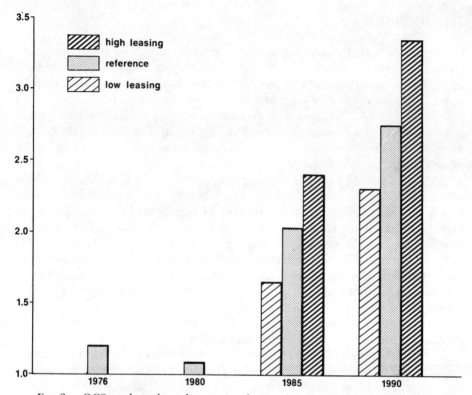

Fig. 2.—OCS crude and condensate production leasing uncertainties in the United States (in million bbl/day). Source: *National Energy Outlook 1977* (Washington, D.C.: Federal Energy Administration, 1977).

however remote the location and however deep the water, will be developed in the foreseeable future. Much of the abyssal depth will be found uneconomic for hydrocarbon exploitation, even if (as seems likely) deposits are one day found in it.

The cost of energy or any other product will remain limited, in the phrase of Henry David Thoreau, by "the amount of what I will call life which is required to be exchanged for it." But, in establishing this balance between a product and its cost in the case of offshore hydrocarbons, the context is one in which life itself may be part of the product being paid for. The value of energy as the underpinning of modern society can hardly be overestimated. Consequently, eagerness to push back the frontiers of offshore oil and gas seems bound to intensify.

BIBLIOGRAPHY

Baldwin, P. L., and Baldwin, M. F. *Onshore Planning for Offshore Oil.* Washington, D.C.: Conservation Foundation, 1975.

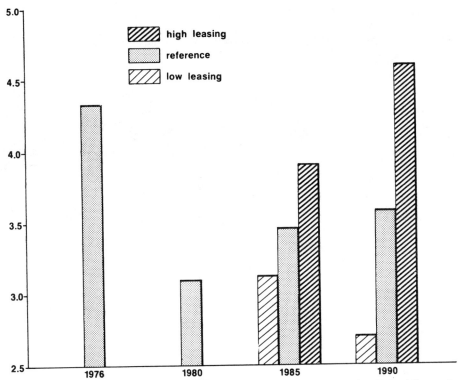

Fig. 3.—OCS natural gas production leasing uncertainties in the United States (in trillion ft³/year). Source: *National Energy Outlook* (Washington, D.C.: Federal Energy Administration, 1977).

Beguery, Michel. *L'Exploitation des océans.* Paris: Presses Universitaires de France, 1976.
Blair, J. M. *The Control of Oil.* New York: Pantheon, 1976.
Bureau of Mines, U.S. Department of the Interior. *International Petroleum Annual 1975.* Washington, D.C.: Department of the Interior, 1977.
Bureau of Mines, U.S. Department of the Interior. *Minerals and Materials.* Washington, D.C.: Department of the Interior, March 1977.
Leipziger, D. M., and Mudge, J. L. *Seabed Mineral Resources and the Economic Interests of Developing Countries.* Cambridge, Mass: Ballinger, 1976.
Library of Congress Congressional Research Service. *Effects of Offshore Oil and Natural Gas Development on the Coastal Zone.* Washington, D.C.: Government Printing Office, 1976.
Symonds, E. "Capital Requirements of Energy Supply." Presented at the Offshore Technology Conference, Stavanger, Norway, 1974. New York: Citibank, 1974.
United Nations. *Economic Significance in Terms of Seabed Mineral Resources of the Various Limits Proposed for National Jurisdiction.* New York: United Nations, 1973.
U.S. Geological Survey. *Geological Estimates of Undiscovered Recoverable Oil and Gas Resources in the United States.* Circular 725. Washington, D.C.: Department of the Interior, 1975.
U.S. Geological Survey. *Mineral Resource Management of the Outer Continental Shelf.* Circular 720. Washington, D.C.: Department of the Interior, 1975.

Whitaker, J. C. *Striking a Balance*. Washington, D.C.: American Institute for Public Policy Research, 1976.

Zuleta, B. "Statement to the Asian-African Legal Consultative Committee." *Law of the Sea Conference Papers*. New York: United Nations, 1977.

Nonliving Resources

Oil and Gas Exploration and Exploitation in the North Sea

Peter R. Odell
Erasmus University, Rotterdam

INTRODUCTION

The North Sea has become the world's most active region of offshore oil and gas developments—with the possible exception of the Gulf of Mexico. Even that is pushed into second place if one takes note of the speed of exploitation, of the degrees of success achieved in terms of the discovery of oil and gas reserves, and of the amount of technological innovation that has been engendered by the exploration and production efforts in the deep water and adverse weather conditions of the North Sea.[1]

What has happened to date in the North Sea over the period from the first offshore activity there in 1964 depends essentially on the international oil companies, for they provide the dominant element in the exploration for and the production of the province's oil and gas resources. Indeed, they have been given—directly and indirectly—well over 70 percent of all the North Sea acreage allocated[2] (including an even higher percentage of the best blocks); they have discovered all the major fields found so far (except two, namely, Brae and Brora) together with most of the minor fields; and they are responsible for almost all the fields in production or under development toward production.

This is so in spite of the interest shown in the North Sea both by new oil companies formed with the North Sea specifically in mind and by non-oil companies in Britain and Norway deciding to diversify into a new field of endeavor. Even the rapid growth of state oil companies—particularly the British National Oil Corporation and Norway's Statoil—seems unlikely to diminish *very much* the role of the international oil companies in determining

---

1. Descriptions of these conditions can be found in K. Chapman, *North Sea Oil and Gas* (North Promfret, Vt.: David & Charles, 1976), and I. L. White et al., *North Sea Oil and Gas: A Study Sponsored by the Council on Environmental Quality* (Norman: University of Oklahoma Press, 1973).

2. In all sectors of the North Sea (see fig. 5), concessions for exploration and/or production are allocated to companies which apply for them. They are not auctioned by the states concerned to the highest bidders. See K. W. Dam, *Oil Resources: Who Gets What How?* (Chicago: University of Chicago Press, 1976), for a description and analysis of the procedures involved.

© 1978 by The University of Chicago. 0-226-06602-9/77/1978-1006$01.79

just how far and how fast the North Sea oil and gas province will be developed. Thus, the motivations of the international oil companies in respect of the North Sea opportunities—together with these companies' responses to the petroleum legislation of the countries surrounding the North Sea—have largely determined the progress toward, and still condition the prospects for, the development of this major new oil and gas province.

THE RESOURCE BASE

Progress and prospects in the North Sea's oil and gas exploitation, though dependent in the final analysis on the "behavior" of the companies (and of the governments), depend in the first instance on the size and wealth of the resource base. Knowledge of this is, in the initial period of exploration, limited and speculative. Thus hypotheses on the likely occurrence of oil and gas are based largely on geological analogies. The size and complexity of the North Sea basin constituted a major difficulty in the early evaluations. Size, though a simple concept, was consistently underrated such that the North Sea came to be seen as a kind of European backyard in which prospects for exploitation were limited and about which the main concern was that false hopes should not be raised over its potential. However, what has already happened in terms of successful exploration and what remains to be done in this respect (and this adds up to an exploration effort that will be several times the magnitude of that already made) should be seen in the context of a potentially petroliferous North Sea province which is almost exactly the same size as the petroliferous region of the Persian Gulf—a comparison which is illustrated in figure 1. As in the Persian Gulf, where there are many different oil and gas "plays" (i.e., many potentially productive horizons in the underlying geological series), so also with the North Sea where the first decade of exploration has also clearly demonstrated that there are many sorts of potential reservoir rocks which are worthy of investigation. This opens up not only the likelihood of exploration in parts of the province which have, hitherto, not been considered worthwhile, but also the desirability of deeper exploration in areas previously investigated only by shallower wells.

Thus, both geographical scale and geological complexities underly the great success to date in the exploration efforts in the North Sea and provide the basis for continuing efforts which should certainly stretch out over the next 25 years. Success to date is pinpointed in table 1, which shows the number of finds of oil and/or gas that had been made in various classes up to the end of 1976. The essential proof of the prolific nature of the basin lies in the fact of over 200 discoveries to date, of which about two-thirds (129) have already been designated as gas or oil and gas fields. Of this number, about 60 percent have had reserve figures declared for them. Moreover, of the 79 fields with declared

Oil and Gas Exploration and Exploitation 141

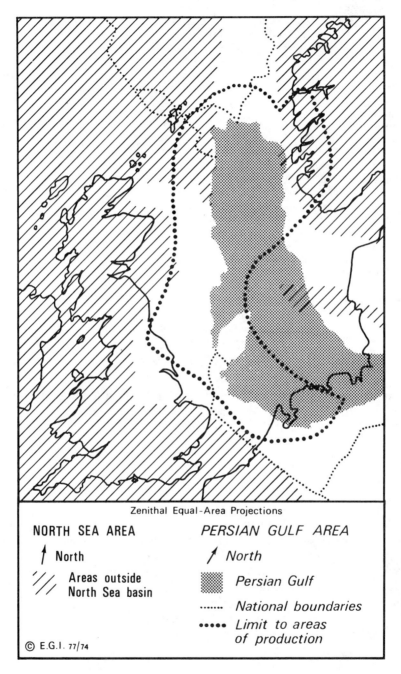

Fig. 1.—A comparison of the size of the potentially petroliferous areas of the Persian Gulf and the North Sea.

TABLE 1.—NORTH SEA OIL AND GAS PROVINCE:
OFFSHORE DISCOVERIES TO JUNE 1977

	Oil and Gas	Gas Only	Not Known	Total
Number of discoveries—total	75	99	34	208
In southern basin	7	88	...	95
In northern basins	68	11	34	113
Designation of the discoveries—total	75	99	34	208
Fields *with* declared reserves	41	38	...	79
Fields *without* reserves declaration	34	16	...	50
Discovery wells (no other information)	...	45	34	79
Size distribution of declared fields—total	41	38	...	79
More than 2×10^9 bbl oil or equivalent	4	1	...	5
$1-2 \times 10^9$ bbl oil or equivalent	9	2	...	11
$0.5-1.0 \times 10^9$ bbl oil or equivalent	12	3	...	15
Smaller fields	16	32	...	48
Production and production plans:				
Number of fields with declared reserves	41	38	...	79
In production	8	9	...	17
Production plans	16	7	...	23
No plans for production	17	22	...	39

reserves, 31 have more than 500 million barrels of oil (or the oil equivalent thereof as natural gas), so making them "giant" fields in usual North American oil industry parlance.

NATURAL GAS

What the discoveries mean in terms of effective overall recoverable reserves is especially difficult to evaluate in the case of natural gas. For example, many of the gas-only fields lie in the Dutch sector of the North Sea (see fig. 5). Here, unfortunately, there is no obligation on either company or government to give any field and reserve information at all to the general public, so eliminating the possibility of depletion policymaking (and, indeed, energy policymaking) which can be seen to be justified by the rate of reserve discovery. Table 2 details the gas fields, discoveries, and reserves positions for the southern basin of the North Sea. In the Dutch sector there are at least 50 offshore discoveries[3] about which virtually nothing has been made known to the public in terms, for

---

3. Ministry of Economic Affairs, *National Gas and Oil in the Netherlands and Its Offshore Area, 1976* (The Hague, 1977).

example, of reserve figures for individual fields. The Dutch government has merely given a total figure for the sector's proven reserves without indicating what is meant by "proven" or even the number of fields whose reserves are included in the total! One can conclude, however, that the officially declared reserves grossly understate the actual position. The Placid field (L10/11) has been forward sold (to Gasunie and to German customers) to the extent of 150 × 10^9 m^3; the Ameland field has been officially announced as containing 55 × 10^9 m^3 (in response to a parliamentary question which asked if the field was a second Groningen with over 2,000 × 10^9 m^3!); and each of the 10 other fields in production or being developed for production must have at least 25 × 10^9 m^3 (a minimum exploitable size) while their average size, based on hints dropped by the companies concerned, seems to be of the 40 × 10^9 m^3.

Thus these 12 fields alone—even on a simple arithmetic addition of their individual reserves—contain between 450 and 600 × 10^9 m^3 of natural gas. However, as shown in table 2, there are at least 40 other gas fields and discoveries in Dutch waters ranging in size from a few milliard up to more than 25 × 10^9 m^3. Assuming, conservatively, an average size of 5 × 10^9 m^3 then there is another 200 × 10^9 m^3 of natural gas to add to the 450–600 × 10^9 m^3 defined above—to give a total, by a simple arithmetic counting-up procedure, of up to 800 × 10^9 m^3. However, in such cases, the simple arithmetic sum of the 90 percent probable reserves from a large number of fields is not reasonable, as the answer given is for a 98–99 percent probability of their being at least that amount. To return to a reasonable 90 percent probability figure for the overall

TABLE 2.—NATURAL GAS RESOURCES OF THE SOUTHERN NORTH SEA BASIN, EXCLUDING ASSOCIATED GAS

	Dutch Sector	British Sector	German Sector
Total no. of gas discoveries	51	31	3
No. of discoveries declared as gas fields	11	15	...
Governments' declarations of remaining proven gas reserves (10^9m^3)	367*	552*	...
No. of fields in production	5	7	...
Other fields with announced production plans	5	...	...
Current (1977) annual production (10^9m^3)	ca. 3	40	...
Estimate of 1980 production from fields currently on production or in development (10^9m^3)	ca. 10	ca. 42	...
Likely remaining reserves† for all fields in each sector at summed 90% probability (10^9m^3)	1,000+	1,050+	50+
1980 production potential with full exploitation‡ of the already discovered reserves (10^9m^3)	40+	50+	2–3

* Arithmetic total of "proven" reserves of declared fields.
† Based on all discoveries made and not just on declared fields.
‡ Based on 20–25-year depletion periods for the fields.

reserves of the group of fields it is necessary to collate the individual probability curves for the size of each field. Though such curves are of a confidential nature in the Dutch information system, a conservative estimate of their shape and a subsequent recalculation of the overall 90 percent reserves from all the fields combined indicates a total of at least $1,000 \times 10^9$ m^3 of recoverable reserves from the fields which have already been discovered in Dutch waters.

More information is, fortunately, available on nonassociated gas reserves in the British part of the southern North Sea basin though even in this case there is no evidence to indicate that all finds made have been announced or that the total figure of remaining reserves of just over 500×10^9 m^3 has been adjusted to give overall probabilities equal to those used for the individual fields (of which only 15 of the 31 in British waters have been declared). Elsewhere in the North Sea the reserves as announced for the Danish sector (by the DUC, the operating company) were for $24-42 \times 10^9$ m^3. The much higher reserve figure of 62×10^9 m^3, as estimated for three fields only by consultants appointed by the Danish government, again points to a continuation of the tradition of understatement of reserves by the companies which have made the discoveries. This is a tradition in Western Europe dating back to the earlier gross understatement by Shell and Esso of the likely reserves of the Groningen field: now known to be the largest gas field in the noncommunist world.

In table 3, therefore, we find a contrast between the proven gas reserves position as it is officially presented and what the situation could be if there were a more reasonable approach to the calculation of reserves. If discoveries of new associated and nonassociated gas reserves continue at the level achieved over the last few years then the estimate of over $8,000 \times 10^9$ m^3 of remaining recoverable reserves of gas by that date in Western Europe—mainly in the North Sea—can now be considered as the *minimum likely* rather than the maximum possible.[4] Moreover, beyond that date one can also be confident that new discoveries—and new reserves in previously known discoveries—will continue to add to the gross available amounts of recoverable natural gas so that, even without the discovery of another major offshore gas province somewhere around the much-indented coastline of Western Europe, an ultimate resource base amounting to at least $20,000 \times 10^9$ m^3 does not seem to be unduly optimistic.

Thus, given this adequate gas resource base and the proximity of a readily available market for as much gas as can and will be produced, one can expect (company and government behavior permitting) a continued rapid contribution of natural gas to the Western European energy economy, particularly, of course, in the countries surrounding the North Sea from which the bulk of the resources will be available for the foreseeable future.

4. In addition, by then, about $2,250 \times 10^9$ m^3 of natural gas produced in Western Europe will have been used.

TABLE 3.—WESTERN EUROPE: ITS CURRENTLY "PROVEN" AND POSSIBLE NATURAL GAS RESOURCES AND AN ESTIMATE OF THEIR DEVELOPMENT BY THE EARLY 1980s

	Remaining Recoverable Reserves			Mid-1980s Annual Production Potential ($\times 10^9 \text{m}^3$)	Millions of Tons of Coal Equivalent* (Approximate)
	Declared "Probable" by 1977 ($\times 10^9 \text{m}^3$)	Probable plus Possible ($\times 10^9 \text{m}^3$)	As Likely by Early 1980s ($\times 10^9 \text{m}^3$)		
Onshore Netherlands	2,030	2,150	2,100	105	120
South North Sea, British sector .	550	725	1,050	50	70
South North Sea, other sectors ..	440	850	1,250	55	75
Onshore West Germany	310	515	450	25	30
Austria, France, Italy, etc.	420	490	600	35	45
Northern North Sea Basin, UK/Norway	900	1,500	2,500	115	135
Rest of European continental shelf (Ireland, Spain, etc.)	50	150	350	20	25
Total	4,700	6,380	8,300	405	500

SOURCE—For 1976 various national and EEC/OECD estimates. Estimates for the 1980s are the author's own.
*Conversion to coal equivalent based on known or estimated calorific values of the various gas supply sources.

OIL

Progress to date in the development of North Sea hydrocarbons thus relates more to natural gas than to oil. However, the even more recent developments (post-1969) in the exploration for oil have produced even more exciting results in terms of the resource base potential than those of the somewhat earlier and longer search for offshore gas.[5] As a result, the prospects for the medium- to longer-term future of oil production from the North Sea now exceed all earlier forecasts and expectations.

Table 4 presents the minimum likely position on reserves by June from already known North Sea oil fields. This shows that ultimately recoverable reserves from discoveries that have already been made are likely to be at least 45,000 million barrels—after making only quite modest allowance for the appreciation of fields with already declared reserves and for fields which have been announced as having been proven but about which no information on reserves has yet been given. It does not, however, make any allowance whatsoever for the more than 60 additional discovery wells which have been drilled (see table 1). Some of these will eventually prove to have recoverable reserves of oil and so push the sum total of discovered reserves to date above the 45×10^9 barrels figure.

Table 5 then puts this real-world reserve development position and prospects for development in the perspective of some of the estimates which have been made over the last few years about North Sea oil. Shell and B.P. both gave earlier estimates for 1976 and 1980 reserves which lie at a level of 50 percent or less of what has occurred, or what can now be seen as most likely to happen in terms of reserve development. These companies' latest estimates of the province's "ultimate" reserves range from 35 to 50×10^9 bbl; but these figures indicate ultimate levels of recoverable reserves which are less than those that are currently known to exist or the existence of which can be confidently extrapolated from present evidence. This is, moreover, is spite of the fact that *most* of the exploration work in the North Sea still remains to be done—including exploration work on some of the largest structures in the Norwegian sector in blocks which have only just been allocated; or even not yet allocated to companies for such work, in accordance with Norwegian conservation policies. Of the oil companies' estimates on the North Sea's reserves only Conoco's 1975 estimate of up to 67,000 million barrels of ultimately recoverable oil remains credible and one is forced to the conclusion that the oil industry's conservatism over the question of evaluating reserves (and/or its wish to protect its vested interests) has once more led it to inappropriate pronouncements on the resource potential of a region whose future economic and even political survival depends on a realistic evaluation of its indigenously available oil.

---

5. This is partly, at least, because the oil companies are more interested in finding oil than gas. In almost all Western European countries gas distribution and sale is a state monopoly and thus subject to more control than the oil companies like.

In this respect it should be noted that table 5 also shows that our simulation model study[6] of the North Sea oil reserves—a model which was generally criticized in the industry for its highly "optimistic" conclusions—is also underpredicting the rate of development of reserves of the province. This, however, does not surprise us, as we described it as a conservative model in which the probabilities of discovery, of size of field, and of recoverability of oil, etc., were oriented to minimum rather than reasonable opportunities. Thus, although it is possible that the 1976 and 1980 results from the model may be underestimates (because of mistakes we made in simulating the *timing* of the exploration and development effort such that the model's results will, within a few years, "catch up" with the real world developments), nevertheless the model's 90 percent probability figure of 78,000 million barrels of ultimately recoverable oil from the North Sea province is beginning to assume a high degree of "reasonableness" when seen in the context of the more than 45,000 million barrels which are likely to be recoverable from known fields and with another 60 discoveries still awaiting evaluation.

THE CONTRASTS IN THE OIL AND GAS RESERVES EVALUATIONS

Why should "officialdom" (governments and the oil companies) have so seriously underestimated the resource potential? It seems to be due largely to an unwillingness to try to quantify the future potential availability of the so-far

TABLE 4.—NORTH SEA OIL RESERVES, DECEMBER 31, 1976

	Million Barrels
As declared	
By simple addition of declared figures for 43 fields	ca. 23,500
After adjusting to a summed 90% probability (from the 39 fields, assuming each is declared at a 90% probability)	≥31,000
On extrapolation:	
With upward revision following production from the 39 fields (average 15%)	≥36,000
Addition of reserves from the so-far undeclared 35 discovered fields* with an assumption that these fields are on average two-thirds smaller than the declared fields	9,500
Minimum total* of North Sea oil reserves discovered to date	>45,000

*In addition there have been, as shown in table 1, over 60 oil (or gas) discovery wells many of which will eventually be declared as oil fields. The 45 x 10^9 barrels of oil shown in this table as already having been discovered is thus a minimum likely figure and not even the most reasonable estimate. See below in table 5 for details on future estimates of North Sea oil reserves.

6. P. R. Odell and K. E. Rosing, *The North Sea Oil Province: An Attempt to Simulate Its Development and Exploitation, 1969–2029* (London: Kogan Page, 1975).

TABLE 5.—NORTH SEA OIL RESERVES: ESTIMATES AND FORECASTS COMPARED

	By 1976	By 1980	Ultimate
As declared:			
By simple addition of declared reserves for 39 fields	22.5	...	....
After adjusting to a summed 90% probability (from the 39 fields, assuming each field is itself declared at a 90% probability)	30+	...	...
Declared plus additional discovered reserves not yet declared:			
From upward revision of reserves in declared fields	...	ca. 35	...
From fields discovered but not yet declared (27 fields)	...	8–9	...
New discoveries from 1976 to 1980	...	±10	...
Totals	30+	53–54	...
As predicted by EGI simulation model (50% probable)	17.8	...	...
As hypothesized by EGI model for 1976 reserves after appreciation plus discoveries 1976–80 (50% probable)*	...	48.6	...
As forecast by EGI model on full development of the province:*			
90% probability	...	...	78
50% probability	...	...	109
As forecast by oil companies:			
By Shell in 1972 (5)†	10–12	17.5	(35)
By B.P. in 1973‡	...	...	38
By B.P. in 1974§	16–18	24–30	44
By Conoco in 1975¶	...	...	45–67
By Shell in 1976#	23	...	35
By B.P. in 1976**	...	...	±50
By Shell in 1977††	...	...	±50

*P. R. Odell and K. E. Rosing, *The North Sea Oil Province: An Attempt to Simulate its Development and Exploitation, 1969–2029* (London: Kogan Page, 1975).
†At E.I.U. International Oil Symposium, London, October 1972 (A. Hols, head, Production Division of Royal/Shell Exploitation and Production Coordination); ultimate figure is 1975 forecast at Tønsberg Conference.
‡At Financial Times North Sea Conference, London, December 1973 (Dr. J. Birks, director, B.P. Trading).
§At meeting Conference of the Society for Underwater Technology, Eastbourne, April 1974 (H. Warman, exploration manager, British Petroleum Co.).
¶At Conference on the Political Implication of North Sea Oil and Gas, Tønsberg, February 1975 (T. D. Eames, Oil Exploration Division, Conoco North Sea, Ltd.).
#Shell Briefing Service, *Offshore Oil and Gas, North West Europe* (London: Shell International Petroleum Co., July 1976).
**At a meeting in London of the European Atlantic Group, December 1976 (P. I. Walters, a managing director of B.P.), reported in *Noroil*, vol. 5, no. 1 (January 1977).
††At a meeting of the Empire Club of Canada, March 1977 (P. B. Baxendell, a managing director of Shell).

undiscovered recoverable resources of the North Sea province—that is, because of an unwillingness to do in Western Europe something which is commonplace in other parts of the world and most notably North America, where the USGS-directed evaluations of undiscovered recoverable oil and gas are used as a prime input for estimating future levels of oil and gas production. Yet this quite reasonable procedure for the United States is even more important for Western Europe, where most of the potential hydrocarbon resources still remain undiscovered, given that most of the petroliferous regions are round the long and much-indented coastline of the continent in off-shore locations in which oil and gas exploration and exploitation has only recently become possible. Figure 2 shows the vast extent of these petroliferous offshore regions. Within the total extent of these regions the North Sea constitutes but a small part.

Within this context, the concept of total potential resource base is very different from the way in which it has been conceptualized at the official European level, where planners seem not to have succeeded in reaching out beyond the severe limitations of "proven" reserves. The familiar resource diagram—figure 3—perhaps helps to clarify the issues involved and is especially important at the present stage in the cycle of petroleum exploration and exploitation in Western Europe, where proven reserves and inferred reserves, which are currently considered to be economic to produce, are very small relative to the total potential which could be developed as knowledge increases, technology improves, politics change, and economics give more emphasis to indigenous resource development.

To return to the North Sea, however, we can simplify the evaluation which is necessary to try to understand how its oil and gas resources are developing. This evaluation is presented in figure 4 with its x- and y-axes representing "discovery" and "recovery"—the complementary aspects of resource development.

The size of the resource base depends, in the first instance, on the number of fields and on their size. Equally obviously, in any province, there are a finite number of fields and how many of these are discovered is a function of the size of the investment in the exploration effort. The more fields that are discovered, the greater the quantity of reserves in the province—a development which is illustrated on the y-axis in figure 4. Increases in reserves figures thus depend first on the continuation of an exploration effort in the province over a long period of time during which the results of the continuing effort are expected to yield knowledge of successively smaller fields. However, in light of specific economic and political conditions, it is possible that the exploration effort will be terminated before all the prospects have been tested so that resources, which are discoverable in a technological sense, will remain undiscovered and so keep the reserves figure of the province at a lower level than would otherwise have been the case.

The degree to which this phenomenon is likely to occur in respect of North Sea exploration remains uncertain. Many companies, however, have indicated

FIG. 2.—Western European regions of offshore oil and gas potential

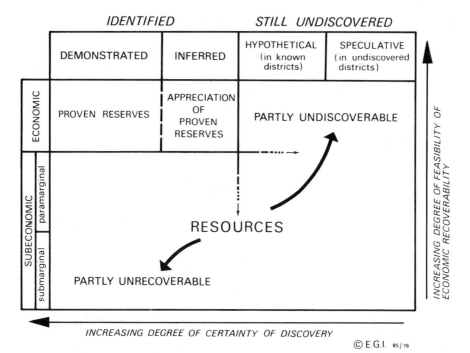

FIG. 3—Resource diagram. This illustrates the categories into which resources (of, e.g., oil and gas) may be divided and the way in which geological and economic factors influence the relationships between the categories. In light of levels of knowledge, technology, and price, the dividing lines between the categories vary over space and time. The proven and inferred reserves of oil and gas in Western Europe at present are relatively small, but conditions have worked against their development to date. With changed conditions—of knowledge, technology, economics, and politics—the resource base can be more effectively explored and developed.

that the much increased physical costs of developing North Sea fields, coupled with the higher share of the revenues which now have to be paid to governments following the renegotiation of the terms of the concession arrangements (and which the companies see, logically enough, as constituting an additional set of costs affecting the viability of their operations), constitute good reasons why exploration should cease when all the larger structures have been tested. This might then exclude the search for all fields which are expected ultimately to yield less than 200 million barrels of oil (or the equivalent thereof as associated or nonassociated natural gas). However, as such reservoirs are unlikely to contain much more than 15–20 percent of the total reserves of the province (based on the usual statistical distribution of reserves by field size in a petroleum province), then the effect of this component in accounting for the difference between the simulated availability

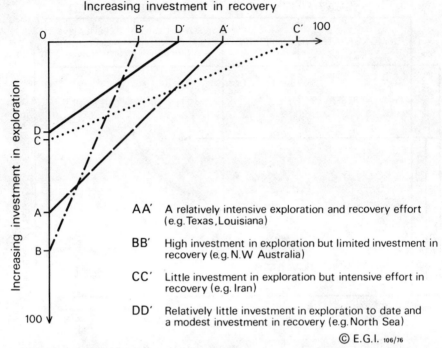

Fig. 4.—Discovery and recovery: complementary aspects of resource base evaluation. The proven oil reserves of any given petroliferous province depend first, on how much investment is put into the exploration effort; i.e., in testing all the possible occurrences of hydrocarbons in different sorts of structures in different horizons in the geological succession. Full exploration is a costly and time-consuming process and most hydrocarbon provinces of the world have, as yet, been explored only to a limited degree. In the case of the North Sea, the exploration effort to date, as shown on the diagram, covers no more than one-third, at most, of the total exploration which is necessary fully to explore the province; and this, if it happens at all, will be spread out over at least the next 2 decades. Second, proven recoverable reserves depend on investment in the facilities needed to recover the discovered oil. Fields may be creamed of their lowest-cost resources (with costs represented almost entirely by investment costs) or developed more intensively so as to push recovery toward the "limit" set by technology. The technology itself, of course, changes over time, as does the commercial viability of recovering more or less oil from a field or group of fields. In this diagram North Sea investment in recovery is shown to be relatively higher than the current level of investment in exploration (line DD'). It is, however, still modest compared with what could be done.

of reserves and the quantities expected by the oil companies will be relatively small, unless, of course, other factors, such as government-imposed limits on the amount of exploration (as in Norway) intervene to inhibit discoveries.

Apart from this influence, however, the much more important component in determining contrasts in estimates of reserves emerges from the other axis (the *x*-axis) in figure 4. This is a component which represents possible variations in the degree to which the resources of oil and gas which have been found in a set of fields are actually exploited. Of the oil and gas in place in any reservoir a certain percentage will be recoverable with a given technology over an economically relevant time period. This percentage figure will be a function of the level of investment made in the oil recovery system such that the more money that is spent, the more oil will be recovered. Moreover, technology also improves over time and so increases the possible recoverability of oil from discoverable fields, so enabling more oil to be recovered with additional expenditure on the development systems. Finally, of course, technological improvement also costs money and investment in this way may also be seen as a means of increasing the recoverability of oil from a field. Overall, therefore, one can hypothesize that, the more investment that is made in developing an offshore field more intensively and extensively, the more oil will be produced, so pushing out to the right in figure 4 the recoverability component. By this means, too, total reserves are increased.

The major difference between our simulation model's predictions of the reserves of the North Sea and the estimates of "officialdom" appear to emerge from the way in which this component in reserves' evaluation is being allowed to work itself out. What it boils down to is a difference between the *possible* and the likely recoverability of oil from a discovered field and this is a question—with both technical and economic aspects—which has not yet been adequately explored in spite of its important policy implications for both companies and governments.

As shown previously (table 1) a large number of oil and gas fields have already been found in the North Sea province. Their distribution, by size and by type at the end of 1976, is illustrated in figure 5. There are almost 30 "giant" fields (using American parlance in which a "giant" is a field with over 500 million barrels of oil or oil equivalent). In the case of each of these fields some part of their technically recoverable reserves will be recovered on the basis of an installed production system, the decision on which will depend essentially on the operating company's evaluation as to how it can earn, at best, maximum and, at worst, sufficient profits. In other words the declared recoverable reserves of an offshore field are not a fixed quantity. On the contrary, they are a highly variable element and are essentially a function of the investment decision which the operating company takes. The initial investment decision is, moreover, one which, in the unique circumstances of the North Sea's physical environment, determines more or less once and for all what percentage of the technically recoverable reserves shall be recovered over the full production life cycle of the field. This is because, given the size, the shape and the deep-water location of the fields, the initial decision on the number of platforms to be put on the field and the number of wells to be associated with them is the critical variable for defining the quanitity of the reserves which will be recovered—that is, for defining the *size of the field* in terms of recoverable reserves.

154 Nonliving Resources

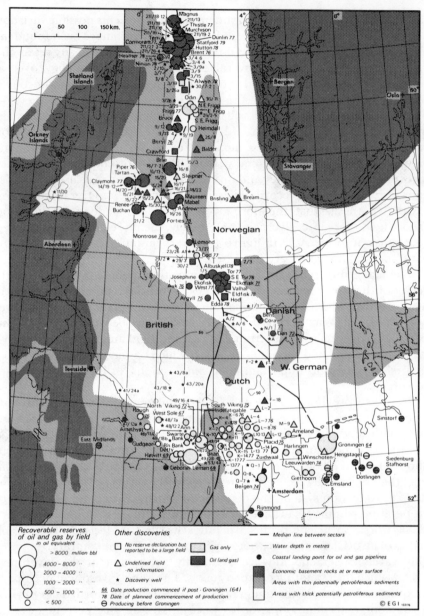

Fig. 5.—Oil and gas fields in the North Sea to the end of 1976

How this works out in practice is illustrated in figure 6 showing how, for a hypothetical field, one, two, or three platforms can be located to deplete the reservoir; or, rather, the oil reserves of part (or of parts) of the reservoir. In economic terms, moreover, each additional platform is less productive than the

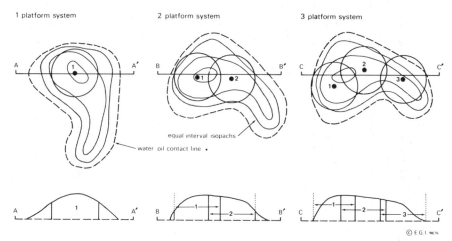

Fig. 6.—Reorganization of platform location with increasing system size. A hypothetical oil field is shown with a one-, a two-, and a three-platform system, respectively, on the three maps of the field. Below each map is an appropriate cross-section diagram. If we compare the one- with the two-platform system and the cross section A-A' with the cross section B-B', we can see that the introduction of the second platform has caused the relocation of the first, resulting in the deepest oil-bearing sands being shared between the two platforms for production purposes. The first platform now produces less oil than the one platform in the one-platform system. Comparing the two- with the three-platform system and sections B-B' and C-C', we can see the same phenomena in respect to the location and productivity of platforms 1 and 2.

previous one (successive platforms produce decreasing quantities of oil but do not cost any less to install or to run) and so, as shown in figure 7, there is a rising average unit investment cost curve as a field is more extensively and/or intensively developed.

Thus, the location and the geographical extent and shape of any North Sea field; its reservoir characteristics; and the economics of the different production systems which can be developed to deplete it have important consequences for the field's unit production costs. Moreover, as the unit revenue curve can be taken to be horizontal (as it is not affected by the production decision), then there are also consequences for the unit profitability of production.

In such circumstances one can argue that there is a high propensity on the part of the operating companies to play safe and thus to take exploitation decisions which mean that the fields are simply "creamed" of their lowest-cost-to-produce reserves. Thus, some, or even most, of the technically recoverable reserves of a field become unrecoverable and so do not constitute recoverable reserves at all in the context of company evaluations on how to achieve acceptable rates of return on investment, when compared with the opportunities which exist for nearly all the companies concerned for investing the same money in other activities or in oil-producing activities elsewhere in the world.

It is this set of facts that forms the principal reason for the contrast between

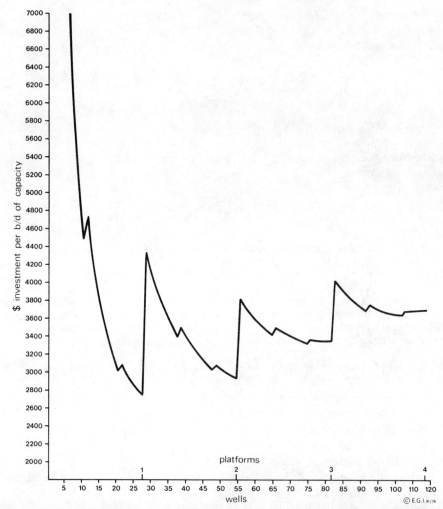

FIG. 7.—Average investment per barrel per day of capacity with a four-platform system. For the installation of this system on the field, four separate platforms each with an ultimate capacity for handling 27 wells are required. Within each platform there is a falling unit cost curve as average productivity increases with the increasing number of wells—except that costs of expanded platform facilities at 12 and 20 wells per platform create the upward kinks in the curves. The installation of each additional platform, however, reduces overall productivity and there is a jump back to a higher unit investment cost—e.g., from the minimum $3,350 when full productivity is achieved from the third platform to $4,000 as the first well on the fourth platform comes into production. Note the steadily increasing unit cost from the most productive situations on platforms 1-4; i.e., from $2,750 to $2,950 to $3,350 to $3,700 with the four platforms.

the simulated level of North Sea reserves (and in which simulation model a primary assumption was that all reserves which are discovered and technically producible will be produced) and the much lower levels which are being defined by the companies. The problem has been fully defined and analyzed in a recently completed study.[7] Here, table 6 offers no more than the conclusion of the study in respect of three North Sea fields (Forties, Piper, and Montrose) which were analyzed in detail in respect of possible alternative production systems; their productivities and associated profit levels and their revenues to government and contributions to foreign exchange earnings. This demonstrates unequivocally the great difference between what reserves the companies have decided to produce from these three fields and what might have been produced if alternative production systems had been installed on the fields with the objective of achieving production levels which would have maximized the economic returns to the United Kingdom. The difference in terms of the reserves figures is of the order of 31.0 percent; the differences from the point of view of the benefits to the U.K. economy are just as significant, ranging from 23.9 percent in terms of the present value of the future flows of tax revenues to over 50 percent in respect of the present value of investments in the production systems; a factor which is of great significance for job creation in the economy in general and in the oil-producing areas in particular.

CONCLUSIONS

Offshore oil and gas exploitation in the North Sea has created a technological frontier—not only in terms of the development of the exploration and production facilities which have been considered in this paper but also in terms of undersea pipelining and maintenance developments. And even this still leaves equally important issues unexamined; issues such as the safety of the installations themselves and of the additional difficulties created for navigation in a busy area; the working conditions and the safety of the men employed on the rigs, the platforms, and the servicing vessels—especially those in the deep-sea diving industry which has had to expand rapidly to cope with the demand for underwater inspections in waters of up to 500 feet in depth and which requirements have already caused a significant loss of life (over 30 deaths to the end of 1976); and, of course, the question of the impact of exploration and exploitation on the marine environments as well as the impact of major

7. P. R. Odell, K. E. Rosing, and H. Vogelaar, *Optimal Development of the North Sea's Oil Fields: A Study in Divergent Company and Government Interests and Their Reconciliation* (London: Kogan Page, 1976).

158 Nonliving Resources

TABLE 6.—FORTIES, MONTROSE, AND PIPER FIELDS: CONTRASTS IN BENEFITS FOR THE U.K. ARISING FROM COMPANY AND COUNTRY OPTIMAL DEVELOPMENTS OF THE FIELDS

Characteristic	Forties		Montrose		Piper		Total of the Three Fields		
	4 Platform	6 Platform	1 Platform	3 Platform	1 Platform	3 Platform	Companies' Optimum	Country's Optimum	% Difference
Quantity (million barrels) of recoverable oil in economically relevant time period	1,935	2,315	159	333	625	903	2,719	3,561	31.0
Production of oil in 1980* (million bbl)	221	255	(ca. 160)	21	42	(642)	71	100	313	397	26.9
Peak year production of oil	1980/81	1981	1979/82	1981/82	1979/82	1981/82	...	...	...
Flow of government revenues in 1980* (million $)	2,138	2,310	140	282	738	983	3,016	3,575	18.4
Foreign exchange value of 1980* oil (million $)	3,210	3,697	320	628	1,071	1,494	4,601	5,819	28.6
Present value of all government revenues (milliard $)	7.04	8.07	.46	.71	2.45	3.54	9.95	12.33	23.9
Present value of total volume of oil produced (milliard $)	13.1	15.1	1.34	2.67	4.82	6.82	19.26	24.59	27.7
Capital investment (million $):									
1974	367	367	42	85	93	113	502	565	12.5
1975	397	443	98	213	217	279	712	935	31.3
1976	361	500	84	215	186	274	632	988	56.5
1977	277	599	28	162	62	191	367	950	158.7
No. of platforms:									
Built by 1976	4	4	1	2	1	2	6	8	33.3
Under construction in 1976	0	2	0	1	0	1	0	4	?

NOTE.—Numbers in parentheses are companies' estimate of oil recoverable.
* 1980 has been selected for illustrating the contrasts in government benefits to show the relatively near future importance of the differences arising from the systems and not because the differences reach their peak in 1980. Indeed, relatively, the gap continues to widen throughout the 1980s but, in terms of the absolute difference between the size of the benefits, the peak is reached in 1984/85.

onshore installations (servicing facilities and terminals especially) in adjacent coastal regions hitherto largely undisturbed by large-scale industrial activity. Such technological and environmental issues are obviously important and raise many hitherto unasked questions concerning the use of the marine environment for oil and gas production.[8]

In this article, however, we have tried to show that the development of the North Sea for oil and gas has also created new frontiers in political and economic issues. These issues are also less than fully understood and/or not put into the context of relationships between national governments anxious for security of energy supply and the oil companies whose justifiable commercial interests seem to indicate a pattern and a type of development of the oil and gas reserves which cannot meet the needs of the communities involved.

ADDITIONAL REFERENCES

Odell, Peter R. *Natural Gas in Western Europe.* Haarlem: Bohn, 1969.
Odell, Peter R. "Indigenous Oil and Gas Developments in Western Europe's Energy Policy Options." *Journal of Energy Policy* 1, no. 1 (June 1973): 47–64.
Odell, Peter R. *The Western European Energy Economy: The Case for Self Sufficiency, 1980–2000.* Leiden: Stenfert Kioese, 1976.
Odell, Peter R., and Rosing, K. E. *Optimal Development of the North Sea's Oil Fields: A Study of Divergent Government and Company Interests and Their Reconciliation.* London: Kogan Page, 1976.
Odell, Peter R.; Rosing, K. E.; and Beke-Vogelaar, H. E. "On the Optimisation of the North Sea Pipeline System." *Journal of Energy Policy* 4, no. 1 (March 1976): 50–56.

---

8. For a survey of these issues—and for references to the specialized literature—see E. deKeyser, ed., *The European Offshore Oil and Gas Yearbook* (London: Kogan Page, 1976).

Nonliving Resources

Other Ocean Resources

R. H. Charlier
Flemish Free University of Brussels and Northeastern Illinois University

WATER

One of the first resources of the ocean is water itself, whether it is used to extract salts, construct electrical power plants, or quench the thirst of man and parched lands.

Combating Volcanic Eruption

The recent eruption in the Icelandic island of Heimaey was the occasion for a major attempt to control a lava flow by use of seawater as a cooling agent. Icelanders are old hands at coping with volcanic phenomena. Over the last 10 yr, four eruptions occurred in this country, and in the year 570 an eruption of the Heimaey volcano was already described in awed terms by the Irish Abbot of Clonfert, Brendan MacFinnloga.

Some 50–200 liters per second of seawater are poured onto the lava flow. A wall of cool rubbly lava is thus created at the flow margin and the flow thickens against it. In places, fractures which were originally deep in the flows are coated with salt left by the seawater that evaporated. While the island's harbor and main city, Vestmannaeyjar, was heavily damaged by eruption and flows, the use of cooling seawater may well have limited further damage. A final assessment of seawater use in such instances awaits further study.

A Source of Freshwater

The shortage of freshwater poses a pressing problem. Water levels are dropping on the continents. Occasionally wells are invaded by brackish waters, artesian wells run dry, and deeper-drilled wells are needed. Kings County, outside New York City, woke up one morning to find its wells put out of use by seepage of Long Island Sound waters. This was a half century ago. Today, New York City and the surrounding communities import their drinking water from the Catskill Mountains, hundreds of kilometers away. Tapping the ocean for

© 1978 by The University of Chicago. 0-226-06602-9/77/1978-1007$03.60

freshwater, an unthinkable luxury not so long ago, is now a necessity. Engineers have examined the possibility of towing icebergs from Antarctica to Australia and to the west coast of the United States. An iceberg with dimensions of 1,000 × 1,000 × 250 m represents 250 million m³ and could be towed in 300 days to near the Atacama desert in Chile. It would lose 86 percent of its water mass but would still represent 35 million m³ of freshwater worth $27 million. The trip would cost $13 million and thus, theoretically, the operation would be profitable. The freshwater available from Antarctic ice alone is staggering.[1]

With 90 percent of the world's ice locked up in cold storage, Antarctica is the world's largest ice mass. Snow and ice have been measured 15,000 m thick at the South Pole, with the average height of the ice sheet above sea level estimated at 2,500 m. In calculating the amount of water locked in ice there, scientists have figured that if the cap should melt, the oceans of the world would rise 260 m, drowning every coastline and wiping out every port, harbor, and coastal city.

Today, there are numerous desalting plants throughout the world providing freshwater for man, cattle, and irrigation purposes. Salt may be removed through a variety of processes such as distillation, the membrane process, congealing, solar distillation, chemical processes, physical and electrical methods, and the flash distillation process used in the San Diego, California, plant. This was the first plant in the United States to use the multiple flash process and is one of the world's largest. The U.S. Office of Saline Water disclosed in 1968 that no less than 627 plants were then in use, being built, or on order, with a production capacity of 800,000 m³ per day. In many cases the energy needed to run the plants is provided by petroleum or gas, an ideal energy source in the Middle East. Kuwait has the lead in freshwater production with 100,000 m³ per day and is building two more plants which together will provide an additional 130,500 m³ of freshwater daily.

The major problem, of course, is the cost of the freshwater produced. Since fuel is no problem in Kuwait, water can be produced at a cost of $0.12 per cubic meter; otherwise, it would cost about $0.30.

The price can probably be cut by planning larger plants, treating at least 100,000 m³ per day, and establishing agricultural-industrial complexes which will recover the waste products. If plans to build nuclear-powered plants go through, then giant desalting factories will further reduce production costs by furnishing between 300,000 and 500,000 m³ per day.

Presently, desalination plant capacity reaches up to 28,000 m³ per day. Since 1968, installed capacity has increased up to 30 percent each year, and it is expected, according to Gould,[2] that by 1978 freshwater production from the

1. W. F. Weeks and W. J. Campbell, "Towed Icebergs—Plausible or Pipedream?" *Journal of the Marine Technology Society* 7, no. 5 (August 1973): 29–32.
2. H. R. Gould, "Minerals from the Sea," in *Oceanography: The Last Frontier*, ed. R. C. Vetter (New York: Basic, 1974), pp. 137–52; see also J. R. Vaillant, *Les Problèmes du dessalement des eaux de mer et des eaux saumâtres* (Paris: Eyrolles, 1970); R. W. Durante, "Economic Desalting Methods," *Ocean Industry* 2, no. 5 (May 1967): 70–72.

sea may well reach 4 million m³ per day, valued currently at at least $250 million per year.

However, there are additional freshwater sources (aquifers and springs) underneath the continental shelf that could, when tapped, provide coastal settlements with a valuable resource. Although much of it remains unused, springs in Argolis Bay (Greece) emit 863,000 m³ of freshwater daily, with a current value of $36 million per year. The spa of Ostende-Thermal in Ostend (Belgium) is located on a flattened dune 50 yd from the high-tide line of the North Sea; Arabian boats, and the island city of Argolis in ancient Phoenicia, tapped submarine springs for freshwater. Brackish water not exceeding a salinity of 2,000 ppm is also available and usable for agricultural, industrial, and, when diluted with freshwater, for domestic purposes.

Sites already located for such springs include several areas of the Mediterranean and possibly the Romanian coast of the Black Sea; in the Pacific, Japan, Australia, Guam, Samoa, and Chile, and the California coast of the United States; in the Atlantic, South Carolina, Florida, several locations in the Gulf of Mexico, the Mexican Yucatan peninsula, the Bahamas, Barbados, and Cuba.

ENERGY ALTERNATIVES FROM THE OCEAN

The steadily increasing shortage and the mounting price of gas and oil have triggered a search for new sources of energy to supplement the conventional fossil fuels—oil, gas, and coal—and the recently developed nuclear, hydroelectric, and geothermal systems. Among such new sources, consideration has been given to oil shales and tar sands and to more sophisticated nuclear reactors, but also to sun, wind, and ocean waters. While at the present it appears that none will solve the energy dilemma, each has the potential to make a substantial contribution to future energy needs. We may be reasonably confident that human ingenuity will overcome technical and physical difficulties, but production of power from nonconventional sources will require time, research, and large capital investments.

Traditional sources of energy include fossil fuels such as petroleum, natural gas and coal, and hydrothermal power. The first are nonrenewable resources which are being depleted at a staggering pace, and the second type accounts for 5×10^{10} w, or about half of 1 percent of current world needs. Doubling hydrothermal power output, the estimated maximal capability would still yield merely 1 percent of man's future needs.

The oceans appear, since they cover 71 percent of the globe, to be a promising source of energy. Power from the ocean encompasses the fossil fuels found beneath the ocean floor, for example, petroleum, gas, methane, coal, and also geothermal reserves, water as a resource, solar radiation, surface air movements, and hydrogen. Yet, there is no unanimous agreement on which alternate sources of energy to develop or on the actual extent of the oil crisis. Woods Hole's K. O. Emery maintains that most undiscovered continental shelf reserves must be located on domains not closely linked to land provinces, and foresees that such reserves will prove to be considerable. Barring large increases in the consumption rate, Emery believes that fossil fuels, namely, petroleum and gas, are likely to continue as major sources of fuel for more than 100 yr.[3]

The list of alternative energy sources is both long and impressive. Calculated in watts, the earth—both land and sea—receives 10^{16} w directly from solar energy. Next is the available ocean heat, totaling some 10^{13} w, followed by the available wind energy (on land and sea) and the precipitations (also on land and sea) which provide, each, 10^{12} w. Ocean currents and tides add another 10^{10} w and so does geothermal energy, on land and sea. There are also waves and surface agitation caused by wind.

In each instance the ocean is a potential major provider of harnessable energy. Thus, the oceans can help solve the energy dilemma in a rather wide variety of ways. But all ocean energy sources do not hold equal promise, and few have actually been tapped, even though we already now, or soon will, possess the needed technology. Most commonly written off is electromagnetic power. Seawater-powered batteries have a long life, high reliability, and are inexpensive and relatively independent of depth, pressure, and salinity factors. However, large power units are bulky and heavy. Salinity "power" includes salinity exchanges and freshwater and saltwater contacts.

TIDES

Tidal energy has fascinated men since classical times: the Greeks tried to tap it; floating tidal mills functioned on the Danube; Dover harbor's entrance had a tide mill; and dozens of such mills tapped ocean energy in England, Wales, and Brittany, some to this day. After some unsuccessful attempts, the French built a large plant on the estuary of the Rance River and the Soviets assembled one on Kislaya Bay. It is claimed that the Soviet Union is now constructing a second one on Mezen Bay to cope with the industrial expansion of the Soviet European Arctic. More ambitious plans have been developed for the Sea of Okhotsk area.

Nevertheless, tidal energy facilities may disrupt local ecological systems. One geophysicist, Vautroys, has even warned of a possible slowing down of the earth's rotation.[4] However, in view of the limited number of economically exploitable sites, tidal power plants are most probably environmentally tolerable on a worldwide scale.

    3. K. O. Emery, "Provinces of Promise," *Oceanus* 17 (Summer 1974): 46–52.
    4. L. Vantroys, "Perturbation apportée au régime des marées par le fonctionnement d'une usine marémotrice," *Houille Blanche* 1 (1957): 188–99.

During the decade 1980–90, Canada plans to build, if then economically competitive, a tidal power plant on the estuary of the Memracook in the upper Fundy Bay with an annual output of 6,500 Gwh, a 2,176-Mw facility. India included in its fifth 5-yr plan the study of a facility in the Rann of Kutch. The four best Australian sites could produce as much as 30,000 Mw; an Argentinian facility in either Golfo San Matias or Golfo Nuevo (Valdes Peninsula) would have a capacity of 600 Mw. The U.S. plant in Maine would have a 1,800-Mw capacity and produce 44 Gwh daily (15,800 annually). The French plant at la Rance has an ultimate capacity of approximately 350 Mw.

Of tidal power in the northeast of the North American continent, the National Petroleum Council (NPC) report said, singling out only a Bay of Fundy plant: "the generation of 50 million kwh annually ... would correspond to 240,000 barrels of oil per day." While maintaining that "potential for annual tidal energy is limited," the NPC admits that "nonetheless, the summation of tidal energy over many years becomes significant," and recognizes that "the ultimate potential in the Passamaquoddy-Fundy area is several times [that of the 50 billion kwh]."[5] A conservative estimate made by Walter Munk of Scripps Oceanographic Institution places the total global tidal energy at 3×10^{12} w, of which some 350 Twh would be harnessable.[6] A National Academy of Sciences report has calculated 13,000 Mw.[7]

Promising sites exist in many areas worldwide, though not always close to the consumer. Besides high capital investments, tidal power also is handicapped in its development by the intermittent nature of the movements and the rhythm of high tides by 50 min a day; hence some of the power produced is used to store up energy for use at peak demand time. Finally, normal tide phenomena may be disturbed by tidal power plants.

Many question whether tidal power plants are really economical. The Electricité de France, ignoring the challenge given their report by the political left, again postponed the construction of the Chausey Island plant near Mont Saint-Michel. The plant could be completed by 1982 and save 8 million tons of oil annually. The price tag killed the British Severn River tidal plant, but today it is freely admitted that, had it been built in 1933 or in 1945, it would have paid off handsomely in 10 yr. In Australia, a single plant built on the Kimberley coast would provide 300,000 kw, or about 50 times the present production of electricity of Australia. The lack of a market led to a negative decision in that case.

Nevertheless, a report presented at the 1975 annual meeting of the American Society of Naval Engineers by Claude Lebarbier[8] provides solid

5. New Energy Forms Task Group, *U.S. Energy Outlook: New Energy Forms* (Washington, D.C.: National Petroleum Council, 1973), pp. 89–94.
6. As told to Elisabeth Borgese by Walter Munk, personal communication.
7. National Research Council, Committee on Resources and Man, *Resources and Man: A Study and Recommendations* (San Francisco: Freeman, 1969).
8. Claude H. Lebarbier, "Power from Tides—the Rance Tidal Power Station," *Naval Engineers Journal* 87, no. 3 (April 1975): 57–71 and no. 4 (June 1975): 57–60.

evidence that existing plants must be acknowledged by the ocean engineering community as full-scale feasibility models. Also, the Canadian Tidal Power Review Board indicated in a September 1974 report that the economic position of tidal power has considerably improved over the last 5 yr. As a result, a $3 million study plan has been funded by the Canadian government, which includes optimization of energy production and electrical power market and transmission assessments.

Numerous sites are suitable for tidal power harnessing (figs. 1 and 2). I have listed 28 major locations.[9] Advanced feasibility studies have been carried out for sites in Wales, the United States, Canada, and Argentina in the Atlantic and Australia in the Pacific. France, the Soviet Union, South Korea, and Norway also report tentative plans for tidal plant construction (fig. 2).

In the world's first actually operating system, the difference of from 9 to 14 m between high and low tide produces over 544,000 kw. Because of its reversible operation, power is tapped in the Rance River plant from the waters as they rush upstream at high tide and as the waters recede toward the sea. The reservoir's level can be raised by pumping, and its storage ability is one of its most valuable assets. The volume of water displaced reaches 718 million m^3. The power plant is 390 × 53 m wide and, when in full operation, 24 generating units are put to work. Started in 1963, in operation since 1966, and costing close to $100 million, the Rance River project required the removal of 1.5 million m^3 of water and the drying up of about 75 ha of the estuary. The dam also accommodates a road which cuts 30 km from the distance between St. Malo and Dinard and eliminates a ferry.

The Soviet Union's small coastal bays hold a power potential exceeding some 3.2 billion kw. The Soviet Union has built an experimental plant on the Kislogubskaia. The plant was completed in December 1968 and has a capacity of 800 kw. Its engineer, Lev Bernstein, foresees the success of tidal power stations in international cooperation: "The tidal energy of the English Channel, linked up with that of Mezen Bay and coordinated, could supply peak power for all of Europe."[10]

The Soviets are now starting commercial plants near Lumbovka on the Kola Gulf and on the Mezen Bay. Capacity of these stations will be, respectively, 300,000 kw and 25 million kw, with tidal amplitudes in these gulfs of 7 and 9 m, respectively. They are also searching for power to industrialize eastern Siberia; by harnessing tides in the Okhotsk Sea bays of Penzhina and Gizhiga, whose amplitudes reach up to 13 m, the Soviets could generate 174 billion kwh. On the

9. See the following studies by R. H. Charlier: "Harnessing the Energies of the Ocean," *Journal of the Marine Technology Society* 3, no. 3 (1969): 13–32 and no. 4 (1969): 59–81; "Harnessing the Energies of the Ocean: A Postscript," *Journal of the Marine Technology Society* 4, no. 2 (1970): 62–66; "The Rance River Tidal Power Station: A Comment," *Naval Engineers Journal* 87, no. 3 (June 1975): 58–59; and "The Energy Dilemma and Ocean Power Potential," in *Symposia of Expo '75* (Okinawa: International Ocean Exposition, 1975), pp. 126–35.

10. *Soviet Life* 160 (January 1970): 37.

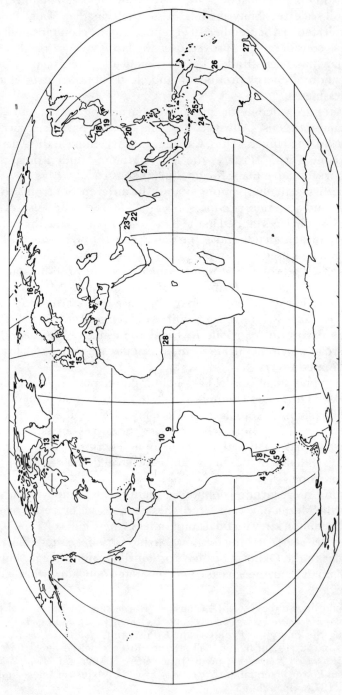

FIG. 1.—Major tidal power plant sites: 1, Cook Inlet; 2, British Columbia; 3, Baja Californa; 4, Chonos Archipelago; 5, Magellan Straits; 6, Gallegos/Sta. Cruz; 7, San Jorge Gulf; 8, San Jose Gulf; 9, Maranhao; 10, Araguaia; 11, Fundy/Quoddy; 12, Ungava Bay; 13, Frobisher Bay; 14, Severn/Solway; 15, Rance; 16, Mezen/Kislaya; 17, Okhotsk Sea; 18, Seoul River; 19, Shanghai; 20, Amoy; 21, Rangoon; 22, Cambay Bay; 23, Cutch Gulf; 24, Kimberleys; 25, Darwin; 26, Broad Sound; 27, Manakau; 28, Abidjan.

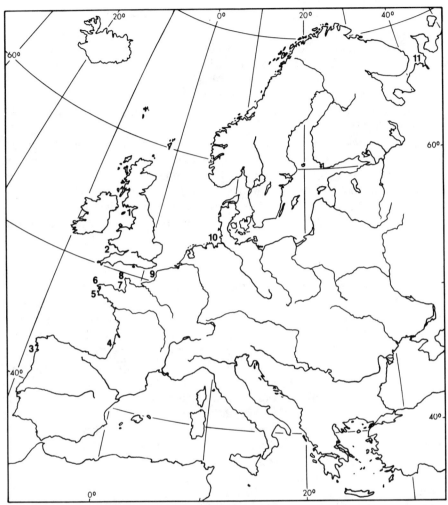

Fig. 2.—Sites of proposed tidal power plants: 1, Solway Firth; 2, Severn River; 3, Chausey Island; 4, Rance River; 5, Aber Benoit; 6, Somme River; 7, Arcachon; 8, Vigo; 9, Mezen; 10, Kislaya; 11, Brest; 12, Busum/Wilhelmshafen.

Canadian/U.S. border, sites have been under consideration (especially in Passamaquoddy Bay) for nearly 50 yr. The Passamaquoddy Bay has tide differences reaching 15 m. The major objection to the tidal energy station project has been the distance between the plant and the eventual consumer. Since transmission is no longer a problem, interest in the project has revived. If ever completed, it would dwarf the Rance River project. The multiple-basin plan at Passamaquoddy actually went into construction in 1935, but work came to a halt when Congress failed to appropriate money. The same attitude prevailed in 1963 and 1964.

Among the main advantages of tidal power is its regularity from year to year, with less than 5 percent annual variation. Though there are substantial disadvantages,[11] such plants permit simultaneous use of the dam for a road or railroad; provide estuary navigation improvements, production of cheap electricity, and a virtually inexhaustible supply of energy; and all are totally pollution free.

Though nuclear power plant proposals are usually tidal power's major competitors, some commentators see the possibility of an ideal "mix"—perhaps a happy marriage—of nuclear and tidal schemes. The nuclear plant could work continuously at full load, hence putting its capital investment to 100 percent use, while the tidal power plant could store and deliver power "with an eye to the value of peak-demand power." Offshore nuclear centrals can be built on land and towed into place, a procedure also followed by the Soviets with their experimental tidal power plant. The Netherlands has been working on a plan to build a huge artificial island in the North Sea which would house a nuclear power plant, a garbage dump, and polluting heavy-industry plants. Tentatively sited some 60 km north of Rotterdam, it could double as ship terminal for vessels reaching 400,000 tons. Oil refineries, iron and steel works, aluminum shelters, and ship-repair facilities could be accommodated. A similar plan exists in Belgium to build an offshore complex some distance from the coast.

Another factor militating against tidal power plants is the strong emphasis placed by some economists on an efficient allocation of resources and their concern for making a choice between competing demands for limited resources. Placing the major importance on discounted cash return, they obscure the real issue behind a veil of possibly faulty projections of interest rates, secondary benefits, transfer functions, exhaustion or increase in cost of alternative power sources, and pollution abatement requirements. Suitable cost-benefit anaylses have shown the economic rationality of such projects, but in capitalistic countries, doubt is being raised, mostly by economists working for investor-owned utilities, as to whether government participation in power development is in the public interest.

THE SUN

The sun may well be the major factor in ocean energy production through radiation. Radiant energy is most efficiently absorbed by the ocean. The energy absorbed by the ocean, excluding ocean-atmosphere storage, amounts to 80 billion Mw daily, or 50,000 times the energy used by man. However, direct use

---

11. See Charlier, "Harnessing the Energies of the Ocean."

of solar radiation, as is done in the Pyrenees and at Valley Forge, Pennsylvania, for instance, is limited by length of insolation and the ensuing need for large plants.

There are basically two types of heliothalasso systems: the first one, already possible, extracts solar energy stored in ocean surface layers; the second system uses concentrating mirrors placed on floating platforms to focus incident solar energy on a boiler. Taking into consideration the fact that bottom waters remain close to 3.9° C whether in tropical or polar areas, while air temperatures in polar zones stay close to 45° C, physicists have suggested that polar waters could be used as a source of warm fluid, while the air could provide the cold fluid. This would yield a temperature amplitude of about 50° C, or double the difference available solely in the oceanic domain. Furthermore, no long water ducts would be necessary. The so-called Barjot system would use butane, a gas that does not mix with water and liquifies at −10° C. Thalassothermal plants based on this principle could be built in Scandinavia, Siberia, and polar areas in general. No environmental or ecological disturbances would result.

Solar energy could be used to overheat ocean surface water covered with a thin oil film in shallow basins; no evaporation would occur, and adduction conduits would not have to reach into considerable depths (300–400 m). Masson[12] studied this possibility near Dakar, Senegal; Hirschmann drew blueprints for a plant that would utilize near-surface water (even as warm as 20° C) and ocean water heated by solar energy to about 70° C in a basin with an electric turbine placed between evaporator and condenser.[13] Hirschmann's scheme is coupled with distilled freshwater production.

Elisabeth Borgese quotes J. Hilbert Anderson and James H. Anderson as estimating construction costs for a thalassothermal energy plant at $200 per kilowatt, as opposed to the needed $700 per kilowatt for a nuclear plant.[14] Even a more pessimistic forecast of $1,100 per kilowatt is believed to make sea thermal power still economically competitive with other power producing systems by the late 1980's. The Andersons estimated that the Florida Straits, often mentioned as an important source of ocean-current power, could provide 20 million Mw by tapping the potential of temperature differences. Problems associated with thalassothermal schemes include corrosion, organism fouling, dissolved gas removal, energy consumed for pumping purposes, and maintenance costs of turbines.

Motions of wind and sea also generate usable energy which is renewable and virtually nonpolluting. These processes are sustained by solar energy provided yearly by the earth's atmosphere (potential energy 5,000 Q [quintillion] Btu; $1 Q = 10^{18}$) and by impact of gravitational fields of sun and moon and

12. H. Masson, "L'Utilisation de l'énergie solaire dans les régions arides," *Annales de mines* 4 (1955): 163.
13. C. Gomella, *La Soif du monde et le dessalement des eaux* (Paris: Colin, 1966).
14. Elisabeth M. Borgese, *The Drama of the Oceans* (New York: Abrams, 1975), p. 142.

the earth's spinning upon water and air (estimated at 65 Q of kinetic energy). In order of magnitude, we have ocean-stored heat, kinetic energy stored in winds, wind-wave stored potential and kinetic energy, and tide and ocean currents stored kinetic energy. In New England, wind power has proved less costly than conventional power, provided that system compenents could be produced on a large scale. Wind power could be produced as well near shore as offshore.[15]

WIND

Wind may thus be tapped as an indirect form of solar radiation, and windmills could be anchored at sea. A deterrent to building such plants is wind-force variability ranging from total calm to gales. Winds also carry air masses loaded with vaporized water from the oceans which, when turning to rain, could provide pollution-free energy; but favorable geographical sites are few. Finally, winds generate waves and currents. Wave power can be tapped directly, or the energy of the breaking wave can be trapped, but the variability of wave height has been a discouraging factor. Problems hampering current energy conversion include the distribution of the energy, the fact that the flow fluctuates throughout the year, and that the current axis moves occasionally.

TEMPERATURE DIFFERENCES

Prospects of tapping ocean temperature differences for energy production are encouraging. Thermal energy results from temperature differences between two water supplies of unlimited discharge; in the tropics, for instance, deep cold waters flowing from the polar regions come into contact, through upwellings, with warm surface waters. Around some Pacific islands the difference between surface and 300-m deep water reaches 20° C. Thalassothermal energy has actually been used in experimental plants. Difficulties with the conduit bringing up the deep waters and the size of the turbines needed led to the eventual abandoning of the scheme. The Société de l'Energie des Mers that had constructed the Abidjan plant went out of business. However, the basic idea is being revived, using an intermediary fluid.

Thalassothermal plants using intermediary fluid as a refrigerant could compete effectively with nuclear plants. The technology exists, developed over the years by researchers such as Claude, Boucherot, and the Andersons. A keen renewed interest has developed using either upwellings, artificial upwellings, or air-sea temperature differentials. The original idea goes back to 1881 when

---

15. W. E. Heronemus, "Using Two Renewables," *Oceanus* 17 (Summer 1974): 20–27.

Arsène d'Arsonval[16] suggested extracting energy from the ocean by using ammonia as a working fluid for a closed-cycle thermal engine. Low-boiling propane could also be used; differences of temperature between deep and surface ocean waters often reach 4.4° C (40° F), which allows thermal fluids to boil and condense; such fluids in the state of pressurized vapor will drive a vapor turbine electric generator. The use of propane would permit turbine diameter reduction from ±7 to 1½ m, and the submersion of condenser (to 50 m) and boiler (to 85 m) would permit wall-size reduction by balancing inside and outside pressure. Needed for an ocean-thermal plant are a partially submerged vessel with a boiler, a turbine, an electric generator, condensers, and auxiliary equipment. There are no fuel costs.

George Claude was the first to use temperature differences of water to produce electricity. To prove the feasibility of his scheme, he used thermal pollution caused by the discharge of manufacturing plants' waste waters in the cooler Meuse River waters near Ougrée (Belgium).[17] He made further attempts off the coast of Brazil and in Cuba. Charles Beau and N. Nizery later founded the Société de l'Energie des Mers which actually built the first full-scale operating plant utilizing, in open cycle, the temperature differences between deep and surface ocean waters.[18] Actually, theirs was merely the culminating effort of a long series of unsuccessful attempts stretching over the 20-yr span separating the two world wars.

A full-scale attempt was initiated in 1942 by the French Minisolonies and the Centre National de la Recherche Scientifique. After 6 yr of study, the construction of a plant was finally undertaken in Abidjan (Ivory Coast). In this project, the thermodynamic cycle consisted of evaporating, under vacuum, part of the warm surface waters at 28° C (82° F) encountered in tropical areas.[19] The steam is taken in by the condenser, cooled by the colder (8° C = 46° F) deep waters, and on its way proceeds through a turbine that drives an electric generator. At that time, sea-thermal energy compared favorably, from an economic viewpoint, with hydroelectric energy, the more so since the energy could be directly used for an evaporating plant for chemical industries. In Abidjan, freshwater was in greater need than electrical energy, and so a project combining both aims was ideal. There, major problems arose with the immersion of the cold-water adduction pipe; repeated experiments with a large-diameter duct all failed. However, when articulated joints for the duct

---

16. Arsène d'Arsonval, "Utilisation des forces naturelles. Avenir de l'électricité. Energie thermique des mers," *Revue Scientifique* 17, no. 9 (September 1881): 370–72.

17. Georges Claude, "Power from the Tropical Sea," *Mechanical Engineering* 52 (1930): 1039–44.

18. See Côte Beau, "L'État actuel des études et travaux en vue de la construction d'une centrale d'énergie thermique des mers en Cote d'Ivoire," *Industries et travaux d'outre-mer* 4 (1955): 222–23.

19. W. O. Skeat, ed., *The Transactions of the 4th World Power Conference* (London, July 11–14, 1950) (London: Lund Humphries, 1952).

and antiwave floaters suspended on cables to hold up the pipe were used, the operation met with success. That the plant nevertheless went out of business is due to the fact that then conventional power plants produced cheaper electricity, that the ducts suffered repeated ruptures, and that turbines required large dimensions. It is noted in this regard that the current hydronautics system calls for no less than six 23-m (78 ft) turbines using temperature differences of 2.2° C (36° F), while the French engineers felt no difference of less than 15° C (59° F) should be considered; hence the hydronautics' system would be suitable for a wider geographic range.

Since Beau and Nizery's scheme, the matter has been reexamined in France by Gouggenheim, Daric (1957), and combination solar-thermal energy schemes were proposed by Gomella (1966).[20] W. Heronemus, following the work of the Andersons, suggested a scheme which envisions a series of such plants spaced in an area 15 mi east to west by 550 mi south to north along the western portion of the Gulf Stream, whose electricity could be transmitted to virtually any U.S. location at competitive cost.[21] The plants would use a closed Rankin cycle with ammonia or propane, for instance, as a cooling intermediate fluid, which could function with a 17° C (62.6° F) temperature difference. The power plant itself would be semisubmerged and contain multiple units. C. Zener, then chief scientist at Westinghouse, already studied in 1965 the practical aspects of tapping ocean thermal power; he and A. Lavi are reexamining the open-cycle option, while the Division of Solar Energy in the government's Energy Research and Development Administration is forging ahead with the closed-cycle system.[22] Their views parallel Lockheed's, though there are divergences where water-pipe and heat-exchanger designs are concerned.

Gerard and Roels explored the possibility of using upwellings in the Virgin Islands as a cold-water source to provide a temperature difference with the surface waters to produce electricity.[23] The study also involved fertilization of the sea. Roels expressed concern with Othmer[24] that the various designs involving vertical suction pipes suspended from vessels or platforms, submerged power cables, and freshwater lines carrying products to shore would augment the difficulties for controlled mariculture, which they propose to link with energy production; they also discussed in detail the engineering design

20. A. Gougenheim, "L'Utilisation de l'énergie des marées," *Calviers océanographiques* 19 (1967): 277–93; G. Daric, "Schéma de fonctionnement d'une centrale sous-marine équipression à fluide auxiliaire," *Houille Blanche* 2 (1957): 694–70; Gomella.
21. See Heronemus.
22. A. Lavi and C. Zener, "Electric Power from the Ocean Thermal Difference," *Naval Engineers Journal* 87, no. 3 (April 1975): 33–46 and no. 4 (June 1975): 47–52.
23. R. B. Gerard and O. A. Roels, "Deep Ocean Water as a Resource," *Journal of the Marine Technology Society* 4, no. 5 (1970): 69–79.
24. R. D. Othmer and O. A. Roels, "Power, Fresh Water, and Food from Cold, Deep Sea Water," *Science* 182 (October 1973): 121–25.

made by Alemco, Inc., for their particular site. The success of thalassothermal plants is no longer dependent on its feasibility but on circumventing some remaining construction, operation, and possibly environmental problems. Engineering firms are now confident that, in addition to saving millions of barrels of fossil fuel, ocean thermal plants can compete with the cost of conventional electricity-producing plants.

The May 1975 Ocean Thermal Energy Conversion Workshop in Houston, Texas, considered the various projects funded by the Energy Research Development Administration (ERDA) and the "Rann" program of the National Science Foundation. Proposals carry suggestions of liquid hydrogen, oxygen, ammonia, methanol, and freshwater liquid nitrogen production, so that even if electricity would be difficult to transmit to a grid, it could provide energy at sea to produce agricultural chemicals and power for the land. The Lockheed system consists of a number of self-contained power-generating modules attached to a semisubmersible structure, and the cold water is piped up. The TRW system is a concrete cylindrical surface vessel with internal power units and uses fiber-reinforced plastic for the adduction pipe; a closed Rankin cycle operates on temperature differences of 4° C (39.3° F); of three fluids tested, ammonia seemed the best, outranking propane and R-12/31. Hydronautics recommended an open cycle plant with the units placed on a platform structure. However, the cold-water adduction duct that is to provide a diagonal structural brace for the platform has not yet been finally designed.

The Andersons, ready with a proposal since 1963, suggest a submerged boiler which reduces pressure differences between working fluid and outside water, using an Allied Chemical refrigerant (R-12/31); the largest part of the floating plant is submerged, sheltering it from the kinds of waves and storms which repeatedly troubled the Abidjan plant.

The Lockheed Aircraft Corporation proposal asserts that its system is superior to fossil-fuel and nuclear-fired power plants. Their baseline design cost is $2,660 per kilowatt, with power costing 36 mills per kilowatt-hour, which compares quite favorably with traditional and nuclear plants. Not only does the Ocean Thermal Energy Conversion power plant (OTEC) add no heat to the atmosphere, it returns its resource to the ocean from where, heated by insolation and cooled by polar surroundings, it can be recycled through the plant; this is not so with other systems and compensates for OTEC's low (3 percent) efficiency. Lockheed's 160-MWe (megawatt electric) (net) baseline design consists of a 260,000-ton concrete cold-water pipe and four 9,200-ton power modules, developing each 60 MWe (gross) electric power.

Sea Solar Power expects to be able to mass produce plants which would provide 100 Mw for about $50 million.[25] The annual income of such a plant

---

25. See, e.g., J. H. Andersen, *Proceedings, Solar Sea Power Plant Conference* (Pittsburgh: Carnegie-Mellon University Press, 1973); see also his article, entitled "Turbines for Sea Solar Power Plants," on pp. 126–52 of that volume.

could amount to $100 million because, besides electricity ($18 million), it can produce a long list of by-products (table 1).

In Hawaii, the construction of an onshore or near-shore ocean thermal energy conversion plant on a lava flow near Keahole Point has been studied. The plant would be a 20-Mw facility. Such a plant, preferably with a capacity of 35 Mw, is feasible in the immediate future on a cost-competitive basis, provided a breakthrough is made in a delivery system from offshore plant to shore. Other U.S. research groups are examining systems with electrical generation capacities of up to 400 Mw. Research involves stations either resting on the ocean floor or floating power stations. Inverted thermal gradients exist in some isolated spots in the oceans, for example, the Red Sea; here the warmer water is on the ocean floor. By rearranging the equipment, such sites are suitable for power tapping and have the added advantage that pumps to return the condensed working fluid to the boiler are unnecessary because it flows by gravity to the boiler.[26]

In conclusion, it appears that transmission problems are no greater than with current traditional power plants. Construction costs are considerable, and so may operational costs be, but returns would be the same as for traditional plants if cost goals for the system can be met. Sea thermal power plants do not pollute or create waste, exact no change of landscape or life-style, or affect the real estate market because they are located in the ocean. Risks involve the danger of sinking, shared with any ship, and that of temperature changes. These would not exceed 1°C, but ecological impact studies should nevertheless be made. The environmental impact could be beneficial: fish will thrive on the nutrients contained in the deep cold water. Water would be used only if the electrolysis process is selected. Plant failure would have no other effect than to curtail electricity production.

According to ERDA, sea thermal power will become feasible if capital costs can be reduced to $1,000 per kilowatt, a figure considered attainable by researchers at all main centers. The ERDA will test an experimental 1,000-kw device in 1979, and it plans systems of, respectively, 5,000, 25,000, and 100,000 kw for 1980, 1981, and 1984. The largest of these plants could service a city of 75,000. If those projects succeed and thalassothermal power becomes competitive with conventional and other sources, then commercial-size plants of 400,000 kw and over could be built by the end of the next decade.

MAGNETISM, CONDUCTION, AND AGITATION

Presently, there is little promise for utilizing either ocean electromagnetic energy, heat flow transmitted by conduction through the ocean bottom, and kinetic energy caused by surface agitation brought about by the wind.

26. A. W. Hagen, *Thermal Energy from the Sea* (Park Ridge, N.J.: Noyes Data, 1975).

TABLE 1.—ESTIMATED VALUE OF BY-PRODUCTS
FROM A SEA THERMAL POWER PLANT (US$ Millions)

By-Product	Value
Hydrogen	34
Carbon dioxide	32
Ammonia	18
Methanol	12
Oxygen	12
Freshwater	11
Nitrogen	6
Food fish	6 (to 78)
Total	131 (to 203)

CURRENTS

The energy of ocean currents is considerable, particularly in lower latitudes. Specifically, near Japan sites exist where a wave-power station activated by the strong winter currents, virtually absent during the summer months, could be coupled with a thalassothermal station using the high summer temperature differentials. Far from having a nefarious environmental impact, such a combination power plant would have the beneficial side effect of stemming coastal erosion.

The kinetic energy available from the sole Florida current segment of the Gulf Stream is equal to the power produced by 25 1,000-Mw conventional power plants. Of these, 2,000 Mw can be readily extracted. The electricity produced in this manner could be competitively priced by 1980 based on contemporary price predictions for current sources.[27] Electricity could also, for instance, be produced by using the thermal difference of the Gulf Stream at the surface and the very cold water about 1,000 m below; the engines would be placed on a float. The temperature gradient between the Gulf Stream and the countercurrent below varies over the 1,000 m from 16° C to 22° C. Several devices have been proposed over the years. Remenieras and Smagghe suggested as early as 1957 designing generators along the same lines as aerogenerators; the matter had been previously approached by Bouteloup and by Romanovsky.[28] An underwater windmill using an open propeller was discussed at a Miami workshop in February 1974 but would require blades 100-m

27. H. B. Stewart, Jr., "Current from the Current," *Oceanus* 17 (Summer 1974): 38–41.
28. G. Remenieras and P. Smagghe, "Sur la possibilité d'utiliser l'énergie des courants marins au moyen de machines à aérogénérateurs," *Houille Blanche* 2 (1957): 532–39; J. Bouteloup, *Vagues, marées, courants marins* (Paris: Presses Universitaires de France, 1950); V. Romanovsky, *La Mer: source d'énergie* (Paris: Presses Universitaires de France, 1950).

long, while blades might not all be in equal-speed water layers. Difficulties arose also with the savonius rotor, used on current meters, and with the Kaplan turbine. The water low-velocity energy converter of Steelman offers an inexpensive system, operated by parachutes which move a no-end belt: open when they move with the current, they collapse on their return journey. Herman Sheets[29] has pointed to the large amounts of energy available and its conversion in plants located close to the areas of demand. By using a vertical axis turbine, a power plant buffered against seasonal variations of flux could be built. For William Heronemus, "a nation that intends to squander almost a billion dollars a year on a breeder reactor program . . . ought to be able to invest a modest sum in demonstration of some energy concepts that would be both socially acceptable and economically prudent."[30] Coastal currents, though not as efficient a source of power as the rather steady ocean currents, could also be tapped where tidal ranges are modest and power small-scale stations. One deterrent to the construction of submerged systems taking advantage of high-energy tidal currents is that they could constitute navigational hazards in heavily traveled areas.

One very modest experiment using "current power" has been attempted in Iceland where, on Hraunsfjord in the Breidafjord, located at 65° north and 23° west, near Malmberg, a small pump powering a mill was driven by tidal and current flow. The experiment was not considered a success, but then, currents are not powerful close to the Icelandic coast.

WAVES

Waves, if sufficiently regular and powerful, could provide an acceptable source of energy. Normally of low height and only a few seconds apart, waves pose few problems; the difficulties are created by storms. The characteristics of ocean waves place constraints upon converting devices. In 1909, Alva Reynolds built a wharf and suspended panels underneath it; the force of the waves was transmitted to a wheel attached to an electric generator. In 1971, his successors tested a wave motor off Pacifica, California. Von Arx would amplify the energy effect of the surf by focusing it on a "horn" designed to accept the surge; a head of tens of meters could thus be created and the water stored to supply hydroelectric turbogenerators.[31] Since a minimum of sediment transport is necessary to absorb incoming wave energy, a large number of such horns could upset the balance.

29. Herman E. Sheets, "Power Generation from Ocean Currents," *Naval Engineers Journal* 87, no. 2 (1975): 47–56.
30. Heronemus (n. 15 above).
31. W. S. von Arx, "Energy: Natural Limits and Abundances," *Oceanus* 17 (Summer 1974): 2–13.

Slow-moving internal waves at the interface between waters of different density or temperature possess considerable power, as evidenced by the "diving ballast and trim" records of submarines; the problem is to concentrate their energy. Von Arx suggested the use of submarine canyons and appropriate bottom topography features as horn or receiver.

Devices for trapping breaking wave energy include converging ramps leading to a natural or man-made reservoir from which the water flows back to the sea after passing through low-pressure turbines and devices that are set in motion by the wave impact itself. Converging channels supplying a basin which constitute the forebay for a conventional low head-power station seem to provide the highest output of any scheme proposed, yet are presently uneconomical. Power Systems conducted successful small-scale tests with a concrete trough parallel to the shorelines in which a pliable strip filled with a hydraulic fluid is secured. Submerged at 7-m depth where wave shape plays no role, the strip breaks the waves, undergoes the hydrostatic pressure of the water mass above, and transmits it to a hydraulic accumulator through the hydraulic fluid. Presure is stored by an accumulator until a specific magnitude is attained, at which point it delivers the energy to a fluid dynamo-connected motor.

Great Britain has one of the world's most favored coastlines for tapping wave power; energy contained in the waves there is variously estimated at from 40 kw/m, to 70 kw/m, depending on location. Fortunately, this energy is available throughout the year and seems to coincide with the seasonal pattern of electricity demand. It has been estimated that 50 percent of the total need for electrical power could be produced by harnessing waves along a 1,000-km ocean stretch.

Among the numerous devices that have been proposed for wave-energy transformation, the British government is selecting four for funded research. The Russell rectifier is a structure with high- and low-level reservoirs that is exposed to waves. Waves drive seawater in the high-level reservoirs and extract it from the low-level reservoirs, separated from the sea by vertical nonreturn flaps. The Salter duck, developed by Salter over many years at Edinburgh University's Bionic Research Laboratory, is an oscillating vane, so shaped that it permits extraction of a high percentage of incident wave energy.[32]

Wavepower Ltd. pursues work with a series of rafts, separated by hydraulic motor-and-pump combinations which convert the raft-motion energy into high pressure in a fluid. The system is designated as "contouring rafts." The British National Engineering Laboratory is involved with the air-pressure ring buoy. A study of floating breakwaters in Japan revealed that wave height could be significantly reduced by shaping breakwaters like inverted boxes, while causing the wave motion in the box to work on air forcing the air in and out of the orifices in the top of the box.

32. S. H. Salter et al., "Characteristics of a Rocking Wave Power Device," *Nature* 254, no. 4 (1975): 504–6.

Salter also considered energy storage.[33] Electrolytic production of hydrogen from seawater appears to him promising in that respect. He suggests self-propelled installations which could move in line ahead, a low-drag condition, into the Atlantic, turn abreast to the waves, and be slowly driven back by wind and wave thrust, storing hydrogen on the way. Once near shore, most of the hydrogen could be unloaded at a terminal, keeping enough in the installation to get out to sea again; among the advantages of this mobility is elimination of mooring problems. The best location for such a wave-power station would be the approaches to the Hebrides.

The Japanese Ryokuseisha TG-2 wave-activated generator uses wave motion to compress an enclosed air supply which rotates an air turbine directly connected to an electric generator; it is currently used on various types of observation towers, as an electricity source for ocean survey instruments, for light houses in isolated sites, and for the beacons of buoys. It could eventually provide a large output by using many pipes or air tubes.

Wave motion already has been tapped for air-column excitation, and bell clapping for navigational aids that whistle or ring. A wave-activated pendulum which drives a spring generator unit and a self-aligning water-wheel generator have been patented. Wave power is pollution free; like wind and sun it is a widely available renewable resource and requires no fuel. Installations could simulaneously provide coast and harbor protection, and power units could be coupled to desalination plants.[34]

GEOTHERMAL RESOURCES

In 1967, Paul H. Jones noted the probable presence of abnormal subsurface pressure zones in neogenic deposits of the northern Gulf of Mexico.[35] High heat flow was associated with the rift and fracture zones and also with volcanically active areas. But the prospect of tapping subsea geothermal energy was in doubt when McKelvey and Wang concluded, in 1969, that "geothermal energy associated with rift zones and volcanic activity had little prospective value in areas far from land because of the difficulty of utilizing it."[36]

In island and coastal situations, however, geothermal power probably

---

33. S. H. Salter, "Wave Power," *Nature* 249, no. 6 (1974): 720–24.
34. J. D. Isaacs et al., "Utilization of the Energy in Ocean Waves," mimeographed (San Diego: Institute of Marine Resources, University of California, 1975).
35. Paul H. Jones, "Geochemical Hydrodynamics—a Possible Key to the Hydrology of Certain Aquifer Systems in the Northern Part of the Gulf of Mexico," in *Proceedings of the 23rd International Geological Congress* (Prague: Academia, 1968), symposium 2, pp. 113–25.
36. V. E. McKelvey and F. F. H. Wang, "World Subsea Mineral Resources," Report, Miscellaneous Geological Investigation Maps I-632, U.S. Geological Survey (Washington, D.C.: Department of the Interior, 1970).

could be tapped to produce electric power, an operation possibly coupled with desalination processes, solution mining of sulfur and potash, repressuring offshore oil fields, and even petroleum recovery. Harnessable geothermal energy is, furthermore, favorably situated because it is likely to be found in zones where petroleum is also present.[37] From an environmental and ecological viewpoint, geothermal power has no unfavorable impact. The Icelanders have tapped their geothermal power for many years, but generally use of this resource is relatively recent.

Only six land-based geothermal plants are fully functional—in Italy, Iceland, Japan, and the United States; Mexico and New Zealand are building plants. The U.S. Geological Survey has conducted various studies dealing with geothermal energy potential, and in a report issued in August 1975, geothermal power tapping on and offshore of Texas' gulf coast is mentioned as a potentially promising energy source. Super-heated water, steam, methane, and natural gas are present in large quantities. The energy potential of the waters in areas of the northern Gulf of Mexico includes both thermal energy and hydraulic energy, in addition to methane. Seemingly, these energy resources are far more promising than geothermal energy from geysers, fumeroles, and hot springs whose worldwide potential is quite modest in comparison to total power demands.

The largest active volcano in the world, on Hawaii, also holds considerable promise. Its steam could be a major energy source for the islands, which are otherwise entirely dependent on fossil fuels. A plant could be in operation before 1980.

SALINITY

Scant attention has been given to tapping the energy which can be derived from the contact of saltwater and freshwater. Recently, some closer observations have been made in La Jolla, California, particularly by Gerald Wick and J. D. Isaacs.[38] Where rivers drain into the ocean, fresh and saltwater come in contact. Isaacs has calculated that the osmotic pressure could represent the energy equivalent of a vertical fall of about 250 m. While a two-fluid battery already has been developed, research has also been pursued to examine possibilities of large-scale use of a salinity gradient. The British have studied the matter and concluded that the power that the Amazon and Brahmaputra release as they enter the ocean is staggering.

37. D. E. White, *Geothermal Energy* (Washington, D.C.: Geological Survey 1969); D. B. Brooks, "Conservation of Minerals and the Environment," in *World Mineral Supplies*, ed. G. J. S. Govett and M. H. Govett (Amsterdam: Elsevier), pp. 268–74.

38. G. Wick and J. D. Isaacs, "Salinity Power," in *Symposia of Expo '75* (Okinawa: International Ocean Exposition, 1975), pp. 153–65.

Salinity-concentration gradients can also be tapped when any two bodies of water, not necessarily salt and fresh, come in contact. The way is open for the direct utilization of osmotic pressure differences, the utilization of electromechanical potential, vapor-pressure differences, and mechanochemical phenomena. Harnessing these energies, however, is not economically feasible at present because of the current state of membrane technology, but new technology or new materials may change this situation rapidly. As a matter of fact, schemes not using membranes may be engineered for the tapping of salinity power from brine more concentrated than seawater. In this instance, the power density is much larger. Batteries using seawater, dry-charge primary batteries, are in common use and were used aboard the *Trieste* at depths of 700 m. Lockheed Electric also developed a low-cost type; several cells have been manufactured but are rather heavy and bulky.

In brief, tapping the ocean's salinity power has been considered, among others, through direct utilization of osmotic pressure difference and utilization of the electrochemical potential, of the vapor pressure difference, and of the mechanochemical potential. Yet, harnessing such power is not economically feasible at the present stage of technology. With the more concentrated brines, however, more immediate application is perhaps possible.[39]

DEUTERIUM

Even less has been said about extracting from the ocean the deuterium fuel necessary for nuclear fission to produce energy. The presence of deuterium reserves may be an additional incentive to locate nuclear power plants offshore in view of growing opposition to land-based nuclear plants. As cited earlier, the Netherlands is contemplating building a huge artificial island, 65 km north of Rotterdam, which would accommodate a nuclear power station, large polluting industries, a chemical waste recycling plant, an industrial and domestic garbage incinerator, and provide a terminal capable of handling 400,000-ton ships. Thermal pollution would be rapidly dissipated, according to project promoters, but that point of view will undoubtedly be challenged by environmentalists.

HYDROGEN

Hydrogen available in huge quantities in the ocean, easily stored and transported, can be extracted using another type of ocean energy, for instance, thermal energy, or conventional energy sources. Harrenstein feels that if the primary energy source for synthetic fuel production is to be petroleum, little will be gained, but an ocean-powered plant could provide the electricity to

39. Ibid.

extract hydrogen from ocean water.[40] The only environmental impact would be that of floating platforms.

Energy delivery is of capital importance because often good thalassothermal sites lie far from large centers of electricity consumers. A solution would be the production of energy forms that can be easily stored and/or transported. Hydrogen answers this need and could be produced by electrolysis. It may become competitive with electricity when adding factors such as storage and transport, and could also be used interchangeably with it, though, as J. K. Dawson points out, there will be competition from liquefied natural or synthesized methane and methanol.[41] Hydrogen can be distributed and transmitted by pipelines and can be stored in huge quantities like natural gas. It thus offers a solution to the problems of multiplication of overhead transmission lines, to the slow development of cryogenic superconducting cables, and to the demand for some form of electrical storage other than reversible hydroelectric stations, which are limited by geographical conditions.

PROGNOSIS

Some years ago little attention was paid to suggestions that we tap the oceans' energy. We were enjoying what seemed to be an inexhaustible and inexpensive petroleum and natural gas economy, and newly discovered nuclear energy made the energy situation seem even brighter. Things have changed considerably since then, and fossil-fuel costs have skyrocketed. Despite large remaining world coal reserves, and especially considering the very seriously endangered environment, it seems that relying on coal and nuclear power to replace petroleum and natural gas is unwise. Furthermore, assuming a growth rate of world fuel consumption of only 4 percent a year, the amount of carbon dioxide in the atmosphere, which has already increased by more than 10 percent since 1900, could double by 2030, triggering ominous changes in the world climates, for example, a temperature increase of the earth's surface of about 2.9° C. The National Oceanic and Atmospheric Administration places the responsibility for the carbon dioxide air content on our relentlessly growing consumption of fossil fuels. Would it not be timely to embark on a rational, well-balanced program of tapping new energy sources, and particularly ocean and solar power as a suppletive source, rather than plunge into a risk-loaded "nuclear economy"? Surely a nation that has spent billions of dollars in space research, in nuclear study, and in sophisticated weaponry can afford to insure its future generations an adequate energy supply by funding research for alternative energy sources.

40. H. P. Harrenstein, "Hydrogen to Burn," *Oceanus* 17 (Summer 1974): 28–29.
41. J. K. Dawson, "Prospects for Hydrogen as an Energy Resource," *Nature* 249, no. 6 (1975): 724–25.

RAW MATERIALS

Marine resources are chemical, geological, or biological. Geological resources can be authigenic, detrital, or organic. They are abundant, although specialists are not sure whether, under present conditions, they can be profitably exploited.

The Economics of Ocean "Mining"

In 1963, an American company was drilling in the Gulf of Mexico at 100 m for petroleum, the Japanese had begun large-scale exploitation of the sand-iron deposits on the bottom of Aiake Bay, and the deposits of the Indian Ocean were being raked for "mineral-containing lumps." It was then estimated that metals could be hauled from the sea at only 50–70 percent of the cost of land mining, a proposition made the more attractive because on land high-grade ore is being rapidly depleted and marine ores are often highly concentrated. Realistic estimates list no less than 60 useful elements in the oceans; but, even though the quantity present is often stunning—such as 10 million tons of gold, over 15 billion tons of manganese, and at least 20 billion tons of uranium—these elements are diluted in billions of tons of water. The ocean floor offers greater concentrations, and the continental shelf is a repository of substances such as bromide, magnesium, tin, iron, phosphorite, sulfur, and even, in spots, diamonds, gas, and petroleum. The abyssal zones are covered with a red clay containing approximately 50 percent silica, 20 percent aluminum oxide, and manganese, cobalt, copper, nickel, and vanadium. Some are mineable (table 2).[42]

Since 1973 and the oil embargo, marine minerals became far more important to the economies of the world. The United States is faced with an increasing dependence on foreign supply sources for 31 minerals, including sand, gravel, copper, nickel, and uranium. Whereas imports cost the United States $6 billion in 1971, they may reach, in steady dollars, $50 billion by 2000.

The world has consumed more minerals over the last 25 yr than were produced in all previous years of recorded history. As could be expected, the largest share was used up by the United States and Western Europe, ranging from about 55 percent for zinc, lead, and tin to about 75 percent for aluminum and copper; petroleum is estimated at 68 percent. However, since 1970 the Soviet Union and Japan have joined the ranks of the large mineral consumers: about 25 percent for aluminum and copper, roughly 31 percent for zinc and lead, and petroleum accounts for 21 percent. Summing it up, this group of

---

42. "Ocean Mining Comes of Age," *Oceanology International* 6, no. 2 (February 1971): 34–41 and no. 12 (December 1971): 34–38; M. J. Cruickchanck et al., "Offshore Mining: Present and Future," *Engineering and Mining Journal* 169, no. 1 (1968): 84–91.

TABLE 2.—RECOVERABLE MINERALS FROM THE MARINE ENVIRONMENT

Mineral	Geographical Location	Water Depth (Feet)
Sand and gravel	Atlantic and Pacific coasts, U.S.	<100
Glass and foundry sand	Atlantic and Pacific coasts, U.S.	<200
Magnetite	Australia; India; Japan; Pacific coast, U.S.	100–400
Glauconite	Pacific coast, U.S.	30–6,000
Rutile	Australia; Atlantic coast, U.S.	<100
Zircon	Australia	<100
Tin	Malaysia; Indonesia; Thailand; Alaska; Great Britain	<400
Silver	Pacific and Alaskan coasts, U.S.	<400
Gold	Pacific and Alaskan coasts, U.S.	<400
Platinum	Pacific and Alaskan coasts, U.S.	<400
Diamonds	Southwest Africa	<200
Manganese	Atlantic and Pacific Oceans; Mediterranean Sea	4,000–18,000
Phosphorite	Atlantic and Pacific Oceans, U.S.; Australia; Africa	100–4,000
Coal	Canada; Great Britain; Japan	<400
Monazite	South India; Sri Lanka	0–200
Shell	Gulf and Pacific coasts, U.S.; Iceland	<100
Sulfur	Gulf coast, U.S.	<100

SOURCE.—J. L. Goodier, "How to Mine Marine Minerals," *World Mining* (July 1967), p. 46.

countries uses nearly all the aluminum and copper and 86 percent of the zinc, lead, and tin. They leave barely 11 percent of the petroleum for the rest of the world.

Though still debated, the need to mine the oceans is with increasing frequency being seen as becoming a reality as early as the next decade. Expansion of ocean mining could be triggered by land-environment considerations, the need to conserve dwindling continental reserves, the ready availability of bulk materials absent or depleted on adjacent lands, higher-quality ores easily reachable in coastal waters, improved techniques to reduce exploitation costs, the rarity of some minerals, and economic considerations such as equilibrating balance-of-payments budgets and conserving hard foreign currency (table 3).

The bedrock beneath the ocean basins may also contain large concentrations of the same suite of metals as the ocean basins themselves, namely, large concentrations of manganese, iron, nickel, copper, and cobalt. It probably contains as well such metals as zinc, mercury, chromium, molybdenum, and the precious metals silver, gold, and platinum. The latter, and some nickel, copper, and lead, could be present in deposits resulting from the differeniation of the molten mantle material from which the basalt itself was derived or from

TABLE 3.—WORLD ANNUAL PRODUCTION OF MINERALS FROM OCEANS AND BEACHES (Estimated Raw-Material Value in US$ Millions)

	Production Value (1972)	% Value from Ocean	Projected Production Value (1980)
Subsurface soluble minerals and fluids:			
Petroleum (oil and gas)	10,300	18	90,000*
Frasch sulfur	25	33	
Salt	.1		
Potash (production expected in 1980s)	None	...	
Geothermal energy	None	...	
Freshwater springs	35†	...	
Surficial deposits:			
Sand and gravel	100	<1	
Lime shells	35	80	
Gold	None	...	2,000‡
Platinum	None	...	
Tin	53	7	
Titanium sands, zircon, and monazite	76	20	
Iron sands	10	<1	
Diamonds (closed down in 1972)	None	...	
Precious coral	7	100	
Barite	1	3	
Manganese nodules (production expected by early 1980s)	...	...	
Phosphorite	None	...	
Subsurface bedrock deposits:			
Coal	335	2	
Iron ore	17	<1	
Extracted from seawater:			
Salt	173	29	
Magnesium	75	61	
Magnesium compounds	41	6	
Bromine§	<20	30	
Freshwater	51	...	2,000‡
Heavy water	27	20	
Others (potassium salts, calcium salts, and sodium sulphate	1	...	
Uranium	None	...	
Total	...	...	94,000

Source.—G. J. S. Govett and M. H. Govett, *World Mineral Supplies* (Amsterdam: Elsevier, 1976).

Note.—Total production value of nonpetroleum commodities = $694 million; total from seawater = $388 million.

*Projections indicate that offshore production by 1980 will probably at least triple the 1972 daily output rate of 9.5 million barrels of oil and 17 billion (10^9) cu ft of gas per day and crude-oil price will probably be stabilized around $10 per barrel.

†Seawater plant at Freeport, Texas, closed down in late 1969 (U.S. Bureau of Mines, *Mineral Year Book 1971* [Washington, D.C.: U.S. Bureau of Mines, 1973], 1: 233).

‡Also assuming an average 30 percent increase of raw minerals.

§More than 200 million gallons of freshwater are recovered per day from submarine springs in Argolis Bay, Greece, but only a small portion of the produced water is utilized.

melting of the crustal plates that were thrust into the mantle.[43] However, there is less variety, though no less quantity, of minerals on the great ocean-basins floor than on the various segments of the continental margin (shelf, slope, rise) and in the small ocean basins.

Ocean mining will be complex, difficult, and costly, but a 1975 report of the National Research Council nevertheless concludes that "marine mining offers enormous potential for becoming independent of foreign counries for some important minerals, including those used as a source of energy."[44] Evidently this reality has not escaped other nations either, and the wild scramble to carve up the ocean is certainly motivated by such considerations.

Careful attention has been paid to potential environmental impact, and according to the 1975 report, marine mining could be undertaken within acceptable limits of environmental risk, although standards must be set before exploitation. It proposes a regulatory system which would prohibit industry from gaining economic advantages by keeping for its private use information it has amassed in prospecting. Instead of area leasing used for offshore oil extraction, work-program proposals would be the basis for allocations. As mining activities must be integrated into coastal zone planning, licensing schedules should be on a 10-yr schedule to insure orderly procedures.

On the continental shelf the probability of mining operations are classified as follows in order of earliest need: sand and gravel, calcium carbonate, titanium and gold placers, and phosphorite. In the United States this would mean mining in the Gulf of Maine, along the Massachusetts coast, and New York–New Jersey bight, the Northwest and Southwest Pacific coasts, the Bering Sea, the Arctic Shelf, and, inland, along the Great Lakes. Based on land experience, a span of at least 7 yr must be forseen between the start of exploration and full production.

Deep ocean mining is entangled in a web of international legal maneuvering. Nevertheless, it is expected that the United States will unilaterally legislate guarantees within the next 2 yr enabling deep-sea mining to start. Much is expected from the mining of deep-sea polymetallic nodules containing high-grade manganese, copper, cobalt, and nickel.

Yet, deep ocean mining on a production scale requires a staggering initial capital investment. In 1975 dollars, between $240 and $900 million would be needed to extract manganese, copper, and cobalt near Hawaii and process 5,000 tons a day in California, while requiring perhaps 20 yr to bring the project on stream. This seems to point to the need of governmental incentives to encourage capital investment in ocean mining. But besides money, further refinement of technology must be encouraged and manpower appropriately trained, the more so since no university offers a formal degree in marine mining.

43. G. J. S. Govett and M. H. Govett, *World Mineral Supplies* (Amsterdam: Elsevier, 1976).

44. National Academy of Sciences, *Annual Report 1975* (Washington, D.C.: Government Printing Office, 1976).

Basic research is steadily conducted but the applied aspects are often embodied in projects. For example, Project FAMCUS (French-American Mid-Ocean Undersea Study), which aimed at further understanding sea-floor spreading and probing the mid-Atlantic rift valley where earth crust is created, will also examine the creation of such minerals as copper, manganese, and chromite. One theory holds that seawater circulating through fractures in the ridge's rock formations may carry off some of these materials and concentrate them elsewhere. Perhaps the study may give a lead to minerals siting in more easily accessible locations, thus increasing the efficiency of prospecting. The submerged extension of the continent off the Atlantic, Pacific, and Gulf coasts of the United States has been claimed by the United States for economic exploitation, the most extensive territory to be added to the country since the Louisiana Purchase in 1803. The Continental Shelf extends, around the United States, from 10 to 300 mi off the coast, including 175 mi off Cape Cod, from 50 to 125 mi off the South Atlantic states; from 50 to 150 mi into the Gulf of Mexico; from 10 to 50 mi off the Pacific Coast; and approximately 300 mi off the Alaskan coast. The Hawaiian islands' shelf extends 10–50 mi offshore.

Already under way in various parts of the world, efforts to extract wealth from beneath the sea include extensive recovery of oil off the shores of the United States; diamond mining off the coast of southwest Africa; iron and coal mining off the Continental Shelf of Japan, and tin off the Malaysian Shelf; and the extraction of magnesium and bromine from the sea at Freeport, Texas, and of tin, aragonite, and sands (see fig. 3).

PRODUCTS

Ten years ago, according to the U.S. Bureau of Mines, the value of mineral offshore production along U.S. coasts alone reached the equivalent of roughly 7.3 billion current dollars. By 1981, seabed production in the free world could account, depending on the mineral, for from 2.5 percent (copper) to 79 percent (cobalt) from all sources, and polymetallic nodules alone could be worth $425 million annually in the 1974 dollars (table 4).[45]

Phosphorite

Clumps of phosphorite appear often near shore, at less than 100 m depth, and mining leases for them have been granted over 12,000 ha of southern California. The fertilizer could have been placed on the market in 1963 at $13.50 a ton, when $15 was the price tag for the imported product. In 1973, the

45. D. B. Johnson and D. F. Logue, "U.S. Economic Interests in Law of the Sea Issues," in *The Law of the Sea: U.S. Interests and Alternatives*, ed. R. C. Amacher and R. J. Sweeney (Washington, D.C.: American Enterprise Institute for Public Policy Research, 1976), pp. 37–76.

Other Ocean Resources 187

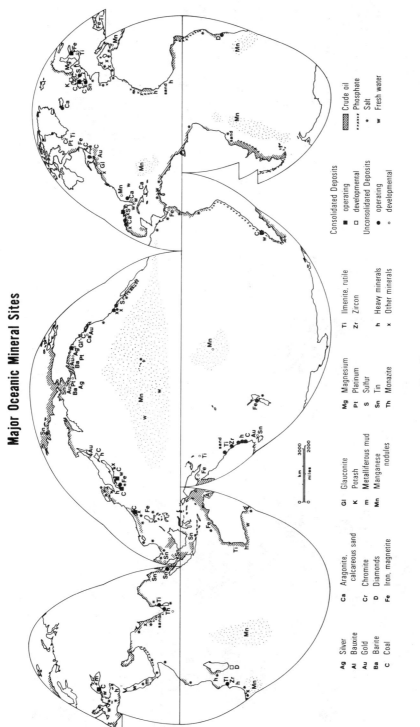

Fig. 3

TABLE 4.—CLASSIFICATION OF MARINE MINERAL RESOURCES

Dissolved	Unconsolidated			Consolidated
	Continental Shelf (0–200 m)	Continental Slope (200–3,500 m)	Deep Sea‡ (3,500–6,000 m)	
Seawater:	Nonmetallics:	Authigenics:	Authigenics:	Disseminated, massive, vein, tabular, or stratified deposits of:
Freshwater*	Sand and gravel†	Phosphorite‡	Ferromanganese nodules and associated	Coal
Metals and salts of:	Lime sands and shells†	Ferromanganese oxides and associated minerals	Cobalt	Ironstone
Magnesium*	Silica sand†	Metalliferous mud with:	Nickel	Limestone
Sodium*	Semiprecious stones†	Zinc	Copper	Sulfur
Calcium*	Industrial sands†	Copper	Sediments:	Tin
Bromine*	Phosphorite	Lead	Red clays	Gold
Potassium	Aragonite	Silver	Calcareous ooze	Metallic sulfides
Sulfur	Barite*		Siliceous ooze	Metallic salts
Strontium	Glauconite			Hydrocarbons
Boron	Heavy minerals:			
Uranium†	Magnetite†			
Other elements	Ilmenite†			
Metalliferous brines:‡	Rutile†			
Concentrations of:	Monazite†			
Zinc	Chromite			
Copper	Zircon†			
Lead	Cassiterite*			
Silver	Rare and precious minerals:			
	Diamonds*			
	Platinum			
	Gold†			
	Native copper			

Source.—M. J. Cruickchank, *Marine Mining: SME Mining Engineering Handbook*, Sec. 20 (New York: Society of Mining Engineers of Aime, 1973).
*Currently recovered commercially offshore.
†Recovered in coastal areas; may include some offshore activity.
‡Under research and development.

phosphorite mined on land in coastal North Africa and Florida was worth $13, and a ton of the marine product costs only $6. With the Arab producer heading a pricing organization, 1 ton was priced at more than $17 in 1976, and the price might still go up.

Phosphorite nodules are also available at great depths, but they are as difficult to retrieve as the polymetallic nodules. Phosphorite is made up of sand, gravel, calcareous organic remains, and fossil phosphorite; it usually contains 30 percent of economically worthwhile material (P_2O_5) and is generally found where other materials brought from land are not accumulated. Exploitable areas include the Pacific coasts of Georgia and Florida, Peru, the west coast of South America, probably the northwest of Australia, Japan, Spain, and the northwest and south of Africa. The Indian geological survey claims to have found deposits near the Andaman Islands.[46] While occurrence of nodules has been known for quite some time off the coast of Guinea and on the Agulhas Plateau, recent investigations revealed deposits on the Chatham Rise and Campbell Plateau, off the coast of east New Zealand, off northern New South Wales on the upper continental slope, and off Quilon (southwest India).

Ocean phosphorite also occurs in other than nodule form. In shallow waters occur phosphatic mudbanks, sands, pellets, and oolites. The mudbanks could be exploited to improve the nearby soils in Malabar (India); the highest content in P_2O_5 occurs in the period between monsoons and may attain as much as 18 percent. Both the muds and the sands are easily exploitable, though they are of rather low grade; sands off Baja California have at best a 12 percent content. Mero has estimated the reserves at 2,000 million tons. Additional deposits are located along the coast of North Carolina near Cape Fear.[47] Consolidated deposits of phosphatic strata are presently too expensive to mine on the seabeds, even though higher in P_2O_5 content. Outcrops also have been sited off Morocco and Rio de Oro near Florida, Georgia, and southern California.

World reserves of phosphorous exceed considerably the projected cumulative demand, estimated at 200 million tons. Even rapid expansion of agricultural demand would not put a great dent in the 21,500 million tons of reserves. Land production, however, contributes to pollution, conflicts with consevation measures, and, because of bulk, results in high transportation costs. These latter considerations are the major factor for looking toward the sea.

46. G. I. Bushinskii, "The Origin of Marine Phosphorites," *Lithology and Mineral Resources* 3 (1966): 292–311; M. C. Manderson, "Commercial Development of Offshore Marine Phosphates" (paper read at the Fourth Offshore Technology Conference, Houston, 1972), to be published in *Proceedings,* vol. 2, pp. 393-98.

47. J. L. Mero, *The Mineral Resources of the Sea* (New York: Elsevier, 1966).

Polymetallic Nodules

Ever since the Challenger reported their presence, nodules containing a high concentration of manganese have been found in many ocean-bottom areas of the Pacific, Atlantic, and Indian Oceans. Pacific nodules contain 30 percent manganese, plus 1.1 percent copper, 0.25 percent cobalt, and 1.25 percent nickel (table 5). Usually less than 20 cm in diameter, one nodule weighing 750 kg and measuring 1 m in diameter was brought up by the British from the Philippine Trench. In 1970, a nodule area covering 390 km^2, in waters of medium depth, was located near Hawaii; heaviest concentrations were found near the north coast of Lihue and south of Kapaa. A French expedition picked nodules off Tuamotu (Tahiti) from depths varying between 1,000 and 1,600 m. Most samples were small, but one specimen weighed 128 kg. Some years ago, John Mero provided an assessment of the reserves of metals found in the Pacific Ocean polymetallic nodules and compared them with land reserves (table 6). Projections for future demand seem to indicate a potential need for nodule extraction (table 7). Six years ago, a team of oceanographers from Columbia University mapped nodule sites, providing a valuable sheet for an economic atlas of the ocean. If Deep Sea Ventures proceeds with its plan of an initial recovery of 1 million tons of nodules yearly, it could recover 252,000 tons of manganese, 11,900 tons of nickel, 10,500 of copper, and 2,400 of cobalt, plus some trace metals such as molybdenum, vanadium, zinc, silver, and platinum. According to U.S. and UN projections, and provided no political complications generate delays, exploitation could commence by 1980, and some 15 million metric tons (16.5 million short tons) of dry nodules —with roughly 50 percent involving U.S. interests—would be produced by 1985.

Polymetalic nodules are seen by some as resulting from precipitation of seawater particles originating from terrestrial and sub-seabed sources and undersea volcanic eruptions. Others view the nodules as organic phenomena, with living organisms as the agents, through which metals are deposited on a nucleus. Still others, while admitting that nodules could occur without bacterial intervention, hold that bacteria produce a catalyzing enzyme helping the reactions occurring at deep-ocean pressure and temperatures. Most likely the sea bottom is enriched in these metals by dissolved and particulate matter coming from the midoceanic ridges.

In addition to Deep Sea Ventures, which had gathered 40 tons of nodules off the Florida and Carolina coasts, Ocean Resources has also been active in this field. So have Canadian, German, Japanese, and French corporations. Bethlehem Steel has also been involved for some time in exploitation studies, as have Summa, Inc., a Howard Hughes subsidiary; Dennicott; Le Nickel; Sumitomo; Metallgesellschaft A.G.; and Preussag A.G. International Nickel has supported research on processing technology and studies in engineering systems.

Ferrous-manganetic concretions have been found in the Sea of Japan, and

TABLE 5.—AVERAGE ANALYSES OF POLYMETALLIC NODULES
(% Dry Weight)

Region	Nickel	Copper	Manganese	Cobalt
North Pacific siliceous ooze	1.28	1.16	24.6	.23
North Pacific red clay	.76	.49	18.2	.25
South Pacific elevations	.41	.13	14.6	.78
South Pacific abyssal plain	.51	.23	15.1	.34
North Atlantic	.38	.15	14.2	.34
South Atlantic	.48	.15	18.0	.31
Indian Ocean	.50	.19	14.7	.28

SOURCE.—United Nations, Committee on the Peaceful Uses of the Seabed and the Ocean Floor beyond the Limits of National Jurisdiction, *Economic Significance, in Terms of Seabed Mineral Resources, of the Various Limits Proposed for National Jurisdiction* (A/AC. 138/87), June 1973.

chromite could be exploited along the Sakhalin coast. According to Mero, the Pacific Ocean alone has reserves of 10^{12} tons of manganese, and a single gathering operation could provide up to 50 percent of the world's production of cobalt. In short, reserves of over 100 billion tons of cobalt, manganese, and nickel probably rest on the bottom of the Pacific Ocean, and less than 1 percent of ocean-bottom reserves would suffice to satisfy current needs in manganese, nickel, copper, and cobalt for 50 yr. According to one estimate by D. B. Brooks,[48] the price of manganese production could drop by 45 percent, that of nickel by 7 percent, and that of cobalt by 30 percent, but other projections see only reductions of 3 percent and 4 percent for manganese and nickel, respectively, though still 27 percent for cobalt.

According to D. S. Cronan of the University of Ottawa, manganese at depths exceeding 3,300 m is mostly todokorite, whereas at lesser depths it is manganese dioxide.[49] Todokorite concentrates nickel and copper which replace the bivalent dioxide that concentrates, instead, cobalt and lead. Todokorite is more common in an oxidation milieu.

Actually, world reserves of manganese can carry us through this century; as an ingredient of steel it may become more valuable if developing nations accelerate their industrialization pace. But the United States imports virtually all of its manganese. Considering that as much as 40 percent of the price paid is

---

48. Brooks, "Conservation of Minerals and the Environment" (n. 37 above), and *Low-Grade and Non-conventional Sources of Manganese* (Washington, D.C.: Resources for the Future, 1966), pp. 116–18. For a recent analysis of effects on future world prices, see Committee on Commerce, U.S. Senate, *The Economic Value of Ocean Resources to the United States* (Washington, D.C.: Government Printing Office, 1974), pp. 25–28.

49. D. S. Cronan, "Regional Geochemistry of Ferromanganese Nodules in the World Ocean," in *Conference on Ferromanganese Deposits on the Ocean Floor*, ed. D. R. Horn (Washington, D.C.: National Science Foundation Office for the International Decade of Ocean Exploration, 1972), pp. 19–30.

192   Nonliving Resources

TABLE 6.—RESERVES OF METALS IN POLYMETALLIC NODULES OF THE PACIFIC OCEAN

Element	Amount of Element in Nodules (Billions of Tons)*	Reserves in Nodules at Consumption Rate of 1960 (Years)†	Approximate World Land Reserves of Elements (Years)‡	Reserves in Nodules/Reserves on Land	U.S. Rate of Consumption of Element in 1960 (Millions of Tons per Year)§	Rate of Accumulation of Element in Nodules (Millions of Tons per Year)	Rate of Accumulation/Rate of U.S. Consumption	World Consumption/U.S. Consumption
Magnesium	25	600,000	L¶	...	.04	.18	4.5	2.5
Aluminum	43	20,000	100	200	2.0	.30	.15	2.0
Titanium	9.9	2,000,000	L¶	...	.30	.069	.23	4.0
Vanadium	.8	400,000	L¶	...	.002	.0056	2.8	4.0
Manganese	358	400,000	100	4,000	.8	2.5	3.0	8.0
Iron	207	2,000	500#	4	100	1.4	4.01	2.5
Cobalt	5.2	200,000	40	5,000	.008	.036	4.5	2.0
Nickel	14.7	150,000	100	1,500	.11	.102	1.0	3.0
Copper	7.9	6,000	40	150	1.2	.055	.05	4.0
Zinc	.7	1,000	100	10	.9	.0048	.005	3.5
Gallium	.015	150,000	...	...	.0001	.0001	1.0	...
Zirconium	.93	+100,000	+100	1,000	.0013	.0065	5.0	...
Molybdenum	.77	30,000	500	60	.25	.0054	.2	2.0
Silver	.001	100	100	1	.006	.00003	.005	...
Lead	1.3	1,000	40	50	1.0	.009	.009	2.5

Source.—John Mero. *The Mineral Resources of the Sea* (New York: American Elsevier, 1966). p. 196.
*All tonnages in metric units.
†Amount available in the nodules divided by the consumption rate.
‡Calculated as the element in metric tons (from U.S. Bureau of Mines Bulletin no. 556).
§Calculated as the element in metric tons.
¶Present reserves so large as to be essentially unlimited at present rates of consumption.
#Including deposits of iron that are at present considered marginal.

TABLE 7.—POLYMETALLIC NODULES: COMMERCIALLY ATTRACTIVE CONSTITUENTS PROJECTIONS FOR FUTURE DEMAND
(Short Tons)

Commodity	U.S. Production (1971)	World Production (1971)	United States (1985)		
			Low Estimate	High Estimate	
Manganese	0	22,792,130*	1,700,000	1,950,000	
Cobalt	Negligible	25,857 (Co content)	14,650	17,050	
Nickel	15,654† (Ni content)	706,069 (Ni content)	356,000	414,000	
Copper	1,522,183 (recoverable)	6,664,079 (Cu content)	3,600,000	4,200,000	

Commodity	Total World (1985)		United States (2000)		Total World (2000)	
	Low Estimate	High Estimate	Low Estimate	High Estimate	Low Estimate	High Estimate
Manganese	13,700,000	17,950,000	2,265,000	2,900,000	18,265,000	23,900,000
Cobalt	25,850	28,250	26,000	34,300	40,550	48,850
Nickel	960,000	1,018,000	632,000	833,000	1,464,000	1,665,000
Copper	11,200,000	13,500,000	6,000,000	7,800,000	15,700,000	20,000,000

SOURCES.—U.S. Commission on Marine Science, Engineering, and Resources, *Marine Resources and Legal-political Arrangements for their Development* (Washington, D.C.: U.S. Commission on Marine Science, Engineering, and Resources, 1969), pp. vii-133, 134, 140, 142; U.S. Bureau of Mines, *Mineral Yearbook, 1971* (see table 3).
*Figure given is gross weight manganese ore (35 percent or more Mn content).
†Includes 2,581 short tons recovered as a by-product of metal refining.

for transportation, the interest of the United States in the nodules is well placed (fig. 4). Some years ago, the cumulative world demand was expected to reach 200 million tons in 1980, with world reserves assessed at 480 million tons; the current U.S. reserves may reach 6 million tons, but the demand is at least three times as much. The total value of manganese-nodule mining activity for 1985 has been placed at $534 million, of which $180 million will be for offshore activity at 1973 prices. From an economic viewpoint, ocean manganese and cobalt can be considered worth at least the value of their land counterparts, which they would replace, and even more if the lower prices were to permit increased consumption.

If only nickel, cobalt, and copper were extracted, the mining operation would still yield a small profit, although cobalt production would become unprofitable if costs of processing nodules exceed $45 per ton.

Only a small percentage of sea bottom has been surveyed, but knowledge is sufficient to start appreciable exploitation, and the United States has the most advanced technology. Most attention, as noted above, has been centered on north Pacific nodules found at 3,000–4,500 m. Four companies have announced they are ready to begin mining.

Copper world reserves amount to 450 million short tons; mine production

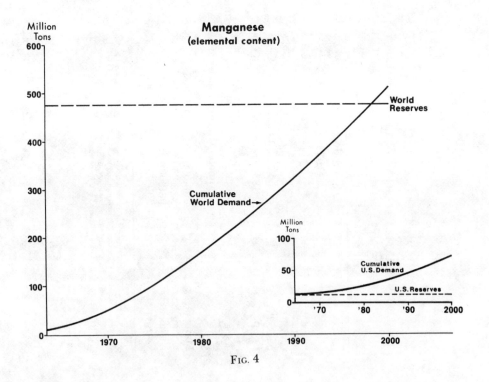

FIG. 4

was over 8 million short tons in 1974. Some additional reserves are hypothesized by the U.S. Bureau of Mines. Probably polymetallic nodule–extracted copper would hardly influence world supplies and prices, representing 1.3 percent of total consumption with its 200,000 tons yield and a projected market value of $261 million in the year 2000 (table 8).

Cobalt, though it has a small market, could be used as a substitute for several other metals, provided it were lower priced. The U.S. projected use increase is about 2.6 percent yearly, but world-demand increase may well reach 8 percent by 1985. Land production is dropping off, though lesser producers may be able to increase their current figures. Nevertheless, with world demand possibly reaching 60,000 tons in 1985, some 30,000 tons from nodules would be usable. Projected market value of the cobalt would be $45 million (table 7).

If the proposed Exclusive Economic Zone (EEZ) is created, polymetallic nodules will be virtually the only mineral resource exploitable in what will be left as international waters within the relatively close future. No significant recovery of nodules can be expected within the EEZ itself, the more so since economics will force oil and gas exploitation to the shallower waters, even though technology would allow deeper-water recovery. Many countries do not need, at this time, some of the metals contained in the nodules: Indonesia, the Philippines, Australia, Canada, the United States, the Dominican Republic, and Guatemala, to name some, have ample supplies of nickel on land. The cost of transporting nodules to processing sites is three to four times as high as the price for lateritic land ores, though such cost differentials could be compensated by copper recovery from the nodules, if copper is not present in lateritic ores. Processing at sea now appears impractical. We must thus consider that 66 percent will be added to the cost of the product to recover the nodules and bring them to a plant. This can be estimated as representing an outlay of capital of $5 per pound per year, or roughly $150 million for 13.6 million kg of nickel (15,000 short tons).

TABLE 8.—1973 PROJECTED MARKET VALUE OF
METALS FROM POLYMETALLIC NODULES
IN 1985 AND 2000

	Price per Ton (1985)	Price per Ton (2000)
Manganese	200	200
Cobalt	1,500	1,500
Nickel	2,900	3,040
Copper	1,660	1,980

SOURCE.—U.S. Congress, Senate, Committee on Commerce, *The Economic Value of the Ocean Resources of the United States*, S. Res. 222 (Washington, D.C.: Government Printing Office, 1974), p. 26.

Iron Ore

The current value of dredged and mined subsea minerals exceeds $600 million annually, but accounts for only 2 percent of world production. The Japanese started retrieving small quantities of iron ore in 1961 from the bottom of Ariake Bay. With the help of two giant dredging units, production rose by 1963 from 1,000 to 30,000 tons a month. Yawata Steel estimated at that time that it had a minimum of 36½ million tons available in reasonably shallow waters.

Construction is under way in the Maritime Province of the Soviet Union of a metallurgical complex which will process manganese, cobalt, nickel, and copper-containing materials of marine origin. Among others, the plants will treat magnetite- and titomagnetite-rich placers mined from the Baltic and Black Seas. Other such deposits have been located in the Sea of Azov and near the Kurile Islands. Iron occurs also in substantial quantities in polymetallic nodules.

Tin

The Soviets also plan to recover tin from the Laptev Sea and the east Siberian coasts and amethyst along the southern littoral of the White Sea. Cassiterite, a tin ore, has been mined from the ocean for some time. In the State of Selangor (Malaysia), tin is mined from 50-m depths, and tin has been mined near Tongkah (Thailand) since 1907.[50] Production in 1973 reached 6,000 tons of tin concentrate, and new deposits of cassiterite spotted in the Andaman Sea near Takuapa have been exploited since 1968. In Indonesia, eluvions and alluvions containing tin ore are dredged from the ocean bottom close to the Singkep, Banka, and Billiton Islands. In 1971, an American concern tested a hydraulic mining system which proved functional at depths down to 1,000 m. The largest tin-ore dredge was built recently by the Japanese. They scrape the ocean bottom, then suck up the material made up of sands and muds. Once aboard the ship, the ore is automatically separated from the bulk of the material, which is then returned to the ocean. The Japanese are also experimenting with a continuous-belt dredge which could perhaps work at depths of up to 4,000 m.

The Soviets reported large tin reserves in Mankina Guba (Yakutia), near Selyakhskaya, stretching from Cape Svyatoi Nos to the Strait of Dmitri Laptev, south of Bolshoi Lyakhov Island, in the harbors of Kuntzeyev and Siahu, and in the Japan Sea. They claim to be ready to extract diamonds, platinum, and especially gold near Kalyma (Lena River), Nakhodka (Okhotsk Sea), and along the northern and southern coasts of the Kamchatka Peninsula. Specially

---

50. P. H. A. Zaalberg, *Offshore Tin Dredging in Indonesia* (London: Institute of Mining and Metallurgy, 1970).

equipped ore-processing ships are being built which will also serve for prospecting and be equipped with a probe and radioisotope sensor which, when put to sea, will reveal on a shipboard indicator the presence of lead, gold, and manganese through gamma-ray absorption.

In the Atlantic Ocean, off the coast of Cornwall, in St. Ives Bay, recovery of tin-containing sands has started, and important deposits have been located off Brest on the coast of Brittany.

Sands and Gravel

Sands and gravels also are extracted from the ocean. Some containing organic remains have been used as building stone (San Marcos Castle, in St. Augustine, is built from coquina). Others have been used for artificial beach building, land fill, cement production, and in the making of prestressed concrete. They are inexpensive and easily transported.[51]

On Ocean Cay, an artificial island built in the Bahamas Archipelago using dredged material, aragonite is sucked up and treated. Production reached 2 million tons during 1971, and reserves are estimated at 50–100 billion tons. Near Muiden, in the Netherlands, sand is dredged from a depth of 75 m under the surface of the former Zuiderzee. Off the British Isles, more than 50 dredges exploit sand and gravel deposits. Recently, high-quality calcareous sands have been located near the Laccadive Islands. Such mining operations might ruin beaches if carried out too close to the coasts, as the Lebanese and the Israelis have found. Yet, the need for marine sands and gravels will increase. Some countries fail to find sufficient quantities on land, and the French foresee that within a decade the Channel will be tapped for the materials needed by Paris, Normandy, and the north of France.

Exploitation is carried on principally in the United States, Japan, Denmark, Sweden, and especially Great Britain, where recovery of sand and gravel is largest and most advanced: 13.5 million tons (then worth $32 million) were extracted in 1971. The most promising U.S. areas are off New York and New Jersey. Occasional operations have also been conducted in Thailand and near Hong Kong. Dredging of sand and gravel already surpasses in value and quantity all other superficial seafloor mining operations.

Calcium carbonate has been dredged as coral sand offshore from Hawaii and Fiji, near Arkanes (Iceland), and in the United States (Gulf Coast and near

51. G. P. Chapman and A. R. Roder, "Sea-dredged Sands and Gravels," *Quarry Managers Journal* (1969), pp. 251–63; M. J. Cruickchank, "Unconsolidated Deposits," in *Marine Mining: SME Mining Engineering Handbook*, sec. 20 (New York: Society of Mining Engineers of Aime, 1973), pp. 20–114; W. F. McIlhenny, "Oceans and Beaches, a Ready Source of Raw Chemical and Mineral Materials," *Offshore* 29, no. 5 (1969): 56–62, 198.

San Francisco) as lime shells and elsewhere as lithothamnium and aragonite mud. Because it is bulky, it may prove economical to exploit local reefs rather than to import the product. It is used in the manufacture of cement and fertilizers, pulp and paper, and in the extraction process of magnesium from seawater.

Recovery of materials offshore is generally more expensive than on land. However, transportation costs may be less and hence prices more competitive when a user is coast or near-coast located. The inland may be served at low cost if materials can be water transported. Furthermore, depletion of land-based supply centers may act as a favorable factor for offshore sites.

Depth is an important element; economically exploited deposits should not lie beneath more than 35 m of water. Emery believes that U.S. Atlantic and Gulf of Mexico continental shelf deposits would satisfy U.S. needs for centuries to come. Some 450,000 million tons of sand are available along the northeastern coast of the United States, but these reserves have been tapped so far only for small landfill and beach-nourishment operations.

Gravel is more difficult to recover because it is often buried under thick sand deposits. In the United States, rich deposits line George Bank and areas off New York City. Calcareous marine shells, abundant along the Gulf of Mexico, have been mined along the coasts of Louisiana, Texas, Florida, and along the Atlantic and Pacific shores, with a yield of 20 million tons a year. Objections have been voiced by environmentalists. Rich deposits exist in Hawaii and the Caribbean.

Some sands contain gold; gravels contain diamonds, and often where gold is found their also is platinum.[52] Goodnews Bay has provided, since 1935, close to 90 percent of the platinum needed by the United States. Extraction is presently carried out on Australian and South African beaches. Here diamonds have been mined from the sea (1962–72) all along the coast from the Orange River mouth to Deay Point. Kimberlite is transported by the river and deposited along the littoral by marine action. The yield is not negligible: in 1964, one company extracted 16,118 carats from a single marine deposit. In Alaska, Inlet Oil exploits petroleum deposits but also has discovered gold deposits in the channel off Bluff; searches in Goodnews Bay for gold, platinum, mercury, and chrome; and in the southeast for gold, silver, copper, zinc, and uranium. Beyond the Burdekin River (Queensland), Australians found gold deposits worth an estimated $100 million.

Off the southeast coast of Greenland, Danish enterprises are prospecting for chromite, rutile, and platinum. Near southeast Alaska, 1,000 tons of barite are mined per day. Heavy minerals have concentrated on the continental

---

52. F. Haber, "Das Gold im Meerwasser," *Zeitschrift für Angewandte Chemie* 40 (1927): 303–16; C. H. Nelson and D. M. Hopkings, *Sedimentary Processes and Distribution of Particulate Gold in the Northern Bering Sea* (Washington, D.C.: U.S. Geological Survey, 1972).

platform at depths averaging 200–300 m. Rutile is currently mined off the Australian east coast, where 95 percent of the world's reserves are located, and 450,000 tons of ilmenite are mined yearly off the west coast. Both are often associated with zirconium and thorium-containing monazite. Zircon and monazite are extracted simulataneously with titanium in Florida, Sri Lanka, and Australia. Though deposits of monazite have been located off India and Alaska, Australia remains the leader with 30 percent of the world's production. The Soviets have mined uranium since 1972 near Liepaja in the Baltic Sea; they also prospect for titanium, ilenite, and rutile. They have detected magnetite and titanium placers in the sands of the northwest coast of the Black Sea, and similar deposits were found near Batumi and in the Sea of Azov. These placers also contain chromite—already mined off the Oregon coast—magnetite, cassiterite, and aragonite. Magnetite was exploited until very recently by the Finns, and the Japanese still extract 40,000 tons per year south of Kyushu Island (table 9).

Though occurrence of titaniferous sands is rather common along beaches and near shores, commercially valuable deposits are, so far, only identified south of Kyushu (Japan), west of Luzon (Philippines), west of New Zealand, and along southern Javanese beaches.

Beach and dune sands in Queensland, New South Wales; New Zealand; India and Sri Lanka; Africa; and Florida and South Carolina have been exploited for some time for rutile, ilmenite, zircon, and monazite. Zircon, rutile, ilmenite, and magnitite could be mined offshore from the Queensland–New South Wales border and along a 1,000-km-long segment of the Australian coast from Fraser Island to Newcastle.

Nickel

According to analysts, world reserves of nickel can satisfy projected cumulative needs to about 2000. Rapid industrialization would not cause a heavier demand because nickel use involves nonferrous alloys and special steel. The U.S. reserves are very modest, but nickel is plentiful in Canada. Nickel can also be retrieved from low-grade sulphide and lateritic ores, but extraction may well be more expensive than from the sea. Oddly, polymetallic nodules may prove more valuable for their nickel than for the manganese because their nickel content is now higher than in land ores (fig. 5).

Based on U.S. Bureau of Mines projections, demand for nickel is to increase at a 3 percent yearly rate.[53] World mine production reached 821,500 short tons in 1975, with worldwide reserves estimated at 59,600,000 short tons, most in currently pro-Western countries. Nickel world reserves are estimated at

53. U.S. Bureau of Mines, *Minerals Yearbook, 1974* (Washington, D.C.: Bureau of Mines, 1974).

TABLE 9.—TYPICAL ONGOING AND POSSIBLE FUTURE MARINE MINING OF UNCONSOLIDATED SURFICIAL DEPOSITS

Commodity	Ore Mineral	Grade of Ore*	Price Range (US$)†
Nonmetallic:			
Silica	Quartz sand	76.93	29–92/st
Lime	Shells and shell sands	1,428.6	1.40/st‡
Magnesite	Magnesite	33.33	60–100/st
Sand and gravel	Various	1,923.1	1.04–7.87/st
Phosphate	Phosphorite nodules and sand	307.7	6.50–1,020
Topaz	Topaz	1 ct/yd³	1–5/ct
Spinel	Spinel	.2 ct/yd³	5–100/ct
Corundum	Corundum	28.55	70–130/st
Heavy mineral sands:			
Beryllium	Beryl	66.67	30–35/st
Titanium	Rutile,	11.43	175/st
	ilmenite	101.8	22–24/lt
Chromium	Chromite	93.33	24–56/lt
Zirconium	Zircon	40.0	56–70/lt
Manganese	Hausmannite, braunite	36.72	61–68/lt
Iron	Magnetite	203.64	11/lt
Thorium	Monazite	12.44	180–200/lt
Columbium	Columbite (10:1)	1.25	1,600–1,700/st
Rare earths	Group of 15 Me oxides	7.15	.14–3,00/lb§
Tin	Cassiterite	.7463	2.80/st¶
Mercury	Cinnabar	.498	152.5/fl(76 lb)
Precious and rare metals:			
Diamonds	Diamond (industrial)	.25 ct/yd³	4–50/ct
Copper	Native metal	2.04	.49–.53/lb
Silver	Native metal	.6369 oz/yd³	1.57/oz
Gold	Native metal	.0203 tr oz/yd³	49.26–49.46/tr oz
Platinum group	Native metal	.0091 tr oz/yd³	110/tr oz
Deepsea authigenes:			
Mn/Fe/Co/Ni/Cu	Manganese nodules	66.84	29.92–56.61/st#
Au/Ag/Cu/Pb/Zn	Metalliferrous oozes	481.93	4.15–12/st**

Source.—Govett and Govett (see table 3), modified from Cruickchank (see table 4).
Note.—st = short ton, lt = long ton, ct = carat, fl = fluid, tr oz = troy ounce.
*Pound per yd³ unless otherwise specified (to convert to kg/m³ multiply by 0.342).
†Low value in price range used; prices from *Engineering and Mining Journal* (May 1972) and U.S. Bureau of Mines, *Minerals Yearbook, 1970* (Washington, D.C.: U.S. Bureau of Mines, 1970.
‡Calculated from per-ton value of calcium in shell at 40 percent of this value, from atomic weights.
§Using $1.50 as the average value, rather than the low price in the price range.
¶Calculated from per-pound value (1972) of tin at 79 percent market value, from atomic weights.
#Varies from average Atlantic to highest Pacific; data from Mero, *The Mineral Resources of the Sea* (see table 6).
**As above, wet value.

30 million tons, and current cumulative world demand exceeds 7 million tons. Nodule exploitation could depress U.S. manganese prices by 40 percent and nickel prices by 10 percent and even more if numerous producers got into action.

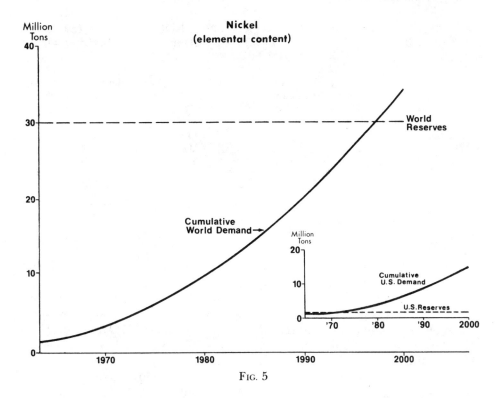

Fig. 5

The UN report (A/Conf. 62/25) estimated that world nickel production from nodules could represent approximately 18 percent of the total demand by 1985 (table 10). These same minerals, taken from the seabed, would have had a total approximate value of $492 million in 1971 (table 11).

Sulfur

World reserves are estimated at 2,200 million tons, and projected current cumulative world demand of 450 million tons can be satisfied until the mid-1990s. United States reserves are ample until after 1980. Substantial sulfur reserves exist in coal deposits, amounting for the United States alone to some 7 billion tons, and the U.S. Clean Air Act may lead to recovery of some of it. Finally, sulfur is abundantly present in gypsum and is economically exploitable (fig. 6).

Yet, prospects for exploitation of marine sulfur are good, and offshore salt domes in the Gulf of Mexico were already exploited in 1965 when sulfur exploitation rights were acquired for $34 million on 29,140 ha at 30–60 m depth. Onshore technology was easily adapted to ocean mining. Some salt domes come close to the ocean surface, and, through dissolving of the halite, bacteria can bring about sulfate reduction and transform nonsoluble matter

TABLE 10.—PROBABLE PRODUCTION FROM NODULES, ESTIMATED WORLD DEMAND, AND ESTIMATED NET IMPORT REQUIREMENT OF INDUSTRIAL COUNTRIES (in Thousand Tons)

	Probable Production from Nodules	Estimated World Demand	Estimated Net Import Requirement of Industrial Countries
Manganese	834	14,878	6,622
Nickel	199	1,106	699
Copper	181	13,517	3,266*
Cobalt	27	54	Not av.

SOURCE.—Third United Nations Conference on the Law of the Sea, *Report of the Secretary-General: Economic Implications of Seabed Mineral Development in the International Area* (A/Conf. 62/25), May 22, 1974.
*Not including centrally planned economies.

into sulfur. Contemporary sea-sulfur production exceeds 10 percent of the total sulfur production in the United States, or about 60,000 tons in 1965. For the peak year of 1968, U.S. production came 20 percent from the ocean. Potential reserves in subsea salt domes appear to be in the millions of tons, based on information obtained from holes bored in the Gulf of Mexico and the Mediterranean Sea. Considering the United States alone, offshore reserves of Frash sulfur exceed 200 million "long" tons, of which 100 million tons are certainly recoverable. Production for the 15-yr span 1960–75 from the outer continental shelf increased 12-fold (table 12). Sulfur can be recovered as a by-product of oil and to a lesser extent of gas extraction. The term "Frash" designates deposits which are recoverable by drilling wells into which hot water is pumped to melt the sulfur; they are found as a by-activity of the oil and gas

TABLE 11.—ESTIMATED VALUE OF MINERAL PRODUCTION (in US$ Millions)

	Co	Cu	Mn	Ni	Total	%
Landbased:						
1971	115	6,125	223	445	6,908	100
1980	130	9,382	310	617	10,413	97
1985	140	11,969	363	825	13,299	96
Seabed:						
1971	0	0	0	0	0	0
1980	70	123	12	135	340	3
1985	120	158	33	181	492	4
Total:						
1971	115	6,125	223	445	6,908	100
1980	200	9,505	322	752	10,779	100
1985	260	12,127	396	1,006	13,789	100

SOURCE.—Third United Nations Conference on the Law of the Sea (see table 10); U.S. Bureau of Mines, *Minerals Yearbook, 1974* (Washington, D.C.: Bureau of Mines, 1974); and various UNCTAD documents.

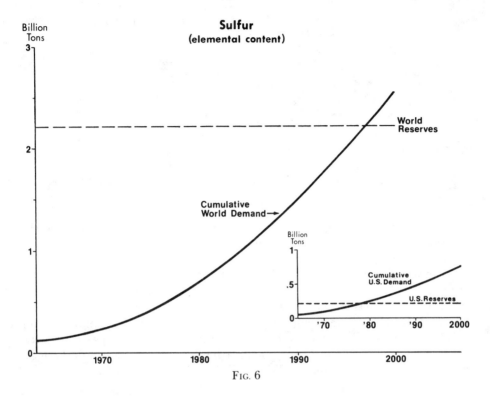

Fig. 6

search. In the United States, offshore sulfur has been exploited—mostly for consumption by fertilizer producers—at Grand Isle and Cominada Pass (Louisiana). Production reaches about 2 million "long" tons a year. A new market may develop as sulfur concrete competes with cement in the construction business.

Copper, Zinc, and Barite

Copper and zinc are mined off Maine, and barite off Alaska, though tides there reach 7 m and wind speeds 170 km/hr, while temperatures fall below −15° C. Barite is also abundant off the California coast. The total production of ocean-mined barite has an economic value of $1 million yearly. The muds needed to drill oil and gas could double demand by 2000.

Muds

Through erosion and alluvial accumulations, muds are deposited at the sea bottom at lesser depths than on the continental platform. Calcareous muds

TABLE 12.—SULFUR, OUTER CONTINENTAL SHELF
OF THE UNITED STATES, 1960–75

Year	Production Tons	Production Value ($)	Royalty Value ($)
1960	98,025	1,762,866	285,784
1965	1,090,950	23,387,005	3,197,532
1970	1,099,584	24,636,736	3,235,874
1975	1,248,134	69,737,571	3,737,784
Through 1975	15,902,662	432,996,588	46,903,057

SOURCE.—U.S. Geological Survey, Conservation Division, "Outer Continental Shelf Statistics, 1953–1975" (Washington, D.C.: Department of the Interior, 1976), p. 57.

originate from shell deposits and can be used in the production of whitewash, as fertilizer, or for its content of calcium, potassium, or barite. Such muds cover about 35 percent of the ocean bottom at depths ranging from 700 to 6,000 m with layers whose thickness is estimated at 400 m on the average. These are spread over 128 million km^2 and accumulate at the rate of 1 billion tons per year, or eight times the quantity yielded by contemporary land operations.

Red muds are argillaceous and constitute a source of aluminum, iron, copper, nickel, cobalt, and vanadium. Some oceanographers believe they cover one-half of the Pacific Ocean and one-fourth of the Atlantic Ocean floors at an average depth of 3,000 m.

Dissolved Substances

Common salt or sodium chloride extraction from seawater represents some 30 percent of the world production for domestic and industrial use. It is exploited off the U.S. Pacific coast, Puerto Rico, France, Sicily, and elsewhere. Bromine and magnesium compounds are simultaneously extracted: 70 percent of the total production of bromine and 60 percent of the magnesium used are of marine origin. Dow, Kaiser, and Merck extract magnesium in Texas, and Dow alone provides, from marine origin, 75 percent of the U.S. needs in bromine. Near Los Angeles, iodine is extracted from the petroleum fields' brackish solutions.

Thick beds of tachydrite, magnesium salt ($CaCl_2 \cdot MgCl_2 \cdot 12H_2O$), have been located along the Brazilian coast (Sergipe salt basin) and in the Congo basin. They probably can be mined by solution methods.

The oceans represent a gigantic storehouse of 50.10^{15} tons of dissolved materials, among which the chloride ion represents 54.8 percent of total salts, the sodium ion 30.4 percent, sulfate 7.5 percent, magnesium 37 percent, calcium 1.2 percent, potassium 1.1 percent, carbonate 0.3 percent, and bromide 0.2 percent. These eight ions account for over 99 percent of the salts. Together chloride and sodium ions account for 85.2 percent of the dissolved salts and are the most easily extracted. Over the last 15 yr salt production from the outer continental shelf has increased fivefold (table 13).

Salt basins have been located in recent years, as a result of offshore petroleum exploration and the Deep Sea Drilling Project activities, in offshore Arctic and eastern Canada, northwest Australia, the Middle East, Africa and the Red Sea, the Mediterranean, and various European sites. Off northeast Yorkshire and southeast Durham, important potash deposits have been located. Such basins contain, besides salt, anydrite and gypsum and probably also potash and magnesium.

Potash deposits in salt basins are large, though not as frequent as salt and gypsum. Reserves of K_2O (potassium oxide) run into tens of billions of tons; large deposits are located under the North Sea.

Crude soda and potash are known to have been extracted from seaweed ashes in Scotland in 1720. Iodine was extracted in the nineteenth century from seaweeds, and magnesia was prepared along the Mediterranean coast. M. G. Seaton reports that in 1923 magnesium chloride and gypsum were produced from the bitterns from solar evaporation of San Francisco Bay water.[54] It is now produced as a by-product of petroleum and other well brines.

Commercial seawater bromine was produced from San Francisco Bay bitterns in 1926; potassium chloride was recovered from Dead Sea waters by evaporation in 1931, and in 1932 bromine was taken from the residual liquors of the potassium plant.

A sudden surge in the demand for ethylene dibromide, a constituent of the antiknock compound of gasoline, occurred in 1933, could not be met by bromine from subterranean plants, and led to the construction of Kure Beach plant (North Carolina) which was producing, 5 yr later, 18,000 tons of bromine a year. A second plant was producing 27,000 tons by 1944 in Freeport (Texas). According to U.S. *Bureau of Mines Bulletin* 667,[55] bromine available from the

TABLE 13.—SALT, OUTER CONTINENTAL SHELF OF THE UNITED STATES, 1960–75

Year	Production Tons	Production Value ($)	Royalty Value ($)
1960	59,794	10,764	1,792
1965	290,894	52,334	8,724
1970	269,691	48,544	8,091
1975	151,032	(54,212)	(8,469)
Through 1975	4,865,209	794,328	132,954

Source.—U.S. Geological Survey, Conservation Division (see table 12), p. 56.

54. M. G. Seaton, "Bromine and Magnesium Compounds Drawn from Western Bays and Hills," *Chemical and Metallurgical Engineering* 38, no. 11 (1931): 638–41.
55. C. L. Klingman, "Bromine," in *Mineral Facts and Problems*, U.S. Bureau of Mines Bulletin no. 667 (Washington, D.C.: U.S. Department of the Interior, 1975).

sea is estimated at 100 trillion tons, dwarfing all other resources. Ocean extraction in the United States was halted in 1969. The Dead Sea alone has reserves of 800 million tons, a sufficient amount to fill current needs for hundreds of years.

Dow Chemical produced the first magnesium metal from seawater from the Gulf of Mexico and Freeport in 1941; within a few years production reached 8,100 tons. The U.S. government completed another such plant at Velasco (Texas) which produced magnesium at 30 percent below land-plant costs; by 1942, it was turning out about 3,300 tons a year.

Although the Kure Beach and Velasco (Texas) plants were closed down as World War II came to an end, the entire virgin magnesium production of the United States is of marine origin and so was at least 75 percent of the bromine. Currently 90 percent of magnesium is produced at Freeport and Velasco, an output of 125,000 tons worth $96 million; bromine now extracted from inland brines could possibly again be extracted from the sea. Magnesium, the lightest structural metal commercially available, is one-fourth as heavy as iron and two-thirds as heavy as aluminum. It has high strength when alloyed with aluminum, zinc, and manganese; is corrosion resistant and easily fabricated; and hence is very valuable in airplane and car construction. Added prior to pouring cast iron, it gives to so-called nodular cast iron strength and ductility comparable to steel.

Whereas bromine production requires deep seawater of high and constant salinity, not organically contaminated, magnesium production requires one-twentieth of the water needed for bromine extraction, and the temperature of the water plays hardly any role. Recovery of magnesium can reach 90 percent. Worldwide capacities exceed 2.16 million short tons (table 14). Presently, eight American companies have a production capacity of 510,000 short tons of magnesium oxide. Plans also exist to complete by 1977 a $50 million seawater magnesia facility in Norway at Bodo with a 220,000 tons/year capacity.

A few years ago hot brines were discovered in the Red Sea. The estimated value of the mineral materials contained in these brines is estimated at more than $15 billion. Layers, occurring at depths of more than 2,100 m, are sometimes 200 m thick and contain, besides precious metals, zinc, lead, iron, copper, and cobalt. Three separate pools of hot brines had been defined in the Red Sea rift valley in 1967. The brines and their associated heavy metals are believed to be periodically discharged from the eastern side of Atlantis II, the largest deep. Seismic reflections are obtained from hot brines and make their location possible; the reflections are due to the differences in density between the hot brine and the overlying Red Sea bottom water. The hot-brine area is located in the middle of the Red Sea rift valley. It is believed that the heat source and some of the heavy metals are probably associated with a local geothermal phenomenon. It is believed that in the Atlantic Deep II alone there are reserves, in the upper 11-m layer, worth $2.3 billion. Such wealth is tempting, but so far, no extraction method has been devised. The Hughes Tool Company and Deep Sea Ventures are conducting research to develop an appropriate system.

TABLE 14.—MAGNESIUM PRODUCTION FROM SEAWATER

Country	Location	Capacity (Short Tons MgO)
Canada	Aguathuna, Newfoundland	30,000
Ireland	Dungarvan, Waterford	75,000
Israel	Arad	50,000
Italy	Syracuse, Sicily	60,000
	Sant'Antioco, Sardinia	120,000
Japan	Hotsu	72,000
	Navetsu	55,000
	Minamata, Onohama, Toyama	187,000
	Ube, Yamaguchi	440,000
Mexico	Cuidad Madero, Tampico	50,000
Norway	Heroya, Oslo Fjord	80,000
People's Republic of China	Liaoning, Manchuria	10,000
Soviet Union	Not av.	100,000
United Kingdom	Hartlepool, County Durham	250,000
United States	Lews, Del.	5,000
	Port St. Joe, Fla.	100,000
	Pascagoula, Miss.	40,000
	Freeport, Tex.	100,000
	Chula Vista, Calif.	5,000
	Cape May, N.J.	100,000
	Moss Landing, Calif.	150,000
	South San Francisco, Calif.	10,000
	Others	80,000
Total		2,169,000

SOURCE.—Adopted from U.S. Bureau of Mines, *Minerals Yearbook, 1974* (see table 11).

The contact between hot waters and cooler oxygenated water above brings about oxidation of the dissolved metals. Chemical and mineralogical analyses led to the division of the deposits into seven bedded and laterally correlative facies: detrital, iron montmorillonite, goethite amorphous, sulfide, manganosiderity, anhydrite, and manganite. Their distribution, unconsolidated nature, and age relations show that the chemistry of the brine has considerably changed with time. The system appears to be an ore body in the formation process.

The Red Sea geothermal system is not unique. Others have been located in the Salton Sea in 1963 and in Chleken (USSR) in 1967.[56] Furthermore, similar enriched sediments exist in a few sites where the heat flow is exceptionally high, such as the crest of the East Pacific Rise, and perhaps also at substantial distance

---

56. See, e.g., J. L. Bischoff, "Red Sea Geothermal Deposits," in *Hot Brines and Recent Heavy Metal Deposits in the Red Sea,* ed. E. T. Degens and D. A. Ross (New York: Springer, 1969), pp. 368–78; E. T. Degens et al., "Red Sea: Detailed Survey of Hot Brines Areas," *Science* 156 (1967): 514–16.

from the active rifts, such as areas adjacent to the Banu Wuhu, a submarine volcano off the coast of Indonesia. The hot brines with salinities of 24 percent and temperatures exceeding 40° C of Atlantic Deep II are estimated to have an in situ value of metals of $2,500 million exclusive of iron, manganese, mercury, and other metals present in the mud. To the value of pools and the muds, one should add that of the geothermal energy which could be tapped.

Seawater analyses have revealed quantities of dissolved copper, cobalt, zinc, gold, lithium, uranium, and deuterium. Concentrations, however, are small; yet these elements could perhaps be recovered as a side operation of desalinization plants. And, if the world were to switch to a hydrogen economy, the sea would naturally be an inexhaustible source of hydrogen. As for deuterium, an essential element of the fusion process, the oceans contain 25 trillion tons of it. Deuterium is most abundant in tropical waters. That amount is the equivalent of 75.10^{26} Btu. Govett and Govett mention that heavy water extraction in Glace Bay (Nova Scotia) utilizes Gulf Stream waters arriving from the equatorial Atlantic.[57] A lithium isotope could be used as well in the fusion process.

Gould estimated in 1974 that the materials presently extracted from seawater account for one-third of the total world production of salt, two-thirds of the magnesium metal production, and one-third of the bromine.[58] Furthermore, it may be repeated that by 1978 freshwater production from the ocean alone will be worth $250 million annually, with an expected production of 4 million m³ per day.

INDIRECT EXPLOITATION

Some activities are not strictly ocean mining, but they are closely linked with the ocean domain, albeit because of the technological problems involved or the location of the zones of extraction.

Coal, etc.

Petroleum companies have drilled since the late nineteenth century off California through the ocean floor at ever increasing depths to reach the petroleum reserves stored under the ocean. Drilling started in the Gulf of Mexico in the last 25 yr. Progress in petroleum extraction in the ocean has raised questions of international law which have not yet been answered.

Even before the Christian era, mines were under exploitation by Greeks

57. Govett and Govett (n. 42 above).
58. Gould (n. 2 above).

beneath the sea at Laurium. Since then, coal, limestone, iron ore, tin, copper, and nickel have been extracted near the British Isles and the Atlantic coast of France; near Greece, Turkey, and Spain in the Mediterranean; Finland in the Baltic Sea; Japan and China in Asia; and on the west coast off Canada, the United States, and Chile. So have gold, mercury, and barium. African concerns such as De Beers, in the Republic of South Africa, have gathered diamonds off the coast of Namibia.

About the year 1530 a shaft sunk in an artificial island off the Scottish coast made coal extraction possible. Today over 100 undersea mines are in operation, with access from land, islands, or artificial islands. Some mines reach depths of 2,400 m, lie below 120 m of water, and are at 8 or more km from shore. Coal, like gas and petroleum, is not extracted from the ocean but rather from under the ocean. Subsea coal mines are operated in China, Japan, Turkey, Great Britain, Nova Scotia and Newfoundland, and Chile. Some 2 million tons were extracted in 1965. The Japanese started subsea mining in 1860 by extending a land-based mine off western Kyushu; new mines have been discovered off eastern Hokkaido; in one mine more than 5,000 are daily at work. Offshore mines accounted, with 10.8 million tons of coals, for 38 percent of the country's total production. Subsea coal accounts for 10 percent of current British coal production. According to McKelvey and others, coal deposits on the continental shelf exist, among other locations, in Alaska, Canada, Brazil, and Argentina and outside the Americas near Australia, Greece, Norway, the Arctic USSR, Spain, and Israel.[59]

By the eighteenth century, the British had already extracted tin off Cornwall. A high-grade barium sulfate has been mined since 1969 off Castle Island (southeast Alaska) by blasting underwater and dredging the ore afterwards. A billion tons of iron are available near Bell Island (Newfoundland), and iron is retrieved near Cockatoo Island (northwest Australia), Jussaro Island (Finland), and Elba (Italy).

While petroleum and gas production is progressing at an accelerating pace in the North Sea, important reserves of methane have been discovered and are also being tapped. Indirect exploitation also includes corals for jewelry (Bahamas), amber (Baltic), and limestones, used for building and cement manufacture.

Among the organic geological resources are shells, petroleum, and natural gas. Gas and petroleum are being depleted at a faster rate than extraction. Wells dug at depths of 250 m are not unusual, and current prospecting is considering deposits more than 400 m under the water surface. Petroleum demand doubles approximately every 10 yr, which has triggered worldwide prospecting.

59. McKelvey and Wang (n. 36 above).

CONCLUSION

The potential mineral wealth of continental rocks is enormous, and submerged continental areas are large enough to contain quantities of minerals which eventually will make important contributions to the world economy. Placer deposits are richer and more extensive in some regions, such as Southeast Asia. The mineral resource potential of the deep ocean, however, is small per unit area as compared to that of continents because there is no thick sialic crust, the seat of ore-producing magmas; no rocks predating the Cretaceous are known to be exposed; and important sedimentary rocks such as coal, evaporites, and bedded phosphate are not known to have formed in the ocean.

Ocean mining may have serious socioeconomic consequences if it displaces minerals that would otherwise be mined on land; places and people may thus be affected. Relationships of marine minerals to onshore minerals must be taken into account. If marine mining presents fewer waste-disposal problems than land operations, then this might be an incentive to turn to the sea, but our knowledge on this topic is still quite scanty.

Current offshore oil and gas production reached a total value of $40 billion in 1974, exceeding all other ocean-derived income combined including shipping. Other ocean mineral production had a total value of $740 million, equal to only about 2 percent of on-land production of the same minerals.

Transportation and Communication

The Organization of Shipping

R. A. Ramsay
UNCTAD

Ocean transportation has three basic elements: the cargoes which create the need for transport, the ships which fulfill those needs, and the carrying arrangements which connect the owner of the cargo with the owner of the ship and regulate the manner in which the cargo is to be shipped.

Cargoes fall into three broad categories: bulk liquids, bulk dry cargo, and liner cargo. Bulk liquids consist mainly of petroleum, which accounts for around 56 percent of the world's cargo movements and which is almost invariably shipped in tankers by the full shipload. Bulk dry cargoes consist largely of five commodities which are known in the shipping industry as the "major bulks": iron ore, coal, grain, bauxite, and phosphate. These five commodities are normally shipped by the full shipload in bulk carriers and account for about 20 percent of the world cargo movements. In addition there are about 16 commodities which are classed as "minor bulk" cargoes.[1] These are shipped sometimes by the full load by bulk carriers, sometimes by full loads in general cargo vessels which are smaller, and sometimes in even smaller consignments which fill no more than part of a general cargo vessel. The shipments of minor bulk commodities which are carried by bulk carriers account for around 4 percent of world cargo movements. Thus, the major part of world cargoes—80 percent in fact—consists of bulk items, liquid and dry, which are handled by tankers or bulk carriers.

One characteristic of these bulk cargoes is that nearly all of the commodities are fluid (i.e., liquid or granular) and can be loaded and discharged by pipelines, suction, augers, or similar equipment. Bulk handling methods are highly efficient, in the technical sense, and economical, notwithstanding the capital expenditure involved in building bulk terminals such as wheat silos. This is one of the main reasons why, as discussed below, bulk shipping is relatively economical in comparison with other forms of shipping.

The remaining 20 percent of world cargoes consists mainly of liner cargoes—mixed cargoes of small consignments of various items which are carried on liner vessels. This percentage also includes shipments of the minor

---

The views expressed are those of the author and do not necessarily reflect those of the United Nations.

1. Manganese ore, mineral sands, raw sugar, scrap, soya beans, pig iron, gypsum, coke, salt, timber, sulfur, cement, iron pyrites, steel, nonferrous ores, fertilizers.

© 1978 by The University of Chicago. 0-226-06602-9/77/1978-1008$00.75

bulk commodities shipped in full loads by general cargo vessels, since it is statistically impossible to separate the two subgroups, but liner cargoes constitute the major share. Unlike bulk cargoes, liner cargoes consist of a mixture of items, most of which have to be handled individually, whether manually or mechanically. Even where the loading and discharging costs are minimized by containerization, liner handling costs are still very high in comparison with handling costs in the bulk trades. This is one of the main reasons why liner services are relatively expensive.

There are four main types of carrying arrangements which link cargoes with ships: industrial carriage, liner shipping, charter parties, and contracts of affreightment.

Industrial carriage means that the ships are owned by the industries which control the cargoes—by exporting companies which sell on c.i.f. terms, by importing companies which buy on f.o.b. terms, or by transnational companies which ship goods from one affiliate to another. Industrial carriage is used mainly by companies which make regular shipments of large volumes of bulk cargoes, but it is also used by some companies with smaller volumes whose products require specialized ships.

In some respects, liner services are comparable to city bus services: vessels run on regular routes at regular intervals according to schedules which are announced in advance, make calls at numerous ports to load and discharge, and cater to the needs of shippers and consignees who ship in small consignments. Such services can be operated by general cargo vessels, container ships, roll-on vessels, LASH (Lighter Aboard Ship) vessels, or even bulk carriers; the distinguishing feature of a liner service is not the type of vessel used, but the type of service offered. In liner services the shipowner accepts responsibility for loading and discharging and charges a freight rate which covers the costs.

Charter parties are basically arrangements for shippers who have full shiploads of cargo, although instances of "part charters" occur. Several types exist to meet the needs of the parties, the most important being voyage charters (whereby a shipper hires a ship for a specified voyage) and time charters (whereby the shippers hire the ship for a specified period of time). Under a voyage charter party the parties can agree, and often do so, that the charterer (i.e., the shipper) will take responsibility for loading and discharging and pay a rate which excludes these costs. This is always the case with time charters. Thus, the rates which are quoted for charter party carriage are not always directly comparable with the rates quoted by liner services; an adjustment must be made to cover handling costs.

Contracts of affreightment are similar to charter parties, but instead of committing a shipowner to hire a particular ship to the shipper, they commit him to transport specified quantities of cargo between specified ports over a specified period, leaving the shipowner with some discretion as to the actual ships he will employ. It is left to the parties to agree which one will accept responsibility for loading and discharging. Such contracts are normally used with respect to large volume bulk movements (of several hundred thousand tons per annum), but contract shipping methods are also employed for lower volume movements of some specialized cargoes.

Industrial carriers are not concerned with rate-fixing problems, since they can adopt their own pricing structure for their own ships, but pricing methods are of vital importance to shippers who must use the services of independent shipowners. Price levels are determined in an entirely different manner in the liner and the nonliner (i.e., the charter and the contract) sectors of shipping. The rates for charter and contract shipping are the subject of bargaining on the open market, so, except in a situation in which there is a world shortage of shipping, bargaining tends to drive rates down toward cost levels. There is comparatively little cartelization in this sector, despite periodic efforts by tanker owners to form rings to maintain price levels, and consequently a high degree of competition between shipowners for available business. In liner shipping, on the other hand, the liner operators traditionally have formed themselves into cartels called "conferences" which fix freight rates and restrain shippers from using nonconference lines by a system of loyalty agreements whereby shippers receive rebates in return for shipping exclusively on conference vessels.

Thus, rate levels in the nonliner sector are relatively low for two reasons: first, this sector deals primarily with bulk items with low handling costs (or with homogeneous cargoes of packed goods which are cheaper to handle than cargoes of mixed items), and, second, excessive profiteering is minimized by the operations of the open market. Conversely, there are two reasons why liner rate levels are relatively high: first, liner cargoes consist of "break-bulk" items which are expensive to handle (handling costs account for 30 percent of freight revenues in many trades), and, second, under normal conditions the conference system enables liner operators to maintain margins over their cost levels. Even in a situation where there is a surplus of shipping, the liner conferences only face limited competition from nonconference lines because the latter cannot usually offer the "bus stop" service which is offered by the conference lines.

This is a broad overall view of the international shipping scene, which is the subject of intense international controversy. What then are the issues which are the subject of controversy?

There are two basic controversies, one between the shipowners of the traditional maritime countries which own most of the world fleet and the shipowners of developing countries who seek a greater share of the fleet; the other between shipowners and ship users who seek more equitable procedures for determining rate levels in the liner trades. Because developing countries tend to be users rather than suppliers of shipping, and because they bear much of the impact of transport costs (as discussed below), this second controversy tends to develop into an issue between the industrialized shipowning countries and the developing countries.

The controversy over participation in shipowning operations has been growing over the past decade. Developing countries generate about 63 percent of world cargo movements, but their fleets amount to only around 7 percent of the world fleet. The United Nations Conference on Trade and Development (UNCTAD) has been examining the reasons for the past failure of these countries to expand their fleets and is investigating means of enabling them to expand further in the future.

The past efforts of developing countries in fleet expansion were largely concentrated in the liner sector for several reasons. Entering the bulk sector would involve trading on the open world market, with all of the consequent uncertainties, and many of the bulk cargoes to and from these countries on a regular basis were controlled by transnational corporations; therefore, access to cargoes appeared easier in the liner trades. Furthermore, the liner services provided by foreign shipowners have always been the subject of complaint, partly on account of the level of rates and partly on account of the dictatorial attitude adopted by many conferences; and, consequently, there has been pressure in many countries to establish national-flag liner services.

These early efforts were often frustrated by the cartel arrangements of the liner conferences, which often treated the established conference member lines as having a monopoly over conference cargoes (i.e., those in respect of which shippers had signed loyalty agreements) and were reluctant to admit new lines, either at all or with acceptable sailing rights. Largely as a result of the shipping committee of UNCTAD, there is now a general acceptance of the rights of a country to participate in the carriage of the cargoes which are generated by its trade, and conferences are now much more willing to admit new lines. Certain countries, notably the United States, have legislated against conference arrangements for closed membership. The UNCTAD Convention on a Code of Conduct for Liner Conferences, which is likely to come into force by 1979, makes provision for trading partner countries each to have the right to carry up to 40 percent of their liner cargoes, leaving the remaining 20 percent for carriage by third-flag vessels (or "cross-traders").

As a result of these developments the general cargo fleets of the developing countries have expanded to 15 percent of the world total, but their share of the total world fleet is still languishing at about 7 percent on account of their lack of expansion in the tanker and bulk carrier fleets, which only stand at 5.7 percent and 5.5 percent of the world tanker and bulk carrier fleets, respectively. At the same time there has been a rapid expansion in the bulk fleets operated under flags of convenience, which grew from 23 percent of the world bulk fleet in 1965 to 32 percent in 1976. Since these flag-of-convenience vessels are understood to be owned in the industrialized countries, this trend would suggest that the "open-registry" countries (which offer flags of convenience) are providing a means whereby the industrialized countries can maintain control over bulk-shipping operations while at the same time gaining all of the advantages of lower-cost labor which only the poorer countries can provide.

Understandably, there is a feeling among developing countries that they should have a greater role in bulk shipping than as mere providers of labor.

At the eighth session of the UNCTAD shipping committee held in Geneva in April 1977, a resolution was adopted which noted the situation with regard to the tanker and bulk carrier fleets of developing countries and requested the UNCTAD secretariat to prepare a set of guidelines to increase competitiveness of the bulk fleets of developing countries. It is expected that this work will be completed by the end of 1978.

Debate on the other controversial issue, the protection of the interests of shippers in dealing with the liner conferences, is also centered in the shipping committee of UNCTAD. After several early attempts to question the justification of liner conferences, the liner conference system is now accepted as a basic necessity in the interest of stability. Attention is now concentrated upon making it work in an equitable manner.

Developing countries are especially concerned with the level of freight rates. Apart from the fact that they are users rather than suppliers of shipping, there is also considerable evidence to suggest that the impact of freight rates is usually borne by exporters of primary products and importers of manufactured goods.[2] It is therefore likely that developing countries, rather than their main trading partner countries, suffer most from high freight levels.

In an effort to insure more equitable freight rates and conditions, about 33 developing countries (as well as the majority of developed market economy countries) have taken steps to establish consultation procedures whereby conferences will confer with a shippers' council before making decisions. However, these efforts have been only partly successful.

The effectiveness of consultation procedures was surveyed by a recent UNCTAD report which examined the reasons for past lack of success and made recommendations for the future.[3] The report concluded that many shipper/shipowner consultations are largely ineffective because the shippers' councils concerned are not truly representative bodies and cannot speak for shipper interests with authority. Furthermore, shippers lack the necessary data, and unless a conference voluntarily provides data on costs and other factors there is often no basis for a meaningful discussion. The report concluded that a truly representative shippers' organization must be based upon an organizational infrastructure of commodity groups and recommended that more emphasis be placed on the organization of shippers of particular commodities, which can in turn become the member organizations of a shippers' council. The report also recommended that shippers' councils should be backed by shipping investigation units which can examine the efficiency of existing shipping methods and investigate whether liner shippers could utilize

---

2. See United Nations, Conference on Trade and Development, Secretariat, *Freight Markets and the Level and Structure of Freight Rates* (E.69.II.D.13), 1969, pp. 71–76.

3. Ibid., *The Effectiveness of Shipper Organizations* (TD/B/C.4/154), November 1976.

alternative methods. These recommendations were endorsed by the recent session of the shipping committee.

The investigation of alternative shipping methods goes right to the heart of the question of strengthening the bargaining power of shippers in dealing with the liner conferences. If shippers have no alternative but to use a conference service, whether they like it or not, they will be unable to exert much influence on conference decisions relating to rate levels.

The key to finding alternative methods of shipment, and hence to strengthening shippers' bargaining power, lies in exploiting the possibilities of potentially bulkable cargo movements. As noted above, liner services basically serve the needs of shippers who ship in small consignments, whereas the nonliner sector of shipping caters primarily to those who ship by the shipload. However, while the consignments of individual liner shippers and consignees may be very small, examination of the overall cargo movements of a liner service often reveals very sizable quantities of particular commodities which, if consolidated, could be shipped much more economically by the shipload. Since bulk shipping methods are most readily applicable to shiploads of homogeneous cargo, it is logical that action to consolidate consignments be initiated by the various commodity trading associations.

Changing shipments from liner to charter or contract methods will involve a certain decrease in sailing frequencies, and it may involve some extra expense in marshaling and distributing the larger consignments. It also involves some surrender of control over shipments by individual shippers, who must transfer cargo booking activities to some form of central agency such as a trade association or a government agency authorized to sign charter parties or contracts of affreightment on their behalves. A commodity group must therefore consider carefully whether there would be a net advantage in adopting bulk methods, keeping in mind the need for delivery frequencies to be compatible with marketing needs. However, whether a commodity group actually takes the step of switching from liner to nonliner shipping methods, it will be in a far stronger position in bargaining with a conference if it places itself in a position to be capable of adopting bulk methods should the need arise.[4]

Thus, the past year has witnessed a change in the manner in which the two most controversial issues in ocean transportation are being tackled: efforts to increase developing countries' share of the world fleet are being directed for the first time toward the bulk sector and efforts to increase shippers' bargaining power vis-à-vis that of the liner conferences are being directed toward strengthening the actions of the commodity groups. The impact of these new approaches can only be assessed in future years.

---

4. For a full discussion of the factors involved in switching from liner to bulk shipping, see ibid., *The Protection of Shipper Interests: Bulking of Cargoes* (TB/B/C.4/127, suppl. 4).

Transportation and Communication

A Safe Voyage to a New World
Thomas Busha and James Dawson

If Christopher Columbus were to return 4 centuries after he left his own world of change, discovery, and the overthrow of ancient certainties, he would find the maritime world of today at a new stage of evolution and again in need of some upturning of established ideologies and of the venturesome spirit.

Despite the occasional recurrence of that spirit, man's understanding of the oceans after millennia of deep-water navigation is not so remarkable as his continuing inability to cope with the sea's vagaries and to understand its demands. He hunts for its living resources in the same way his remotest ancestors did, and as a seafarer he is scarcely less a victim of the oceans than they were. Along with many other uses to which the salt-water surface of the world is put, navigation today poses more problems than the current level of social and technical intelligence seems capable of solving.

As the end of the twentieth century comes nearer, it seems possible to predict that a new view—indeed a revolutionary concept—of the legal and technical means of ensuring ships' safety will in time impose itself on the maritime world. The years 1976 and 1977 have, in particular, seen events of importance in the development of ocean uses, and it is appropriate that the advent of the *Ocean Yearbook* should come at such a crucial time. In those two years there has been a discernible coalescence of opinion about future ocean management, and in this the factor of future measures for safe navigation has important significance.

More than a million tons of shipping were lost *each year* in the years from 1972 to 1976.[1] One million tons of shipping represents a greater fleet of ships than the individual fleets of nearly 100 nations of the world. In the first quarter of 1976, some 317,738 tons of shipping were lost and in the second quarter 177,833, mainly (according to the intelligence department of Lloyd's) to fire, explosion, and collision.

Why, in circumstances of modern technology, should vessels be lost in 6

The views expressed in this article are those of the coauthors writing in a personal capacity and do not represent any official view of IMCO or its Secretariat or reflect any opinion other than the private views of the coauthors.

1. Michael Baily, shipping correspondent, reporting in the *Times* (London), on figures compiled by the Liverpool Underwriters Association which revealed also that four States (Cyprus, Greece, Liberia, and Panama) led the world in ships' casualties in the beginning of 1977 (February 4, 1977).

© 1978 by The University of Chicago. 0-226-06602-9/77/1978-1009$1.79

months of 1976 at a cost of more than $200 million to the insurance market? The safety of ships at sea is the concern of an elaborate apparatus of law and regulation, both national and international, daily implemented by navigators everywhere. Therefore, why this extravagant expenditure of life and property in the transport of men and goods by sea?

The major perils of the sea are known even to landlubbers; perhaps nothing more compelling to the romantic imagination than seafaring exists or has existed from before Homer to the present day. Those perils being natural do not decline in frequency, and the causes of many phenomena dangerous to shipping are still unknown.

Man-made perils also abound. It would be more common today than in the days of Joseph Conrad for a ship to encounter a derelict "ghost ship" on an ocean lane. The *Duarte*, a Colombian steamer which disappeared 26 years ago, has been sighted nine times since, and the *Dunmore*, abandoned in 1908 after her cargo shifted in a mid-Atlantic storm, has been seen adrift on even more occasions than the *Duarte*. A French ore carrier, the *Montlucon*, was adrift from 1951, when she was abandoned in a storm, until 1973, when she was sighted in the Bay of Bengal and taken in tow to the wrecker's yard. These ghost ships and other drifting objects occasionally collide with other vessels, as happened in 1974 when the cargo vessel *Herakos* was rammed in the North Sea by a drifting derelict which then disappeared.[2] More recently, a supply ship disappeared in the same area without trace, and was thought by many to have hit a drifting buoy uprooted by bad weather, at night.

Often less visible than a hulk abandoned with enough buoyancy for continued drifting are the objects lost at sea, in particular drifting lengths of metal pipeline, containers, and other deck cargo. The lengths of pipe, sometimes 100 feet long and 10 feet in diameter, and cargo containers floating upright with 5–6 feet of freeboard and 30–50 feet long, painted black, may present an unseen hazard and remain afloat for a considerable time. Timber and buoyant tank containers lost from the decks of cargo vessels also continue to constitute a hazard, and such objects as a pontoon, a "possible rocket head," a cable reel, and a weather monitor buoy, in addition to the mines frequently sighted, figure among the detritus which presents a hazard to vessels at sea.

The man-made obstacles to safe navigation are added to the natural dangers of the oceans which can suddenly overwhelm a ship. Immense waves which can be expected to develop over the world's continental shelves are the "devils" known to ancient mariners and able today to cripple or sink any ship. These episodic waves occur, for example, in the southern Indian Ocean off the east coast of Africa.[3]

    2. James Dawson, "Ghost Ships Are a Menace to World's Ocean Lanes," *American Marine Engineer* (a house journal of the Todd Shipyards Corp., no longer published) (September 1974).

    3. James Dawson, "Freak Ocean Waves Are Episodic," *New Scientist* 73, no. 1033 (January 6, 1977): 7–9.

The Agulhas current, sweeping down the southeast coast of Africa, is favored by tanker owners and masters who can use this current with the object of saving fuel. When in these waters the south-bound 5-knot current encounters much-traveled storm waves from Antarctic waters, there is a backup of waves caused by the undersea terrain and the tangential opposition of the stormy seas. That backup can produce a wave disastrous to the biggest ships, indeed may be more fatal to a 300,000-ton tanker than a 68-foot schooner.

In the case of the *Texaco Oklahoma*, the same sea conditions off Cape Henry, south of New York, in March 1971 broke her in half. The liner *Queen Elizabeth* on a trans-Atlantic troop-carrying voyage in 1943 was passing over Greenland's continental shelf when an episodic wave of massive proportions lifted her like a cork and plunged her into a vast trough, smashing the toughened glass on the bridge like a windowpane and obliterating the heavy gear fixed to the forecastle. At least one vessel, the *Waratah*, on her second voyage vanished off southeast Africa, undoubtedly the victim of an episodic wave, and none of her 119-man crew or 92 passengers was ever seen again.

Enough unnecessary mystery has been projected into the documented losses of vessels in certain areas of the oceans. It is now time to activate available systems and begin to mount the massive oceanographic programs needed for the anticipation of the causes of these ships' casualties, which are weather related and occur in certain similar conditions. While some studies have, for example, been done by the South African Council for Scientific and Industrial Research, documentation of the precise location where such waves have caused damage or destruction is very difficult to establish.

Surface objects and heavy weather are primary hazards often augmented by the human element in ships' casualties. The common collision or stranding or other accident frequently owes its origin to that element.[4]

The English Channel is the busiest shipping lane in the world and has been a scene of many casualties of the same sort that arise from "dangerous driving" on land. In 1971, the *Texas Caribbean* ran aground on the Varne Bank. The *Texas Caribbean* was a tanker, and efforts to inform other ships that she was aground did not succeed in averting a collision (perhaps owing to her great length) in which the *Brandenburg*, having collided with the tanker, sank and took the *Texas Caribbean* with her. Several days later a third ship, the *Niki*, hit the wrecks. Altogether 51 people perished. The crew members of the lightship sent to warn shipping of the sunken ships were terrified at night as other vessels bore down on the site.

A casualty which included the element of accidental oil pollution occurred 3 years later in another strait. *Metula*, a tanker of 206,719 deadweight tons carrying oil from the Persian Gulf to Chile, was transitting the Magellan Strait

---

4. U.S. Department of Commerce, *Human Error in Merchant Marine Safety* (Washington, D.C.: Maritime Transportation Research Board, National Technical Information Service, June 1976); see also n. 20 below.

on August 9, 1974. Having passed into the first narrows westbound, she grounded on Satellite Bank, striking bottom at nearly her full speed of 14.5 knots and coming to a stop in about 260 feet—"with an action described as 'like a shock wave.'"[5] Having 194,000 long tons of light Arabian crude oil aboard (which spreads and pollutes more quickly than heavier oils), the breach of *Metula's* forward compartments caused an initial loss of 6,000 tons of oil, mounting between August 19 and 20 to 40,000 tons, and on September 19 to 50,000 tons, a total of 54,000 tons being released.

There was variable weather, and strong currents were hitting *Metula* at the time of grounding. The Strait is about 310 miles in length, and the difficulties and dangers in navigating it are the same as in narrow channels and close harbors elsewhere, "But if the weather is thick, as is likely to be the case for more or less protracted periods, the navigation of Magellan Strait is both difficult and dangerous because of incomplete surveys, the lack of aids to navigation, the great distance between anchorages, the strong currents, and the narrow limits for the manoeuvring of vessels."[6]

The reader might well ask why the very large crude carriers (VLCCs) are found in such areas; the answer is speed and competition. It is not seen as dicing with disaster for ships of a big tanker fleet with superior navigational aids and better officer-training schemes to navigate in dangerous waters, but the smaller owners and the single-ship owners may allow or encourage hazardous routes and ignore risks of weather in order to remain competitive.

It may be surmised that the level of seamanship and the appreciation of wind, tide, and wave were of a higher order in the past than in the present day of automatic pilot and digital computer. Even with the available hydrographic charts, there is reason to think that commercial interests have not made the best use of automated systems at sea. There is evidence that the *Torrey Canyon* carried no Admiralty Pilot Book (which must be used in conjunction with nautical charts) of the area where she stranded on the Seven Stones Reef, northeast of the Scilly Isles on March 18, 1967. The *Arrow*, which grounded with a cargo of oil in Chedabucto Bay in Nova Scotia on February 4, 1970, had no coastal chart aboard, and her navigational aids were not functioning. This resulted in Canadian government legislation making the carrying of Canadian charts in Canadian waters mandatory.

The massive pollution which resulted from two groundings of the *Urquiola* at the entrance to the harbor of La Coruña, Spain, in May 1976 was caused in part by ignorance of the depth of the channel. The Spanish charts showed a minimum channel depth of 21 meters. The draft of *Urquiola* was 18 meters

---

5. U.S. Coast Guard, "Report of the VLCC *Metula* Grounding, Pollution and Re-floating in the Strait of Magellan in 1974" (unclassified, unpublished report submitted by Rear Admiral R. I. Price, Chief, Office of Marine Environment and Systems, U.S. Coast Guard, 1975).

6. Ibid., enclosure 15.

after she first went aground; yet she ran hard aground again in the same passage when trying to gain the sea, and her bow tanks were again ruptured.

The significance of these incidents takes on a dramatic quality if one contrasts them with commerce by air. It is unimaginable that travel through that medium—perhaps less fraught with danger in the present day than the oceans—should be undertaken today in a jumbo jet with no more charts or navigational equipment than were available to Charles Lindbergh. Yet the seas are daily navigated by ships whose charts have not been updated since the days of steam and even sail, and the *Igara*, a new ore carrier valued at about $25,000,000, struck an "uncharted" rock on March 1973 off Singapore. The ship was navigating in waters initially "explored" in 1862 by Britain and charted from Netherlands surveys by lead and line done between 1881 and 1907.[7]

One of the world's several areas of restricted navigation which caused difficulties for the mariner, the politician and the international lawyer is the Straits of Malacca and Singapore. Big-ship operators are confronted with complicated navigational instructions and devious routes through a passage much used by such ships. A very large crude carrier on passage from the Persian Gulf to Japan, if made to use the Lombok Strait instead of the Malacca, adds 1,200 miles to the voyage, or over 3 days at 15 knots—a costly deviation. In 1974 in the Straits of Malacca between Sumatra and the Malay Peninsula, when the *Showa Maru* went aground on such a voyage, her cargo endangered the coasts and fisheries of three states.

Those states—Indonesia, Malaysia, and Singapore—are alive to the danger. In February 1977 the foreign ministers of the three states met in Manila to consider measures to enhance safety of navigation and to coordinate their antipollution policies and measures in the Straits of Malacca and Singapore. They adopted a dozen recommendations, and a traffic separation scheme for three critical areas of the Straits has since been prepared by those states, acting together.

The hazards which are encountered by the VLCCs are not only those of safe navigation. The 350,000 tonner has about 3 million square feet of plates and a vulnerability both to weather and to explosion by reason of gas trapped in empty tanks of enormous capacity. The static from a rayon shirt can lead to an explosion which will rip open hundreds of yards of metal like a can opener.

Other hazards on the voyages of the VLCCs include the results of boredom of crew members. Nöel Mostert's dramatic account of the voyage of one of these giant ships was much criticized on points of detail but undoubtedly served to direct attention to matters such as the dangers of boredom aboard ship which continue to justify that attention.[8]

It is the factor of public awareness which greatly contributes to a

---

7. James Dawson, "Antiquated Charting Perils Shipping," *Marine Technology Society Journal* 8, no. 2 (February 1974): 15–17.

8. Noël Mostert, *Supership* (New York: Knopf, 1974).

heightened responsiveness by governments acting cooperatively to deal with maritime safety. Relatively speaking, the organized international involvement in other aspects of shipping is weak. Unlike international air commerce, where states became involved from the earliest days by reason of the legal fiction of their control of all airspace over their territory, the shipping community has resisted with success until recent years almost all encroachment by governments on the transnational freedom they enjoy in their well-insulated area of the international economy.

However, the world of shipping makes relatively little distinction between the governmental and the nongovernmental in operational matters of some kinds. The existence of the Tankers Owners Voluntary Agreement concerning Liability for Oil Pollution (TOVALOP) and the Contract Regarding an Interim Supplement to Tanker Liability for Oil Pollution (CRISTAL) owes itself to an alert, international point of view and a desire to "put its house in order" on the part of the oil industry, as do the Load-on-Top system and the "Clean Seas Code," another voluntary response of the oil industry. These policies also owe themselves to the less altruistic fact that autoregulation in an industry is a weapon of the strong to squeeze out the weak operator by economic means.

Turning to the intergovernmental level, it is evident that despite centuries of interconnection between nations through commerce in ships the beginning of concerted and institutional cooperation by states on the international plane dates from the post–World War II fervor for international organizations. Having its roots in the coordinated shipping policies of the allies in that war, a scheme for states' consultation in maritime matters was created in 1948 and languished for a decade because those states did not immediately ratify the treaty which established a new organization. The Inter-Governmental Maritime Consultative Organization (IMCO) was established in London in 1959 and it arrived on the scene not a moment too soon.

The 1960s revealed that such a maritime organization was both a political and functional necessity in the United Nations family of states and organizations. Moreover, the issues of primary concentration were not those of flag discrimination and restrictive practices in shipping (the so-called commercial and economic issues which the "maritime nations" did not then wish aired in a world forum), but rather the technical concerns which had over the years occupied the attention of diplomatic conferences convened from time to time (generally by the United Kingdom) to deal with safety of life at sea (1929, 1948, 1960), the load lines of ships (1930), and pollution of the sea by oil (1954).

What could only be predicted in IMCO's infancy was the rapid expansion of shipping throughout the world; the problems it would raise came sporadically and generally called for urgent *post facto* solution rather than being anticipated.

In general terms, the growth of the world's fleet into the sixties was about 8 percent per annum, and this growth was reflected in part by a 7.5 percent per annum increase in the traffic through the Suez Canal between 1955 and 1966. In the latter year, the last prior to its closure, 14 percent of the world's seaborne trade traversed the canal (73 percent of it oil and 27 percent dry cargo).[9] Taking 1968 as a revealing statistical year, the tonnage total of the world's merchant fleet rose by 12,052,000 from the year before, Liberian tonnage being the world's largest with a rise of 3,122,000 tons and the United Kingdom second. Norway, Japan, and the USSR were rapidly adding to their fleets, and only the United States and Greece were scrapping old tonnage and actually declining in fleet size. In that year, the world's total of tanker tonnage rose to almost 70,000,000 tons and represented 33.6 percent of all vessels (in the two previous years it had been an even higher percentage).[10]

The growth of tanker tonnage was accompanied by the phenomenon of bigger vessels. In 1968 ships of 50,000 gross tonnage and over rose from 118 to 178 (oil tankers from 105 to 151). The Suez Canal had been closed in June 1967 when one-third of oil imports by sea to Europe were normally taking the Suez route. This phenomenon of bigger ships was to continue during the following decade, owing to the closure of the canal and to the technological breakthrough which made economies of unrestricted scale possible when building tankers. The maximum size (deadweight) of a tanker traversing the Suez Canal was 60,000 tons.[11] The period from June 1967 to the end of 1974 saw the heyday of the VLCC. In those years the ships of 200,000 tons and above rose from one to 478, that is, 46 percent of the world's tanker fleet. In 1975 there were 30 ships of over 300,000 tons deadweight in service and 151 on order: the supply of tanker tonnage in 1975 was increasing more rapidly (3.8 times since 1963) than the trade in oil (2.8 times).[12]

What was not foreseen by the states most active in IMCO in the early sixties was a thicket of problems arising from both the older and the newer vessels sailing the seas. The organization had been entrusted with a treaty dealing with oil pollution (the International Convention for the Prevention of the Pollution of the Sea by Oil, 1954) long before the word "environment" acquired daily currency. The states acting through the machinery of IMCO dealt with the gradual improvement of that treaty (including amending it in 1962) as a form of environmental treaty limited by many technical practicalities then known to the mariner, in part because the treaty dealt with operational pollution, not accidental pollution of the *Torrey Canyon* kind.

More important to IMCO in those years than the 1954 treaty was the 1960 International Convention for the Safety of Life at Sea, the implementation and improvement of which was the stock-in-trade of the organization during its formative years in London. It was in respect of this convention that the first incident of a recurring pattern of international activity at IMCO came about. This pattern has relevance to future ocean management.

The pattern begins with a self-protective response by a coastal state to a

9. Hans Laurin, "Suez and World Shipping," *Lloyd's List* (May 8, 1975).
10. "12M Ton Rise in World Merchant Fleet," *Lloyd's List* (November 13, 1968).
11. Laurin.
12. Ibid.

maritime casualty or series of casualties affecting its populace and their interests. Impelled in large measure by the demands of a public outcry in the media—and the political reactions thereto—the state concerned turns to IMCO not only for the endorsement and sanction of measures in its self-interest, but for the necessary extension onto the international plane of any lawmaking which vitally affects its own and other states' maritime commerce.

From the point of first enlisting the aid of friendly states and setting forth its program for rapid improvement of international safety standards, the state in question must balance its domestic compulsions against the multipartite process of international rule making in IMCO. The urgency of a situation and a clamor in the press may require immediate and forthright measures of questionable legality (e.g., the destruction by the United Kingdom government of the *Torrey Canyon* on the high seas) or recourse to some degree of "unilateral action" on the regulatory plane (e.g., in the United States the introduction of congressional bills and the issuance of Coast Guard Notices of Proposed Rulemaking following the incidents of the *Yarmouth Castle* and the *Argo Merchant*).

The first of the major casualties of the sixties which gave impetus to such initiatives in IMCO for improved safety measures was the *Yarmouth Castle*.

The Panamanian SS *Yarmouth Castle* was nearly 40 years old when in the night of November 13, 1965, she took fire on one of her biweekly cruises from Miami to Nassau and 5 hours later capsized and sank with the loss of nearly 100 passengers and crew.

The Marine Board of Investigation convened to investigate the casualty recommended efforts "to upgrade and amend the International Convention for the Safety of Life at Sea, 1960, with respect to passenger vessels that contain large amounts of combustible material in their construction."[13] These efforts were undertaken by the United States in May 1966; Panama stood aside. An aroused public and a vigilant Congress in the United States brought the matter of passenger safety to a level not usually preoccupied with the technical rulemaking of IMCO's Maritime Safety Committee, and no less a personage than former Ambassador Averell Harriman was sent to London to enounce the United States government's urgent interest in fire protection for ships.[14]

For IMCO, the reaction of the United States government to the *Yarmouth Castle* and the fire-protection amendments to the 1960 Safety Convention which resulted represented the first significant stirrings of "unilateral action" as a factor in international regulatory development. In attaining the expected

---

13. Admiral E. J. Roland, "Commandants' Action concerning Recommendation of the Board," 5943/YARMOUTH CASTLE (February 24, 1966).

14. After the *Argo Merchant* and other casualties, it was the U.S. secretary of transportation, Mr. Brock Adams, who on May 23, 1977, took the role which Ambassador Harriman had earlier filled. He explained the U.S. government's urgent interest in oil-tanker safety measures to the IMCO Council, remarking tht "President Carter stressed ... that he prefers international solutions to problems which spring from international concerns" (see nn. 18 and 21 below).

results, the United States government was obliged to consider the interests of other states, but both in 1966 and 10 years later it proved a complicated matter to resolve the conflicts between domestic interests and those of the United States in the wider international context. Different perspectives as between the Congress, the Department of State, and the Coast Guard are sometimes the result of which organ of government is more alert to the possibilities of international retaliation.

Four bills were introduced in the United States Congress before it finally enacted Public Law 89-777 in November 1966, containing a relatively harmless program (from the aspect of the international relations of the United States) for protecting American passengers on cruise ships.

These bills when consolidated were steadfastly opposed in the House Committee on Merchant Marine and Fisheries by the Department of State and the Coast Guard. The influential special interests bent on using the sovereign power of the United States were indifferent to the conflicts with the commitments of the United States in relation to other states which were posed in the bills. The specter of using congressional power to override treaty obligations was raised briefly. Admiral Paul E. Trimble of the United States Coast Guard and Mr. William K. Miller of the Department of State pointed to the possible contravention by the United States of its treaty obligations, both in the 1960 convention and various treaties of friendship, commerce, and navigation. In the background of the laudable and humanitarian desire to protect human lives, the possibility of retaliation by foreign shipping in the cargo sector loomed over the deliberations.[15] This arose in part from the fact that the states whose ill will was aroused at the time were carrying not only 30 percent of the cruise trade out of the United States but some 45 percent of the exports and imports of the United States, with, in some cases, even higher percentages of their own trade.[16]

To some of the interested persons who viewed IMCO's response to the United States' initiatives, the response was unsatisfactory. Congressman Maillard's accusations against the organization for "condoning" shipboard hazards was a stricture which would be echoed on other occasions by those who expected a forum of states to foresee events which the states themselves did not anticipate, or preferred to ignore.

In the event, IMCO went to work on fire safety measures for existing ships in 1966 and adopted amendments to the 1960 Safety Convention, turning to the problem of future passenger ships' fire protection in 1967.

On a matter of delayed enforceability, however, the congressman was

---

15. A detailed exposition of the events of 1965 and 1966 in their legislative context is found in Thomas A. Clingan, Jr., "Legislative Flotsam and International Action in the *Yarmouth Castle*'s Wake," *George Washington Law Review* 35, no. 4 (May 1967): 675–97, from which this brief account is drawn. The discussion is relevant to events 10 years later.

16. Ibid., p. 692.

accurate when he predicted that it "would take upward of 5 years" for the IMCO standards to become effective.[17] In fact, nearly 12 years have passed and the 1966 amendments are not yet in force, nor the 1967 amendments. Like many unenforceable results of international consensus in IMCO, the fact that they have been adopted has, however, had its own salutary impact. The older passenger ships, including the *Queen Mary* and the *Queen Elizabeth*, were retired from service and a new era of cruise ships, better protected against fire, is well under way, irrespective of the status of the amendments but certainly in knowledge of their existence and significance.

The *Torrey Canyon* incident in 1967 followed close upon the *Yarmouth Castle*, in this case provoking even more urgent and far-reaching activity at IMCO. The dramatis personae in this next enactment of the "crash program" pattern were the same maritime states, but the leading role was taken again by the victim-state, the United Kingdom, closely followed by a fellow victim, France.

Barely 2 weeks after the stranding on March 18, 1967, of this 120,000-deadweight-ton tanker with a resulting "black tide" spreading from the Seven Stones Reef to pollute southern England, Brittany, and the Channel Isles, the United Kingdom government called for an urgent meeting of the IMCO Council to discuss the problems brought to light by that casualty. On May 4 and 5 an extraordinary session of the council laid down an 18-point program of work, both technical and legal, which was launched without delay and which in some measure was completed by 1976–77 but is also a continuing part of IMCO's expanding environmental concerns.

The significance of this event as part of a pattern of state behavior at IMCO was, as mentioned above, that it was another crash program for the amelioration of existing treaties and other regulatory regimes as well as the creation of new ones. It was seen that greater emphasis would have to be put on establishing sea-lanes incorporating traffic separation schemes and that port advisory services and pilotage would have to be improved. The use of shipborne aids to navigation including radar and electronic position-fixing equipment was to be studied and encouraged.

An incident in the *Torrey Canyon* drama was that her master, having found himself in trouble, also found that his ship could not be turned because the wheel was disengaged, the ship was on automatic pilot, and the helmsman did not know how to change over to manual steering. By the time the captain had helped the helmsman to turn the wheel hard to port it was too late. At IMCO, the Sub-Committee on Safety of Navigation in 1967 stressed that the possibility of instantaneously securing human control over a ship's movements should always be assured.

Moreover, the need for up-to-date charts and other publications necessary for safe navigation was emphasized, and navigational lights and signals, shore guidance of ships, and standards of training of mariners were to be taken up

---

17. Ibid., p. 693.

internationally for study and improvement. In addition, wide-ranging programs of measures for abatement of pollution and for legal matters were inaugurated.

Most of these programs have progressed at IMCO to the point where a vast body of regulatory prescription exists in the form of treaties, codes, recommendations, guidelines, and resolutions of many kinds concerned with safety of navigation and pollution prevention which have issued from IMCO as a result of the *Torrey Canyon*. Yet the oceans become more crowded and dangerous.

The pattern of urgent programs of work at the behest of a member state is being repeated as a result of the grounding, on December 15, 1976, off the shore of Nantucket Island, of the tanker *Argo Merchant*, from which some 25,000 tons of oil escaped. This was followed, in the same month, by a series of casualties, including the explosion of the *Sansinena* (December 17), the puncture of an oil-filled hold of the *Olympic Games* in the Delaware River (December 27), the grounding (without pollution) of the *Daphne* in a harbor of Puerto Rico (December 28). In January 1977 the melancholy series continued: there were groundings in Delaware Bay *(Universe Leader)* and in Texas *(Barcola)*; a lost ship *(Grand Zenith)* off Nova Scotia; an explosion during tank cleaning *(Mary Ann)* off Norfolk, Virginia; and, on January 18, the *Irenes Challenger* broke in half southeast of Midway Island in the Pacific.

Public opinion, as in the days of the *Yarmouth Castle* and *Torrey Canyon*, was aroused and hostile to the operators of the tankers involved. The Delaware River casualty led to the arrest of the ship's master. The president of the United States took action and called upon a task force under the secretary of transportation to draw up new safety standards. A spate of bills was introduced in Congress. As in the days of the *Yarmouth Castle*, Coast Guard Notices of Proposed Rulemaking were issued and hearings held by the Coast Guard in San Diego and Washington.

Fears of "unilateral action" were raised as in 1965 and 1966.[18] The United States appeared to have set course on irreversible national policies which might crystallize before negotiations with other states had begun or had concluded. Nearly 1,000 vessels were examined by the Coast Guard on arrival in United States ports (about half were found deficient) after warnings that U.S. and foreign tankers calling at those port would be boarded and examined at least once a year.

Although the president also proposed "international initiatives,"[19] a conflict arose again between the angry proponents of national interests and the spokesmen for international realism. In the spring of 1977, as in the days of the *Yarmouth Castle*, "Congress was laboring to generate a legislative proposal that

---

18. Office of the White House Press Secretary, "Fact Sheet—Actions to Reduce Maritime Oil Pollution" (March 16, 1977).

19. Ibid.

would be responsive to diverse domestic interests, while the executive branch moved forward on the international level toward a solution acceptable to foreign shippers."[20]

On May 23, 1977, the secretary of transportation, Brock Adams, told the IMCO Council that the organization would have time to consider an international approach to tanker problems, but that the United States would go ahead with its national legislative program. He explained that the United States would want provision for all oil tankers of 20,000 deadweight tons or more using its ports to have segregated ballast tanks (which are exclusively for seawater ballast and which therefore cannot be contaminated with oil which would be discharged along with ballast water from oil cargo tanks used for ballast), inert gas systems (for safe cleaning of oil tanks), backup radar with collision avoidance equipment, and improved emergency steering arrangements. Finally, new vessels would require double bottoms.

The last of these proposals is the most controversial, and the costs of segregated ballast retrofitting and inert gas systems in existing tankers seems likely to send many of these tankers to the wrecking yards if the proposals are to be adopted.

The opponents of some of the more costly "hardware" solutions to ships' safety point to the element of human error as the most significant in avoiding casualties, but the factor of crew standards and training was not ignored by President Carter or the U.S. authorities.[21]

In addition to the U.S. proposals, the United Kingdom and other states are contributing ideas to a special group set up in IMCO to prepare for a diplomatic conference which IMCO has scheduled for February 1978. The United Kingdom proposal is for setting up a marine safety corps consisting of ships' surveyors and other experts from the more advanced maritime countries to assist states in improving their inspection and certification procedures.

The new *démarches* at IMCO point to yet more treaty making and urgent regulatory draftsmanship by maritime technicians and lawyers. An unspoken premise of the exercise is that whatever IMCO does will have some *cachet* of international authority, and that even if a "standard" is not in force as between all the states that adopted it a government may proceed to enforce it on its own ships or even on foreign ships in its own ports.[22]

20. Clingan, p. 677.
21. Secretary Adams pointed out at IMCO that 80–85 percent of tanker accidents are caused by human error. His address to the IMCO Council is summarized in IMCO document C XXXVIII/SR.1 (May 23, 1977). An International Conference on the Training and Certification of Seafarers is scheduled for the spring of 1978 by IMCO.
22. The shipping industry, which tends once again to regard the U.S. tanker safety issue as a kind of "ultimatum" by that country, has submitted to a multitude of new international regulations since 1967, and it still reserves its strongest complaints not so much for the substance of new proposals as for the unpredictability of regulations with which a shipowner may be required to comply in the various states. The fear of U.S. action is that ships may arrive in ports to find that they do not meet the standards of the

This recurring pattern of ad hoc domestic and international rule making raises one of the major questions which will continue to confront states: the efficacy and necessity of treaties. This device of creating new norms of international behavior, and codifying old ones, is becoming ever more cumbersome and arcane. It is a commonplace of international treaty negotiations to speak of "standards," but the precise meaning of the term is elusive. Ambassador Beesley of Canada put his finger on the problem: "When does an international standard become an international standard? When it is reflected in a Convention that is being drafted and approved but has not been ratified? When it comes into force—and, in that event, how does it affect those countries that have not ratified? At what point does it become *ius cogens* under the Law of Treaties—binding on everyone whether they ratified it or not?"[23]

The nature of the IMCO constitution and its continuing use of diplomatic conferences to enshrine in separate maritime treaty instruments all the most important obligations of state behavior has affixed it to a Procrustean bed, shakily upheld by old treaties long superseded but still on the books, newer treaties in force or not in force, and constellations of amendments, in force or not in force, to existing treaties in force. In some cases even amendments to treaties not yet in force have been adopted in anticipation of the entry into force of both treaty and amendments.

Whether in these technical areas of interest the existence or nonexistence of enforceable contracts between states is still vitally significant to the promotion of their shared interests in maritime safety might be gauged by the number of times those states go to court to assert a breach of those contracts. That question of international law may be left to treaty lawyers and futurists, but it seems plain that with all the complex efforts to create and bring into force a host of treaties and their amendments on maritime matters these instruments betray ever more sadly their medieval origins in a world greatly in need of rational ocean management for a single world community. Moreover, treaties are like all legislation: they do not prevent human failings or stop the mishandling of ships by ill-trained mariners.

A glance at the case of a nontreaty which has become a treaty may illustrate the international preoccupation with form (as well as its possible irrelevance) or, at least, serve to emphasize Ambassador Beesley's point.

On July 15, 1977, at 12:00 hours zone time, the International Convention on the International Regulations for Preventing Collisions at Sea, 1972, came into force. One historical significance of this incident possibly was unremarked

---

authorities which inspect them. This feature of the American program may give rise to the "international discord" warned against recently in such journals as *Fairplay International Shipping Weekly* ("IMCO's Time of Crisis on Tanker Safety Issue" [June 2, 1977]).

23. "Pacem in Maribus VI," in *Symposia of Expo '75*, Official Report of the Japan Association for the International Ocean Exposition (Okinawa: Dai Nippon, 1975), p. 55.

even by mariners, for a phenomenon primarily of law gives it a stamp of uniqueness.

The oddity is "that the Rules of the Road" (as the collision regulations are known), although part of the "common law of the sea, adopted by the common consent of States,"[24] had never before 1977 been incorporated in a treaty. Rather, they were a set of rules adopted by most states in their national law and beginning as rules published in 1840 by Trinity House and recognized as enforceable by the British Admiralty Court.[25] In 1862, in Great Britain the "Regulations for Preventing Collisions at Sea" were appended to the Merchant Shipping Amendment Act,[26] and these regulations were taken up within the decade following by more states (including the United States of America in 1864) until they became, with successive changes, a worldwide expression of rules for safe navigation. Conferences of states held in 1929, 1948, 1954, and 1960 examined, revised and readopted these Rules of the Road, but never made them into a treaty. In 1972 they were finally given the formal character of written obligations between states, although it might be argued that they have long been part of the binding law of nations irrespective of not being treaty law.

The genesis of the Rules of the Road as well as the IMCO work on fire protection and prevention of pollution point to the leadership which a single state such as the United Kingdom or a small group of states may exercise in the maritime world with consequences great and small. Historically, Great Britain led the way in opening the seas to free navigation after a period of exclusivity and efforts by some sovereigns (including the English kings) to close off large areas of sea from merchant and naval vessels of other states (described by Chief Justice Cockburn as "vain and extravagant pretensions").[27] Great Britain was also in the forefront of the movement to survey and chart the seas, beginning with the setting up of the Surveying Service and Hydrographic Department of the Admiralty in 1759, and to create a common Code of Signals (first published in Great Britain in 1857).

If the hydrographic services of the United Kingdom have contributed more than a single nation's share to the world of shipping, the Corporation of Trinity House—a body chartered in 1514 as a semireligious guild—is probably the most prestigious of the 70-odd constituents of the International Association of Lighthouse Authorities (IALA), although it is, somewhat anomalously, free

---

24. *The Scotia* 81 U.S. (14 Wall) 170 (1871). In this famous appellate action in the United States Supreme Court, the defense against an allegation of fault in a collision case was that the American ship which collided with the *Scotia* was carrying improper lights. Although the prescription of lights to be used on the high seas originated in a British enactment, the court held that the rules were not those of a British or an American statute but were an obligation of the general maritime law, created by "common consent of States."
25. The Duke of Sussex [1841] 1 Wm Rob. 274.
26. 25–26 Vict. C. 63.
27. Regina v. Keyn [1876] 2 Exch. Div. 63, p. 175.

of United Kingdom government control. It is the principal pilotage and lighthouse authority of Britain, responsible for buoyage and marking and disposal of wrecks, and active under statutory authority to safeguard shipping in British waters.

The present British government has expressed a wish to create a central governmental board to reorganize pilotage, but a "takeover" of Trinity House has not been a popular idea, since for all of its antiquity it works extremely well. Moreover, its influence is not solely national; it trains students from abroad and has sent experts to such countries as Bangladesh, the Ivory Coast, and Mexico. Australia, Belgium, and Canada use its research facilities and expertise, and it deploys the latest equipment, such as the "Lanby" (Large Automatic Navigation Buoy) from its modest budget for lights and other navigation aids. Trinity House is a seafarers corporation, run by mariners, with an unbureaucratic administration of three independent departments supervised by men called Elder Brethren and Younger Brethren,[28] nine of the former being "active" and with the experience of many years at sea. It is a body which represents not only the organic and pragmatic growth of agencies of consensus and action in the maritime world but also the kind of institutional link between the practical and the policymaking levels so badly needed on the international level.[29]

Such a link is in part being provided by IALA and IMCO in the attempt to deal with buoyage. The existence of some 30 buoyage systems in the world (nine in Europe) was a reason for attempts at uniformity which have been under way since the days of the League of Nations. A worldwide system has not been created, but the plethora of systems has now been reduced to two, and in 1977 Trinity House was able to set out some new buoys in the English Channel based upon a new system. The 6 years of work since the *Texas Caribbean*, *Brandenburg*, and *Niki* were sunk have now resulted in better marking of wrecks and warnings of danger to ships navigating in those congested waters.[30]

Responsibility for working out System A and System B fell to the IALA which had reached agreement on the first of these (a combination of what are called the "cardinal" and "lateral" systems) and sought the blessing of IMCO. In May 1976 the Maritime Safety Committee having approved System A, the

28. Royalty and other distinguished persons figure among the nine "honorary" Elder Brethren, and the master is the Duke of Edinburgh. Winston Churchill was an Elder Brother of Trinity House and is reported to have replied in his fearless French when asked in that language about the significance of his attire—the Trinity House frock coat with gold lace: "Je suis frère aîné de la Trinité." "Quel influence!" was said to have been the reply.

29. "I believe that a drastically new concept of management is vital—one which leads in the opposite direction from overblown, centralized, permanent superbureaucracies with decision-making authority concentrated at the top. Indeed, I believe that we are fast approaching and in some cases may have exceeded, the effective limits of centralized bureaucratic structure, especially in public institutions" (Maurice F. Strong, "One Year after Stockholm," *Foreign Affairs* 51, no. 4 (July 1973): 690–707).

30. "New 'Road Signs' for Sailors," *Observer* (London) (April 17, 1977).

scheme was circulated to governments by IMCO.[31] System B (a lateral system) is still under study in the two bodies. At least one reason for the slowness with which such schemes are implemented is the difficulties which they present to hard-pressed hydrography services which must set about changing charts and issuing notices to mariners.

Buoyage is an essential element in navigation, and it has for some years been evident that the routeing of ships is also. What IMCO refers to as "a complex of measures concerning routes followed by ships and aimed at reducing the risk of casualties" is rapidly becoming ever more necessary as the convergence of large ships in narrow and dangerous waters increases.

Ships have an almost magnetic inducement to collide in such waters, especially when rapid maneuvers are called for, volunteered, or induced by ignorance or panic. The most popular areas are in such waters as the Delaware River, the English Channel, and the Straits of Malacca and Singapore. One adds to these hazards such factors as (1) the practice of arranging, when loading in the Persian Gulf, for the ship's load line to be exactly correct when entering, say, the port of New York; (2) the effects of "squat" in slow-moving ships; and (3) the increase in draft when a square-bottomed ship heels.[32]

These and many other reasons for changes in the way ships navigate in approach waters and other converging areas are known to ships' masters, many of whom would be glad to see such changes or even the adoption of compulsory routeing. But the pressures on shipowners contribute to their seeing things in a different light from society at large, underwriters, or ecologists. Long experience also shows that, in matters of navigation, information is often lacking, communications poor, and clarity of purpose badly needed. Remarkably, the voice of the mariner himself is not often heard, that of the merchant banker who stands behind the shipowner, never, and hard-working hydrographers hampered by dwindling funds, far too seldom.

Serious consideration should now be given to studying whether special-purpose carriers and deep-draught ships might be subject to compulsory routeing as well as to travel-to-schedule throughout their voyages, much in the manner of aircraft.[33]

With the entry into force of the new Regulations for Preventing Collisions at Sea on July 15, 1977, routeing schemes for various parts of the world became institutionalized, and obligations will be imposed on vessels navigating in or near such routeing schemes. The problem of providing for the rapid

31. IMCO, FN/Circ. 80 (21 May 1976).
32. Dawson, "Antiquated Charting," p. 17; and Captain A. F. Dickson, "Underkeel Clearance," *Proceedings of the Marine Safety Council, United States Coast Guard* 34, no. 3 (March 1977): 51–57.
33. On this subject, see Commander R. B. Richardson, F.R.I.N. (a Port of London Authority havenmaster), "Confusion of Arts," *IALA Bulletin*, no. 62 (January 1975), pp. 28–31; and J. S. McKenzie, "The Routeing of Ships," *Journal of Transport Economics and Policy* 5, no. 2 (May 1971): 201–15, from which much of what follows is gratefully drawn.

dissemination of the details of such new obligations arises. Also, frequent changes or inadequate or confusing information will have adverse effects on the mariner and the safety of ships.

Rule 10 of the new collision regulations applies only to schemes "adopted by the Organization" and prescribes the conduct of vessels navigating in or near traffic separation schemes. Compliance therefore presupposes knowledge by mariners.

The other presupposition is that the coastal state must be involved when schemes are adopted for approaches to ports. No "jurisdictional" difficulties in this regard have been encountered by IMCO. The authorities of coastal states are generally happy to promulgate the schemes and put them to IMCO which then examines and approves them and makes them known. For an example of the first stage of the process, traffic lanes from Port Conception, California, through the Santa Barbara Channel to Santa Monica Bay were first made effective on January 1, 1966, after being decided on by a committee which consisted of representatives of United States government agencies, shipping interests, and the oil industry. Similar schemes were introduced even earlier for New York, Delaware Bay, and San Francisco, but some (e.g., New York) have not yet been endorsed by IMCO under the new collision regulations. Endorsement by IMCO is a later stage of the process generally beginning with presentation of a scheme to the Sub-Committee on Safety of Navigation. In the case of the Straits of Malacca and Singapore this was done jointly by Indonesia, Malaysia, and Singapore in September 1977.

Realistic solutions to the growing traffic problem—over and above the strivings of IMCO[34]—are hard to come by. "Governments and the shipping industry in so many countries seem loath to step in any positive direction, being so sensitive to the lobbies and interests that support them."[35] Additionally, the solutions to the problem are not sought outside the narrow concerns of national jurisdiction and one-industry policy, even though it is as much a matter of transport economics as of the safety of the voyage. One system worthy of wider notice and greater use is found in Sweden. It consists of cables set out in approach waters which guide ships safely to berth through their picking up signals electronically from these cables.

34. Various institutes of navigation began work on traffic separation immediately after the 1960 Safety Conference in London, and priority was given to the Strait of Dover for which a scheme was completed in 1967. Recommended routeing systems for areas outside national jurisdiction called for international sanction of the kind IMCO was prepared to provide. In 1967 at the fifteenth session of the Maritime Safety Committee, it was unanimously decided that "IMCO was the only international body responsible for establishing and recommending measures concerning the separation of traffic in congested areas on an international level." The assembly of IMCO approved this conclusion, and since 1967 over 200 schemes have been approved as IMCO recommendations. On July 15, 1977, the element of obligation was introduced by the entry into force of the new collision regulations.

35. Richardson, p. 30.

Air transport began without air traffic control. "Its introduction became necessary, not so much because of the increase in aircraft speed, but because of the increase in density at focal points, e.g., in the approaches to an airport and at the crossway point of airways."[36] While ships differ in speed and mass from aircraft, the first purpose of directing both is to avoid collisions; the second is for reasons of economy.

The profound difference between maritime and air commerce is that in the former one finds a random distribution of arrivals, departures, allocation, and use of port services (berths, pilots tugs, bunker facilities, etc.) without any but the most rudimentary shore management or even knowledge of the whereabouts of individual ships. This individualism is part of the orthodoxy of the sea; it is the enemy of modern technology and ships' safety.[37]

The cost-effectiveness of modern communications should be reviewed in this context. The use of existing facilities and the addition of satellite communication for long-range anticipation of bad weather and for navigational purposes can certainly provide the infrastructure for better management of maritime traffic.[38] Commander Richardson, in a paper delivered to the International Congress of European and American Institutes of Navigation in Hanover in 1973, expressed his view about traffic management by putting the objective in these terms: "For this end we require organizational compulsion of a sort, so that those conducting the ship vehicle can know that they are working in an environment where others are at least morally required to base their decisions on the same intelligence and where the whole system is designed to keep them all fully informed all the time. Although a mandatory requirement with provisions for penalties does not guarantee 100% participation, nevertheless the probability of achieving such conformity is far greater than if left on a voluntary basis. There will be fewer impurities left in the traffic flow."[39] The "mandatory requirement" should not be unduly delayed. But if the lack of up-to-date hydrographic and other technical knowledge and the reluctance of governments to fashion "a newer culture for the conduct of our ships at sea in close waters"[40] render it impossible to achieve by the means at hand in intergovernmental bodies such as IMCO, it might be done through the

---

36. McKenzie, p. 201.
37. Richardson, p. 30.
38. In November 1976 a conference convened by IMCO adopted the Convention of the International Maritime Satellite Organization (INMARSAT) to improve maritime safety systems, among other purposes. That convention is included as an appendix to this volume. On July 14, 1977, the *Journal of Commerce,* London, reported on a new geostationary meteorological satellite which will "keep open a weather eye" for typhoons and other disturbances in the Pacific Ocean. The program is part of the World Weather Watch.
39. Richardson, p. 31.
40. Ibid.

toughening of insurance requirements, touching the shipping industry where it hurts most—in its pockets.[41] The hydrographers, the maritime telecommunications experts, and the insurance industry should no longer assume that ships are solely the concern of mariners.

"In seamanship and its associated arts we find some of the oldest skills known to mankind. So old, in reality, that its practitioners are loath to admit the possibility that any other people are entitled to proclaim with any knowledge at all the ways of ships. Inevitably a lore has passed down for a few thousand years about the ways of shipmen and their guilds, their helpers and lodesmen—now pilots—and other longshore activities, which is all accorded a sanctity that tends to be embalmed in the sheer romance of the whole sea saga. This may have been sound in the days of the Master-Owner, who sailed and traded his own ship but nowadays, when the concern is more that of conflict between ships rather than the total dominance of each individual ship, more and more people are coming to recognize that traffic in close waters is no longer a matter solelh within the competence of the expert mariner."[42]

The economics of shipping would also profit from orderly control and scheduling of the arrival of vessels, since delay in waiting to berth would be reduced, to the profit of both shipowners and port authorities. Tugs, pilot, water, and bunkers could be arranged in advance, as similar services are arranged for the arrival of aircraft. Manpower would be saved, and the ordering of the arrivals at a port would obviate the need for ships to travel at full speed (at heavy cost in bunker fuel) solely to find a place in the queue—and then possibly to wait for days to berth.[43]

One factor in scheduling which may be more complex than in air traffic is weather. This will, however, be subject not only to more reliable forecasting as the use of satellites and computers expands the knowledge of meteorologists but also to new technologies of weather management which, although now in the "what if" stage,[44] may be more possible technologically than on the legal and political plane. Again, the problems confronted are primarily a matter of surmounting inertia and developing the will to move forward on the international level.

Deep-ocean routeing will greatly reduce collision risk of the sort which

41. Dawson, "Antiquated Charting," pp. 15, 16.
42. Richardson, p. 29.
43. McKenzie has set out in his admirable article (n. 33 above) a statistical projection of berth occupancy and turnaround time in graph and tabular form. A simulation was used for average rates of arrival and turnaround and from this with a computer program it was possible to establish the relationship between random arrivals and the queues which can form of ships wanting to berth.
44. Wendell Mordy, "Weather Modification, as a Non-polluting Technology Capable of Changing the Marine Environment," *Symposia of Expo '75*, p. 53, n. 19.

occurred to the *Andrea Doria* and the *Stockholm*. There is no reason why a central control and coordinated routeing center, with subordinate area centers and dependent port systems, could not pinpoint the location on the globe of every merchant ship on a long voyage. This is already done in the implementation of rescue and salvage operations for vessels in distress.[45]

Container transport already has led to rapid advances in the techniques of multimodal transport in which sea, road, rail, and air are together regarded as integrated means of moving goods transnationally. The development of scheduling and certainty in sea transport would result in savings in rolling stock and other forms of transport optimization,[46] to the benefit of shippers and consumers. Standardized equipment and procedures in ports (a long-established feature of air commerce) would also effect appreciable savings.

McKenzie has asked a pertinent question with regard to routeing: "If the position of every ship is known for certain, what is the point of having defined lanes of shipping?"[47] He asserts that it would be more logical for ships to use all available sea room and have their routes chosen by reference to destination, size, and speed. This would avoid the dangers inherent in the present traffic flow where coasters and VLCCs use the same sea-lane; it would also dispense with the need for ships' detours, and it would give opportunities for some ships to travel more safely at greater speed and with less fear of collision or stranding.[48]

Undoubtedly, the development of hundreds of traffic separation schemes by governments, using the consultative machinery of IMCO, is the first and essential step toward achieving much wider measures of ocean management in this crucial segment of future ocean navigation. That process is under way. The problems arise in how and when the next step will be taken.

The intensity and complexity of the post–World War II development of ocean uses will seem modest when the accelerated activities of the coming decades unfold, whether in the investigation or exploitation of ocean space and whether on the high seas or the many exclusive economic zones and other special areas. That future, as Dr. Pardo and Mrs. Borgese remind us in their writings on ocean management, has already begun. Moreover, when it became evident that the third United Nations Conference on the Law of the Sea would be a protracted exercise in accommodating the separate state interests of the world community, it equally became clear that only those activities which *compel a community approach to ocean uses* will succeed in frustrating to some degree the new colonization of the seas in which "he who controls the area imposes the law."[49]

45. McKenzie, p. 207.
46. Ibid.
47. Ibid., p. 211.
48. Ibid.
49. Oliver S. Schroeder, Jr., "The Law and the Sea: An Introductory Comment," *Case Western Reserve Journal of International Law* 8, no. 1 (Winter 1976): 5–12.

The two fields in which such imperatives of community development are most evident were pointed to by the late Professor Wolfgang Friedmann in these words: "The freedom of the seas cannot remain a *laissez-faire* freedom. In our overcrowded world, navigation, as well as the exploitation of the living and mineral resources of the sea, must be the subject of planning and regulation for the common benefit of mankind."[50]

Such uses of the word "must" are regarded cynically by some, but the time will surely come when repeated ad hoc efforts to impose order on many of the maverick activities of the shipping and fishing industries will be proved to be impotent.

The degradation of the ocean environment is already compelling new international measures of pollution prevention with each passing year. In the quite near future substandard ships will be driven from the seas.[51] The days of the VLCC may, for several reasons, be numbered. The quiescent animosity toward flags of convenience again has come to life, and while governments the world over have vacillated between condemnation and condonation, the elements of this practice which concern standards of safety and manning competence will undoubtedly be subjected to ever-increasing regulation.

However, patterns of international behavior may have hardened, particularly where industrial and governmental conservatism are most evident, as they are in the field of maritime commerce. The words of Friedmann pronounce a warning which he gave some 6 years ago: "It is much easier to build up an international régime from scratch than to transform an established system. But for the same reason, every year that passes without effective measures diminishes the prospect of international ocean control."[52]

This may yet prove to be a sad truth, but it is in any case generally impossible to start from scratch where international shipping legislation is concerned.

What must now be done by the nations of the world is for governments to move within the existing institutions of international cooperation toward those attainable but far-reaching measures for safe navigation which these states, themselves, should propose. No enormous institutional structure is required,

---

50. Wolfgang Friedmann, *The Future of the Oceans* (London: Dobson, 1972), preface.

51. 1976 and 1977 also saw IMCO moving ahead with programs for improved enforcement of conventions, in particular to identify and deal with ships that do not comply with international standards. For all the appeal to some shipowners (and many seafarers) of operating under a flag of convenience and the substantial efforts of Liberia to dispel a reputation for indifference to such standards, the poor safety record of the flags of convenience is a major reason for the current insistence on "port State jurisdiction," which most shipowners resist. There is also evidence that some shipowners are bowing to the pressures to leave the convenience flags (see Michael Grey, "Flags of Convenience—the Pressure Grows," *Fairplay International Shipping Weekly* [July 14, 1977], pp. 49–51).

52. Friedmann, p. 118.

and these measures can be made viable and applicable by unfettering the lawmaking functions of the United Nations and allowing representatives of states to legislate for the world community as technicians charged with harmonizing and promulgating the laws of a large community rather than solely as servants of the traditional local experiences and interests. This process of transforming international forums could be assisted by developing countries which *are* able in some degree to start from scratch nationally with novel shipping legislation—a process which is accelerating with the aid of the United Nations Development Programme and IMCO. Sound ocean management need be neither unattainable nor very difficult to administer with existing institutions, regional and global.

An example of an effective national practice which could be extended to international use without great difficulty, and to the benefit of safe navigation, is the "Sea Court." Such courts have been functioning in Germany since 1871.

These Sea Courts are established for the simple and rational purpose of hearing cases of maritime character within a time frame which accommodates human memories and the mobility of vessels.[53] Within hours of a collision in a German coastal area the court, consisting of a civil justice, a state attorney, and four naval or merchant marine assessors, with a clerk, convene to hear all necessary testimony. A system of this character, with whatever refinements of further adjudication and enforcement of judgment which might be devised, could in time be appropriate as an international measure and helpful to environmental protection as well as to the safety of life and property at sea.

Advanced maritime technical expertise is available to a specialized agency such as IMCO in abundant supply. The contributions of many enlightened representatives of states and the well-founded policies which those states direct their representatives to negotiate and refine in sessions of the major organs, subcommittees, and working groups of IMCO are positive features of international cooperation leading to sound decisions and broadening their application.

In August 1977 IMCO was well on the way toward preparing the basic documentation for a conference in February 1978 on safety of tankers. In New York, the sixth session of the Third United Nations Conference on the Law of the Sea had concluded, and an Informal Composite Negotiating Text, reproduced as an appendix to this volume, was the result. In this document the mariner will find much with which he is familiar: Article 87 proclaims the freedom of the high seas, comprising, *inter alia*, freedom of navigation; Article 90 confers the right of every state to sail ships under its flag on the high seas; other rights and duties and many other ocean activities are dealt with in more than 300 provisions of the basic treaty.

The exigencies of safe navigation simply will not conform with the

---

53. James Dawson, "Sea Courts and the U.S. Navy," *U.S. Naval Institute Proceedings* 99, no. 1/839 (January 1973): 91–93.

fragmentation of the oceans by a new colonialism. If sovereignty remains the rule, it must be transformed into functional sovereignty: "interweaving of national and international jurisdiction within the same territorial space."[54] And if the world is to achieve a rational management of shipping to maintain the safest movement of vessels in all parts of the world's oceans, that institution of the world community entrusted with this important aspect of ocean affairs must be released from the straitjacket of formalism engendered by national reluctance to delegate decision-making power on matters of technology which require global policy and swift implementation of it.

54. Arvid Pardo and Elisabeth Mann Borgese, *The New International Economic Order and the Law of the Sea*, IOI Occasional Paper, no. 4 (Malta: Royal University of Malta, 1976).

International Maritime Satellite-Communication System: History and Principles Governing Its Functioning

M. E. Volosov and A. L. Kolodkin
Research Institute of Maritime Transport, Moscow

Y. M. Kolosov
USSR Ministry of Foreign Affairs

On September 3, 1976, the signing of the constituting acts in London crowned the efforts of the parties concerned to establish an intergovernmental organization—INMARSAT.[1] The purpose of the new organization is to develop, put into commission, and operate a maritime satellite-communication system. This step was conditioned by a number of serious circumstances arising from the history of seafaring over a period of many years.

HISTORICAL, ECONOMIC, OPERATIONAL, AND TECHNICAL PRECONDITIONS

It is generally recognized that from the earliest times merchant shipping has played a big role in the development of society by helping establish, maintain, and advance economic, political, cultural, and other ties between peoples, countries, and continents. Mankind has also used seafaring as a means for obtaining from the world ocean valuable biological and mineral resources and geographical, geological, and oceanographic information. However, though the seafarers were engaged with these important problems, they found themselves in a paradoxical position; they were always confronted with the extremely difficult problem of maintaining regular communication between ships at sea or in foreign ports and various fleet services of their respective nations, and between ships, on one hand, and shipowners, cargo owners, insurance agents, charterers, etc., on the other. It should be noted that the possibility of establishing shore-to-ship, ship-to-shore, and ship-to-ship communication is an essential condition for improved performance of the complex marine-transport system and the other sectors of the economy connected with

---

1. Abbreviation of "International Maritime Satellites." The text of the INMARSAT Convention will be found in Appendix B of this issue.

© 1978 by The University of Chicago. 0-226-06602-9/77/1978-1010 $02.42

navigation. The function of marine communication is to ensure safety of navigation, protection of human lives at sea, rational control of shipping, increased carrying capacity of shipping, and, thus, to improve the profitability and performance of the fleet.

In the context of the current scientific and technological revolution, reliable communication, operating around the clock and accessible to sailors in all regions of the world ocean, or at least on the most frequented international routes, is more important than ever. In this connection, it would be appropriate to mention one point. In analyzing the development of the industrial revolution of the late eighteenth and early nineteenth centuries, Karl Marx proposed an important law-governed feature characterizing this process. It consists in the fact that a revolution in the sphere of industrial production inevitably sparks off a revolution in the means of communication and transport.[2] Indeed, the achievements of the current scientific and technological revolution which are being widely introduced into the merchant cargo and passenger, fishing, research, and other fleets are entailing far-reaching revolutionary changes in all the elements of marine transport, in all of its structural, quantitative, and qualitative indicators. They are opening more and more possibilities for the progressive growth of the rate of marine shipments[3]—international shipments in particular, because the latter promote in the greatest degree the exchange of products, international division of labor, and economic integration of the various countries.[4] These achievements open access to an ever-wider assortment of sea and ocean resources which are vital for the economy. Finally, they make it possible to conduct on a global scale the comprehensive scientific research without which it is impossible to improve the functioning of sea transport.

The use of advanced techniques and new construction materials in shipbuilding today, installation in the ships of more powerful and more economical propulsion plants, improved propulsive devices, and auto-navigators, and the introduction of highly efficient cargo-handling means in both ships and ports have directly increased the overall number of ships and have contributed to the construction of larger ships with a deadweight of 150,000–500,000 tons (and subsequently, perhaps, to one million tons)[5] and

    2. Karl Marx and Friedrich Engels, *Sochinenya*, 2d ed. (Moscow: Gosudarstvennoe izdatelstvo politicheskoy literatury, 1960), 23:395.
    3. W. J. Golovin, *Prognoz razvitiya morskogo transportnogo flota mira na 1971–1990 gody* (Moscow: Centralnoe buro nauchno-teschnicheskoy informacii Ministerstva morskogo flota SSSR, 1972), pp. 10, 14, 15.
    4. S. A. Wyschnepolsky, *Mirovye morskie puty i sudoschodstvo. Ocherky* (Moscow: Gosudarstvennoe izdatelstvo geograficheskoy literatury, 1953), p. 282; S. M. Baew and E. D. Rodin, *Vajneyschie tendentzii v razvitii sovremennogo morskogo sudoschodstva* (Moscow: Centralnoe buro nauchnoteschnicheskoy informacii Ministerstva morskogo flota SSSR, 1975), pp. 3, 4, 15.
    5. *Zosen* 3 (1969): 14–37; *World Ships on Order* 27 (1970): 2; H. H. Beinhauer and E. Schmacke, *Fahrplan in die Zukunft. Digest internationaler Prognosen* (Düsseldorf: Droste, 1970), p. 20.

with speeds of up to 25–35 and even to 55 knots[6] which require smaller crews. Besides, the new ships spend less time in ports.[7] Such cardinal changes have made navigation increasingly difficult and the operation of fleets more complicated in general.

All these factors and developments confront the sailors, fleet adminstrators, and support personnel with difficult tasks and requirements intended to improve the safety of shipping, raise its competitive capacity, and guarantee maximum returns on funds invested in marine transport and related sectors of the economy. This aim will only be secured if the losses in vessels are considerably decreased or ruled out completely, that is, if losses caused by accidents, inefficient traffic supervision, idle time, empty runs, and other factors are cut down or eliminated altogether. Success in this sphere largely depends on whether marine transport is outfitted with advanced means of communication and whether these means are efficiently employed.

As to the conventional means of communication employed in sea shipping, namely, medium-wave, short-wave, and very-high-frequency (i.e., within the range of direct visibility) radio stations, cable, and radio-relay lines, it should be stated that, in the opinion of specialists,[8] these no longer meet the modern requirements. In the immediate future, that is, by about 1980, the number of ships in the world ocean will increase 50 percent as compared with the present, and the intensity of radio traffic will triple in some cases.[9] When this happens, the present means of communication will be unable to meet the actual requirements of the fleet. This is conditioned by a whole range of objective causes. Among these is the relatively low traffic capacity of the existing communication channels. The radio-frequency bands allocated for the marine-communication services[10] are totally inadequate for the traffic going through them. Other causes are that the performance of the high-frequency radio and telephone channels depends on industrial, atmospheric, and other physical factors,[11] that it takes a long time to establish radio contact (from 5 to 6

---

6. *Shipping World and Shipbuilder* 37/51 (1965): 48–50; Beinhauer and Schmacke; W. J. Golovin, pp. 172–74.

7. *Zosen* 9 (1969): 55; *Motor Ship*, vol. 583 (1969); Beinhauer and Schmacke, p. 146; Golovin, pp. 22, 178.

8. A. D. Koval, G. R. Uspensky, and V. P. Yasnov, *Cosmos cheloveku* (Moscow: Izdatelstvo "Maschinostroenie," 1971), p. 177; I. V. Charyk, "Future Prospects of Satellite Telecommunication Systems," *Telecommunication Journal* 38 (May 1971): 296–300; Ch. Dorian, "Application of Space Communications to the Maritime Mobile Service," *Telecommunication Journal* 38 (May 1971): 338–47.

9. Ph. J. Klass, "Maritime Satellite Economic Studies," *Aviation Week and Space Technology* (May 1, 1972), pp. 52–53.

10. IMCO, "Determination of the Requirements of Marine Services on Space Technology Use." Report on 22d Session of IMCO Maritime Safety Committee (MSC XXII/22), October 16, 1970, sec. 4-a (V).

11. L. N. Korsunsky, *Rasprostranenie radiovoln pri sviazi s iskusstvennimi sputnikami Zemly* (Moscow: Izdatelstvo "Sovetskoe radio," 1971), chap. 1.

hr to establish unilateral contact),[12] and that the cost of construction and operation of cable and radio-relay lines[13] is high (as a result, the cost of marine-communication services is extremely high, too). The situation is further complicated by the fact that the improvement and expansion of conventional communication systems calls for great capital investments, with the consequence that they are unable to keep up with the rapidly growing demand for communication facilities. This is acutely felt when the demand of marine merchant shipping for such services soars.

The launching of Sputnik I, the world's first artificial Earth satellite, by the USSR on October 4, 1957, opened much broader possibilities for mankind in organizing communications. Sputnik I carried two radio transmitting devices which operated simultaneously.[14] Since then, the multiaspect employment of artificial satellites, including for purposes of long-distance communications, has been extended and improved. Today, communication satellites ensure highly reliable, economical, and expedient transmission of a large volume of information varied in form and character. This information can be transmitted by correspondents separated by great distances. Thus, communication satellites can meet the most exacting requirements of various users. In this connection, the statement of Kurt Waldheim, ex-chairman of the UN Committee for Exploration and Use of Outer Space for Peaceful Purposes and the present UN Secretary General, made in 1968, is noteworthy. He said that the real possibilities of a satellite global communication system were such that there was hardly a field of activity that would not benefit from it, either directly or indirectly.[15]

Bearing in mind the tremendous potentialities of satellite equipment for the organization of global communication, it is easy to see the reason for the growing interest in it on the part of the mobile-transport services, including the marine-communication service, which can be reorganized on a fundamentally new basis. Indeed, experimental data and the practical experience which has been accumulated in the use of space equipment in this field have shown that artificial satellites can make communications readily accessible to ships at sea regardless of the distance to the nearest shore, that is, in the whole of the world ocean from 82° north latitude to 82° south latitude.[16] The reliability of marine communications increases immeasurably. Artificial satellites are practically immune to all sorts of interference. Besides, the time needed to establish contact ("waiting" or "delay") is reduced to only a few minutes.[17] The number

12. Dorian; Klass, p. 52.
13. A. D. Ursul, *Osvoenie cosmosa (philocophsko-metodologicheskie i socialnie problemy)* (Moscow: Izdatelstvo "Misl," 1967), p. 68.
14. *Pravda* (October 5, 1957).
15. United Nations, General Assembly, (A/AC. 105/L. 44), p. 33.
16. Dorian.
17. Ibid.

of working channels in satellite communications[18] is steadily increasing. This helps remove the load from the cable and radio-relay lines and the strain from the radio-frequency spectrum which is now in use. Satellite communication facilities enable the marine-communication service to extend the number of services made available to clients through the use of the telephone, teletype, phototelegraph, and high-speed data-transmission techniques (i.e., weather, navigational data, and information on market conditions, etc.). The transmission of such information is essential for the normal functioning of the fleet. More than that, these services help intensify the performance of the fleet.

The possibilities offered by space equipment enable the ship at sea to use it not only as a means of communication, but also as an important navigational aid. What is meant is that satellites outfitted with radio beacons or satellites serving as passive retransmitters (as long as they are in a geostationary or geosynchronous orbit) may be employed as constant markers for position finding, that is, radio position fixing at sea.[19] With the help of satellites, it is possible to conduct these operations with practically no interruption and with a much higher degree of accuracy than before. Position fixing can be carried out regardless of the hour of the day or the weather.[20] This is of special importance not only to cargo vessels, but also to research ships and fishing fleets.

It is necessary to point out that satellite communications which are characterized by the above possibilities could play a great part in advancing the safety of navigation. They can be used directly for preventing distress at sea (to provide urgent information on the situation), for location of ships in distress, for organization of rescue thereof, and for coping with such situations.[21] Finally, the introduction of satellite communications into the merchant service will help improve the services made available to passengers and the working conditions of the crews by satisfying in maximum degree their cultural and everyday requirements. This, in turn, will ultimately promote the solution of the acute problem of fluctuation of crews in the fleet.[22]

Thus, the use of space equipment for the needs of marine shipping will help effectively meet the most exacting requirements of the fleet-administration service and the service of safety of navigation and improve the

---

18. In the beginning of 1969 the number of such channels over the Pacific, Indian, and Atlantic Oceans was 571; in February 1970, 1,423 (*Telecommunication Reports* 36, no. 34. [1970]: 4–6); by December 1970, 9,000 (Charyk). In the opinion of experts, this quantity may be increased 10 times (Charyk).

19. G. Gleadle, "Satellite Communication in Maritime Service," *Telecommunication Journal* 38 (May 1971): 348–53.

20. R. B. Kershner, *Status of Navy Navigation Satellite System* (Baltimore: Johns Hopkins University, 1969).

21. D. G. Pope, "Possible Earth Station Techniques for a Survival Craft Distress Service by Satellite" (paper presented at the IEE Conference on Earth Station Technology, October 1970); W. R. Crawford, "Satellite Alarm Rescue" (paper presented at the North Atlantic SAR Seminar, New York, October 1970).

22. Klass, p. 53.

performance of marine transport, raising it to a qualitatively new stage.[23] This will lead to a revolution not only in marine communications, but above all in the merchant service and other sea services.

However, it is possible to realize the idea of employing artificial Earth satellites in the interests of shipping only with the establishment of an efficient international service. The organization of such a service will require many states to contribute to the effort by pooling the available technical and economic (including financial) resources. This will call for close cooperation in all spheres of interstate relations wherever it may prove necessary.[24] And, of course, to solve this problem it will be essential from the outset to conduct scientific research and then to tackle a whole complex of legal problems, including above all international legal and organizational problems.

PREPARATORY MEASURES

As soon as the space age set in, efforts were begun to study and solve the organizational, legal, technical, and operational problems created by the need to employ artificial Earth satellites for long-distance communications in general and marine communications in particular. They have been undertaken at both national and international levels. The measures and decisions taken by the United Nations and its specialized agencies, such as the Inter-governmental Maritime Consultative Organisation (IMCO) and the International Telecommunications Union (ITU),[25] or under their aegis, have played and are still playing a large positive role.

Thus, way back on December 12, 1959, the UN General Assembly passed Resolution 1472 (XIV) on the establishment of a Committee for Space, which has contributed to the elaboration of international legal rules applicable to marine satellite communications, too. In Resolution 1721 (XVI), adopted on December 20, 1961, the General Assembly pointed to the need to make satellite communications available to all nations of the world on a nondiscriminatory global basis. Finally, in Resolution 1802 (XVII) of December 14, 1962, the General Assembly admitted that " ... the use of scientific and technical achievements in outer space, in particular in the field of meteorology and means of communication, may give mankind tremendous advantages and promote economic and social progress." It further stated that "satellite

---

23. Koval et al., pp. 78–80, 176–80.
24. M. E. Volosov, A. L. Kolodkin, and N. D. Smirnov, "IMCO i nekotorie pravovie voprosi sozdaniya mejdunarodnoy slujbi morskoy sputnikovoy sviazi," *Materiali po morskomu pravu i torgovomu moreplavaniyu (Normativnie aktii praktika, kommentarii)*, viousk 6 (Moscow: Izdatelstvo "Transport," 1974), pp. 24–25.
25. A. H. Abdel-Gani, "The Role of the United Nations in the Field of Space Communications," *Telecommunication Journal* 38 (May 1971): 393–96.

communications would extend transmissions with the help of the radio, telephone and television."[26]

The ITU took a series of decisions and carried out a number of measures which were of great importance to the problem. Their purpose was to support and advance international cooperation in the rational employment of all types of long-distance communications, including marine communications, and to promote the development of means of communication and more efficient operation thereof in this specific field. In 1967, ITU held a World Administrative Conference on Radio Communications (WACR-67), which appealed to the governments of the member countries of the union to determine the tentative operational requirements for the maritime-communication services which would employ space-communication means. The WACR-67 adopted a special recommendation addressed to IMCO, requesting the latter to organize investigations with a view to establishing the requirements for communication satellites to ensure the safety of shipping at sea and improve navigation.[27] In 1971, the World Administrative Conference on Radio Communications (WACR-71) paid special attention to the employment of space-communication facilities by the maritime services. The WACR-71 accepted a special recommendation proposing that the international organizations concerned, including IMCO, and the national telecommunication administrations should elaborate the requirements for the mobile maritime-communication service and study the question of distribution of frequencies to meet these requirements.[28]

Proceeding from these decisions and in pursuance of these measures which were adopted or carried out by the United Nations and ITU, several countries have set up national satellite-communication systems. Corresponding international organizations—INTELSAT[29] and INTERSPUTNIK[30]—were founded. Though the aforementioned and other measures created a precedent, and though the above organizations could serve as prototypes in the solution of organizational and legal aspects of the problem of employment of artificial satellites for purposes of navigation, the problem of ensuring regular communication for marine shipping remained unsolved. It was necessary to solve it.

As an organization specially authorized to "ensure the mechanism of cooperation between the governments" in the sphere of legal regulation and

---

26. U.N. General Assembly Resolutions adopted at its seventeenth session, New York, 1963, pp. 6–8. Roman numerals in parentheses indicate session number.

27. Resolution of WACR-67, no. MAR 3.

28. Resolution of WACR-71, no. SPA 11.

29. Abbreviation of "International Telecommunication Satellites." See J. Johnson, "The INTELSAT System," *Telecommunication Journal* 38 (May 1971): 270–78.

30. Abbreviation of "Internazionalnie Sputniki." See W. S. Vereschetin, *Cosmos. Sotrudnichestvo. Pravo* (Moscow: Izdatelstvo "Nauka," 1974), pp. 109–14.

adoption of measures bearing on all sorts of technical matters in international merchant shipping,[31] IMCO engaged in this work without limiting itself solely to the study of the technical aspects of the problem. Thus, IMCO undertook a thorough study of the possibilities and conditions for the creation of a worldwide system of marine satellite communications. At its twenty-third session, in 1977, the IMCO Committee for Safety at Sea passed a decision obliging IMCO from the outset to play an active part in the creation and commissioning of a maritime satellite system. In pursuit of this task, IMCO was to work out an international agreement.[32]

Subsequently, the committee pointed to the international character of the mobile maritime services and, therefore, to the need for all-around cooperation in this field among all interested states taking part in it.[33] Proceeding from this, the committee passed a decision on the elaboration of an organizational plan for commissioning an international maritime-communication satellite system.[34] The purpose of the contemplated measures was to prepare a draft of a formal international agreement on the establishment and functioning of a maritime-communication system with the use of artificial Earth satellites and on the founding of an international organization to operate it.

In keeping with the plan, the IMCO Subcommittee on Radio Communications formed a special working group in 1972, which was later transformed into the Group of Experts on Maritime Satellites. In addition to conducting a thorough all-around investigation of the technical, economic, and operational matters, the chief purpose of the group was to elaborate in detail the organization and legal aspects of the problem, that is, questions bearing on the establishment of a new international mechanism or the employment of an existing mechanism for the creation, development, financial backing, operation, and subsequent improvement of a maritime-communication satellite system.[35]

Acting on the opinion of the overwhelming majority of the specialists represented in it, the group of experts concentrated its attention chiefly on the creation of a new international organization, formulation of the principles of representation of interested states in it, and finding of financial resources for its functioning.[36]

In November 1972, the group of experts directly proceeded with the elaboration of the organizational and legal fundamentals of the future system and the international organization to be placed in control over it. The Soviet specialists had drafted a document entitled "Tentative Principles for the

31. See Article 1(a) of the Convention of IMCO.
32. IMCO, (MSC XXIII/19), par. 75.
33. IMCO, (MSC XXIV/5/2), par. 65.
34. IMCO, (MSC XXIV/WP. 4), par. 39.
35. IMCO, (COM IX/12), par. 30.
36. IMCO, (MARSAT 1/7), sec. 7, pars. 23–24.

Establishment of an International Organisation for a Marine Satellite Communication Service and Radio Position Finding (INMARSAT),"[37] which they submitted to the group for consideration and which the group accepted as a basis for discussion. Most of the cardinal provisions of the document were accepted as "Draft Articles of a Convention"[38] and later as the "Convention on the Establishment of an International Marine Satellite Communication System." This document reflected the basic questions bearing on the practical management of the technical services for the proposed system and coordination of the work of the future organization with the measures conducted by IMCO, ITU, and other organizations functioning in related fields. Among the most important provisions of the Soviet document are the articles on the international legal fundamentals of the new organization, the status of the ground-based and space elements of the system, functions and structure of its organs, rights and duties of its members, etc. In conformity with the tentative principles, INMARSAT was to (1) comply in its work with the universally recognized principles and rules of international law, including the corresponding provisions of the UN Charter, multilateral agreements relating to the exploration and use of space, the Geneva Conventions of 1958 on the Law of the Sea, etc.; (2) contribute to ensuring safety at sea and improvement of methods of fleet control, while ruling out the possibility of employment of the available technical means for military purposes; (3) act on the basis of the widest possible exchange of information and with account of the universally accepted technical requirements, conditions, criteria, and standards approved and recommended by the corresponding international organizations, such as ITU, with respect to long-distance communications; (4) have the right to conclude agreements and contracts, acquire, rent, and alienate property, undertake pleadings, and bear international responsibility for its commitments; (5) accept members regardless of the social, political, economic, racial, or other features characterizing one interested state or another; and (6) grant access to all forms of services it provides to ships of all nations, regardless of whether the states whose flag they are flying are members of INMARSAT or not.

The world's first socialist state has been and is pursuing a Leninist foreign policy. As L. I. Brezhnev pointed out, the Soviet state "cannot stand aloof from the solution of . . . problems affecting the interests of all mankind."[39] Thus, as a reflection of the general line of such a foreign policy, the said Soviet initiative has again convincingly demonstrated the USSR's readiness for broad cooperation with all states on principles of equality and mutual advantage. At the same time, this initiative has revealed the Soviet Union's firmness and consistency in

37. IMCO, (MARSAT ES. 1/36).
38. IMCO, (MARSAT V/6).
39. See *Materialy XXV sezda KPSS* (Moscow: Izdatelstvo politicheskoy literaturi, 1976), p. 56.

steering a course ensuring that all participants in the international association observe progressive and democratic principles and standards of present-day international law, not only in the settlement of political, but also, in equal degree, of economic matters in international life.

It would be no exaggeration to say that this initiative largely determined the general line and progressive political comtent of the work both of the Group of Experts on Maritime Satellites[40] and subsequently of the Conference on the Establishment of an International Maritime Satellite Communication System. The draft convention on INMARSAT produced by the group of experts at its six sessions offered a positive solution to the task of setting up an independent international organization of interested states, an organization that would not be subject to the influence of any international or national monopoly association. The draft accepted the system of INMARSAT organs proposed in the Soviet document. In the opinion of the group, the system should comprise a principal representative body—an assembly which should include all the INMARSAT members; the other principal body—a council consisting of the limited number of members of the future organization shouldering the main financial burden in the realization of measures planned and executed by the organization; and a responsible executive body—a directorate to be headed by the director general of INMARSAT.

The documents adopted by the group of experts have not introduced any major changes into the Soviet proposals on the functions of these agencies. The fundamental functions are: formulation of the general policy of the new organization (assembly), elaboration and approval of the plans and measures for the creation, development, acquisition or leasing, and operation of the elements of the communication system and adoption of decisions on financial matters (council), and practical implementation of the policy and realization of all measures to be carried out by the organization (director general).

The draft convention on INMARSAT that the group of experts has drawn up contains other provisions of the aforementioned Soviet document, namely, on the universal and nondiscriminatory character of membership in INMARSAT and on the accessibility of the services it grants to users of all nationalities, with the financial burden to be shouldered by the member states depending on the degree of use of services. On general agreement, the draft contains a provision in keeping with which INMARSAT could own a space segment, that is, the satellites and technical facilities essential for the normal functioning of the system, or could lease it from states having such facilities. As to space-communication stations based on the land ("Earth land stations") which

---

40. A. L. Kolodkin and M. E. Volosov, "Some International Law Questions on the Establishment of an International Maritime Satellite Communication," *Proceedings of the Seventeenth Colloquium on the Law of Outer Space. International Astronautical Federation* (New York: Davis, 1975), pp. 221–23.

would be capable of using INMARSAT satellites, the draft stipulates that they should be owned by the member states of the future organization. Finally, it was recognized that the work of the marine satellite-communication service and the international mechanism administrating it should be closely coordinated with the measures, decisions, recommendations, standards, and criteria adopted or executed by IMCO and ITU.

However, it should be mentioned that, though the general result of the work of the group is positive, there have been a few negative points too. Thus, the draft convention on INMARSAT did not obtain the unanimous approval of the members of the group of experts on account of the position of the U.S. delegation. Referring to its internal legislation, in conformity with which communications belong to the sphere of private enterprise, the U.S. delegation refused to accept on behalf of its country any financial commitments arising from the agreement on INMARSAT. This, in short, was its position. That was why the U.S. delegation proposed to transfer the function of control over the marine satellite-communication system to INTELSAT, which was actually under the control of Comsat General,[41] a national U.S. corporation. Another solution the U.S. delegation was prepared to accept consisted in the establishment of a specialized international consortium with the participation of both national and international organizations, including purely commercial organizations.[42] It is obvious that such an organization, which would be known by the noteworthy name of INMARCOMSAT,[43] could give rise to a real danger of monopolization of marine satellite communications in the whole world by private companies, U.S. companies above all. It appears that this is the reason why the majority of the participants in the group of experts did not support the idea advanced by the U.S. delegation.

The general course of the studies of the concrete questions and the positive prospects which became evident in the work of the group of experts gave them reason to make an optimistic appraisal, after the first 3 yr of work, of the practical possibilities for solving the problem of creating an international marine satellite-communication system and a corresponding organization to run it. Proceeding from this, the IMCO Assembly adopted in November 1973 a decision on the convocation in 1975 of a special diplomatic conference which was to look into this question from all angles and to take final decision on it.[44]

41. *Aviation Week and Space Technology* (April 21, 1975).
42. IMCO (MARSAT/CONF/3), October 30, 1974, p. 93.
43. Abbreviation of "International Maritime Communication Satellites," formed under the name of the American space-communications corporation, "Comsat."
44. IMCO/ Resolution A.305 (VIII) (November 23, 1973).

CONSIDERATION OF THE QUESTION ON THE ESTABLISHMENT OF INMARSAT AT A CONFERENCE OF GOVERNMENTS AND ADOPTION OF CONSTITUTING ACTS

The conference for the establishment of INMARSAT, convened under the auspices of IMCO, opened in London on April 23, 1975. Delegations of 54 states took part in the work of the conference. The representatives of 16 international organizations whose work is connected with the use of the world ocean or outer space, including representatives of ITU, the World Meteorological Organization, the European Space Agency, INTELSAT, and the International Air Transport Association, attended the conference as observers. The conference held three sessions: in April–May 1975, February–March 1976, and September 1976. They lasted a period of about 40 days and ended in the adoption of a Convention on the International Maritime Satellite Communications Organization (INMARSAT) and an Operational Agreement on the International Maritime Satellite Communications Organization (INMARSAT).

The main result of the first session of the conference was that it adopted a fundamental resolution on the need to create a marine satellite-communication system which is to be administered by a newly established special intergovernmental organization.[45] The session revealed disagreements between the various groups of participants in the conference on the organizational principles of the future system and the international mechanism for its control. It has been pointed out above that these contradictions were already apparent while the IMCO Group of Experts on Maritime Satellites was at work. As a result, the first session failed to reach agreement on the concrete wording of the articles of the draft convention on INMARSAT. To achieve a compromise and to reformulate the wording of the constituent act of INMARSAT, the conference formed an Intersession Working Group.

The main task of the intersession working group was to make compatible the vital organizational principle of the new organization supported by the majority of the interested states (including the socialist and developing countries), which was to set up INMARSAT as an independent organization of governments, and the principle which would permit, in addition to governments, the participation of national operational competent organizations (including private organizations)—entities designated by their respective governments.

After a series of consultations held both within the framework of the intersession working group and at unofficial meetings of representatives of a number of states, it was decided that fundamental questions of the work of the maritime satellite-communication system should be settled by the states—

---

45. Final act of first session of Conference on INMARSAT.

members of the future organization—whereas the practical matters of financial, technical, and operational character should be solved in INMARSAT by the representatives designated by competent entities. Owing to this, the question of responsibility arising from participation in INMARSAT with respect to the former and latter should be decided differentially. This is essential to relieve the governments of this responsibility in certain cases. At the same time, each government shall play the role of chief guarantor exercising control over the work of the organizations it designates. Thus, it was made binding on the government of every state—"Party to the Convention on INMARSAT"—to issue to the organization it designates pertinent instructions which would be compatible with the domestic legislation of the given country. This condition is necessary to ensure the fulfillment of commitments accepted by the designated organization as the participant—the signatory—in the special constituting document of the future organization—the Operating Agreement on INMARSAT. Besides, the government of every member state was obliged to see that in its work the organization designated to act as participant in the operational agreement should not violate the requirements confronting the state concerned as party to the convention on INMARSAT and as cosignatory of other international agreements, the UN Charter above all, and also the Treaty on Principles Governing the Activities of States in the Exploration and Use of Outer Space, including the Moon and Other Celestial Bodies, of January 27, 1967, the Convention on International Liability for Damage Caused by Space Objects of March 29, 1972, the Geneva Conventions on the Law of the Sea of 1958 now in force, and other international agreements concerning navigation, etc.

It should be pointed out that the above provisions, later consolidated in Article 4 of the Convention on INMARSAT, reflect the universally accepted principles of contemporary international law. In keeping with these principles, the states shall bear international responsibility for national activities in outer space, regardless of whether they are conducted by governmental or nongovernmental bodies of the corresponding country. More than that, it is binding on the states to exercise constant control over the work of these bodies in outer space. These principles (as is known) were proclaimed in the declaration on the basic principles governing the activities of states in the exploration and use of outer space adopted by the UN General Assembly on December 5, 1963, and in the aforementioned Outer Space Treaty of 1967.

In compliance with its mandate,[46] the intersessional working group was to carry out the aforementioned tasks, and, in addition to this, it was to draw up two new constituting documents—a Convention on INMARSAT and an Operational Agreement on INMARSAT—proceeding from an earlier draft convention. It was considered that only states would be subjects of ("parties to") the convention. At the sime time, either the states themselves or national

---

46. IMCO, (MARSAT/CONF/10), May 9, 1975.

organizations designated by the governments concerned ("signatories") would be subjects of the operational agreement.

Thus, the main aim formulated by the first session of the conference was hereby achieved, namely, to make compatible the different approaches to the question of the participants in INMARSAT. The purpose was ensured by introducing a special principle in keeping with which both states and organizations designated by them could be subjects of the newly formed organization. If it is a state that is a participant in the operational agreement, it will voluntarily assume the duties normally performed by an organization which acts as a legal person. It will bear liability for the fulfillment of commitments arising from the operational agareement and will agree in advance to have disputes (which may crop up in connection with INMARSAT activities) referred to arbitration, even if the plaintiff happens to be a government-designated entity. If an originally designated organization withdraws from the operational agreement, the government which authorized it shall designate another national competent entity to act as participant in its stead or shall itself act in that capacity. The government concerned shall thus meet the requirements of the said principle. If the government fails to take these measures accordingly, it will be considered to have withdrawn from INMARSAT.[47]

The intersessional working group included the above provisions in the draft convention and draft operational agreement on INMARSAT in the form of concrete articles. Upon completion, the documents were submitted to the conference for consideration.

The second session of the conference adopted all the articles of the operational agreement and nearly all the articles of the convention. The session failed to reach an agreement on the text of three articles: on the weight carried by the votes of the representatives of the signatories in the INMARSAT Council, on the official and working languages of the organization, and on the possibility (or impossibility) of making reservations with respect to the convention. That was why the final adoption of the constituting documents on INMARSAT was to be made by the third session of the conference which, as pointed out above, was held in London over September 1–3, 1976, lasting only 3 days. The third session ended in the adoption of the Convention on INMARSAT and the Operational Agreement on INMARSAT.

The text of the Convention on INMARSAT had been done in the Russian, English, Spanish, and French languages. As of May 1, 1977, the representatives of 14 states had affixed their signatures to the convention, those of the USSR, the Ukrainian Soviet Socialist Republic, the Byelorussian Soviet Socialist Republic, Poland, Bulgaria, the United Kingdom, Norway, Greece, Australia, Japan, Liberia, Iran, Kuwait, and Cameroon.

Since then, it has been necessary to create INMARSAT in fact. This calls

47. IMCO, (MARSAT/CONF/27 1/2), May 1, 1975.

for the execution of a series of practical measures after the adoption of the constituting acts till they acquire legal force. This task was to be fulfilled by a special preparatory committee, formed in pursuit of Resolution 2 which was adopted by the second session of the conference.[48]

The representatives of the governments which have signed the convention and operational agreement and the representatives of designated entities which have signed the operational agreement can take part in the work of the preparatory committee. The representatives of states or designated entities could also be permitted to participate in the work of the committee, provided the governments of the corresponding countries have started to carry out the procedural measures with a view to joining INMARSAT. To take part in the work of the preparatory committee, the state concerned shall forward an application in writing to the secretary general of IMCO.

The program of works[49] to be carried out by the committee includes studies in standards and technical requirements to ship-based and ground-based stations on the Earth, possible requirements to and versions of the space segment and its technological and operational parameters, appraisal of the potential demand for marine satellite-communication facilities, drawing up of proposals on the structure, functions, and powers of the directorate, drafting rules of procedure of the assembly and council, establishing contacts with the country of the seat of the organization, and drawing up of the draft protocol on the privileges and immunities to be enjoyed by INMARSAT.

The first session of the preparatory committee was held in January 1977 and the second in May 1977 at the IMCO headquarters in London.

LEGAL STATUS OF INMARSAT AND ITS ORGANS

Aims and Principles Governing the Work of INMARSAT

Under the terms of the convention, the purpose of INMARSAT is to make provision for the space segment necessary for improving maritime communications, thereby assisting in improving distress and safety-of-life-at-sea communications, efficiency and management of ships, maritime public corresponding services, and radiodetermination capabilities (Article 3).

The convention states that INMARSAT will seek to provide service in all areas of the world where there is need for maritime communications. The organization shall act exclusively for peaceful purposes. However, this does not mean that in peacetime warships will not be able to avail themselves of the services of INMARSAT's space segment.

48. Articles 29-3 and 30-6 of the Convention on INMARSAT.
49. IMCO, (MARSAT/CONF/27), Annex 1.

The organization can acquire a space segment of its own or rent one. The space segment of INMARSAT will be available for use by ships of all states, regardless of whether they participate in the organization or not.

The Earth land stations shall be owned by the states which are parties to the Convention on INMARSAT and shall be located within their respective territories.

The fund for the acquisition of the space segment and maintenance of the organization, initial capital amounting to at least U.S. $200 million, shall consist of contributions made by the signatories to the operational agreement. The contribution to be made in money shall be determined on the basis of the share of the traffic put through the space segment by the ships and the shore stations of the state concerned.

The INMARSAT shall operate on a sound economic and financial basis, with regard to accepted commercial principles. To this end, the organization will charge for the services it grants at rates set by the INMARSAT Council. The overall sum of the revenues derived from the charges should cover the operating costs and adminstrative expenses of the organization, costs of technical maintenance of the facilities, replenishment of operating assets, repayment to the participants of their contributions to the capital, and the interest on them. It follows that INMARSAT will operate on the basis of self-recoupment. It is quite possible that the organization shall derive a profit on invested capital.

However, this does not mean that INMARSAT will cease to be an organization which is intended to promote progress in marine transport in order to become a profitable commercial enterprise with profit and superprofit being ends in themselves. Such an interpretation of the aims of INMARSAT does not in any way coincide with the concrete provisions of the organization's constituting documents. In particular, this is in contradiction to the provisions of Article 8 of the operational agreement. In keeping with this article, the signatories to INMARSAT are entitled to "remuneration for use of their capital," and not to "profit" on invested capital. Besides, this remuneration will be paid only after the revenues accruing to the account of the organization have covered all its expenses which are made for the humane purposes set forth above. Therefore, it is hardly possible to accept the description of INMARSAT as "an international enterprise whose activities are directed at acquiring profit"[50] or to agree to attempts to identify it with some of the purely commercial international organizations, such as the INTELSAT consortium.[51]

---

50. R. F. Preuschen, "INMARSAT—eine weltweite Satellitenorganisation für die Schiffert à la INTELSAT," *Zeitschrift für Luft- und Weltraumrecht* 3 (1976): 1–6.
51. Ibid.

Questions Bearing on Financial Backing and Participation in INMARSAT

The conference decided to fix in percent the initial contribution each signatory to the organization should make. The distribution was based on information on the volume of traffic currently handled with the help of conventional means. These data being approximate, the delegations declared that their governments were prepared to contribute a certain share expressed in percent to provide initial financial backing for the organization. This apportionment was laid down in the Annex to the Operational Agreement and is regarded as its integral part. In keeping with this annex, the initial capital of INMARSAT will be made up of contributions from 42 signatories, the share of the United States being 17 percent, the United Kingdom, 12 percent, the joint share of the USSR, the Ukraine, and Byelorussia, 11 percent, etc. It has been further established that the share of any signatory cannot be less than 0.05 percent of the total investment shares.

Initially, the council will consist of the representatives of 23 signatories, including 19 on the basis of the principle the biggest shares contributed and four on the principle of just geographical representation. This composition is temporary, as will be explained below.

The Convention on INMARSAT can come into force only if adequate financial backing is provided for the organization. In conformity with Article 33, the convention will become valid 60 days after the states concerned contribute 95 percent of the investment shares. (The states jointly contributing this sum will acquire the status of parties.) This means that the minimum initial investment in the organization should be U.S. $190 million. If it is impossible to collect this sum in the course of 3 yr after the convention is open for signing (i.e., by September 3, 1979), the convention shall not come into force. The purpose of this provision is to prevent the freezing of the capital of the parties and participants who intend to take part in the convention but will not be able to provide the total sum consisting of separate contributions that will be needed to create the proposed system.

All interested states can become parties to the convention. To do so, a state has to sign the convention, to sign it upon its ratification, adoption, or approval, or, finally, to join it after it comes into force (Article 32). It should be mentioned that no state can become a party to the convention until the state concerned or the organization designated by the government of the said state has signed the operational agreement (Article 32, par. 4). At the same time, under Article 17 of the operational agreement the latter will come into force for the participant the day the convention becomes valid for the party concerned under Article 33 of the convention.

Leading Bodies of INMARSAT

Under the terms of the convention, an assembly, council, and directorate shall be set up to effect guidance over the organization and the marine satellite-communication system.

The assembly consists of all the parties to the convention. Each party has only one vote in the assembly. The functions of the assembly are to lay down the general policy and aims of the organization, to work out recommendations to the Council of INMARSAT, and to adopt decisions on the recommendations the council submits to it on matters bearing on the relations between INMARSAT, the states, and international organizations and on other questions of principle. The assembly shall hold sessions once in 2 yr.

As a rule, the Council of INMARSAT shall consist of 22 signatories in the operational agreement. Eighteen representatives are members of the council because they have made the biggest financial contributions to INMARSAT. Four other representatives are elected by the assembly in pursuit of just geographical representation in the council. (This principle has been reflected in the Convention on INMARSAT on the initiative of the USSR delegation.)[52]

There can be more members of the council, if two or more participants claim a vacancy because they have made an equal contribution to the financial fund of the organization. Precisely such a situation is taking shape in the council which has been formed for the first time. It has already been stated that this council will have 19 participants represented on it on the principle of the biggest financial contribution, the contribution of the eighteenth and nineteenth participants being equal. In later years, such a situation will take shape only in an exceptional case, because the financial contribution of every participant will be based on an accurate estimate of the actual use of INMARSAT communication channels by the ships of each nation. That these indicators may coincide is hardly probable.

These calculations will be made every year on the basis of actual information on the use of services made available by the system during the preceding year. It follows that some of the members of the council included in it on the principle of the biggest financial contribution may automatically change every year. As to the four members elected in addition to the above, they can change only once in 2 yr, because they are elected at the regular sessions of the assembly. (Though extraordinary sessions of the assembly may possibly be held, they will not conduct elections to the council.)

It has already been said that, initially, the council will have 19 members. According to the table of initial contributions, the members of the council will be the representatives of the United States, the United Kingdom, the USSR,

---

52. IMCO, (MARSAT/CONF/C. 1/WP.11); (MARSAT/CONF/5/5).

Norway, Japan, Italy, France, the FRG, Greece, the Netherlands, Canada, Spain, Sweden, Denmark, Australia, India, Brazil, Kuwait, and Poland. The first session of the assembly will elect four more members. The membership of the council shall remain unchanged for 2 yr, till the next regular session of the assembly is held. At this session, the assembly will reelect these four members. These representatives will remain members of the council indefinitely. The reason for this is that under paragraph 4, Article 5, of the operational agreement, the shares to be contributed will be determined for the first time only 3 yr after the INMARSAT space segment for the Atlantic, Pacific, and Indian Oceans has been put into operation.

Before the space segment for the Atlantic, Pacific, and Indian Oceans has been put into operation, the participants shall shoulder the expense of maintaining the organization in keeping with the above apportionment of initial contributions. However, the first membership of the council may change long before that. This may occur if a group of participants not represented on the council pools their resources to have one common representative on the council. If the joint contribution of such a group is greater than that of some individual member of the council (this refers only to those members of the council who have entered it on the basis of the biggest financial contributions), the latter will be compelled to leave the council and cede his seat to a group representative. Since an application on group representation may be made at any moment, the composition of the council may change even before it holds its first meeting.

The functions of the council are essential to the practical work of INMARSAT. Thus, the council is responsible for the functioning of the space segment, which is vital for the realization of the aims of the organization, in the most economical, effective, and expedient way. It is also authorized to formulate the requirements to the marine satellite-communication system, to decide on the criteria and procedure for the issue of licenses for access to the space segment, and to determine the financial policy of the organization. The INMARSAT Council will hold sessions at least three times a year.

The procedure for the adoption of decisions in the council is rather complicated. When the pertinent provisions of the convention were drawn up, the delegations of the Western countries defended a view of their own. They maintained that the organization should be managed by those who provide the financial backing, that is, by the members that are most concerned with the success of INMARSAT's financial policy. The principle of so-called weighted voting was advanced in the council. In accordance with this principle, the votes of the participants in the council should be distributed among them in proportion to their financial contributions. The Soviet delegation maintained that membership in the council of participants who made the biggest contribution to the organization's fund would adequately ensure this goal. Therefore, the Soviet delegation thought that it was not necessary to subdivide further the rights of the members of the council in proportion to their respective contributions. The USSR representatives officially proposed that every member of the council should have one vote.[53] At a plenary meeting of the conference held on April 25, 1975, they stated that when the council passes a decision the principle of equality of all its members should be applied.[54]

It is a pity that this proposal failed to obtain adequate support. Seeking to achieve a constructive solution of the problem of establishment of INMARSAT for the early employment of satellite facilities for navigation, the Soviet delegation thought it advisable to accept a compromise in this matter. The Soviet representatives said later that, though in principle the USSR upholds the principle of equality of all members of international organizations, in this special case they were prepared to accept the proposal on weighted voting in the council, bearing in mind the specificity of the work of INMARSAT. However, they declared that this should not be regarded as a precedent.[55] To avoid the situation that has taken shape in INTELSAT, which is an international consortium where more than one-third of the votes in the leading body belonged to the U.S. representative and where, as a result, no decision requiring a competent majority vote can be taken without his consent, the delegations of the USSR, the Ukraine, Byelorussia, Bulgaria, Cuba, Czechoslovakia, the GDR, Hungary, and Poland submitted a proposal to the effect that one representative in the INMARSAT Council cannot control more than 20 percent of the votes.[56]

This proposal was received with satisfaction and supported by the representatives of the developing countries and several West European states. As a result, 15 delegations voted for the proposal, 15 abstained, and only 13 voted against it. The U.S. delegation pointedly insisted on the participants in the council wielding an unlimited share of votes. As a compromise, it proposed to set the limit at 30 percent.

The West European delegations dared not vote against this proposal, because they were afraid to lose a big contributor to the fund of the new organization. If the United States should withdraw, the remaining contributors would have to put in an extra $34 million. At the same time, they were not in a position to oppose the proposal of the socialist countries, because the latter were prepared to invest in INMARSAT 14 percent of the funds the organization needed. Besides, the West European countries were counting on the USSR's support, because its position as a big space power would save them in INMARSAT from the role of unequal partner they were playing in INTELSAT. That was why this group of countries used their votes at the

53. IMCO, (MARSAT/CONF/5/5).
54. IMCO, (MARSAT/CONF/SR.5), p. 3.
55. IMCO, (MARSAT/CONF/SR.27), p. 25.
56. Weighted voting is used effectively in other international organizations, such as in the nongovernmental Baltic and International Maritime Consultative Organization (BIMCO), each member of which has a number of votes depending on the amount of its investment share.

conference to torpedo both proposals. As a result, the second session of the conference failed to reach agreement on this question.

A series of consultations was held in preparation for the third session of the conference. At these consultations, a compromise formula was agreed upon. It stated that the weight of a vote of a single representative in the council from any participant should not exceed 25 percent of the total number of votes. If the share of any representative happens to exceed that limit, he can invite the other participants to buy out that surplus, or part of it, to increase the weight of their votes. If the other participants refuse to take the offered surplus or agree to take only a part of it, the weight of the vote wielded by the holder of the surplus will be allowed to surpass the set limit in proportion to the undistributed part. Paragraph 3 of Article 14 of the convention was included in it in this wording.

The directorate of INMARSAT is the executive body of the new organization. It is headed by the director general, who is appointed for a term of 6 yr. Though this matter was hotly debated at the conference, this provision of the convention may be considered acceptable because it does not prevent the governments of the INMARSAT member countries from expressing their attitude toward the nominee for the post of director general. The director general, appointed by the council, will be considered approved only if more than one-third of the INMARSAT members raise no objection to the candidacy in the course of 60 days after the Depositary of the Convention has notified them of the appointment.

The Space Segment, Technical Recommendations, and Use of Services Rendered by INMARSAT

In actual fact, the INMARSAT system will start functioning as soon as the space segment has been created. The convention defines the space segment of INMARSAT as satellites, tracking installations, telemetry, telecontrol, control, observation, and other equipment and facilities connected therewith and essential for the functioning of communication satellites either owned by the organization or operated by it on lease.

Under the Convention on INMARSAT it is the council that is responsible for ensuring the functioning of the space segment. In executing its duties, it must take into account the opinions and recommendations of the assembly. It shall deal with all matters bearing on design, development, construction, housing, acquisition (through purchase or hire), operation, technical maintenance of the space segment, and acquisition of all services necessary for the launching of satellites. When the organization finds it advantageous, it can conclude contracts for the fulfillment of technical and operational functions. These functions shall be performed by contractors with the corresponding capabilities.

The provisions of the convention on the acquisition by INMARSAT of property and services (Article 20) proved to be one of the most acute issues debated at the conference. The interests of the industrial and other business circles of the United States clashed on this point with those of Western European countries. The West European delegations tried to prevent the United States from establishing a monopoly in the delivery of equipment and provision of services needed by the organization. They did their utmost to sell INMARSAT their own products. The West European countries had invested heavily in the manufacture of these products.[57] Upholding common positions, they managed to ensure a favorable wording on the matter.

Thus, it is laid down in the convention that in acquiring property and services the council shall pursue a policy of encouraging worldwide competition in the interests of the organization. To this end, property and services shall be acquired through purchase or hire on contracts concluded at open international tenders. The U.S. delegation, in turn, got the conference to include in the convention the provision which states that such contracts shall be awarded to bidders offering the best combination of quality, price, and favorable delivery time.

This stipulation was counterbalanced by a paragraph included in the convention which states that, if there are several tenders containing comparable combinations of the above criteria, the council shall conclude a contract with account of the interests of the organization to conduct a policy of encouraging worldwide competition. The idea behind this formula is that the council shall purchase property and services from various firms in turn, even if some supplier already cooperating with the organization happens to offer property and services that are equally good as those offered by the new supplier.

In some cases, the council will be permitted to acquire property and services without tenders, provided it encourages competition in the interests of the organization. These cases include the conclusion of contracts to a sum of no more than U.S. $50,000, urgent contracts under emergency circumstances, and contracts for products and services coming from limited sources of supply.

The U.S. delegation has not given up the desire to create a space segment for INMARSAT on the basis of its multipurpose satellite of the INTELSAT series. With this end in view, the U.S. delegation persuaded the conference to adopt a recommendation on the study of organizational, financial, technical, and operational consequences that may arise from the employment by INMARSAT of multipurpose satellites suitable both for navigation and aeronautics.[58] The response to this American move was that the recommendation now points to the need to conduct the study in a way that would not adversely affect the programs for planning and creating a marine satellite-communication system.

57. IMCO, (MARSAT/CONF/SR. 21), p. 12.
58. W. S. Vereschetin, pp. 65–80; Y. M. Kolosov, *Borba za mirny cosmos* (Moscow: Izdatelstvo "Mejdunarodnie otnoscheniya," 1968), pp. 10, 110.

In addition to the above functions, the INMARSAT Council will also decide on the criteria entitling earth stations (i.e., ship and shore stations) to use the space segment of INMARSAT (Article 15 of the convention). Several technical recommendations of this kind were accepted at the second session of the conference. But these were of a general character. Thus, it was recognized that general standardization of specifications for ships' stations should ensure considerable economic, operational, and technical advantages. Therefore, it was recommended that all states—parties to INMARSAT—establish with the help of pertinent international bodies the lowest world technical and performance standards for equipment which should be agreed upon and accepted as a basis for the specifications of ships' stations using INMARSAT services. Another recommendation drew attention to the need to impart a global character to marine satellite communications without discrimination against any nation. This recommendation urged all the parties to the convention to undertake joint steps to establish world technical and performance standards to make possible effective telecommunications between ships' stations, fixed Earth-based stations, and subscribers on the shore. The conference further proposed that the states should look into the question of permitting ships' stations to work on such frequencies as 1535–1542.5 Mc/sec and 1636–1644 Mc/sec while the ships are in waters within the national jurisdiction of foreign states. Finally, the conference proposed to study the organizational, financial, technical, and operational consequences that may arise from the use by INMARSAT of multipurpose satellites capable of meeting the needs of sea and air navigation.

The recommendations adopted by the conference are of optional character. However, the technical criteria to be adopted by the council shall be obligatory. In this particular case, the standards adopted by an international organization shall be obligatory for the states. Initially the technical standards and requirements for the space segment and for Earth-based stations of the INMARSAT system will be worked out by a preparatory committee and then will be transferred to the council.

It has been pointed out that the space segment of INMARSAT shall be open for use by ships of all nations on conditions to be determined by the council. The council shall not discriminate among ships on the basis of nationality (Article 7 of the convention). By "ship" the convention implies vessels of any type operating in the marine environment, including inter alia hydrofoil boats, air-cushion vehicles, submersibles, flotation craft, and platforms not permanently moored.

Scientific and technological progress has led to the creation of various sea installations which, though not sea-going vessels, are operated at sea and are therefore in need of communication services.[59] Such installations may use

---

59. *Sovremennoe mejdunarodnoe morskoe pravo* (Moscow: Izdatelstvo "Nauka," 1974), pp. 283–84.

INMARSAT satellites only with the council's permission. The council will grant such permission in every case only if the work of their stations shall not significantly affect the provision of service to ships.

The Earth land stations through which the ships communicate with the shore shall be established only within the territories of states which are parties to the convention. Other states will be able to have such stations only with the permission of the council.

Under Article 32 of the convention, every state has the right to notify the depositary of the convention on its application of definite registers of ships operated under authority of that state. This does not imply that there will be several registers in a given state.[60] This provision was included in the convention at the request of the British delegation. It was an alternative to another provision the initial draft contained. This provision aroused objections on the part of several delegations (the said provision was on the right of states to make declarations on the applicability of the convention to territories for which these states are responsible with respect to their international relations). The formula which was found does not contradict the principle of granting independence to colonies and peoples and at the same time makes it possible to extend the provisions of the convention to ships listed in the registers. This only points to the highly unpopular character of all manifestations of colonialism in international affairs and in international law.

The question of the right of states—parties to INMARSAT—to create space-segment facilities other than the space segment of INMARSAT gave rise to difficulties during the discussion. The delegations that were eager to make the INMARSAT system a paying concern tried to impose a ban on the members of the organization that would prevent them from creating similar systems without preliminary consultations with the council. In the opinion of the Soviet Union, such a ban would provide legal justification for inteference in the internal affairs of sovereign states. Therefore, the USSR delegation did the right thing when it opposed this ban. In doing so, it won the support of a considerable number of other representatives. Finally it was decided that to ensure technical compatibility and to prevent inflicting any major economic damage to the organization every state which is a party to INMARSAT shall notify it about the intention to create within its jurisdiction independent space-segment means or to use an existing space segment for the same purposes as INMARSAT. The opinion on technical compatibility is to be expressed by the council in the form of an optional recommendation. As to the opinion on the possibility of economic damages, it should be expressed by the assembly, also in the form of a nonobligatory recommendation.

The need to notify INMARSAT in combination with the proposed recommendations is intended to restrict attempts to create competitive marine satellite-communication systems. However, the form in which this condition

60. IMCO, (MARSAT/CONF/SR. 23), pp. 14–15.

has been formulated in Article 8 of the convention will hardly create a serious obstacle to the formation of such systems, first, because this obligation does not concern systems designed to ensure national security and because it does not affect systems created before the Convention on INMARSAT comes into force. Thus, there is nothing to stop the interested countries from using the services in commercial marine communications offered by the INTERSPUTNIK or INTELSAT side by side with INMARSAT. The second reason why this condition presents no obstacle is that it is possible to create new systems, as long as they inflict no economic damage on the organization as competitors in granting paid services in telecommunications.

Questions of Responsibility

It has been mentioned above that in the constituting acts of INMARSAT the questions of responsibility were settled on the basis of universally accepted principles formulated, in particular, in the Treaty on Principles Governing the Activities of States in the Exploration and Use of Outer Space, including the Moon and Other Celestial Bodies, of January 27, 1967, and the Convention on International Liability for Damage Caused by Space Objects of March 29, 1972. In conformity with these principles, the system of liabilities for the failure to fulfill the commitments arising from membership in INMARSAT is formed in such a way that in the final count it is the government of the state that bears the political responsibility and the signatories in the operational agreement the material liability.

Article 30 of the convention deals with the question of the suspension of rights and forceful discontinuation of membership in the organization. In actual fact, Article 30 establishes sanctions. If a party fails to carry out its commitments, the directorate may be duly informed. The convention does not say who shall make these notifications. It should be assumed that these notifications may come both from the INMARSAT Council and from any party. However, they cannot come from the assembly, because it is authorized to take decision on such matters, or from signatories of the operational agreement, because it can hardly be expected that a national organization shall be authorized to file a complaint on the actions of a government of a foreign power in failing to observe its commitments.

In keeping with paragraph h of Article 12 of the convention, such matters as discontinuation of membership in the organization and adoption of pertinent decisions are designated within the competence of the assembly. Besides, the assembly is authorized to take decision on the official relations of the organization with the states (par. f of Article 12). Proceeding from this, the directorate should forward such notifications to the assembly for consideration. If the assembly establishes the failure of a party to live up to its commitments arising from the convention, and if this has impaired the efficient functioning of the organization, the assembly may adopt a decision on the discontinuation of membership of the party concerned.

Discontinuation of membership under the convention automatically entails the cessation of activities in the operational agreement. If a signatory of the operational agreement fails to live up to a commitment arising from membership in the organization, the matter is brought before the INMARSAT Council for consideration. The council will adopt a resolution confirming the failure to fulfill the commitment, and the signatory concerned is notified accordingly. If the signatory continues in the same fashion for 3 mo more, the council may suspend the rights of the said signatory. Finally, if this goes on for another 3 mo, the council will adopt a recommendation addressed to the assembly, requesting it to discontinue the membership of the signatory in question. The assembly is empowered to take such a decision.

Proposals were advanced on automatic suspension of signatories' rights. However, it was the Soviet delegation that proposed the more democratic procedure which has been accepted (this procedure has been described above). Under the convention, the membership can be automatically discontinued in one case only, that is, if the signatory fails to pay its contribution in the course of 4 mo after payment is due. If the payment is delayed for another 3 mo, the council can take a decision on discontinuing the membership of the said signatory. In this case, the sanction imposed by the council may refer to a government, if it acts in the capacity of a signatory.

It has already been mentioned that INMARSAT is a legal person and bears responsibility for its actions and for its commitments. This provision has been consolidated in Article 25 of the convention.

Although the English text of the document uses the term "responsible," the British delegation made a special statement, emphasizing that in the context of the provisions on responsibility contained in the operational agreement the term should be understood to mean "liable," that is, materially responsible. If this interpretation is accepted as the correct one, it will be necessary to draw the conclusion to the effect that the question of political responsibility of INMARSAT has been evaded in the documents of the organization. Therefore, it can be settled in keeping with conventional international usage.

The parties will not be held liable for actions on commitments of the organization. The only exception to this rule are the nonmember states. This means that the parties will bear responsibility to nonmember states, including responsibility for damage incurred to natural and legal persons of the latter, if such responsibility arises from pertinent international treaties (Article 22). As this article was worked out the lawyers—members of the delegations to the conference—interpreted this provision as recognition of the fact that the parties to the Convention on INMARSAT should bear responsibility in keeping with the provisions of the abovementioned Convention on International Liability for Damage Caused by Space Objects, 1972. This means that the parties shall bear liability only if damage has been caused to an INMARSAT member-state or to natural or legal persons of these states. In conformity with this interpretation, an INMARSAT member who has suffered damage from a space object another INMARSAT member has launched at the request of and for the organization can claim compensation for such damage from the organization, but not directly from the launching state. In our opinion, this interpretation does not quite conform to the provisions of the aforementioned Convention of 1972. Article 22 of this convention states that, if an intergovernmental organization is responsible for damage caused by a space object, it shall shoulder joint liability with the member states. It was not fortuitous that the Japanese delegation had made a reservation in connection with Article 22 of the Convention on INMARSAT to the effect that this provision should not rule out reference to the Convention of 1972. This logically followed from paragraph 2 of Article 22 of the Convention of 1972. The paragraph reads that the states can conclude international agreements confirming, supplementing, or extending its provision. It is obvious that Article 22 of the Convention on INMARSAT introduces a change into the commitments arising from the provisions of the Convention of 1972. The cosignatory states of the Convention of 1972 will have to be guided in their mutual relations by its provisions, despite Article 22 of the Convention on INMARSAT, if they are faced with claims in connection with damage incurred by space objects.

If the organization is required to meet liabilities, the participants in the operational agreement will have to shoulder the expenses in proportion to their shares as of the date of the liability (Article 11 of the operational agreement). If a signatory has to meet liabilities arising from the activities of the organization, the latter will compensate the signatory for the losses it has suffered. The other signatories shall bear the burden of expense in proportion to their shares.

In conclusion to this brief statement on the problem of liabilities in INMARSAT, it should be said that neither the signatories nor the organization (Article 12 of the operational agreement) shall be liable for loss or damage sustained by reason of any unavailability, delay, or faultiness of telecommunications services provided or to be provided pursuant to the convention or the operating agreement.

Settlement of Disputes

The system for the settlement of disputes is of great importance to an international organization of the type of INMARSAT which is engaged in regular economic and operational activities. Such disputes may arise between the members of the organization or between the members, on one hand, and the organization, on the other hand. The delegations of the Western countries at the conference supported the establishment of an obligatory procedure for the settlement of disputes through arbitration.

Disputes concerning the rights and obligations under the Convention on INMARSAT should be settled through negotiation between the parties to the dispute. If a dispute is not settled by this means in the course of 1 yr, the parties concerned can come to agreement on another method of settlement, including the arbitration procedure (Article 31). Disputes arising between the organization and a party (parties) to the convention, such as disputes over agreements they have concluded, should be referred at the request of one of the sides to an arbitration, unless they can be resolved through direct negotiation in the course of 1 yr or unless conciliation is reached through other means.

If one of the sides in the dispute is a party to the convention and the other is a signatory of the operational agreement, the dispute shall be referred to an arbitration tribunal only with the consent of all sides involved. Disputes between signatories or between signatories and the organization arising from the convention or the operational agreement shall be settled through negotiation. Should they fail to come to terms in the course of 1 yr or to agreement on another means, the dispute shall be referred to an arbitration tribunal at the demand of one of the sides in the dispute (Article 16 of the operational agreement).

Should a dispute occur between a signatory and the organization on an agreement they have concluded, it should be referred to an arbitration tribunal in the course of 1 yr, unless agreement is reached on another course of action.

Thus, the member states of INMARSAT shall settle all disputes either through talks or through another procedure on which mutual agreement has been reached. The arbitration procedure is optional. Disputes between states, on one hand, and INMARSAT, on the other, on agreements concluded between them are an exception to this rule. Such disputes may be submitted to an arbitration tribunal at the request of one of the parties to the dispute. The signatories shall settle their disputes with the help of an obligatory arbitration procedure (at the request of one of the sides in the dispute), with the exception of disputes between them, on the one hand, and states—parties to the convention—on the other. Disputes between the organization, on the one hand, and parties or signatories on the other shall be settled through obligatory arbitration, disputes between the organization and the states (parties) over matters bearing on the Convention on INMARSAT being an exception to the rule. Parties, signatories, and the organization may be involved in disputes over the convention in any combination. It appears, however, that signatories should not be parties in disputes over the rights and obligations arising from the Convention on INMARSAT. In any case, if such disputes should crop up and if the signatories should be involved in them, they should not be referred to obligatory arbitration. This opinion inevitably suggests itself because, in this case, any signatory, even if it is a designated entity, obtains the opportunity to interfere in the competence of a state, whereas the interpretation and application of the provisions of the convention are an inalienable and exclusive prerogative of such a state as a sovereign participant in international intercourse. Therefore, it would be more correct to interpret paragraph 1 of

Article 16 of the operational agreement as follows: disputes between the signatories over the operational agreement and disputes between the signatories, on the one hand, and the organization, on the other, over the convention and operational agreement should be settled through compulsory arbitration.

It should be emphasized that the arbitration procedure cannot be used to reverse the decisions of the assembly on the discontinuation of membership of a party in the organization (par. 1 of Article 31 of the convention).

The procedure for the settlement of disputes through arbitration within the framework of the new organization has been laid down in the Annex to the Convention on INMARSAT—"Procedures for the Settlement of Disputes Referred to in Article 31 of the Convention and Article 16 of the Operational Agreement." According to this document, when a dispute is referred to arbitration, the plaintiff shall forward an official memorandum to the directorate and the respondent, laying down the essence of the dispute, stating the measures he expects the arbitration tribunal to take, setting forth the reasons why the dispute is being referred to arbitration and why a settlement has not been reached through other means, the proof of violation of commitments, and giving the names of members of the arbitration tribunal which is to be nominated by the respondent within 60 days. In the course of the next 30 days, both arbitrators must reach agreement on the third member of the arbitration tribunal, who will be the president of the arbitration tribunal. If a member of the arbitration tribunal is not appointed within the established period, any party to the dispute may request the President of the International Court of Justice to appoint one (Article 3 of the "Procedures").

The arbitration tribunal shall examine all cases behind closed doors. However, representatives of the organization and the parties to the Convention on INMARSAT who have been appointed as litigants can take part in the proceedings. If the organization itself is a party in the case, all the parties and signatories may take part in the sittings of the arbitration tribunal.

All decisions of the arbitration tribunal shall be in accordance with international law and be based on the provisions of the Convention on INMARSAT, the operational agreement, and generally accepted principles of law. The decision is binding on the sides in the dispute. The arbitration costs are borne by all sides in equal degree.

Thus, under the Convention on INMARSAT no permanent arbitration tribunal has been set up; it is therefore an ad hoc arbitration tribunal.

INMARSAT AND INTERNATIONAL LAW

In finishing the review of the historical, organizational, legal, operational, and technical aspects bearing on the establishment of the International Maritime Satellite Organization, it should be mentioned that INMARSAT is a new form of international cooperation in the use of outer space and the latest equipment and that the basic principles of space law are, beyond doubt, fully applicable to it.

The INMARSAT is to create and then administer a communication system. Some of the aspects of its activity will come under international law on telecommunications. The new organization is intended to provide services to marine activities. Therefore, its work cannot be fulfilled independently of and in isolation from the corresponding principles of the law of the sea. Finally, the peculiar features of INMARSAT from the standpoint of the composition of its membership are bound to be of interest to the law of international organizations.

The draft drawn up by the Group of Experts on Maritime Satellites contains a preamble to the Convention on INMARSAT, with a provision to the effect that the employment of marine satellite communications should correspond to the most effective and expedient use of the radio-frequency spectrum of orbital space. Until now, international documents normally used the expression "outer space." The introduction of a new term—"orbital space"[61]—might have been interpreted as a possibility for dividing outer space into spheres subject to different legal treatment. It should be noted that the opposite follows from the Treaty on the Principles of Activity of States in the Exploration and Use of Outer Space, including the Moon and Other Celestial Bodies, of January 27, 1967. The principles consolidated in this universal multilateral agreement cover the whole of outer space, thereby constituting the basis for the establishment of uniform legal treatment governing the use of all of its spheres. At the Conference on the Establishment of INMARSAT, the Soviet representatives[62] drew attention to this circumstance. As a result, the term "orbital space" was deleted from the preamble. On a proposal advanced by the U.S. delegates, it was replaced by the expression "satellite orbits."[63] This helped emphasize effective and expedient use of certain orbits only—in particular, the geostationary orbit—and not outer space in general.

However, the above does not rule out the extension of the generally accepted principles of international space law to the sphere of activity of INMARSAT, as stated above.

In speaking about the importance of the creation of INMARSAT, it is necessary to say that this is a major step in pursuit of the agreements reached as a result of the signing of the Final Act of the Conference on Security and Cooperation in Europe on August 1, 1975. This historic document shows that the cosignatories intend to "encourage the development of transport and the

---

61. See *Mejdunarodnoe cosmicheskoe pravo* (Moscow: Izdatelstvo "Mejdunarodnie otnoscheniya," 1974), pp. 5–11.
62. IMCO, (MARSAT/CONF/SR. 21), p. 6.
63. Ibid.

solution of existing problems by employing appropriate national and international means."[64] As a new intergovernmental organization, INMARSAT will serve to realize this agreement.

It should be mentioned specially that the organizational foundations of INMARSAT correspond on the whole to the basic principles and rules of international law of the contemporary period. This is confirmed, for instance, by the fact that INMARSAT is an international organization that is open to all states without discrimination. This is also confirmed by the fact that INMARSAT has been created for peaceful purposes only. An analysis of the constituent acts of the organization shows that they are strictly oriented on the observance of the principle of cooperation between states, as consolidated in the aforementioned Final Act of the European Conference. The cosignatories of the Final Act declared they would endeavor, "... in developing their cooperation, to improve the well-being of peoples and contribute to the fulfillment of their aspirations through, inter alia, the benefits resulting from increased mutual knowledge and from progress and achievement in the economic, scientific, technological, social, cultural and humanitarian fields. They will take steps to promote conditions favourable to *making these benefits available to all;* they will *take into account the interest of all in the narrowing of differences in the levels of economic development* and in particular the interest of developing countries throughout the world."[65] (italics added)

We are witnessing that the achievements of the scientific and technological revolution are helping more and more to bring about a rapprochement between the peoples. These achievements have considerably increased the potentialities of modern transport and communications. It is possible, in turn, to improve the work of transport on the basis of the latest equipment and techniques, to promote effective joint activity of states with different social systems, mutually advantageous cooperation between them in the technical, operational, economic, financial, and other spheres, only if the necessary conditions exist in the sphere of international relations, only if this is accompanied by profound detente.

The cooperation of states in the improvement of marine communications, the establishment of INMARSAT, is a convincing example of this development.

---

64. *Pravda* (August 2, 1975).
65. Ibid.

Marine Science and Technology

Perspectives on the Sciences of the Sea

Lord Ritchie-Calder
Center for the Study of Democratic Institutions

MARINE SCIENCE AS A GROWTH INDUSTRY

In 1950, midway through the twentieth century, 750 oceanographers were identified in 48 countries, or (a ludicrous statistic) two qualified scientists for each million cubic miles of ocean space. By 1970, 5,740 marine scientists were located in 91 countries, and by 1975, 12,000 in 130 countries. This increase in scientific manpower was supplied both by academic departments of oceanography and by scientists and graduate engineers in related fields who acquired their specialized marine knowledge through extended experience and through special training courses. This growth took place primarily in developed countries with their adaptable systems of higher education. Most developing countries, including the coastal states claiming (at the UN Conference on Law of the Sea) extended territorial waters and responsibilities for their exclusive economic zones, still lack an adequate marine science base; they need teaching and research facilities, libraries, research ships, but, above all, trained scientists and engineers.

THE BEGINNINGS OF OCEANOGRAPHY

Oceanography, which H. U. Sverdrup defined as "the application of all the sciences to the study of the sea," can be said to have begun on December 2, 1872, when the *Challenger*, a British warship converted into a research vessel with laboratories, dredges, winches, sounding ropes, and nets, sailed from Portsmouth on an expedition organized by Charles Wyville Thomson, professor of natural history at Edinburgh University. The voyage lasted 3½ yr, traversing the Atlantic, the Pacific, and the Indian Oceans and penetrating the Antarctic. The report filled 52 vols. So many deep soundings were taken that the main contours of the ocean basins were established. (The deepest spot plumbed was 26,850 ft off the Marianas in the North Pacific.) The small team of six scientists made physical, chemical, and geological observations, discovered

© 1978 by The University of Chicago. 0-226-06602-9/77/1978-1011$01.79

4,717 new species of sea creatures and 717 new genera, and brought back quantities of interesting pebbles dredged up from the depths of the Atlantic, the Pacific, and the Indian Oceans. These were to remain museum curiosities until their economic significance as ferromanganese nodules was recognized in the 1950s.

Marine biology, on the other hand, dates back to Aristotle (384–322 B.C.). He made remarkable observations of many marine animals, both anatomically and in their natural habitats. His objective studies meant extensive surveys and careful dissections, and if his scientific methods had not been neglected in the intervening 2,000 years, the subject would have been more advanced and more authoritative than it is today.

Much later, with the navigators and explorers who led the "expansion of Europe," came mapmaking and charting, providing a functional picture by which ships could get from one place to another. The old navigators knew quite a lot about prevailing winds and something about currents, although their sailing records produced little that could be called scientific studies of such currents. When the Royal Society of London was founded by Charles II in 1660 its "heads of inquiry" laid emphasis on navigation, but this preoccupation was diminished by the increasing specialization of the exact sciences. Benjamin Franklin was able to draw up a passable chart of the Gulf Stream, based partly on the ships' logs of American sailing captains, which he first published in the *Transactions of the American Philosophical Society* in 1786. On his several Atlantic crossings he made his own observations of water temperatures by scooping up bucketfuls. Captain James Cook also gathered scattered oceanographic information on his three voyages of exploration, and, in spite of his unacademic, before-the-mast origins, was given scientific recognition when the Royal Society made him a fellow and awarded him the prestigious Copley Medal. Mainly to deal with the meticulous bathymetric charts and detailed information accumulated by Cook on his voyages, an office was established in the British Admiralty, and the first official hydrographer was appointed in 1795. From then on, a pretty thorough record of the seabed on the regular shipping routes was built up from naval and mercantile soundings.

This aggregation of nautical knowledge began to assume the attributes of a scholarly science through the pioneer work of an American naval officer, Matthew Fontaine Maury (1806–73). He collected all the information he could find about winds, currents, and temperatures from thousands of ships' logs and compiled the world's first textbook on ocean physics—*Physical Geography of the Sea* (1855). At the same time, an adventurous and scientifically scrupulous school of marine biologists was growing up. Scholars were going to sea and recovering specimens from deep and distant waters. Profound insights were gained, like Charles Darwin's from the voyage of *The Beagle,* but scientists were supercargo until the *Challenger* expedition took a faculty to sea and systematized research.

A PLANETARY SCIENCE

From that time on, oceanography became increasingly a planetary science, applying physics, chemistry, biology and mathematics to the study of the oceans in space and time. It belonged with meteorology, the study of the earth's atmosphere; with geology, the study of the lithosphere, the solid part of the earth; and with astronomy, the study of bodies outside the earth. Like those others, it is concerned not only with present manifestations but with events in the distant past—a long history which is decipherable only with great difficulty. As in any other branch of natural science, oceanographers can build hypotheses and theories in explanation of observed phenomena. An oceanographic theory, while abstracting and schematizing processes relying on abstract propositions in physics and chemistry, and using mathematical analyses, can only be a consensus of views, subject to long-term observations with no means (as in physics) of experimental confirmation.

How could it be otherwise? The medium itself is never still. Every drop of seawater is constantly in motion, from the surface upheavals of the waves to the slow movements of the deep currents; from the flow of ocean-rivers, greater than any rivers on land—like the Gulf Stream in the Atlantic, the Kuro Shio off Japan, or the Humboldt Current of the Southern Pacific—to the ocean rapids—like tidal bores or straits' races. Much of the motion is swirling and irregular, like smoke curling in the air, but in general water particles move in horizontal paths, stratified into relatively narrow layers, moving independently of each other like the traffic flow on the overpass and underpass. Though, in general, the average direction and volume of the major currents near the sea surface are familiar, changes in the currents from season to season cannot be defined with certainty. The Gulf Stream, for instance, meanders in its course, shrinks and expands from time to time, and changes speed and direction. In Franklin's day such information would have been of practical interest to navigators. Today it is of critical importance to peoples and nations and governments because on the vagaries of such currents depend not only the behavior of fish populations but the productivity of land agriculture as well. In the interrelationship between the sea and the atmosphere, a deviation of an ocean current can affect not only temperatures and growing time but also the rain-bearing winds, deflecting them from precarious areas like the Sahel, or upsetting the monsoons of southern Asia on which depend the food of hundreds of millions. There is not yet any way of controlling such planetary effects, but knowledge of them is essential to manage resources wisely.

Even maps of the ocean floor, while exciting imagination and speculation, are comparable in accuracy and detail only with the maps of the land surface published 250 yr ago. Fifty years ago, although the Wegener theory of continental drift had its exponents, most geologists thought of the oceans and the continents as features of the earth's surface which had existed in their present form long before the beginning of the geological record written in the rocks. Their origins were not a matter for serious scientific investigation by responsible scientists. The ocean floor was generally regarded as a featureless and uninteresting abyssal plain, covered with mud that had accumulated slowly and continuously over billions of years from the silt scoured off the continents.

PLATE TECTONICS

The activities of submarines marked a technological breakthrough that led to a remarkable increase in knowledge. First, naval submarines had to maneuver in three dimensions in the high seas, and second, their hunters had to try to locate them. This led to sonic radar by which sound pulses, traveling through the water, are reflected back to the ship sending them, reporting direction and, by time interval, distance. In the 1950s echo sounding revolutionized the mapping of the ocean bed. Ships, bouncing echoes off the floor, could make a continuous record of its profile. By this method, British, Danish, German, and U.S. investigators were able to show that a great ridge extended along the axis of the North and South Atlantic Oceans. Another was shown in the center of the Indian Ocean and yet another in the eastern Pacific. Since 1964 techniques have been so improved that the continuous profile can be defined with a margin of error of only 1–2 m. This has been further refined by the use of satellites by which the precise location of the ship taking the soundings can be determined. In 1956 it was postulated and subsequently proved by survey expeditions that the ridges form a continuous system, winding 60,000 km through all the world's oceans. The system extends the whole length of the Atlantic Ocean, passes midway between Africa and Antarctica, and turns northward into the Indian Ocean in the center of which it branches, the main ridge continuing midway between Australia, New Zealand, and Antarctica, to the east side of the Pacific Basin running all the way to the Gulf of California. There are lateral systems branching out from the main "spine," and many of them are seismicaly active—a submarine volcanic system.

The pattern of the ridges throughout the oceans is so regular and consistent that it cannot be explained by geographicaly distributed coincidences; there must be a common, all-embracing explanation. That explanation is now so convincing that is has not only silenced the critics of Wegener's theory of continental drift but has subsumed it. That explanation is now termed "plate tectonics." "Tectonics" means "construction," and the new theory maintains that the geological features of the Earth were structured by the action of so-called plates. The outer 50 km of the planet, the lithosphere, is a brittle shell. This shell has been broken into six large and perhaps a dozen smaller plates, like the bits of a chocolate Easter egg. Convection currents, upswellings from the deep interior of the globe, shift those plates in relation to each other—parting, colliding, and riding over or plunging under each other. Between the plates the upwelling of molten rock from the interior spills over to create the ridges and a new ocean floor. Very little change occurs in the middle of a plate. The reconstruction is all done at the edges, and these edges can usually be identified by looking at the map to see where earthquakes occur frequently. No gaps can occur between two plates, because hot rocks rise from below to act as a weld. Nor can plates overlap to any great extent. If plates are moving toward each other, one of them dips underneath, subducts, and reenters the earth's interior. The subducting plate creates an ocean trench. The process causes earthquakes and volcanoes on the far side of the trench. Thus the ocean floor margins keep changing.

Studies of the sediments and rocks beneath the deep sea have provided revolutionary insights into the history of the planet. The earth's crust is relatively thin under the oceans, and the great mass of rock which forms the mantle and which makes up most of the substance of the planet and encloses the molten core is easier to study at sea than on land. By mapping the shape of the seafloor; by drilling into the seabed and taking samples; by measuring the thickness, nature, and distribution of sediments; by checking the variations in the forces of gravity in the earth's magnetic field and in the heat coming from the interior, a convincing picture is being built up of the physical properties of the planet.

In the course of these developments, oceanography, from being a poor relation to be patronized by the exact sciences, has become a pacemaker for them. It is setting out lines of inquiry and insisting on exact answers from physicists, chemists, and mathematicians. Moreover, marine biologists, from being seagoing naturalists who brought back interesting specimens, are now regarded as the custodians of a threatened ecosystem. Biological oceanography, concerned with the interrelationship between the plants and creatures, the waters, the currents and the sediments of 71 percent of the earth's surface, has now become of paramount importance to mankind.

THE WEALTH OF THE OCEAN

The point of departure for contemporary oceanography was the International Geophysical Year (IGY), 1957–58. This, the year of the sun-spot maximum, was made the occasion for a combined operation of 30,000 scientists of 100 nations. This network of scientists studied Planet Earth, its composition, its surface, its atmosphere, and the cosmic forces which impinge on it. The studies included the aurora and airglow, geomagnetism, gravity, glaciology, ionospheric physics, seismology, refinements of longitudes and latitudes, cosmic rays, and the depths of the seas. Apart from the academic data collected by the IGY, two events of immense political and economic importance occurred. One was the launching of the space programs and the other was the discovery of commercial possibilities of ferromanganese nodules. The Soviet Union and the

United States had agreed to contribute instrumented satellites which would go into orbit outside the earth's atmosphere to "look" inward and outward and to intercept cosmic rays and measure forces. In the first event, the Soviet Union was first into space with Sputnik in an orbit which had required a powerful launcher. Militarily, it was claimed, this showed a "missile gap" between the Soviet Union and the United States, and the intensive "space race" followed, including "man on the moon by 1970." At the same time, survey vessels of many nations traversed the oceans sampling and charting the ocean depths. Their sweepings showed the widespread existence of ferromanganese nodules, with a particularly big haul by the Scripps Institution expedition on the Tuamotu escarpment, just east of Tahiti. By scanning the deposits with television and still cameras, the various expeditions revealed densities like a pebbled causeway. At a depth of 3,700 m in a stretch of the Pacific the concentration was calculated at 45,000 tons per square mile. The neglected sea bottom had become a desirable piece of real estate.

The oceans had long been recognized as a "liquid mine" because all the elements were present in solution or suspension. Abortive attempts to recover gold had been made in the 1920s, but the extraction of magnesium and bromine by treatment of enormous quantities of seawater—1 million gal an hour for magnesium and twice as much for bromine—made the oceans the main source of those minerals. Nature has its own alchemy. Ferromanganese nodules are so called because iron and manganese have the capacity to attract and concentrate other elements. Colloids of manganese and iron oxides forming in seawater, increasing in size and sinking downward, collect the other elements in proportions varying according to their availability in different parts of the oceans. They bear an electric charge and go on growing on the sea bottom as potato-sized black-to-brown lumps. In some locations the concentration of nodules can be 100,000 tons per square mile, containing as much as 2.5 percent copper, 2.0 percent nickel, 0.2 percent cobalt, and 35 percent manganese. Such proportions would be regarded as high-grade ores if found on land. Even at a depth of 6,000 m cable buckets or hydraulic dredges should be capable of mining the nodules at rates from 10,000 to 15,000 tons per day.

CONTINENTAL SHELF

At about the same time (1958), the Geneva Conference on the Law of the Sea was drafting a convention (to take effect in 1964) on the continental shelf. The jurisprudential transformation of this geological feature derived from a proclamation by President Truman in 1945. This proclamation declared that since modern technology (i.e., offshore drilling for oil) was capable of exploiting the resources of the continental shelf, since recognition over such resources was necessary, and since the exercise of such jurisdiction by the contiguous state was just and reasonable, the United States regarded the continental shelf as appertaining to it and subject to its jurisdiction and control. In other words, the continental shelf was the submerged extension of the coastal state. At a stroke, the United States had added 860,000 sq mi to its territory, an area larger than the Louisiana Purchase which had doubled the size of the then United States. The continental shelf was not defined in the proclamation, but the State Department set its seaward limit as the 200-m isobath.

The continental shelf can be defined as that area of the sea or ocean floor between mean low waterline and a sharp change in the inclination of the floor that indicates the edge of the "cliff" which is the continental slope. This drop can be at varying depths, usually around 130–150 m (so 200 m was generous), and the width of the shelf ranges from less than 1 mi to up to 800 mi.

The Geneva Convention on the Continental Shelf confirmed the rights of a coastal state to exercise sovereign rights over this submerged province and defined the shelf as:

"(a) the seabed and subsoil of the submarine area adjacent to the coast but outside the area of the territorial sea, to the depth of two hundred metres or, *beyond that limit, to where the depth of the superadjacent waters admits the exploitation of the resources of the said areas;* (b) the seabed and subsoil of marine areas adjacent to the coasts of islands."

That "or" was to become one of the main reasons for the long-drawn-out Third Law of the Sea Conference beginning in June 1974. When the oceanographers had revealed the potential mineral wealth in the nodules, the UN Committee on the Peaceful Uses of the Seabed and the Ocean Floor had been set up in 1967 to consider the management of those resources. There was concurrence with the principle of the "common heritage of mankind" proposed by Arvid Pardo, then the ambassador of Malta. This applied to "the ocean floor beyond the limits of national jurisdiction." But how could the extent of the common heritage be determined if a nation's jurisdiction could extend as far as its technology could take it? Moreover, coastal nations which had narrow continental shelves had a grievance because they had no seaward submarine territory over which to claim jurisdiction. So, instead, they claimed a "patrimonial sea." This later was to become, in the Law of the Sea Conference, the "exclusive economic zone."

LIVING RESOURCES

Since the Second World War, intensified exploration and research have taken place to identify and quantify marine biota. Long-range fishing vessels and factory ships ply the seven seas. In their search for commercial stocks, they carry marine biologists and research equipment and supplement the work of the research vessels. The North Atlantic, east and west, has been constantly investigated, and the North Sea has been studied for nearly a century. An international expedition, the Guinea Trawling Survey, has explored 2,000 mi of the coast of West Africa. The western Pacific from the East China Sea to the Bering Straits has been searched avidly for exploitable stocks.

Fish will be found most readily in the upwelling areas (where the cold underwaters come to the surface) in the tropical and subtropical seas. Three main upwelling areas are off Peru, North Africa, and Southwest Africa. In the Indian Ocean upwelling areas lie off Somaliland, Arabia, the Malabar Coast, the Orissa coast of India, Ceylon, and the Andaman Islands. In the western Pacific there are upwelling areas off northwest Australia and in the Banda and Flores Seas, but in all those areas while the inshore and oceanic stocks are known those of the continental shelf are unfamiliar, and since it is on the edges of the continental shelf that the upwellings occur and where the productivity is highest, this is a serious lacuna. There are poorly known areas in the southern parts of the Atlantic and the Pacific, in the Arctic, and, especially, in the Antarctic.

Factors apart from breeding and feeding affect fish stocks. The physical conditions in the sea can produce dramatic effects. For example, the distribution of cod appears to be bounded by the 2°C isotherm on the bottom; when the cod get into colder waters they are affected physiologically, and they tend to remain on the warm side of the boundary. Their distribution in Arctic waters is controlled by this phenomenon. In the western Indian Ocean, in the late 1950s, an estimated 20 million tons of dead fish were seen floating at the surface. It is believed that this holocaust was caused by an oxygen-poor layer rising close to the surface. The pulsation of warming and cooling in the eastern Pacific produces the recurring phenomenon of El Niño (The Christ Child, so called because it occurs at Christmas time). The effect is the spreading of warm tropical water in a thin layer over the normally cold, upwelled water of the Humboldt Current, off the Peruvian coast. The region is the most intense upwelling area in the world, maintaining a population of anchovies to the amount of 20 million tons. In addition it supports yellowfin, skipjack tuna, hake, and bonito as well as a huge bird population the on-shore droppings of which have produced perennial supplies of guano as agricultural fertilizer. El Niño brings disaster to the anchovies, affects the stocks of other fish, and diverts the birds to more southerly waters, where fish are more readily accessible to them, leaving the nesting birds to die.

DRUGS FROM THE SEA

There is an increasing interest in the substances of therapeutic importance which are to be found in the 500,000 species of marine life. Since the chemistry of life began in the sea, it is sensible to go there to look for basic processes.

Marine bacteria, fungi and phytoplanktonic, and higher algae have been shown to contain antibacterial, antifungal, and antiviral properties. In a study of antibiotic activity of microorganisms isolated from various depths of the ocean down to 3,500 m, EN Krasil'nikova found that 124 of 362 cultures were active against *Staphylococcus aureus, Escherichia coli, Myobacterium luteum* and

Saccharomyces cerevisiae. One had a broad antibacterial spectrum.[1] The culture fluid of a fungus, isolated from a sewage outfall, was toxic to a wide range of bacteria. Subsequently Florey (the Nobel Prize winner for the discovery of penicillin) and his colleagues Abraham and Newton showed that this fungus produced a number of antibiotics, notably *Cephalosporin C* which was toxic to a wide range of bacteria while resistant to inactivization by the enzyme which, in some cases, rendered penicillin ineffective. The polychete, *Thelepus setosus,* produces a potent fungicide. Kainic acid, an active principle of red algae, *Digenia simplex,* is an effective vermifuge (worm remover). One of the most potent poisons known to toxicologists is tetrodotoxin, from the puffer fish, precatiously enjoyed by the Japanese as a table delicacy, *fugu.* At low dosage levels tetrodotoxin is used clinically as a muscle relaxant and as a pain killer. Apart from venoms and poisons, extracts of marine organisms have effects which are antitumor, neuroactive, cardioactive, and cardiovascular. They produce hormones (including insulin) analogous to those of humans.

NEW WAYS OF FINDING OUT

There are several approaches to increasing the essential knowledge of fish distribution: underwater sound; surveys of fish eggs and larvae; the collection of fish scales in stratified bottom sediments and remote sensory detection. The most useful method of exploration is an acoustic one combined with capture. Echo surveys, which can be automated, can be used to make relative estimates of abundance. Single fish can be counted separately from shoals. Counts per cubic meter can be made down to depths as great as 600 m. Collection of pelagic fish eggs and larvae can be done with simple ships and simple plankton nets. High-speed nets (towed at about 5 knots) are an effective way of catching larvae. The bottom sediments, with fish scales, ear drums, and other remains, provide a valuable time scale. The deposit of a decade or even a year may be distinguishable and can contain sufficient identifiable fish remains to define the persistence of a fish species, show the interrelations of different species, and reveal gross changes in the composition of marine fauna. Sediment cores representing 1,000 yr have shown that in the inshore waters off California hake and anchovy have dominated, whereas sardines have come and gone. Such historical evidence can be important in considering management of contemporary fish populations.

Airplanes and helicopters have been used to assist fishing fleets in detecting and pursuing shoals and for research purposes. With the development and use of air-dropped bathythermographs for detecting and relaying temperature information, data could be obtained about subsurface layers as well as information about surface phenomena. In the North Pacific expendable

1. E. N. Krasil'nikova, *Microbiology* 30 (1961): 545–49.

bathygraphs were dropped every 25 mi along a great circle line 1,800 mi long to study temperature as a function of position, depth, and time. Since the rate at which sound waves pass through ocean water varies with temperature, a sound-velocity chart, checked at monthly intervals, was provided for a vast area of the ocean.[2]

Spacecraft also have opened up new opportunities for oceanographers. In common with the meteorologists they could have photographs of planet Earth and every part of it. Satellite navigation systems could pinpoint the exact positions of their research ships, giving their conventional observations an accuracy which they could never guarantee before. With color photography, across the spectrum from ultraviolet to infrared, they could get revealing pictures of the oceans. They could distinguish the fertile and desert areas of the oceans. Off the coast of Peru they could see an area of green pastures as productive for animal life as is the blackest soil of the Ukraine for grain. In the central regions of the North Atlantic and North Pacific Oceans, they could see the purplish-blue color of waters as barren as the Sahara. With a special radiometer, tried out on high-flying aircraft and operating on four wavelengths (466, 525, 550, and 600 nm), it was possible to get measurements of the albedo (the ratio of the amount of radiation reflected by a surface to the amount falling upon it) of the oceans.[3] Further refinements produced a method of measuring the chlorophyll content and the turbidity of seawater, independent of cloudiness, surface waves, and sky radiation.

SCANNING THE DEPTHS

The sea is opaque. Light and radio waves travel only a short distance before becoming scattered (like headlights in a fog) so that images cannot be recognized. Thus the observational satellites cannot really see what is happening under the surface. Hence the military importance of the oceans in which nuclear submarines, with their hydra-headed weapons, can escape space surveillance.

On the other hand, sound waves, due to their small absorption in water, can propagate over great distances, although absorption is greater at high frequencies. There is scarcely a phenomenon in the ocean's depths which does not leave an identifiable trace in the sound pattern. Refinements, and high sensitivity, in acoustical methods are such that a fish only centimeters long can be detected at a range of several kilometers due to resonance scattering of sound by its swimming bladder. The study of the reflection and scattering of

2. *Encyclopedia Britannica*, 15th ed., s.v. "Undersea Exploration," Macropaedia vol. 18, p. 847.

3. June–July 1975 over the Gulf of Guinea. Reported by P. Y. Deschamps, P. Lecomte, and M. Viollier to the Joint Oceanographic Assembly, Edinburgh 1976.

sound waves can determine the spectrum of surface waves and the concentration and dimensions of air bubbles beneath the surface. The use of echo sounders and acoustic side scanners makes the mapping of the sea bottom relatively simple, while minor irregularities of the sea bottom are revealed by the sound scattering at the water-bottom interface. Beyond that, in examining the geological structure of the crust down to the mantle, seismic and seismoacoustic methods are effectively employed.

Propagation of sound along fixed paths of different lengths makes it possible to study turbulence and internal waves and measure sea currents. The use of ultrasonics can sharpen up the acoustical image of an underwater object and, by converting sound signals into electromagnetic signals, produce a visible image on a screen.

The combination of radio and acoustic techniques has produced remote sensing which can overcome the blindfold effect of turbid, nontransparent water. Apart from underwater television cameras and conventional hydrophones "wired" to surface vessels, instruments can be installed on the sea bottom or suspended in the water column and linked to surface buoys, unmanned installations which can record and store the information. These ocean data banks can be triggered off when necessary and the information transmitted to, or through, satellite systems to land-based computers.

The advanced technologies for the investigation and exploration of the ocean now operate at all levels—instrumented spacecraft and aircraft, with remote sensors; oceanographic buoys; research submersibles; offshore observatories and undersea habitats.

MAN AT DEPTH

Man, like all other creatures, began in the oceans. In his evolution as a respiratory being, with lungs instead of gills to extract oxygen from the atmosphere instead of from water, he biologically more or less excluded himself, except as a surface traveler, from the seven-tenths of his planet which is ocean. Nevertheless, the ambition to return not as a sailor or a swimmer but as an explorer at depth dates back a long time. Alexander the Great (365–323 B.C.) is reputed to have gone down to the seafloor in a diving bell such as his tutor, Aristotle, described. Leonardo da Vinci devised diving equipment akin to modern versions. The diving suit consisted of a sealed waterproof tunic with an inner vest of coiled armor plate to protect the chest against compression; the helmet had glass eye panes and a bifurcated mouthpiece with pliable tubes attached to a snorkel apparatus, the air intake of which was supported at the surface by disks of cork; the nether garment included webbed flippers like those of the modern frogmen. For longer endurance and greater mobility, he provided the equivalent of modern gas cylinders, in the form of a skin bag with, in his notes, a hint of compressed air, but he supressed his ideas of submarine vessels "by reason of the evil nature of man who would use them as a means of destruction at the bottom of the sea."

Purposeful research diving at great depths belongs mostly to the second half of the twentieth century, beginning with the construction of the bathyspheres and continuing with the conversion of military submarines to research needs. A formidable fleet of deep-sea submersibles is in commission. Two of the best known are the *Aluminaut,* launched in 1964, made of aluminum alloy (its cylindrical hull is rounded off at either end with hemispheres equipped with portholes and, designed to reach a depth of 4,500 m, it weighs 73 tons and cruises at 2.5 knots), and the *Alvin* operated by the Woods Hole Oceanographic Institution.

The U.S. Deep Diver series consists of 8-ton submarines used to convoy divers to an exact location and to stand by and pick them up as soon as they complete the assignment, so that their decompression period can begin immediately. The Beaver type of research submarine can carry passengers to a depth of 600 m. The Deepstar series was designed to be capable of submerging to 6000 m, and the Dolphin series, with a capacity of 22 men, to operate at 1,000 m.

Bathyscaphes, used to explore the deepest parts of the oceans (e.g., the record descent of 6.78 mi by Jacques Piccard and D. Walsh in the *Trieste*) have a life-support system similar to the small submersibles. Dependent on battery power, they perform short missions, are not heated, and maintain a pressure equal to that at sea level. They use oxygen from cylinders, and for carbon dioxide removal they use Baralyme or lithium scrubbers.

Jacques Piccard devised a pioneer life-support system for the research submersible *Ben Franklin* which was designed to drift in an unpowered state, remaining submerged for weeks at a time. The features included the passive removal of carbon dioxide by lithium hydroxide absorption; low temperature oxygen storage; preheated stored hot water; silica gel humidity control, and the use of the sea to absorb excess heat radiated from the vessel. The sea-level air environment in the *Ben Franklin* was maintained by the use of the original atmospheric nitrogen and oxygen from cylinders. For the crew of six (on the Gulf Stream drift mission) the oxygen consumption was 680 g per man per day, the equivalent of 2,250 calories. Temperature control was passive, and no additional heat was supplied other than that from the operation of the lights and instruments and the body heat of the crew. At the end of 30 days, the crew was in good shape and conditions still tolerable.

HABITATS

Historically, the diving bell, like an inverted tumbler with entrapped air, was used as a life-support system. Today, it has returned to favor. In modern practice it is sealed at the surface and lowered to the work site, retaining the surface pressure inside. After preliminary inspection the diver compresses the bell to the pressure of the surrounding depth and, in his diving gear, opens the hatch, leaves the bell, does what is necessary, reenters the bell, seals the hatch, and signals to be raised to the surface. Gas is vented from the bell at the controlled rate called for in the decompression tables. On the ship deck or platform the pressure in the decompression chamber is brought up to that of the bell and the two are locked together. The hatches between them are opened, and the diver then transfers to the more comfortable decompression chamber to await his return to normal. The bell is also used as part of the "hookah" system in which the helmeted diver can draw, through a tube, from the bell's breathing facilities.

With existing materials capable of withstanding pressures at great depths; with life-support systems which could encapsulate a living environment; with communication systems and with means of supply, it is theoretically possible to establish at almost any depth installations for life and work. Captain Jacques-Yves Cousteau operated his Conshelf I habitat at 10 m below the Mediterranean Sea for 7 days. In 1963 five oceanauts lived for 1 mo in Conshelf II at a depth of 11 m, and in 1964 two men remained at 132 m in the Caribbean for over 48 hr in an inflated rubber habitat and were able to swim and work near their habitat by means of a "hookah" rebreather. This pumped gas from the habitat to the diver and returned the expired gas for recycling. In the same year the U.S. Navy started a series of experiments (Sealab I) keeping four men at 58.8 m for 9 days. Sealab II was set in 62.5 m of water off the coast of California; it was a cylindrical tank 17.4 m long and 3.6 m in diameter. Twenty-eight divers manned three 15-day tours, one staying 30 consecutive days. The divers had free access to the sea. There were no enduring ill effects.

Free diving with the aqualung, the portable life-support system, has progressed very rapidly. It has been accompanied by careful physiological studies in France, Monaco, the United States, the USSR, and the U.K. which have demonstrated the possibility of diving and remaining at great depths under saturation conditions. A major problem of deep diving is the compulsory release of accumulated gas from solution in the blood and other body fluids which must be achieved by decompression. This procedure usually requires a series of stops on ascent to allow gases to escape from the body slowly enough to prevent the formation of bubbles in the blood (like the bubbles in champagne). Standard decompression tables can be used for short periods and at depths not greater than 120 m. When the required operating depths are in excess of 120 m the decompression time is so great that diving cannot be carried out safely without special decompression equipment. For example, a dive to 200 m for 1 hr of work will require about 24 hr of decompression. A submersible decompression chamber can help divers to regain the surface in the best possible decompression conditions. Saturation diving, however, can open the way to extensive human exploration and exploitation of the seabed. Once the diver's body is saturated with inert gas his decompression time does not increase and he can do long spells of work (usually limited by cold tolerance). In practice divers are compressed slowly to working pressure, generally in a compression chamber on deck, and are then transferred to a sealed diving bell and lowered to the site. There is a saturation point for any given depth.

SURFACE OBSERVATORIES

There is a large proportion of oceanographers who take a conservative attitude to the ideas of underwater research laboratories. They still maintain that practically everything that is required in terms of observations and measurements can be done by surface research ships and surface installations.

Since 1950 there has been an increase in the number of research ships specially designed and built for specific types of inquiry—unlike the converted naval and mercantile vessels with which oceanographers had to make do. The international listing of such research vessels exceeds 500. All ocean vessels usually over 70 m in length are operated by large academic institutions or government agencies. Smaller vessels for coastal work are operated by practically all oceanographic groups, including commercial contractors. Apart from the essential features (the oceanographic winch, the boom, and the crane), a research vessel requires several laboratory spaces. A deck laboratory is necessary for instrument preparation and operation. Other laboratories are needed for performing chemical, biological, and geological analyses, for deploying electronic equipment, and for photographic development and printing.

For long-term observations at selected sites, research platforms, akin to oil platforms, are installed on fixed towers standing on the seabed. Large unmanned oceanographic buoys and their sensory auxiliaries are classified as platforms. Some of them can measure and record 100 separate channels of data, sending the data on command by radio to shore stations thousands of miles away. The information can be stored for periods as long as 1 yr, and the master buoy acts as a clearinghouse for strings of moored buoys, collecting their measurements and transmitting them ashore. Another important advance is the submerged buoy providing stable support for underwater instruments at depths up to 6,000 m.

Both the United States and the USSR have made extensive use of ice islands. Prefabricated laboratories are imbedded in the ice with facilities equal to a well-equipped land institute. One such ice-island station drifted 5,000 mi in 4 yr from Point Barrow in Alaska, through the Arctic almost to the North Pole, and was evacuated off Iceland. Using a nuclear-powered supply ship, the USSR established an ice-island station at the North Pole itself.

Typical of another generation of research installations is FLIP (Floating Laboratory Instrument Platform). Designed for the U.S. Navy in 1962, it met the requirements for an extremely stable and yet mobile platform from which ultraprecise acoustical measurements could be made at sea. A FLIP-type vessel is like a knobkerrie. The operational part is in the knob; the stick—two thirds of the vessel—consists of ballast tanks. The vessel is towed to the research area and 100 m of ballast tankage is flooded. The stern sinks, the prow rises, and the whole vessel stands to attention, with about four stories of accomodation above the water. The deep draft makes the installation stable. Its only propulsion power is that required for the orientation propellors which rotate the vessel around its axis. The whole process can be reversed. With the evacuation of the ballast tanks the ship reverts to horizontal to be towed elsewhere.

Indispensable, however, are the long-range oceangoing research ships in the tradition of the *Challenger* and the *Albatross* but embodying present-day advanced technology. These include ships like the *Glomar Challenger, Glomar Explorer,* and *Atlantis II*. The *Glomar Challenger* is a 10,500-ton vessel built on the basis of the technology available in the 1960s. It can maintain an exact position for days and weeks at a time by means of an acoustical beacon dropped to the seafloor; hydrophone reception of the beacon signals; and a shipboard computer to make calculations and adjustments for special propulsion units. The ship is stabilized so that the drillers can effectively control, even in high seas, 23,000 ft of drill pipe. A drilling derrick operates through the bottom of the ship, extracting cores from depths of thousands of meters—cores which tell the history of oceans: climates, mountain building, volcanic eruptions, life and evolution, and the movements of the seafloor and continents. It uses the latest navigation aids, including satellite observations and electric magnetic systems. This gives it precision of positioning so that a bit drilling the sea floor can be withdrawn and reinserted—like threading a needle through 16,000 ft of water. The *Glomar Explorer,* with the technology of the 1970s, is even more remarkable. *Atlantis II* was responsible for the discovery of the metalliferous brines in the Red Sea.

THE MECHANISMS OF CLIMATE

The oceans are mainly responsible for the weather. The interactions between the hydrosphere and the atmosphere determine climatic changes. The major part of the energy of storms and winds is transmitted from the sun to the atmosphere through the ocean. Air, heated by contact with the warm ocean surface, carries water vapor alóft. As the rising air cools and contracts, the vapor condenses, releasing its latent heat, and the air expands again and rises still further. The density distribution in the atmosphere is thus greatly perturbed. Enormous amounts of energy enter the air through this mechanism of evaporation at the sea surface and condensation aloft. Under the constraints exerted by the rotation of the earth, this energy contributes to the formation of the hurricanes of the tropics and the cyclonic storms of the midlatitudes. The winds, in turn, drive the surface currents of the sea, thereby determining the location of the warm water masses that are the principal regions of evaporation and hence of energy transfer from sea to air. In this way, the ocean and the atmosphere form a feedback system on a local and global scale. Meteorological studies have shown that changes in the hemispheric weather patterns over periods of weeks to years are related to the changes in temperature distribution of the water layers near the surface of the sea. To study those changes and the accompanying atmospheric processes requires measurements at many points over vast areas. No nation by itself has sufficient research ships or oceanographers to obtain all the needed data.

This is not just a question of academic curiosity. The economic implications and indeed the issues of human survival which are involved make such studies of practical and urgent importance.

Because the sea behaves more sluggishly than the air, improvements in long-range weather forecasting can be made through the studies of the interactions between sea and air. The present accuracy of long-range forecasting is low, and great economic benefits would follow any improvements. The planting and harvesting of crops could be better regulated in terms of what and when to plant. Flood damage could be reduced by management of flood-control structures—for example, by lowering the water levels in reservoirs prior to periods of heavy precipitation and snow melt. Economies could be made in energy production and supply if the public utilities and fuel producers could plan production, transportation, and storage on the basis of reliable forecasts of warm or cold winters or hot or cool summers. The costs of construction of building highways, telephone and telegraph lines, dams, and public utilities could be lowered if scheduling of labor and machinery could be planned to take advantage of good weather. The potential savings are in the range of billions of dollars, many orders of magnitude greater than the research investments required to acquire the data, especially when the costs and the efforts are shared internationally. This is recognized (but still not adequately) in World Weather Watch, organized by the World Meteorological Organization and the Integrated Global Ocean Station system initiated by the Intergovernmental Oceanographic Commission (IOC).

INTERNATIONAL COOPERATION

Just as the earth's atmosphere is continuous with people everywhere breathing essentially the same air, so too are the ocean waters indivisible. (We speak of the "Seven Seas," but there is only one ocean.) Thus international scientific cooperation is not only desirable but necessary if human understanding of the ocean is to keep pace with human needs.

Experience has shown at least 10 ways in which scientific benefits can be gained by international cooperation in the study of the world ocean: (i) speeding up of exploration of little-known regions, (ii) intercalibrations of methods, (iii) synoptic studies of air-sea interaction, (iv) studies of the fluctuations in sea level, (v) biological censuses in the ocean, (vi) studies of phenomena in special areas (coral atolls, deep-sea trenches, the Arctic and Antarctic; monsoon changes of winds and ocean currents in the Indian Ocean and the South China Sea, etc.), (vii) mapping of the deep seafloor, (viii) data exchange, (ix) the use of navigational aids, and (x) exchanges among individual scientists.

The International Indian Ocean Expedition (IIOE) and the International Cooperative Investigations of the Tropical Atlantic (ICITA) are examples of how combined operations can quicken and intensify the exploration of little-known parts of the ocean. There are still large and critically important areas where the ocean waters, their populations of living things, and the underlying seafloor are very little known. The new tools of oceanic exploration have barely been used in some parts. A concentrated, task-force approach to these unknown regions by modern oceanographic ships of several countries would yield large returns.

Within the United Nations, the principal oceanographic body is the IOC, founded in 1961, based at UNESCO, Paris (UNESCO also has an Office of Oceanography which provides staff support for the commission and has operational functions). These bodies work in conjunction with the World Meteorological Organization (WMO) on climatic problems, with the Food and Agriculture Organization (FAO) on marine fisheries, and with the Inter-Governmental Maritime Consultative Organization on environmental and other questions.

In 1968, the United Nations General Assembly endorsed the concept of LEPOR, the long-term and expanded program of oceanographic research and exploration "to increase the knowledge of the ocean, its contents, and the contents of its subsoil, and its interfaces with the land, the atmosphere, and the ocean floor and to improve understanding of processes operating in or affecting the marine environment, with the goal of enhanced utilization of the ocean and its resources for the benefit of mankind."

Six months before, President Lyndon Johnson on behalf of the United States had proposed an International Decade of Ocean Exploration (The Decade) (1970–80). Member states of IOC were encouraged to cooperate and to present new projects to the commission for endorsement as components of The Decade. The commission laid down the criteria that (1) all programs must be multinational in character, (2) the data collected must be channeled through the World Data Centre, (3) all programs must be of exclusively peaceful purpose, (4) they must be scheduled to take place during the 1971–80 period, (5) the projects must involve such scientific substance and international cooperation as to produce results more rapidly than could be achieved by separate action, and (6) scientists from other nations must be involved from the start of the programs. A second "Decade" is projected for the 1980s.

Part of the grand design of the long-term program is IGOSS, the Integrated Global Ocean Station System. This, in conjunction with World Weather Watch of WMO, includes the modern technological means for observations, radio communication and data processing. It is intended to provide synchronous undelayed information from the whole ocean.

The idea of a General Bathymetric Chart of the Oceans (GEBCO) was suggested at the Seventh International Geographical Congress in Berlin in 1899 and became a reality when Prince Albert I of Monaco assembled a group of scientists to work on it in 1903. It was later handled by the International Hydrographic Bureau. With the invention of continuous sounding, the flood of data became overwhelming. In 1974 the workload had become so heavy that the (now) International Hydrographic Oganization agreed to cooperate with the IOC of UNESCO. Following a study of the Scientific Committee on Oceanic Research (SCOR), it was decided to develop an entirely new form of presentation consisting of 18 charts covering the world on a scale of 1:10 million. (Sheets are obtainable from the Hydrographic Chart Distribution Office, Dept. Fisheries and the Environment, 1875 Russell Road, P.O. Box 8080 Ottawa, Ontario, Canada K1G 3H6.)

The revelation of the topography of the ocean floor is immensely exciting—a picture of the world drowned beneath the sea. But in addition to mapping the sea bottom, it is necessary to find out what is beneath the seafloor. Seismic refraction has provided some information about the thickness of sediments, but refinements in the technique of seismic-reflection profiling permits a continuous cross section of sediment stratification from a vessel traveling at a few knots. A promising new technique lies in the use of multichannel seismic reflection for recording oblique reflections from the crust and mantle. Another is the use of seismic receivers in the deep sea drilled holes.

RESEARCH AND HUMAN NEEDS

The new knowledge being revealed by this extensive and intensive ocean research opens up immense possibilities of the use of the resources of the sea for human needs, including the industrialization of the seabed not only on the continental shelf but also of the ocean depths.

The oceans, however, are not an inexhaustible resource store. We extract vast quantities of oil from beneath the seafloor, but these will foreseeably run out. Fish have been harvested, by the relatively primitive methods of hunting, to such an extent that the normal cyclical replenishment has been seriously impaired. In many regions excessive damage has been done to coastal areas where some 60 percent of the human population lives and where the heaviest demands for harbors, industry, tourism, and recreation are felt.

A major concern with implications not only for the present but for posterity relates to the pollution of the sea. The oceans serve as a common, uncontrolled sink for the world's waste products drained from the land and as a dustpan for vast quantities of windblown pollution. Sewage, oil-cargo residues, domestic, industrial, radioactive, and toxic wastes are being indiscriminately dumped in the sea or in rivers which carry them to the sea. It has been estimated, for example, that on an annual basis 4,000–5,000 tons of toxic mercury are discarded and the amounts of petroleum hydrocarbons annually entering the ocean have been reckoned as high as 6,113,000 tons.

The wastes and poisons already released to the marine environment have resulted in a whole host of well-documented adverse effects. Contamination of beaches with inadequately treated sewage has led to bacterial and virus infections of epidemic proportions as well as to reducing the amenities. Oil spills have contaminated beaches, killed sea birds, and spoiled the flavors of fish and shellfish. Methods of dealing with such spills have introduced additional pollutants which themselves are toxic to marine life. The accumulation of insecticide residues and insecticides, such as halogenated hydrocarbons, in fish, sea birds, and seals provides serious health hazards. The reproductive processes of birds have been reduced by reason of pesticide ingestion. Dissolved copper, zinc, and mercury, which are rare in seawater but common in industrial waste, affect filter-feeding invertebrates, resulting in the poisoning of consumers.

There is proper popular and official concern about the nature and extent of marine pollution, but there is also a vast ignorance which only intensive and highly organized research can remove. For most pollutants, the acceptable degree of exposure of man or other organisms is either unknown or poorly quantified. No quantitative statement can be made on acceptable levels of exposure without some estimate of the rate of incidence of harmful effect resulting from a given degree of exposure. Resort has to be made, in the case of man, to epidemiological studies and the exposure of organisms under experimental conditions. There is an urgent need to develop methods of the assessment of the effects of chronic exposure on individual organisms, populations, and ecosystems.

It is also necessary to study the physical, chemical, and biological processes in which the pollutants take part. For instance, a substance originates from a source on the land or in the atmosphere and passes through the oceanic reservoir, which consists not only of water but of the metabolism of its inhabitants, until it eventually sinks to the ocean-bottom sediment. The time scale of this progression may vary from near-instantaneous to millions of years. During this time the pollutant takes part in many processes, and it is the study of these processes that will provide an indication of the relationship between the inputs and the distributions.

One of the ultimate aims of this kind of investigation is to develop a capacity to predict future change in the marine environment. Only when the constituent processes are thoroughly understood will it be possible to construct a model of such changes. A "comprehensive plan for the global investigation of pollution in the marine environment (GIPME) provides an international framework within which national and regional programs on various aspects of marine pollution may be coordinated to contribute to an understanding of global problems. It is under the aegis of the IOC of UNESCO.

RESTRICTIONS ON RESEARCH

Few nations are today capable of coping with the problems raised by the marine environment. Out of 113 countries with coastlines 49 have coastlines longer than 1,000 km; 19 countries have continental shelves equal to or larger than their land area. The problem is how to assist nations, many of which are developing and lack the educational base from which to undertake the scientific and technological responsibilities involved. Without access to the necessary knowledge they will not be able to fulfill the international obligations incumbent upon them in assuming extended jurisdiction.

It cannot be repeated often enough that the ocean system is a single system. What happens in one part of that system can have consequences (often of catastrophic dimensions) in localities or in subsystems (biologically speaking) far removed. There are no fences nor physical boundaries in the marine ecosystem. What happens in estuaries, in coastal waters, and on the continental shelf can determine the fate of whole species. A coastal state wanting to exercise protection over its commercial fisheries, even to the limit of 200 mi, will find that what happens in some other country's jurisdiction or in the high seas will affect the fish population. The classical example is that of El Niño which recurrently brings disaster to the coastal fisheries of western Latin America. A 200-mi exclusive economic zone with the right to control over fishing would not avert a failure of the anchovy fishing of Peru. The upwelling of the cold Antarctic waters of the Humboldt Current in this case is overlaid by a warm current originating thousands of miles away in the eastern Pacific. The studies of the marine ecology, now imperative, require global cooperation. Denial of access of cooperation in international research must redound to the disadvantage of the intransigent nation.

The 200-mi economic zones proposed at UNCLOS (United Nations Conference on the Law of the Sea) encompass approximately 37 percent of the ocean area, and together they represent regions where perhaps 90 percent of the oceanic biological activity occurs; where the interaction of the atmosphere and the ocean in determining climate is most effective; where most of the world's undersea earthquakes occur; and where answers to some of the most complex and challenging problems concerning the history of the earth are to be found. The concern of the scientific community about the seemingly antiscientific policies being formulated at the Law of the Sea Conference is therefore understandable. The oceanographers are afraid that they will be severely handicapped. Already there have been complaints about the restrictive practices of many coastal states. The fleet of academic research ships which make application to governments for the right of access to waters over which they claim control have found that about half their scheduled cruises have had to be canceled because requests were denied or undue hindrance disabled the intentions. Some requests are never acknowledged, or the approval is delayed until it is too late for the program to be conducted successfully.

In spite of the assurances by the academic researchers that nationals of a coastal state would be allowed to participate and that the results of the findings would be published in reputable scientific journals and made available to the states, suspicion prevails. The less developed countries now exercising, or claiming, a jurisdiction over extended waters have misgivings about what they perceive to be military or commercial rather than scientific intentions. They are also self-conscious about their inadequate scientific resources; some of them would be hard put to it to find any qualified person to participate on shipboard in the survey cruise. They are also at a disadvantage in handling the research material when it is available. Potential exploiters, on the other hand, either have or can buy the expertise necessary to realize the potentials of the surveys and studies.

Scientists naturally resent the distrust of their intentions. They are impatient to follow up the exciting leads now being given in marine research. The answer can only be more research and more countries effectively involved in that research and more people, in the less developed countries, capable of evaluating that research. That is the difficulty: poor nations do not have the educational infrastructure from which to develop their cadres of marine experts. Here is a question of genuinely sharing knowledge and skills and coaching the men of affairs in the developing countries in what is the true nature of marine research and the best use which can be made of it.

This predicament, which is not only a present complication but a danger to posterity, is well understood by the intergovernmental and nongovernmental bodies involved in the acquisition of scientific knowledge on which proper ocean management should be based. The need is to convince governments that their best interests lie in cooperative, disinterested research on a global scale—a scale as big as the ocean itself.

BIBLIOGRAPHY

Borgese, Elisabeth Mann. *The Drama of the Oceans.* New York: Abrams, 1975.
Buzzati-Traverso, Adriano. *Perspectives in Marine Biology.* Berkeley: University of California Press, 1958.
Calder, Nigel. *The Restless Earth.* New York: Viking, 1972.
Deacon, George, ed. *Seas, Maps, and Men.* 2d ed. London: Hamlyn, 1968.
Knauss, John A. "Marine Science and the 1974 Law of the Sea Conference." *Science* 184 (June 28, 1974): 1335–41.
Knauss, J.; Frye, P.; and Wooster, W. "The Marine Scientific Research Issue in the Law of the Sea Negotiations." *Science* 197 (July 15, 1977): 230–33.
Linklater, Eric. *The Voyage of the Challenger.* London: Murray; New York: Doubleday, 1972.
Loftas, Tony. *The Last Resource.* London: Hamilton, 1969.
NAS, Ocean Policy Committee, Commission on International Relations. *Marine Scientific Research and the Third Law of the Sea Conference: 2nd Substantive Session* (report prepared by the Freedom of Ocean Research Group). Washington, D.C.: National Academy of Sciences, 1976.

Ritchie-Calder, Lord. *The Pollution of the Mediterranean.* Bern: Lang, 1972.
Schaeffer, M. B. "Freedom of Scientific Research and Exploration in the Sea." *Stanford Journal of International Studies* 4 (June 1969): 46–70.
Wooster, Warren S., ed. *Freedom of Oceanic Research.* New York: Crane, Russak, 1973.

Environment

The Oceans: Health and Prognosis
Peter S. Thacher and Nikki Meith-Avcin
UN Environment Programme

How "sick" are the oceans? To what extent have human activities altered the natural structure of marine communities, by means direct and indirect, obvious and obscure? To what extent are the more dramatic consequences of pollution—brought to mind by such names as Minamata and Torrey Canyon—really indicative of the long-term and perhaps irreversible damage we may be doing to our ocean environment? To answer such questions we must first decide where to look, what criteria to use, and how to define so ambiguous and relative a term as "pollution."

The obvious place to look for pollution is where we expect to find it in its greatest concentrations—in coastal waters, estuaries and areas where population and industry are concentrated. Here we can observe directly the physical transport of a polluting material, its entry into and accumulation within local food webs, its effects on marine organisms, communities, and sometimes on humans themselves. The next place to look is where we *least* expect to find evidence of man's polluting activities—far-removed areas such as the polar ice caps, the deep sea, the upper layers of the atmosphere. When a substance such as DDT is ubiquitous and can be found in the tissues of Antarctic organisms (marine invertebrates, fish, and penguins), in the livers of bottom-dwelling ocean fish, and in the atmosphere over the Sargasso Sea, there is indisputable cause for concern.

Which criteria we choose becomes a key question when trying to determine the state of health of the sea. We must look at a local situation dynamically—to see not only what is there but how it is likely to change with time: what the sources of pollution are, how the rate is increasing or decreasing, where the polluting material is transported and what its ultimate fate is.

Much recent ecological debate has centered around the question of the validity of making predictions about what will happen to an ecosystem as the result of environmental stress (e.g., introduction or increase in pollution load), based on inferences from what is happening in similar systems elsewhere or from what is known about the history of the system in question. Although these are interesting theoretical considerations, there simply has not been enough basic study to allow them a place in formulation of international environmental policy. Whether or not a particular change (e.g., population fluctuation) in a community is natural or pollution induced is a question that can be approached

© 1978 by The University of Chicago. 0-226-06602-9/77/1978-1012$03.54

only through the most controlled experimental conditions, but what we *can* say, with some confidence, is that the introduction of (1) chemicals which do not naturally occur in the environment (such as DDT and PCBs) and (2) materials which do occur but are released to the sea in unprecedented quantities (such as heavy metals, trace elements, hydrocarbons, and nutrients) can have stress effects on natural systems which may be irreversible.

Defining pollution in an uncircular and workable manner can be difficult. Perhaps it is best to consider pollution as simply one type of environmental stress—the introduction by man of materials or energy into a system which causes unexpected or unpredictable change to take place, relative to what we consider the normal course of events. A legal definition of pollution, one that is acceptable to countries concerned with direct effects of pollutants on the health and well-being of its human residents, is necessarily anthropocentric. For this reason, the definition of pollution employed by the United Nations at the 1972 Stockholm conference on the human environment reads as follows: "Pollution means the introduction by man, directly or indirectly, of substances or energy into the marine environment resulting in such deleterious effects as harm to living resources, hazards to human health, hindrance to marine activities including fishing, impairment of quality for use of sea water and reduction of amenities." We find this definition sufficiently broad and well suited to its legal purposes.

The objectives of this paper are to describe the most obvious examples of man-induced pollution stress as applied to marine systems and the international policy which has been devised in response as one means of countering the threat to a shared vital resource. We shall first consider briefly the pollutants which man has introduced into the oceans, their observed levels, predicted increase, and effects on marine life. We shall then consider the history of international programs and legislation dealing with pollution, with emphasis on those with which the United Nations Environment Programme (UNEP) has been associated in its 5 years of existence. Our treatment will be broad and unavoidably sketchy but will perhaps serve as a satisfactory introduction to the subject. Future editions of the *Ocean Yearbook* will provide adequate opportunity to supplement and revise this general picture of mankind's varied approaches to one of its greatest problems.

OCEAN POLLUTANTS: SOURCES, LEVELS, AND EFFECTS

Petroleum Hydrocarbons

Concern over the increasing additions of petroleum hydrocarbons to the ocean environment has been prompted largely by the visible results—soiling of beaches, occurrence of oil films and tar balls in surface waters, lethal and sublethal effects on marine organisms. Although these localized effects are easily observed and described, realistic assessment of the distribution of the many petroleum compounds which are released to the environment is technically and analytically very difficult. Thus we have relatively little descriptive data on the petroleum content of waters, sediments, or organisms.

There are three general source categories of petroleum-containing hydrocarbons: man-generated substances in the form of crude oil and refined petroleum products, organism-produced alkanes and alkenes, and petroleum from natural seepage through the ocean floor.[1] The quantitative separation of oceanic hydrocarbons from these various sources is difficult, although some guidelines for the differentiation of petroleum and biogenic hydrocarbons have been indicated by the National Academy of Sciences (NAS) and one technique has been devised for sediments by Blumer and Sass.[2] Crude oil consists of thousands of hydrocarbon compounds; trace amounts of metals (such as nickel, vanadium, and iron) complexed with organic chelates; porphyrins and other compounds which contain sulphur, nitrogen, and oxygen. The four principle classes of hydrocarbons found in crude oil are alkanes (paraffins), cycloalkanes, aromatic hydrocarbons, and olefins (in trace amounts).[3] Pollution by crude oil is caused primarily by tanker accidents, deballasting operations, and tank washing, as well as from natural seepages and losses from off-shore production.[4]

The NAS estimate of the total input of petroleum hydrocarbons to the oceans is approximately 6 million metric tons annually. About 2 million tons of this is attributable to transportation of petroleum by sea,[5] and it is predicted that this will increase to 6 million tons by 1980.[6] The remainder is delivered by a variety of sources (table 1). The total burden of oceanic hydrocarbons may be as high as 14 billion tons,[7] but whatever the correct value, a 4 percent annual increase in petroleum transport guarantees its rise unless precautions are taken.[8]

Due to the varying physical and chemical properties of the compounds

1. Edward D. Goldberg, *The Health of the Oceans* (Paris: UNESCO, 1976).
2. NAS, *Petroleum in the Marine Environment* (Washington, D.C.: National Academy of Sciences, 1975); M. Blumer and J. Sass, "Indigenous and Petroleum Derived Hydrocarbons in a Polluted Sediment," *Marine Pollution Bulletin* 3 (June 1972): 92–93.
3. Edward P. Myers and Charles G. Gunnerson, *Hydrocarbons in the Ocean*, NOAA/MESA Special Report (Washington, D.C.: Government Printing Office, 1976).
4. IMCO/FAO/UNESCO/WMO/WHO/IAEA/UN Joint Group of Experts on the Scientific Aspects of Marine Pollution (GESAMP), *Impact of Oil on the Marine Environment*, Reports and Studies, no. 6 (Rome: FAO, 1976).
5. NAS.
6. IOC/WMO, *Report of the Second IOC/WMO Workshop on Marine Pollution (Petroleum) Monitoring*, IOC Workshop Report, no. 10 (Paris: IOC, UNESCO, 1976).
7. Goldberg.
8. B. W. Halstead, "Toxicity of Marine Organisms Caused by Pollutants," in *Marine Pollution and Sea Life*, ed. Mario Ruvio (London: Fishing News [Books], 1972), pp. 584–94.

TABLE 1.—ESTIMATED INPUTS OF PETROLEUM HYDROCARBONS
ENTERING THE OCEAN ANNUALLY, CIRCA 1969–71 (NAS 1975)

Source	Millions of Tons (Metric) per Annum
Marine transportation	2.133
Offshore oil production	.08
Coastal oil refineries	.2
Industrial waste	.3
Municipal waste	.3
Urban runoff	.3
River runoff	1.6
Natural seeps	.6
Atmosphere	.6
Total	6.113

SOURCE.—NAS, *Petroleum in the Marine Environment* (Washington, D.C.: National Academy of Sciences, 1975).

which comprise crude oil and its derivatives, several environmental fates are possible:

1. Evaporation. Low molecular weight hydrocarbons (12 or less carbon atoms per molecule) are lost rapidly from an oil slick, and the rougher the sea, the faster the loss.

2. Solution. Low molecular weight and high polarity contribute to relative solubility of hydrocarbons. Inorganic and microbial degradation processes can produce compounds of increased solubility.

3. Emulsification. The formation of emulsions of water-in-oil and oil-in-water depend on turbulence, and the dissolution of oil components proceeds most rapidly when a dispersion of oil-in-water is formed due to the dependence of the process on surface area. Water-in-oil emulsifications are thought to be the precursors of beach tars and tar lumps after chemical alteration.

4. Sedimentation. When hydrocarbons become aggregated into lumps, emulsions, or residues, or when they adhere to sinking particles, sedimentation can take place. The latter process can be accelerated by the turbulent suspension of sediments which often occurs in shallow coastal waters.

5. Oxidation. Photochemically or thermally induced oxidative reactions occur at various rates depending on the nature of hydrocarbon components and their location. Surface slicks are more susceptible to such destruction than emulsions or tar balls.

6. Microbial degradation. Over 90 species of microorganisms are capable of degrading many constituents of oils, and different species of bacteria metabolize different components. Such factors as water temperature, availability of nutrients, and the presence of microbial predators can influence rates of

degradation. Molecular configuration is also an important factor influencing hydrocarbon biodegradability, with alkanes the most rapidly attacked compounds.[9]

7. Biological incorporation. Marine organisms take up petroleum hydrocarbons directly from the water by absorption or ingestion of particles or by predation on other organisms which contain such compounds. Once incorporated, the petroleum is metabolized, stored, or excreted.

Measurements of hydrocarbons in ocean water throughout the world have been compiled in a comprehensive report by MESA and NOAA (Marine Ecosystems Analysis Program and National Oceanic Atmospheric Administration, U.S. Department of Commerce) which finds that most surface and near-surface waters have from 1 to 10 ppb total hydrocarbons, and both biogenic and petroleum hydrocarbons appear to be ubiquitous, the latter somewhat concentrated in coastal shipping lanes.[10]

Physical states of petroleum in sea water can be divided into three categories:

1. Dissolved and particulate phases. In 1975 NAS estimated the oceanic burden of these phases at 400 million tons. Values from a variety of sources (comprehensively reviewed in the NOAA/MESA report) vary widely and are generally found to be depth dependent, with highest concentrations occurring at the surface. The study on which the NAS estimate was based found 6 μg/liter in the upper 10 m from a profile taken off the coast of Bermuda.

2. Sea surface microlayer. Wade and Quinn report 14–559 ppb total hydrocarbons in the surface microlayer of the Sargasso Sea, with a median value of 105 ppb.[11] This contrasts with a median value of 60 ppb only 20–30 cm below the surface. Surface films from the eastern Mediterranean were found to contain extremely high concentrations of organics (40–230 mg/m^2), only 5 percent of which were of natural origin.[12]

3. Tar balls. Butler estimates the tar content of oceanic surface waters at about 700,000 tons.[13] Analysis suggest that these come from tanker ballast waters. In 1975 NAS summarized numerous tar-ball analyses from worldwide samples (table 2). Due to their varying size and density, tar balls tend to be distributed throughout the water column.[14]

Hydrocarbon concentrations in marine sediments are extremely variable with values of 1–4 ppm (dry weight) in open-ocean sediments and up to 12 ppt (parts per thousand) in highly polluted areas.

9. GESAMP.
10. Myers and Gunnerson.
11. T. L. Wade and J. G. Quinn, "Hydrocarbons in the Sargasso Sea Surface Micro-Layer," *Marine Pollution Bulletin* 6 (April 1975): 54–57.
12. R. J. Morris, "Lipid Composition of Surface Films and Zooplankton from the Eastern Mediterranean," *Marine Pollution Bulletin* 5 (May 1974): 105–9.
13. J. N. Butler, "Pelagic Tar," *Scientific American* 232 (June 1975): 90–97.
14. GESAMP.

TABLE 2.—TAR DENSITIES IN THE WORLD OCEANS

Location	Area (10^{12} m^2)	Tar (mg/m^2) Maximum	Mean	Total Tar (10^3 Tons)
Northwest Atlantic marginal sea	2	2.4	1	2
East coast continental shelf	1	10	.2	.2
Caribbean	2	1.2	.6	1.2
Gulf of Mexico	2	3.5	.8	1.6
Gulf Stream	8	10	2.2	18
Sargasso Sea	7	40	10	70
Canary and north equatorial current	3	1,000?	?	?(large)
Mediterranean	2.5	540	20	50
Indian Ocean	75	...	?	?(large)
Southwest Pacific	45	...	< .01	< .5
Southeast Pacific	45	...	?	?
Kuroshio system	10	14	3.8	38
Northeast Pacific	40	3	.4	16

SOURCE.—NAS, *Petroleum in the Marine Environment* (Washington, D.C.: National Academy of Science, 1975).

Data from hydrocarbon analysis in marine organisms from several areas are compiled in table 3. Values were consistently found to be in the parts per million level, even in open-ocean species. Such widespread occurrence may be due in part to the fact that hydrocarbons, after ingestion and incorporation by organisms, are chemically very stable as they pass through the food chain.[15]

Pollution by crude oil and oil fractions have been found to have the following damaging effects on marine ecosystems: (1) direct kill of organisms through coating and asphyxiation; (2) direct kill through contact poisoning of organisms; (3) direct kill through exposure to the water-soluble toxic components of oil; (4) destruction of the more sensitive juvenile forms of organisms; (5) destruction of the food sources of higher species; (6) incorporation of sublethal amounts of oil and oil products into organisms, resulting in reduced resistance to infection and other stresses (the principle cause of death in birds surviving the immediate exposure to oil); (7) destruction of food values through the incorporation of oil and oil products into fisheries resources; (8) incorporation of carcinogens into the marine food chain and human food sources; (9) low level effects that may interrupt any of the numerous events necessary for the propagation of marine species and for the survival of those species which stand higher in the marine food web.[16] Damage occurs to organisms through such mechanisms as mechanical clogging or blanketing by

---

15. M. Blumer, "Oil Contamination and the Living Resources of the Sea," in Ruvio, pp. 476–81.
16. Ibid.

TABLE 3.—PETROLEUM HYDROCARBON LEVELS
IN MARINE MACROORGANISMS

Organisms	Area Type*	Hydrocarbon Type	Estimated HC Amount (μg/g)
Macroalgae:			
Fucus	4	Bunker C†	40 dry
Enteromorpha	4	No. 2 fuel oil	429 wet
Sargassum	1	C_{14-30} range	1–5 wet
Higher plants:			
Spartina	4	No. 2 fuel oil	15 wet
Molluscs			
Modiolus, mussel	4	No. 2 fuel oil	218 wet
Mytilus, mussel	4	No. 2 fuel oil†	36 dry
Mytilus	4	Bunker C†	10 dry
Mytilus	4	Bunker C, aromatics	74–100 wet
Mytilus	3	n-C_{4-37}†	9 dry
Mya, clam	4	No. 2 fuel oil	26 wet
Pecten, scallop	4	No. 2 fuel oil	7 wet
Littorina, snail	4	Bunker C, aromatics	46–220 wet
Mercenaria, clam	3	C_{16-32} range	160 dry
Crassostrea, oyster	2	Polycyclic aromatics	1 wet
Crustacea:			
Hemigrapsus, crab	4	Bunker C†	8 dry
Mitella, barnacle	4	Bunker C†	8 dry
Lady crab	3	C_{14-30}	4 wet
Plankton	2	Benzopyrene	.4 wet
Sargassum shrimp	1	C_{14-30}	3 wet
Lepas, barnacle	1	C_{14-30}	6 wet
Portunas, crab	1	C_{14-30}	34 wet
Planes, crab	1	C_{14-30}	11 wet
Fish:			
Fundulus, minnow	4	No. 2 fuel oil	75 wet
Anguilla liver, eel	4	No. 2 fuel oil	85 wet
Smelt	3	Benzopyrene	0–5 dry
Flatfish	2	C_{14-20}	4 wet
Flying fish	1	C_{14-20}	.3 wet
Sargassum fish	1	C_{14-20}	1.6 wet
Pipefish	1	C_{14-20}	8.8 wet
Triggerfish	1	C_{14-20}	1.7 wet
Birds:			
Herring gull, muscle	4	No. 2 fuel oil	535 wet
Echinoderms:			
Asterias, starfish	4	Bunker C, aromatics	20–147 wet
Luidia, starfish	2	C_{14-20}	3.5 wet

SOURCE.—NAS, *Petroleum in the Marine Environment* (Washington, D.C.: National Academy of Science, 1975).
*1, oceanic; 2, chronic pollution, coastal; 3, chronic pollution, harbor; 4, single spill.
†n-alkanes only.

absorbed oil, irritation of mucous membranes and other respiratory surfaces and transport membranes, and interference with chemoreception and neurological processes.[17]

In general, the low-boiling fractions of petroleum (e.g., benzene, toluene, xylene) produce anesthesia and narcosis in a wide variety of lower animals at low concentrations, leading to death as concentrations are raised. These compounds are also toxic to humans. The higher-boiling fractions contain multiring aromatic compounds known to be carcinogenic. They can also interfere with nutrition and communication in marine animals.[18]

Toxic effects of oil have been determined for phytoplankton, zooplankton, certain sensitive invertebrates, and fish larvae from the Black Sea.[19] Fish larvae are quite susceptible to oil toxicity and are unable to avoid oil-contaminated water.[20] When oil-spill effects on marshland were observed, it was found that plants varied in their susceptibility and that the application of emulsifiers to "clean up" the spill had more toxic effects than the oil itself.[21] Although "tainting" effects of oil on fish and shellfish are well known, only recently has it been determined that the oil passes essentially unchanged through the intestinal barrier to stabilize in the lipid pool of the organisms, much in the same manner as other persistent chemicals.[22]

Effects of oil spills on higher animals are perhaps most notable in the case of sea birds, thousands of which are killed annually by being coated with oil in the English Channel and North Sea.[23] The 1969 oil spill in Santa Barbara Channel resulted in very little mortality to organisms due to toxic effects, but smothering and coating resulted in high mortality in pelagic birds and a species of intertidal barnacle. The relatively low commercial fishery landings in the 6 months following the spill could not be attributed directly to it, and mammal surveys did not indicate abnormal mortality rates.[24]

Laboratory studies have determined that thick layers of oil can interfere with gaseous exchange across the air/sea interface, but so far no areas have been identified where an oil slick has been associated with oxygen depletion.

---

17. Goldberg.
18. Blumer.
19. O. G. Mironov, "Effect of Oil Pollution on Flora and Fauna of the Black Sea," in Ruvio, pp. 222–24.
20. W. W. Kühnhold, "The Influence of Crude Oils on Fish Fry," in Ruvio, pp. 315–18.
21. E. B. Cowell, J. M. Baker, and G. B. Crapp, "The Biological Effect of Oil Pollution and Oil-cleaning Materials on Littoral Communities, including Salt Marshes," in Ruvio, pp. 359–64.
22. M. Blumer, G. Souza, and J. Sass, "Hydrocarbon Pollution of Edible Shellfish by an Oil Spill," *Marine Biology* 5, no. 3 (1970): 195–202.
23. Ilmo Hela, "Marine Productivity and Pollution," in *The Environmental Future*, ed. Nicholas Polunin (New York: Barnes & Noble, 1972), pp. 250–72.
24. D. Straughan, "Biological Effects of Oil Pollution on the Santa Barbara Channel," in Ruvio, pp. 355–59.

Likewise, the theory that surface slicks accumulate chlorinated hydrocarbons and mercury has not been substantiated.[25]

It appears that large-scale environmental damage resulting from such spills as Santa Barbara and Torrey Canyon has not yet occurred. However, we have insufficient knowledge of long-term and low-term effects of petroleum hydrocarbon pollution on organisms and communities to discount its importance. It is clear that natural systems can handle a limited overload of these compounds, but the rate of addition must not exceed the rate of recovery. Unfortunately, recovery rates are difficult to determine.

Halogenated Hydrocarbons

Modern technology has given man the capability of synthesizing a wide variety of compounds which do not occur in nature. Of such synthetics, the halogenated hydrocarbons have received most attention, one reason being that their presence in the environment is indisputably due to human activities. They are generally persistent, biologically active chemicals, and these and other properties have caused halogen-containing organic compounds to become the most widely used synthetics.

DDT and PCBs
DDT and its metabolites and the polychlorinated biphenyls (PCBs) are ubiquitous in the marine environment. They are found in plankton collected worldwide and in the tissues of Arctic seals and Antarctic penguins. They have been discovered far from their obvious sources, and it now appears that their major mode of transport from the continents to the oceans is atmospheric.

In 1971 the NAS estimated that 2.0×10^{12} grams of DDT had been manufactured since production began, most of it by the United States.[26] By one estimate, in 1969 the United States exported 37,000 tons of DDT, 23,370 of which were used in antimalarial programs, and 13,930 for agriculture and other purposes.[27] After entering the environment, DDT is partially degraded to DDE and other metabolites by photochemical or biological reactions. DDT residues then enter the atmosphere through vaporization processes and are washed into the marine system by rainfall. In 1971 NAS estimated that 2.4×10^{10} grams of DDT are injected per year into the marine system through rainfall alone, calculated from a measured 80 ppt mean concentration in rainfall over the United Kingdom in July 1967. This value has since been shown to be a reasonable estimate for periods of high DDT usage in the Northern

25. GESAMP.
26. NAS, *Chlorinated Hydrocarbons in the Marine Environment* (Washington, D.C.,: National Academy of Sciences, 1971).
27. Goldberg.

Hemisphere. Once in the sea, DDT residues are found concentrated in the surface monolayer (15 μ), probably due to absorption by organic chemical slicks.[28]

PCBs are used widely in hydraulic fluids, paints, plastics, paper products, and in "closed-system" electric and heat-transfer equipment. There are an estimated 80 PCB compounds used in industry, and each has a different environmental behavior. They are released to the environment through leakage, spillage, breakage of containers, and combustion, and are less readily degraded than DDT. PCB production in several industrialized countries for 1971 is given in table 4.

Like DDT, PCBs are found in the atmosphere and were measured in the marine air over the Sargasso Sea.[29] Although they tend to concentrate in the surface monolayer of the water, they appear to have but a short residence time in surface waters, perhaps due to association with falling particles.[30]

DDT and PCBs occur in marine organisms according to their exposure history. A study by Williams and Holden of marine plankton off the Scottish Coast demonstrated a marked gradient according to distance from shore.[31] From the Firth of Clyde to an open-ocean station 400 miles west, DDT residues decreased by a factor of 10 and PCBs by a factor of 12.

Table 5 shows levels for DDT residues and PCBs from a number of

TABLE 4.—PRODUCTION OF PCBs IN CERTAIN COUNTRIES, 1971 (in Tons)

Country	Quantity
United States	18,000*
Federal Republic of Germany	8,000
France	7,600
United Kingdom	5,000
Japan	6,800†
Italy	1,500
Spain	1,500‡
OECD countries§	48,000

SOURCE.—Reproduced in Edward D. Goldberg, *The Health of the Oceans* (Paris: UNESCO, 1976), from OECD, *Polychlorinated Biphenyls, Their Use and Control* (Paris: OECD, 1973).
*38,000 tons in 1970.
†11,000 tons in 1970.
‡Estimate.
§Including the above.

28. Ibid.
29. T. F. Bidleman and C. E. Olney, "Chlorinated Hydrocarbons in the Sargasso Sea Atmosphere and Surface Water," *Science* 183 (February 8, 1974): 516–18.
30. G. R. Harvey, W. G. Steinhauer, and H. P. Miklas, "Decline of PCB Concentrations in North Atlantic Surface Water," *Nature* 252 (November 29, 1974): 387–88.
31. R. Williams and A. Holden, "Organochlorine Residues from Plankton," *Marine Pollution Bulletin* 4 (July 1973): 109–11.

TABLE 5.—DDT RESIDUES AND PCBs IN
A VARIETY OF PLANKTON SAMPLES

Area	No. of Samples	DDT Residues	PCBs
Gulf of St. Lawrence	9	...	2,000–93,000
Scottish coast	26	<3–107	10–2,200
Sargasso Sea	4	.7	7–450
South Atlantic	4	.2–2.6	19–638
Northeast Atlantic	22	2–26	10–110
Clyde, Scotland	15	6–130	40–230
California	...	.2–206	.7–30
California	...	...	100–1,300
Iceland (phytoplankton)	1	...	1,500
North and South Atlantic	53	...	.2

Source.—Edward D. Goldberg, *The Health of the Oceans* (Paris: UNESCO, 1976).
Note.—DDT residues and PCBs in parts per billion wet weight.

plankton samples. A long-term survey of DDT and its metabolites in molluscs from coasts of 15 states of the United States found measurable levels in 63 percent of the samples with a general peak concentration in 1968.[32]

In 1969–71 a study of DDT levels in mussels from several European countries and North America found highest concentrations in samples from the United Kingdom and the Mediterranean Coast of Spain, with values even up to 30 ppm wet weight. Levels in herring often exceeded even this, and were highest in the Baltic.[33] Such results are typical and not too surprising, given the proximity of monitored organisms to sources of pollution from industry and agriculture. Perhaps more significant is the detection of chlorinated hydrocarbons in several species of Antarctic fauna and plankton, in bottom-dwelling fish from deep Atlantic waters, Arctic ringed seals, and Pacific whales.[34]

Biological effects of DDT residues in nature are well known in the case of birds which feed upon marine organisms. Reproductive failures due to

---

32. P. A. Butler, "Residues in Fish, Wildlife and Estuaries," *Pesticide Monitoring Journal* 6 (March 1973): 238–362.

33. A. V. Holden, "International Co-operative Study of Organochlorine and Mercury Residues in Wildlife, 1969–1971," *Pesticide Monitoring Journal* 7 (June 1973): 37–52.

34. C. S. Giam, R. L. Richardson, M. K. Wong, and W. M. Sackett, "Polychlorinated Biphenyls in Antarctic Biota," *Antarctic Journal* 8 (1973): 303–5, cited in Goldberg; N. Meith-Avcin, Stanley M. Warlen, and Richard T. Barber, "Organochlorine Insecticide Residues in a Bathyl-Demersal Fish from 2,500 Meters," *Environmental Letters* 5 (Fall 1973): 215–21; A. V. Holden, "Monitoring Organochlorine Contamination of the Marine Environment by the Analysis of Residues in Seals," in Ruvio, pp. 266–72; A. A. Wolman and A. J. Wilson, "Occurrence of Pesticides in Whales," *Pesticide Monitoring Journal* 4 (June 1970): 8–10.

eggshell thinning in pelicans have been reported.[35] DDT and PCBs have also been implicated in abortions of sea-lion pups.[36] Crustaceans have been shown to be especially sensitive to these compounds, and an accidental discharge of PCBs in Escambia Bay, Florida, led to a large mortality of shrimps. Laboratory experiments have indicated possible influence of organochlorines on species composition of phytoplankton communities.[37] Butler, Childress, and Wilson suggested that DDT residues from local agriculture use were responsible for observed losses in marine productivity off the south Texas coast.[38]

No harmful effects of DDT on humans have been observed, but accidental ingestion of PCBs was associated with the occurrence of skin eruptions, increased pigmentation, and possibly some deaths.[39]

Other Synthetic Chemicals

The environmental significance of some other synthetic chemicals may be described as follows:

1. HCB (C_6Cl_6). Used as a grain fungicide, it is also a contaminant of certain pesticides. Agricultural uses allow this compound, which has a very high vapor pressure, to be totally dispersed. It is extremely unreactive, and therefore persistent in the environment, and has been found in many marine organisms.[40] Health effects of HCB have been observed in human inhabitants of Turkey who consumed HCB-coated grain and subsequently developed a metabolic disease associated with a disturbance in heme synthesis.[41]

2. Mirex. This is an organochlorine compound used against the fire ant in the southeastern United States and as a flame-retarding additive in plastics. In 1971 it was found in measurable levels in nine of 77 samples of marine invertebrates and fish collected from the Gulf of Mexico and southeastern coastal areas.[42] Laboratory studies have indicated potential hazards from its

35. L. J. Blus, C. D. Gish, A. A. Belisle, and R.M. Prouty, "Logarithmic Relationship of DDT Residues to Eggshell Thinning," *Nature* 235 (February 18, 1972): 376–77.

36. R. L. Delong, W. G. Gilmartin, and J. G. Simpson, "Premature Births in California Sea Lions: Association with High Organochlorine Pollutant Residue Levels," *Science* 181 (September 21, 1973): 1168–69.

37. J. L. Mosser, N. S. Fisher, and C. F. Wurster, "Polychlorinated Biphenyls and DDT Alterations of Species Composition in Mixed Cultures of Algae," *Science* 176 (May 5, 1972): 533–35.

38. P. A. Butler, R. Childress, and A. J. Wilson, "The Association of DDT Residues with Losses in Marine Productivity," in Ruvio, pp. 262–66.

39. M. Kuratsune, T. Yosimura, J. Matsuzaka, and A. Yamaguchi, "Epidemiologic Study on Yosho, a Poisoning Caused by Ingestion of Rice Oil Contaminated with a Commercial Brand of Polychlorinated Biphenyls," *Environmental Health Perspectives* 1 (April 1972): 119–28.

40. NAS, *Assessing Potential Oceanic Pollutants* (Washington, D.C.: National Academy of Sciences, 1975).

41. Goldberg.

42. G. P. Markin, J. C. Hawthorne, H. L. Collins, and J. H. Ford, "Levels of Mirex and Some Other Organochlorine Residues in Seafood from the Atlantic and Gulf Coastal States," *Pesticide Monitoring Journal* 7 (March 1974): 139–43.

increased use, since high concentrations (0.01–10 ppb in seawater) have lethal effects on crab larvae[43] and juvenile crayfish.[44]

3. Low molecular weight hydrocarbons. Ubiquitous in the atmosphere and oceans, this group of compounds includes the chlorofluorocarbons (trichlorofluoromethane and dichlorofluoromethane) used as propellant solvents and refrigerants; tri- and tetrachloroethane, widely used as cleaning solvents; and substances which are probably of natural origin, such as methyl iodide, chloroform, and carbon tetrachloride. Very little is known about effects of these compounds in the marine environment, but the suspected effects of chlorofluorocarbons on the tropospheric ozone layer are the subject of widespread debate. Their uses as propellants are entirely dispersive, and thus the annual world production (estimated at 1 million tons per year) is assumed to quickly enter the environment.

4. EDC tars. Disposal of wastes from the production of the plastic polyvinyl chloride, primarily in the form of EDC (1,2-dichloroethane) tars, has affected some fish in the North Sea.[45] Most of the EDC tar formed to date has been dumped into the oceans,[46] and due to its physical nature it tends to be retained in the upper layers of the ocean.

Heavy Metals

Heavy metals enter the marine environment naturally through processes of weathering, but in some cases the amount of a particular metal being mobilized in the environment as a consequence of human activities rivals that from natural processes. Two major sources of metals are cement production, which releases great amounts of mercury, and the burning of fossil fuels, responsible for the release of a billion grams of zinc per year into the environment.[47] Other sources of metallic wastes are chemical, food, mining, metallurgical, photographic, plating, printing, textile, and tanning industries. Nonindustrial sources include agriculture (pesticides and fertilizers) and water-treatment

---

43. C. G. Bookhout, A. J. Wilson, T. W. Duke, and J. I. Lowe, "Effects of Mirex on the Larval Development of Two Crabs," *Water, Air, Soil Pollution* 1 (April 1972): 165–80.

44. J. L. Ludke, M. T. Finley, and C. Lusk, "Toxicity of Mires to Crayfish, *Procambarus blandingi*," *Bulletin of Environmental Contamination and Toxicology* 6 (January-February 1971): 89–96.

45. S. Jensen, A. Jernelöv, R. Lang, and K. H. Palmork, "Chlorinated Byproducts from Vinyl Chloride Production. A New Source of Marine Pollution," in *Proceedings. FAO Technical Conference on Marine Pollution and Its Effects on Living Resources and Fishing* (Rome: FAO, December 1970).

46. A. Jernelöv, "Heavy Metals, Metalloids and Synthetic Organics," in *The Sea*, ed. E. D. Goldberg (New York: Wiley Interscience, 1974), vol. 5 cited in Goldberg (n. 1 above).

47. Goldberg.

processes.[48] Extraction of minerals from the sea bed is a possible source of metal pollution, the potential dangers of which are reviewed by Portmann.[49]

Modes of entry of most metals into the ocean include rivers, industrial and domestic sewage discharges, and direct dumping of wastes. Lead, however, seems to enter largely via the atmosphere, following its release from internal-combustion engines using antiknock fuels. About 1 kg/day is the estimated discharge from cars per capita in the United States.[50] The total amount of lead entering the oceans from atmospheric washout is thought to approximately equal the 20,000 tons discharged annually from rivers.[51]

Only in the case of lead do we have any indication of a significant increase in concentration in coastal ocean waters due to human activities. Although many reported levels are thought to be too high due to sample contamination, the entry of lead aerosols into the coastal water of the Pacific, Atlantic, and Mediterranean has significantly raised surface seawater concentrations over those found before the widespread use of lead alkyls in fuel.[52] In the case of mercury, no increase has been seen in seawater concentration or in tissue concentrations of bottom-dwelling fish.[53] However, the release of an estimated 5,000 tons of this highly reactive and toxic element into the coastal sea each year has profound implications for commercial fisheries and marine ecosystems.[54]

Reactive elements such as heavy metals are rapidly removed from seawater to the sediments in organic precipitates or solid biological phases due to their participation in intense biological and chemical activities. Metals have been found concentrated in recently deposited sediments in the Baltic Sea and off Southern California.[55] In both cases, enhanced concentrations of lead, cadmium, zinc, and copper were found in surface layers. Radiocarbon dating determined that the increases corresponded generally with the onset of the

48. W. J. North, G. C. Stephens, and B. B. North, "Marine Algae and Their Relation to Pollution Problems," in Ruvio, pp. 330–40.

49. J. E. Portmann, "Possible Dangers of Marine Pollution as a Result of Mining Operations for Metal Ores," in Ruvio, pp. 343–46.

50. H. A. Schroeder, *The Poisons around Us* (Bloomington: Indiana University Press, 1974).

51. FAO, *Seminar on Methods of Detection, Measurement and Monitoring of Pollutants in the Marine Environment* (Rome: FAO, 1970).

52. T. J. Chow and C. C. Patterson, "Concentration Profiles of Barium and Lead in Atlantic Waters off Bermuda," *Earth and Planetary Science Letters* 1 (1966): 397–400, cited in Goldberg.

53. R. T. Barber, A. Vijayakumar, and F. A. Cross, "Mercury Concentrations in Recent and Ninety-Year-Old Benthopelagic Fish," *Science* 178 (November 10, 1972): 636–38.

54. Hela.

55. H. Erlenkeuser, E. Seuss, and H. Willkomm, "Industrialization Affects Heavy Metal and Carbon Isotope Concentrations in Recent Baltic Sea Sediments," *Geochimica Cosmochimica Acta* 38 (1974): 823–42, cited in Goldberg; K. W. Bruland, K. K. Bertine, M. Koide, and E. D. Goldberg, "History of Metal Pollution in Southern California Coastal Zone," *Environmental Science and Technology* 8 (May 1974): 425–32.

Industrial Revolution. A study of four basins off the coast of North America compared recent rates of metal accumulation with those of 100 years ago (table 6). In these areas the anthropogenic fluxes of lead are three times higher than natural fluxes, and yet for nickel, cobalt, manganese, and iron there are no such changes. For other metals—chromium, zinc, copper, silver, vanadium, cadmium and molybdenum—fluxes due to technological activity approximate those of natural origin. A rough comparison of various transport pathways in the area, derived from independent estimates, is shown in table 7.

Once in the ocean, metals may be transformed into various chemical states. In a pH-dependent series of reactions, speciation of metals may occur among ionic, particulate, or complexed states. Certain elements, notably mercury, tin, platinum, gold and thallium, can be methylated in the environment by microorganisms into potentially toxic compounds.[56]

Attention has been focused recently on heavy metal pollution due to instances of human poisoning by mercury (the renowned Minamata Bay tragedy) and cadmium (the *itai-itai* epidemic in Japan following consumption of contaminated rice).

Metals are toxic due to their interference with vital biochemical processes. Lead is an enzyme inhibitor which also impairs cell metabolism in human beings,[57] although no deleterious effects have thus far been observed in marine organisms. Mercury is a highly reactive and toxic metal, and its biological effect depends strongly on its concentration, chemical form, and species of target organism.[58] Cadmium is a potential threat to humans due to its high bioaccumulation factor. Zinc, although of low oral toxicity to man, can have deleterious effects on marine larvae and has high concentration factors in certain marine organisms.[59]

Due to their persistence, toxicity, and tendency to accumulate to high levels in organisms, the heavy metals should be monitored carefully in the marine environment, especially in the coastal waters where their potential for biological damage may be greatest.

Sewage and Fertilizers

Microorganisms of human and animal origin find their way into coastal waters by direct disposal, rivers, storm-water runoff, and even through the atmosphere. The primary dangers from such contamination are a direct risk to public health and possible effects on the quality and safety of sea food.

56. J. M. Wood, "Biological Cycles for Toxic Elements in the Environment," *Science* 183 (March 15, 1974): 1049–52.
57. Hela.
58. S. Keckes and J. K. Miettinen, "Mercury as a Marine Pollutant," in Ruvio, pp. 276–89.
59. Halstead.

TABLE 6.—FLUXES OF HEAVY METALS INTO SEDIMENTS
OF CALIFORNIA COASTAL BASINS

Element and Flux	Fluxes (in $\mu g/cm^2/Year$)				
	San Pedro	Santa Monica	Santa Barbara	Soledad	Average
Pb:					
Anthropogenic	1.7	.9	2.1	...	1.6
Natural	.26	.24	1.0	.23	.5
Rainfall	...	...	...	...	...
Cr:					
Anthropogenic	3.1	2.6	2.9	...	2.9
Natural	2.8	2.1	10.7	4.6	5.2
Zn:					
Anthropogenic	1.9	2.1	2.2	...	2.1
Natural	3.1	2.8	9.7	2.8	5.2
Rainfall	...	...	...	...	...
Cu:					
Anthropogenic	1.4	1.1	1.4	...	1.3
Natural	1.2	1.0	2.6	1.4	1.6
Rainfall	...	...	...	...	...
Ag:					
Anthropogenic	.09	.09	.10	...	.09
Natural	.05	.03	.11	.08	.06
V:					
Anthropogenic	1.5	2.6	7.8	...	4.0
Natural	3.5	3.4	13.6	4.6	6.8
Cd:					
Anthropogenic	...	...	.07	...	.07
Natural	...	...	.14	...	.14
Mo:					
Anthropogenic	...	.8	...	...	.8
Natural	...	.08	...	...	.08
Ni:					
Natural	1.6	1.3	4.1	2.3	2.3
Rainfall	...	...	...	...	...
Co:					
Natural	.33	.26	1.0	.17	.53
Mn:					
Natural	13	8	24	7	15
Rainfall	...	...	...	...	...
Fe:					
Natural	1,260	1,200	3,060	840	1,800
Al:					
Natural	1,740	1,630	4,860	1,280	2,700

SOURCE.— K. W. Bruland, K. K. Bertine, M. Koide, and E. D. Goldberg, "History of Metal Pollution in Southern California Coastal Zone," *Environmental Science and Technology* 8 (May 1974): 1049–52.

TABLE 7.—FLUXES OF MATERIALS TO SOUTHERN CALIFORNIA
COASTAL REGION (in Tons/Year/12,000 km^2)

Element	Flux of Anthropogenic Components to Sediments	Waste Waters	Storm Water Plus Dry-Weather Flow	Washout Fluxes
Pb	190	213	90	156
Cr	350	649	25	12
Zn	250	1,680	101	550
Cu	160	567	18	60
Ag	11	15	1	5
V	480	...	...	17
Cd	8	54	1	48
Mo	100	...	...	24

SOURCE.—K. W. Bruland, K. K. Bertine, M. Koide, and E. D. Goldberg, "History of Metal Pollution in Southern California Coastal Zone," *Environmental Science and Technology* 8 (May 1974): 1049–52.

The most common pathogenic organisms from fecal discharge of warm-blooded animals are strains of *Salmonella, Shigella, Leptospira,* enteropathogenic *Escherichia coli, Pasteurella, Vibrio, Mycobacterium,* human enteric viruses, cysts of *Endamoeba hystolytica,* and hookworm larvae. Many of the bacterial, parasitic, and viral diseases of man are transmitted by fishery products, and a summary of the major of these is presented in table 8.[60]

In estuarine areas, lagoons, and archipelagoes a great number of bathing beaches have been closed due to sewage contamination, and in eastern Canada many of the molluscan shellfish-growing areas have been closed.[61] The Mediterranean receives a great deal of untreated sewage, especially in its northwestern basin, and seldom is it discharged far enough from shore to avoid affecting beaches.[62]

Microorganisms eventually die, and seawater is thus said to be "self-cleansing." Therefore, risks from sewage discharges remain localized and tend to cause problems at or close to the site of contamination. But with the rapid increase in population growth and urbanization the situation is likely to worsen unless particular care is exercised to prevent contamination of beaches and fishing grounds.[63]

60. GESAMP, *Principles for Developing Coastal Water Quality Criteria,* Reports and Studies, no. 5 (Rome: FAO, 1976).
61. C. M. Blackwood, "Canadian Experience on Sewage Pollution of Coastal Waters: Effect on Fish-Plant Water Supplies," in *FAO Seminar on Methods of Detection, Measurement and Monitoring of Pollutants in the Marine Environment* (Rome: FAO, 1970).
62. GFCM/ICSEM Group of Experts on Marine Pollution, "Review of the State of Pollution in the Mediterranean Sea," in Ruvio, pp. 28–32.
63. Hela.

TABLE 8.—CHARACTERISTICS OF DISEASES IN MAN

A. PRINCIPAL BACTERIAL AND VIRAL FISH- AND SHELLFISH-BORNE DISEASES

Disease and Etiological Agent	Principal Aquatic Food Animals Involved as Source of Infection	Sources of Infection or Aquatic Food Animal	Pathogenicity for Aquatic Food Animal	Mode of Transmission to Man	Disease in Man and Most Common Manifestations
Bacterial infection: Salmonella spp.: a) S. typhi, S. paratyphi b) Other species (e.g., S. typhimurium, S. enteritidis)	Fish or shellfish secondarily contaminated through polluted waters or through improper handling	a) Human feces and waters contaminated by human feces b) Human and animal feces, polluted waters	Nonpathogenic	Ingestion of raw or insufficiently cooked contaminated fish or shellfish	a) Typhoid and paratyphoid fever, septicemia b) Salmonellosis gastroenteritis
Vibrio parahaemolyticus	Marine fish and shellfish	Organism occurs naturally in the marine environment	May cause death of shrimps and crabs; experimentally pathogenic for fish	Usually through consumption of raw or inadequately cooked fish or shellfish that has not been properly refrigerated	Diarrhea, abdominal pain
Bacterial intoxication: Clostridium botulinum	Fermented, salted, and smoked fish	Sediment, water, animal feces	Toxin can kill fish	Ingestion of improperly processed fish or shellfish	Botulism: neurological symptoms with high case-fatality rate
Staphylococcus aureus	Fish or shellfish secondarily contaminated through improper handling	Man—nose and throat discharges, skin lesions	Nonpathogenic	Ingestion of fish or shellfish cross-contaminated after cooking	Staphylococcal intoxication: nausea, vomiting, abdominal pain, prostration

Bacterial intravital intoxication:* *Clostridium perfringens*	Fish or shellfish secondarily contaminated through polluted waters or through improper handling	Polluted water, human and animal feces, sediment	Nonpathogenic	Ingestion of cooked fish or shellfish that has not been properly refrigerated	Diarrhea, abdominal pain
Bacterial skin infection: *Erysipelothrix insidiosa*	Fish, particularly spiny ones (e.g., searobins, redfish) —organism is present in fish slime and meat	...	Nonpathogenic	Through skin lesions—usually an occupational disease	Erysipeloid—severe inflammation of superficial cutaneous wounds
Viral infection: Virus of infectious hepatitis	Shellfish	Human feces and water polluted by human feces	Nonpathogenic	Ingestion of raw or inadequately cooked contaminated shellfish	Infectious hepatitis

TABLE 8. Continued

B. PRINCIPAL PARASITIC FISH- AND SHELLFISH-BORNE DISEASES

Disease and Etiological Agent	Principal Aquatic Food Animals Involved as Source of Infection	Life Cycle of Parasite	Pathogenicity for Aquatic Food Animal	Mode of Transmission to Man	Disease in Man and Most Common Manifestations
Parasitic infection: Trematodes:					
Clonorchis sinensis (Chinese liver fluke)	Freshwater fish— Cyprinidae family (e.g., carp, roach, dace)	1st intermediate host: snail; 2nd intermediate host: fish; definitive host: man, dog, cat, other fish-eating mammals	Muscle cyst infection	Ingestion of raw or insufficiently cooked, infected fish (dried, salted, or pickled fish may be involved)	Clonorchiasis: signs and symptoms related to liver damage
Opisthorchis felineus, O. viverrini	Freshwater fish— Cyprinidae family (e.g., whitefish, carp, tench, bream, barbel)	1st int. host: snail; 2nd int. host: fish; def. host: man, dog, fox, cat, other fish-eating mammals	Muscle and subcutaneous cyst infection	Ingestion of raw or insufficiently cooked, infected fish	Opisthorchiasis: cirrhosis of the liver
Heterophyes heterophyes	Freshwater or brackish-water fish	1st int. host: snail; 2nd int. host: fish; def. host: man, dog, cat, fish-eating birds	Encyst in muscles and skin	Ingestion of raw or insufficiently cooked, infected fish (frequently salted or dried fish)	Heterophyiasis: abdominal pain, mucous diarrhea; eggs may be carried to the brain, heart, etc., causing atypical signs
Metagonimus yokogawai	Freshwater fish (e.g., trout, sweetfish, dace, whitebait)	1st int. host: snail; 2nd int. host: fish; def. host: man, dog, pig, cat, fish-eating birds	Encyst in gills, fin, or tail	Ingestion of raw or insufficiently cooked, infected fish	Metagonimiasis: usually mild diarrhea

The Oceans: Health and Prognosis 313

Organism	Source	Location/Larvae	Mode of Infection	Disease
Paragonimus westermani, P. ringeri (Oriental lung fluke)	Freshwater crab and crayfish	1st int. host: snail; 2nd int. host: crab, crayfish; def. host: man, dog, pig, wild carnivores	Encyst in gills, muscles, heart, liver	Ingestion of raw or insufficiently cooked, infected crabs or crayfish, or ingestion of water contaminated by metacercariae that have escaped from a crab or crayfish	Paragonimiasis: usually chronic cough and hemoptysis from flukes localized in the lungs; flukes may invade other organs
Cestodes: *Diphyllobothrium latum*	Freshwater fish (e.g., pike, trout, turbot)	1st int. host: copepod; 2nd int. host: fish; def. host: man, dog, cat, pig, fox, polar bear, other fish-eating animals	Plerocercoid larvae infection of muscles and other organs	Ingestion of raw or insufficiently cooked, fish (frequently inadequately pickled fish)	Diphyllobothriasis: disease may be mild or inapparent; may see signs of gastroenteritis, anemia, weakness
Nematodes: *Anisakis matina*	Marine fish (e.g., cod, herring, mackerel)	...	Internal larvae infection	Usually from ingestion of raw or partially cooked, pickled, or smoked herring	Aniakliasis: eosinophilic enteritis
Anglostrongylus cantonensis	Freshwater shrimp, land crab, possibly certain marine fish	1st int. host: slug, land snail; def. host: rat; paratenic host: shrimp, land crab	...	Ingestion of raw or inadequately cooked shrimp or crabs (sometimes pickled)	Eosinophilic meningitis

*Intoxication by toxin produced in the body by bacteria present in heavily contaminated foods.

The effect of nutrient enrichment of seawater from agricultural land runoff and sewage disposal is not necessarily harmful to the marine environment. But in coastal areas with restricted water exchange, discharge of organic wastes can cause a chain reaction beginning with the oxidation of the organic matter and leading eventually to eutrophication of the surface waters. Although in some cases this may seem beneficial due to increased fish yield in the area, the buildup of decaying organic matter sometimes can lead to total oxygen deficiency which can destroy the benthic biota. Wherever stagnant basins act as nutrient traps, large amounts of nutrient salts can collect at the bottom. When rich deep waters are circulated to the surface, plankton blooms occur, producing more organics to sink and decay and beginning a fertilization cycle which will recur again and again. Recovery of the system from such conditions is difficult.[64]

Radionuclides

There are three categories of radioactive materials which man introduces into the sea: (1) nuclear fuels such as uranium-235 and plutonium-238, which may be released from nuclear-powered ships and satellites, (2) fission products, such as strontium-90, cesium-137, and barium-140 from nuclear detonations and energy production, and (3) activation products such as zinc-65 and iron-55 from nuclear reactors and weapons.[65]

After the first nuclear detonations began in the late 1940s, intensive testing of nuclear weapons in the atmosphere gave rise to global radioactive pollution of land and oceans. Since 1962, however, the rate of atmospheric testing has decreased so considerably that the physical decay of fission products in the environment has outpaced new inputs, and the inventories of such nuclides as ^{90}Sr and ^{137}Cs have fallen significantly.[66]

Other man-made inputs to the radioactivity of the oceans have resulted from the burnup on reentry into the atmosphere of a spacecraft carrying a ^{238}Pu power generator in 1964 and from the crash of a spacecraft carrying atomic weapons near Thule, Greenland, in 1966. The ^{238}Pu and ^{239}Pu activity that these accidents added to the oceans did not contribute significantly to human exposure, and their interest is now purely historical.[67]

The nuclear fuel cycle (mining and milling of uranium ore, fuel fabrication, reactor operation, fuel reprocessing, and waste management) inevitably

---

64. S. H. Fonselius, "On Eutrophication and Pollution in the Baltic Sea," in Ruvio, pp. 23–28.
65. Goldberg.
66. Francesco Sella, personal communication (May 1977).
67. Ibid.

involves releases of radioactive material to the environment. These have been kept within strict limits so that the resulting doses have so far been kept well below the dose limits for humans recommended by the International Commission on Radiological Protection (ICRP).

The importance of radionuclides released into the ocean is due to the ionizing radiation they emit. These may affect adversely all biological systems in proportion to the dose of radiation that the systems absorb. In living marine systems we know only the effects of radiation doses much higher than those due to the current inputs of man-made radionuclides, and these consist of increased mortality of invertebrates and fish in various larval stages.

A study on the cycling of three biologically active radionuclides (manganese, iron, and zinc) in the estuarine environment identified several estuarine organisms which tend to concentrate radionuclides.[68] Although such studies point to possible risks to humans from radioactive seafood products, even at the time of intensive weapons testing the contribution of marine products to the human dietary intake of the artificial radionuclides of main concern for human health was only a fraction of the total dietary intake.

Much more information is available on the effects of ionizing radiation on man, obtained through direct observations on irradiated individuals or by extrapolation of data in experimental animals. The most important long-term effects are the induction of mutations or chromosomal changes, which can be transmitted from the exposed individual to his descendants and may manifest themselves as defects of various degrees of intensity even generations after the exposure, and the induction of malignancies in the irradiated individuals themselves, the site depending on which tissues have been exposed. Both categories of effects are nonspecific and can be due to exposure to other environmental agents. Both can be assumed to occur with frequencies proportional to dose, at least at low doses.

At the doses to which we are now exposed from the current radioactive pollution of the environment, no increase in the frequencies of these effects can be detected.[69] However, the planned expansion of the nuclear industry calls for very strict regulations and surveillance of all releases of radioactivity to the environment, including the oceans. International legal instruments, such as the London Convention for the Prevention of Marine Pollution by Dumping of Wastes and Other Matter (see Appendix, pp. 692–96), are likely to become increasingly necessary if pollution of the oceans by radioactive materials is to be kept within limits consistent with the safety of man and his environment.

68. T. R. Rice, J. P. Baptist, F. A. Cross, and T. W. Duke, "Potential Hazards from Radioactive Pollution of the Estuary," in Ruvio, pp. 272–76.
69. Sella.

Solid Wastes

Disposal of solid wastes can cause somewhat more serious effects than aesthetic assault. The millions of tons of solid materials entering the sea each year (table 9) can interfere with shipping by becoming entangled in ships' propellers, clogging intake systems, or even blocking channels. Rubbish from packaging materials—plastic, metal, cloth, glass, or wood products—often becomes deposited on the bottom with adverse effects on benthic communities. Disturbance of substrate can affect larval settlement, covering of the bottom can lead to anoxia at the sediment surface, and the debris can interfere physically with the movement and growth of benthic organisms.

Damage to intertidal communities from floating materials can be readily observed in areas of intense logging activities, where organisms are scraped repeatedly from rocks and sediment substrates are denuded of grass and invertebrate fauna.

The littering of the marine environment is a serious problem in only a few localized coastal areas. In general this sort of pollution is easily brought under control, and since much of the litter comes from ships, effective control of vessel discharge would reduce the input significantly.

Conclusion

In spite of the ocean's immensity and the eventual dilution and dissemination of polluting substances which enter it, we cannot continue limiting our pattern of response to the pollution threat to mere reaction to individual polluting accidents. Poisonous, persistent, man-made chemicals are in the ocean—in the food web, in the deep-sea sediments, in the water column. We know they are there even when they occur at undetectable levels because we put them there. Once they reach detectable and measurable levels in the ocean system, it is too late to prevent whatever effects they may have on the resident biota.

There is an indisputable need for further basic research on the improvement of analytical techniques for detection of persistent chemicals and on the environmental behaviors, fates, and effects of polluting substances in general. However, we already know quite enough to conclude that mere reaction is an inadequate approach—we must try to prevent further environmental deterioration.

Evidence is beginning to accumulate showing widespread effects of polluting materials on reproductive behavior and physiology of marine organisms and the viability of their offspring. There is increasing risk of species extinctions, which in turn poses the danger of absolute and irreversible alteration of marine communities. Buildup of toxic chemicals in sediments and initiation of fertilization cycles have caused deterioration of certain near-shore

TABLE 9.—TOTAL LITTER ESTIMATES

Source	Amount of Litter Generated (10^6 Tons per Year)
Passenger vessels	.028
Merchant shipping:	
Crew	.110
Cargo	5.600
Recreational boating	.103
Commercial fishing:	
Crew	.340
Gear	.001
Military	.074
Oil drilling and platforms	.004
Catastrophies	.100
Total	6.360

SOURCE.—NAS, *Petroleum in the Marine Environment* (Washington, D.C.: National Academy of Science, 1975).

ecosystems to the point where only massive efforts and long-term energy expenditure can restore them to their previous state. Prevention of such deterioration in the first place is less costly in every way, and we are rapidly developing the technological means for this. But pollution is rarely a local problem in any strict sense, and increased cooperation between neighbors— whether individuals, communities, countries or continents—is essential.

INTERNATIONAL EFFORTS TO PROTECT THE MARINE ENVIRONMENT

Concern for environmental protection can unite even the most economically and politically diversified countries. This has been clearly demonstrated in the few years since the beginning of international efforts to promote cooperation among countries sharing the use and benefits of the oceans.

The general basis for this cooperative effort was furnished by the adoption of the Action Plan for the Human Environment at the 1972 United Nations Conference on the Human Environment at Stockholm and the subsequent formation of the United Nations Environment Programme as mobilizer and coordinator of efforts at all levels to protect and preserve the biosphere from unnatural stresses due to human activities.

The Stockholm Action Plan was the first attempt to develop a comprehensive program for international action in the field of environment. Since it was considered impractical to define the activities of UNEP in terms of a single, totally comprehensive "super-program," efforts were made from the beginning to build up programs in carefully selected priority areas: human settlements and human health; terrestrial ecosystems, their management and control; environment and development; oceans; energy; and natural disasters.

The Stockholm conference decided that an approach to the problems of oceanic pollution requires two categories of action—assessment and control. The assessment category includes (1) the identification of high-priority pollutants; (2) the identification of the sources of marine pollutants and an evaluation of the risks they pose; (3) basic research and monitoring; and (4) the collection, storage, retrieval, and dissemination of information. The control category includes (1) the establishment of maximum acceptable levels of pollutants; (2) limitation of the discharge of pollutants; and (3) promotion and enforcement of compliance. The categories are interrelated in that research can help to identify problems for which control is needed and monitoring may be used, not as an enforcement measure but to check the effectiveness of control measures.

At its third session, UNEP's Governing Council endorsed the following general strategy for its oceans program: (1) promotion of international and regional conventions, guidelines, and action for the control of marine pollution and for the protection and management of aquatic resources; (2) assessment of the state of pollution and of living resources; (3) monitoring marine pollution and aquatic resources.

Due to the variability of the nature of specific pollution problems from one region to another, and the differing political and economic realities which must be accommodated in efforts to solve them, it was decided that a regional application of this strategy was most practicable. Accordingly, governments of countries bordering endangered seas would be encouraged to agree on a common approach to the pollution problems of their area and to propose regional projects to be implemented with technical, financial, and advisory support from cooperating specialized agencies of the UN system. The scientific aspects of each regional program—monitoring and assessment of pollution and determination of its nature, sources, and impact—would be carried out primarily by existing national research institutions.

The regional seas selected as initial concentration areas were the Mediterranean, the Persian Gulf, the Caribbean, and the West African Coast. To these were added East Asian Waters (including the Malacca Straits) and the Red Sea.

The Mediterranean

The Mediterranean Action Plan was adopted at the Intergovernmental Meeting on the Protection of the Mediterranean at Barcelona in early 1975 (see Appendix, pp. 702–33), and its success to date has rendered it a useful model for the other programs.

The Mediterranean region was an ideal place to begin since the Mediterranean Sea is subjected to pollution stress of every variety. It is a relatively well-known sea, scientifically speaking, and has great importance in human terms due to the dependence of millions of its shores' inhabitants on its resources. In 1974, 1,300,000 tons of fish worth $700 million were caught in the Mediterranean. More than 100 million tourists migrate to its coasts every year.

The sea's average depth is 1,500 m, with maxima of 5,000 m; its area is 3.5 million km^2, and its east-west extent is 3,800 km. The sill which separates the Mediterranean from the Atlantic at the Strait of Gibraltar is 365 m deep and only 15 km wide, which makes the Mediterranean nearly an inland sea. The renewal period of its generally hypersaline waters (38 ppt) is estimated at 80 years.[70]

The main pollution problems of the Mediterranean coastal zones are caused by a general lack of adequate treatment of domestic sewage and industrial discharges, the use of pesticides in agriculture, and oil pollution from accidental and operational discharge from vessels. These factors contribute to a general pollution load of toxic chemicals in sediments and biota, an overload of nutrients in certain areas with a resulting increase in BOD (Biological Oxygen Demand), and occurrence of pathogenic organisms in waters and shellfish. The state of open waters is not yet critical, but many coastal zones are considered badly polluted.

[EDITORS' NOTE.—The pollution issue was recognized early by Pacem in Maribus, which in 1971 instituted a project on pollution in the Mediterranean. The results are reported in Lord Ritchie-Calder's *Pollution in the Mediterranean* (Bern: Lang, 1972). In 1972, a major conference on the Mediterranean was held in Split also under PIM (International Ocean Institute) auspices, the proceedings of which appeared as Norton Ginsburg, Sidney Holt, and William Murdoch, eds. *The Mediterranean Marine Environment and the Development of the Region* (Msida: International Ocean Institute, 1974), well in time to have some effect on the terms of the Barcelona Convention.

Parenthetically, the International Ocean Institute also has supported a Caribbean project (see pp. 329–31 below) which began in 1973 with a conference on regional maritime problems held in Kingston, Jamaica, under the auspices of the Government of Jamaica. Several papers from this conference were later published in E. M. Borgese and D. Krieger, *The Tides of Change* (New York: Mason/Charter, 1975), and cooperation continues with CEESTEM and other regional organizations.]

The Barcelona Action Plan's frontal attack on the above problems consists of four components: scientific assessment, legal, integrated planning, and institutional and financial arrangements. We will consider the first three of these.

70. UNDP, *Management of the Fresh-Water Resources of the Mediterranean Basin and Protection of their Quality* (Geneva: UNEP, 1976).

Scientific Assessment

In order to take environmental considerations into account in national planning and decision making, it is necessary that governments have an accurate assessment of the presence and likely effects of pollutants in their region. In the Mediterranean region, institutions exist in every country which have at least the potential capability of making such assessments, but many of these institutions need governmental and international support to help them in the collection of relevant data according to standardized methods, a generally expensive and demanding task.

In September 1974, under UNEP sponsorship, the Intergovernmental Oceanographic Commission (IOC), the General Fisheries Council for the Mediterranean (GFCM) of FAO, and the International Commission for Scientific Exploration of the Mediterranean (ICSEM) convened an International Workshop on Marine Pollution in the Mediterranean, at which 40 scientists from Mediterranean marine research centers reviewed the state of Mediterranean pollution and the facilities already available throughout the region for its assessment. On the basis of the recommendations emanating from this workshop, the 1975 Barcelona meeting approved a Co-ordinated Mediterranean Pollution Monitoring and Research Programme, to be executed by UNEP in collaboration with several specialized UN agencies (the General Fisheries Research Council [GFCM] of the Food and Agriculture Organization [FAO], the Intergovernmental Oceanographic Commission [IOC] of the United Nations Educational, Scientific and Cultural Organization [UNESCO], the World Health Organization [WHO], the World Meteorological Organization [WMO], and the International Atomic Energy Agency [IAEA]). The program consists of seven pilot projects, described as follows:

1. Joint IOC/WMO/UNEP Pilot Project on Baseline Studies and Monitoring of Oil and Petroleum Hydrocarbons in Marine Waters. Despite the concern so often expressed at high levels about the problem of the oil and petroleum hydrocarbon pollution which threatens coastal recreational areas, we know little about actual levels and effects on marine communities. Participants in this pilot project will make visual observations of oil slicks and other floating pollutants from suitably located observation points, sample tar balls from ships and research vessels and, pending the solution to certain methodological and analytical problems, undertake direct seawater sampling and analysis.

2. Joint FAO(GFCM)/UNEP Pilot Project on Baseline Studies and Monitoring of Metals, Particularly Mercury and Cadmium, in Marine Organisms. Due to the potential toxicity to man of accumulated and concentrated heavy metals, this project concentrates its monitoring activities on mercury and cadmium levels in various species of edible marine organisms from different habitats and trophic levels (striped mullet, Mediterranean mussel, bluefin tuna). The monitoring of copper, lead, manganese, selenium and zinc was also recommended.

3. Joint FAO(GFCM)/UNEP Pilot Project on Baseline Studies and

Monitoring of DDT, PCBs and Other Chlorinated Hydrocarbons in Marine Organisms. According to this project, levels of persistent organochlorine compounds will be measured seasonally in selected organisms (striped mullet, Mediterranean mussel, pink shrimp) which are nearly ubiquitous in the Mediterranean and of great economic importance to regional fisheries.

4. Joint FAO(GFCM)/UNEP Pilot Project on Research on the Effects of Pollutants on Marine Organisms and Their Populations. This project deals primarily with the investigations of sublethal effects of potential pollutants on marine organisms and populations of organisms, with attention to functional as well as morphological changes. Research will involve such aspects as effects on sensitive developmental stages, specific mechanisms involved in accumulation and physiological effects of specific pollutants, and damage to genetic material. The overall aim is to provide a general assessment of the relative hazards of polluting chemicals to marine biota in order to develop water quality criteria and background information for future monitoring activities.

5. Joint FAO(GFCM)/UNEP Pilot Project on Research on the Effects of Pollutants on Marine Communities and Ecosystems. The project will focus on the most vulnerable marine communities—those found in coastal waters, lagoons, and brackish coastal lakes. Use will be made of areas which have been studied in the past in order to take natural successional processes into account and to trace the historical response of particular communities to increasing pollution stress. Polluted areas will be compared with clean but physically equivalent ones. Analysis of community structure and determination of functional indices (such as diversity) and body burden of pollutants will be among the most common methods of describing pollution stress effects on target communities.

6. Joint IOC/UNEP Pilot Project on Problems of Coastal Transport of Pollutants. Water circulation in localized coastal areas will be emphasized in this project, since this is where our knowledge of Mediterranean circulation patterns is most deficient. Since most pollutants entering the sea do so via rivers, study of circulation patterns near river outlets and water exchange between coastal and offshore regions will show how and where particular pollutants enter the region and move to other areas.

7. Joint WHO/UNEP Pilot Project on Coastal Water Quality Control. The objective of this project is to produce an adequate system of assessing the present level of coastal pollution which is known to have adverse effect on human health. The population in many areas of the Mediterranean is threatened by salmonellosis, dysentery, viral hepatitis, poliomyelitis, and even cholera. Good correlation data relating disease and water pollution is needed, as well as a program of sanitary and health surveillance of coastal areas used for recreation and aquaculture.

Work began on these projects in late 1975. In spring and summer of 1977 midterm reviews of the pilot projects were held in order to assess their progress. A total of $690,000 worth of equipment and $130,000 in training has already been made available to participating Mediterranean research institutions which lacked adequate facilities or specialized know-how to carry out the program. An intercalibration center helps to insure the comparability of acquired data, and a maintenance service is provided to all recipients of analytical instruments.

The participation of Mediterranean countries in the pilot project, as of April 1977, is summarized in table 10. Figure 1 shows the location of the 73 research institutions participating at that time and the regional activity centers responsible for scientific coordination of the respective projects.

Legal

Provision of international environmental legislation is an essential and obvious first step to pollution control in so politically diverse an area as the Mediterranean. Initial drafts of three international agreements were presented at the 1975 Barcelona meeting, and in February 1976 these were approved in their revised form by 16 of 18 Mediterranean States at the Barcelona Conference of

TABLE 10.—NUMBER OF RESEARCH INSTITUTIONS IN EACH MEDITERRANEAN COUNTRY PARTICIPATING IN THE MEDITERRANEAN POLLUTION MONITORING AND RESEARCH PROGRAM'S SEVEN PILOT PROJECTS

Country	Pilot Project						
	1	2	3	4	5	6	7
Albania	...	...	...	...	...	...	...
Algeria	...	1	...	...	1	...	...
Cyprus	1	1	1	1	1	1	...
Egypt	1	1	1	1	1	1	...
France	9	8	6	4	3	5	...
Greece	3	5	4	4	3	2	2
Israel	1	1	1	1	1	1	4
Italy	...	5	1	1	1	3	8
Lebanon	1	1	1	1	1	1	1
Libya	...	...	...	...	...	...	...
Malta	1	1	1	1	...	1	3
Monaco	...	...	...	...	...	1	1
Morocco	...	2	2	2	1	...	...
Spain	3	2	2	2	2	3	...
Syria	...	...	...	...	...	...	...
Tunisia	1	1	1	...	...	...	...
Turkey	2	5	3	3	2	2	1
Yugoslavia	3	4	4	3	4	2	3
Total	26	38	28	24	21	23	23

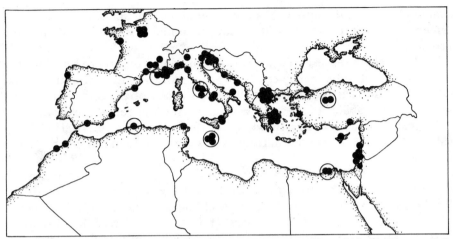

FIG. 1.—Research institutions participating in the Co-ordinated Mediterranean Pollution Monitoring and Research Programme, as of April 1, 1977. Large circles indicate locations of the seven regional activities centers.

Plenipotentiaries of the Coastal States of the Mediterranean Region on the Protection of the Mediterranean Sea (text can be found in the Appendix on pp. 702–33 of this issue).

The first of these legal instruments was the Convention for the Protection of the Mediterranean Sea against Pollution. This convention sets up a general framework for cooperation between governments and the development of additional legal protocols, institutional arrangements, and procedures for the settlement of disputes. By April 1977 it had been signed by Cyprus, Egypt, France, Greece, Israel, Italy, Lebanon, Libya, Malta, Monaco, Morocco, Spain, Tunisia, Turkey, Yugoslavia and the European Economic Commission (EEC).

The second agreement was the Protocol for the Prevention of Pollution of the Mediterranean Sea by Dumping from Ships and Aircraft, which included an annex listing prohibited substances such as organohalogens, organosilicons, mercury, cadmium, persistent synthetic materials, crude oil and petroleum derivatives, radioactive wastes, acid and alkaline compounds, and any form of materials used for biological or chemical warfare. A special permit is required for the dumping of such materials as arsenic, lead, copper, zinc, other metals, pesticides, cyanides, fluorides, and synthetic chemicals. The dumping of other wastes would require a general permit issued by competent national authorities. All of the above-mentioned states and the EEC have signed this protocol.

The third agreement is the Protocol concerning Co-operation in Combating Pollution in the Mediterranean Sea by Oil and Other Harmful Substances in Cases of Emergency. In this document great importance was attributed to communication and mutual assistance among coastal states and a regional center. By April 1977 all of the above-mentioned states had signed the protocol.

The 16 states represented at the 1976 Barcelona conference called for the establishment of a regional oil-combating center in Malta to serve as the central information service and to advise those concerned of the occurrence of oil spills, the availability of cleanup assistance, the danger of environmental damage, and similar matters. In December 1976, through efforts of UNEP and IMCO (Inter-governmental Maritime Consultative Organization) the center was inaugurated.

In February 1977 UNEP convened an intergovernmental meeting in Athens to discuss principles for a Draft Protocol for the Protection of the Mediterranean from Land-based Sources of Pollution. Also, UNEP is cooperating with several other UN agencies (WHO, FAO, UNESCO, Economic Commission for Europe [ECE], United Nations Industrial Development Organization [UNIDO], and IAEA) on a related project which will provide information on the type and quantity of pollution from major land-based sources which enters the sea via rivers or directly from off-shore man-made structures or coastal dumping. The project will also examine the status of waste discharge and water pollution management practices in the region.

Integrated Planning
The purpose of the integrated planning component of the Mediterranean Action Plan is to encourage continuing economic and social development in the Mediterranean on the basis of environmental principles. Continuing development is both desirable and necessary, but it must—and can—take place without further environmental degradation.

Work preparatory to the intergovernmental meeting held in Split, Yugoslavia, at the beginning of 1977 disclosed the necessity of two lines of approach to the problems of environmentally sound development in the Mediterranean region.

First, there must be in-depth study to identify the imminent risks to the environment accompanying the processes of development throughout the region. This study program, titled the Blue Plan, will lead to recommendations on means of integrating environmental considerations into development plans, policies, and decisions. The objective is to place at the disposal of decision makers and planners in the Mediterranean countries information enabling them to formulate methods for optimum socioeconomic development on a sustainable basis—without environmental degradation. Centralized activities will include conducting a survey of major development activities now being carried out in the region and their environmental consequences, compilation of a directory of regional institutions and experts available for participation in the plan, and the provision of technical assistance and training in such fields as industrialization and urbanization, energy production, fisheries, agriculture, coastal zone management, and water resource management. Priority for such assistance will be assigned to developing countries.

A parallel line of approach is to initiate cooperative efforts by coastal states regarding the adoption of appropriate environmental management practices in selected priority areas. Termed the Priority Actions Programme, this approach will entail the identification of alternatives for environmentally sound development in the six initial fields recommended by delegates at the Split meeting. These are (1) protection of soil, (2) management of water resources, (3) marine living resources: management of fisheries and aquaculture, (4) human settlements, (5) tourism, and (6) soft technologies for energy, including solar energy.

The progress made in each of the four components of the Mediterranean Action Plan will be reported to governments at a general review meeting to be held in Monaco in late 1977.

East Asian Waters

East Asia (fig. 2) is an area of extreme geographical, cultural, and environmental complexity. It includes both temperate and tropical ecosystems which are threatened by a wide variety of man's activities (table 11).[71] Little work has been done to assess the effects of these activities on the particular marine communities of the region.

In April of 1976, IOC of UNESCO, FAO, and UNEP held a scientific workshop in Penang, Malaysia, on marine pollution in East Asian waters in an attempt to set up an action plan for the region. Regional and extraregional experts identified sources of pollutants, determined research priorities, and proposed pilot projects based on the particular problems of each subregion. Table 12 summarizes the priority grouping of categories of pollution and their regional aspects.[72]

The workshop recommended the following regional pilot projects:

1. Mangrove systems as sewage and sediment buffer zones. Mangrove forests are fragile systems which are being destroyed on a large scale. They are important spawning and feeding grounds for economically important aquatic species. The aim of this project is to provide a basis for their preservation and rational use.

2. Comparative study of metals using oysters as indicators. Oysters occur naturally throughout the region and are cultured commercially in many areas. Since they have been shown to accumulate heavy metals and to harbor human pathogens they are considered valuable bioindicators of pollution. The project

---

71. IOC, *Report of the IOC/FAO(IPFC)/UNEP International Workshop on Marine Pollution in East Asia,* IOC Workshop Report, no. 8 (Paris: UNESCO, 1976).
72. Ibid.

FIG. 2.—East Asian waters and the subregions for which pilot projects have been proposed.

will aim at the establishment of regional contamination levels and guidelines for standards of hygiene.

3. Studies of red tides in East Asian Waters. Red tides, which occur globally, have been increasing in the region and may be a consequence of pollution. Through study of the species responsible for these phenomena, their size, concentration, and toxicity to marine organisms, scientists hope to determine their cause and relation to other occurrences.

4. A study of physical dispersal processes in coastal waters. The effects of pollution from sewage outfalls from populated or industrialized areas are largely dependent on the natural dispersal, mixing, and transport of the pollutants. Knowledge of the dynamic physical processes of the sea in the area receiving the effluent is necessary before appropriate disposal criteria can be determined.

TABLE 11.—MAIN HUMAN ACTIVITIES AND THE MAIN POLLUTANTS ASSOCIATED WITH THEM IN EAST ASIA

Types of Activity	Type of Pollutant of Environmental Problem
Urbanization	Domestic sewage; earth-moving and land reclamation; dredging of harbors, etc.; industrial wastes
Agriculture, forestry	Pesticides, particularly chlorinated hydrocarbons, organophosphates, carbamates, fertilizers (hypertrophication); silt
Oil extraction, refinery and transport, dispersants	Oil and oil dispersants
Mining (including seabed mining)	Metals and metalloids (tin, copper, nickel arsenic; silt; destruction of corals by silt or direct mining for builing materials
Metallurgy	Metals, especially copper, zinc, nickel, cadmium
Cellulose	Organochlorine compounds from chlorine bleaching; mercury from production of caustic soda and chlorine used in cellulose treatment, organic slimicides
Plastics	By-products from vinyl chloride production, monomers, cadmium and other stabilizers; plastic litter
Power generation	Heat; radioactive waste
Desalination	Heat; salt

Source.—IOC, *Report of the IOC/FAO(IPFC)/UNEP International Workshop on Marine Pollution in East Asia*, IOC Workshop Report, no. 8 (Paris: UNESCO, 1976).

In addition to the above, there were six subregions for which pilot projects were proposed to deal with specific local problems.

1. Bay of Bengal. This is an area of increasing industrial and agricultural development, resulting in a large influx of pesticides, metals, and untreated wastes. These factors, as well as the heavy siltation from river erosion and drilling, mining, and filling operations, threaten the fisheries resources and commercial seaweed production of the area, not to mention the health of local inhabitants. The proposed pilot project will assess levels of DDT and heavy metals in the bay and its organisms, the hazards to human health from sewage disposal practices, and the effect of siltation on fisheries.

2. Malacca Straits. Due to extremely heavy navigation through the Straits, including annual transport of more than 300 million tons of crude oil and its derivatives, a project on oil-pollution assessment and its impact on living resources was proposed. Other projects deal with effects of sedimentation due to river discharge and mining, land-clearing and dredging activities; the monitoring of tin, arsenic, and lead in certain indicator organisms; and the effects of untreated sewage on human health.

3. Gulf of Thailand. Several countries in this subregion are planning nuclear power plants whose discharges may cause severe thermal stress effects on tropical marine organisms. A long-term research program on the effects of

TABLE 12.—MAJOR MARINE POLLUTANTS IN EAST ASIA IN APPROXIMATE ORDER OF PRIORITY BY GROUPS AND THE COUNTRIES HAVING SPECIAL CONCERN WITH THEM OR BEING AFFECTED BY THEM

Priority Grouping and Pollutant	Country and/or Area Having Special Concern or Being Affected*
1:	
Oil	*Malacca Straits,* South China Seas, Seto Inland Sea, Tokyo Bay, Hong Kong, *Singapore,* Ulsan (Korea), Manila Bay
DDT, pesticides, organochlorines, etc.	India,† Philippines, Vietnam, Korea, Malaysia, Indonesia, Japan, Thailand, Australia (North)
Heavy metals	India, Philippines, Japan, Hong Kong, Indonesia, Malaysia
Organic and biological pollutants, fertilizers	Singapore, Thailand, Korea, Hong Kong, India, Indonesia, Japan, Malaysia, Philippines, Australia (North; fertilizers)
Silt	*Malacca Straits,* Marinduque and *Cebu* (Philippines), *Phuket* (Thailand), India, Hong Kong, Kalimantan and Java (Indonesia), Sarawak and Sabah (Malaysia), Korea, Seto Inland Sea (Japan), Australia (North)
2:	
Heat	*East coast of Gulf of Thailand,* Bintulu (Sarawak), Bagac Bay (West coast of Luzon, Philippines), Straits of Johore, Hong Kong
Metalloids	Japan, Indonesia, Malaysia, India, Philippines
3:	
Plastics	Hong Kong, Japan
Radioactive wastes	Japan
Salt (desalination)	*Hong Kong,* Japan, Singapore

SOURCE.—IOC, *Report of the IOC/FAO(IPFC)/UNEP International Workshop on Marine Pollution in East Asia,* IOC Workshop Report, no. 8 (Paris: UNESCO, 1976).
*Based on available information. Italicized areas were those specifically identified as being high priority areas.
†Only eastern India was considered.

thermal effluents on regional marine organisms was considered necessary prior to the establishment of disposal criteria. Other projects in this region deal with the effects on coastal ecosystems of the increased organic load from agro-industrial wastes (primarily from palm-oil, rubber, and tapioca-starch industries), and with the assimilation capacity of local waters for disposed wastes.

4. South China Sea. This is an area of mangrove forests which provide resources vital to aquaculture and inshore fishing industries. A project on the effects of local pollutants on the integrity and productivity of this ecosystem will be initiated, along with others on the levels and effects of toxic metals, siltation, and oil on marine biota and coastal resources of the subregion.

5. Sea of Japan, Yellow Sea, East China Sea. This subregion, comprised of

both coastal and open-water systems, will be investigated to determine general pollution trends. Other projects will deal with the distribution of heavy metals and organochlorines in organisms, sediments, and water; the degree of shellfish contamination and self-purification; and eutrophication effects with emphasis on the Red Tides common in the area.

6. Eastern Archipelago. The high population density of the area makes sewage disposal a particular threat to the fragile tropical ecosystems found here. It is thought that mangrove communities might be used as disposal areas, but thorough investigation of the ability of these areas to assimilate wastes must be undertaken before this disposal method can be adopted. Coral reefs abound in the region and are vital as spawning grounds, nurseries, shelters, and feeding grounds for tropical species. The reefs are threatened by siltation and coral harvesting, and a project was proposed on the effects of these factors on reefs and adjacent communities.

Since the Penang workshop, efforts at local and international levels have focused on the establishment of effective research network and monitoring systems so that these projects may be implemented as quickly as possible. The opening of lines of communication in this politically complex region is of primary importance to the success of all aspects of the proposed action plan.

The Caribbean

The Caribbean region comprises an assortment of complex and fragile tropical and subtropical ecosystems which extend throughout the Caribbean Sea and Gulf of Mexico west of the Greater and Lesser Antilles (fig. 3). The total area of this region is 4.31×10^6 km^2, and its mean depth is 2,174 m.[73] A great deal of basic oceanographic research has been carried out in the region by the numerous institutions on its coasts, and much is already known about the hydrographic characteristics of particular importance in the accumulation and transport of pollutants. The International Workshop on Marine Pollution in the Caribbean and Adjacent Regions was held in Port-of-Spain, Trinidad, in December 1976 by IOC, FAO, and UNEP. In relation to this meeting, UNEP had begun developing a regional action plan for the wider Caribbean region comprised of the following subjects: protection of the marine environment; promotion of environmental health; human settlements; tourism; industrial, technical and natural resources development; natural disasters. The Caribbean Action Plan will be carried out in cooperation with the Economic Commission for Latin America (ECLA), with emphasis on the management of coastal development problems. The work will be undertaken almost exclusively by the

73. FAO/WECAFC Secretariat, *Preliminary Review of Problems of Marine Pollution in the Caribbean and Adjacent Regions*, doc. IOC/FAO/UNEP/IWMPCAR/6 (Rome: FAO, 1976).

Fig. 3.—The Caribbean region and coastal states

experts and institutions of the region, supported by extensive training and other means of strengthening the capabilities of the national institutions.

After determining the nature of regional marine pollution problems, the Trinidad workshop recommended the following projects:

1. Sources, effects, and fates of petroleum and petroleum products in the Caribbean, Gulf of Mexico, and adjacent regions. This project should help identify the impact of oil production and transportation in the region on tourism, recreation, and food production by examining the pathways by which petroleum hydrocarbons enter the sea, their effect on coastal marine communities, and their ultimate fate.

2. Health aspects of the disposal of human wastes into the marine environment. By considering only those pollutants known to enter the Caribbean marine environment and to be related to human disease, the project should help to guide the development of water quality criteria for recreational waters and areas where aquaculture is practiced and recommend means of improving methods of sewage treatment and disposal.

3. Investigation of the hydrological regime as it affects the transport and fate of pollutants in coastal lagoons and estuaries. The transition areas between freshwater and marine environments are generally productive areas which serve as habitat, breeding grounds, and nurseries of many marine species. The project should provide information important for the preservation of these areas.

4. The effect of medium-scale eddies in the transfer and mixing of pollutants. Eddies can trap pollutants in coastal regions and greatly influence their transport. A study of their structure, persistence, and renewal should lead to better understanding of the distribution and fate of pollutants—where they are likely to concentrate and have their greatest impact.

5. The effect of pollutants, especially those from domestic and industrial sewage, on tropical ecosystems of economic importance. Through bioassay and toxicity tests on the resident species of such systems as mangrove swamps and coral reefs, the project should provide data helpful to the establishment of water quality criteria.

6. Baseline and monitoring studies of persistent chemicals in the Caribbean, Gulf of Mexico and adjacent regions. Due to population pressures and agrarian economies on many Caribbean countries, the input of organochlorine insecticides and PCBs is increasing rapidly. Information on which regulatory measures can be based should be provided by regional monitoring activities.

7. Controlled experiments on the effects of pollutants on tropical marine organisms and ecological communities. Tropical ecosystems beg study since most research in community ecology has been carried out in temperate regions. The objective of this project is to assess the effects of pollutants on tropical systems through controlled field experiments.

Another international environmental organization which has been active in the Caribbean is the International Union for the Conservation of Nature (IUCN). As part of the IUCN Caribbean Regional Programme, in cooperation with UNEP and the Tropical Agricultural Research and Training Centre (CATIE), a regional survey of national parks and reserves in the Caribbean was begun in 1976. A major component of the regional program is the Caribbean activity of the IUCN/World Wildlife Fund (WWF) Marine Programme, which will involve such activities as the establishment of reserves and sanctuaries for endangered species and mapping surveys of threatened ecosystems (mangrove forests, coral reefs, etc.). Implementation of the program should take place in late 1977 or early 1978.

The Persian Gulf

The Persian Gulf (fig. 4) is a shallow, marginal sea 1,000 km in length and 200–300 km in width, with a surface area of about 224,000 m^2 and an average depth of 35 m. The waters are warm (14° C average in winter, up to 35° C in summer) and highly saline (over 40 ppt) due to the high evaporation rates which also cause a relative lowering of sea level and a strong surface current flowing into the gulf through the Straits of Hormuz. A compensating outflow of high salinity water occurs at deeper levels.[74]

74. UNESCO, *Marine Sciences in the Gulf Area*, UNESCO Technical Papers in Marine Science, no. 26 (Paris: UNESCO, 1976).

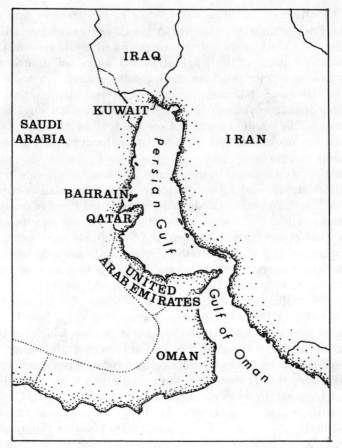

Fig. 4.—Countries which are considering an action plan for the Persian Gulf region

The coasts of this body of water are undergoing the world's most rapid development. Unprecedented oil development, industrial growth, and diversification and urbanization are taking place in a fragile natural environment. The sea has a low purging capacity for the increasing industrial and urban sewage and the huge brine output from numerous desalinization plants. Accidental oil spills or deliberate discharge could be extremely hazardous to local marine communities.

In March, April, and May of 1976, as a cooperative effort of UNEP and ESA (Department of Economic and Social Affairs of the UN Secretariat), a seven-man interagency field mission was sent to the coastal states of the Persian Gulf to collect information, define problems, and draw up a statement on the technical and scientific capabilities of the area preliminary to the formulation of an action plan for the region. A draft action plan was proposed at a consultation of experts in Kuwait, December 1976, and governments of the region met in Baghdad in June 1977 to consider the proposed action plan before its presentation to an intergovernmental meeting in Kuwait, October 1977. The plan consists of the following components:

1. Environmental assessment. With human health and well-being as their ultimate goal, a number of measures were recommended, including identification of factors currently influencing the quality of the marine environment in the gulf, compilation of a regional directory of institutions and experts available for participation, and development of coordinated regional marine science and meteorological programs.

2. Development and environmental management. Activities for which regional management policies should be formulated include coastal area development, rational exploitation of fishery resources and development of aquaculture, contingency planning for combating pollution, and establishment of water quality standards. In addition, an extensive training program of local experts was recommended in areas such as food protection and sanitary inspection, pollution control in urban areas, land use, and industrial plant siting.

3. Legal aspects. Legal experts met in Bahrain in January 1977 to prepare a draft convention which would provide the basis for regional cooperation to protect the marine environment. This convention, the Kuwait Regional Convention for Co-operation on the Protection of the Marine Environment from Pollution, as well as an additional protocol on cooperation in pollution emergencies, was then considered at the October Kuwait meeting. It was proposed that protocols on land-based sources, seabed exploration, and scientific and technical cooperation be developed in the future.

The West African Coast

The Gulf of Guinea is a regional sea which comprises both tropical and subtropical systems. It is generally a shallow and variable region due to upwelling and shifts of the Benguela current. Its coasts are witnessing highly localized and specialized development which results in somewhat unique problems for their human inhabitants.

For nine weeks beginning in April 1976, a UNEP exploratory mission surveyed pollution problems in West African countries which border on the Gulf of Guinea. The mission visited 14 coastal countries and two island states, indicated in figure 5. The report of the mission described the following types and sources of pollution in the region:[75]

1. Oil pollution. Nearly every country visited expressed concern about the

---

75. M. P. Angot and D. Kaniaru, *Marine Pollution Problems of the West African Coastal Countries of the Gulf of Guinea* (Nairobi: UNEP, 1976).

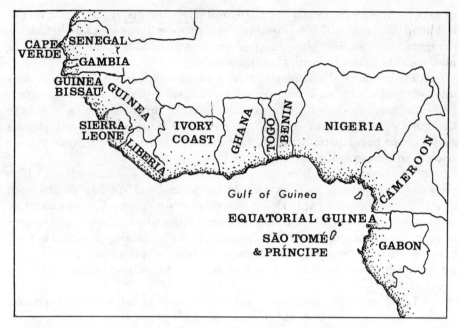

Fig. 5.—The coastal states of West Africa visited by the UNEP exploratory mission

problem of oil polluting the beaches, primarily as a result of tankers flushing their tanks at sea rather than at appropriate shore facilities.

2. Sewage. There are no treatment plants for waste in any of the cities of the region, and few have adequate sewage systems. In some cases untreated sewage has been discharged into lagoons, killing most of the organisms in the worst case, reducing fishery resources in another.

3. Industrial waste. Nearly all the countries surveyed have oil refineries or plan to establish them in the future, and the existing ones have little or no treatment of waste discharge. Other industries which threaten to pollute the region are phosphate, sugar, paper and pulp factories, and chemical, cement, and fertilizer manufacturers. Agricultural development will also add to the chemical load of the coastal sea, due to increased use of fertilizers and pesticides.

4. Logs. One potentially serious hazard is that of logs floating at sea or obstructing beaches and waterways. Due to the increasing value of the wood lost or discarded, the problems caused by such carelessness may disappear without the need for legislation.

5. Coastal erosion. Although some erosion is natural, it appears that much is the result of human activities in this region, such as the building of harbors and breakwaters which interrupt normal coastal current patterns. Beaches and roadbeds are being badly eroded in some areas, and fishing villages occasionally have been displaced.

On the basis of these findings, the mission recommended the development of a convention to protect the marine environment against these forms of pollution and recommended guidelines for their abatement.

In November 1976 UNEP convened an interagency consultation to consider draft principles for a possible regional convention. In addition, this meeting proposed the development of a comprehensive action plan for the region, including a legal component which was discussed at a meeting of government experts in late 1977. The scientific input to the action plan for this region will be formulated by an experts' workshop on marine pollution, organized by IOC in cooperation with UNEP.

All West African countries contacted so far have demonstrated support and willingness to cooperate in efforts to abate pollution in their region, and it is expected that a regional convention will materialize in the near future.

The Red Sea

The Red Sea (fig. 6) is 2,000 km long and has an average depth of about 500 m. Due to high evaporation rates its waters reach a salinity of 40–42 ppt as they flow northward, eventually cooling and sinking to create a deep countercurrent which flows over southern sill through the Gulf of Aden and into the Indian Ocean. The circulation pattern reverses from June to September, and intermediate waters from the Gulf of Aden carry nutrients into the relatively nutrient-poor sea. There is little fishing of any kind, and rates of primary production appear to be low.[76]

The Red Sea is of unique geological and biological interest. It seems to be a new ocean in the first stages of formation and contains the northernmost coral reefs in the world. The possible effect of its elevated temperatures and salinities on its marine inhabitants, the presence of endemic species, the relatively low species diversity compared with the Indian Ocean, and the possible effect of the reopening of the Suez Canal on species composition on either side are all questions which render the sea especially intriguing to scientists.

To date the Red Sea is relatively unpolluted due to low population densities and scarcity of industry on its shores. Pollution, where it exists, is extremely localized. But like the Persian Gulf, the Red Sea is an area of rapid increase in oil exploration, production, and shipping, and the environmental impact is expected to be considerable. Pollution by oil from spills and intentional discharge as well as the physical disturbance of construction and dredging operations are all potentially stressful to local benthic communities.

In October 1974 UNESCO responded to a request by the Arab League Educational, Cultural and Scientific Organization (ALESCO) to convene a scientists' Workshop on Marine Sciences Programmes for the Red Sea, which

76. UNESCO, *Marine Science Programme for the Red Sea*, UNESCO Technical Papers in Marine Science, no. 25 (Paris: UNESCO, 1976).

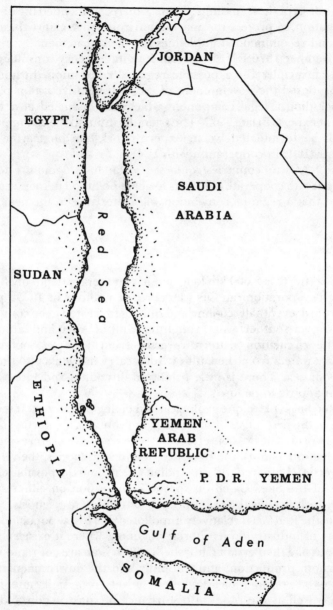

Fig. 6.—The eight countries which participated in the 1976 ALESCO conference at Jeddah.

was held in Bremerhaven, Germany. The workshop report provided the scientific basis for the development of an action plan for the region at two subsequent expert meetings in Jeddah, Saudi Arabia, in late 1974 and early 1976. The plan, adopted at the 1976 conference, includes a draft convention and protocols and specific proposals for a scientific research and monitoring program. These proposals include development of both regional and national research institutions and the establishment of training courses and symposia on marine science teaching. Follow-up activities to these proposals are being coordinated by ALESCO, and a third Jeddah conference is planned for 1978. At the request of ALESCO, UNEP has agreed to participate in this program, contributing to the costs of training scientists and technologists who will operate the marine research and monitoring stations and to the provision of equipment, and carrying out a survey of living marine resources in the Red Sea, especially turtles and dugongs, with a view to restoring their populations for sustained productivity.

Also, IUCN has been active in the Red Sea region, urging the establishment of marine reserves, in cooperation with UNESCO's Man and the Biosphere Programme, for the purpose of nature conservation, tourism, education, and scientific research.

Other Regions of Interest

Besides the six priority regions in which UNEP has concentrated its recent efforts, there are other areas where international cooperation for environmental protection has taken place or is envisaged.

The Baltic
One of the best examples of regional cooperation among countries for the purpose of protecting a shared body of water is that of the Baltic (fig. 7). The Baltic is among the most polluted seas in the world, due to its restricted water exchange and long shore lines. It is a partially stagnant basin, with low nutrient content in its surface waters and accumulation in deep water. Chemical pollution is also a problem, and up to 5 mg mercury per kilogram has been found in fish from the Swedish coast. Baltic organisms contain about 8–10 percent higher concentrations of PCBs and DDT than those from the Swedish west coast, and incidents of ships dumping their oil occur daily.[77]

In 1973 a meeting of experts from the Baltic States took place in Helsinki to identify and discuss the Baltic Sea's unique features and the potential problems from pollution and the encroachment of human activities on species' habitats. In March 1974 the seven Baltic countries signed the convention on the Protection of the Marine Environment of the Baltic Sea Area. The Interim Baltic Marine Environment Commission first met in November 1974 and established a scientific-technological working group which would examine environmental problems and scientific activities in the region and make recommendations concerning possibilities of cooperation and coordination among regional research institutions.

77. Fonselius.

Fig. 7.—The Baltic States

By November 1976 the third Baltic commission meeting was acting on recommendations of the working group concerning, among other matters, criteria and standards for discharge of noxious substances, assessment of conditions of the marine environment, and development of cooperative relations among Baltic countries and international organizations.

The North Sea

The North Sea region provides another example of extensive cooperation between states in the protection of the marine environment. Although there is no single, comprehensive plan for this area, governments of the region have been responsible for the development of a number of legal instruments concerning marine pollution, including (1) Agreement concerning Cooperation in Dealing with Pollution of the North Sea by Oil (1969), (2)

Convention for the Prevention of Marine Pollution by Dumping from Ships and Aircraft (1972), (3) Convention for the Prevention of Marine Pollution from Land-based Sources (1974), and (4) Convention on Civil Liability for Oil Pollution Damage from Offshore Operations (1975).

The South Pacific
Yet another area of increasing international activities relating to the marine environment is the South Pacific, where, at the request of southeast Pacific countries, IOC and FAO have organized a workshop on the scientific aspects of marine pollution in this region, in Santiago, Chile.

The action plans developed for each of the above regions are designed to contribute to development within the region in a way that will meet basic human needs on a sustainable, environmentally sound basis. At the international level, extensive mutual consultations and interactions are required in order that the resources available throughout the UN system can be put at the disposal of the regional governments.

The role of UNEP in these regions is to act as a catalyst for the initiation of regional activities to protect against a commonly perceived danger. Acting in concert with the other UN agencies, it provides a forum—a common meeting ground—for the governments of a region and offers general guidance in the formulation of an interdisciplinary, action-oriented approach appropriate to the particular character of the local environment. When necessary, UNEP also provides a portion of the financial support needed to get the action started.

CONCLUSIONS

Although we are finally learning how to prevent rather than just react to cases of environmental degradation due to pollution stress, we must go far beyond mere methodology to actually accomplish this goal. The global effort to "protect and enhance the present and future quality of the environment for human life and well-being," which is the stated objective of the Stockholm Action Plan, requires enlightened public opinion, political support, and engagement of the skills of the scientific community on a worldwide basis.

Where once environmentalists were pessimistic about saving the oceans—as we know them—from pollution sickness and slow death, we now believe that efforts in the last few years have changed the prognosis. Where before there was ignorance and apathy, there is now increased individual awareness of the natural environment, increased willingness to respond to the warnings inherent in scientists' findings, and determination to affect the course of environmental events. We hope that UNEP and the many other agencies concerned with environmental matters have done their share to coordinate and consolidate the good intentions and energetic efforts of those who wish to maintain the health of their good Earth.

Environment

Radioactive Waste Disposal in the Oceans

Robert A. Frosch, Charles D. Hollister, and David A. Deese
Woods Hole Oceanographic Institution

I. THE PROBLEM

One of the prime issues in deciding whether the United States and other countries should rely on nuclear power as a major energy source arises from the nature of the wastes produced in the commercial nuclear fuel cycle. High-level nuclear wastes are extremely toxic, with some fission product radionuclides having effective lifetimes of more than 1 million yr.

While some think that this long-lived toxicity poses a unique problem, many substances in common use also have long toxic lifetimes. For example, arsenic and heavy metals, such as lead, are indefinitely toxic. Usually, we are not faced with having to dispose of large quantities of these materials, although we do find them in our dwellings and elsewhere in amounts that possibly could be fatal if swallowed or otherwise put into the human body. These poisons, though, do not pose a "hazard from mere proximity" as some radioactive materials do. Contact with lead, for example, is not dangerous, but even short contact with cesium 137 can be. We are only beginning to appreciate the dangers of these nonradioactive poisons.

As we delve into this problem, we should keep in mind that nearly all energy systems produce wastes in one form or another that pose a danger to man and his environment. We are thus faced with a complex system of trade-offs among various sources of energy that all carry some penalty for use. The comparison with other toxins and energy systems, however, does not change the problem created by dangerous commercial nuclear waste. This material must be stored or disposed of in a manner that will be safe from a human and environmental point of view.

Failure to provide convincing plans for the management of this waste resulted in a July 1976 decision by the U.S. Court of Appeals in the *Vermont Yankee* case. The court decision had the effect of barring the Nuclear Regulatory Commission (NRC) from licensing commercial nuclear power

---

Based on three articles by Frosch, Hollister, and Deese, in *Oceanus*, vol. 20 (Winter 1977). The three sections of this report were prepared separately by Frosch, Hollister, and Deese.

© 1978 by The University of Chicago. 0-226-06602-9/77/1978-1013$00.95

plants in the United States until the court is convinced that the NRC has adequately examined the question of disposition of waste.

In its decision, the court noted: "Once a series of reactors is operating, it is too late to consider whether the wastes they generate should have been produced, no matter how costly and impractical reprocessing and waste disposal turn out to be; all that remain are engineering details to make the best of the situation which has been created." It also commented that the decisions to license nuclear reactors were "a paradigm of irreversible and irretrievable commitments of resources" that must receive "detailed" analysis under the National Environmental Policy Act.

The United States is not alone in this problem. The question of inadequate waste management recently became a public issue in Britain after the Royal Commission on Environmental Pollution issued a report that stated: "There should be no commitment to a large programme of nuclear power until it has been demonstrated beyond reasonable doubt that a method exists to ensure the safe containment of long-lived highly radioactive waste for the indefinite future." Other countries, too, are examining this question closely.

The growing international awareness of the dangers inherent in commercial nuclear wastes undoubtedly has been influenced by the fact that, after 30 yr of manufacturing plutonium for nuclear weapons, the United States has not firmly established management plans for the final disposal of millions of gallons of high-level wastes produced for military programs.

Estimates vary on the time when waste from commercial reactors in the United States would equal that produced in our weapons program. The precise time depends on the rate of expansion of commercial nuclear power and what happens in the future in regard to the production of weapons plutonium. The "crossover" dates also vary, depending on whether the estimate is based on the volume or the activity of waste. Reactors that produce plutonium for weapons yield wastes with lower radioactive intensities per unit volume than do commercial power reactors. The amount of a radioactive material is usually stated in curies (a curie is a quantity of radioactive material that undergoes 37 billion disintegrations/sec, which is equivalent to the radiation intensity of 1 g of radium).

Since shortly after the beginning of the nuclear era in 1942, the majority of military waste has been stored at government sites at Hanford in the state of Washington and at Savannah River, North Carolina. In addition, high-level wastes from nuclear-powered vessels have been kept at Idaho Falls, Idaho.

These waste materials are stored in several forms. At Hanford, 50 million curies of strontium 90 and cesium 137, the most intensely radioactive elements (many curies per gram) and those generating the most heat, have been separated out and are stored in heavy steel canisters in water pools. This represents about 80–90 percent of the strontium and cesium in the waste. The remaining materials — residual fission products, trivalent actinides, and a small amount of unextractable plutonium — are stored in large tanks as liquids, sludge, and sodium nitrate salt cake. The current inventory is about 75 million gal of liquids, sludge, and salt cake. Of this amount, about 22 million gal of liquid and 29.5 million gal of salt and sludge are stored at Hanford. At Savannah River, about 25 million gal of unseparated alkaline liquids, salt cake, and sludges are stored in double-walled steel tanks provided with cooling coils to remove the heat produced by the strontium and cesium.

The total amount of nuclear waste expected to be in storage at Hanford sometime after 1980 has been estimated to be 360 million curies. The total military waste at all sites in the United States during this period will probably be about 500 million curies, the exact amount depending upon details of future weapons production programs. According to the Energy Research and Development Administration (ERDA), the accumulated solidified high-level wastes from commercial Light-Water Reactors (LWRs) at a federal repository would reach 500 million curies between 1988 and 1999. This is based on an estimate of 254,000 mw installed LWR electrical capacity by 1988. Since this is two-thirds of the total nuclear capacity estimated to be installed by then, the crossover date on this estimate — when commercial wastes will exist in greater quantities than military wastes — comes somewhat earlier in the 1980s.

The failure to plan satisfactorily for the long-term disposal of military waste has led to skepticism in many quarters that the United States can manage the waste from civilian power reactors. While the fact that proper action was not taken in the past does not mean that it cannot be taken now, the history of past management leads to a perhaps justified lack of confidence that proper planning and management will be done, or can be done.

It should be noted, however, that no one has been harmed by our stored military wastes. In principle, they could, with care and vigilance, continue to be stored in tanks for a long time. There have been several leaks from the tanks, which have left radionuclides in the soil in the immediate vicinity. We must face the fact that no metal tank left in the open, or in shallow burial, can be expected to last without corrosion and related problems for hundreds of years. Thus, continued care, maintenance, and replacement of tanks is a necessity. Indefinite use of such a system poses a hazard to the workers involved and requires a commitment of almost endless human attention.

In any case, the very existence of the military high-level waste means that we already have a management problem that must be solved whether or not we commit ourselves to a major expansion of nuclear power.

The long-lived nature of the radioactive wastes (involving periods up to and beyond 1 million yr) also raises a number of new questions about social responsibility. We are quite used to worrying about the effect of policy on ourselves, our children, and our grandchildren, but our concern gets more difficult to define and deal with when it extends beyond that stage. How are we to think about the problem of management of radioactive materials when the danger may last for times similar to the archaeological history of man? Can we devise a management plan that will somehow continue to operate longer than any known human system, longer than some geological and climatological times? Should we even try to do this? Or should we assume that generations to come will somehow improve upon the actions that we take?

Other major questions of social import arise, too. The largest is connected with the nonproliferation of nuclear weapons in which we all have a stake. In the commercial energy cycle, spent nuclear fuel rods used to power reactors can be reprocessed to gain more fuel. In this process, plutonium, central to the manufacture of nuclear explosives, can be extracted. The further sale or dissemination of reprocessing technology thus could lead to a larger number of nations owning nuclear weapons. There is, of course, the possibility that these weapons could fall into the hands of terrorists. It has been suggested that the extra energy made available from these reprocessing units would not be worth the potential danger. The United States, in fact, has been trying to slow what many feel is the inevitable worldwide growth of these facilities with their attendant waste-management problems.

It is in the context of these dilemmas that we must consider the technical, scientific, and social means for the management of nuclear waste.

Three Basic Types of Reactors

There are three basic possible designs for reactors. The first produces heat at peak efficiency, with plutonium a by-product (present commercial LWRs); the second is designed to produce plutonium, with heat a by-product (military reactors used for weapons); and the third is a combination of the first two that will produce both plutonium and heat. This is known as a "breeder" reactor. These basic reactors can in turn be designed with different heat exchange materials (coolants) other than water, including gases and liquid metals. The advantage of the breeder reactor is that by producing both heat (from the fission of uranium 235) and plutonium (from the transmutation of uranium 238) it provides power, while at the same time increasing the total amount of reactor fuel available over that available from using only natural uranium 235. A disadvantage is that highly purified plutonium can be made into nuclear explosives without the expensive and difficult physical separation processes necessary to get uranium 235 in a form pure enough to be used for fission weapons. As noted earlier, this gives rise to concern about the proliferation of nuclear weapons.

Other reactor systems, notably those based on the heavy element thorium, are possible. However, they will not be discussed because at the present time the most active planning is based on uranium and plutonium cycles.

Reprocessing

After the processes we have described to date have taken place in the core of a reactor of any design, the fuel then gradually changes it character, consisting more and more of fragments from the fission of uranium 235 or plutonium and of transuranic elements. Thus, at a certain point, the effectiveness of the rod in producing energy begins to decrease, the fuel in the rod having been "burned," leaving many nonfissile elements present to interfere with new fission reactions. At this point, it is removed and replaced by a fresh fuel rod.

There are two basic possibilities for handling the spent fuel rods: they may be reprocessed so that the reusable uranium and plutonium can be extracted, or they may be discarded as waste (this is known as the "throw-away fuel cycle").

If the choice is to reprocess the spent fuel rods, uranium and plutonium will then be extracted and made into new fuel elements to be burned further in the fission process. Fuel elements may be made from uranium compounds only, plutonium compounds, or from a combination of the two (mixed-oxide fuel). The remaining materials — a mixture of fission products, generally highly radioactive, with some residual plutonium, uranium, and other transuranic elements — are the high-level wastes that must be dealt with in some manner. In a breeder cycle, the plutonium from the spent fuel rod would be almost entirely separated out (99.5 percent) for further use as fuel; but, if it were decided not to use plutonium as a fuel, this material would be included in the waste faction.

Other types of radioactive waste also arise from the nuclear fuel cycle. These include the materials of the reactor itself that may have to be changed in the course of maintenance or repair and that have become radioactive by contact with the core. Also, there are materials that are involved in purifying the water used in the reactor and ensuring that nothing from the core goes out as effluent, etc. These materials are mostly of much lesser radioactive content and toxicity, but their safe handling and disposal does pose problems.

The Nature of the Waste

Most of the materials in the waste are radioactive. Radioactive elements are described in terms of their half-lives and the energy and kind of radiation emitted. The half-life is the interval of time that must elapse during which a half of the nuclei will undergo radioactive decay — the time in which the radioactivity of the material in question decreases to half its original value. Thus, in two half-lives the intensity of the radiation emitted by material will be a quarter of what it was when measurements began; in three half-lives an eighth as intense, and so on.

The half-lives of the radionuclides in high-level wastes vary from a fraction of 1 yr up to periods of more than 1 million yr. Because of the complexity of their chemical behavior and the numerous biological effects that these radionuclides may have, it is difficult to describe their toxicity or effects in terms of simple indices. One such "hazard index" gives the amount of water (in cubic meters) required to dilute the material to the "maximum permissible concentration" in public water supplies as allowed by current federal government guidelines.

The quantity of wastes estimated to be produced in the commercial nuclear fuel cycle in the United States varies, of course, with predictions of how much electrical power will be produced by nuclear reactors. At the moment, there are some 60 commercial reactors licensed to operate in the United States, with triple that number foreseen by the end of the century. The reactors today provide approximately 9 percent of the nation's electricity. By 1985 the figure is expected to rise to 26 percent. One estimate by ERDA projects that by the end of the year 2000 there will have accumulated at federal repositories about 2,500 m^3 of solidified waste (a cube 45 ft on a side), containing 10.9 billion curies of radioactivity. At that time, this will be accumulating at a rate of 350 m^3 and 1.8 billion curies of radioactivity per year.

In large quantities, nuclear fuel waste is extremely dangerous. Even in very small amounts, some of its constituents, including iodine, strontium, cesium, and plutonium, can be carcinogenic or lethal if ingested by human beings, animals, or other living organisms.

There are large variations in the biological effects of radioactive materials. Cesium 137, for example, is extremely dangerous because it is a bone-seeking element that will do severe damage to living tissue. Plutonium, on the other hand, is relatively unlikely to cause damage from mere contact, or even ingestion, since it will be eliminated by the digestive tract. However, it is highly toxic and carcinogenic if inhaled because particles can get stuck in the lungs, or if it gets into the bloodstream, since it is then deposited in the bones.

How Do We Dispose of the Waste?

How shall we finally dispose of the high-level radioactive wastes we have created over the last 30 yr and will produce more of in the future? There are three principal alternatives, each of which has numerous subalternatives. We can disperse the material, we can store it and guard it, or we can put it somewhere with very difficult access so that we can leave it there safely without continuous concern. A combination of these alternatives also is possible.

II. THE SEABED OPTION

Oceanographic data acquired since the late 1960s suggest that the ocean floor is continually built and destroyed by dynamic processes of crustal movement. This process, once called continental drift, has come to be known as plate tectonics or seafloor spreading. The globe is made up of a number of solid-rock or lithospheric plates composed of oceanic and continental crusts. These plates move in predictable directions at predictable speeds. They collide in regions of seismically active deep-sea trenches or of mountain building. Plate boundaries thus can be areas of crustal destruction where new crust, if the earth's diameter is to remain constant, is made at a rate equal to the destruction rate. Such growth takes place at the center of the Mid-Oceanic Ridge, a globe-circling and spreading welt of about 40,000 km in length. Along this active volcanic line new molten basalt is constantly being injected into the ridge, which widens at a rate of 2–20 cm/yr.

Rejected Options

Why not put the high-level waste into the deep-sea trenches instead of the mid-plate, mid-gyre region? Because the deep-sea trenches, popular press accounts to the contrary, are unpredictable (material from their bottoms has been thrust up onto the continent in the past) and unstable. In addition, they are usually near continents—and, therefore, man—and often lie beneath biologically productive ocean waters. Another, perhaps minor, consideration is that at present we do not have the technology for penetrating crustal rock at trench depths.

The only direct data we have about the structure and composition of crustal rock have come from a few holes drilled with great difficulty through approximately ½ km of basalt in shallow Mid-Oceanic Ridge crest areas. Core samples taken on the Mid-Oceanic Ridge suggest that this rock is broken up and badly fractured, with perhaps very high bulk permeability. None of the data so far suggest that shallow ocean crustal rock is monolithic. These considerations lead us, at least for the present, to the thought that emplacement of wastes in the crustal rock, at least at shallow depths, would not be prudent. An abyssal midplate region should be drilled, however, before we finally abandon this disposal option, as it is conceivable that crustal rock is effectively healed and sealed — leading to a very low permeability — by the time it reaches midplate depths of 5 km (a journey requiring at least 50 million yr of seafloor spreading).

Placing the high-level waste on top of the seafloor simply by kicking a canister off the fantail effectively puts the waste directly into the biosphere, as it is difficult to conceive of making a canister that would survive without leaking for hundreds of thousands of years in the corrosive marine environment. Any leak, either during the disposal operation or after, would inject radioactive material into the marine ecosystem. From samples, photographs, and current meter data, we know that the energetics of the biological and physical processes of the sediment/water interface (benthic boundary layer) can be very high and very unpredictable.

Another suggested disposal option is to dilute the high-level waste by dispersing it into the ocean waters. This method, favored by some Western European countries, has been and is being used to dispose of low-level wastes.

Calculations show that the waters of the ocean are not vast enough to take all of the waste from all of the military and industrial sources without being contaminated beyond safe limits within the next few decades. Dispersion and concentration mechanisms — biological, physical and chemical — are so poorly known that researchers are not yet ready to predict possible pathways and rates of transfer from ocean bottom to man's food chain.

The MPG Seabed

Given the serious defects in most ocean disposal options, the geologic formations beneath the seafloor should be assessed with a view toward establishing some site selection criteria. Initially, it seems prudent to avoid areas where earthquakes had been recorded. This leaves the central portions of large plates, some of which are thousands of miles from areas of crustal destruction. Here the seafloor is apt to be covered with a blanket of soft, sticky, chocolate-colored oxidized clays. These clays have certain chemical and physical properties that, if taken together, might conceivably provide a suitable waste isolation medium, even assuming total failure of the canister after emplacement — obviously the worst possible case. The ion-retention and permeability characteristics of these clays might be adequate to chemically and physically contain the waste for the periods needed.

Future Plans

We are still trying to prove the adequacy of the sediment barrier to waste migration. A next step is to identify the best possible sediment with respect to the retention of radionuclides, whether it be oxidized red clay, reduced hemipelagic clay, or biogenic ooze. Sediments that have adequate containment properties will have to be studied at sea to determine whether they can be found in sufficient thickness in MPG-type settings. If so, we will determine whether the sediment is uniform over large areas. Finally, once a barrier is proven from chemical and permeability measurements, we will determine in situ the physical and dynamic response to emplacement, to establish if the sediments do in fact fully close above an emplaced canister. If this does not occur, we will have to design a technique for permanent hole closure.

In summary, after an initial 3 yr of research it appears that the relatively impermeable, highly sorptive clayey sediment like that found in MPG areas has the potential to isolate high-level radioactive wastes from the ocean and from man. If we continue to find the concept to be scientifically sound, we must present our data as soon as possible, so that the seabed can be fully examined by disinterested experts and compared with other disposal alternatives. Certainly, there appears to be no scientific or technical reason to abandon the seabed disposal concept at this time.

III. THE LAW OF THE SEA

No assessment of a subseabed option for the disposal of high-level radioactive wastes would be complete without an analysis of the parallel political, legal, and institutional implications of such a concept. These considerations could prove to be at least as complex as the scientific endeavors. The path to a subseabed option divides in several directions through a thicket of conflicting national and international interests. Which way the United States might ultimately decide to turn would be dictated largely by the extent to which the country decided it must rely on a subseabed option for future nuclear waste disposal.

Where do we stand at present on the national and international paths toward a subseabed option? We can say that the option is scientifically and technically plausible, but that it could turn out to be unworkable from a legal and political point of view. On the other hand, the national and international mechanisms exist that could, if judiciously set in motion, make it the most viable of all alternatives. To understand the complex international and national procedures that would be necessary to adopt this option, it is first necessary to briefly call to attention the recent efforts to dispose of high-level radioactive wastes.

For many years, the waste-management component of nuclear energy programs in this country and elsewhere received very low priority treatment. However, the large increase in the United States waste-management budget for 1976–77 is evidence of a very recent transition to a high-interest program with considerable institutional and financial support.

One reason for this growing interest is that radioactive waste management has become one of the two or three major issues raised by opponents of nuclear energy. Many groups are now focusing on high-level waste disposal as the primary defect in increased reliance on this power source. Severe pressure to demonstrate waste-disposal technologies has developed from environmental and political opposition to the construction and operation of more nuclear power plants, the reprocessing of spent fuels, and the overall expansion of nuclear energy, including the risks of further proliferation of nuclear weapons.

Every country operating nuclear reactors for energy has, or will have, highly radioactive spent-fuel bundles in storage. These must either be finally disposed of, creating a throwaway cycle, or reprocessed, creating a large variety of waste streams for further management. Even if reprocessing were to be done externally, the country of origin might well have to take the wastes back.

All nations with major energy programs agree that it is necessary to solidify high-level wastes after an initial cooling period. But the technology required for the most effective solidification is not available to all nations. For example, France and the Soviet Union, both well advanced in this field, are reluctant to divulge the technologies.

Other countries are exploring alternate disposal concepts, but there are no common international criteria or standards to guide individual national efforts. And, as with the high-level waste-solidification process, any effective technology that may be developed is likely to be kept secret.

Most nuclear countries are studying geologic formations as their prime option for final high-level waste disposal. At the moment, only West Germany has an operational repository for wastes at the Asse Salt Mine. However, this complex will only serve as a test facility for high-level wastes. Aside from the West German focus on salt and American plans to use salt and rock formations eventually, no countries have made commitments to specific land disposal sites for high-level wastes.

Some countries, such as Japan, Switzerland, Belgium, Britain, and the Netherlands, have serious geographic, demographic, geologic, or hydrologic restrictions. Most nations do not have the vast land areas that are available to the United States, the Soviet Union, and Canada. Added to these physical constraints is the growing environmental, consumer, and political opposition to nuclear energy. This opposition has highlighted two key factors; (1) the sensitivity of national groups to the prospect of offering territory that would be used as a repository for another nation's high-level wastes, and (2) the highly interdependent nature of decision making in the international system.

Thus, some nuclear powers, such as Britain, may have to find disposal sites outside their borders, perhaps in countries without nuclear programs or in specially designated international areas (islands, seabed, etc.). All told, it is possible that many countries without major nuclear energy programs will become directly involved through international organizations in the final disposal of high-level wastes.

Delays in the construction and licensing of facilities abroad, just as in the United States, are translated into higher energy costs and revised construction plans. The result has been increasing interest among the highly industrialized countries in an acceptable high-level waste-disposal solution. This has led to a significant, but very reticent, expansion of the International Atomic Energy Agency's (IAEA) program on waste management. (The IAEA, based in Vienna, is an autonomous agency of the United Nations that is charged with both promoting the peaceful uses of nuclear energy and preventing its application to military purposes among its 110 members.) However, disposal programs in the major nuclear nations are still in the very early stages of development, and serious efforts by the IAEA to solve this problem are just beginning.

Coastal Management

Coastal Area Development and Management and Marine and Coastal Technology

INTRODUCTION

A. Background

1. The present report has been prepared in accordance with Economic and Social Council resolution 1970 (LIX) of 30 July 1975. In paragraph 8 of the resolution, the Council requested the Secretary-General to report to the Council at its sixty-third session on progress made in implementing the resolution, the key operative provisions of which may be summarized as follows:

a) Endorsement of the Secretary-General's programme of activities in the field of coastal area development, as outlined in paragraph 56 of his report on coastal area management and development (E/5648), namely, preparation of a manual on coastal area development, organization of seminars, and broadening of the scope of research institutes;

b) A request to the Secretary-General to submit to the Council, on a regular and continuing basis, pertinent information on the uses of the sea;

c) A request to the Secretary-General to continue the process of identifying various regions or subregions specially amenable to coastal area management and development and, at the request of the Governments concerned, to assist in the elaboration of comprehensive plans of action for such an approach;

d) A further request to the Secretary-General to take effective measures for the promotion of a better and wider application of marine technology suitable for developing countries in their endeavors relating to coastal areas, including the provision of assistance in the matter of training, institution-building and the acquisition and implantation of appropriate technology;

e) A request to the organizations of the United Nations system concerned, including the regional commissions, to give full support to the Secretary-General in implementing programmes for coastal area development as a joint endeavour and to further their existing close collaboration, using whenever appropriate established mechanisms of co-ordination.

2. The Council will recall that in its resolution 1802 (LV), section II, of 7

Progress report of the United Nations Secretary-General, May 18, 1977 (UN Document E/5971).

Copyright is not claimed for this document.

August 1973, entitled "Marine co-operation," it requested certain actions on the part of the Secretary-General, in close co-operation with the organizations of the United Nations system concerned, including the regional economic commissions, which provided the basis for the development of a United Nations programme in coastal development. The Secretary-General subsequently undertook a comprehensive interdisciplinary study to identify and review the problems of coastal area development, using for this purpose the expertise of the entire United Nations system in technical and scientific matters, and also in development planning. This study was submitted to the Council at its fifty-ninth session (E/5648).

3. In response, the Council adopted resolution 1970 (LIX), which set out the operative elements summarized earlier, thereby providing a refinement of directives previously given in resolution 1802 (LV) and further guidance to the Secretary-General in developing the United Nations coastal area programme, particularly in the field of marine and coastal technology.

B. CO-ORDINATION

4. United Nations activities in coastal area development have been envisaged and implemented as joint undertakings with the competent organizations of the United Nations system. This is in accordance with the Economic and Social Council's directives to this end and in keeping with the inherently multidisciplinary and intersectoral nature of integrated coastal area development.

5. In directing its attention to "integrated coastal area development" and "a comprehensive approach to coastal area development" the Council introduced a new field for the application of co-ordination. However, the Council also provided the conceptual framework—an integrated approach; a planning framework (with physical referents)—coastal areas; and the organizational framework—regional or subregional—within which co-ordination could be carried out.

6. It is clear that in this context the main thrust of co-ordination is less towards avoiding or minimizing duplication of work among United Nations organizations than towards ensuring that the full range of expertise available within the system is assembled and integrated in the implementation of inherently co-operative activities. The emphasis therefore is less on the delimitation of competencies, though this is still necessary, than on the synthesis of competencies.

7. The co-operation within the United Nations system called for by the Economic and Social Council has been carried out under the auspices of the ACC Sub-Committee on Marine Affairs (formerly the Sub-Committee on Marine Science and its Applications), which, with the new terms of reference recommended by the Sub-Committee at its eighteenth session, held from 10 to 14 January 1977 (see CO-ORDINATION/R.1199, annex III), as approved by

the Preparatory Committee of ACC at its eighty-fifth session, promises to be a more effective mechanism in promoting interagency collaboration in integral coastal area development as in other common endeavours. Within the United Nations, an Intra-Departmental Task Force on Coastal Area Development has also been established in the Department of Economic and Social Affairs to ensure a United Nations input that takes account of the interests and expertise of the various relevant programmes within the Department.

I. MANUAL ON COASTAL AREA DEVELOPMENT: A SUMMARY[1]

A. BACKGROUND

8. The concept of coastal area management and development is a very broad one, requiring inputs from the pure, applied and social sciences. At present a number of intergovernmental bodies, professional societies, non-governmental organizations, universities and scientific expert groups are carrying out work in the field of coastal area management.[2] Such work, however, deals with particular scientific and technical questions and does not attempt to relate the economic issues of coastal management to the general development planning process. Coastal management, as recognized by the Economic and Social Council in resolution 1802 (LV), is an integral part of national development planning. The Manual, which will form a "bridge" between conceptual work and operational activities, breaks new ground in the sense that it puts the diverse work done by other groups into one coherent framework, whose

---

1. The summary presented in this section is based on the first full draft of the Manual, which may require revision in the light of any comments and suggestions made by United Nations organizations and outside experts to whom the draft was circulated.

2. Several thematic papers dealing with coastal management were submitted to the United Nations Water Conference, e.g., "The coastal zone: a challenge to environmental engineering" (Canada) (E/CONF.70/TP.154); "Coastal zone management" (United States of America) (E/CONF.70/TP.190); and "Environmental study of the Tejo Estuary" (Portugal) (E/CONF.70/TP.211).

The Institute of Biology of the Federal University of Bahía (Salvador, Brazil) held an International Symposium on Coastal problems: Planning, Pollution and Productivity, in December 1976.

The International Federation of Landscape Architects, at its fifteenth Congress, held at Istanbul from 6 to 9 September 1976, discussed the role of the landscape architect in coastal development.

The Joint Group of Experts on the Scientific Aspects of Marine Pollution (GESAMP) has recently established a working group on the pollution implications of sea-bed exploitation and coastal development.

The International Union for the Conservation of Nature is launching a marine programme which deals with coast conservation.

emphasis is on the economic factors that impinge on coastal development.[3] The objectives of the Manual are:

a) To provide planners and decision-makers with detailed guidelines for establishing a national programme of coastal management;

b) Within these guidelines, to focus attention on the kinds of information required for coastal management and on the problem of presenting multidisciplinary information in a unified format;

c) To provide university researchers, government planners and policy makers with a guide to the international technical literature on coastal management issues.[4] The Manual will contain over 600 carefully selected internationally available references to literature on such fields as: coastal geomorphology, processes and engineering; environmental management techniques for the coastal area; coastal mapping and charting; coastal management in diverse countries; economic analysis of coastal problems; monitoring and control of coastal and marine pollution; conservation and marine reserves; land-use planning in the coastal area. Extensive cross-referencing in the text will allow readers to pursue points discussed in the text more fully.

9. Development of the coastal area, in the context of the Manual, includes the following economic activities:

a) Fishing and mariculture;
b) Mining of minerals (including hydrocarbons);
c) Transport and shipping;
d) Foreign and domestic tourism;
e) Manufacturing;
f) Agriculture;
g) Forestry;
h) Settlement.

These activities enter directly into the flow of national income and product. In addition to these activities, conservation, waste disposal and scientific research are also relevant to the coastal area, but do not directly enter into the development process. Since each of these development activities has several associated disciplines, they cannot be discussed in the Manual. Rather, the objective of the publication is to treat systematically the important economic and technical

---

3. Social factors, levels of technology and political organization are, of course, important determinants of the level and type of coastal development. A number of continuing programmes within the United Nations system—unified approach to social and economic planning, rural development and ecodevelopment (UNEP) are treating these issues.

4. The International Working Group on Project 5 of the UNESCO Programme on Man and the Biosphere (MAB), meeting in Paris from 13 to 17 May 1974, in its final report recognized the need for a "compilation and review of the existing socio-economic literature on coastal zones," which it felt to be "essential for the formulation of guidelines." (See UNESCO, MAB Report Series No. 21, 1974, p. 77.)

elements of coastal development, with a view to establishing some basic features of a national programme to manage coastal development.

B. The Concept of Coastal Area Development and Management

10. Coastal area development and management is a hybrid concept, combining elements of regional analysis[5] and environment management. The coastal area is a band of varying width, which includes a land component and a seaward component. Within the national context, it is a region which contains a certain proportion of national capital, income and employment. The question of how to allocate these aggregates among the various national regions is one of the traditional problems of development planning. Within the coastal area, there is the problem of how to allocate physical capital, e.g., settlements, roads, industrial plants, etc., in space. This is the normal function of the physical planning process, including landscape architecture. In respect of the regional planning and physical planning decisions, coastal management and development is intimately related to over-all development planning. It should be noted, however, that the normal application of planning processes to the coastal area is complicated by the fact that many national government agencies may each have only partial jurisdiction in the coastal area.

11. Environmental management[6] is particularly important in the coastal area, as estuaries and coastal zones are among the biologically most productive areas on earth. As such, development activities can have particularly strong interactions with each other and with the coastal environment.

12. These two elements—regional analysis and environmental management—form the concept of coastal management discussed in the Manual. A national programme can combine these two elements in different proportions, depending on the country's size, resource base, administrative structure and level of development. Higher-income industrial countries, for example, tend to weight their coastal programmes towards environmental management and protection.

13. The Manual reviews the national programmes of coastal management in a cross-section of countries. On the basis of this review, the diversity of emphasis among countries is apparent. It is concluded that there is no model coastal management programme existing that could be adopted by developing countries. All of the national programmes reviewed do, however, contain certain common features, and the experience of these countries will be useful as a background for developing countries.

---

5. Also referred to as "regional science" or "regional economics."
6. The process of monitoring, protecting and conserving environmental values (e.g., air and water quality and aesthetics) by suitable political, legal and economic mechanisms.

C. The Nature of the Management Problem

14. Coastal management problems are not fundamentally different from land use and environmental management problems in arctic regions, arid zones, river basins and inland areas. Every society must strike a balance between the growth of its material output and the level of environmental quality. This fundamental problem of economic choice applies to the coastal area.

15. Although not fundamentally different from other management problems on land, coastal management is a decidedly more complex process. This is partly because of the relatively rudimentary state of knowledge of the coastal environment, compared to that of terrestrial ecosystems. The coastal area is an ecotone, a transitional environment between two distinct systems (land and marine); management of such a system is therefore more difficult than terrestrial management.

16. Three major problems of coastal management are identified in the Manual.

17. The first problem involves interactions among uses of the coast. Two or more uses of the coast can physically conflict, owing to the limited space; an example of such a conflict would be where a dredge (for mining or for channel clearing) cuts submarine pipelines or cables. Conflicts can occur between groups of users of coastal facilities. An example would be the case of a fishing harbour which is chosen to become a supply base for off-shore oil activities.

18. The destruction or deleterious modification of valuable environments is the second problem. The coastal areas of the world contain a number of low-energy environments which are valuable as breeding grounds for water fowl, as nurseries for fish and as shoreline stablilizers. These environments include wetlands,[7] salt marshes, coastal peat and mangroves. There is always a strong pressure to develop these areas for commercial and residential purposes.[8]

19. The third problem to be dealt with is the pollution of coastal and estuarine waters. The pollution of coastal waters, and the resulting harm to human health, amenities and fisheries, represents a particular class of what economists call external effects or "externalities." It is in the coastal area that the critical problems stemming from marine pollution exist. Pollution, however, is a by-product of economic growth, and no society can afford simply to move to a state of "zero discharge" of all pollutant materials. This would be costly, inefficient and wasteful. The problem is further complicated by the fact

---

7. "Wetland" is a commonly used expression, although it is not a standard scientific term. It applies to all intertidal coastal lands subject to periodic (diurnal to annual) marine inundation and largely overgrown by halophytic vegetation.

8. E. Robes Piquer, "The threat of fast development on the coastal areas in developing countries" (paper presented to the fifteenth Congress of the International Federation of Landscape Architects, held at Istanbul in September 1976).

that "What is rampant pollution in one society may be favourably received in another."[9]

20. Therefore, a trading off of some degree of economic growth must occur in exchange for an increase in environmental quality. No society can maximize both of these desirable goals. In the Manual, the analysis of the technical problems of pollution management is done in economic terms.

D. THE UNDERLYING CAUSES OF COASTAL AREA MANAGEMENT PROBLEMS

21. The fundamental reason identified in the Manual for most existing coastal problems, particularly in the developing countries, is the lack of an explicit government policy for the development of the coast and the management of its resources. The reasons for the lack of a national policy are manifold, namely:

   a) National economic problems;
   b) Cultural differences in the valuation of environmental quality;
   c) Lack of awareness of potential interactions between coastal development activities;
   d) Costs associated with a policy which are perceived to outweigh the benefits.

22. In the West African coastal region, for example, a recent expert mission[10] found that logs from local timber operations were an obstacle to tourism (when they washed onto beaches) and a potential hazard to navigation (when they were floating). In the case of pollution from floating or washed-up logs, there may not be an awareness of the potential interactions with coastal tourism and navigation. More fundamentally, there is no economic incentive for the timber industries to recover these logs, since they are generally low-grade lumber, and there is no discernible economic loss, at the present time, from allowing them to drift seaward via rivers. The lack of a national policy for coastal development is the fundamental obstacle to effective management of the area and its resources.

23. A second cause of the management problem related to the lack of a national policy is the fact that numerous government departments and bodies have a jurisdiction in the coastal area. This problem was identified in the 1975 report of the Secretary-General on coastal area management and development (E/5648). As a result of the multiple and sometimes conflicting divisions of responsibility among government bodies, it is difficult to build up an integrated information base for coastal management, and implementation of large-scale projects may be delayed as a result of interagency difficulties.

---

9. See UNESCO, MAP Report Series No. 21, p. 77.
10. UNEP exploratory mission to study marine pollution problems of the West African coastal countries of the Gulf of Guinea (25 April–2 July 1976).

E. Establishing a National Programme in Coastal Area Development and Management: A Systems Engineering Approach

24. UNESCO, in its publication *Marine Sciences in the Gulf Area*,[11] cites the need for a "systems engineering or operational research approach to marine planning and development" in this region. Chapter IV of the Manual specifies the necessary data gathering, research, evaluation and organizational requirements for a national programme of coastal management; these activities are grouped together according to a systems engineering framework.

25. In the planning phase of an engineering approach to coastal management programme, the objectives, goals and performance standards of the programme are formulated. In this process, the broad macro-economic and social objectives are taken as given. During the planning phase, one of the most important tasks is the definition of the management boundaries, namely the extent of the coastal area.

26. There are four possible criteria for defining the coastal area, each of which is discussed in detail in the Manual. Each criterion has certain virtues and drawbacks, and the choice of criteria will be influenced by many factors, including the administrative capacity of the country. The coastal area may be defined according to:

a) Physical criteria. For example, the landward extent of the coastal area may be defined by a drainage basin, while the seaward extent may be defined by the continental shelf.

b) Administrative boundaries. The coastal area could correspond, under this criterion, to a county or district boundary (landward component).

c) Arbitrary distances. The landward component of the coastal area may be defined as being, for example, 1,000 feet inland from mean high-water mark.

d) Selected biophysical units. The coastal management area may, in some cases, be composed of certain important biophysical units, e.g., estuaries, lagoons, deltas, bays, etc.

27. The application of each of these criteria is reviewed in the Manual, using examples from international experience. Whatever criteria are used for defining the management boundary (coastal area), the operational definition should satisfy certain guidelines, e.g.:

a) The definition should be clearly stated and easily understood;

b) The coastal area should be susceptible to uniform cartographic representation;

c) The coastal management boundaries should respect, as far as possible, existing social and political subdivisions;

d) The coastal area should be of economic significance.

It is recommended in the Manual that the operational definition of the coastal

---

11. UNESCO, *Technical Papers in Marine Science,* No. 26 (1976).

area be made on the basis of natural features, with additional provisos so as to meet the above guidelines.

28. A technical annex to chapter IV of the Manual provides a comprehensive listing of natural coastal features which will assist developing countries in the task of defining coastal management boundaries.

29. The second phase of a systems engineering approach to coastal management is the operating phase. The work necessary to accomplish the objectives of the coastal management programme is defined, and the work is divided into manageable components and functions. These functions are grouped into an orderly organizational structure, and systems and procedures for performing the work are devised.

30. The Manual focuses on one of the most important functions in the operating phase, namely the building up of an information base for management. A developing country will have a number of technical services, which conduct surveys and investigations independently of coastal management. A number of these services will have acquired broad regional data that are important for coastal management: oceanographic and hydrographic service; tidal-gauge service; geological and mineral resources survey; cartographic service;[12] soil science or pedology division; biology service;[13] fisheries division; meteorological service, including port meteorology.

31. Where existing data on coastal features and processes are inadequate, certain special surveys and investigations may have to be commissioned. In addition to the biophysical data, socio-economic data on the households, settlement and infrastructure are also required. Since no survey—no geological survey, for example—is ever "completed," the information base must be revised and up-dated periodically.

32. As the information base is developed, one very important long-term exercise should be a national classification of coastal areas. There are four types of classifications: (i) descriptive; (ii) genetic; (iii) dynamic; and (iv) geotectonic. A coastal classification scheme is not an academic exercise. It is essential both for development and for conservation of the coastal area. According to a 1975 study of the International Union for the Conservation of Nature:

> A classification scheme is the basis for the establishment of a system of preserves by means of which marine ecosystems will eventually be conserved, studied and monitored. ...A classification is at the heart of the work of marine conservation, research and monitoring.

Guidelines are presented for the development of a national system of coastal

---

12. Sometimes known as the lands department, cadastral survey or geographical institute.

13. In some countries, this service may be divided into botany and zoology; it may also be called an ecological service.

classification based on the "composite resource" concept, which is developed in the Manual. A technical annex also describes important physical coastal processes.

33. Management of the coastal area requires that the nature and location of land and water users be controlled so as to ensure that development is compatible with the objectives of the management programme, e.g., to ensure the continuing productivity of the coastal ecosystems. Ideally, the entire national territory—including the coastal area—should be mapped on a uniform geographical reference system, according to certain basic characteristics, e.g., topography, hydrology, geology, land cover, etc.[14] Similarly, water quality criteria should be developed for all national surface waters (rivers, lakes, estuaries and coastal waters). The coastal management process would then function within the over-all national programme of land and water use.

34. In most developing countries, however, there is no national plan for land and water use. Therefore, a coastal management programme may have to have a somewhat restricted scope to begin with. To that end, it is recommended that:

a) The coastal area be divided into broad zones, which would indicate fragile areas, areas of recreational or scientific interest, zones of preservation, and zone of development. There is a discussion in the Manual of the techniques and limitations of suitability analysis, along with a review of country experience with these techniques.

b) Coastal development projects should be submitted to an evaluation at the feasibility stage, which concentrates on:

  i) the social profitability of the project;

  ii) its environmental impact, including the question of site suitability.

There is a detailed review of current methods of project appraisal and environmental impact assessment.

c) A coastal co-ordinating council should be established to act as the policy-making and management body. It should be composed of representatives from government departments, and it should allow for industrial, academic and public participation.

d) Suitable legislation should be drafted to allow the desired degree of control over coastal development activities.

F. Conclusion

35. The requirements for a national programme of coastal management can be grouped into three broad categories:

  a) An information base and a management information system;

---

14. Mexico's Comisión de Estudios del Territorio Nacional (National Land Survey Commission) is building up a system of this type.

b) A legal and administrative framework for operation of the programme;

c) A planning and evaluation procedure for choosing the kinds of development and their location in the coastal area.

With respect to these needs, the Manual specifies the kinds of scientific information required for management decision-making, proposes a possible solution to the problem of multiple jurisdiction in the coastal area[15] and critically reviews current international progress in ecological modelling, environmental impact assessment and project appraisal. Guidelines for applications of these techniques are presented.

36. The technical papers presented to the United Nations Water Conference made it clear that "efforts to undertake planning for geographically homogeneous regions like coastal zones . . . need to be encouraged."[16] In addition, the Environment Committee of OECD has formulated several recommendations for international action on coastal management.[17] The Manual is also responsive to these initiatives.

II. PROGRAMME ACTIVITIES: COASTAL AREA DEVELOPMENT AND MANAGEMENT

A. REGIONAL AND SUBREGIONAL ACTIVITIES

37. Following the convening of a Group of Experts on Coastal Area Development at United Nations Headquarters from 11 to 15 November 1974 (see E/5648), regional and subregional activities have been implemented in the four coastal regions initially selected for specific attention.

38. In each of these regions, activities are being developed and implemented in close consultation with the regional commissions, whose central role in such endeavors was fully acknowledged by the Economic and Social Council.

1. PERSIAN GULF

39. The region of the Persian Gulf was designated as one of four to receive priority examination under the coastal area programme undertaken by the

---

15. The subject of legislation for coastal management, although mentioned in the Manual, will be treated at length in a separate volume that is under preparation. The first part, which will contain a survey of existing national legislation pertaining to coastal activities, will seek to determine the coverage of national laws and to identify common characteristics and gaps in legislation. The second part of the volume will provide basic rules and guidelines for carrying out a number of specific activities.
16. E/CONF.70/10, p. 42.
17. *The OECD Observer*, no. 83 (September/October 1976).

Secretary-General in response to Economic and Social Council resolution 1802 (LV). A plan for a multidisciplinary expert mission was designed at the meeting of the Group of Experts on Coastal Area Development.

40. The project concept was subsequently endorsed by UNDP and included among the regional projects of the Regional Bureau for Europe, Mediterranean and the Middle East. The competent agencies and organizations of the United Nations system were consulted and the integrated nature of the project was ensured by their agreement to take part.

41. The document drafting and recruitment of experts for the final project took place during 1975, when it appeared that funds would be accessible. At the end of 1975 the UNDP funding situation was less promising and a request was made to UNEP to undertake a large part of the financing for the mission.

42. In early 1976, UNEP agreed to undertake a major role in project financing and a number of alterations were made in the mission's terms of reference to reflect a greater emphasis of marine environment problems. The team of experts was assembled for substantive briefings by United Nations staff members at Headquarters in March 1976.

43. Following briefings at Headquarters and at Geneva, the mission spent two months in the region, visiting each of the eight participating countries. The working papers were compiled and edited by the project manager and circulated to agencies for comment in June 1976. A summary report of the working papers, incorporating agency comments and suggestions, was prepared by the project manager and UNEP in August 1976, along with a draft action plan to be submitted to an expert meeting on coastal area development and protection of the marine environment jointly sponsored by the Department of Economic and Social Affairs of the United Nations Secretariat and UNEP. This meeting was held in Kuwait from 6 to 9 December 1976 and was attended from all eight countries as well as organizations in the United Nations system and by the project manager.

44. As a result of this meeting, a revised draft action plan was prepared for submission to an intergovernmental meeting on the protection of the marine environment.

2. CARIBBEAN

45. At its first session, held from 31 October to 4 November 1975 the Caribbean Development and Co-operation Committee (CDCC) of the Economic Commission for Latin America adopted a work programme that included a section on coastal area development. In response to this expression of interest by Governments in the Caribbean subregion, which took full account of the recommendations of the Group of Experts on Coastal Area Development concerning the region, the Department of Economic and Social Affairs, initially in conjunction with a parallel undertaking by UNEP, prepared a draft

background study entitled "Development and environment in the Caribbean: coastal and marine aspects," in consultation with the United Nations organizations and agencies concerned. The study was intended to provide a comprehensive review of the current situation in the Caribbean with respect to marine resource development and potential, uses of coastal and marine space and the state of the marine environment. A considerably abridged version of the lengthy study was subsequently prepared by the Secretariat, covering the main problem areas and policy issues relating to coastal area development and marine-related activities in the Caribbean. Its aim was to facilitate deliberations at an Interagency Meeting on Co-ordination for Implementation of the Work Programme of CDCC, convened at United Nations Headquarters from 28 to 30 June 1976. The full study was also made available at the meeting.

46. A variant of the abridged study was also prepared by the Secretariat and submitted to the first session of the Intergovernmental Oceanographic Commission's Association for the Caribbean and Adjacent Regions (IOCARIBE), which held its first session at Caracas from 19 to 23 July 1976. IOCARIBE is the successor body in the region to the IOC Co-operative Investigation of the Caribbean and Adjacent Regions. IOCARIBE, in its recommendation I-6, took note of the "development programme of the ... Ocean Economics and Technology Office [of the Department of Economic and Social Affairs] in the Caribbean, particularly as it relates to training and education, coastal area development and management and the IOCARIBE project on environmental geology and evolution of the continental and insular margins of the Caribbean and Gulf of Mexico area."

47. IOCARIBE further considered it essential that it work closely with ... [the Ocean Economics and Technology Office and UNEP] and requested "the IOCARIBE secretariat to establish open channels of communication particularly with these two bodies...."

48. The Department of Economic and Social Affairs was also represented at the second session of CDCC, held at Santo Domingo in March 1977, when specific needs and problems related to coastal area development to which the United Nations could respond were identified.

49. Plans are also under way to prepare model maps of representative coastal areas in the Caribbean region for the purpose of demonstrating the potential of the integrated mapping of information for marine environmental and coastal area management.

3. WEST AFRICA

50. As a follow-up to the meeting of the Group of Experts on Coastal Area Development, the Department of Economic and Social Affairs prepared a paper entitled "Preliminary study on the major uses of the coastal area, Gulf of Guinea—a proposal for a regional coastal area development scheme," which

was sent to UNDP and the Economic Commission for Africa (ECA) for comment at the end of 1975.

51. Subsequently, the United Nations Secretariat consulted with ECA officials on the matter of establishing a Marine Technology and Science Centre in West Africa and on the mineral resources in the medium-term plan for the period 1978–1981, including the proposal for the establishment of coordinating off-shore prospecting committees in various African coastal regions.[18]

52. The United Nations Secretariat is also carrying out consultations with officials of several West African countries on the need for and feasibility of developing a project for the prevention of coastal erosion.

4. SOUTH-EAST ASIA

53. A proposal for an economic background study to be linked with the results of the International Workshop on Marine Pollution in East Asian Waters, held at Penang, Malaysia, from 7 to 13 April 1976 and to be followed by a regional symposium, was formulated by the Department of Economic and Social Affairs in co-operation with the Committee for Co-ordination of Joint Prospecting for Mineral Resources in Asian Offshore Areas late in 1975. The proposal has been sent to Governments in the region and was favourably received at an Interagency Meeting on Regional Seas, convened by the Executive Director of UNEP in Paris from 16 to 18 June 1976.

B. SEMINARS

54. The Department of Economic and Social Affairs and the German Foundation for International Development convened jointly an Interregional Seminar on Development and Management of Resources of Coastal Areas at Berlin (West), Hamburg, Kiel and Cuxhaven from 31 May to 14 June 1976.

55. Representatives from some 30 developing countries, assisted by resource specialists from the United Nations system as well as from industry and educational institutions, examined the problems and opportunities associated with coastal resource development and management with a view to developing concrete action proposals that would receive expression in follow-up activities at the national, regional or subregional levels.

56. The format of the Seminar was structured in such a way as to encourage action-oriented results within an integrated planning framework. Thus the

---

18. *Official Records of the General Assembly, Thirty-first Session, Supplement No. 6A* (A/31/6/Add.1), paras. 1255–1256.

participants first were exposed to a full week of lectures and discussion, ranging from the need for integrated planning to artificial beaches and coastal mapping from satellite altitudes. They were then organized into concurrent workshops that focused on specific priority themes and problems identified on the basis of presentations of national experience.

57. As a follow up to the Interregional Seminar, regional and subregional seminars and workshops are expected to be organized to address needs and problems related to coastal area development experienced by countries in specific geographical contexts. The first such workshop is being planned for Central America in 1978.

C. TRAINING

58. In response to concern expressed by the Economic and Social Council, the Department of Economic and Social Affairs, through its Ocean Economics and Technology Office, has considerably expanded its support for and involvement in training activities in marine and coastal affairs over the past two years.

59. In addition to organizing seminars that served a training purpose, such as the interregional seminar described in paragraphs 54 to 56 above, as well as the planned regional seminars in the planning stage, the Secretariat late in 1976 published a *Register of Courses and Training Programmes in Marine Affairs*.[19] Prepared by the Department of Economic and Social Affairs (Ocean Economics and Technology Office) with the collaboration of the Intergovernmental Oceanographic Commission (IOC), the *Register* is intended to inform interested persons, within and outside programmes offered by institutions throughout the world in marine related fields. It is hoped that the *Register* will help the developing countries, in particular, in their efforts to develop an infrastructure of trained manpower in the marine sector and will thereby contribute to the development of national marine capabilities.

60. Particularly with respect to training problems and needs related to coastal area development and management, the Secretariat will play an active role at the second session of the IOC Working Committee for Training, Education and Mutual Assistance (TEMA), which will be held at Headquarters in July 1977.

61. The Department of Economic and Social Affairs recognizes TEMA as the most appropriate and effective framework within which diverse marine-related training activities within the United Nations system may be organized, and accordingly it prepared for submission to the second session of the Working Committee a document suggesting the incorporation within the scope of TEMA of certain activities supportive of integrated coastal area development that would complement the "science"-oriented activities that are at present the

---

19. United Nations publication, Sales No. E.77.II.A.2.

focus of TEMA concern. As follow-up to the second session of the Working Committee for TEMA, it is expected that training programmes related to coastal area development will be organized and implemented in co-operation with several organizations of the United Nations system.

III. THE APPLICATION OF MARINE AND COASTAL TECHNOLOGY IN DEVELOPING COUNTRIES

A. BACKGROUND

1. SCOPE OF THE PROGRAMME

62. The Economic and Social Council affirmed in resolution 1970 (LIX) that the development of coastal areas was predicated on the better and wider application of appropriate technology and it requested the Secretary-General to take effective measures to that end.

63. In keeping with the Council's emphasis on utilizing the expertise of the entire United Nations system, the United Nations programme in coastal area development has been implemented across traditional sectoral and disciplinary lines, relying upon effective co-ordination and co-operation to unite all sectors of the system in joint undertakings. The positive acceptance of this approach appears to point to its feasibility in a multidisciplinary programme in the acquisition and application of technologies employed in coastal area development and marine-related activities.

64. Consequently, the Secretary-General is seeking to establish a concerted approach, in close co-operation with the competent organizations of the United Nations system. Although in any programme related to the oceans the use of technology is always an integral element, there has been little recognition of the fact that the separate sectoral activities under way constitute an over-all marine and coastal technology resource base that can be applied systematically to multisectoral marine and coastal-related development goals.

2. ROLE OF A MARINE AND COASTAL TECHNOLOGY (MACTECH) PROGRAMME WITHIN THE UNITED NATIONS SYSTEM

65. The United Nations system is deeply involved in and committed to furthering the transfer of technology to developing countries. Emphasis has been placed on this subject by the General Assembly at the sixth and seventh special sessions and its importance is also reflected in the decision by the General Assembly in resolution 31/184 of 21 December 1976 to convene in 1979 a United Nations Conference on Science and Technology for Development. Within this over-all context, considerable importance has been attached to the

need for technological information, as in the case, for example, in the Lima Declaration and Plan of Action on Industrial Development and Co-operation[20] and in resolutions of the General Assembly and the Economic and Social Council.[21]

66. Within this wider framework of concerns and activities, the objective of the United Nations in the specific field of marine and coastal technology is first of all to help Governments to evaluate their needs and capabilities and consequently to assist them according to their own priorities, in (i) acquiring and applying the appropriate technologies, and (ii) developing such technologies in the perspective of greater collective self-reliance and co-operation.

67. Several organizations of the United Nations system are dealing with the full range of problems relating to the acquisition and application of technology *per se*, while others approach the subject more restrictively, either by function or by sector. A programme that seeks to deal with a wide range of problems or functional issues but in relation to a single broad technology sector—e.g., marine and coastal area development—would appear to combine the merits of both approaches and promote the in-depth treatment of an area of recognized importance. The programme components are summarized in the table at the end of this section.

68. The first step towards the application of technologies is the acquisition of information conducive to planning, evaluation and development. For the practical purpose of informing developing countries of all aspects of applied coastal and marine technology, it is essential to recognize that, while many organizations of the United Nations system deal with particular types of this technology, no single programme deals with marine and coastal technology as a whole. Such a programme would not supplant, bypass or infringe upon any programme of a United Nations organization but, on the contrary, would increase its accessibility to developing countries through new publications and other tools for the dissemination of information. When certain activities fall wholly within the terms of reference or competence of a given United Nations

---

20. Adopted by the United Nations Industrial Development Organization at its Second General Conference, held at Lima from 12 to 26 March 1974 (A/10112, chap. IV).

21. The General Assembly, on 15 December 1975, adopted resolution 3507 (XXX) in which it reaffirmed "the importance of wider dissemination of scientific and technological information, ... and the need to enable developing countries to select technologies which meet their requirements." The Secretary-General was requested to "establish an interagency task force ... with a view to the preparation of a plan for the establishment of a network for the exchange of technological information." Similarly, in para. 1 of resolution 1902 (LVII) the Economic and Social Council requested the Secretary-General "to undertake ... a feasibility study on the progressive establishment of an international information exchange system for the transfer and assessment of technology."

body or agency, then the latter will be relied upon to handle all relevant aspects in so far as they are covered.[22]

3. MARINE AND COASTAL TECHNOLOGY INFORMATION SERVICE (MACTIS)

69. The Marine and Coastal Technology Information Service (MACTIS) is being developed by the Secretary-General initially to serve as a focal point for activities within the United Nations system dealing with the collection, evaluation and dissemination of information. It is operated within the Department of Economic and Social Affairs in such a manner as to be compatible with other activities in the area and has access to the facilities of the Electronic Data Processing and Information Systems Service of the United Nations Secretariat. The functions subsumed under MACTIS are being implemented or are planned to be implemented in two phases, as described in the following sections.

B. PROGRAMME DEVELOPMENT: FIRST PHASE

1. IDENTIFICATION OF THE TECHNOLOGY NEEDS OF DEVELOPING COUNTRIES

70. The principal criterion for the selection of the technologies to be covered in the programme will be their relevance to coastal area development and management activities in developing countries. Many programmes in technology transfer have begun with a technology that is mostly in use in developed countries and have attempted to find applications for it in developing countries. A more responsive approach is to examine the needs of developing countries in order to determine which of these needs can be met with current technology or through its adaptation. Thus, the preferred approach will be to match carefully identified needs, either, when feasible, with available technologies, or with appropriate adaptations of those technologies, rather than to redefine needs and problems so that they can be accommodated by existing technologies.

22. Some of the competencies in marine and coastal technology are fairly clearly established and in this regard mention might be made, *inter alia,* of the Department of Economic and Social Affairs of the United Nations Secretariat in non-living resources; FAO in fisheries, other living resources and fish processing; UNCTAD in shipping and port and harbour development, including dredging; UNESCO in marine science; UNIDO in industries located in coastal areas and industrial pollution; UNEP in marine and coastal pollution; IMCO in shipping safety and navigation; WMO in marine meteorology; WHO in sanitary engineering, environmental health and water quality; IAEA in the disposal of radio-active wastes; and ITU in marine and coastal telecommunications.

71. A number of seminars, expert meetings, background reports and other forums have examined or discussed subjects directly related to the technology required for coastal area development. A careful examination and analysis of the statements, discussions and conclusions contained in the relevant documents is currently under way and a preliminary list of priority activity areas and technology needs in developing countries is being drawn up.

Coastal Area Development Meetings
72. At each of the following meetings, representatives from developing countries discussed their problems and priorities:

   a) Meeting of the Group of Experts on Coastal Area Development, held at United Nations Headquarters from 11 to 15 November 1974 (see para. 37 above);

   b) Interregional Seminar on the Development and Management of Resources of Coastal Areas, held at Berlin (West), Kiel and Cuxhaven from 31 May to 14 June 1976 (see paras. 54–56 above);

   c) Joint United Nations/UNEP Consultation on the Protection and Development of the Marine Environment and Coastal Area, held in Kuwait from 6 to 9 December 1976 (see para. 43 above).

Other United Nations Meetings
73. Many other meetings have dealt with subjects of significance in determining the priority needs of developing countries in coastal area development and for the technologies required to meet these needs.

74. These meetings include the *ad hoc* regional and other meetings on training, education and mutual assistance (TEMA) held under the auspices of IOC, the Conference of Ministers of Arab States Responsible for the Application of Science and Technology to Development (CASTARAB) and other UNESCO meetings on marine science and technology, UNCTAD meetings on ports and related activities, meetings of CDCC, meetings of the Joint Group of Experts on the Scientific Aspects of Marine Pollution (GESAMP) on marine pollution, UNEP meetings on environmental management and various meetings of the United Nations Institute for Training and Research on related matters.

75. In addition, relevant discussions at the sixth and seventh special sessions of the General Assembly, the Economic and Social Council, the Governing Councils of UNDP and UNEP, IOC and IMCO have also given indications of the major needs and interests of developing countries.

76. It is expected that the regional commissions and the various meetings held under their auspices will also be rich sources of pertinent information, as will the various regional meetings on technical co-operation among developing countries (TCDC) convened in preparation for the Conference on Technical Co-operation among Developing Countries, to be held at Buenos Aires from 27 March to 7 April 1978.

Surveys and Background Papers
77. A number of field surveys have been carried out under the United Nations coastal area development programme. These have covered the countries of the Gulf of Guinea and the countries of the Persian Gulf. The reports provide valuable information on the facilities and needs of developing countries. In addition, a number of background papers for such regions as the Caribbean and West African have been prepared, dealing with coastal area development concerns. These too form an input into the preliminary identification of the most urgent technological needs of developing countries.

2. UNITED NATIONS REFERRAL SERVICE

Interagency Co-operation
78. The first steps toward implementing the marine and coastal technology (MACTECH) programme have been taken by the Secretary-General under the mandates of the Economic and Social Council resolutions 1802 (LV) and 1970 (LIX). In order to make a comprehensive survey of the availability of information on MACTECH within the United Nations system and the capabilities of the system to respond to inquiries, the specialized agencies, together with other competent organizations, and particularly the regional commissions, have been requested to join the United Nations in the establishment of a central referral service. The United Nations Secretariat would then undertake the role of referral centre for all requests which can be dealt with by the various organizations of the United Nations system. This referral service will be part of the over-all MACTIS and in the first instance, by informing each organization of the activities and capabilities of others, it would also promote the exchange of information within the system and thereby assist in avoiding unnecessary duplication of effort.

Guidebook to Sources of Information within the United Nations System
79. The responses from United Nations organizations will be compiled into an index guide to information sources in MACTECH within the United Nations system. It will include a summary of the capabilities of the United Nations in MACTECH as well as a listing of relevant information systems. The guide, which will be made available in published form, should thereby facilitate access by developing countries to the resources, expertise, information and services available within the system in this field.

Gaps in United Nations Coverage and Reference to Sources outside the United Nations System
80. A comparison of the priority MACTECH needs with the capabilities of the United Nations system may lead to the identification of concerns not adequately dealt with at present. The gaps so identified will require recourse to

information sources outside the United Nations and their incorporation into MACTIS. The identification of sources of information outside the United Nations system has begun with the circulation of a note verbale to every coastal State Member of the United Nations. In addition, the participants from developing countries at the Interregional Seminar on the Development and Management of the Resources of Coastal Areas have been requested to supply sources of technical information within their regions.

81. Further identification of sources of information in developed and developing countries will provide the basis for a supplemental publication, a guide to information sources in the MACTECH industries. Such guides have been produced for numerous industries by UNIDO and it is proposed that the publication be prepared jointly with UNIDO, thereby taking advantage of that organization's expertise and experience. Its preparation and updating, as needed, will require the establishment of links and continuing working relationships with information sources and referral systems outside the United Nations.

C. PROGRAMME DEVELOPMENT: SECOND PHASE

1. EXPANSION OF INFORMATION SERVICES

82. The referral centre and other components of MACTIS will operate through existing technological information systems such as the Aquatic Sciences and Fisheries Information System (ASFIS), the Science and Technology Policies Information Exchange System (SPINES), the Industrial Inquiry System (IIS) etc. Through co-operation with the Universal System for Information in Science and Technology (UNISIST) and the Inter-organization Board for Information Systems and Related Activities of ACC, compatibility with existing systems will be assured. With respect to ASFIS, for example, with which co-operation is expected to be close, information outputs initially prepared under the programme (e.g., the guides above) will be put under the umbrella of the system for revision and updating on a regular basis. With respect to the information services provided by the Centre for Natural Resources, Energy and Transport of the United Nations Secretariat in the field of land-based mineral, energy and freshwater resources, MACTIS will serve as a substantive and co-ordinating link with other marine-oriented systems, such as ASFIS. In practice, this would mean that information relating to marine subjects within the United Nations competence generated by or collected by the Department of Economic and Social Affairs would under the MACTECH programme and specifically through its MACTIS component, be disseminated in the form of suitable outputs through the most appropriate channel, whether ASFIS services or United Nations services. By operating at the area of interface and interaction between the Department of Economic and Social Affairs and

other information systems and services of the United Nations system, MACTIS would help to ensure proper co-ordination of related information activities and the avoidance of unnecessary duplication of work.

Technology Reviews
83. For specific components of coastal area activities alternative techniques and equipment are available. A series of publications is planned that will describe not only the kinds of equipment available but also the requirements for their application. These reviews will be prepared with the assistance of experts who have worked in or are from developing countries and are aware of the need, in relation to certain types of problems, to adapt sophisticated technologies of developed countries to situations where labour and time may be less expensive; United Nations offices and organizations, including the regional commissions, that have particular competence in the subject area concerned will also co-operate closely in the preparation of the reviews. It is envisaged that the reviews will provide a guide to the selection of appropriate technology and guidelines for its application. Included will be questions of environmental impact studies and cost/benefit analysis in both the short and the long term. Subjects under consideration include:
   a) Coastal protection technology;
   b) Beach and near-shore mining technology;
   c) Selected aspects of off-shore oil and gas exploitation technology;
   d) Selected aspects of coastal transport technology;
   e) Mapping in the coastal area;
   f) Remote sensing technology for the inventory and evaluation of coastal area resources.

Inventories of Products and Services
84. Concurrently with research for the technology reviews, lists will be compiled of products and services relevant to the technology in question. These lists will also provide an input to MACTIS and will include and emphasize sources in developing countries.
   85. In response to a note verbale circulated to Governments early in 1976, a preliminary listing has been compiled of marine-related industries in member countries. This listing reflects the difficulty of defining the components of the MACTECH industry and obtaining information on existing capabilities. The experience gained in this first compilation has influenced the programme by indicating that a breakdown of activities into components would have been desirable at an earlier state.

Marine and Coastal Technology Abstracts (MACTA)
86. An important method of disseminating information on MACTECH goes beyond the compilation of information sources to include summaries and outlines of particular subjects. These abstracts will be similar to the aquatic

sciences and fisheries abstracts (ASFA), which are a product of ASFIS; they will contain excerpts and analytical summaries of articles in current literature dealing with those marine and coastal technologies not covered by the programmes of the United Nations system of organizations. The eventual integration of these abstracts into ASFIS will depend upon suitable arrangements with agencies and upon the possibility of a United Nations financial input.

2. ACQUISITION AND ADAPTATION OF TECHNOLOGY

Triangular Linkage
87. The interposition of the United Nations as a link or go-between with respect to the consumers and suppliers of MACTECH would aim at making developing countries aware of the costs and benefits of alternative technologies while making producers and technicians aware of the specific problems of developing countries, such as availability of trained personnel and research and administrative support, as well as physical, social and economic factors.

88. An understanding of the complexity of MACTECH, coupled with an appreciation of the opportunities it can offer, will help a developing country to assess technological needs and options. A knowledge of variations in such factors as the availability of labour in different countries may, it is hoped, influence the supplier technologist to adapt his technology systems to varying operating environments.

89. An early activity in the linkage programme will be the convening of a meeting of a group of experts, with representatives of industry and developing countries interacting. The aim of this meeting will be to prepare a series of guidelines to be followed by suppliers wishing to be responsive to the needs of developing countries and guidelines for developing countries who will be users of technology and should be aware of their options and the costs and benefits of technology acquisition. The emphasis of the first meeting will be on identifying technological needs.

3. PROMOTION OF TECHNICAL CO-OPERATION AMONG DEVELOPING COUNTRIES (TCDC)

90. The TCDC programme undertaken by UNDP with the co-operation of the United Nations system provides an excellent means for promoting the sharing of technical resources, skills and capabilities among developing countries for mutual development. The suppliers of an appropriate technology may, in many instances, be from developing countries themselves. A regional approach to TCDC in the coastal and marine field might offer special potential, as would appear to have been demonstrated by the results of the several *ad hoc* regional meetings on training, education and mutual assistance (TEMA), held under

the auspices of IOC, particularly the meeting convened at Manila, at which a wide range of opportunities for intra-regional co-operation were identified.

91. Therefore, a major activity to be undertaken will be a detailed survey of the existing capabilities within developing countries that could be made available to other developing countries. The survey will include private firms and consultants, government experts, universities, laboratories and research facilities and will conclude with an analysis of the opportunities and limitations of TCDC in coastal and marine affairs. The information obtained together with the information on suppliers from developed countries, will form inputs into the referral capability of MACTIS. It is anticipated that the TCDC component of this referral capability would become part of the UNDP Information Referral System within the TCDC programme, with the responsibility of responding to inquiries concerning MACTECH assistance from developing countries. It would also contain references to all relevant information on MACTECH that has been gathered by regional centres, projects and research programmes.

92. The TCDC programme will also seek to encourage countries to share their experiences in adapting technologies for use in developing countries. The exchange and training of technical personnel during project implementation is in line with the excellent example set by the Project for Off-shore Prospecting in East Asia, which is supported by UNDP and uses experts, training and equipment from developing countries.

93. Any comment and guidance that the Council may wish to provide concerning the development and direction of the programme of activities described above would be most welcome.

APPENDIX

MARINE AND COASTAL TECHNOLOGY PROGRAMME

Components of the Programme	Activities	Specific Output
First phase:		
1. Definition of needs	United Nations meetings Agency meetings Seminars Surveys	Internal listing of subjects to be covered
2. Establishment of a referral system	Interagency co-operation through ACC Sub-Committee Uses of existing computer facilities	Users' guide to United Nations sources of information in MACTECH
3. Coverage of gaps in the United Nations system by reference to outside sources and establishment of list of these sources	Analysis of United Nations capabilities Compilation of replies to note verbale to coastal States Working relationship with outside sources	Users' guide to outside sources of information (with UNIDO)
Second phase:		
4. Expansion of information services	Analysis of specific technologies Review of selected outside publications	1. Technology reviews 2. Inventory of products and services 3. Publishers of abstracts (MACTA)
5. Acquisition and adaptation of appropriate technology	Linkage of producers and consumers through groups of experts and *ad hoc* meetings	Publication of guidelines for users and consumers in marine and coastal technology

6. Promotion of TCD............	Survey of the capabilities of developing countries Analysis of opportunities and limitations	Guide to the capabilities of developing countries Input to the over-all referral system for TCDC (with UNDP), including technical assistance

Military Activities

Strategic Submarines and Antisubmarine Warfare

Frank Barnaby
Stockholm International Peace Research Institute

Antisubmarine warfare (ASW) involves the detection, localization, and destruction of hostile submarines, both tactical and strategic ones. The consequences of strategic ASW technology have considerable significance for world security, because a breakthrough in this field would seriously undermine the invulnerability of American and Soviet strategic offensive nuclear forces and threaten the strategic nuclear balance between these powers. By the mid-1980s the invulnerability of land-based ICBMs may be in great doubt because of the development of highly accurate missiles, and consequently submarine-launched ballistic missiles (SLBMs) may then be the main guarantee against a first-strike capability. But such considerable resources are being devoted to ASW, particularly by the United States and the USSR, that a breakthrough cannot be ruled out, even though the task of detecting enemy nuclear strategic submarines is an exceedingly difficult one. What makes the ASW task particularly difficult is that the entire enemy strategic submarine fleet (or at least the vast majority of it) would have to be destroyed within such a short time that no submarine (or very few) could fire its missiles. A single strategic submarine can carry many dozens of nuclear warheads. Both the United States and the USSR have a relatively large number of strategic submarines.

AMERICAN AND SOVIET STRATEGIC SUBMARINE FLEETS

The Unites States currently has 41 strategic nuclear submarines—11 Polaris and 30 Poseidon (see Appendix G, table 9G, p. 851). Each Polaris submarine is equipped with 16 Polaris A-3 SLBMs and each Poseidon submarine with 16 Poseidon C-3 SLBMs. The American fleet is, therefore, armed with 656 SLBMs.

The development of a new U.S. strategic submarine, called the Trident, is under way. The first Trident will probably be operational in 1978, and it is planned that 10 Tridents will join the fleet by the mid-1980s; each one will carry 24 Trident-1 SLBMs.

The Polaris A-3 SLBM has a range of about 2,500 nautical mi and can carry three multiple reentry vehicles (MRVs), each with a yield of about 200,000 tons

© 1978 by The University of Chicago. 0-226-06602-9/77/1978-1015$00.75

(200 kt) of TNT. The Poseidon C-3 SLBM can carry up to 14 multiple independently targetable reentry vehicles (MIRVs), each with a yield of about 40 kt, over a range of 2,500 nautical mi. The Trident-1 SLBM will have a range of about 4,000 nautical mi and will carry up to 8 MIRVs. A Trident-2 SLBM is now under development, designed to carry up to 14 MIRVs over a range of nearly 5,500 nautical mi.

Currently, the U.S. strategic nuclear submarine fleet probably carries a total of about 5,000 independently targetable nuclear warheads. If the planned deployments of Trident SLBMs are carried through, this number may double during the 1980s.

The accuracy of ballistic missile warheads is measured by the circular error probable (CEP), which is the radius of the circle centered on the target within which 50 percent of a large number of warheads fired at the target will fall. The Polaris A-3 SLBM has a CEP of about 0.5 nautical miles and the Poseidon C-3 a CEP of about 0.3 nautical miles (see Appendix G, table 10G, p. 852). The Trident-1 should have about the same accuracy as Poseidon at twice the range.

Future SLBMs may have very high accuracies. Research is underway in the United States on a new SLBM warhead, the Mark 500, which will have the capability to maneuver away from approaching defensive missiles. If this warhead were provided with terminal guidance, in which lasers and radar would guide it right on to its target, its CEP could be less than 100 m. Such a warhead would be capable of destroying relatively small hardened targets, including land-based ICBMs in their silos, and would certainly be a first-strike weapon.

The Soviet strategic nuclear submarine fleet consists of eight Hotel class submarines, each equipped with three SS-N-5 SLBMs; 34 Yankee class, each with 16 SS-N-6 SLBMs; 11 Delta I class, each with 12 SS-N-8 SLBMs; and seven Delta II class, each with 16 SS-N-8 SLBMs. None of these Soviet SLBMs carries MIRVs, and each is armed with a warhead of about 1,000 kt (1 mt) yield. Some SS-N-6s, however, may carry three MRVs, each with a 200-kt warhead.

The range of the SS-N-5 is about 700 nautical miles, that of the SS-N-6 about 1,500 nautical mi, and that of the SS-N-8 about 4,200 nautical mi. Soviet missiles are less accurate than their American counterparts. The CEPs of the SS-N-6 and SS-N-8 are thought to be about one nautical mile. But the technological gap is closing. Thus, last November, the Soviet Union flight-tested from a submarine in the White Sea a new SLBM—the SS-N-18—designed to carry three MIRVs. This missile, which will probably replace the SS-N-8 on Delta class submarines, is thought to have a range exceeding 5,000 nautical mi.

Soviet strategic submarine forces are currently equipped with nearly 800 missile warheads, a number which will considerably increase as MIRVs are installed.

ANTISUBMARINE WARFARE

Antisubmarine warfare, which dates back to World War I when the British effectively used listening devices (hydrophones) and aircraft to detect German U-boats, and depth charges and mines to sink them, is today a very complex activity. It is often more of an art than a science, relying on judgments based on probabilities rather than certainties.

That ASW is complex is indicated by the fact that in the United States at least six different ASW weapon systems are deployed or being developed. These are based on land-based (Orion P-3C) and sea-based (Viking S-3A) ASW patrol aircraft, the antisubmarine guided-missile frigate program, the destroyer ASW program, the guided-missile cruiser program, and the hunter-killer submarine. Continuous refinements are being made in the equipment for these programs and in the communication systems associated with them.

The Soviet ASW program is less advanced than the American one and is based mainly on naval helicopter carriers and long-range, land-based aircraft. The helicopters rely on very sophisticated electronic equipment to detect and track enemy submarines and armed helicopters to destroy them. Two Soviet Moskva-class helicopter carriers and one Kiev-class carrier are in service and are used for fleet defence. The Soviet Union also operates a number of ASW cruisers and destroyers.

Several types of Soviet long-range aircraft, equipped with high-resolution radar and magnetic anomaly detection equipment, are designed to find and sink U.S. strategic submarines. Soviet ASW activities are generally confined to areas close to Soviet territory and to Soviet fleets. The U.S. ASW system, however, is better able to provide a long-range capability.

The ASW techniques using aircraft to detect submerged submarines by active or passive sonobuoys (hydrophones suspended by cable from a floating buoy) are relatively effective. Underwater sounds picked up by a hydrophone are relayed by a radio on the floating buoy to the aircraft. If a submarine is detected, the aircraft can release several more buoys to localize it by comparing the intensity of the signals from the various buoys. The exact position of the submarine can then be found by using more precise equipment, such as the magnetic anomaly detector which measures local distortions of the geomagnetic field caused by the submarine's hull. A homing acoustic torpedo released by the aircraft will seek out the submarine and destroy it with either a conventional or a nuclear explosion.

The U.S. Orion P-3C is an interesting example of an ASW aircraft. It has a crew of 12, a cruising speed of about 200 knots, and can patrol for nearly 12 hr. The Orion carries radar, TV, and magnetic anomaly detection equipment as well as passive-directional and active sonobuoys. The aircraft carries a total of about 90 sonobuoys, which can detect submarines at distances of up to about 10 km. The information from the detection equipment is analyzed automatically by an on-board computer which evaluates the data and launches antisubmarine weapons.

But a more effective ASW weapon against strategic nuclear submarines is another nuclear submarine—the hunter-killer—equipped with sonar and other ASW equipment, underwater communications, a computer to analyze data from the detectors and to fire antisubmarine weapons—such as torpedoes with active and passive acoustic terminal guidance and nuclear warheads. Both the United States and the USSR operate large hunter-killer fleets, each consisting of several dozen nuclear submarines. A hunter-killer submarine is designed to find and then to follow its quarry until the order comes for its destruction. In the case of nuclear war, the aim would be to destroy as many as possible of the enemy strategic submarines simultaneously.

Extensive ASW arrays of fixed hydrophones, such as the U.S. array in the Gulf of Mexico, involving a large number of devices moored to the bottom of the oceans, are used for long-range surveillance. The ultimate objective is the real-time surveillance of all the oceans by transmitting data from such ASW arrays to central computers by satellites.

Progress in ASW will have important (perhaps even fatal) consequences for humankind. The more progress there is, the greater is the chance of general nuclear war.

Military Activities

The ASW Problem:
ASW Detection and Weapons Systems[1]

The 1969/70 SIPRI *Yearbook* made a detailed study of ASW. The following is a summary of part of this study.

1. STATEMENT OF THE ASW PROBLEM

The ASW problem can be stated in the following general way:

Detection: The first element in ASW is the determination that there is a submarine in the water. This can be a most difficult task since the enemy submarines will do their utmost to conceal themselves. Nuclear submarines, with their capacity to remain submerged for very long periods, can hide anywhere in the upper 10 percent of the ocean volume. Even modern conventional submarines using snorkels are very difficult to find.

Classification: When a suspected submarine has been detected, it must be properly classified. Many "false targets" in the ocean react to ASW detection methods: schools of fish, whales, certain water layers, etc. Friendly submarines must be sorted from enemy submarines and the latter classified according to type. The last point is important since it determines the kind of countermeasures used.

Localization: The accurate position of the detected submarine must be established. The submarine must then be tracked continuously until the attack has been delivered. The submarine may, of course, use the time lag between detection and attack for evasive movements.

Attacking: The submarine must be attacked by a weapon that can reach it while it is diving into deeper water.

---

1. The subject of ASW has been extensively dealt with in the following SIPRI publications:
   Stockholm International Peace Research Institute, *SIPRI Yearbook of World Armaments and Disarmament 1969/70* (Stockholm: Almqvist & Wiksell, 1970).
   Stockholm International Peace Research Institute, *World Armaments and Disarmament, SIPRI Yearbook 1974* (Stockholm: Almqvist & Wiksell, 1974).
   Stockholm International Peace Research Institute, *Tactical and Strategic Antisubmarine Warfare* (Stockholm: Almqvist & Wiksell, 1974).
   This section is summarized by Frank Barnaby from the first of these publications.

© 1978 by The University of Chicago. 0-226-06602-9/77/1978-1016$00.75

Destruction: The weapon must actually hit the submarine or make such an indirect impact on it that it either is destroyed immediately or is forced to surface where it can be destroyed.

The ASW strategy most likely to be used in a war, at least by the United States, is an offensive forward strategy. This means, in particular, to attack enemy submarines as near their home bases as possible, before they reach their stations in the open ocean and while they may be passing through straits or other geographical barriers.

2. ASW MEANS OF DETECTION

Essentially three means of detection are available: electromagnetic, acoustic, and magnetic. The *Yearbook* discusses each of these. A summary of this discussion follows.

Electromagnetic Detection

Electromagnetic detection includes the optical field, radar, infrared, and laser. Optical identification and radar are, of course, used by surface ships, aircraft, and helicopters whenever a submarine surfaces. The radar systems are also efficient against snorkels and floating antennae. Infrared detection may be employed by aircraft, helicopters, or satellites to trace the heat emitted by submarines, which to some extent shows up in surface water movements; but this is also a near-surface method. All these methods are handicapped by the fact that the penetration of electromagnetic energy in water is low.

Acoustic Detection

Acoustic detection is the primary method of finding submarines below the surface. The most important term in acoustic detection is "sonar," which stands for "sound, navigation, and ranging." There are two broad categories of sonar techniques, passive and active sonar. Passive sonar devices simply listen to sounds created by the submarine's propulsion machinery or by its movement through the water. They can detect a submarine over fairly long distances; they cannot, however, determine the distance to it with any accuracy.

Active sonar systems send out high-energy sound waves which strike underwater objects and return echoes to listening instruments. Active sonar possesses three advantages over passive sonar: it does not depend on the target to generate noise, the distance as well as the direction of the target can be measured, and a comparison of the acoustic frequencies of the echo and the transmitted pulse will produce a so-called Doppler signal which indicates whether targets are moving toward or away from the sonar. Important disadvantages of active sonar are: the transmissions reveal the presence of the sonar transmitter over an area far greater than its own zone of detection, and it receives echoes from a wide variety of objects difficult to distinguish from submarines. Two additional problems are that for geometrical reasons the signal received on reflection from the target varies as the fourth power of the range, that is, if the range is doubled the echo received will diminish by 15/16ths; and that using stronger transmitting signals may cause "cavitation," that is, an air-vapor pocket in front of the transmitter. The attenuation of sound in water may be partially overcome by using low-frequency sound waves; however, these are more difficult to generate and are less suitable for target discrimination, being too long. The problems of sound generation under water, however, can be overcome by high-energy transducers that can generate megawatts of acoustic power rather than a few tens of kilowatts.

Sophisticated passive sonars now in use can receive sounds almost 100 miles away. The ranges in the active mode are usually much shorter, generally 10–15 mi. However, ranges in excess of 30 mi have been achieved by U.S. active sonars using the bottom-bouncing technique.

Sonars can be used in many different ways, some of which the *Yearbook* describes:

Hull-mounted on surface ships, usually destroyers.—Modern destroyers often have combined passive/active sonars. The newest types are very large and high powered. The output of the sonars controls the firing of the ship's antisubmarine weapons.

Hull mounted on submarines.—One example is the AN/BQQ-2 on U.S. nuclear attack submarines. It has both a passive acquisition subsystem and an active/passive tracking subsystem. Since concealment in the water is the biggest advantage of submarines, they use the passive mode as much as possible and put on the active sonar immediately before attacking.

Variable depth sonar (VDS).—This sonar is dragged through the water by cable from a surface ship. The main advantage of VDS sonar is that it can be used to sense targets under the thermocline.

Helicopter-dipped sonar.—Helicopters may be sent out from surface ships and, while hovering, lower a cable with a sonar at its end to listen for submarines.

Sonar buoys.—Essentially there are two sorts: those dropped from aircraft in order to obtain a "fix" on a suspected submarine, and those established as long-term listening posts. Both kinds of buoys listen for sounds under water and report by radio to aircraft, satellites, or surface ships. Sonobuoys may be, and probably are, used for establishing surface barrier detection systems for submarines.

The principal drawback with existing passive sonobuoys has been the necessity to drop a whole string of them, together with explosive charges that pump sound into the water at preselected depths, in order to detect and localize a submarine. The system requires a great deal of electronics analysis in the aircraft. The biggest breakthrough in sonobuoy technology is said to be the AQA-7, DIFAR, which uses directional sensing: Fewer buoys need to be dropped and fewer data processed on board the aircraft. Sonobuoys are expendable; they eventually turn themselves off and sink.

Bottom sonars.—Fixed bottom sonars have been deployed by the Americans on the U.S. continental shelf since the 1950s. The Sonar Surveillance Systems (SOSUS) basically consists of series of passive sonars (i.e., hydrophones) connected by cable to land. Bottom sonars are also used at present for barrier control purposes. Since they can be installed at fixed positions, bottom sonars are much more efficient than other forms of sonars for accurately measuring the distance to a target submarine. Because of the need to use very base lines for submarine detection, bottom sonars are as a rule installed in widely separated pairs—at greater distance from each other than is possible with shipborne sonars.

Despite their limitations due to the difficulty of predicting sound propagation in water, sonar systems are likely to remain the chief means of detecting submarines. As submarines become quieter—their chief countermeasure—more advanced and expensive technology will go into the sonars.

Magnetic Anomaly Detection

A magnetic anomaly detector (MAD) is a sensitive magnetometer which senses local variations in the earth's magnetic field. On land natural anomalies are common, but at sea local anomalies are rare because bodies of ore are too remote from the surface for detection. However, when a metallic body, such as a submarine, nears the surface, the magnetic field disturbance can be measured. The favorable features of this sensor system are that it is unaffected by the surface and is invulnerable to jamming. MAD's severe range limitation of less than 3,000 feet is its great drawback. This limitation allows the system to be used only for final confirmation of contact and localization for destruction, not for searching large expanses of ocean.

MAD devices are mostly used on ASW aircraft and helicopters. Another use of magnetic anomaly detection is in a cable system on the seabed in shallow water. The cables will register all magnetic objects that pass over them and alert ASW forces.

3. ASW WEAPONS SYSTEMS

Antisubmarine warfare usually requires the cooperation of several systems—unmanned surveillance systems, aircraft, surface ships, and submarines—against a single submarine. The ASW forces of the U.S. Navy, for instance, operate in task forces generally consisting of an ASW carrier with specialized aircraft, destroyers, and attack submarines (hunter-killer groups).

ASW carriers.—ASW carriers are as a rule smaller than attack carriers. Only the Western navies have ASW carriers; the Soviet Union has none besides two new helicopter carriers.

Destroyers and helicopters.—The newest destroyers have powerful sonars that control the firing of antisubmarine torpedoes and missiles. Some destroyers have also been equipped with helicopters for submarine hunting. Relatively speaking, the importance of destroyers as a means of detection in ASW seems to have diminished in recent years in favor of aircraft and antisubmarine submarines. But operating together with other sensor systems the destroyer retains part of its importance as a weapons platform against enemy submarines.

Land-based ASW aircraft.—Land-based ASW aircraft are considered an efficient means of looking for submarines on, or not too deep below, the surface. With their long endurance—more than 10-hr patrol times—they can scan vast expanses of the ocean. The advantages of ASW aircraft are summed up thus. With their sensors they can detect the presence of submarines on or near the surface and direct surface ships, antisubmarine submarines, or other aircraft to the place for attack. Equipped with sonobuoys, depth charges, and torpedoes, they may attack and destroy the submarine. If possible they will operate with a screen of destroyers around the suspected position of the enemy submarine in order to prevent it from escaping. The aircraft is itself invulnerable to attack from the submarine. The aircraft's main disadvantages are that its sensors cannot penetrate deep below the surface where the submarine may be hiding and that, because of its speed, it has difficulty maintaining the position of the submarine.

Antisubmarine submarines.—The antisubmarine submarine has many advantages as an ASW hunter. It uses the same medium as the target and can therefore do whatever it can do—for instance, remain undetected for long periods, listen quietly over long distances, dive deeper, attack with the same weapons. This applies particularly to the new nuclear attack submarines developed as a countermeasure to the nuclear-powered ballistic-missile submarines. Antisubmarine submarines are, however, expensive and, consequently, few in number; they have the same communication and navigation problems as other submarines.

In the forward ASW tactics deployed by the U.S. Navy, the antisubmarine submarines are usually assigned patrol sections along barrier lines, such as the lines between Greenland–Iceland–Faroe Islands–Scotland or between the Aleutian Islands and the northern Japanese islands in the northern Pacific.

ASW MEANS OF ATTACK

The main categories of ASW means of attack are torpedoes, missiles, depth charges, and mines.

Torpedoes.—Torpedoes are the classical weapon of the submarine. Several advanced types are in existence or being developed: (1) Homing torpedoes that use acoustic ranging to locate their target and in that way seek to overcome their main restriction—limited speed (modern high-speed torpedoes may reach a maximum speed of about 60 knots, but that is not enough against submarine targets which may move at 30–35 knots); (2) wire-guided torpedoes that are continuously fed new guidance orders from the submarine that fired them; (3) torpedoes with nuclear warheads that do not have to hit the target in order to destroy it; (4) torpedoes launched from surface ships or dropped from aircraft.

Antisubmarine rockets.—One way of overcoming the limited range and speed of conventional torpedoes is to fire them by rockets in a ballistic trajectory through the air. The U.S. Navy has developed two types:

ASROC is fired from surface ships; the rocket usually carries a torpedo which is released by parachute before the end of the ballistic trajectory. Once in the water, the torpedo homes in on the target. The rocket may also carry a depth charge, nuclear or conventional, which descends without parachute to predetermined depths and detonates.

SUBROC, a nuclear depth charge launched by a submarine underwater in a ballistic trajectory against a target submarine.

Air-dropped depth charges.—Several air-dropped depth charges are in existence.

Mines.—Existing mines include: (1) moored conventional mines that depend on physical contact for their activation; (2) cable-controlled bottom mines; (3) pressure mines, a ground mine for use in shallow waters (not more than 180 feet). Lying on the bottom, the mine detonates when an underwater structure passes over it, causing a pressure differential to be registered in the mine. Pressure mines are very difficult to sweep and therefore constitute a particularly potent antisubmarine weapon; (4) magnetic mines, activated by the magnetic field created by a submarine; they may be either moored or ground mines, and their range is limited to a few hundred feet; (5) acoustic mines, which react to the noise created by a submarine, may be either moored or lying on the bottom, and presumably have a longer range than pressure or magnetic mines; (6) torpedo mines. The U.S. Navy is developing one deep ocean mine, CAPTOR, that is really an encapsulated torpedo: Upon activation it can seek out the target and destroy it. CAPTOR has a large radius and probably a nuclear warhead.

Examples of the above systems are given in Appendix G. (pp. 835–71).

Military Activities

The Seabed Treaty

Jozef Goldblat
Stockholm International Peace Research Institute

Besides the bilateral agreement of 1972 on the prevention of incidents on and over the high seas between the ships of the armed forces of the United States and the Soviet Union, four multilateral treaties have been concluded in the post–World War II period restricting the military uses of the seas.

The Antarctic Treaty, effective since June 23, 1961, declared the Antarctic as an area to be used for peaceful purposes only. This declaration was reinforced by the prohibition of any measures of a military nature, such as the establishment of military bases and fortifications, the carrying out of military maneuvers, and the testing of weapons. Nuclear explosions of any kind and the disposal of radioactive waste material have also been prohibited. The provisions of the treaty apply south of the 60° south latitude. Thereby, not only the Antarctic continent but also the ice shelves which surround the continent and occupy an area estimated at 800,000 square miles have been put under the demilitarized regime.

Next came the treaty banning nuclear-weapon tests in three environments, called the Partial Test Ban Treaty (PTBT), which entered into force on October 10, 1963. One of the environments prohibited for testing is the underwater environment. To stress the comprehensiveness of the prohibition, the text of the treaty enumerates, for illustrative purposes, territorial waters and high seas. High seas have been singled out to remove the possibility of an argument being put forward that these parts of the seas were not covered by the prohibition because they are not under the "jurisdiction or control" of any party, as stipulated elsewhere in the treaty. And once the high seas had been mentioned, it was found expedient to denote them, at least by implication, as a direct extension of the territorial waters (irrespective of the breadth of the latter) so as not to leave gaps in the banned environment.

The treaty prohibiting nuclear weapons in Latin America (Treaty of Tlatelolco) was signed in Mexico City on February 14, 1967, and has been in force since then for each state which has ratified it and waived the requirements specified in one of its articles. The zone of application of the Treaty of Tlatelolco is the whole of the territories for which the treaty is in force, but the term "territory" is defined as including the territorial sea, air space, and any other space over which the state exercises sovereignty "in accordance with its

© 1978 by The University of Chicago. 0-226-06602-9/77/1978-1017$02.00

own legislation." In view of the existing legislation in Latin American countries, the zone may encompass large portions of the Atlantic and Pacific Oceans, which are considered by most other states as high seas. Indeed, the treaty provides that the zone eventually will cover an area between 150° west longitude and 20° west longitude and thus will extend hundreds of kilometers off the coast of the states party to the treaty. The United States, which is party to Additional Protocol II of the Treaty of Tlatelolco and has undertaken to respect the statute of military denuclearization of Latin America, stated that its ratification could not be regarded as implying recognition of any legislation which did not, in its view, comply with the relevant rules of international law. A statement in similar terms was made by the United Kingdom and France. These three powers are, therefore, not bound by the treaty stipulations extending beyond the limits which they consider as defining the territorial sea "under international law." But China made no reservation on the extent of the zone; it has thereby accepted limitations on its military activities in the same broad area of the high seas that the Latin American countries designated as free of nuclear weapons.

The latest addition to arms-control measures in the seas is the so-called seabed treaty, imposing some limitations on the use of the seabed, the ocean floor, and the subsoil thereof. It was signed on February 11, 1971 (Appendix A). The following section contains an account of the negotiations at the Geneva Disarmament Committee and the UN General Assembly which led to the signing of the seabed treaty, an analysis of its text, and an assessment of its significance. In conclusion, the status of the implementation of the treaty is reviewed, and a few suggestions are made for possible further action.

THE NEGOTIATING HISTORY

Arms-control measures concerning the seabed began to receive attention in 1967, when the United Nations decided to engage in a thorough examination of the possibility of reserving exclusively for peaceful purposes the seabed, the ocean floor, and the subsoil thereof underlying the high seas beyond the limits of present national jurisdiction and the use of their resources in the interest of mankind.[1]

In 1968, the Soviet Union suggested that the Eighteen-Nation Committee on Disarmament (ENDC) should consider the prohibition of the use for

---

1. United Nations, General Assembly, *Examination of the Question of the Reservation Exclusively for Peaceful Purposes of the Seabed and the Ocean Floor, and the Subsoil Thereof, Underlying the High Seas beyond the Limits of Present National Jurisdiction, and the Use of Their Resources in the Interests of Mankind* (hereafter referred to as A/RES/2340 [XXII]), December 18, 1967.

military purposes of the seabed beyond the limits of the territorial waters.[2] The United States proposed to take up the question of arms limitation on the seabed with a view to preventing the use of this environment for the emplacement of weapons of mass destruction.[3]

Negotiations at the ENDC began in the spring of 1969 with the presentation by the Soviet Union of a draft treaty which provided for total demilitarization of the seabed and ocean floor.[4] The United States, in opposing the Soviet comprehensive approach, submitted its own draft calling only for denuclearization of the seabed environment.[5]

It soon became evident that no comprehensive ban on the military use of the seabed would be achieved in the foreseeable future. On October 7, 1969, following major concessions by the Soviet Union, the two great powers, cochairmen of the Conference of the Committee on Disarmament (CCD),[6] tabled a joint draft treaty under which the parties would undertake "not to emplant or emplace on the seabed and the ocean floor and the subsoil thereof beyond the maximum contiguous zone provided for in the 1958 Geneva Convention on the Territorial Sea and the Contiguous Zone any objects with nuclear weapons or any other types of weapons of mass destruction, as well as structures, launching installations or any other facilities, specifically designed for storing, testing or using such weapons."[7]

The majority of nations found the text of the joint draft treaty inadequate. Criticism concerned mainly the scope of the prohibition, the area of the seabed to which the prohibition should apply, the methods for verifying compliance

---

2. United Nations, General Assembly, Ad Hoc Committee to Study the Peaceful Use of the Seabed and the Ocean Floor beyond the Limits of National Jurisdiction, 3d Session, *Summary Records of the Thirteenth to Twenty-sixth Meetings* (hereafter referred to as A/AC. 135/S.R. 13–26), 1968, pp. 98–108. This is the summary document that contains reference cited (A/AC. 135/20). See also United Nations, General Assembly, *Letter Dated 5 July 1968 from the Permanent Representative of the Union of Soviet Socialist Republics to the United Nations, Addressed to the Secretary General,* July 8, 1968.

3. A/AC. 135/S.R. 13–26, pp. 157–71.

4. Conference of the Eighteen-Nation Committee on Disarmament, Union of Soviet Socialist Republics, *Draft Treaty on Prohibition of the Use for Military Purposes of the Seabed and the Ocean Floor and the Subsoil Thereof* (hereafter referred to as ENDC/240), March 18, 1969.

5. Conference of the Eighteen-Nation Committee on Disarmament, United States of America, *Draft Treaty Prohibiting the Emplacement of Nuclear Weapons and Other Weapons of Mass Destruction on the Seabed and Ocean Floor* (hereafter referred to as ENDC/249), May 22, 1969.

6. In the summer of 1969, the membership of the Eighteen-Nation Committee on Disarmament was enlarged, and it was decided that the new name of the conference would be the Conference of the Committee on Disarmament (CCD).

7. Conference of the Committee on Disarmament, Union of Soviet Socialist Republics and United States of America, *Draft Treaty on the Prohibition of the Emplacement of Nuclear Weapons and Other Weapons of Mass Destruction on the Seabed and the Ocean Floor and in the Subsoil Thereof* (hereafter referred to as CCD/269), October 7, 1969.

with the obligations assumed, and the procedure for amending the treaty. It was strongly urged that the formulations used ensure that the interests of all coastal states would be safeguarded. Requests were also put forward for entrusting the United Nations with the task of securing the observance of the treaty and for periodic reviews of the operation of the treaty. Several delegations to the CCD prepared working papers with specific proposals for changes.

On October 30, 1969, a revised joint draft was submitted by the United States and the Soviet Union; it included a few amendments on which the cochairmen agreed.[8] The new draft clarified the status of the zone lying between the outer limit of territorial seas narrower than 12 miles and the outer limit of the maximum contiguous zone, it provided for the right of recourse of the parties to the United Nations Security Council in the event of serious doubts concerning the fulfillment of the treaty obligations, and it included a provision for a review conference and established an equal voice for all parties in deciding which amendments should be introduced in the future.

The text was discussed at the twenty-fourth UN General Assembly session. It still proved unsatisfactory to most delegations, which claimed that their fundamental concerns had not been met. A series of new proposals was made. The General Assembly called upon the CCD to take those into account and to continue its work on the subject.[9]

On April 23, 1970, a third version of the joint U.S.-Soviet draft treaty was issued by the cochairmen of the CCD.[10] It incorporated suggestions made by different delegations, particularly by Argentina, Brazil, Canada, and Mexico. The verification provisions were elaborated in greater detail than in the previous texts; the concept of a "seabed zone" was used in place of the earlier references to the "maximum contiguous zone"; the so-called disclaimer clause, dealing with the relationship of the obligations assumed under the treaty and other international obligations of the states parties to the treaty, was expanded and appeared as a separate article; and a provision was included to the effect that the treaty would not affect international agreements concerning the establishment of nuclear-free zones.

In the opinion of a number of countries, the draft still required improvement. Demands were put forward for a binding commitment to continue negotiations on further measures prohibiting the military use of the seabed, recognition of the principle of international responsibility for verification procedures, and full respect for the sovereign rights of coastal states.

These demands were partially met in the fourth consecutive version of the

---

8. Ibid., rev. 1.
9. United Nations, General Assembly, *Resolution Adopted by the United Nations General Assembly (on the Report of the First Committee A/7902), Question of General and Complete Disarmament* (A/RES/2602 [XXIV]), January 21, 1970.
10. CCD/269, rev. 2, April 23, 1970.

draft treaty submitted by the Soviet Union and the United States on September 1, 1970.[11] The resulting text (see Appendix A) was judged acceptable by the Disarmament Committee and was commended by the twenty-fifth UN General Assembly in a resolution of December 7, 1970.[12] However, in agreeing that the treaty should be opened for signature and ratification, many UN representatives made it clear that their affirmative votes for the resolution did not prejudge the positions which their governments would eventually adopt. The reasons for this reserve were explained in the course of the UN debate.

Some nations, including France,[13] considered the scope of the treaty prohibition to be too narrow, restricted as it was to activities of little military interest. The treaty was also found insufficient as a denuclearization measure, since it exempted from the ban a seabed zone 12 miles wide. Peru considered the exemption to be "unjustifiable discrimination" in favor of the nuclear powers.[14] An amendment to enlarge the geographical extent of the prohibition to cover the entire seabed and ocean floor[15] was not accepted, but, significantly enough, out of 99 delegations participating in the vote on the amendment, as many as 39 (i.e., over a third) abstained.[16]

The provisions on verification, though rather elaborate, proved unsatisfactory to France and Pakistan,[17] in that no genuinely international control system was established.

Even more objectionable than these omissions were the imputed "sins of commission." It was charged, especially by Latin American countries, that the reference in the treaty to the 1958 Geneva Convention on the Territorial Sea and the Contiguous Zone was meant to set an international precedent in support of the 12-mile limit of territorial waters, which most of these countries opposed, and that this was the real purpose of the great powers rather than to promote disarmament. The language of the treaty was criticized for lack of precision; Ecuador did not hesitate to call it a network of ambiguities and errors.[18]

11. Ibid, rev. 3; and CCD/317, annex A, 1970.
12. United Nations, General Assembly, *Resolution Adopted by the General Assembly (on the Report of the First Committee A/8918), Treaty on the Prohibition of the Emplacement of Nuclear Weapons and Other Weapons of Mass Destruction on the Seabed and the Ocean Floor and in the Subsoil Thereof* (A/RES/2660 [XXV]), February 9, 1971.
13. United Nations, General Assembly, *Meeting Records, 25th Session* (hereafter referred to as A/C.1/PV. 1754), November 9, 1970.
14. United Nations, General Assembly, *Meeting Records, 25th Session* (hereafter referred to as A/C.1/PV. 1763), November 17, 1970.
15. United Nations, General Assembly, *Limited Series, Question of General and Complete Disarmament: Report of the Conference of the Committee on Disarmament* (A/C.1/L.528), November 10, 1970.
16. United Nations, General Assembly, *Question of General and Complete Disarmament* (A/8198), December 4, 1970.
17. A/C.1/PV. 1754.
18. A/C.1/PV. 1763. See also United Nations, General Assembly, *Meeting Records, 25th Session* (A/C.1/PV. 1764), November 18, 1970.

Nevertheless, since the misgivings of coastal states were allayed by a disclaimer clause, many of them saw no harm in endorsing an agreement covering an environment of marginal military importance. It was also appreciated that the treaty, which in essence represented a bilateral U.S.-Soviet self-limitation, had been negotiated and finalized multilaterally. Consequently, and as a result of pressure exercised by the United States and the Soviet Union, only two delegations (El Salvador and Peru) opposed the UN General Assembly resolution commending the seabed treaty, and two abstained (Ecuador and France), while 19 delegations were absent during the voting.

SCOPE OF THE SEABED PROHIBITIONS

United Nations resolutions calling for the reservation of the seabed and ocean floor and the subsoil thereof exclusively for peaceful purposes[19] formed a framework for possible arms-control measures. However, a controversy arose over the meaning of the phrase "exclusively for peaceful purposes."

The nonaligned countries contended that the United Nations had invariably understood the use of a given environment for exclusively peaceful purposes to mean the prohibition of all military activities whatever their purpose, and that there should be no departure from this approach in the case of the seabed. Some of them reasoned that since the seabed must be used for the benefit of all states (as stated in the above resolutions), any military use of it represented an unjustified territorial usurpation hampering peaceful exploitation of the environment.

The Soviet Union also equated "peaceful purposes" with "nonmilitary purposes." Its approach was similar to that applied to Antarctica under the Antarctic Treaty (see above). Accordingly, the first Soviet draft treaty aimed at demilitarizing completely the seabed and the ocean floor as well as the subsoil thereof.[20]

The United States interpreted the phrase "peaceful purposes" as not barring military activities generally. It argued that specific limitations of certain military activities would require detailed agreements, and that activities not precluded by such agreements would continue to be conducted in accordance with the principle of the freedom of the seas. It saw the analogy with the 1967 Outer Space Treaty, which does not provide for the use of outer space exclusively for peaceful purposes but specifically prohibits to place in orbit

---

19. A/RES/2340 (XXII). See also United Nations, General Assembly, *Resolutions Adopted by the General Assembly (on the Report of the First Committee A/7477), Examination of the Question of the Reservation Exclusively for Peaceful Purposes of the Seabed and the Ocean Floor, and the Subsoil Thereof, Underlying the High Seas beyond the Limits of Present National Jurisdiction, and the Use of Their Resources in the Interests of Mankind* (hereafter referred to as A/RES/2467 [XXIII]), January 14, 1969.

20. ENDC/240.

around the earth objects carrying nuclear weapons or other kinds of weapons of mass destruction. Accordingly, the United States proposed that states undertake not to emplant or emplace fixed nuclear weapons or other weapons of mass destruction or associated fixed launching platforms on, within, or beneath the seabed and ocean floor.[21]

In advocating denuclearization, the United States asserted that only weapons of mass destruction could have enough significance militarily to warrant the expense of their stationing on the seabed. It expressed the belief that realistic possibilities did not and would not soon exist for such conventional military uses of the seabed as would be threatening to the territories of states. Some nonnuclear but clearly military uses of the seabed (e.g., devices for detection and surveillance of submarines) were essential to the security of states and therefore indispensable. In the opinion of the United States, complete demilitarization would, moreover, raise verification problems by imposing a task of deciding whether each object or installation emplaced on the seabed was of a military nature. In any event, the United States, being a major naval power, was not prepared to accept a ban on all military activities on the seabed.

Compromise suggestions ranged from a general ban, subject to exception for devices and activities not of a directly military nature or of a passive defensive character, to a ban comprising weapons of mass destruction as well as conventional weapons to be agreed upon in a list, to a prohibition of weapons of mass destruction in a first stage to be followed by a ban on conventional weapons at a later stage.

The text eventually agreed upon provides for an undertaking by states parties to the treaty not to emplant or emplace on the seabed and the ocean floor and in the subsoil thereof any nuclear weapons or any other types of weapons of mass destruction as well as structures, launching installations, or any other facilities specifically designed for storing, testing, or using such weapons. The parties also undertake not to assist, encourage, or induce any state to carry out activities prohibited by the treaty and not to participate in any other way in such actions (article 1).

When asked for more precision, the sponsors of the treaty explained that the treaty prohibits, inter alia, nuclear mines anchored to or emplaced on the seabed. It does not apply to facilities for research or for commercial exploitation not specifically designed for storing, testing, or using weapons of mass destruction, but facilities specifically designed for using such weapons could not be exempted from the prohibitions of the treaty on the grounds that they could also use conventional weapons. The prohibitions are not intended to affect the application of nuclear reactors or other nonweapons applications of nuclear energy, consistent with the treaty obligations. It was also explained that while submersible vehicles able to navigate in the water above the seabed would be viewed as any other ships and would not be violating the treaty when

21. ENDC/249.

anchored to or resting on the bottom, bottom-crawling vehicles which could navigate only when in contact with the seabed and which were specifically designed to use nuclear weapons would be banned. Since the prohibition embraces not only fixed (as was originally provided for in the U.S. draft) but also certain mobile facilities, the use of the terms "emplant or emplace" does not seem to be entirely compatible with the scope of the treaty.

GEOGRAPHICAL EXTENT OF THE AREA COVERED BY THE PROHIBITIONS

There was a general understanding that seabed disarmament measures were to include the area reserved exclusively for peaceful purposes. The latter was defined in UN resolutions[22] as the area underlying the high seas beyond the limits of national jurisdiction. The vague language of the definition reflected the lack of agreement as to where these limits actually lay. However, a view prevailed that a precise boundary, devised specifically for arms-control purposes and expressed in terms of distance from the coast, should be agreed upon.

Under the Soviet draft treaty the prohibition was to cover an area beyond a 12-mile maritime zone of the coastal states.[23] The U.S. draft provided for a prohibition beyond a 3-mile band adjacent to the coast.[24] Eventually, the area of prohibition was defined (articles 1 and 2) as lying beyond the outer limit of a seabed zone coterminous with the 12-mile outer limit of the zone referred to in part 2 of the Convention on the Territorial Sea and the Contiguous Zone, signed at Geneva on April 29, 1958.

The undertakings by states parties to the treaty are applicable also to the 12-mile seabed zone, except that within such a zone they shall not apply either to the coastal state or to the seabed beneath its territorial waters. In other words, since the treaty does not contain an absolute prohibition on the placement of weapons of mass destruction beyond the parties' own seabed zone, and since an exception has been made with regard to territorial waters, states have the right, according to the language of the treaty, to install weapons of mass destruction on the seabed beneath the territorial waters within the 12-mile seabed zone of other states, obviously with the consent and authorization of the states concerned ("allied option"). This would not be permitted in the band between

---

22. A/RES/2340 (XXII); A/RES/2467 (XXIII); and United Nations, General Assembly, *Resolutions Adopted by the General Assembly (on the Report of the First Committee A/7834), Question of the Reservation Exclusively for Peaceful Purposes of the Seabed and the Ocean Floor, and the Subsoil Thereof, Underlying the High Seas beyond the Limits of Present National Jurisdiction, and the Use of Their Resources in the Interests of Mankind* (A/RES/2574 [XXIV]), January 15, 1970.
23. ENDC/240.
24. ENDC/249.

the outer limit of the territorial sea and the 12-mile limit of the seabed zone in cases where the breadth of the territorial waters is narrower than 12 miles. The United States and the Soviet Union pointed out that the exemption with respect to the seabed beneath the territorial waters within the seabed zone left unaffected the sovereign authority and control of the coastal state within the territorial sea.[25] A call for a voluntary abstention from placing nuclear weapons on the seabed under the territorial seas until such time as that area, too, was covered by the treaty went unheeded.

The strongest objections, however, were related to the reference in the seabed treaty to the 1958 Geneva Convention on the Territorial Sea and the Contiguous Zone. It was contended that the reference was misleading and gave rise to arguments of a legal and practical nature.

The contiguous zone is a surface criterion applying to superjacent waters. There seems to be no relationship between its characteristics and a prohibition on the employment of weapons on the seabed. The linking of the limits of the zone exempted from the seabed treaty prohibitions within the limits of the maximum contiguous zone, as provided in the Geneva Convention, also poses a question of what would happen to the seabed treaty if and when the convention were to be amended specifically with regard to those points that were taken as reference points for the relevant articles of the treaty or replaced altogether by another law of the sea convention. Only a minority of nations have ratified the Geneva Convention (see Appendix B). Some states qualified its provisions as highly controversial, narrow, and antiquated and thought it inappropriate to invite nonparties to an agreement to accept its formulations in defining new obligations.

Indeed, there was no necessity to make even an indirect reference to the 1958 Geneva Convention. A simple and self-sufficient formula could be used, stating the extent of the zone and how it should be measured. This would have distinctly separated the regime of the seabed treaty from the general regime of the law of the sea.

The treaty contains a clause to the effect that nothing in it shall be interpreted as supporting or prejudicing the position of any state party with respect to existing international conventions, including the 1958 Convention on the Territorial Sea and the Contiguous Zone, or with respect to rights or claims which such party may assert or with respect to recognition or nonrecognition of rights or claims asserted by any other state related to waters off its coast, including inter alia territorial seas and contiguous zones, or to the seabed and the ocean floor, including continental shelves (article 4). But such a sweeping disclaimer, if taken literally, may contradict the very sense of the treaty. To have meaning, a disarmament or a nonarmament measure must

---

25. Conference of the Committee on Disarmament, *Final Verbatim Record of the Four Hundred and Ninety-second Meeting, Held at the Palais des Nations, Geneva, on Tuesday, 1 September 1970* (CCD/PV. 492), September 1, 1970.

restrict, at least in some degree, the freedom of action as well as certain rights or claims the states may have asserted hitherto.

VERIFICATION OF THE PROHIBITIONS

The problem most extensively discussed was how to verify compliance with the prohibitions. In the first drafts the question was dealt with rather cursorily. The Soviet Union proposed that all installations and structures on the seabed and the ocean floor and the subsoil thereof should be open on the basis of reciprocity to representatives of other states parties to the treaty for verification of the fulfillment of the obligations.[26] The United States proposed that the parties should be free to observe activities of other states on the seabed and ocean floor, without interfering with the activities or otherwise infringing rights recognized under international law, including the freedom of the high seas.[27]

Neither proposal proved acceptable. It was pointed out that while verification must protect the interests of all, a mere proclamation of the right to verify would be meaningless for states with less-developed undersea technology—that is, for the overwhelming majority of states. Such countries would be unable to exercise the right of verification even if they suspected that they were threatened by weapons or military installations in adjacent areas of the seabed, unless they were guaranteed assistance in carrying out the necessary operations by the technologically more advanced states. On the other hand, some countries felt that in seeking direct aid from one or another nuclear-weapon power they would compromise their policy of nonalignment.

The United States and the Soviet Union refused to commit themselves formally to assisting any complaining state in the verification. The United States said that, given the present state of technology, heavy expenses as well as hazards were involved in performing major underwater searches, the equipment and personnel for these specialized activities being in short supply. Besides, varying political relations among countries that might become parties to the treaty made it impossible for the United States to accept a firm obligation in this respect. A clause dealing with this subject was, nevertheless, inserted in the treaty, stipulating that verification may be undertaken by any state party using its own means or with the full or partial assistance of any other party (article 3, par. 5).

The reluctance to resort to the optional aid of the technologically advanced states and to rely for security on such uncertain factors as the good will, availability of equipment, or the changing circumstances of the world situation brought about demands from many states for the internationalization of

26. ENDC/240.
27. ENDC/249.

control. Some urged the setting up of a special body responsible for the observance of the seabed treaty prohibitions. Others envisaged the use of existing international organizations for channeling verification requests. Canada suggested recourse to good offices, including those of the UN Secretary General, in identifying the state responsible for activities giving rise to concern relating to compliance with the treaty, as well as in arranging assistance in carrying out verification procedures.[28] Still others understood internationalization only as a possibility of calling upon the United Nations Security Council to settle disputes over verification.

The United States and the Soviet Union considered special international arrangements for carrying out control or the turning of verification functions over to the United Nations as needless and, in any event, premature and wasteful of resources. They believed that reliance should be placed on consultation and cooperation, but admitted that if consultation and cooperation had not removed the doubts and there remained a serious question concerning fulfillment of the obligations assumed under the treaty, a state party may, in accordance with the provisions in the UN charter, refer the matter to the Security Council for action in accordance with the charter (article 3, par. 4).

Nevertheless, the nonaligned countries continued to insist on the inclusion in the text of at least some reference to the possibility of international verification. This insistence was probably not directly related to the requirements of the seabed treaty. No one expected that there would really be much need for verification. What mattered was the establishment of a principle to be followed in future disarmament measures of greater importance. Finally, it was agreed that verification may be undertaken also "through appropriate international procedures within the framework of the United Nations and in accordance with its Charter" (article 3, par. 5).

As regards the nature of verification, the right of each state to observe was taken for granted, but it was felt that the right to verify would be deprived of substance if it were limited to observation. A number of countries insisted on access without restriction, so that dubious installations may not only be "looked at" but also "looked into." They also asked that the parties should be obliged to disclose their activities.

The Soviet Union was in favor of access to seabed facilities, similar to that provided under the Antarctic Treaty. The United States maintained that a right to go into a facility emplaced on the seabed or to open up equipment for the purpose of verifying whether nuclear weapons had not been installed there

---

28. United Nations, General Assembly, *Question of General and Complete Disarmament: Report of the Conference Committee on Disarmament, Canada, Working Paper on the Provisions of Article III of the Draft Treaty on the Prohibition of the Placement of Nuclear Weapons and Other Weapons of Mass Destruction on the Seabed and Ocean Floor and Subsoil Thereof (Annex A, Doc, A/7741)* (hereafter referred to as A/C.1/992), November 27, 1969.

would be difficult to exercise and unnecessary. Under the freedom of the high seas, parties could approach the area of a facility or an object, so long as there were no interference with the activities of the states concerned. Emplacements for nuclear weapons on the scale required to be of significant military value would be difficult to build without the knowledge of other countries. The placement of such installations would require a great deal of sophisticated material, unusual engineering activities, and a highly visible support effort also on the surface of the sea. In addition, the deploying country would obviously try to develop security systems to protect the military secrets of such installations. All this would attract attention of other maritime countries. The United States held that the configuration and operation of facilities specifically designed for nuclear weapons and other weapons of mass destruction would be conspicuous and identifiable.

Following compromise proposals made by Brazil and Canada,[29] as well as some other countries, the United States and the Soviet Union agreed on and wrote into the treaty the following procedure: Each party shall have the right to verify through observation the activities of other parties on the seabed and the ocean floor and in the subsoil thereof beyond the seabed zone, provided that observation does not interfere with such activities. If after such observation reasonable doubts remain concerning the fulfillment of the obligations assumed under the treaty, the party having such doubts and the party responsible for the activities giving rise to the doubts shall consult with a view to removing them. If the doubts persist, the party having such doubts shall notify other parties, and the parties concerned shall cooperate on such further procedures for verification as may be agreed, including appropriate inspection of objects, structures, installations, or other facilities that reasonably may be expected to be of a kind prohibited by the treaty. If the state responsible for the activities giving rise to the reasonable doubts is not identifiable by observation of the object, structure, installation, or other facility, the party having such doubts shall notify and make appropriate inquiries of parties in the region of the activities and of any other party. If it is ascertained through these inquiries that a particular party is responsible for the activities, that party shall consult and cooperate with other parties provided above. If the identity of the state responsible for the activities cannot be ascertained through these inquiries, then further verification procedures, including inspection, may be undertaken by the inquiring party (article 3, pars. 1, 2, and 3).

The extent of inspection envisaged in the treaty has not been spelled out. The formula employed poses a problem of how to proceed if the identity of the

---

29. United Nations, General Assembly, *Question of General and Complete Disarmament: Report of the Conference Committee on Disarmament, Brazil, Revised Working Paper on the Provisions of Article III of the Draft Treaty on the Prohibition of the Emplacement of Nuclear Weapons and Other Weapons of Mass Destruction on the Seabed and Ocean Floor and the Subsoil Thereof (Annex A, Doc. A/7741)* (A/C.1/993/rev.1/corr. 1), December 12, 1969. See also A/C.1/992.

state responsible for the activities giving rise to doubts became known only after the verification procedures had been initiated, and if the state in question proved to be a nonparty to the treaty.

In the course of the debate, some countries, especially those claiming extensive continental shelves, expressed concern that a hostile state might, under the guise of activities authorized by the treaty, collect information about conventional armaments or indulge in industrial espionage of facilities installed for peaceful exploration or even exploit the resources of the seabed belonging to another state.

To safeguard the interests of coastal states, it was decided to include a proviso in the treaty that verification activities shall not interfere with activities of other parties and shall be conducted with due regard for rights recognized under international law, including the freedoms of the high seas and the rights of coastal states with respect to the exploration and exploitation of their continental shelves (article 3, par. 6).

It is also stipulated that parties in the region of the activities giving rise to doubts concerning the fulfillment of the obligations under the treaty, including any coastal state and any other party so requesting, shall be entitled to participate in consultation with a view to removing the doubts and in cooperation on further procedures for verification. Also, whenever the identity of the state responsible for such activities cannot be ascertained through inquiries, the inquiring party shall invite the participation of the parties in the region, including any coastal state, and of any other party desiring to cooperate, in undertaking verification procedures, including inspection (article 3, pars. 2, 3).

SIGNIFICANCE OF THE SEABED TREATY

The seabed treaty binds, in fact, only the superpowers. A prohibition on placing nuclear weapons on the bottom of the sea cannot be a restraint on the military policies of countries which have formally renounced acquisition of such weapons under the Nonproliferation Treaty, or of those others including France, the United Kingdom, and China which, while having nuclear weapons, may not have the means to conduct the banned activities on a significant scale for a long time to come. One issue that directly engages nonnuclear-weapon states is the question of nonnuclear military installations on the continental shelves. However, this question has not been covered by the treaty.

The arms-control value of the seabed treaty is low. Nuclear installations on the seabed, once considered a possibility, have proved unattractive to the military. Because of the requirement of invulnerability, mobile systems are always favored over fixed systems. The treaty has banned something which did not exist and which, even without a prohibition, was not likely to develop. In view of its limited scope it is much less important as a preventive or nonarmament measure than the Antarctic Treaty, the Treaty on Outer Space, or the Treaty of Tlatelolco. And since it permits the use of the seabed for facilities servicing free-swimming nuclear weapon systems, it is no obstacle to a nuclear arms race in the whole of the sea environment.

IMPLEMENTATION OF THE SEABED TREATY

The seabed treaty entered into force on May 18, 1972, but on December 31, 1976, it had no more than 62 adherents (see Appendix C). The low number of parties is in consonance with a weak interest in this agreement which, with all the appearances of an arms-control measure, is devoid of real content.

Some states made reservations upon signing or ratification to make sure that their rights under the law of the sea were not adversely affected or to reiterate the points of view which had not been taken into account during the negotiating process. Thus Argentina stated that, in its understanding, the reference to the rights of exploration and exploitation by coastal states over their continental shelves (article 3, par. 6) was included solely because those could be the rights most frequently affected by verification procedures; it precluded any possibility of strengthening, through the seabed treaty, certain positions concerning continental shelves to the detriment of others based on different criteria. Brazil stressed that nothing in the treaty should be interpreted as prejudicing in any way its sovereign rights in the area of the sea, the seabed, and the subsoil thereof adjacent to its coasts; in its understanding, the word "observation," appearing in connection with verification (article 3, par. 1), refers only to observation that is incidental to the normal course of navigation in accordance with international law. By December 31, 1976, neither Argentina nor Brazil had ratified the treaty.

Yugoslavia, a party to the treaty, made an interpretative declaration a few months after the deposit of its instrument of ratification. In its view, the article which deals with verification through observation (article 3, par. 1) should be interpreted in such a way that a state exercising its right under this article is obliged to notify in advance the coastal state, insofar as its observations are to be carried out "within the stretch of the sea extending above the continental shelf of the said state." The United States and the United Kingdom rejected this interpretation and placed on record their formal objection to the Yugoslav reservation on the ground that it was incompatible with the object and purpose of the seabed treaty.

In joining the treaty, Canada and India reserved the right to verify, inspect, or remove any weapon, installation, facility, or device implanted or emplaced by other countries on their continental shelves.

Italy pointed out that in the case of agreement on further measures in the field of disarmament relating to the seabed, the question of the delimitation of the area within which these measures would find application shall have to be examined and solved in each instance in accordance with the nature of the measures to be adopted. This was a reference to the proposal which had been put forward by Italy during the negotiations on the seabed treaty but was not accepted—namely, that weapon-emplacement prohibitions should apply beyond a curve corresponding to a depth of 200 meters. (This bathymetric approach was one of the criteria applied in determining the continental shelf under the Geneva Convention of 1958.)

The parties to the seabed treaty have undertaken to continue negotiations concerning further measures in the field of disarmament for the prevention of an arms race on the seabed, the ocean floor, and the subsoil thereof (article 5). No such negotiations have as yet taken place. The reason for this inaction, put forward mainly by the great powers, is that the possibility of reaching a new agreement depends on the outcome of the Third United Nations Conference on the Law of the Sea. This conference deals with a broad range of issues, including those concerning the regime of the high seas, the continental shelf, the territorial sea (including the question of its breadth and the question of international straits), fishing and conservation of the living resources of the high seas, the establishment of an international regime and machinery to govern the exploration and exploitation of resources of the seabed beyond national jurisdiction, the preservation of the marine environment, and scientific research. There has been no discussion of measures related to arms control on the seabed. However, the seabed treaty envisages conferences of parties to review the operation of the treaty and to assure that its purposes and provisions are being realized (article 7). These conferences may generate new ideas about arms-limitation undertakings.

SUGGESTIONS FOR FURTHER ACTION

In 1969, Canada advanced a concept of a 200-mile zone extending from the outer limits of a 12-mile coastal band in which only the coastal state, or another state acting with its explicit consent, would be able to perform the defensive activities not prohibited under the seabed treaty. Nigeria suggested a 50-mile zone for similar purposes. In the UN seabed committee in 1973, a few Latin American states suggested that the emplacement of any kind of facilities on the seabed of the "adjacent area" should be subject to authorization and regulation by the coastal state, and at the 1974 Law of the Sea Conference in Caracas, Kenya and Mexico, supported by some other countries, formally proposed that no state should be entitled to construct, maintain, deploy, or operate on or over the continental shelf of another state any military installations or devices or any other installations for whatever purposes without the consent of the coastal state.

Considering the ban on the emplacement of nuclear weapons and other weapons of mass destruction beyond a 12-mile seabed zone under the 1971 seabed treaty, and the much older restrictions on laying mines in peacetime outside territorial waters, it is difficult to see what kind of conventional weapons could be safely emplanted on the continental shelf of other states. Shore-bombardment weapons and installations from which manned incursions could be mounted against a coastal state were mentioned in the debate on the seabed treaty. But the military value of such costly offensive systems would be doubtful, given their detectability and the need to protect them. More useful would appear to be devices monitoring communications of the coastal state and/or capable of disrupting them, submarine navigation systems, devices monitoring the entrance or exit of submarines to and from harbors, as well as instruments designed to render ineffective the surveillance and defenses of another state. These devices and instruments would be more autonomous than weapon systems but would still pose problems of information transmission and supply of power. They would also be sensitive to possible countermeasures. Nevertheless, technical difficulties can be overcome, and the risks inherent in operations conducted far from the shores of the emplanting state might perhaps be found worthwhile taking under certain circumstances.

It will be noted that the trend toward a coastal state's wider jurisdiction over the marine resources including protection of the marine environment, as expressed in the concept of a 200-mile economic zone, may imply the right for the coastal state to impose certain regulations and, in particular, to restrict certain military activities as well as research for military purposes (insofar as the nature of research can be unambiguously determined) in an area often even wider than its continental shelf. Coastal states may claim that installation of military devices by another state in a zone of their economic activities would interfere with such activities. Indeed, exclusive or preferential economic rights could not be effectively exercised without the right to prevent emplantation of undesirable objects on the seabed or the use of peaceful facilities for nonpeaceful aims. In possible competing uses, peaceful applications must have priority over military applications. What matters most is the status of the shelf because this is a more convenient place to emplant the devices in question than the outlying areas.

Only a few states have the capability of carrying out, with required sophistication, significant submerged operations of military importance. This is what they actually do on their own continental shelf. But it would be unjust to leave them the right to use for military purposes the continental shelf of others, a right which in most cases could not be reciprocated. Military installations in the proximity of other states cannot be justified on the grounds that they serve the defense interests of the state emplacing them, even if the installations are not of a patently offensive nature. Moreover, the question of neutral rights and duties under the law of war could arise, if these activities were directed against a third state. It would seem useful, therefore, to establish a seabed security zone adjacent to the coast, in which the coastal state would have the exclusive right to mount military equipment or other devices for military purposes (without obstructing international navigation) as well as to conduct research for such purposes. The zone would have to be sufficiently large to promote a sense of security among smaller nations; preferably, it should cover the whole continental shelf. (The legal status of the superjacent waters must not be affected by the rules governing the seabed.)

Under the existing rules, consent of the coastal state is necessary even for peaceful ventures on the continental shelf, namely, for research concerning the shelf and undertaken there. A consent regime for military ventures would certainly also be in order. In principle, the coastal state must have the right to allow another state to use its continental shelf for military purposes in the exercise of collective self-defense. In practice, however, certain countries would probably never use this right.

APPENDIX A

TREATY ON THE PROHIBITION OF THE EMPLACEMENT OF NUCLEAR WEAPONS AND OTHER WEAPONS OF MASS DESTRUCTION ON THE SEABED AND THE OCEAN FLOOR AND IN THE SUBSOIL THEREOF[30]

The States Parties to this Treaty,
Recognizing the common interest of mankind in the progress of the exploration and use of the sea-bed and the ocean floor for peaceful purposes,
Considering that the prevention of a nuclear arms race on the sea-bed and the ocean floor serves the interests of maintaining world peace, reduces international tensions and strengthens friendly relations among States,
Convinced that this Treaty constitutes a step towards the exclusion of the sea-bed, the ocean floor and the subsoil thereof from the arms race,
Convinced that this Treaty constitutes a step towards a treaty on general and complete disarmament under strict and effective international control, and determined to continue negotiations to this end,
Convinced that this Treaty will further the purposes and principles of the Charter of the United Nations, in a manner consistent with the principles of international law and without infringing the freedoms of the high seas,
Have agreed as follows:

Article 1

1. The States Parties to this Treaty undertake not to emplant or emplace on the sea-bed and the ocean floor and in the subsoil thereof beyond the outer limit of a sea-bed zone, as defined in Article II, any nuclear weapons or any other types of weapons of mass destruction as well as structures, launching installations or any other facilities specifically designed for storing, testing or using such weapons.
2. The undertakings of paragraph 1 of this Article shall also apply to the sea-bed

30. United Nations, General Assembly (A/8198), February 9, 1971.

zone referred to in the same paragraph, except that within such sea-bed zone, they shall not apply either to the coastal State or to the sea-bed beneath its territorial waters.

3. The States Parties to this Treaty undertake not to assist, encourage or induce any State to carry out activities referred to in paragraph 1 of this Article and not to participate in any other way in such actions.

Article 2

For the purpose of this Treaty, the outer limit of the sea-bed zone referred to in Article 1 shall be coterminous with the twelve-mile outer limit of the zone referred to in Part II of the Convention on the Territorial Sea and the Contiguous Zone, signed at Geneva on 29 April 1958, and shall be measured in accordance with the provisions of Part I, Section II, of that Convention and in accordance with international law.

Article 3

1. In order to promote the objectives of and ensure compliance with the provisions of this Treaty, each State Party to the Treaty shall have the right to verify through observation the activities of other States Parties to the Treaty on the sea-bed and the ocean floor and in the subsoil thereof beyond the zone referred to in Article 1, provided that observation does not interfere with such activities.

2. If after such observation reasonable doubts remain concerning the fulfilment of the obligations assumed under the Treaty, the State Party having such doubts and the State Party that is responsible for the activities giving rise to the doubts shall consult with a view to removing the doubts. If the doubts persist, the State Party having such doubts shall notify the other States Parties, and the Parties concerned shall co-operate on such further procedures for verification as may be agreed, including appropriate inspection of objects, structures, installations or other facilities that reasonably may be expected to be of a kind described in Article 1. The Parties in the region of the activities, including any coastal State, and any other Party so requesting, shall be entitled to participate in such consultation and co-operation. After completion of the further procedures for verification, an appropriate report shall be circulated to other Parties by the Party that initiated such procedures.

3. If the State responsible for the activities giving rise to the reasonable doubts is not identifiable by observation of the object, structure, installation or other facility, the State Party having such doubts shall notify and make appropriate inquiries of States Parties in the region of the activities and of any other State Party. If it is ascertained through these inquiries that a particular State Party is responsible for the activities, that State Party shall consult and co-operate with other Parties as provided in paragraph 2 of this Article. If the identity of the State responsible for the activities cannot be ascertained through these inquiries, then further verification procedures, including inspection, may be undertaken by the inquiring State Party, which shall invite the participation of the Parties in the region of the activities, including any coastal State, and of any other Party desiring to co-operate.

4. If consultation and co-operation pursuant to paragraphs 2 and 3 of this Article have not removed the doubts concerning the activities and there remains a serious question concerning fulfilment of the obligations assumed under this Treaty, a State Party may, in accordance with the provisions of the Charter of the United Nations, refer the matter to the Security Council, which may take action in accordance with the Charter.

5. Verification pursuant to this Article may be undertaken by any State Party using

its own means, or with the full or partial assistance of any other State Party, or through appropriate international procedures within the framework of the United Nations and in accordance with its Charter.

6. Verification activities pursuant to this Treaty shall not interfere with activities of other States Parties and shall be conducted with due regard for rights recognized under international law, including the freedoms of the high seas and the rights of coastal States with respect to the exploration and exploitation of their continental shelves.

Article 4

Nothing in this Treaty shall be interpreted as supporting or prejudicing the position of any State Party with respect to existing international conventions, including the 1958 Convention on the Territorial Sea and the Contiguous Zone, or with respect to rights or claims which such State Party may assert, or with respect to recognition or non-recognition of rights or claims asserted by any other State, related to waters off its coasts, including, *inter alia,* territorial seas and contiguous zones, or to the sea-bed and the ocean floor, including continental shelves.

Article 5

The Parties to this Treaty undertake to continue negotiations in good faith concerning further measures in the field of disarmament for the prevention of an arms race on the sea-bed, the ocean floor and the subsoil thereof.

Article 6

Any State Party may propose amendments to this Treaty. Amendments shall enter into force for each State Party accepting the amendments upon their acceptance by a majority of the States Parties to the Treaty and, thereafter, for each remaining State Party on the date of acceptance by it.

Article 7

Five years after the entry into force of this Treaty, a conference of Parties to the Treaty shall be held at Geneva, Switzerland, in order to review the operation of this Treaty with a view to assuring that the purposes of the preamble and the provisions of the Treaty are being realised. Such review shall take into account any relevant technological developments. The review conference shall determine, in accordance with the views of a majority of those Parties attending, whether and when an additional review conference shall be convened.

Article 8

Each State Party to this Treaty shall in exercising its national sovereignty have the right to withdraw from this Treaty if it decides that extraordinary events related to the subject-matter of this Treaty have jeopardised the supreme interests of its country. It shall give notice of such withdrawal to all other States Parties to the Treaty and to the United Nations Security Council three months in advance. Such notice shall include a statement of the extraordinary events it considers to have jeopardised its supreme interests.

Article 9

The provisions of this Treaty shall in no way affect the obligations assumed by States Parties to the Treaty under international instruments establishing zones free from nuclear weapons.

Article 10

1. This Treaty shall be open for signature to all States. Any State which does not sign the Treaty before its entry into force in accordance with paragraph 3 of this Article may accede to it at any time.
2. This Treaty shall be subject to ratification by signatory States. Instruments of ratification and of accession shall be deposited with the Governments of the United Kingdom of Great Britain and Northern Ireland, the Union of Soviet Socialist Republics and the United States of America, which are hereby designated the Depositary Governments.
3. This Treaty shall enter into force after the deposit of instruments of ratification by twenty-two Governments, including the Governments designated as Depositary Governments of this Treaty.
4. For States whose instruments of ratification or accession are deposited after the entry into force of this Treaty, it shall enter into force on the date of the deposit of their instruments of ratification or accession.
5. The Depositary Governments shall promptly inform the Governments of all signatory and acceding States of the date of each signature, of the date of deposit of each instrument of ratification or of accession, of the date of the entry into force of this Treaty, and of the receipt of other notices.
6. This Treaty shall be registered by the Depositary Governments pursuant to Article 102 of the Charter of the United Nations.

Article 11

This Treaty, the Chinese, English, French, Russian and Spanish texts of which are equally authentic, shall be deposited in the archives of the Depositary Governments. Duly certified copies of this Treaty shall be transmitted by the Depositary Governments to the Governments of the States signatory and acceding thereto.

In witness whereof the undersigned, being duly authorized thereto, have signed this Treaty.

APPENDIX B

COUNTRIES WHICH HAVE SIGNED, RATIFIED, ACCEDED, OR SUCCEEDED TO THE CONVENTION ON THE TERRITORIAL SEA AND THE CONTIGUOUS ZONE AS OF DECEMBER 31, 1975

Country	Signed	Ratified*
Afghanistan	Oct. 30, 1958	...
Argentina	Apr. 29, 1958	...
Australia	Oct. 30, 1958	May 14, 1963
Austria	Oct. 27, 1958	...
Belgium	...	Jan. 6, 1972
Bolivia	Oct. 17, 1958	...
Bulgaria	Oct. 31, 1958	Aug. 31, 1962
Byelorussian SSR	Oct. 30, 1958	Feb. 27, 1961
Cambodia	...	Mar. 18, 1960
Canada	Apr. 29, 1958	...
Colombia	Apr. 29, 1958	...
Costa Rica	Apr. 29, 1958	...
Cuba	Apr. 29, 1958	...
Czechoslovakia	Oct. 30, 1958	Aug. 31, 1961
Denmark	Apr. 29, 1958	Sept. 26, 1958
Dominican Republic	Apr. 29, 1958	Aug. 11, 1964
Fiji	...	Mar. 25, 1971
Finland	Oct. 27, 1958	Feb. 16, 1965
German Democratic Republic	...	Dec. 27, 1973
Ghana	Apr. 29, 1958	...
Guatemala	Apr. 29, 1958	...
Haiti	Apr. 29, 1958	Mar. 29, 1960
Holy See	Apr. 30, 1958	...
Hungary	Oct. 31, 1958	Dec. 6, 1961
Iceland	Apr. 29, 1958	...
Iran	May 28, 1958	...
Ireland	Oct. 2, 1958	...
Israel	Apr. 29, 1958	Sept. 6, 1961
Italy	...	Dec. 17, 1964
Jamaica	...	Oct. 8, 1965
Japan	...	June 10, 1968
Kenya	...	June 20, 1969
Lesotho	...	Oct. 23, 1973
Liberia	May 27, 1958	...
Madagascar	...	July 31, 1962
Malawi	...	Nov. 3, 1965
Malaysia	...	Dec. 21, 1960
Malta	...	May 19, 1966
Mauritius	...	Oct. 5, 1970
Mexico	...	Aug. 2, 1966
Nepal	Apr. 29, 1958	...
Netherlands	Oct. 31, 1958	Feb. 18, 1966
New Zealand	Oct. 29, 1958	...

APPENDIX B. Continued

Country	Signed	Ratified*
Nigeria	...	June 26, 1961
Pakistan	Oct. 31, 1958	...
Panama	May 2, 1958	...
Portugal	Oct. 28, 1958	Jan. 8, 1963
Republic of China (Taiwan)	Apr. 29, 1958	...
Romania	Oct. 31, 1958	Dec. 12, 1961
Senegal†	...	Apr. 25, 1961
Sierra Leone	...	Mar. 13, 1962
South Africa	...	Apr. 9, 1963
Spain	...	Feb. 25, 1971
Sri Lanka	Oct. 30, 1958	...
Swaziland	...	Oct. 16, 1970
Switzerland	Oct. 22, 1958	May 18, 1966
Thailand	Apr. 29, 1958	July 2, 1968
Tonga	...	June 29, 1971
Trinidad and Tobago	...	Apr. 11, 1966
Tunisia	Oct. 30, 1958	...
Uganda	...	Sept. 14, 1964
Ukrainian SSR	Oct. 30, 1958	Jan. 12, 1961
Union of Soviet Socialist Republics	Oct. 30, 1958	Nov. 22, 1960
United Kingdom	Sept. 9, 1958	Mar. 14, 1960
United States	Sept. 15, 1958	Apr. 12, 1961
Uruguay	Apr. 29, 1958	...
Venezuela	Oct. 30, 1958	Aug. 15, 1961
Yugoslavia	Apr. 29, 1958	Jan. 28, 1966

*Ratification refers to deposit of instruments of ratification or accession, or of notification of succession, with the Secretary-General of the United Nations.

†The Secretary-General received a communication on June 9, 1971, from the government of Senegal denouncing this convention as well as the Convention on the Living Resources of the High Seas and specifying that the denunciation would take effect on the thirtieth day from its receipt. The said communication, as well as the related exchange of correspondence between the secretariat and the government of Senegal, was circulated by the Secretary-General to all states entitled to become parties to the conventions concerned under their respective clauses. In this connexion, a communication from the government of the United Kingdom was received by the Secretary-General on January 2, 1973, stating inter alia: "... As regards the notification by the Government of Senegal purporting to denounce the two Conventions of 1958, the Government of the United Kingdom wish to place on record that in their view those Conventions are not susceptible to unilateral denunciation by a State which is a party to them and they therefore cannot accept the validity or effectiveness of the purported denunciation by the Government of Senegal. Accordingly, the Government of the United Kingdom regard the Government of Senegal as still bound by the obligations which they assumed when they became a party to those Conventions and the Government of the United Kingdom fully reserve all their rights under them as well as their rights and the rights of their nationals in respect of any action which the Government of Senegal have taken or may take as a consequence of the said purported denunciation."

APPENDIX C

COUNTRIES WHICH HAVE SIGNED, RATIFIED, OR ACCEDED TO THE SEABED TREATY AS OF DECEMBER 31, 1976

Country	Signed	Ratified*
Afghanistan	Feb. 11, 1971 (L,M,W)	Apr. 22, 1971 (M)
		Apr. 23, 1971 (L)
		May 21, 1971 (W)
Argentina†	Sept. 3, 1971 (L,M,W,)	...
Australia	Feb. 11, 1971 (L,M,W)	Jan. 23, 1973 (L,M,W)
Austria	Feb. 11, 1971 (L,M,W)	Aug. 10, 1972 (L,M,W)
Belgium	Feb. 11, 1971 (L,M,W)	Nov. 20, 1972 (L,M,W)
Benin (Dahomey)	Mar. 18, 1971 (W)	...
Bolivia	Feb. 11, 1971 (L,M,W)	...
Botswana	Feb. 11, 1971 (W)	Nov. 10, 1972 (W)
Brazil‡	Sept. 3, 1971 (L,M,W)	...
Bulgaria	Feb. 11, 1971(L,M,W)	Apr. 16, 1971 (M)
		May 7, 1971 (W)
		May 26, 1971 (L)
Burma	Feb. 11, 1971(L,M,W)	...
Burundi	Feb. 11, 1971 (M,W)	...
Byelorussian SSR	Mar. 3, 1971 (M)	Sept. 14, 1971 (M)
Cambodia	Feb. 11, 1971 (W)	...
Canada§	Feb. 11, 1971 (L,M,W)	May 17, 1972 (L,M,W)
Central African Republic	Feb. 11, 1971 (W)	...
Colombia	Feb. 11, 1971 (W)	...
Costa Rica	Feb. 11, 1971 (W)	...
Cyprus	Feb. 11, 1971 (L,M,W)	Nov. 17, 1971 (L,M)
		Dec. 30, 1971 (W)
Czechoslovakia	Feb. 11, 1971 (L,M,W)	Jan. 11, 1972 (L,M,W)
Democratic Yemen	Feb. 23, 1971 (M)	...
Denmark	Feb. 11, 1971 (L,M,W)	June 15, 1971 (L,M,W)
Dominican Republic	Feb. 11, 1971 (W)	Feb. 11, 1972 (W)
Equatorial Guinea	June 4, 1971 (W)	...
Ethiopia	Feb. 11, 1971 (L,M,W)	...
Finland	Feb. 11, 1971 (L,M,W)	June 3, 1971 (L,M,W)
Gambia	May 18, 1971 (L)	...
	May 21, 1971 (M)	
	Oct. 29, 1971 (W)	
German Democratic Republic¶	Feb. 11, 1971 (M)	...
	July 27, 1971 (M)	
Germany, Federal Republic of #	June 8, 1971 (L,M,W)	Nov. 18, 1975 (L,W)
Ghana	Feb. 11, 1971 (L,M,W)	Aug. 9, 1972 (W)
Greece	Feb. 11, 1971 (M)	...
	Feb. 12, 1971 (W)	
Guatemala	Feb. 11, 1971 (W)	...
Guinea	Feb. 11, 1971 (M,W)	...
Guinea-Bissau	...	Aug. 20, 1976 (M)
Honduras	Feb. 11, 1971 (W)	...
Hungary	Feb. 11, 1971 (L,M,W)	Aug. 13, 1971 (L,M,W)
Iceland	Feb. 11, 1971 (L,M,W)	May 30, 1972 (L,M,W)

APPENDIX C. *Continued*

Country	Signed	Ratified*
India**	...	July 20, 1973 (L,M,W)
Iran	Feb. 11, 1971 (L,M,W)	Aug. 26, 1971 (L,W)
		Sept. 6, 1972 (M)
Iraq††	Feb. 22, 1971 (M)	Sept. 13, 1972 (M)
Ireland	Feb. 11, 1971 (L,W)	Aug. 19, 1971 (L,W)
Italy‡‡	Feb. 11, 1971 (L,M,W)	Sept. 3, 1974 (L,M,W)
Ivory Coast	...	Jan. 14, 1972 (W)
Jamaica	Oct. 11, 1971 (L,W)	...
	Oct. 14, 1971 (M)	
Japan	Feb. 11, 1971 (L,M,W)	June 21, 1971 (L,M,W)
Jordan	Feb. 11, 1971 (L,M,W)	Aug. 17, 1971 (W)
		Aug. 30, 1971 (M)
		Nov. 1, 1971 (L)
Laos	Feb. 11, 1971 (L,W)	Oct. 19, 1971 (L)
	Feb. 15, 1971 (M)	Oct. 22, 1971 (M)
		Nov. 3, 1971 (W)
Lebanon	Feb. 11, 1971 (L,M,W)	...
Lesotho	Sept. 8, 1971 (W)	Apr. 3, 1973 (W)
Liberia	Feb. 11, 1971 (W)	...
Luxembourg	Feb. 11, 1971 (L,M,W)	...
Madagascar	Feb. 14, 1971 (W)	...
Malaysia	May 20, 1971 (L,M,W)	June 21, 1972 (L,M,W)
Mali	Feb. 11, 1971 (W)	...
	Feb. 15, 1971 (M)	
Malta	Feb. 11, 1971 (L,W)	May 4, 1971 (W)
Mauritius	Feb. 11, 1971 (W)	Apr. 23, 1971 (W)
		May 3, 1971 (L)
		May 18, 1971 (M)
Mongolia	Feb. 11, 1971 (L,M)	Oct. 8, 1971 (M)
		Nov. 15, 1971 (L)
Morocco	Feb. 11, 1971 (M,W)	July 26, 1971 (L)
	Feb. 18, 1971 (L)	Aug. 5, 1971 (W)
		Jan. 18, 1972 (M)
Nepal	Feb. 11, 1971 (M,W)	July 6, 1971 (L)
	Feb. 24, 1971 (L)	July 29, 1971 (M)
		Aug. 9, 1971 (W)
Netherlands§§	Feb. 11, 1971 (L,M,W)	Jan. 14, 1976 (L,M,W)
New Zealand	Feb. 11, 1971 (L,M,W)	Feb. 24, 1972 (L,M,W)
Nicaragua	Feb. 11, 1971 (W)	Feb. 7, 1973 (W)
Niger	Feb. 11, 1971 (W)	Aug. 9, 1971 (W)
Norway	Feb. 11, 1971 (L,M,W)	June 28, 1971 (L,M)
		June 29, 1971 (W)
Panama	Feb. 11, 1971 (W)	Mar. 20, 1974 (W)
Paraguay	Feb. 23, 1971 (W)	...
Poland	Feb. 11, 1971 (L,M,W)	Nov. 15, 1971 (L,M,W)
Portugal	...	June 24, 1975 (L,M,W)
Qatar	...	Nov. 12, 1974 (L)
Republic of China (Taiwan)	Feb. 11, 1971 (W)	Feb. 22, 1972 (W)

APPENDIX C. *Continued*

Country	Signed	Ratified*
Romania¶¶	Feb. 11, 1971 (L,M,W)	July 10, 1972 (L,M,W)
Rwanda	Feb. 11, 1971 (W)	May 20, 1975 (L,M,W)
Saudi Arabia	Jan. 7, 1972 (W)	June 23, 1972 (W)
Senegal	Mar. 17, 1971 (W)	...
Sierra Leone	Feb. 11, 1971 (L)	...
	Feb. 12, 1971 (M)	
	Feb. 24, 1971 (W)	
Singapore	May 5, 1971 (L,M,W)	Sept. 10, 1976 (L,M,W)
South Africa	Feb. 11, 1971 (W)	Nov. 14, 1973 (W)
		Nov. 26, 1973 (L)
South Korea	Feb. 11, 1971 (L,W)	...
Sudan	Feb. 11, 1971 (L)	...
	Feb. 12, 1971 (M)	
Swaziland	Feb. 11, 1971 (W)	Aug. 9, 1971 (W)
Sweden	Feb. 11, 1971 (L,M,W)	Apr. 28, 1972 (L,M,W)
Switzerland	Feb. 11, 1971 (L,M,W)	May 4, 1976 (L,M,W)
Togo	Apr. 2, 1971 (W)	June 28, 1971 (W)
Tunisia	Feb. 11, 1971 (L,M,W)	Oct. 22, 1971 (M)
		Oct. 28, 1971 (L)
		Oct. 29, 1971 (W)
Turkey	Feb. 25, 1971 (L,M,W)	Oct. 19, 1972 (W)
		Oct. 25, 1972 (L)
		Oct. 30, 1972 (M)
Ukrainian SSR	Mar. 3, 1971 (M)	Sept. 3, 1971 (M)
Union of Soviet Socialist Republics	Feb. 11, 1971 (L,M,W)	May 18, 1972 (L,M,W)
United Kingdom##	Feb. 11, 1971 (L,M,W)	May 18, 1972 (L,M,W)
United Republic of Cameroon	Nov. 11, 1971 (M)	...
United Republic of Tanzania	Feb. 11, 1971 (W)	...
United States	Feb. 11, 1971 (L,M,W)	May 18, 1972 (L,M,W)
Uruguay	Feb. 11, 1971 (W)	...
Viet Nam	Feb. 11, 1971 (W)	...
Yemen	Feb. 23, 1971 (M)	...
Yugoslavia***	Mar. 2, 1971 (L,M,W)	Oct. 25, 1973 (L,M,W)
Zambia	...	Oct. 9, 1972 (L)
		Nov. 1, 1972 (W)
		Nov. 2, 1972 (M)

NOTE.—L = London, M = Moscow, W = Washington (place of signature and/or deposit of the instrument of ratification or accession).

*Deposit of instruments of ratification or accession.

†On signing the treaty, Argentina made a declaration. It stated that it interprets the references to the freedoms of the high seas as in no way implying a pronouncement or judgement on the different positions relating to questions connected with international maritime law. It understands that the reference to the rights of exploration and exploitation by coastal states over their continental shelves was included solely because those could be the rights most frequently affected by verification procedures. Argentina precludes any possibility of strengthening, through this treaty, certain positions concerning continental shelves to the detriment of others based on different criteria.

‡On signing the treaty, Brazil states that nothing in the treaty shall be interpreted as prejudicing in any way the sovereign rights of Brazil in the area of the sea, the seabed, and the subsoil thereof adjacent to its coasts. It is the understanding of the Brazilian government that the word "observation," as it appears in par. 1 of article 3 of the treaty, refers only to observation that is incidental to the normal course of navigation in accordance with international law.

§In depositing the instrument of ratification, Canada declared that article 1, par. 1 cannot be interpreted as indicating that any state has a right to implant or emplace any weapons not prohibited under article 1, par. 1 on the seabed and ocean floor and in the subsoil thereof beyond the limits of national jurisdiction, or as constituting any limitation on the principle that this area of the seabed and ocean floor and the subsoil thereof shall be reserved for exclusively peaceful purposes. In the view of Canada, article 1–3 cannot be interpreted as indicating that any state but the coastal state has any right to implant or emplace any weapon not prohibited under article 1, par. 1 on the continental shelf or the subsoil thereof appertaining to that coastal state, beyond the outer limit of the seabed zone referred to in article 1 and defined in article 2; article 3 cannot be interpreted as indicating any restrictions or limitation upon the rights of the coastal state, consistent with its exclusive sovereign rights with respect to the continental shelf, to verify, inspect, or effect the removal of any weapon, structure, installation, facility, or device implanted or emplaced on the continental shelf or the subsoil thereof appertaining to that coastal state, beyond the outer limit of the seabed zone referred to in article 1 and defined in article 2. On April 12, 1976, the Federal Republic of Germany stated that the declaration by Canada is not of a nature to confer on the government of this country more far-reaching rights than those to which it is entitled under current international law, and that all rights existing under current international law which are not covered by the prohibitions are left intact by the treaty.

¶The United States has not accepted the notification of signature by the German Democratic Republic.

#On signing the treaty, the Federal Republic of Germany stated that its signature does not imply recognition of the German Democratic Republic under international law. On ratifying the treaty, the Federal Republic of Germany declared that the treaty will apply to Berlin (West).

**On the occasion of its accession to the treaty, the Government of India stated that as a coastal state, India has, and always had, full and exclusive sovereign rights over the continental shelf adjoining its territory and beyond its territorial waters and the subsoil thereof. It is the considered view of India that other countries cannot use its continental shelf for military purposes. There cannot, therefore, be any restriction on or limitation of the sovereign rights of India as a coastal state to verify, inspect, remove, or destroy any weapon, device, structure, installation, or facility which might be implanted or emplaced on or beneath its continental shelf by any other country or to take such other steps as may be considered necessary to safeguard its security. The accession by the government of India to the seabed treaty is based on this position. In response to the Indian statement, the U.S. government expressed the view that under existing international law, the rights of coastal states over their continental shelves are exclusive only for purposes of exploration and exploitation of natural resources and are otherwise limited by the 1958 Convention on the Continental Shelf and other principles of international law. On April 12, 1976, the Federal Republic of Germany stated that the declaration by India is not of a nature to confer on the government of this country more far-reaching rights than those to which it is entitled under current international law, and that all rights existing under current international law which are not covered by the prohibitions are left intact by the treaty.

††A statement was made containing a disclaimer regarding the recognition of states party to the treaty.

‡‡On signing the treaty, Italy stated, inter alia, that in the case of agreements of further measures in the field of disarmament to prevent an arms race on the seabed and ocean floor and in their subsoil, the question of the delimitation of the area within which these measures would find application shall have to be examined and solved in each instance in accordance with the nature of the measures to be adopted. The statement was repeated at the time of ratification.

§§The ratification covers the Netherlands Antilles.

¶¶Romania stated that it considered null and void the ratification of the treaty by the Taiwan authorities.

##The instrument of ratification states that the treaty is ratified in respect of the United Kingdom of Great Britain and Northern Ireland, the Associated States (Antigua, Dominica, Grenada, Saint Christopher-Nevis-Anguilla, Saint Lucia, and Saint Vincent) and territories under the territorial sovereignty of the United Kingdom, as well as the state of Brunei and the British Solomon Islands Protectorate. The United Kingdom recalled its view that if a regime is not recognized as the government of a state, neither signature nor the deposit of any instrument by it nor notification of any of those acts will bring about recognition of that regime by any other state. Grenada became independent on February 7, 1974.

***On February 25, 1974, the ambassador of Yugoslavia transmitted to the U.S. secretary of state a note stating that in the view of the Yugoslav government, article 3, par. 1 of the treaty should be interpreted in such a way that a state exercising its right under this article shall be obliged to notify in advance the coastal state, insofar as its observations are to be carried out "within the stretch of the sea extending above the continental shelf of the said state." On January 16, 1975, the U.S. secretary of state presented the view of the United States concerning the Yugoslav note as follows: "Insofar as the note is intended to be interpretative of the treaty, the United States cannot accept it as a valid interpretation. In addition, the United States does not consider that it can have any effect on the existing law of the sea." Insofar as the note was intended to be a reservation to the treaty, the United States placed on record its formal objection to it on the grounds that it was incompatible with the object and purpose of the treaty. The United States also drew attention to the fact that the note was submitted too late to be legally effective as a reservation. A similar exchange of notes took place between Yugoslavia and the United Kingdom. On April 12, 1976, the Federal Republic of Germany stated that the declaration by Yugoslavia is not of a nature to confer on the government of this country more far-reaching rights than those to which it is entitled under current international law, and that all rights existing under current international law which are not covered by the prohibitions are left intact by the treaty.

Military Activities

Naval Forces

Ronald Huisken
Stockholm International Peace Research Institute

INTRODUCTION

Historically, navies have played a major role in shaping and supporting the international order, and there has been a consistently high interest in, and expenditure on, naval forces. In neither respect has the stature of naval forces waned in the 30 years since World War II. For a period during the 1950s regard for conventional forces, including navies, declined because the predominant scenario was nuclear war. It was generally conceded that once nuclear weapons were extensively used any war would be over before conventional forces could play any significant role.

This view was short-lived. As the Soviet Union expanded and diversified its nuclear arsenal, it became apparent that nuclear weapons were essentially unusable except as a last resort. This, of course, expanded the range and scale of conflicts that could conceivably occur without recourse to these weapons, and, accordingly, regard for conventional forces flourished.

At about the same time, the introduction of nuclear-powered submarines capable of launching long-range missiles while submerged brought the most important and prestigious military function of the postwar era, strategic deterrence, firmly into the naval fold. Even in its infancy the ballistic-missile submarine was recognized as a weapon system ideally suited to the task of strategic deterrence, and its relative fitness for this role has grown continuously as technological developments have made land-based missiles and bombers increasingly vulnerable.

Conventional naval forces retain important wartime missions, influenced predominantly by the pivotal role played by navies in World War II. For most countries this is simply deterring or defending against attack from the sea, although several countries have sufficient naval superiority to contemplate offensive naval operations in a regional context. For the major Western powers the central scenario is a conflict in Europe. A sustained conventional defense effort by the European NATO countries is contingent upon reinforcement from the United States on a scale that is only feasible with sea transport. The protection of convoys in this scenario is what basically determines the sea control capability that the NATO alliance seeks to have.

The projection of power continues to command significant resources, as

© 1978 by The University of Chicago. 0-226-06602-9/77/1978-1018$01.93

evidenced by the U.S. Marine Corps. It is worth pointing out, however, that while a continuing role is seen for seaborne air power in some projection operations, the opportunities for amphibious assault are declining, one reason being the worldwide proliferation of advanced weaponry.

Apart from strategic deterrence and wartime missions, a third dimension of the utility of naval forces is their deployment for political ends in peacetime. Considerable importance is attached to this function. To quote one observer: "This effectiveness [of a navy] short of war is difficult to characterize but is nevertheless pervasive and may well comprise the most significant benefit a nation derives from its naval investment."[1] Perhaps as a result of the fact that both the United States and the Soviet Union now have naval forces capable of sustained operations in distant areas, the interest in and importance attached to this function appears to be on the increase.

Navies are more or less unique among the armed services in having an active peacetime role. The key to this distinction is flexibility. Compared with armies and air forces, naval forces are uniquely flexible. All states have exclusive rights to their territory and the air space above it but, except for a few miles out from the coast, oceans and seas are international. Moreover, the majority of states in the world border on at least one ocean or sea. The impact of the widespread adoption of 200-mile maritime zones remains to be seen.

This flexibility gives the political leadership of a naval power a comparatively wide range of nonviolent options in any attempt to influence the course of events on land. In the extreme case a very substantial military capability can be deployed in such a way that it can be brought to bear in a short space of time and yet, at least in a legal sense, remain totally uncommitted. More generally, naval deployment in a given region can be regular or periodic, the composition of the deployed force can be varied, and it can be reinforced, scaled down, and maneuvered in a variety of ways in an endeavor to convey appropriate diplomatic signals.

The relative suitability of the navy in this role is reflected in a recent study of the use of the armed forces as a political instrument by the United States over the period 1945–75. Under a five-point definition[2] this study identified 215 instances in which the United States used its armed forces as a political

---

1. L. W. Martin, *The Sea in Modern Strategy* (London: Institute for Strategic Studies, 1967), p. 133.
2. These five points were as follows: (1) there must have been a physical change in the disposition of at least part of the armed forces, (2) there had to be a conscious purpose behind this activity, (3) the aim must have been to attain objectives by influence rather than imposition, (4) the intention must have been to avoid significant violence, and (5) some specific behavior had to have been desired in the target states or actors. To be included an incident had to meet all five criteria (B. M. Blechman and S. S. Kaplan, "The Use of Armed Forces as a Political Instrument" [Washington, D.C.: Brookings Institution, December 1976]).

instrument. In 177 (or 82 percent) cases naval units were used, and in 100 cases only naval units were used. Moreover, there was an observable trend toward relatively more use of the navy. Since 1955 naval units have been involved in more than nine out of every 10 incidents.

The same study includes a review of Soviet activity in this regard. Although this review is less detailed than that for the United States, it is apparent that the Soviet Union also finds its navy increasingly useful as a political instrument. Thus, of the 83 instances over the period 1946–68 in which the Soviet Union used its armed forces for political purposes, only 18 appear to have involved naval units. In contrast, over the period 1969–75 Soviet naval units were involved in at least 19 of the 32 identified incidents.

Since it is likely that the United States and the USSR will more often than not have contradictory political objectives, the ambition of one to be relatively more free than the other to use naval forces for political gain (or to prevent political loss) will add fuel to the other pressures tending to produce naval competition.

The prospect of a naval arms race between the United States and the USSR is not the only factor of potential significance for naval developments in the foreseeable future, although it is certainly the main one. In recent years there has been a widespread surge of interest in naval forces. In some areas this is part of a general military rivalry between states, but the more general explanation is probably the growing appreciation of the importance of the oceans, particularly the growing accessibility of oil and gas on the continental shelves and minerals on the deep-ocean floor. In this context the expectation over the past several years that coastal states would acquire jurisdiction over ocean resources out to 200 miles has led these states to undertake the acquisition of naval forces appropriate for the protection and policing of these new interests.

MEASURING NAVAL STOCKS

Given this general background, it is of some interest to measure the trend in the volume of resources embodied in the world's naval forces. Such an exercise faces both methodological and practical problems. Very few nations provide a sufficiently detailed breakdown of their expenditures on naval forces to permit a separation of the resources devoted to ship construction and any subsequent refits and modernizations.

The measures traditionally used to compare naval strengths—numbers of vessels and displacement or tonnage—cannot be regarded as adequate surrogate measures of the volume of resources absorbed in naval ships. Even for the purpose of comparing naval strength, these measures have lost much of their validity. Naval vessels vary enormously in size, armament, speed, and range, making numerical comparisons hazardous. It is also becoming increasingly difficult to divide the navies of different countries into comparable categories. As the major naval powers have pursued their individual philosophies regarding the function and design of warships, traditional categories such as cruisers, destroyers, frigates, and, most recently, aircraft carriers have lost much of their classifying capacity.

Another, and most important, reason to question the validity of these traditional measures is technological change. The postwar period has been characterized by rapid changes in technology, most particularly military technology. From a military point of view this puts a major limitation on numbers or tonnage as measures of relative naval strength. A difference of 10, or even 5, years in the construction dates of two outwardly similar warships can imply a militarily significant difference in their capabilities, as the later vessel will normally incorporate improvements in armaments (offensive and defensive), hull design, propulsion, radar, sonar, navigation, communication, fire control, and so on.

Similarly, the heavy and continuous emphasis on technological improvement across the board means that tonnage has become an increasingly partial measure of the resource input in warship construction. A major part of the cost of a modern warship is the electronic equipment deployed on it. Indeed, it has reached the point where in many cases the shipyard has ceased to be the major contractor in a construction program. This role has passed to the principal supplier of electronic equipment and/or the concern responsible for the complex task of integrating hull, armament, and equipment into a weapons system.

As a result, any measure of the resources devoted to naval construction must take account of the technological factor in addition to numbers and tonnage. Specifically, account must be taken of the sophistication of a ship's armament and related equipment and, owing to the continuous improvement in technology, the date of its construction.

The method used here is essentially a cost index which incorporates all three elements: numbers, tonnage, and technology. There is enough information in open sources to obtain construction costs for naval vessels of different sizes and weapons and equipment fits, although these data are exclusively of Western origin. From these data a basic list was prepared of the cost per ton (in 1973 prices) of six categories of warships. These were further subdivided on grounds of sophistication as follows: (1) aircraft carriers (attack; antisubmarine/amphibious assault; escort/utility); (2) submarines (strategic nuclear [12–16 missiles]; strategic nuclear [three missiles]; strategic conventional; nuclear, cruise missile; nuclear, other; conventional cruise missile; conventional, other); (3) cruisers (missile armed; conventionally armed); (4) destroyers, frigates, escorts (missile armed; conventionally armed); (5) patrol boats (missile armed; torpedo boats; other); and (6) battleships.

Technology was represented by an improvement factor. On naval advice this was assumed to be 3.5 percent annually.[3] This provided a set of comparable value per ton figures for the 6 years selected. These values were then combined with tonnage. All separate categories and subclasses within categories of ships were calculated at their respective displacements.

To reduce the calculations to manageable proportions, several limitations were imposed. First, the calculation was restricted to fighting vessels, strictly defined. Second, the world's navies were observed at just six points over the postwar period: 1950, 1955, 1960, 1965, 1970, and 1976. Third, it was assumed that naval authorities in the major navies insist on exacting standards of performance for ships included in the active fleet: this would be particularly true for the frontline ships included here. Technical improvements are incorporated in the programs of modernization and refitting. Thus, in the first 10 years of its life, a ship is assumed to benefit fully from the incorporation of new technology: that is, its value will rise by 3.5 percent a year for 10 years as a result of the improvement factor. After 10 years the aging of the ship offsets any incorporation of further technical improvements.[4] It is assumed that ships held in reserves do not benefit from technical improvements.

The assumption about modernization and refitting in the first 10 years was made for developed countries only. For underdeveloped countries it was assumed that ships were not regularly modernized and refitted this way. These ships, therefore, remained fixed at their date-of-birth valuation unless there was specific evidence to the contrary.

The method obviously has serious limitations. In particular, the figures arrived at do not in any way reflect the fighting efficiency of different fleets. However, as an index of the resources committed to naval forces, it is, at least in principle, superior to either numbers or tonnage, and from the point of view of exposition it has the very considerable advantage of obviating the need to compare physically different naval units.

---

3. A study prepared by the U.S. Department of Defense concluded that over the period 1945–75 technological advances resulted in an average annual increase in the real cost of major ships, submarines, and aircraft carriers of 4.5 percent (U.S. Congress, House, Committee on Armed Services, *Department of Defense Authorization for Appropriations for Fiscal Year 1976: Hearings on Military Posture and H.R. 3689/H.R. 6674,* 94th Cong. 1st sess., pt. 1, February, March, April, and May 1975, pp. 1826–29).

4. If there is a major conversion to missile armaments, then the value of the vessel is raised to the appropriate value per ton for missile-equipped vessels, and it is treated as a new vessel from the date of its major conversion. The old destroyers converted under the Fleet Rehabilitation and Modernization Program in the United States were treated as a special case. They were valued at 1955 values per ton before conversion; after conversion the values were raised to 1960 values per ton and were kept at those figures in subsequent years.

TRENDS IN THE ESTIMATED VALUE OF WORLD NAVAL STOCK

Over the period 1950–76, the estimated value of the world stock of fighting ships increased more than threefold, roughly the same as total world military expenditure in constant prices. The upward trend was sharply interrupted between 1955 and 1960 (table 1). Between 1960 and 1970, however, there was a marked acceleration in the rate of growth compared with the previous decade (table 2). In recent years the rate of growth has moderated somewhat, but the average annual rate of increase of 6.4 percent since 1960 is still almost double that for the period 1950–60.

To a large extent this acceleration in the rate of growth after 1960 is attributable to a single type of ship—the strategic nuclear submarine. The share of the world naval stock accounted for by these vessels increased from 1.8 percent in 1960 to 28.2 percent in 1976. However, even if strategic nuclear submarines are excluded, the naval stock has more than doubled over the postwar period; and the average annual rate of increase since 1960, 3.8 percent, is still slightly faster than that for the period 1950–60.

It should be borne in mind, however, that the numerical increase of strategic nuclear submarines does not constitute the whole influence that these vessels have had on the world naval stock. Their existence has been accompanied by a large increase in the resources devoted to antisubmarine warfare; the primary function of nuclear hunter-killer submarines, for example, is to track down and destroy strategic submarines.

Not only has the rate of growth of the world naval stock varied over time, but rates of growth for different countries and regions have varied significantly. As a result there have been marked changes in the distribution of the world naval stock (table 3). By far the most significant change has been the emergence of the Soviet Union as a major naval power. Over the period 1950–76 the Soviet Union's naval stock has, on the average, increased twice as fast as the world total and more than three times as fast as that of NATO.

TABLE 1.—WORLD NAVAL STOCK,
1950–75 (U.S. $ Billion)

Year	A	B
1950	35.1	35.1
1955	48.7	48.7
1960	48.2	47.3
1965	74.1	64.8
1970	98.2	80.5
1976	129.1	92.7

NOTE.—A = total; B = excluding nuclear-powered ballistic-missile submarines.

TABLE 2.—WORLD STOCK OF FIGHTING SHIPS: ESTIMATED GROWTH RATES IN VALUE OF STOCK

Country/Region	Percentage of Total World Stock in 1976	Average Annual (%) Growth Rates			
		1950–60	1960–70	1970–76	1950–76
World total	100	3.3	7.4	4.6	5.1
Total NATO	48.2	1.2	6.3	.8	3.0
United States	34.0	1.4	6.4	–.1	2.9
Other NATO	14.2	.5	6.1	3.3	3.3
Total WTO	39.3	12.1	10.9	9.9	11.1
USSR	38.7	11.9	11.0	10.3	11.2
Other WTO	.6	17.4	9.4	–6.1	8.5
Other Europe	1.6	5.4	2.3	3.9	3.9
Other developed	2.4	7.7	7.2	6.0	7.1
China	2.5	18.3	10.9	16.0	14.9
Total Third World	5.9	5.4	5.3	9.9	7.2
Far East	1.6	7.2	7.7	5.3	7.0
Middle East	1.0	8.0	9.5	11.9	9.5
South Asia	.8	8.7	6.1	8.5	7.6
South America	1.8	4.4	.9	7.8	3.8
Central America	.4	–.1	4.9	6.0	3.2
Africa	.4	...	35.0	11.5	...

SOURCE.—Stockholm International Peace Research Institute (SIPRI) worksheets.

In assessing this information it is important to bear in mind that in 1950 the value of the Soviet naval stock was very small, making it comparatively easy to achieve a high percentage rate of increase. For the same reason all new naval construction in the Soviet Union, for the first 20 years or so, meant an equivalent increase in the absolute value of the naval stock. While new naval construction in NATO has, broadly speaking, easily matched that in the Soviet Union, the existing naval stock was very much larger, and, also, new

TABLE 3.—DISTRIBUTION OF THE WORLD STOCK OF FIGHTING SHIPS (%)

Year	NATO	WTO	Other European, Other Developed Countries, China	Third World
1950	81.2	9.2	4.6	5.0
1955	75.7	14.0	5.1	5.2
1960	66.2	22.0	6.9	4.9
1965	63.9	24.6	6.6	4.9
1970	60.1	28.8	6.2	4.9
1976	48.2	39.3	6.5	5.9

construction has been offset by the disposal of large numbers of old vessels. Over the next few years these positions will essentially be reversed. The United States, with over 70 percent of NATO's naval stock in 1976, has disposed of nearly all its naval units constructed during World War II. The U.S. fleet, though currently numerically smaller than the Soviet, is on the average considerably younger and programmed to increase steadily at least through 1980. The Soviet Union has been steadily retiring some of its older units, but the available data suggest that the impact of the block obsolescence of units constructed in the 1940s and early 1950s has yet to be felt.[5] Moreover, if rates of new construction in recent years are not accelerated, the numerical size of the Soviet navy will continue to fall. Thus, while up to the present time the disparate rates of growth have markedly redistributed the world's naval stock in favor of the Soviet Union, its share is likely to stabilize and perhaps decline over the next few years.

REGIONAL ANALYSIS

The countries of NATO and the WTO account for the lion's share of the world's military resources, and naval forces are no exception. In 1976 some 87 percent of the world's naval stock was owned by the countries of these two alliances and an even higher concentration than their share of total world military expenditure (71 percent in 1976). The United States and the USSR dominate these respective alliances, each maintaining a navy several times larger than that of any other nation.

The United States and the USSR

In 1976 the estimated value of the Soviet fighting fleet was some 14 percent greater than that of the United States. Although this aggregate index says very little about the ability of the two navies to carry out their respective tasks, it nonetheless reflects a remarkable transformation. In 1950 the U.S. advantage, in terms of estimated value, was 6.5:1 and even in 1970 it was 1.6:1. Thus, it is only in the last decade or so that the United States has lost its status as the world's unrivaled naval power, although it is probably still the strongest and certainly the most flexible naval power.

The structure of the U.S. and Soviet navies are markedly different. These

---

5. It is entirely possible that the rate of retirement of naval vessels has been faster than that shown by our principle source, *Janes Fighting Ships*. For example, our estimate of the total number of submarines in the Soviet navy at the end of 1976 is 390. An estimate provided by the U.S. secretary of the navy on February 9, 1976, was 325. Unfortunately, the latter figure was not broken down by class of vessel.

differences presumably reflect differences in the functions expected of the navy in each country or, more generally, in naval strategies. Since it takes a considerable period of time to design and built ships, particularly in the numbers required to change the structure of a navy significantly, the existing structure may reflect past rather than current strategy. This is a caveat that must be borne in mind in the following discussion.

The navies of the United States and the Soviet Union have at least one function in common—strategic deterrence. In both countries, nuclear-powered ballistic-missile submarines constitute a major part—in terms of deliverable warheads, a growing part—of the strategic nuclear forces. And in both countries these vessels account for a large share of the estimated value of the naval stock, 30 and 41 percent for the United States and the USSR, respectively, in 1976 (table 4).

The significance of the Soviet lead in numbers (and value) of strategic submarines (55 vs. 41 in 1976) is difficult to assess. If one accepts that the purpose of these boats, together with land-based missiles and bombers, is to deter nuclear attack, then both countries have more than enough already, so that a numerical inequality means very little.[6] Indeed, if the index used is the number of targets that the missiles in these submarines can threaten, then the advantage is sharply reversed, with the United States having a lead of about 7:1 in 1976. By the end of 1976, 29 of the 41 U.S. boats carried the Poseidon missile with 14 independently targetable warheads, while none of the Soviet submarine-launched missiles had this capability.[7] On this basis, therefore, the U.S. vessels should be assigned a much higher value than those of the Soviet Union. To estimate an appropriate value differential based on the number of

TABLE 4.—PERCENTAGE OF U.S. AND SOVIET NAVAL STOCK IN STRATEGIC SUBMARINES

	1960	1965	1970	1976
United States	2.2	18.4	24.5	30.0
USSR	6.6	16.0	21.7	40.7

Source.—See table 2.

6. Perhaps the only situation in which a numerical imbalance is significant would be if the breakthrough in ASW techniques dramatically raised the possibility of detecting and destroying these vessels. At the present time this probability is essentially zero. Moreover, the increasing range of submarine-launched ballistic missiles permits these vessels to remain in or close to their homelands, making ASW operations by the opponent exceedingly hazardous.

7. However, two submarine-launched ballistic missiles, with the U.S. designations SS-NX-17 and SS-NX-18 and believed to have this capability, are under development in the Soviet Union. It is estimated that the SS-NX-18 can deliver three independently targetable warheads.

deliverable warheads and other factors, such as missile accuracy and quietness of the submarine, would, however, be a complex exercise even if the data were available.

For these reasons and because these vessels have such a unique role, all subsequent comparisons of the naval forces in these two countries in terms of numbers and value will exclude them. When this is done, the two fleets, in terms of estimated value, are essentially equivalent, although major structural differences remain.

The principal functions assigned to the U.S. Navy, and their relative priority, have remained essentially constant over the postwar period. The primary task of the U.S. Navy is control of the sea. Experts argue at length about the precise meaning of this phrase, but for the layman an acceptable definition is that one's own shipping—both naval and merchant—can use the oceans in time of war without incurring unacceptable losses and that the enemy is denied this flexibility. For the United States this means primarily the ability to resupply NATO in the event of a war in Europe and to permit naval forces to strike shore targets wherever necessary. The second function, then, is power projection, the ability to deliver ground and air strikes from naval platforms in distant countries. The United States possesses by far the most powerful and sophisticated capability for this function with 15 attack aircraft carriers and the 200,000-man Marine Corps with its own specialized equipment for amphibious operations, including seven 18,000-ton amphibious assault ships and five 39,300-ton units coming into service. The final element is an overseas presence, that is, the ability to deploy naval vessels periodically or continuously in distant oceans to "show the flag" and, on occassion, to endeavor to influence events on land in a favorable direction.

To support this strategy the United States maintains a fleet of aircraft carriers, surface ships, and submarines. The aircraft carrier, capable of being configured for both sea control and power-projection missions, is central to the fleet structure (table 5). The majority of surface ships, for example, are oriented toward antisubmarine and/or antiair warfare to serve in the carrier protection role.

The structure of the Soviet naval stock, shown in table 6, is significantly different. This is to be expected from the absence, until 1976, of aircraft

TABLE 5.—STRUCTURE OF U.S. NAVAL STOCK, EXCLUDING STRATEGIC SUBMARINES (%)

Type of Vessel	1950	1955	1960	1965	1970	1976
Aircraft carrier	30.7	35.1	38.1	37.7	33.3	22.0
Submarine, conventional	8.8	8.3	6.1	3.9	2.2	.6
Submarine, nuclear	...	...	2.7	6.6	14.5	24.0
Surface ships	60.0	56.1	52.6	51.3	49.5	52.9
Patrol boats	.5	.5	.5	.5	.5	.5

TABLE 6.—STRUCTURE OF SOVIET NAVAL STOCK,
EXCLUDING STRATEGIC SUBMARINES (%)

Type of Vessel	1950	1955	1960	1965	1970	1976
Aircraft carrier	...	...	...	...	...	1.0
Submarines, conventional	27.2	27.2	25.9	21.4	17.4	13.0
Submarines, nuclear	...	...	2.2	15.2	30.8	36.9
Surface ships	52.2	58.2	51.8	44.2	40.6	38.5
Patrol boats	20.6	14.8	20.1	19.2	11.2	10.5

carriers, but there are also other significant differences, notably the emphasis on submarines.

There are two partially divergent schools of thought on Soviet naval strategy. One school argues that the primary function of the Soviet navy throughout the postwar period has been to defend the homeland against attack from the sea. This school maintains that all the major developments in the Soviet navy, the increasing size of surface ships and trends in their main armament, their wider deployment, and the persistent emphasis on submarines are all essentially reactions to the changing nature of the threat from the sea.

The second school maintains that while defense remains an important function the Soviet navy has gradually assumed more positive roles, in particular the protection and promotion of Soviet interests throughout the world. We will not go into this debate here except to mention that considerable evidence can be assembled to support both schools of thought. It can also be pointed out that, insofar as the mere presence of naval forces in distant waters can be exploited for political ends, there is no incompatibility between this function and the wider deployment of naval units for reasons of defense.

Aggregate naval strength

As mentioned, in terms of the estimated value of naval ships excluding strategic submarines, the United States and the Soviet Union were essentially in parity in 1976. There were, however, offsetting differences. In 1976 the United States retained a 1.5:1 advantage in major surface ships excluding aircraft carriers, or 2:1 if these units are included.[8] The Soviet Union, on the other hand, had a 1.9:1 advantage in nonstrategic submarines.

A full discussion of the major weapon systems of these two navies would take us too far afield. There is, however, an essential difference in the offensive weaponry of these two navies that has been the subject of considerable debate

---

8. This is true despite the Soviet numerical superiority in this category. Nearly 100 of the 274 major surface vessels in the Soviet navy in 1976 displace only 850–950 tons, too large to be classified as patrol boats, but not contributing very heavily to the Soviet stock of major warships.

and discussion. In the U.S. Navy the principal tactical offensive weapon is the carrier-borne aircraft. The 13 attack aircraft carriers currently operational in the U.S. navy carry about 800 aircraft with an air-to-surface strike capability using bombs and a variety of air-to-surface missiles. The Soviet Union, on the other hand, has concentrated on surface-to-surface cruise missiles. In 1976 the Soviet Union had 44 surface ships and 69 submarines armed with a total of 242 and 430 launchers, respectively, for surface-to-surface missiles.[9] Of these submarines, 41 were nuclear powered and 14 equipped with the short-range SS-N-7 missile, which can be launched while the submarine is submerged; the others carried the older but longer-range SS-N-3 missile, which can only be launched while the submarine is on the surface.

Thus, the U.S. Navy has a large number of offensive aircraft, but these are concentrated on only 13 ships while the Soviet Union has 108 ships and submarines with an offensive capability against surface ships. This imbalance has been the subject of much criticism within the U.S. Navy, but it is expected that this will be remedied in the near future. The U.S. Harpoon surface-to-surface missile, due to become operational in 1977, will be installed on many surface ships; and a version is also being developed that can be fired from submerged submarines. At the present time, over 70 U.S. ships are armed with the Standard 1 surface-to-air missile modified to provide a horizon-limited surface-to-surface capability.

NATO and the WTO

If we now widen the comparison to include all the countries in NATO and the WTO, respectively, the margin of superiority widens significantly in favor of NATO. In terms of the estimated value of naval ships, NATO had a superiority of 1.2:1 in 1976. If strategic submarines are excluded, the margin widens to 1.6:1.

The members of NATO (excluding the United States) include the third- and fourth-ranking navies in the world (the United Kingdom and France, respectively) and a number of other countries with comparatively large navies (table 7). In 1976 these countries accounted for 29 percent of the estimated value of the NATO's naval stock. The non-Soviet countries of the WTO, on the other hand, accounted for less than 2 percent of the estimated value of the WTO's naval stock.

Although there are many contingencies in which the United States cannot

---

9. It was suggested recently that the missile carrying the designation SSN-10 is, in fact, an antisubmarine rather than antiship weapon. If this is the case, the number of major surface units with antishipping missiles drops to 19 and the number of launchers on these ships to 90. The possibility must be considered, however, that the cannisters associated with SSN-10 are capable of launching two types of missiles, one antisubmarine and the other antiship.

TABLE 7.—ESTIMATED VALUE OF NAVAL STOCK: SELECTED NATO COUNTRIES (U.S. $Million)

Country	1950	1955	1960	1965	1970	1976
United Kingdom	5,110	6,610	4,110	4,135	5,277	7,359
France	1,194	1,666	1,444	1,721	2,610	4,326
F.R. Germany	...	50	277	832	1,165	1,450
Italy	416	694	472	890	1,070	1,184

rely on the support of its NATO allies, as the events of October 1973 demonstrated, for example, this imbalance is possibly of great significance. If the WTO were to try to match NATO's naval strength, this would require a Soviet fleet considerably larger than that of the United States, and it is unlikely that the latter would permit this to happen.

As mentioned in the introduction, the stock of fighting ships possessed by a particular country is a rather narrow indicator of that country's naval capacity. Apart from fighting ships, support ships, naval bases, and geographic location make important contributions to naval capability, but the assessment of these factors is probably even more complex than the capabilities of fighting ships.

The one point on which all observers agree is that Soviet naval flexibility is severely handicapped by geographical factors. The Soviet Union maintains four geographically separate fleets. Two of these, the Baltic and Black Sea fleets, are almost totally blocked in.[10] But even the egress routes for its Northern and Far Eastern fleets are, with modern detection devices, considered relatively narrow. These limitations, together with the Soviet Union's enormous geographic size and comparative self-sufficiency, contribute to the atmosphere of alarm surrounding its naval expansion. Some observers feel that it is both extremely difficult and unnecessary for the Soviet Union to be a strong naval power.[11] The fact that it has become one, therefore, has given rise to considerable suspicion as to its motives. The Soviet Union can hardly be criticized for seeking to defend itself against strikes by carrier-borne aircraft or for endeavoring, however vainly, to counter the threat posed by ballistic-missile submarines. Undoubtedly, the Western powers, particularly the United States, are piqued by the fact that they have lost their monopoly of naval power. At the same time the Soviet Union undoubtedly regards the breaking up of a Western monopoly as an attractive objective in itself.

10. The canal linking the Baltic and White Seas has been recently enlarged to permit transit by vessels displacing up to approximately 5,000 tons.
11. The commander-in-chief of the Soviet navy, Admiral Gorshkov, agrees with the latter observation (*Naval Digest* [March 1972], p. 22).

Other Areas

The fact that NATO and the WTO account for nearly 90 percent of the world naval stock does not mean that naval developments outside these two alliances are of no consequence. Naval power is a relative concept, and there are many navies throughout the world which—though miniscule compared with those of the United States or the USSR—have an important bearing on regional military balances.

In Europe (table 8) substantial navies are maintained by Spain, Sweden, and Yugoslavia, in that order. The Spanish naval stock has increased erratically, actually falling between 1955 and 1960 and again between 1965 and 1970. Since 1970, however, it has increased by 50 percent with the addition of destroyers, frigates, and submarines, both newly built and secondhand vessels from the United States. In Yugoslavia the naval stock increased rapidly between 1950 and 1965 and then declined by some 10 percent in the ensuing decade. However, local construction of submarines and missile-armed patrol boats is under way, so that this trend may be reversed.

In other developed countries (table 8) the Japanese naval stock has increased at an average annual rate of 7.2 percent, and in 1976 it ranked sixth in the world. In China it is apparent that the rate of increase in the naval stock has accelerated continuously over the postwar period. In 1976 it ranked fifth in the world in terms of the estimated value of the naval stock. The bulk of the Chinese stock is made up of coastal units, particularly missile-armed patrol boats. But series production of ocean-going units—patrol submarines and destroyers—was begun around 1970.

The estimated value of the stock of fighting ships in Third World countries—taken as a whole—has increased somewhat faster than the world average over the period 1950–76, 7.2 percent annually against 5.1 percent. For the period 1970–76, however, the divergence has been much greater, with the naval stock in the Third World growing twice as fast as the world total. To a greater or lesser degree this is a general phenomena with all the major regions contributing (table 9).

In South America the average annual rate of growth of the naval stock was comparatively modest through 1970. Since then, however, the stock has increased by 57 percent. All the countries in the region have participated in this

TABLE 8.—NAVAL STOCK, MISCELLANEOUS (U.S. $Million)

Country	1950	1955	1960	1965	1970	1976
Other European	779	1,121	1,314	1,608	1,645	2,072
Other developed	511	907	1,076	1,509	2,146	3,049
China	89	162	481	722	1,352	3,301

TABLE 9.—ESTIMATED VALUE OF NAVAL STOCK:
THIRD WORLD COUNTRIES (U.S. $Million)

Area	1950	1955	1960	1965	1970	1976
Middle East	124	171	269	375	666	1,310
South Asia	147	215	339	506	611	994
Far East	358	505	718	1,157	1,508	2,053
Africa	...	...	13	88	274	527
Central America	207	208	205	318	331	470
South America	867	1,175	1,333	1,460	1,452	2,280

buildup through the commissioning of newly built ships and submarines, the modernization of existing vessels, and the acquisition of secondhand units from the United States.

In the Far East the estimated value of the naval stock has increased by 36 percent since 1970. Again, this upward trend is a general phenomenon, with the single exception of Indonesia. Indonesia received over 100 vessels from the Soviet Union and other WTO countries between 1958 and 1965. By the end of 1976 the fleet had been disposed of almost entirely, but a program to rebuild the navy is under way with frigates and missile-armed patrol boats on order from the Netherlands and South Korea, respectively.

In the Middle East, as might be expected, the naval stock has been rising rapidly. The successive Arab-Israeli conflicts, though primarily contests between armies and air forces, have increasingly taken on a naval dimension. In the Persian Gulf the Iranian naval stock has increased more than fourfold since 1970, and further major acquisitions are in the pipeline, including four 7,800-ton *Spurance*-class destroyers from the United States.

In South Asia the Indian naval stock has increased consistently since 1950 at an average annual rate of 9.6 percent. In Pakistan, on the other hand, the stock has been relatively constant since 1960, with a slight decline after 1970 due to losses suffered during the war with India in 1971. However, reports in December 1976 that Pakistan had taken delivery of submarines and destroyers from China, if true, would probably reinstate a modest upward trend for the period 1970-76.

TRENDS IN THE MAJOR TYPES OF WARSHIPS

The general postwar trend has been away from large vessels and toward small ones. There are several reasons for this, but the two main ones are probably the enormous rise in the cost of weapons and the increasing vulnerability of large platforms of any kind to relatively inexpensive counterweapons. At the same time technological developments have made it possible to install highly effective weapons systems on comparatively small platforms.

Aircraft Carriers

The number of attack aircraft carriers in operation around the world has fallen steadily over the postwar period from 44 in 1950 to 21—or 19 if the two U.S. *Hancock*-class carriers in reserve are excluded—in 1976 (table 10). This trend appears to be a fairly clear-cut case of prohibitive cost on the one hand and increasing vulnerability on the other. The attack aircraft carrier is unquestionably a flexible and immensely powerful weapons system. But few nations have been able to support the cost of building and operating up-to-date versions of these vessels. Only the U.S. Navy has sustained an uncompromising commitment to the aircraft-carrier concept. The latest product of this commitment is the nuclear-powered *Nimitz*, a vessel displacing 93,400 tons when ready for combat and constructed at a cost of $1,881 million (in 1976 dollars). Two similar ships are under construction and a fourth has been approved. The cost of each of these three is expected to exceed $2,000 million (in 1976 dollars).

There has been considerable debate in recent years in U.S. naval circles regarding the wisdom of investing so much in a single unit and skepticism regarding the willingness of Congress to appropriate the funds required to maintain the existing carrier force level. At the same time there is still widespread appreciation of the effectiveness of the aircraft—both fixed-wing and rotary—as a weapon against surface ships and submarines. This situation has focused attention on alternative ways of taking air power to sea.

In July 1975 the then U.S. secretary of defense, James Schlesinger, directed the navy to conduct feasibility studies of medium-sized carriers

TABLE 10.—ATTACK AIRCRAFT CARRIERS

Country	1950	1955	1960	1964	1970	1976
United States	27	19	14	16	15	15*
United Kingdom	12	16	8	5	4	1
France	2	4	3	3	2	2
Australia	1	2	2	1	1	1
Canada	1	1	1	1	...	...
Netherlands	1	1	1	1	...	...
Argentina	...	...	1	1	2	1
India	...	...	...	1	1	1
Total	44	43	30	29	25	21

SOURCE.—See table 2.
*Only 13 were operational in 1976.

displacing some 20,000-25,000 tons less than the *Nimitz* class. These studies apparently indicated that this reduction in size involved a loss in aircraft-carrying capacity disproportionate to the anticipated financial saving. A far more significant compromise was the U.S. Navy proposal, advanced a few years ago, for a fleet of Sea Control Ships. These vessels would have displaced just 14,300 tons fully loaded and been equipped primarily with helicopters for antisubmarine warfare. However, they would also have carried three VSTOL fixed-wing aircraft for air defense and maritime strike. This proposal was also rejected as having too little capability with relation to cost.

To overcome this objection the Navy has offered a concept known as the VSTOL Support Ship (VSS). These would displace either 22,000 tons or 32,800 tons fully loaded. The smaller version would carry 22 helicopters and four VSTOL aircraft, while the larger may have steam catapults to allow the operation of a small number of conventional fixed-wing aircraft as well as VSTOL aircraft and helicopters.

Operations with the British Harrier have demonstrated the viability of VSTOL aircraft as a weapons system, and this has permitted far more flexibility in the design of future aircraft carriers. The first vessel to become operational whose overall capability depends heavily on VSTOL aircraft was the Soviet Union's 40,000-ton *Kiev* antisubmarine cruiser. This vessel, deployed into the Mediterranean in July 1976, is capable of carrying a mixed force of about 35 VSTOL aircraft and helicopters. A second vessel is almost complete, and at least one more is under construction. The United Kingdom is moving in a similar direction with its "Through Deck Cruisers," 19,500-ton vessels designed to carry 10 ASW helicopters and five Sea Harriers. The United States, as we have noted, is preparing the VSS design, and France remains committed to aircraft carriers with its proposed 18,400-ton nuclear-powered helicopter carrier.

In sum, although the number of large aircraft carriers will probably continue to fall, modified versions of the aircraft-carrier concept are likely to become more numerous. It is also likely that an increasing number of countries will acquire aircraft carriers of some kind. For example, a proposal that has attracted considerable attention is the Vosper Thornycroft Harrier carrier, a 4,000–5,000-ton vessel carrying eight Harriers and two helicopters. The likely cost of a vessel of this size would put it within reach of many navies.

Another noteworthy development in aircraft carriers is the large and essentially self-contained amphibious assault ship. The United States currently has nine of these ships. The *Iwo Jima* class (seven units) is capable of landing (by helicopter) about 1,000 troops together with artillery, vehicles, and other equipment. The *Tarawa* class (two units operational and three under construction) is about twice as large and, in addition to helicopters, is provided with an internal docking well for four landing craft, each of which can transport 170 tons of equipment. The proposed VSS ships also have a significant, though secondary, amphibious assault capability. The United Kingdom, having converted two aircaft carriers for this role, is the only other country with specialized ships of this kind, although one was placed in reserve in 1976 and the other is now employed primarily in the antisubmarine role. Obviously any country with helicopter carriers for antisubmarine warfare can transform them readily into effective assault ships simply by changing the type of helicopter embarked. The primary difference, however, is the length of time for which such a ship—with an embarked amphibious assault force—could be deployed in the vicinity of the potential assault point. Large, specifically designed ships have far greater endurance in this respect, which confers prolonged capability for mounting an effective assault.[12]

Submarines

The capabilities of virtually all weapons have dramatically increased in the postwar period, but this is perhaps more true of the submarine than any other. At the end of World War II, antisubmarine weapons and techniques were proving highly effective against the then existing types of submarines, but since then the advantage has swung decisively in favor of the submarine. The key to this change was the advent of nuclear power, which obviated the need for the submarine ever to come to the surface and gave it a sustainable underwater speed equal to or greater than that of any surface ship.

The nuclear-powered ballistic-missile submarine, because of its invulnerability, is by far the most prized weapons system for strategic deterrence, and in 1976 over 100 of these vessels were in operation in four countries (table 11). The submarine, particularly when nuclear powered, has proven to be a most effective antisubmarine as well as antisurface ship weapon. The number of

TABLE 11.—NUCLEAR-POWERED BALLISTIC-MISSILE SUBMARINES

Country	1960	1964	1970	1976
United States	3	33	41	41
USSR	4	9	24	55
United Kingdom	...	...	4	4
France	...	...	1	4
Total	7	42	70	104

Source.—See table 2.

12. A large number of navies, of course, have ships and landing craft suitable for more limited amphibious operations. The Turkish invasion of Cyprus in July 1974 demonstrated that, given suitable conditions, vessels of this kind are of great importance.

nuclear-powered attack submarines, armed with sophisticated acoustic detection devices, long-range homing torpedoes, and—for some Soviet vessels—cruise missiles, is increasing (table 12). The primary task of these submarines is to locate and destroy (it is necessary to add, if possible) the strategic submarines, but their antishipping capabilities are also formidable. It is estimated that a force of 30 of these submarines could sink some 50 million tons of shipping a year, which is more than three times the maximum U.S. shipbuilding effort during World War II.[13] In addition, there is the possibility that hunter-killer submarines will acquire a strategic retaliatory strike capability by being equipped with long-range cruise missiles fired from standard torpedo tubes.

Similarly, the speed and range of the nuclear-powered submarine, together with developments in the range of weapons that it can carry and in sensor technology, have led some experts to suggest that it should be considered for a wider range of tasks than is presently the case. Specifically, nuclear-powered submarines could be used as escorts for surface-ship task forces (U.S. submarines already are used sometimes to escort carrier task forces) and to protect convoys. It is suggested that smaller and less sophisticated nuclear-powered submarines would be a cost-effective weapons system for the latter task.

Nuclear power has not, however, meant the demise of the conventionally powered submarine. It is true that the number of conventionally powered patrol submarines (defined as those displacing 700 tons or more) has been falling since 1965, but the total number in 1976 was still substantially higher than in 1950 (table 13). Moreover, in the Third World and in other European countries the number has increased continuously. It is worth noting that nearly all the submarines recently acquired, even by Third World countries, are newly constructed.

Technological developments also have enhanced significantly the capabilities of the conventionally powered submarine. Current types of these vessels can remain submerged for several days without great difficulty and can achieve submerged speeds of about 20 knots, though only for short periods.

TABLE 12.—NUCLEAR-POWERED ATTACK SUBMARINES (N)

Area	1950	1955	1960	1965	1970	1976
USSR	...	...	4	26	58	80
United States	...	1	11	22	46	68
United Kingdom	...	...	...	...	4	9
World total	...	1	15	48	108	157

13. Sir Arthur Hezlet, *The Submarine and Sea Power* (New York: Stein & Day, 1967), p. 259.

TABLE 13.—CONVENTIONALLY POWERED PATROL SUBMARINES (N)

Area	1950	1955	1960	1965	1970	1976
Developed countries	351	527	502	516	459	365
Third World, including China	4	5	33	60	76	133
World total	355	532	535	576	535	498

They are also comparatively quiet in operation and are designed and constructed so as to reduce the risks of detection by radar when using the periscope or snorkel or by magnetic detectors in overflying aircraft. In addition, sophisticated weaponry and sensor equipment can, of course, be installed as readily in a conventionally powered vessel as in a nuclear-powered one. It is apparent that the conventionally powered submarine remains a popular naval weapon system (see appendices to this volume).

Other Major Surface Warships

The number of major surface warships—cruisers, destroyers, frigates, and escorts—has declined continuously since 1955. There was a large diminution between 1955 and 1960 when, owing to obsolescence and changing naval requirements, many World War II vessels were removed from the active fleets, particularly in the NATO countries. A second major drop occurred between 1970 and 1976, primarily due to cuts in the U.S. fleet.

Two constituent opposing trends of this general reduction are worth pointing out. First, the number of vessels armed with missiles for defense against aircraft, missiles, and submarines is increasing rapidly—from 18 in 1960 to 383 in 1976. Second, the number of major surface ships in the Third World, taken as a whole, has increased steadily (table 14).

It is sometimes suggested that the large surface ship is an obsolescent

TABLE 14.—MAJOR SURFACE WARSHIPS

Area	1950	1955	1960	1965	1970	1976
United States	817	837	719	712	555	206
Other NATO	520	582	402	349	351	298
USSR	150	256	261	251	245	274
Third World, including China	183	231	269	276	288	304
World total	1,783	2,044	1,807	1,761	1,605	1,206

NOTE.—Ships include cruisers, destroyers, frigates, and escorts.

weapons system. The capabilities of the principal opponents of the surface ship, namely, aircraft carriers and submarines, have increased far more than its own defensive capabilities. Moreover, defensive-weapon systems have taken up a growing share of the available space, leaving little room for any offensive armament. A surface ship, to be capable of independent operations in the present environment, would require missiles for defense against aircraft, cruise missiles, and submarines. Large naval guns have been abandoned in favor of smaller, rapid-fire weapons which can supplement the antiaircraft missiles. Additional gun systems are regarded as desirable as a second line of defense against cruise missiles. For the ships to have an offensive capability, an additional cruise-missile system would have to be added. Ships of this kind would obviously be large and complex and therefore expensive both in construction and in operation. Nuclear propulsion, so significant for the submarine, provides only a marginal increase in the performance of a surface ship and adds substantially to its acquisition cost. Some observers argue, therefore, that, insofar as fighting capability is concerned, large surface warships are no longer cost-effective.

However, even if the surface ship is at a comparative disadvantage, this does not mean that it is totally helpless. Missiles can engage attacking aircraft at great distances: in 1970 two North Vietnamese jet fighters were shot down from a distance of 105 km by U.S. Talos shipborne surface-to-air missiles. The range of antisubmarine weapons has also increased. The U.S. ASROC missile transports a torpedo or a nuclear depth charge to a maximum range of 10 km; the Australian Ikara system can be used to the maximum range of the ship's sonar. The inherent limitations of the surface ship as a sonar platform (because of the noise it generates) have been largely overcome by the use of high-endurance helicopters with dipping sonars and, more recently, by sonar devices towed at an appropriate distance behind the ship.

Nevertheless, the relative vulnerability of the large surface ship is undoubtedly a factor in the decline in the number of these ships in the world's principal navies. But so long as navies are regarded as useful in situations short of war, major surface ships will remain an important component because they are flexibile and visible, two important characteristics for the peacetime utilization of naval forces.

Finally, a major technological development is in the offing which would enhance significantly the capabilities of the surface ship. In this development the bulk of the weight of a ship is supported by an artificially created air bubble, permitting speeds two or three times greater than present surface ships. The surface ship would, therefore, regain a decisive speed advantage over the submarine—even the nuclear-propelled type—and would be less vulnerable to the torpedo, which is still the main armament on most of the world's submarines.

In sum, the world's principal navies continue to regard the major surface ship as a viable weapons system, although the future size, weaponry, and ultimately even general appearance of these ships can be expected to change markedly.

Patrol Boats, Torpedo Boats, and Gunboats

The number of these minor fighting vessels has increased very rapidly over the postwar period. Being relatively cheap and suitable for coastal waters, they have proved particularly popular in Third World countries.

The missile-armed patrol boat is perhaps the most fashionable naval weapons system at the present time. The sinking in 1967 of an Israeli destroyer by surface-to-surface missiles fired from a Soviet-supplied Egyptian patrol boat provoked a number of countries—including the United States, France, Italy, and Israel—to develop or to accelerate the development of comparable missile systems. As a result there are now six basic types of antiship missile systems on the market outside the Soviet Union. These are Exocet (France), Otomat (France/Italy), Sea Killer (Italy), Penguin (Norway), Gabriel (Israel), and Harpoon (United States). In addition, a number of existing missiles have been "navalized" (for example, the French SS-12) or adapted to provide secondary antiship capability (for example, the U.S. Standard ARM). Most of these missile systems are sufficiently compact to be installed on patrol boats, providing these small, fast vessels with a highly destructive capacity. Some observers have forecast major changes in naval tactics and in the world balance of naval power as a result of the proliferation of missile-armed patrol boats. The possession of these vessels by many countries—and they are cheap enough to be within reach of even the smallest states— "can affect the tactical and strategic employment of the seapower of the major navies. Large warships have much to fear from an unexpected attack from these small missile boats. It is an ideal weapon for guerilla warfare at sea."[14] And the proliferation of these vessels has indeed been rapid. In 1960 only the Soviet Union possessed missile-armed patrol boats, but by 1976 there were over 530 of these vessels in 37 countries (table 15).

It is interesting to note, however, that the missile-armed patrol boat is moving away from the original concept of a small, inexpensive, exclusively

TABLE 15.—MISSILE-ARMED PATROL BOATS (N)

Year	Countries	Boats
1960	1	5
1965	7	141
1970	17	282
1976	37	533

14. L. I. Smith, "New Naval Tactics," *Ordnance* (November–December 1972).

offensive weapons system. The pioneering unit, the Soviet Union's *Komar* class, displaced just 70 tons and carried two launchers for the SSN-2 Styx missile. In all countries designing and building these boats, there has been a conspicuous upward trend in size and complexity to provide a more stable firing platform and to permit the carriage of more launchers, more powerful radars, and more ECM equipment.

CONCLUSIONS

There can be little doubt that a major naval buildup is under way around the world and will persist if construction programs maintain their present momentum. As mentioned in the Introduction, the most plausible general explanation for this development is the emerging capability to exploit systematically the resources in and under the oceans plus the expectation that coastal states will acquire a degree of sovereignty over large areas of the oceans. In view of this it seems inevitable that for the majority of states ocean resources will be, for some time at least, a net liability rather than an asset.

In addition, persisting military rivalries in many parts of the world have long had or are acquiring a naval dimension, for example, in South Asia and the Middle East. Clearly, however, the most important determinant of naval developments over the foreseeable future is the manner in which the United States and the Soviet Union manage their increasing interactions on the high seas.

Whether or not a naval arms race is under way between these two countries is a moot point. The more alarmist Westen statements of recent years on the existence or imminence of Soviet naval superiority have given way to more sober, and reassuring, assessments. Nevertheless, the Western powers and particularly the United States remain acutely sensitive to Soviet naval rivalry, and the evidence clearly indicates that they will endeavor to resist the challenge. In 1976 the outgoing Ford administration in the United States proposed a $28 billion naval construction program for the 5 years 1977–81.

Constructing and maintaining a modern blue-water navy is an enormously expensive undertaking. The life-cycle costs of a single *Nimitz*-class aircraft carrier and its embarked airwing have been estimated at $13.3 billion; 20 conventionally powered guided-missile destroyers (DDG-47) would cost $24 billion over their effective lifetime.[15] For most of the postwar period naval forces have not been an arena of intense competition between the United States and the Soviet Union. Should it become one the costs will be extremely large.

A naval arms race between these two countries that has as a principal

---

15. U.S. Congress, Congressional Budget Office, "Planning U.S. General Purpose Forces: The Navy," Budget Issue Paper (Washington, D.C.: Government Printing Office, 1976).

objective the ability to exploit sea power for diplomatic and political ends would be highly dangerous. Because they each possess massive nuclear arsenals, the greatest stress has been placed on preventing any confrontation anywhere between their armed forces. With land and air forces this has been done, but at sea it will be far more difficult. The international status of the oceans and seas and the apparent determination of both powers to exploit this status in their respective and usually contradictory interests must make confrontations at sea a distinct possibility.

It follows that there is an urgent need, at least, to establish a set of rules to minimize the risk of naval clashes. The existing U.S.-Soviet agreement on the prevention of incidents on and over the high seas, which entered into force on May 1972, could serve as a basis. Far better, of course, would be the commencement of negotiations on the limitation of naval armaments. This could save both countries enormous expense and perhaps forestall a contest that involves a high risk of confrontation.

Military Impact on Ocean Ecology

Arthur H. Westing
Stockholm International Peace Research Institute

OCEAN ENVIRONMENT AND ECOLOGY

The world ocean has a volume of perhaps 1.37×10^{18} m^3. This vast amount of water has a total surface area of about 361×10^{12} m^2, which represents 71 percent of the earth's surface (61 percent in the northern hemisphere and 80 percent in the southern). Approximately 36×10^{12} m^2 (or 10 percent) of the ocean area overlies the continental shelves of the world, these having an average width of about 78 km. Between 18×10^{12} m^2 and 27×10^{12} m^2 (or 5 percent to 7 percent) of the total ocean surface is covered at any one time by a layer of ice that averages about 2 m in thickness. Much of this ice-covered water coincides with the Arctic Ocean, and the remainder fringes the Antarctic landmass.

Wind action, the earth's rotation (Coriolis forces), density differences owing to temperature and salinity differences, and tidal action brought about by the gravitational pull of moon and sun combine to produce not only surface waves, but also a complex system of great ocean currents within and among the various ocean basins. The ceaseless ocean tides and currents serve to keep the ocean waters of the world more or less well mixed and uniform.

The worldwide hydrological cycle plays an indispensable role in the biosphere, its series of pathways serving to link all of the ecosystems on earth. The ocean is the great reservoir for this cycle, representing over 97 percent of the total earthly supply of water and more than 99 percent of its supply of liquid water. It is thus crucially involved in the global water balance, and thereby in both global mineral nutrient cycling and global climate.

The ocean provides suitable niches for an amazing diversity of living things, plant and animal, large and small, sessile and mobile. Taxonomic diversity is substantially greater in the ocean than on land. Although some life can be found in almost any area of the ocean and at any depth, by far most of the marine biomass occurs within the shallow regions that overlie the continental shelves, or in the vicinity of islands. Some ocean habitats, especially the nutrient-enriched ones, are highly productive. These include certain close-inshore areas endowed with a continuing supply of nutrients from terrestrial runoff (such littoral zones as estuaries and especially mangrove swamps) as well

© 1978 by The University of Chicago. 0-226-06602-9/77/1978-1019$02.42

as regions of upwelling, where sunken nutrients are brought up from the lower reaches of the ocean. Continental-shelf waters are over twice as productive as the open ocean, regions of upwelling more than five times as productive, and estuaries and mangrove swamps about ten to twenty times as productive.

Estuaries and mangrove swamps represent a minute fraction of the total ocean habitat. They are important, however, not only for their high productivity, but also because at least three-quarters of all species of marine fishes require this inshore habitat during at least one critical stage in their life cycle. Moreover, estuaries and mangrove swamps are a crucial habitat for numerous sorts of birds for at least some part of the year.

Ocean Use

Civil
Humans make continuing use of the ocean in an enormous and increasingly necessary variety of ways. A number of these uses are quite benign in their long-term impact, but many are damaging, often avoidably so. In considering the many marine resources, it will be useful to distinguish between nonextractive and extractive resources, the latter category in turn divisible into nonliving and living.

First of all, the ocean serves as a separator or buffer of greater or lesser importance among nations. Second, and of prime importance, the ocean is used for transportation, both coastal and intercontinental. Some 23,100 "large" plus an additional 42,800 "small" merchant ships currently ply the seas (see Appendix A). Among the large ones, five are nuclear powered and at least 32 can carry over 400×10^3 m^3 of oil. Transportation is, or at least can be, a relatively benign use of the ocean with the following notable exception. The development of port facilities is frequently at the expense of estuaries and similar habitats so important to marine ecology.

Other nonextractive uses of the ocean include the running of communication cables and oil pipelines, the former quite harmless but the latter potentially less so. The ocean is also used for a variety of recreational, scientific, and aesthetic purposes, most of which are relatively benign. However, the dumping of wastes and the creation of polderlands (e.g., some 40 percent of the Netherlands) must be listed among the detrimental nonextractive uses of the ocean.

The nonliving extractive resources of the ocean are assuming an ever greater share of the world economy. This has resulted from a combination of factors, among them the haphazard terrestrial distribution of minerals vis-à-vis national boundaries, the ever increasing quantities of minerals being utilized, and the increasingly exotic demands of industry (both civil and military) that range ever more widely over the periodic chart of elements. Important ocean extractives include sodium chloride (30 percent of world production), bromine (70 percent of world production), magnesium (60 percent of world and 95 percent of U.S. production), thorium (30 percent of world production), platinum (90 percent of U.S. production), sulphur (10 percent of U.S. production), and pearls (almost 100 percent of world production). By far the most important mineral claimed from the ocean today is oil. The several thousand producing wells in the continental shelves of the world account for about 20 percent of current world production (and the associated natural gas for perhaps 10 percent of world production).

The living ocean resources of commercial importance include both flora and fauna. The marine fishery—which includes true fish (finfish), crustaceans and mollusks (shellfish), and such marine mammals as whales and seals—is an indispensable source of food and other products. The annual world fish catch seems to have at least temporarily leveled off at about 70×10^9 kg fresh weight (ca. 23×10^9 kg dry weight; ca. 10×10^9 kg protein).[1] The marine fishery thus provides the world's human population with about 17 percent of its annual animal protein intake (and about 5 percent of its annual total protein intake). In Japan fish protein represents about 50 percent of the animal protein intake, in the USSR about 20 percent, and in the United States about 3 percent. Between 80 percent and 90 percent of the commercial fish catch of the world occurs over the continental shelves.

With respect to marine plants, approximately 2.6×10^9 kg fresh weight of giant kelp (*Macrocystis*, Laminariaceae) and other large algae are harvested annually throughout the world for fertilizer, food, feed, and various chemical constituents.[2] This annual harvest has more than doubled during the past 15 yr and continues to rise.

Military

The military use of the ocean is rapidly growing in importance and complexity.[3] It can be either defensive or offensive, in territorial waters or on the high seas, on the surface or beneath it. Military use of the ocean by a nation is generally in support of its terrestrial interests, although an increasing fraction is in support of its direct ocean interests.

The navies of the world together maintain about 1,130 "large" ships plus an additional 1,330 "small" ships on the high seas (Appendix A). Among the large ones more than 260 are nuclear powered. Although there are some 51 navies in the world, those of the United States and USSR dwarf all of the others (table 1). The relative military importance the United States attaches to its navy is suggested by the facts that 25 percent of its total number of armed forces is

---

1. *U.N. Statistical Yearbook, 1975*, p. 160; D. Pimental et al., "Energy and Land Constraints in Food Protein Production," *Science* 190 (1975): 754-61; ibid. 193 (1976): 1070, 1073-76.
2. "Seaweed and Seaweed Products," *Ceres* 9, no. 6 (1976): 7, 9.
3. *SIPRI Yearbook, 1969/70*, pp. 92-184.

TABLE 1.—NAVIES OF THE WORLD

Nation	"Large" Surface Ships	"Large" Submarines	Total "Large" Ships	Naval Personnel (10^3)
USSR	195	186	381	450
United States ...	181	106	287	525
Great Britain ...	47	13	60	76
France	38	5	43	70
China	20	3	23	275
46 others	340	0	340	858
Total	821	313	1,134	2,254

SOURCES.—(a) Numbers of "large" surface ships are from *Jane's Fighting Ships, 1976–77*, being a summation of battleships, aircraft carriers, cruisers, destroyers, and half the frigates. Numbers of "large" submarines are from the same source, being a summation of ballistic-missile, cruise-missile, and fleet submarines (see Appendix A). (b) Numbers of naval personnel are from *Military Balance, 1976–77* (London: International Institute for Strategic Studies, 1976), with estimates (totaling 36 × 10^3) being supplied for the six nations missing from among the 46 miscellaneous nations. (c) See also *SIPRI Yearbook, 1975*, pp. 255–307; *SIPRI Yearbook, 1976*, pp. 212, 228–31; and SIPRI, *Tactical and Strategic Antisubmarine Warfare* (Stockholm: Almqvist & Wiksell, 1974), pp. 102–9.

naval[4] and that 33 percent of its total military budget is directly devoted to the navy.[5]

The missions of a navy continue to be the protection of a nation's shores, the safeguarding of its merchant fleet on the high seas, and the intimidation of other, especially smaller, nations (so-called naval presence or showing of the flag). Another naval mission that has taken on a new dimension of importance is the establishment and subsequent protection of a nation's interests in the extractive resources of the ocean, both living and nonliving. Other standard naval missions include the protection of a nation's military troop and supply carriers and the denial of the use of the ocean to an enemy (including blockading). The primary targets of a navy continue to be the military (naval) and civil (merchant) ships and coastal facilities of enemy nations.

Superimposed upon all of these traditional naval functions have been two strategic missions of vast significance. The first of these is strategic deterrence (or attack), made possible over the past 15 yr or so by the so far essentially invulnerable missile-armed nuclear-powered submarines having an intercontinental nuclear second-strike (or first-strike) capability.[6] The second is the counterability, that is, antisubmarine warfare (ASW), a capability, however, not

---

4. *Military Balance, 1976–1977* (London: International Institute for Strategic Studies, 1976), p. 5.
5. D. H. Rumsfeld, *Annual Defense Department Report FY 1978* (Washington, D.C.: Department of Defense, 1977), p. Al.
6. H. Scoville, Jr., "Missile Submarines and National Security," *Scientific American* 226, no. 6 (1972): 15–27, 136; R. L. Garwin, "Antisubmarine Warfare and National Security," *Scientific American* 227, no. 1 (1972): 14–25, 122.

as yet developed to any significant degree.[7] Most by far of the extensive oceanographic and much of the marine biological research in the world is carried out (either directly or via financial support) by the several major navies of the world. The fruits of these efforts, while they have great civil value, are, of course, meant to enhance naval capabilities, largely with reference to antisubmarine warfare.[8]

Ocean Abuse

Civil

The ocean continues to be abused, despite its obvious importance to human welfare and despite an enormous literature of concern.[9] Those estuaries which are not filled are dredged, and all of them are befouled with a continuing stream of the noxious refuse of society. Indeed, the ocean has always been either the direct or ultimate sump, sewer, and cesspit for most human wastes.

Marine pollution reaches the ocean by diverse avenues. There is the discharge of sewage, industrial wastes, and agricultural wastes (pesticides, fertilizers, food-processing wastes) into the ocean either directly or via streams. There are fallout or washout of volatile compounds and particulate matter from the atmosphere, the disruption and pollution associated with the extraction of seabed minerals, the accidental release or intentional dumping of noxious materials or cargoes from ships, and the sinking of the ships themselves. Wastes too complicated to dispose of on land, either because of the bulk involved or because of their intrinsic danger, are routinely dumped into the sea. Oil appears to be the marine pollutant of greatest current concern, not only because of the large quantities that are regularly introduced into the ocean, but also because of the increasing potential for even greater levels of such contamination, as noted in greater detail below.

Ocean wastes exert their adverse influence by covering up littoral and other benthic (seabed) habitats, by direct toxicity, by stimulating the overgrowth of bacteria (which in turn use up the dissolved oxygen necessary for the survival of marine fauna), and in other ways. Fortunately, however, the ocean is immense, continuous, interlaced with currents, and the home of a wide array of microorganisms. Thus, considering the ocean as a whole, the continuing influx of pollutants into this vast and complex system has so far been rendered more

---

7. SIPRI, *Tactical and Strategic Antisubmarine Warfare* (Stockholm: Almqvist & Wiksell, 1974).

8. W. W. Behrens, Jr., "Environmental Considerations in Naval Operations," *Naval War College Review* 24, no. 1 (1971–72): 70–77.

9. E. D. Goldberg, *Health of the Oceans* (Paris: UNESCO, 1976); M. Ruivo, ed., *Pollution: An International Problem for Fisheries*, World Food Problems Report no. 4 (Rome: Food and Agriculture Organization of the United Nations, 1971).

or less innocuous by dilution and decomposition (both abiotic and biotic). On the other hand, local areas subjected to especially high levels of continuing discharge or whose waters are partially isolated from the rest of the ocean (e.g., the New York City bight, the Baltic Sea) have not fared so well.[10] Moreover, the deep-sea habitat is particularly slow to recover from contamination[11] or other disturbance.[12] Therefore, with worldwide gross national product increasing annually[13] and with it the production of pollutants, there is little room left for complacency about the future health of the ocean.

Military

General.—A number of the military abuses of the ocean are comparable to the civil abuses discussed above and even merge into them, whereas others are more readily separable from civil abuses. An overview of the military abuses is provided in the present section, and a number of special ones will be singled out for discussion in subsequent sections, namely, underwater explosions; contamination with radioactive isotopes, chemical-warfare agents, and oil; coastal (littoral) disruption; and the cutting of canals.

To begin with, some fraction of the pollutants that are continually introduced into the ocean via stream flow and atmospheric fallout has its origin in munition factories and other military facilities and activities of all sorts. Since about 6 percent of the combined gross national products of the world is devoted to military activities,[14] one can make a first rough approximation that this same fraction of the routine input of ocean pollution is of military origin. Regarding the abuse of the ocean attributable specifically to its use by ships, one can see that 3.0 percent of the world's "small" ocean-going ships and 4.7 percent of its "large" ones are naval (Appendix A).

It might be suggested that the military fraction of the overall ocean pollution is somewhat less than the above percentages suggest. This could be the case since environmental standards might, on average, be somewhat more stringently established and more rigidly enforced at military than civil installations. This, in fact, appears to be the case in the United States.[15] A more substantial objection to attributing a fraction of ocean pollution to the global military sector of the economy comparable to its share of the world gross national product is

---

10. M. G. Gross, "Pollution of the Coastal Ocean and the Great Lakes," *U.S. Naval Institute Proceedings* 97, no. 819 (1971): 228–43.

11. H. W. Janasch et al., "Microbial Degradation of Organic Matter in the Deep Sea," *Science* 171 (1971): 672–75.

12. J. F. Grassle, "Slow Recolonisation of Deep-Sea Sediment," *Nature* 265 (1977): 618–19.

13. *World Military Expenditures and Arms Transfers, 1966–1975*, Publication no. 90 (Washington, D.C.: Arms Control and Disarmament Agency, 1976), p. 14.

14. Ibid.

15. U. S. Department of Defense, "Maintaining Defense Efficiency with Minimal Impact on Man and Environment," *Commanders Digest* 19, no. 23 (1976): 3–8.

that military reductions would probably lead to almost equivalent civil increases and thus lead to little net environmental gain.[16]

Although the specific military activities singled out for discussion in the subsequent sections are all detrimental to marine life, there can be some beneficial corollaries as well. The denial of large areas of the ocean to commercial exploitation for extended periods owing to wartime activities permits the buildup of fishery populations. For example, fish catches on the Atlantic continental shelf of Europe were remarkably better—by three times or more, depending upon the species—immediately after World War II than before it.[17] The Newfoundland seal (Phocidae) fishery improved as a result of World War II.[18] However, fishing can be substantially hampered during a postwar period owing to the residuum of military artifacts. The sea mines, sea-mine anchors, and sunken ships remaining from World War II continue to this day to prevent access to large fishing areas in Swedish waters.[19] Similarly, large quantities of barbed wire were dumped off the Pacific coast of Canada after World War II and still create a local hazard to fishnets.[20]

Finally, one might mention a rather specialized military abuse of the ocean, that of employing marine mammals for military purposes.[21] Although very little is known about some aspects of such naval operations, it has been suggested that the U.S. Navy is training dolphins (Dolphinidae) and sea lions (Otariidae) for various missions, among them the suicidal one of delivering warheads.[22]

Underwater explosions.—Underwater explosions associated with military activities can occur: (1) in conjunction with undersea warfare and other hostile actions, (2) during military training exercises, (3) when sea mines are set off by unintentional and other civil actions, and (4) when unwanted explosive munitions are disposed of at sea. Adverse effects on marine life can stem both from the shock (blast) wave generated and from the toxic or radioactive properties of the chemicals released.

The size of the lethal zone of an underwater explosion depends upon

---

16. M. E. Chacko et al., *Economic and Social Consequences of the Arms Race and of Military Expenditures* (New York: United Nations, 1972), p. 48.

17. R. S. Clark, ed., "Effect of the War on the Stocks of Commercial Food Fishes," *Conseil Permanent International pour l'Exploration de la Mer Rapports et Procès-Verbaux des Réunions* 122 (1947): 1–62; R. S. Wimpenny, *Plaice* (London: Edward Arnold, 1953), pp. 75–78.

18. J. S. Colman, "Newfoundland Seal Fishery and the Second World War," *Journal of Animal Ecology* 18 (1949): 40–46.

19. B. Anderberg, Swedish Army, personal communication (February 10, 1977).

20. R. O. Brinkhurst, Canada Institute of Ocean Sciences, personal communication (January 25, 1977).

21. F. G. Wood, *Marine Mammals and Man: The Navy's Porpoises and Sea Lions* (Washington, D.C.: Luce, 1973).

22. B. Wallace, "Conscription at Sea," *Saturday Review of Sciences* 1, no. 2 (1973): 44–45.

numerous factors, among them: (1) the type of explosive, (2) the magnitude of the explosion, (3) the dimensions of the body of water, that is, its areal extent and its depth, (4) the nature of the bottom, (5) the depth at which the charge is set off, and (6) the organisms in question.

Explosives are either conventional or nuclear.[23] If the latter, they can be either atomic or hydrogen bombs (see Appendix B). If the former, they can be one of quite a number of explosive formulations. The two most commonly employed basic ingredients in conventional munitions are 2,4,6-trinitrotoluene (TNT) and hexahydro-1,3,5-trinitro-s-triazine (also referred to as cyclonite or RDX or hexogen). Nitramine (also known as tetryl) is perhaps the most used detonator. One commonly employed bomb filler is a 4:1 mixture of TNT and powdered aluminum, called tritonal. Another is a 9:6:4:1 mixture of cyclonite, TNT, powdered aluminum, and ammonium picrate (the ammonium salt of 2,4,6-trinitrophenol) (the code name for this formulation is H6). A common cannon-shell filler is a 3:2 mixture of cyclonite and TNT. The explosive filler in some depth charges is an 8:1 mixture of ammonium picrate and powdered aluminum. If the magnitude of the shock (blast) wave of TNT is taken to be 1, then that of cyclonite is 1.2 and that of ammonium picrate plus powdered aluminum 1.5. The various formulations used in conventional munitions probably all fall within the range of 1 and 1.5.

The overpressure of the shock wave of an underwater explosion travels outward in all directions and can therefore be expected (under ideal conditions) to diminish as an inverse function of the cube of the distance traveled. Thus, if the overpressure at a radius of 1 m is taken to be 100 percent, it will diminish by 90 percent at a radius of 2.15 m, by 99 percent at 4.64 m, and by 99.9 percent at 10 m. The overpressure at a radius of 100 m will be one-millionth of that at 1 m. It also follows from this relationship that in order to double the blast overpressure at any given distance from an explosion, one must increase the size of the charge by a factor of 8 (i.e., by a factor of 2^3); to triple it, by a factor of 27 (i.e., 3^3); and so forth. It should also be noted here that shock waves which reach the seabed reflect back to a greater or lesser extent, depending on the nature of the floor. Shock waves that reach the surface also reflect back or, if of sufficient magnitude, break through to the atmosphere. Finally, the overpressure wave is followed by a rarefaction or negative pressure wave which can also contribute to any damage done.

The amount of overpressure required to kill a marine organism—and thus also the size of lethal zone from any given charge—varies considerably from one type of organism to another. Broadly speaking, the marine organisms of interest here can be divided into the true fish, the crustaceans, the mollusks, and the warm-blooded vertebrates (mammals and birds). (In the marine

23. J. Wilcke, "Unterwasser-Explosion: Vorgänge, Auswirkung, Probleme," *Soldat und Technik* 14 (1971): 74–77; S. Glasstone, ed., *Effects of Nuclear Weapons*, rev. ed. (Washington, D.C.: Atomic Energy Commission, 1964).

fishery, the true fish are referred to as the "finfish," and the crustaceans and mollusks together as the "shellfish.")

By far most of the true marine fishes fall into one of two groups, the cartilaginous fish (e.g., sharks, rays, skates) and the ray-finned bony fish. The latter group accounts for perhaps 95 percent of all true fish and for all of the commercially important ones. Among other things, these ray-finned bony fish are characterized by a large thin-walled sac in the body cavity known as the swim or air bladder. This organ regulates the animals' state of buoyancy and has other functions as well.

It turns out that the air bladder is readily ruptured by an underwater explosion. For example, Aplin[24] studied representatives of nine genera of Pacific Ocean fish off the coast of California, five with air bladders and four without. Under similar blast conditions (ca. 16 m from a ca. 4 kg charge of dynamite), all four of the genera without air bladders survived unhurt, whereas four of the five with air bladders were killed. On average, according to data of the U.S. Department of Defense, marine animals possessing air bladders are 64 times as vulnerable to blast damage as those without.[25]

Of the crustaceans that have been studied, shrimp and lobsters appear to be considerably more resistant to blast damage than air-bladder fish, but crabs only somewhat more so. Gowanloch and McDougall[26] observed in the Gulf of Mexico that blast overpressures which killed air-bladder fish left shrimp uninjured (their data suggested that the shrimp were at least 43 times as resistant as the air-bladder fish). Aplin, in conjunction with the work noted above, made equivalent and confirmatory observations with lobsters. Trials by the Chesapeake Biological Laboratory[27] in Atlantic coastal waters suggest (on the basis of distance to attain 50 percent mortality from a 14 kg charge of TNT) that crabs are about 11 times as resistant as air-bladder fish.

Of the mollusks for which there is information, Aplin observed that a benthic (bottom-dwelling) gastropod, the abalone (*Haliotus,* Haliotidae), appeared to be almost as sensitive as the air-bladder fish he studied. Studies by the Chesapeake Biological Laboratory suggest that oysters are perhaps 12 times as resistant as air-bladder fish. Those by Gowanloch and McDougall suggest an even greater relative resistance.

Observations on marine mammals and birds are exceedingly limited. Fitch

24. J. A. Aplin, "Effect of Explosives on Marine Life," *California Fish and Game* 33 (1947): 23–30.
25. *Ocean Dumping: A National Policy* (Washington, D.C.: Council on Environmental Quality, 1970), p. 15.
26. J. N. Gowanloch and J. E. McDougall, "Biological Effects on Fish, Shrimp, and Oysters of the Underwater Explosion of Heavy Charges of Dynamite," *Transactions of the North American Wildlife Conference* 11 (1946): 213–19.
27. Chesapeake Biological Laboratory, *Effects of Underwater Explosions on Oysters, Crabs and Fish,* Chesapeake Biological Laboratory Publication no. 70 (Solomons Island: Maryland Board of Natural Resources, 1948).

and Young[28] noted off the coast of California that sea lions (*Zalophus*, Otariidae) were killed by blast overpressures that seemed not to damage gray whales (*Rhachianectes*, Eschrichtidae). These same authors noted that cormorants (*Phalacrocorax*, Phalacrocoracidae) appeared to be especially sensitive since they were invariably killed when diving in the general area of an underwater explosion. Pelicans (*Pelecanus*, Pelecanidae) would also die if they dipped their heads beneath the surface at the time of such an explosion.

As noted earlier, the actual dimensions of the lethal zone of an underwater explosion are determined by the size of the charge in conjunction with several other factors. The explosion of a typical depth charge could be expected to be lethal to most marine animals within a radius of 77 m, within an area of 1.9 ha and a volume of 1.9×10^6 m^3. For fish possessing air bladders these values would have to be multiplied by 4, 16, and 64, respectively (table 2).

In addition to the depth charge just described, underwater military explosions are likely to be associated with the use of torpedoes, sea mines, and bombs (both conventional and nuclear), as well as with the disposal of unwanted munitions (table 2). Indeed, the disposal of obsolete or otherwise undesirable munitions may result in huge underwater explosions. For example, during 1965–70 the United States on at least 11 occasions loaded unwanted explosive munitions aboard ships that were then scuttled beyond the continental shelf (in both the Atlantic and Pacific Oceans) and their cargo blown up at depths of between 300 m and 1,200 m.[29] The actual amounts of explosive involved reported for eight of these events ranged from 370×10^3 kg to 1.95×10^6 kg, the average being 920×10^3 kg. A charge of 920×10^3 kg has a radius lethal to most marine animals of about 1,600 m and thus a lethal area of 820 ha and a lethal volume of 18×10^9 m^3—and, as above, for data pertaining to air-bladder fish these figures must be multiplied by 4, 16, and 64, respectively. Only the nuclear explosives provide more impressive figures than these (table 2).

The frequency of underwater explosions in space and time depends on the military situation. It does appear clear, however, that the combined lethal zone of a number of underwater explosions is close to a simple function of the number of nonoverlapping explosion zones. This is the case because sessile forms cannot vacate a danger area and because neither fish[30] nor whales[31] are frightened out of an area by explosions. In fact, certain carnivorous fish are known to be attracted to such an area by the casualties present.[32]

---

28. J. E. Fitch and P. H. Young, "Use and Effect of Explosives in California Coastal Waters," *California Fish and Game* 34 (1948): 53–70.
29. *Ocean Dumping*, p. 7.
30. Aplin; C. M. Coker and E. H. Hollis, "Fish Mortality Caused by a Series of Heavy Explosions in Chesapeake Bay," *Journal of Wildlife Management* 14 (1950): 435–44; Fitch and Young.
31. Fitch and Young.
32. Ibid.

TABLE 2.—LETHAL BLAST ZONES RESULTING FROM UNDERWATER EXPLOSIONS

Munition and Explosive Content	Lethal Radius (m)	Lethal Area (ha)	Lethal Volume (m^3)
Depth charge or rocket (100 kg):			
Most fauna	77.1	1.87	1.92×10^6
Most fishes	309.	29.9	$123. \times 10^6$
Torpedo or moored sea mine (250 kg):			
Most fauna	105.	3.44	4.81×10^6
Most fishes	419.	55.1	$308. \times 10^6$
Seabed sea mine (750 kg):			
Most fauna	151.	7.16	14.4×10^6
Most fishes	604.	115.	$923. \times 10^6$
Munition disposal (10^6 kg):			
Most fauna	1,660.	868.	19.2×10^9
Most fishes	6,650.	13,900.	1.23×10^{12}
20 KT atomic bomb ("9.07×10^6" kg):			
Most fauna	3,470.	3,780.	$175. \times 10^9$
Most fishes	13,900.	60,400.	11.2×10^{12}
1 MT hydrogen bomb ("494×10^6" kg):			
Most fauna	13,100.	54,300.	9.51×10^{12}
Most fishes	52,600.	868,000.	$609. \times 10^{12}$
10 MT hydrogen bomb ("4.94×10^9" kg):			
Most fauna	28,300.	252,000.	95.1×10^{12}
Most fishes	113,000.	4.03×10^6	6.09×10^{15}

Sources.—(a) Munitions and explosive contents: The average or likely quantity of explosive to be found in depth charges, rockets, torpedoes, moored (floating) sea mines, and seabed sea mines is from J. Wilcke, "Unterwasser-Explosion: Vorgänge, Auswirkung, Probleme," in *Soldat und Technik* 14 (1971): 74–77; see also SIPRI, *Tactical and Strategic Antisubmarine Warfare* (Stockholm: Almqvist & Wiksell, 1974), pp. 85–95. Depth charges up to 10 times as large have been used. Torpedoes usually fall within the range of 200 kg and 300 kg (*Jane's Weapons Systems, 1976*, pp. 188–96). Moored (floating) sea mines can go as high as 350 kg, according to J. Marriott, "Mine Warfare," in *NATO's Fifteen Nations* 19, no. 5 (1974–75): 44–50. Seabed sea mines can vary between 500 kg and 1,000 kg. A likely quantity of munition disposal is estimated from the examples given in *Ocean Dumping: A National Policy* (Washington, D.C.: Council on Environmental Quality, 1970), pp. 6–7. The equivalent quantity for the 20 KT atomic bomb is 50 percent of its yield and for the 1 MT and 10 MT hydrogen bomb it is 54.5 percent of their yields, as estimates of the proportion of total energy expended in the blast wave (Appendix B). The modest differences in energy yield between 2,4,6-trinitrotoluene (TNT) and other military explosives are not taken into consideration here. (b) Radii and lethal overpressures: Radii are derived via cuberoot scaling from U.S. Department of Defense values for a charge weighing 907×10^3 kg, reported to be lethal to most marine organisms for a radius of 1,610 m, and to fish possessing air bladders, i.e., to about 95 percent of all fish, for a radius of 6,440 m, according to the Council on Environmental Quality (*Ocean Dumping*, p. 15). Based on graphed data of recorded underwater overpressures from TNT by the Chesapeake Biological Laboratory (in *Effects of Underwater Explosion on Oysters, Crabs and Fish*, Chesapeake Biological Laboratory Publication no. 70 [Solomons Island: Maryland Board of Natural Resources, 1948], p. 15), the lethal overpressure necessary to kill most marine organisms is about 2,000 kPa and that for air-bladder fish about 750 kPa. (c) Areas and volumes: The area and volume figures presented are arithmetic calculations of a circle and sphere, respectively. Water dimensions of sufficient magnitude are assumed. A seabed explosion, for example, would reduce a given lethal volume by roughly half (actually somewhat less than half because of reflection).

The actual number of fish and other marine animals killed by an underwater explosion depends, of course, entirely upon the number present within the lethal zone. This density, in turn, depends upon location, depth, season, and so forth. However, as already noted, biological damage may not significantly depend upon the immediate past local explosion history. The data of Coker and Hollis[33] provide an example of the sorts of variations that can occur. These authors observed a series of 21 explosions at intervals between May and August of 1 yr, set off at a particular site in Chesapeake Bay on the Maryland coast. These were charges of HBX (a mixture of TNT, cyclonite, and powdered aluminum) weighing an average of 324 kg and detonated at an average depth of 22 m in water about 48 m deep. The numbers of fish killed and surfacing per charge fluctuated markedly, ranging from a low of zero to a high of 8,035 (average, 1,555); or, in terms of fresh weight, from zero to 2,497 kg (average, 407 kg). It should be noted here that a small number of dead (or soon dead) fish do not float to the surface, including especially those not possessing air bladders. On the basis of three underwater postexplosion surveys, Fitch and Young determined that 7.8 percent of the total kill sinks to the bottom.[34]

The toxic properties of the chemical constituents of underwater munitions, as well as the damaging radioactivity also associated with nuclear munitions (see next section), provide further possible sources of stress to marine biota. These substances include not only the explosive fillers as modified by the detonation, but also the substances themselves following partial or nondetonation, the various breakdown products (via alkaline hydrolysis, etc.), and the substances making up the munition casings and mechanisms. The danger to marine life from the introduced chemicals hinges upon a number of factors, among them: (1) the intrinsic toxicity of the substance as well as its ability to gain entry into the organism, (2) its solubility and density, (3) the rapidity of its breakdown to innocuous substances, and (4) the amount of water movement and thus the rapidity of its dilution.

One might use TNT as the first example of a chemical contaminant of munition origin.[35] This material is sparingly soluble in seawater (94 g/m^3) and is heavier as well (1,654 kg/m^3 as opposed to 1,025 kg/m^3). It thus sinks largely to the bottom, where it dissolves slowly. Moreover, it is highly stable in seawater, with its pH of 8.2 (there being no measurable decomposition through at least 108 days), is known to kill or inhibit the growth of a number of freshwater microorganisms, and is lethal to several species of freshwater fish at concentrations of less than 5 g/m^3. Moreover, it is acutely toxic to various mammals at less than 200 mg/kg of body weight.

33. Coker and Hollis.
34. Fitch and Young, p. 66.
35. J. C. Dacre and D. H. Rosenblatt, *Mammalian Toxicology and Toxicity to Aquatic Organisms of Four Important Types of Waterborne Munitions Pollutants: An Extensive Literature Evaluation*, Technical Report no. 7403 (Aberdeen Proving Ground, Md.: U.S. Army Medical Research and Development Command, 1974), pp. 94–166; J. C. Hoffsommer and J. M. Rosen, "Hydrolysis of Explosives in Seawater," *Bulletin of Environmental Contamination and Toxicology* 10 (1973): 78–79; W. D. Won, L. H. DiSalvo, and J. Ng, "Toxicity and Mutagenicity of 2,4,6-trinitrotoluene and Its Microbial Metabolites," *Applied and Environmental Microbiology* 31 (1976): 576–580; *Merck Index, 1976*, Compound no. 9397.

Cyclonite, another of the important military explosives, is a dangerous mammalian nerve poison, sometimes used commercially as a rat killer.[36] Its solubility in seawater is 56 g/m^3 and its weight is 1,820 kg/m^3. The half-life in seawater of cyclonite can be calculated to be about 630 days.

Nitramine, a favored detonator, provides one additional example.[37] Its solubility in seawater is 26 g/m^3, and its weight is 1,570 kg/m^3. The half-life in seawater of nitramine can be calculated to be about 33 days. The major decay product is picric acid (2,4,6-trinitrophenol), which is highly soluble (13 kg/m^3). Both nitramine and picric acid are considered to be highly toxic. The lethal concentration to the Pacific coral-reef damsel fish (*Dascyllus*, Pomacentridae) (i.e., the 96 hr LC$_{50}$ of ammonium picrate) is about 95 g/m^3.[38]

The sinking or scuttling of munition-laden ships can provide a large source of explosive contamination. During 1964–67, the United States scuttled at least four ships loaded with unwanted explosive munitions (in addition to the 11 mentioned earlier), which were not blown up.[39] To provide a different example, a Japanese freighter carrying a cargo of about 45×10^3 kg of depth charges sank in about 34 m of water in the Truk lagoon, Eastern Carolines, during World War II.[40] The casings began to deteriorate about 3 decades later, and the leaking explosive (ammonium picrate plus powdered aluminum) threatened fish in the lagoon. It was decided to blow up the cargo to get rid of the toxic chemical, but the detonation was incomplete and some subsequent local fish mortality was attributed directly to the toxic releases.

Contamination with radioactive isotopes.—Radioactive contamination of the ocean associated with military activities occurs as a result of: (1) nuclear weapon manufacture, (2) nuclear weapon testing, (3) the hostile use of nuclear weapons, (4) routine emissions from nuclear-powered naval vessels, (5) the accidental or intentional destruction of nuclear-powered ships and military satellites, and (6) accidental introductions of all sorts. Indeed, almost all of the present global burden of radioactive pollution is of military origin.

Atomic bombs produce about 56.7 g/KT of mixed fission products and hydrogen bombs about 28.4 kg/MT (Appendix B). In the only two instances to date of the hostile use of nuclear weapons, the United States during World War II detonated atomic bombs over Hiroshima and Nagasaki in August 1945. These two bombs—14.6 KT and 23.0 KT in size, respectively—produced a total of about 2.13 kg of miscellaneous fission products (disintegrating, after 100 days at the rate of about 1.32 MCi), some small fraction of which found its

---

36. Hoffsommer and Rosen; *Merck Index, 1976,* Compound no. 2741.
37. Hoffsommer and Rosen; *Merck Index, 1976,* Compound no. 6389.
38. S. C. Jameson, "Toxic Effect of the Explosive Depth Charge Chemicals from the Ship SANKISAN MARU on the Coral Reef Fish *Dascyllus aruanus* (L)," *Micronesia* 11, no. 1 (1975): 109–13.
39. *Ocean Dumping,* p. 7.
40. Jameson.

way into the ocean. The World War II experience is barely suggestive of what a nuclear war of the future might entail.[41]

Of the various sources of radioactive pollution so far, nuclear weapon testing has been by far the worst offender. During the 3 decades since the first experimental device was set off at Alamogordo, New Mexico, in July 1945, the United States, the USSR, and several other nations have detonated almost 1,000 additional "small" (<1 MT) nuclear bombs as well as over 100 immense (>1 MT) hydrogen bombs.[42] Of the overall total, six were exploded beneath the surface of the ocean, 35 on its surface, and some 373 in the atmosphere (table 3).

The six underwater detonations to date have had a combined yield of about 2.1 MT and thus produced about 56.5 kg of mixed fission products (table 3), virtually all of which contaminated the ocean. The 35 water-surface detonations to date have had a combined yield of 63.6 MT and thus produced about 1,710 kg of mixed fission products. Fission products may be classified as fast-decaying or short-lived products (whose half-lives are measurable in seconds or days) or as slow-decaying, long-lived ones (whose half-lives are measurable in years). Perhaps 50 percent (855 kg) of the short-lived component of these fission products entered the ocean from the water-surface detonations. Of that minute but important fraction of long-lived isotopes, an estimated 90 percent eventually entered the ocean (i.e., from about 1,540 kg of original fission products).

The 373 other atmospheric detonations—both air bursts and land-surface bursts—had a combined yield estimated at about 230 MT and thus produced about 6,190 kg of mixed fission products. One can assume that virtually none of the short-lived component entered the ocean. Of the long-lived component approximately 80 percent originating from the air bursts and 40 percent originating from the land-surface bursts may have reached the ocean. Of the 6,190 kg total, about 3,900 kg (63 percent) derived from air bursts and the remaining 2,290 kg (37 percent) from land-surface bursts. From these data it can thus be calculated that the long-lived isotopes originating from 4,030 kg (65 percent) of the originally produced mixed fission products eventually entered the ocean.

To summarize, of the estimated 414 nuclear detonations that occurred either in or on the ocean or in the atmosphere during 1945–76, about 912 kg of short-lived fission products entered the ocean (disintegrating, after 100 days, at the rate of about 565 MCi). The long-lived fission-product component from about 5,630 kg of initially produced fission products must be added to this total.

---

41. M. A. Vellodi et al., *Effects of the Possible Use of Nuclear Weapons and the Security and Economic Implications for States of the Acquisition and Further Development of These Weapons* (New York: United Nations, 1968).

42. B. -M. Tygård, *Hagfors Seismic Array Station* (Stockholm: Swedish National Defence Research Institute, 1977), pp. 5–6.

450   Military Activities

TABLE 3.—OCEANIC AND ATMOSPHERIC NUCLEAR EXPLOSIONS

Year	Underwater			Water Surface			Other Atmospheric						Overall Total
	USA	USSR	Total	USA	Gt Br	Total	USA	USSR	France	Gt Br	China	Total	
1945	...	...	...	...	...	...	3	...	...	...	...	3	3
1946	1	...	1	...	...	...	1	...	...	...	...	1	2
1947	...	...	...	...	...	...	...	...	...	...	...	...	...
1948	...	...	...	...	...	...	3	...	...	...	...	3	3
1949	...	...	...	...	...	...	...	1	...	...	...	1	1
1950	...	...	...	...	...	...	...	...	...	...	...	...	...
1951	...	...	...	...	...	...	15	2	...	...	...	17	17
1952	...	...	...	...	1	1	10	...	...	...	...	10	11
1953	...	...	...	...	...	...	11	2	...	2	...	15	15
1954	...	...	...	4	...	4	2	1	...	...	...	3	7
1955	1	...	1	...	...	...	13	4	...	...	...	17	18
1956	...	...	...	6	...	6	8	7	...	6	...	21	27
1957	...	...	...	...	...	...	27	13	...	7	...	47	47
1958	2	...	2	24	...	24	33	25	...	5	...	63	89
1959	...	...	...	...	...	...	...	...	...	...	...	...	...
1960	...	...	...	...	...	...	...	...	3	...	...	3	3
1961	1	...	1	...	...	...	...	30	1	...	...	31	32
1962	1	...	1	...	...	...	40	39	...	...	...	79	80
1963	...	...	...	...	...	...	...	...	...	...	...	...	...
1964	...	...	...	...	...	...	...	...	...	...	1	1	1
1965	...	...	...	...	...	...	...	...	...	...	...	...	...
1966	...	...	...	...	...	...	...	...	5	...	3	8	8
1967	...	...	...	...	...	...	...	...	3	...	2	5	5
1968	...	...	...	...	...	...	...	...	5	...	1	6	6
1969	...	...	...	...	...	...	...	...	...	...	1	1	1
1970	...	...	...	...	...	...	...	...	8	...	1	9	9
1971	...	...	...	...	...	...	...	...	5	...	1	6	6
1972	...	...	...	...	...	...	...	...	3	...	2	5	5
1973	...	...	...	...	...	...	...	...	5	...	1	6	6

1974	...	...	...	...	...	...	...	...	1	...	...	...	...	8
1975	...	...	...	...	...	...	...	...	...	...	...	3	...	3
1976	...	...	...	...	...	...	...	...	...	...	...	3	...	3
Total	5	1	6	34	...	...	1	35	166	124	45	20	18	373	414
Yield (MT)	...	...	2.1	...	...	...	...	63.6	...	...	...	...	...	230	296

SOURCES.—(a) The data for 1945–62 are compiled from those of S. Glasstone, ed., *Effects of Nuclear Weapons*, rev. ed. (Washington, D.C.: Atomic Energy Commission, 1964), pp. 671–81b. It is possible that up to several of these reported detonations are misplaced as to year. Moreover, a number of the detonations during this period have gone unreported (perhaps in the neighborhood of 30). No more reliable data appear to be available for these years. The data for 1963–76 are from *SIPRI Yearbook, 1977*, p. 403, and are considered reliable; they were derived in part from those of J. Zander and R. Araskog, *Nuclear Explosions 1945–1972: Basic Data*, Report no. A 4505-A1 (Stockholm: Swedish National Defence Research Institute, 1973). (b) Underwater detonations: The five U.S. underwater detonations were all in the Pacific Ocean, three in the western Pacific (one near Bikini and two near Eniwetok) and two in the eastern Pacific. The depth of four of these ranged from 27 m to 610 m (average, 209 m). Three had yields of 30 KT or less each. At least three of them vented. The Soviet detonation occurred in the Barents Sea (near Novaya Zemlya) and had a yield of less than 20 KT. An estimate for the total yield of the six underwater detonations is 2.1 MT. (c) Water-surface detonations: The 34 U.S. water-surface detonations were all in the Western Pacific (19 near Bikini and 15 near Eniwetok). The British detonation was in the Indian Ocean (near northwestern Australia). The 35 water-surface detonations had a total yield of 63.6 MT, according to Glasstone, pp. 483–40. (d) Other atmospheric detonations: The 373 detonations in this category include those that occurred in the air (63 percent by yield), on the land surface (37 percent by yield), and underground but sufficiently close to the surface to have vented (0.02 percent by yield). For locations of most (and yield estimates of some) of those during 1945–62, see Glasstone, pp. 671–81b. For locations of most (and some yield estimates) of the detonations during 1972–74, see *SIPRI Yearbook, 1975*, pp. 506–11; for 1975–76, see *SIPRI Yearbook, 1977*, pp. 400–402. The 201 atmospheric tests, the 40 remaining U.S. detonations had a total yield of 27.4 MT. Based on average yield of all past Soviet atmospheric tests, the 69 remaining Soviet detonations had a total yield of 62.5 MT. Based on an extrapolation to the present from the data of SIPRI (*French Nuclear Tests in the Atmosphere: The Question of Legality* [Stockholm: SIPRI, 1974], pp. 33–34), total French yield comes to 10.6 MT. Based on an extrapolation to the present from the same data (p. 37), total Chinese yield comes to 18.1 MT. The sum is thus 228.8 MT, which rounds to 230 MT. (The range was from substantially less than 1 KT to 58 MT.) (e) All of the detonations listed in the table were for military purposes, two for hostile purposes and the remainder for weapon testing and related activities. (Some half-dozen of the many underground tests are claimed to have been for civil purposes.) (f) The overall total number of nuclear detonations during 1945–76 has been estimated by the Swedish National Defense Research Institute (FOA) to be 1,081, dividing among the nations responsible as follows: United States (614), USSR (354), France (64), Great Britain (27), China (21), and India (1), according to B.-M. Tygård, *Hagfors Seismic Array Station* (Stockholm: Swedish National Defense Research Institute, 1977), p. 5. Most of the 667 detonations not accounted for in the table were underground. (g) Of the estimated 1,081 nuclear detonations that have occurred to date, about 977 have been atomic (fission) bombs and thus only about 104 (or 10 percent) have been hydrogen (fission/fusion) bombs. On the other hand, the hydrogen bombs represent about 95 percent of the combined yield of the 1,081 detonations (this actually being the percentage that can be calculated from the atmospheric detonations that occurred during 1945–58, 1964–73 on the basis of the data of Glasstone (pp. 483–84) and SIPRI, *French Nuclear Tests in the Atmosphere*, pp. 33–37. A fission-product estimate for this mix of bombs is thus 26.9 g/KT (i.e., 47.5 percent of 56.7) (see Appendix B).

Nuclear-powered ships also contribute to the radioactive contamination of the ocean. There are in operation today about 259 large nuclear submarines with such propulsion plus an additional seven naval surface vessels (table 4), as well as five nuclear-powered merchant ships. The routine releases from a nuclear-powered ship have been estimated to average 250 μCi/yr,[43] which, assuming 100 days of "cooling," represents 403 ng/yr of mixed fission products. Thus, the 266 naval ships might contribute about 107 μg of mixed fission products (disintegrating, after 100 days, at the rate of about 65 mCi) to the ocean each year.

Of far greater significance is the potential for massive radioactive releases that might be associated with an accidental sinking (a number of such cases are referred to below). And even more serious would be the intentional destruction during wartime of enemy nuclear-powered ships, both military and civil. In that case, it would be less likely that an intact reactor could be retrieved and more likely that it was ruptured in the first place.

A nuclear-powered submarine might have a 35 MW(th) reactor, and a nuclear-powered cruiser one of 140 MW(th) (Appendix C). Their reactors would thus be producing about 22.4 g and 89.7 g, respectively, of mixed fission products during each day of operation. The short-lived component of this conglomerate decays almost as fast as it is being produced (the equilibrium value, soon reached, being about 108 percent of daily production). The long-lived component, on the other hand, builds up in almost direct proportion to the number of days of operation. Under peacetime conditions a nuclear submarine might be on patrol for 60 days out of every 180 days.[44] Under wartime

TABLE 4.—NUCLEAR-POWERED SHIPS OF THE WORLD

Nation	Military (Naval)		Civil (Merchant)
	Submarine	Surface	Surface
USSR	135	...	2
United States	106	7	1
Great Britain	13	...	...
France	4	...	...
China	1?	...	...
West Germany	...	...	1
Japan	...	...	1
Total	259	7	5

Sources.—Numbers of military ships are from *Jane's Fighting Ships, 1976–77.* Numbers of civil ships are from the *World Almanac and Book of Facts, 1977,* p. 147.

43. C. L. Comar et al., *Effects on Populations of Exposure to Low Levels of Ionizing Radiation* (Washington, D.C.: National Academy of Sciences, 1972), p. 18.

44. *SIPRI Yearbook, 1975,* p. 64.

conditions it might well be in continuous operation for many months at a stretch.

Thus, if a nuclear-powered submarine were destroyed, it would at essentially any time release about 23.8 g of mixed fission products (disintegrating, after 100 days, at the rate of 14.7 kCi). These fission products, comprised primarily of short-lived isotopes, are equivalent in amount to the release of a 0.42 KT atomic bomb. The long-lived component, on the other hand, would be derived from an additional 673 g of original fission products for every 30 days at sea. As a result, after 51 days of operation they would build up to the equivalent of what a 20 KT atomic bomb would release.

The manufacture of nuclear weapons and of the fuel for naval vessels results in some modest amount of routine radioactive air and stream pollution that eventually finds its way into the ocean. A somewhat more substantial amount of such contamination has, at least in the past, resulted from ocean dumping of radioactive wastes arising from these operations. The United States, for example, during 1946–60 dumped into the ocean relatively large quantities of both liquid and solid radioactive wastes, primarily of military origin. These amounted to a reported total of 93.7 kCi at the time of disposal, divided between the Atlantic and Pacific Oceans,[45] which, assuming 100 days of "cooling," is the equivalent of 151 g of original fission products. The United States has virtually ended such dumping, having added only another 1.0 kCi (1.6 g) during 1961–70. Great Britain is said to have similarly disposed of some 40 kCi (65 g) into the Atlantic Ocean during 1951–66.[46]

Accidents involving nuclear weapons and other military items containing radioactive isotopes can also lead to ocean contamination. The hydrogen bomb that dropped into the Mediterranean Sea off Palomares on the southeast coast of Spain in January 1966 was a well-publicized event. That bomb was eventually retrieved, but there have been a number of similar losses with less benign outcomes.[47] For example, an estimated 25 Ci (390 g) of ^{239}Pu was introduced into the Atlantic Ocean 11 km off the coast of Thule, Greenland, in January 1968 as the result of an accident involving four nuclear bombs.[48] Marine organisms over a considerable area became contaminated with this material.

At least two military navigation satellites equipped with nuclear generators are known to have accidentally contaminated the ocean with their radioactive contents.[49] In one instance, in April 1970, a satellite reentered the atmosphere and simply fell into and was lost in the Pacific Ocean. In the other instance, in

45. *Ocean Dumping*, pp. 6–7.
46. N. Turner, "Nuclear Waste Drop in the Ocean," *New Scientist* 68 (1975): 290.
47. *SIPRI Yearbook, 1977*, pp. 52–85.
48. A. Aarkrog, "Radioecological Investigations of Plutonium in an Arctic Marine Environment," *Health Physics* 20 (1971): 31–47; UNSCEAR, *Ionizing Radiation: Levels and Effects* (New York: United Nations, 1972), p. 54.
49. E. P. Hardy, P. W. Krey, and H. L. Volchok, "Global Inventory and Distribution of Fallout Plutonium," *Nature* 241 (1973): 444–45; UNSCEAR, p. 54.

April 1964, the satellite burned up at an altitude of about 50 km and thereby injected almost 1 kg of ^{238}Pu (ca. 17 kCi) into the stratosphere, roughly three-quarters of which has by now entered the ocean.

Regarding nuclear-powered ships, the United States has now lost at least two such submarines, the *Thresher* in the western Atlantic in April 1963 and the *Scorpion*, also in the Atlantic, off the Azores in May 1968. The USSR may have lost as many as four during 1968–71, two in the Atlantic Ocean and one each in the Mediterranean Sea and Pacific Ocean.[50]

Aside from the cataclysmic nearby effects of a nuclear explosion, the ecological impact of radioactive contamination varies according to a blast's magnitude and areal extent.[51] Based upon observations on the effects of the bomb trials at the Bikini and Eniwetok test sites in the western Pacific, it appears that the general structure and dynamics of marine communities recover in due course.[52] Moreover, regarding the ocean as a whole, one would expect to observe little if any overall impact from the level of radioactive contaminants so far introduced. As suggested earlier, all of the nuclear testing to date has introduced into the ocean an estimated total of 912 kg of fission products. One tentative way of evaluating the significance of this introduction is to compare it with the natural level of radioactivity present in the ocean. Thus, as a crude approximation one can compare the activity from these fission products (which represents most of the activity of military origin) with that originating from ^{40}K (which represents most of the activity of natural origin). As ^{40}K has a half-life of 1.28×10^9 yr, it disintegrates at the rate of 7.27 mCi/kg (via β^- emission). The K in nature includes 118 mg/kg of ^{40}K, and K is present in the ocean at 390 g/m^3. Thus, since the ocean contains 1.37×10^{18} m^3 of water, the total continuing activity emanating from the ^{40}K can be estimated to come to 458 GCi. The continuing activity from the natural ^{40}K is thus about 800 times that of the 100 days activity from the 912 kg of fission products (i.e., 565 MCi) and about 8,000 times that of the 1-yr activity (i.e., 55.7 MCi).

On the other hand, an examination of the overall radioactivity of the above sort overlooks the possibility of adverse biological effects arising from the long-lived isotopes arising from such tests. A number of these are of particular concern because they mimic certain of the elements essential to life and are taken up for that reason, and others because they are taken up anyway. Having once gained entry into living organisms, they travel up food chains (some of which culminate in man) and also continue to cycle within the marine ecosystems until they eventually are dispersed (diluted) or decay to insignificance.

50. *SIPRI Yearbook, 1977*, p. 74.
51. Goldberg, pp. 79–95; UNSCEAR; Vellodi et al.; A. O. C. Nier et al., *Long-Term Worldwide Effects of Multiple Nuclear-Weapons Detonations* (Washington, D.C.: National Academy of Sciences, 1975), pp. 102–61; S. H. Small, ed., *Nuclear Detonations and Marine Radioactivity* (Kjeller: Norwegian Defence Research Establishment, 1963).
52. N. O. Hines, *Proving Ground: An Account of the Radiobiological Studies in the Pacific, 1946–1961* (Seattle: University of Washington Press, 1962).

Among those of particular concern are ^{90}SR and ^{137}Cs, because they mimic Ca and K, respectively (table 5). ^{14}C is of concern because it mimics ^{12}C in CO_2, and ^{3}H is also because it mimics ^{1}H in H_2O. The problem with ^{3}H, however, is mitigated somewhat by the natural abundance of the isotope it mimics (table 5). Also of concern are ^{239}Pu and ^{241}Am because of their radioactive and chemical toxicities and because they are taken up by marine organisms, both plant and animal, attaining levels that can be 1,000 to even more than 10,000 times more concentrated than in ambient water. The amounts of these isotopes arising from nuclear bombs, both atomic and hydrogen (table 6), and from destroyed nuclear-powered submarine reactors (table 7) are formidable.

Contamination with chemical-warfare agents.—Chemical-warfare agents have been introduced into the ocean both accidentally and intentionally, although in the latter case not for hostile purposes.

Intentional introductions have been made on a large scale for the purpose of disposing of obsolete or otherwise unwanted chemical munitions. For example, large amounts of unwanted German World War II munitions of all sorts were dumped into the southern Baltic Sea east of the island of Bornholm during 1945–48. Among these was an unknown fraction containing the lethal dermal (blister) agent "mustard gas" (HD; bis [2-chloroethyl] sulphide). In a recent 4½-yr period Danish fishermen netted at least 16 of these mustard-gas bombs, some of which contaminated their catches and went on to result in human illness.[53]

Obsolete British chemical munitions were dumped into the Atlantic Ocean off the coasts of Scotland and Ireland at various times during 1945–56. These contained either mustard gas as above or else the lethal lung (choking) agent

TABLE 5.—CHARACTERISTICS OF SELECTED FISSION PRODUCTS

Isotope	Half-life (yr)	Disintegration Rate (kCi/kg)	Type of Decay	Mimic of	Present at (g/m³)
^{90}Sr	28.1	147.	β	Ca	410
^{137}Cs	30.23	89.8	β, γ	K	390
^{239}Pu	24,400.	0.0638	α	...	...
^{241}Am	458.	3.37	α, γ	...	...
^{3}H	12.26	10,100.	β	^{1}H	114,000
^{14}C	5,730.	4.64	β	^{12}C	ca. 60

Sources.—(a) The half-lives and types of decay are from the *Handbook of Chemistry and Physics, 1974–1975*, p. B248 ff. Disintegration rates were calculated on the basis of individual mass numbers and half-lives. (b) Seawater concentrations of Ca and K are from the *Handbook of Chemistry and Physics, 1974–1975*, p. F190, based on a seawater density of 1,025 kg/m³; the concentration of ^{1}H is based upon the H_2O molecule; the concentration of ^{12}C is composed of ca. 29 g/m³ of inorganic solids in solution plus ca. 2 g/m³ of organic solids in solution plus ca. 27 g/m³ from the CO_2 in solution.

53. F. Garner, "Mustard Oil on Troubled Waters," *Environment* 15, no. 2 (1973): 4–5.

TABLE 6.—QUANTITIES AND DECAY RATES OF SELECTED NUCLEAR-BOMB FISSION PRODUCTS

A. Atomic (Fission) Bombs

Isotope	Quantity or Activity Remaining After					
	1 Hr (g/KT)	1 Hr (Ci/KT)	100 Days (Ci/KT)	1 Yr (Ci/KT)	10 Yr (Ci/KT)	25 Yr (Ci/KT)
^{90}Sr	.680	100.	99.3	97.6	78.1	54.0
^{137}Cs	1.78	160.	159.	156.	127.	90.2
^{239}Pu	31.3	2.00	2.00	2.00	2.00	2.00
^{241}Am	.237	.800	1.10	1.89	2.98	2.93
^{14}C	9.49	44.0	44.0	44.0	43.9	43.9
^{3}H	>0	>0	>0	>0	>0	>0

B. Hydrogen (Fission/Fusion) Bombs

Isotope	Quantity or Activity Remaining After					
	1 Hr (kg/MT)	1 Hr (kCi/MT)	100 Days (kCi/MT)	1 Yr (kCi/MT)	10 Yr (kCi/MT)	25 Yr (kCi/MT)
^{90}Sr	.340	50.0	49.7	48.8	39.1	27.0
^{137}Cs	.890	80.0	79.5	78.2	63.6	45.1
^{239}Pu	15.7	1.00	1.00	1.00	1.00	1.00
^{241}Am	.119	.400	.548	.944	1.49	1.47
^{14}C	4.74	22.0	22.0	22.0	22.0	21.9
^{3}H	.0874	885.	871.	836.	503.	215.

SOURCES AND NOTES.—(a) For information on nuclear bombs, see Appendix B; for basic characteristics of the selected fission products, see table 5; (b) The 1-hr weight values are calculated from the 1-hr disintegration values, on the basis of the disintegration rates provided in table 5. (c) One-hour disintegration values: The value for ^{90}Sr is from Glasstone, p. 484. The values for ^{137}Cs, ^{239}Pu, ^{241}Am, and ^{14}C are factors of the value for ^{90}Sr. See UNSCEAR, *Ionizing Radiation: Levels and Effects* (New York: United Nations, 1972), pp. 52, 54, 57; and A.O.C. Nier et al., *Long-Term Worldwide Effects of Multiple Nuclear-Weapons Detonations* (Washington, D.C.: National Academy of Sciences, 1975), p. 104. The value for ^{3}H is calculated from data of UNSCEAR, p. 57. See also Appendix B. (d) Subsequent disintegration values: These are calculated from the individual half-lives provided in table 5, with the exception of ^{241}Am, the values of which are derived from the data of G. Netzén et al. in *Använt Kärnbränsle och Radioaktivt Avfall. II*, Report no. SOU-1976: 31 (Stockholm: Swedish Ministry of Industry, 1976), pp. 36, 39. The unexpected amounts of ^{241}Am are the result of neutron capture by ^{239}Pu. (e) To convert the given data to values in terms of the amounts of original fission products generated—i.e., to g/kg or Ci/kg, as the case may be—divide the values in table 6, A by 0.0567.

"phosgene" (carbonyl chloride). Some of the mustard-gas bombs washed ashore along the Welsh coast during 1976.[54]

Obsolete U.S. chemical munitions containing the lethal nerve (anticholinesterase) agent "sarin" (GB; isopropyl methylphosphonofluoridate) have been disposed of in the ocean at one or more sites on at least several occasions. In three known cases (in 1967, 1968, and 1970) these were imbedded in concrete within steel vaults and placed aboard ships that were subsequently scuttled several hundred kilometers out to sea at depths of several thousand

---

54. "World War II Poisons," *Marine Pollution Bulletin* 7 (1976): 179.

TABLE 7.—QUANTITIES AND DECAY RATES OF SELECTED
SUBMARINE-REACTOR FISSION PRODUCTS AFTER A MONTH AT SEA

Isotope	Quantity or Activity Remaining After					
	1 Hr (g/sub)	1 Hr (Ci/sub)	100 Days (Ci/sub)	1 Yr (Ci/sub)	10 Yr (Ci/sub)	25 Yr (Ci/sub)
^{90}Sr	10.8	1,590.	1,580.	1,550.	1,240.	857.
^{137}Cs	25.2	2,260.	2,250.	2,210.	1,800.	1,270.
^{239}Pu	113.	7.21	7.21	7.21	7.21	7.20
^{241}Am	.898	3.03	4.15	7.15	11.3	11.1
^{14}C	.00207	.00961	.00961	.00961	.00960	.00958
^{3}H	.00167	16.9	16.6	15.9	9.58	4.10

SOURCES AND NOTES.—(a) For information on submarine reactors, see Appendix C; for basic characteristics of the selected fission products, see table 5. (b) The 1-hr weight values are calculated from the 1-hr disintegration values, on the basis of the disintegration rates provided in table 5. (c) The disintegration values are calculated from the 30-day disintegration values of Netzén et al., pp. 36–37, on the basis of the half-lives provided in table 5, with the exception of ^{241}Am, the values of which are derived from the data of Netzén et al., pp. 36, 39. The unexpected amounts of ^{241}Am are the result of neutron capture by ^{239}Pu. (d) For additional months of operation, the given data can be multiplied by that number; to make the data applicable to a nuclear-powered cruiser, the given data can be multiplied by 4.

meters. For example, the 1970 disposal site, code named "CHASE X," was some 400 km east of Florida at a depth of almost 5,000 m.[55] The ship scuttled in 1970 contained 61 × 10³ kg of sarin (2.5 percent of total cargo weight). Based on the 1970 data one can estimate that the ship scuttled in 1967 contained 204 x 10³ kg of sarin and the one in 1968 176 × 10³ kg (based on total cargo weights from the Council on Environmental Quality.[56] The United States also dumped a small quantity of the lethal nerve agent 'VX' (S-[2-diisopropylaminoethyl] O-ethyl methyl phosphonothiolate) as a part of these operations[57] and perhaps some mustard gas as well.[58]

Turning now to accidental releases of chemical agents, the most spectacular and tragic incident occurred in the Adriatic Sea along the coast of Italy during World War II.[59] One evening in December 1943 the Germans carried out a major bombing attack against Allied Ships in the harbor of Bari. Of the two dozen ships destroyed, one was a U.S. freighter with a cargo of about 100 × 10³ kg of mustard-gas bombs. As a result, much of the mustard gas was released into the water, some of which dissolved into the floating oil. More than 1,000 people were killed by the raid; of these deaths, more than 100 were directly attributable to mustard-gas poisoning and many more to indirectly associated reasons such as disablement followed by drowning.

55. K. M. Ferer, *Fifth Post-Dump Survey of the CHASE X Disposable Site,* Memorandum Report no. 2996 (Washington, D.C.: Naval Research Laboratory, 1975).
56. *Ocean Dumping,* pp. 6–7.
57. Ferer.
58. C. L. Poor, "Disposal of Chemical Agents," *Ordnance* 54, no. 295 (1969): 30, 32.
59. D. N. Saunders, "Bari Incident," *U.S. Naval Institute Proceedings* 93, no. 9 (1967): 35–39; G. B. Infield, *Disaster at Bari* (New York: Macmillan, 1971).

The likelihood of ecological damage resulting from the release into the ocean of chemical-warfare agents depends on the various factors already noted, in conjunction with the release of explosive chemicals. Sarin, for example, is completely miscible with water, a medium in which it breaks down rapidly.[60] Its half-life in seawater is of the order of 1 hr, and its decomposition products (isopropyl alcohol, etc.) are relatively harmless. Sarin is exceedingly toxic to both mammals and insects and thus perhaps to many additional sorts of vertebrates and arthropods. One could thus expect considerable mortality for a brief time in the immediate area of release.

Mustard gas is an oily liquid that solidifies at 14°C.[61] It is very sparingly soluble in water and sinks to the bottom since it is heavier than water (liquid weight, 1,274 kg/m^3; solid weight, 1,338 kg/m^3). Its rate of hydrolysis is "very slow" and its intermediate breakdown products (2,2'-thiodiethanol, etc.) are also somewhat toxic. In mammals at least, mustard gas is a "cellular" poison (with an LD_{50} in laboratory rodents of about 6 mg/kg of body weight). Mustard gas lying on the seabed would provide a continuing local source of poison for some time.

Contamination with oil.—The ocean can be contaminated with oil as the result of military activities in at least two ways: (1) oil tankers can be sunk by hostile actions, and (2) offshore oil facilities can be similarly destroyed.

Of all the "large" merchant ships currently afloat, about 5,350 (or 23 percent) are oil tankers (and of the "small" ones an additional, though relatively unimportant, 1,670) (Appendix A). These 5,350 tankers have cargo capacities that range from about 2×10^3 m^3 to 653×10^3 m^3, the average being 70.8×10^3 m^3. At least 32 of the oil tankers currently in use have capacities in excess of 400 $\times 10^3$ m^3, of which seven surpass 500×10^3 m^3. Together, all of these tankers transport about 1.5×10^9 m^3/yr of oil across the ocean, that is, roughly half of annual world production, which stood at 3.1×10^9 m^3 in 1975.[62]

In time of war, enemy oil tankers could become prime naval targets of opportunity owing to the crucial importance of this commodity to the functioning of most sorts of sustained military effort. During World War II, for example, 674 "large" U.S. merchant ships were sunk by hostile actions; 152 were oil tankers.[63] The combined "gross registered tonnage" of these 152 sunken tankers was 1,235,097. They thus had a total oil-carrying capacity of perhaps

---

60. W. A. Adams, "Nerve Gas—Isopropyl Methylphosphonofluoridate (GB)—Decomposition and Hydrostatic Pressure on the Ocean Floor," *Environmental Science and Technology* 6 (1972): 928; *Military Chemistry and Chemical Agents,* Technical Manual no. 3-215 (Washington, D.C.: Department of the Army, 1967), pp. 15–16; *Merck Index, 1976,* Compound no. 8127.

61. *Military Chemistry and Chemical Agents,* pp. 21–22; *Merck Index, 1976,* Compound no. 6142.

62. *World Almanac and Book of Facts, 1977,* p. 134.

63. M. M. Stephens, *Vulnerability of Total Petroleum Systems* (Washington, D.C.: Department of the Interior, Office of Oil and Gas, 1973), pp. 42–44.

2.73×10^6 m³ (and an average one of 18.0×10^3 m³) (Appendix A). Revelle's data[64] suggest that U.S. tanker losses during World War II represented one-quarter of the total, and half can be assumed to have been loaded when sunk. Therefore an estimated 300 tankers released an estimated 5.5×10^6 m³ of oil into the ocean during World War II, that is, approximately 1.5×10^6 m³/yr for a period of about 3½ yr. The far larger oil tankers in service today (and the even larger ones envisioned for the future) will enormously exacerbate contamination associated with naval warfare in years to come.

Turning now to offshore oil production, several thousand wells drilled into the continental shelves of the world currently produce about 600×10^6 m³/yr, that is, almost 20 percent of annual world production, and more wells are in the offing. Although offshore oil facilities have not as yet been subjected to military attack, their vulnerability in this regard is obvious. Despite their vulnerability, however, their target value would in many instances be overshadowed by that of refineries and terminal areas, which are fed by a multiplicity of offshore and other facilities.

Were an offshore oil production facility in fact to be destroyed in war, it would probably release relatively little oil into the ocean. This is the case owing to the variety of shut-off devices that are now routinely built into such systems in order to protect against calamities of natural (or human) origin, something that cannot be done for an oil tanker.

To put the magnitude of possible military introductions of oil into the ocean into some sort of perspective, it is useful to note that an estimated 700×10^3 m³/yr of oil enters the ocean from natural sources.[65] Moreover, current anthropogenic additions from civil activities have been estimated to be about 4×10^6 m³/yr[66] to 6×10^6 m³/yr[67] (i.e., about 0.16 percent of production). Moreover, about 95 percent of the annual anthropogenic introductions are routine and more or less deliberate, and thus only about 5 percent are accidental (sinkings and so forth). The worst offshore accident to date was the Ekofisk blowout in the North Sea between Scotland and Denmark in April 1977, which released about 24×10^3 m³ of oil into the ocean. The worst tanker accident to date was the breakup of the *Torrey Canyon* off the southwest coast of England in March 1967, which released about 120×10^3 m³.[68]

 64. R. Revelle et al., "Ocean Pollution by Petroleum Hydrocarbons," in *Man's Impact on Terrestrial and Ocean Ecosystems*, ed. W. H. Matthews, F. E. Smith, and E. D. Goldberg (Cambridge, Mass.: M.I.T. Press, 1971), pp. 303–4.
 65. R. D. Wilson et al., "Natural Marine Oil Seepage," *Science* 184 (1974): 857–65.
 66. W. B. Travers and P. R. Luney, "Drilling, Tankers, and Oil Spills on the Atlantic Outer Continental Shelf," *Science* 194 (1976): 791–96.
 67. Goldberg, p. 122.
 68. D. J. Bellamy et al., "Effects of Pollution from the Torrey Canyon on Littoral and Sublittoral Ecosystems," *Nature* 216 (1967): 1170–73; A. J. O'Sullivan and A. J. Richardson, "Torrey Canyon Disaster and Intertidal Marine Life," *Nature* 214 (1967): 448, 541–42. In March 1978 the break-up of the *Amoco Cadiz* off northwest France released 260×10^3 m³.

A huge literature has accumulated on the fate and ecological impact of oil that is introduced into the ocean.[69] In brief, of the oil that is introduced into the ocean the volatile fraction, perhaps 25 percent of it, evaporates away within several days (although this takes rather longer under arctic conditions). Photo- and bacterial degradation and decomposition account for an additional 60 percent or so within several months. The remaining 15 percent forms into small more or less dense clumps of asphaltic substance (1 mm to 100 mm in diameter) with a more extended life span.[70] Some of these so-called tar balls float on the surface, some wash ashore, and some sink to the bottom.

A massive oil spill of the sort that might result from the sinking of an enemy oil tanker could have a dramatic local ecological impact, especially if it occurred along the coast or in some other biotically rich area and even more especially if the area were partially isolated from the main body of the ocean.[71] One of the most immediately obvious effects of such a spill is the tarring of marine avifauna. Some 3,700 dead birds were counted following the Santa Barbara blowout, an estimated 10 percent of the actual number killed and a substantial fraction of the entire local sea-bird population in residence at the time. (One might guess that the zone of contamination posing a threat to birds could be of the order of 10 ha/10^3 m^3.) Bird mortality is attributable primarily to feather fouling.

It appears that the fish present in a spill zone experience no serious increase in mortality. There is, however, some modest increase observable among the benthic (seabed) mollusks and crustaceans. Moreover, commercial finfish and shellfish present in the area become "tainted" and unusable, a condition that persists for perhaps as long as 6 mo to a year. Various of the intertidal lower flora and fauna are killed by a coating of oil, partly via poisoning and partly via oxygen deprivation. Substantial ecological recovery of a locally oil-decimated area takes of the order of several years, perhaps as many as 6.

The foregoing suggests that the ecological impact of oil spills of military origin could have serious local consequences of several years' duration. Regarding the ocean as a whole, however, such incidents need not be a cause for major concern. This is concluded on the basis of the current levels of continuing civil anthropogenic introductions which, though at least an order of magnitude greater than those of natural or potential military origin, have resulted in as yet no demonstrable impact on the overall marine ecology. There is one potential

---

69. D. F. Boesch, C. H. Hershner, and J. H. Milgram, *Oil Spills and the Marine Environment* (Cambridge, Mass.: Ballinger, 1974); Goldberg, pp. 117–36; Ruivo, pp. 15–21.

70. M. H. Horn, J. M. Teal, and R. H. Backus, "Petroleum Lumps on the Surface of the Sea," *Science* 168 (1970): 245–46; C. S. Wong, D. R. Green, and W. J. Cretney, "Quantitative Tar and Plastic Waste Distributions in the Pacific Ocean," *Nature* 247 (1974): 30–32.

71. M. Blumer and J. Sass, "Oil Pollution: Persistence and Degradation of Spilled Fuel Oil," *Science* 176 (1972): 1120–22.

problem, however, that requires special mention. It has been suggested with sufficient authority to provide cause for concern that a major oil spill in the Arctic Ocean could initiate a chain of events leading to extensive ice melting of long duration.[72] Such an occurrence would in turn modify global albedo, perhaps sufficiently to substantially modify the global climate.

Coastal (littoral) disruption.—Near-shore ocean habitats are, as was noted earlier, especially important ones from an ecological standpoint. Such littoral habitats are also the marine ecosystems most often disrupted by human activity. Military activities routinely contribute to such littoral disruption through the development and use of naval bases and other port facilities and in a number of other ways, several of which are alluded to below.

To begin with, military landing operations disrupt the near-shore benthic (seabed) fauna. For example, Tolmer[73] has described the severe local damage that occurred to the littoral (coastal) fauna as a result of the heavy landing traffic along the coast of Normandy on D-Day in June 1944, still in evidence several years later.

The mangrove habitat in South Vietnam suffered especially heavy damage during the Second Indochina War.[74] An estimated 124×10^3 ha of true mangrove (41 percent of that entire subtype) plus another 27×10^3 ha of rear (back) mangrove (13 percent of that subtype) were subjected to military herbicide spraying. The result was virtual annihilation of the vegetation and severe coastal erosion. Sylva and Michel[75] noted significant decreases in the number and variety of planktonic and benthic forms (diatoms, copepods, etc.) as well as in fish eggs; and Davis[76] described reductions in molluskan populations. Moreover, declines in the offshore fishery, involving both finfish and shellfish, have been attributed to the disruption of these breeding and nursery grounds.[77] It is expected that substantial habitat recovery will take a century or more.[78]

And finally it must be noted here that under special conditions underwater

72. W. J. Campbell and S. Martin, "Oil and Ice in the Arctic Ocean: Possible Large-Scale Interactions," *Science* 181 (1973): 56–58; ibid. 186 (1974): 843–46; R. O. Ramseier, "Oil on Ice," *Environment* 16, no. 4 (1974): 6–14.

73. L. Tolmer, "Faune maritime du Calvados et le débarquement du 6 Juin 1944," *Feuille naturalistes,* n.s., 2 (1947): 90.

74. SIPRI, *Ecological Consequences of the Second Indochina War* (Stockholm: Almqvist & Wiksell, 1976), pp. 38–40.

75. D. P. de Sylva and H. B. Michel, *Effects of Herbicides in South Vietnam. B [15], Effects of Mangrove Defoliation on the Estuarine Ecology and Fisheries of South Vietnam* (Washington, D.C.: National Academy of Sciences, 1974).

76. G. M. Davis, *Effects of Herbicides in South Vietnam. B. [6], Mollusks as Indicators of the Effect of Herbicides on Mangroves in South Vietnam* (Washington, D.C.: National Academy of Sciences, 1974).

77. K. D. Brouillard, *Fishery Development Survey: South Vietnam* (Saigon: U.S. Agency for International Development, 1970).

78. H. T. Odum et al., *Effects of Herbicides in South Vietnam. B [13], Models of Herbicide, Mangroves, and the War in Vietnam* (Washington, D.C.: National Academy of Sciences, 1974), p. 289.

nuclear explosions might generate a tsunami that could, in turn, bring about substantial damage to a coastal habitat.[79]

Cutting of canals.—A number of major interoceanic straits have been dug, the two most notable being the Suez and Panama canals. The construction of at least the latter was motivated in large measure by military considerations, namely, to facilitate the transfer of U.S. naval ships between the Atlantic and Pacific Oceans.[80] The continuing U.S. interest in this canal is similarly motivated, an interest that could even lead to its rebuilding or replacement.[81]

The ecological impact of linking more or less isolated bodies of seawater has been a source of some concern.[82] The breaching of the Isthmus of Suez in 1869 reestablished a ready sea-level connection between the Mediterranean and Red Seas that had not existed since the late Miocene epoch, perhaps 12 million yr ago. During the century or so following the opening of the canal some two dozen species of fish are known to have emigrated from the Red Sea to the Mediterranean, although none appears to have done so in the other direction.[83] Moreover, some of the immigrants seem to be supplanting indigenous species. This pattern of events has been attributed largely to a sparse fish fauna in the Mediterranean.

The breaching of the Isthmus of Panama in 1914 reestablished a near-equatorial connection between the Atlantic and Pacific Oceans that had not existed since the mid-Pliocene epoch, perhaps 4 million yr ago. During the half century or so following the opening of this canal only one species of fish is known to have managed to successfully emigrate from the Atlantic to the Pacific, and apparently none has done so in the other direction.[84] The low level of immigration can be attributed to the barrier to such movement provided by the locks and especially to that provided by the lengthy included stretch of inhospitable fresh water (i.e., Lake Gatun). Whether or not the differing habitat conditions between the oceans have additionally contributed to the existing state of affairs will not become evident until such time as a sea-level canal is dug. It is, however, an established tenet of plant and animal geography that unsatisfied habitat requirements are the only certain preventive to range extensions.

It could be noted here that small marine invertebrates are known to be

---

79. W. H. Clark, "Chemical and Thermonuclear Explosives," *Bulletin of the Atomic Scientists* 17 (1961): 356–60.

80. I. Cameron, *Impossible Dream: The Building of the Panama Canal* (New York: Morrow, 1972), p. 106.

81. V. P. McDonald, "Panama Canal for Panamanians: The Implications for the United States," *Military Review* 55, no. 12 (1975): 7–16; J. P. Speller, *Panama Canal: Heart of America's Security* (New York: Speller, 1972).

82. I. Rubinoff, "Central American Sea-Level Canal: Possible Biological Effects," *Science* 161 (1968): 857–61; I. Rubinoff, "Sea-Level Canal Controversy," *Biological Conservation* 3 (1970–71): 33–36; R. W. Topp, "Interoceanic Sea-Level Canal: Effects on the Fish Faunas," *Science* 165 (1969): 1324–27.

83. Topp.

84. R. W. Rubinoff and I. Rubinoff, "Interoceanic Colonization of a Marine Goby through the Panama Canal," *Nature* 217 (1968): 476–78.

transported through the Panama Canal either attached to ships' hulls (the so-called fouling organisms) or in seawater being used as ballast, some of which might survive such passage.[85] On the other hand, such ocean life is constantly being carried from ocean to ocean anyway by ships that continue to ply the high seas.

One can conclude that the construction of sea-level canals which create links between previously isolated seas and oceans could lead to at least some successful invasions into newly available habitats. Such colonization may, in turn, cause perturbations in the established biotic communities, perhaps even leading to species replacement and local extirpations. The possibility of interspecific hybridizations has also been suggested, especially if the isthmus being breached is of relatively recent geological origin and the isolation of some of the species of common origin has not been sufficiently well established.

CONCLUSION

The world ocean is an inextricable and indispensable component of the global system of nature—of the world ecosystem. It is, moreover, an essential and increasingly important source of both living and nonliving natural resources for man, but the ocean also is the immediate or ultimate sump for most human wastes, wastes that continue to grow in both quantity and complexity.

Despite the immensity and ecological resiliency of the world ocean, there is a growing concern that it is being abused to the point where natural self-renewal will not be able to match the anthropogenic insults.

Of the order of 6 percent of all human activity is devoted to military matters, and one concludes as a first approximation that this proportion also represents the fraction of ocean pollution attributable to the military sector. No aspect of ocean abuse is unique to military activities. However, the present radioactive contamination of the ocean is largely of military origin and could, moreover, increase to catastrophic proportions in time of war.

It is becoming increasingly urgent for man to curtail his abuses of the ocean. If such curtailment involved those of military origin, the benefits would not be limited to the fishes of the sea.

APPENDIX A

SHIP NUMBERS AND SIZES

The numbers of merchant ships used in the text refer to June 30, 1976, and are from *Lloyd's Register of Shipping* via the Shipbuilders Council of America (Washington), the Institut für Seeverkehrswirtschaft (Bremen), and the *World Almanac and Book of Facts*,

85. I. Rubinoff, "Sea-Level Canal Controversy."

1977 (pp. 146–47). The numbers of naval ships refer to mid-1976 and are from *Jane's Fighting Ships, 1976-77* (pp. 804–5).

For merchant ships, the designation "small" here refers to ships with a "gross registered tonnage" of between 100 and 1,000, and the designation "large" to those of 1,000 or more. The "gross registered ton" is a measure of volume in which 1 ton represents 2.832 m³ of permanently enclosed space.

For naval ships, the designation "small" here refers to corvettes, the smaller frigates, and patrol submarines; and the designation "large" to battleships, aircraft carriers, cruisers, destroyers, the larger frigates, ballistic-missile submarines, cruise-missile submarines, and fleet submarines.

Naval ships are measured in terms of "standard displacement tonnage," a measure of weight in which 1 ton represents 1,016 kg of total ship weight. There is no ready means of interconversion between "standard displacement tonnage" and "gross registered tonnage." The approximation used here (based on data provided in the British Admiralty *Manual of Navigation, 1964* (1:61–63) is to multiply gross registered tonnage by 2.21 in order to arrive at standard displacement tonnage. The designations "small" and "large" as used in the text are on this basis comparable for merchant and naval ships.

Merchant ships are also measured in terms of their "deadweight tonnage," which is a measure of weight in which 1 ton represents 1,016 kg of cargo-carrying capacity. Using a conversion factor that is based on summations of all the world's "large" merchant ships, with the exception of oil tankers, the approximate deadweight tonnage of a ship (other than an oil tanker) is obtained by multiplying its gross registered tonnage by 1.51. For an oil tanker the equivalent factor comes to 1.86. Moreover, the average weight of oil can be taken to be 855.4 kg/m³ (see, e.g., *U.N. Statistical Yearbook, 1975* [pp. 191–92]).

APPENDIX B

NUCLEAR BOMB CHARACTERISTICS

Nuclear bombs are rated according to their total energy yield in terms of the weight of 2,4,6-trinitrotoluene (TNT) that would produce an equivalent energy yield. The energy yield of so-called standard TNT is taken to be 4.615 MJ/kg, according to G. F. Kinney.[86] The total energy yield of a nuclear bomb is thus 4.187×10^{12} J per "kiloton" (KT) or 4.187×10^{15} J per "megaton" (MT) (the "ton" referred to here weighs 907.2 kg).

Two basic categories of nuclear bomb exist: (1) the atomic bomb, which relies on the complete fission of about 58.8 g/KT of ^{235}U and/or ^{239}Pu; and (2) the hydrogen bomb, which relies on fission to trigger the fusion of H isotopes (for purposes of approximation, the yield of this latter bomb being considered to be half from fission and half from fusion). A bomb relying on the fission of ^{235}U is about 5 percent efficient and thus contains about 1.18 kg/KT of fission yield, whereas one relying on ^{239}Pu is about 15 percent efficient and thus contains about 392 g/KT of fission yield.[87]

The atomic bombs are usually in the "kiloton" range, whereas the hydrogen bombs are usually in the "megaton" but can also be in the "kiloton" range. About 50 percent of the energy yield of atomic bombs is released in the form of a blast (shock) wave, about 35 percent as thermal radiation, and the remaining 15 percent as nuclear radiation. The

86. G. F. Kinney, *Explosive Shocks in Air* (New York: Macmillan, 1962), p. 2.
87. Vellodi et al., pp. 54–55.

comparable values for hydrogen bombs are about 54.5 percent, 38 percent, and 7.5 percent, respectively.

The fission reaction of a nuclear bomb produces about 290×10^{21} fission fragments per kiloton of fission yield,[88] weighing about 56.7 g.[89] These two values, it is assumed, refer to the situation 1 hr after detonation. On the basis of Avogadro's number (602×10^{21} particles /mol), the fission products per kiloton of fission bomb after 1 hr thus represent the equivalent of 482 mmol of a conglomerate of radioactive isotopes having an average mass number of 118. This mixture (the result of both fission and neutron activation) contains some 200 isotopes of about three dozen elements.

The fusion reaction of a hydrogen bomb results in an excess of between 100×10^{21} and 1×10^{24} atoms of ^3H per kiloton of fusion yield.[90] Using the geometric mean of these two extremes (i.e., 316×10^{21}) and dividing this value by Avogadro's number as well as by the mass number (i.e., 3), one arrives at 175 mg of ^3H per kiloton of fusion yield.

To recapitulate, a 1 KT atomic (fission) bomb produces about 56.7 g of mixed fission products (i.e., as determined 1 hr after detonation). A 1 MT hydrogen bomb (half fission, half fusion) produces about 28.35 kg of mixed fission products plus about 87.4 g of ^3H.

The numerous fission products of a nuclear bomb decay at greatly varying rates. One hour after the detonation of an atomic (fission) bomb there are said to be 400 MCi/KT of fission yield (i.e., 7,055 MCi/kg of mixed fission products).[91] (One curie equals 37×10^9 disintegrations per second, the nominal rate of decay of 1 g of ^{226}Ra.) The ^3H disseminated by a fusion reaction (with its half-life of 12.26 yr) has a decay rate of 10.12 MCi/kg. Thus, 1 hr after the detonation of a hydrogen bomb (½ fission, ½ fusion) there would be 200,000 MCi/MT from the mixed fission products plus about 885 kCi/MT from the ^3H.

Most of the fission products are extremely short-lived. Their combined decay rate during the initial half year approximates a log/log linear curve having a slope of -1.2; thereafter the overall decay rate becomes even more rapid, following a new log/log linear curve of slope -2.3.[92] In other words, 50 percent of the fission products present at 1 hr after detonation have disappeared about 1.78 hr after detonation, 90 percent about 6.81 hr after detonation, 99 percent about 46.4 hr after detonation, and 99.9 percent about 316 hr (13.2 days) after detonation. About 42.6 parts per million (ppm) remain after half a year and about 8.66 ppm after 1 yr. One hundred days after detonation is sometimes used as a reference point. If what remains at 100 days (i.e., 87.85 ppm of the 1-hr value) is taken as 100 percent, then 48.5 percent remains ½ yr after detonation and 9.86 percent remains 1 yr after detonation.

These decay parameters can also be expressed more directly in relation to the nuclear bombs themselves. For a fission bomb, the 400 MCi/KT that obtains at 1 hr has decayed to 35.1 kCi/KT after 100 days and to 3.46 kCi/KT after 1 yr. (Thus, at 100 days there are 620 kCi/kg of original fission products.) For a hydrogen bomb, the 200,000

---

88. From figures in UNSCEAR, p. 57.
89. Glasstone, p. 417.
90. UNSCEAR, p. 57.
91. See C. S. Shapiro, "Effects on Humans of World-wide Stratospheric Fall-out in a Nuclear War," *Bulletin of Peace Proposals* 5 (1974): 186–90.
92. See Glasstone, p. 420.

MCi/MT of mixed fission products obtaining at 1 hr have decayed to 17.6 MCi/MT after 100 days and to 1.73 MCi/MT after 1 yr. The latter bomb's 885 kCi/MT of ^{3}H present at 1 hr decays to 871 kCi/MT after 100 days and to 836 kCi/MT after 1 yr. On the other hand, a small but biologically important group of fission products is rather long-lived (i.e., with half-lives measurable in years), including ^{90}Sr and ^{137}Cs (see table 6).

APPENDIX C

REACTOR CHARACTERISTICS OF NUCLEAR-POWERED NAVAL VESSELS

The nuclear reactor of naval vessels (both submarine and surface) is taken to be a pressurized light-water cooled and moderated reactor, as is the case where such information has been provided by *Jane's Fighting Ships, 1976–77*. Sizes have been estimated as follows: A number of nuclear-powered submarines are recorded in *Jane's Fighting Ships, 1976–77* as having shaft horsepowers of 15,000, and a number of such cruisers are rated at 60,000 shaft h.p. (1 h.p. = 746 W). The efficiency of conversion of heat to electricity is taken to be 33.3 percent (see below). Such a nuclear submarine would thus have a reactor of about 33.6 MW(th), here rounded to 35, and a nuclear cruiser one of about 134.3 MW(th), here rounded to 140.

A pressurized light-water reactor can be expected to use up about 513 mg of ^{235}U per MW(th) per day of operation. It converts heat to electricity at an efficiency of about 33.7 percent, here rounded to 33.3 percent. These values are based on the six most recent civil reactors of this sort to have become operational in the United States (all those during 1975–76), being the average of U.S. reactor numbers 312, 315, 317, 334, 336, and 341.[93]

For each kilogram of ^{235}U used up in a nuclear reactor (these data based on a boiling light-water reactor) about 1.25 kg of a large variety of fission products is produced,[94] that is, about 641 mg/MW(th)day. For a 35 MW(th) submarine that has operated for 30 days, the total amount thus comes to 673 g and for a 140 MW(th) cruiser to 2.69 kg.

As was the case with the fission products of an atomic bomb (see Appendix B), the overall decay rate of this conglomerate is exceedingly rapid, for the first half year again following a log/log linear course having a slope of −1.2 and then decaying even more rapidly. However, these products are also constantly being added to during operation. The net effect of these two opposing trends is that within a week the cumulative amount is 105 percent of the daily production, within 2 weeks it is 106 percent, within 1 mo it is 107 percent, and within 2 mo 108 percent. The equilibrium value, reached after half a year or so, is almost 109 percent.

As soon as the reactor ceases to operate, the short-lived fission products (those with half-lives measurable in seconds or days) diminish at a precipitous rate. On the other hand, the small fraction of long-lived fission products produced (those with half-lives measurable in years) build up almost arithmetically during the months of operation and then, when operation ceases, disappear only slowly (see table 7).

93. From the International Atomic Energy Agency, *Power Reactors in Member States* (Vienna: International Atomic Energy Agency, 1976).
94. According to G. Netzén et al., *Använt Karnbränsle och Radioaktivt Avfall. II*, Report no. SOU-1976: 31 (Stockholm: Swedish Ministry of Industry, 1976), p. 36.

Regional Developments

The Southern Ocean

G. L. Kesteven
Food and Agriculture Organization

INTRODUCTION

Of the oceans generally, the *RIO (Reshaping the International Order)* report to the Club of Rome says,

> The oceans are the last and, in some respects, the greatest resource of our planet. They are also a battlefield of conflicting interests. They are also a "common heritage" and as such "belong" neither to the countries with the capability to exploit them nor exclusively to coastal states or powerful vested marine interests. As a common heritage, all mankind must benefit from their exploration, exploitation and conservation. That this cannot be guaranteed by the traditional law of the high seas or by unilateral national policies has become patently obvious. It is also clear that no attempt to fashion a new international order can afford to exclude the oceans. This question represents a historic opportunity, not only to ensure that all will benefit from the exploitation of the oceans, but also to develop new forms of international cooperation.[1]

The foregoing might well have been written expressly about the Southern Ocean, of which much has yet to be discovered, about which much has yet to be done, and from which much is to be gained, not least in the field of international relations. The Southern Ocean is large and most of it international. It contains immense resources, and its space accommodates oceanographic processes of worldwide effect. These resources can be utilized only through development of an advanced technology. The devising and application of that technology will call for considerable managerial skill and the application of concepts and procedures in international relations whose evolution will have far-reaching effect. Scientific research which must be carried out with regard to the Southern Ocean will be expensive, but the task of procuring funds for research will be simpler than that of establishing international agreement as to its conduct and the application of its results. Still more difficult is the matter of distributing the benefits from utilization of resources: resolution of this issue will put the new law of the sea to severe test.

Thus the Southern Ocean may well become a testing ground for procedures in

---

1. J. Tinbergen and A. J. Dolman, eds., *Reshaping the International Order: A Report to the Club of Rome* (New York: Dutton, 1976), p. 41.

© 1978 by The University of Chicago. 0-226-06602-9/77/1978-1020$2.49

all these fields. This gives it supreme importance now, because it is the example par excellence of the common heritage of mankind, and because the rewards it offers in raw materials and energy, power, and prestige are so great that the temptations they present will test the morality of international negotiation severely.

The Southern Ocean cannot be considered in isolation. Although the significance of its living resources, minerals, and energy derives from the uses to which they can be put elsewhere, their nature is determined in part by the operation of processes in other systems on the Antarctic continent and in the waters to the north of it. Similarly, its political status will be affected by what happens with regard to the Antarctic continent. The Southern Ocean and the Antarctic continent are politically similar: ostensibly they are both common heritage and hence not subject to appropriation. Yet the temptation to secure the benefits of the massive mineral resources of the Antarctic by establishing territorial claims is already considerable and, if anything, is reinforced by the difficulties of setting up alternative international arrangements on behalf of the world community. Nevertheless, a reversion in the Antarctic to past methods of seizure of territory could jeopardize the whole concept of common heritage, with only partially predictable consequences for the resources of the seas.

At present, stock is being taken of knowledge of the area's living resources, under the auspices of SCAR (the Scientific Committee on Antarctic Research, International Council of Scientific Unions [ICSU]), SCOR (the Scientific Committee on Oceanic Research, ICSU), the FAO, and the UNDP; the FAO is also reviewing current activities directed at development of the technologies of krill extraction and utilization. At the same time, discussions are taking place about effective resource utilization in the area and the institutional arrangements that might be appropriate for the management of operations in the area. It is hoped that these several activities will encourage an international policy and plan of action for the southern seas. Completion of a comprehensive stocktaking is an indispensable preliminary to those moves.

A necessary first step of this stocktaking is a discussion of the name "Southern Ocean" itself. Several names have been given to the southern circumpolar seas, and several areas have been signified by these same names. Although "Southern Ocean" is very widely but not universally accepted and is the name adopted here, some difference of opinion exists as to the area to which it should refer.

THE CONCEPT AND NAMING OF THE SOUTHERN OCEAN

Specification of the limits of a particular natural earth feature is often very difficult, especially if the limits can change, as is the case for the edge of the sea. Across broad reaches of the seas, the task is even more difficult, not only because the position of surface-defining features such as convergences may shift, but also because there is no vertical bounding curtain below a demarcation line on the surface to separate one ocean from another; and this is important because the significance of an ocean is in its depth, its three-dimensionality.

In an article on the Southern Ocean in the encyclopedic work *The Sea,* Deacon noted that "Southern Ocean, Antarctic Ocean, Southern Seas and South Polar Seas are some of the names used" to refer to the whole of the circumpolar ring of ocean around the Antarctic continent. "It is not a name that needs to be defined rigidly, but one that can be used without much risk of ambiguity to refer to the ring of ocean. It is a remarkably uniform ocean: differences between one sector and another are small compared with differences between one latitude and another."[2]

On the other hand, Sverdrup, Johnson, and Fleming remarked, "It is difficult to assign a northern boundary to the Antarctic Ocean because it is in open communication with the three major oceans: the Atlantic, the Indian, and the Pacific Oceans. In some instances the Antarctic waters are dealt with as parts of the adjacent oceans and are designated the Atlantic Antarctic Ocean, or the Indian or Pacific Antarctic Ocean, whereas in other instances the Antarctic waters must be considered an integral part of all oceans."[3]

In this essay the subject is not so much which of several names given to some part of the southern seas is most appropriate for that entity as a consideration of the several entities to which the name "Southern Ocean" has been or could be given.

The problem is simply stated by Foster, who has written, "There does not seem to be any general agreement as to the extent of the Southern Ocean; some take the northern boundary as the Antarctic Convergence, but most polar oceanographers prefer a less rigid definition and include most of the ocean up to the southern tips of Africa and Australia."[4] For other purposes, the Antarctic Treaty set 60° south latitude as the northern limit, whereas the FAO and the UNDP in the Southern Ocean Programme declared an interest in waters south of 45° south, but neither of these boundaries corresponds to what oceanographers would designate for the Southern Ocean. Agreement seems to hold among oceanographers that an observable oceanic structure can be designated the northern boundary, and there appears to be little debate about the main hydrographic features of the southern seas. Deacon discussed six features. These are *(a)* the warm deep water, *(b)* Antarctic bottom water, *(c)* Antarctic surface water, *(d)* the Antarctic Convergence, *(e)* the sub-Antarctic zone, and *(f)* the subtropical zone.[5] These features are represented in figures 1 and 2, which also represent the Antarctic Divergence.[6] As remarked by Sverdrup, Johnson, and Fleming, "It is evident that one cannot consider the Antarctic Circumpolar Water mass as a body of water which circulates around and around the Antarctic Continent without renewal. On the contrary, one has to bear in mind

---

2. G. E. R. Deacon, "The Southern Ocean," in *The Sea,* ed. N.M. Hill (New York: Interscience, 1963), 2:281.

3. H. U. Sverdrup, Martin W. Johnson, and Richard H. Fleming, *The Oceans:* (New York: Prentice-Hall, 1946), p. 605.

4. Theodore D. Foster, "The Physical Oceanography of the Southern Ocean: Key to Understanding Its Biology" (paper presented at the SCOR/SCAR meeting, Woods Hole, Massachusetts, August 17–21, 1976.

5. N. 2 above.

6. G. A. Knox, "Antarctic Marine Ecosystems," in *Antarctic Ecology,* ed. M. W. Holdgate (London: Academic Press, 1970).

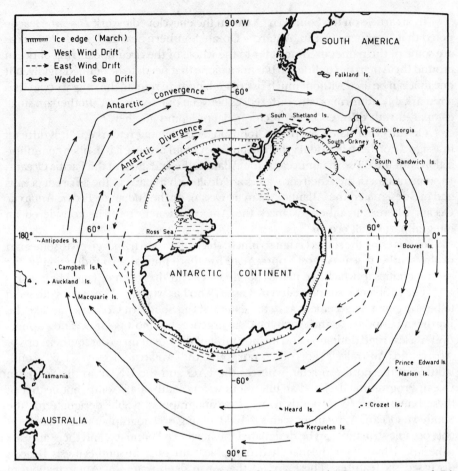

Fig. 1.—Antarctica, showing the position of the Antarctic Convergence and Divergence and surface currents. (With permission from G. A. Knox, "Antarctic Marine Ecosystems," in *Antarctic Ecology*, ed. M. W. Holdgate [London: Academic Press, 1970], p. 70. Copyright by Academic Press Inc. [London] Ltd.)

that water from the Antarctic region is carried towards the north and out of the region both near the surface and near the bottom, and that deep water from lower latitudes is drawn into the system in order to replace the lost portions." They also observe that "from the oceanographic point of view it is logical to consider an Antarctic Ocean extending from the Antarctic Continent to the Subtropical Convergence except in the eastern South Pacific, where an arbitrary limit has to be established."[7]

The view that "the Antarctic Ocean is not an isolated entity but an integral part

7. Sverdrup et al., p. 620.

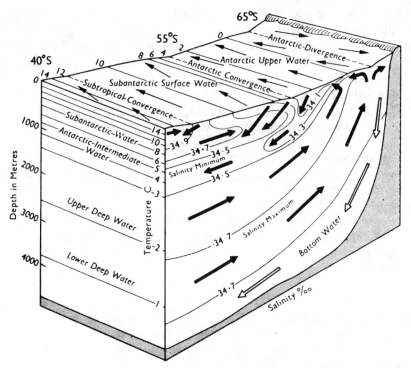

Fig. 2.—Schematic diagram of the meridonial and zonal flow in the Southern Ocean. The diagram represents summer conditions; average position of Convergence and Divergence are shown. The Upper Deep Water is best developed in the Atlantic sector. The south-going component in the Lower Deep Water is weak or reversed in the Pacific. (With permission from G. A. Knox, "Antarctic Marine Ecosystems," in *Antarctic Ecology*, ed. M. W. Holdgate [London: Academic Press, 1970], p. 71. Copyright by Academic Press Inc. [London] Ltd.)

of the world oceanic system,"[8] which could scarcely give rise to controversy, has been well substantiated.

Accounts show two major circumpolar currents: one of water flowing westerly along the continental margin, the Antarctic Coastal Current or East Wind Drift, driven by predominantly easterly winds near the continent; the other of water flowing easterly north of about 60° south, the Antarctic Circumpolar Current or West Wind Drift, driven by westerly winds. Between these move other currents of much irregularity. From the south "cold water and ice spread outwards as a surface current about 100 to 150 m thick in summer, and this upper layer is of fairly low salinity and at temperatures close to freezing point. The water has a northward and eastward movement, except close to the continent south of about 63° S. There is also

8. R. I. Currie, "Environmental Features in the Ecology of Antarctic Seas," in *Biologie antarctique*, ed. R. Carrick, M. Holdgate, and J. Prévost (Paris: Hermann, 1964), cited by Holdgate, p. 9.

a northward flowing cold saline layer forming the Antarctic bottom current. Between these two layers there is a compensating southward movement of relatively warm and saline water and this is drawn from beneath the waters of the world ocean, some of it being traceable from sources as far away as the Mediterranean or the North Atlantic."[9]

Although the processes that effect exchanges between the Antarctic Ocean and other parts of the world oceanic system are of considerable importance, the constancy of the main hydrographic features of this region ought not to be discounted. The Antarctic Convergence is remarkably steady in its location, although the Sub-Antarctic Convergence is less so. A notable uniformity of temperature and salinity prevails between the Antarctic Convergence and the continent. On the other hand, the sea ice cover in this zone varies considerably from one season to another, the outer limit of pack ice lying not far from the continent in summer but extending out to about 60° south in winter. At the same time, mention must also be made of the considerable range of light in this zone, from virtually nil in winter to an abundance in summer.

Thus the southern seas are divided into clearly recognizable zones: first, that which lies between the Antarctic continent and the Antarctic Convergence; second, that which lies between the two convergences. The northern bound of the first is well marked and relatively constant in location, whereas the second is less well marked, less constant, and not recognizable in the eastern South Pacific. This two-zoned system draws water from both north and south and contributes water to the world ocean.

The hydrographic characteristics of these seas, some substantially autonomous and conservative, others forming part of a larger system, are reflected in the pattern of other processes, some internal to these seas and others part of other systems. Exemplifying the former are primary production and phytoplankton communities, sedimentation, and the benthic communities; but, insofar as these examples are biotic systems, they also participate in wider-ranging exchanges with the Antarctic continent on one hand and the world ocean on the other. Migratory animals, notably marine mammals, are participants in this exchange, but birds and seals on the continental margin and fishes and cephalopods that cross from one zone of these seas to another or go beyond the subtropical convergence also participate. Given also that human activities remove material in the exploitation of resources and introduce material by pollution, it remains impossible to consider these seas in isolation. It is then necessary to seek a demarcation line which will specify a Southern Ocean appropriate to institutional arrangements for management of exploitation of its resources.

9. M. W. Holdgate, "The Antarctic Ecosystem," *Philosophical Transactions of the Royal Society of London* 252B (1967): 363–83.

Considering only superficial features, there are three areas that might be designated "Southern Ocean": (1) the zone lying between the Antarctic Convergence and the Antarctic continent, (2) the zone from the subtropical convergence south to the Antarctic continent, and (3) a zone from the Antarctic continent north to an outer boundary set north of the subtropical convergence. In the first two cases at least a problem exists in deciding whether the southern or northern edge or some intermediate line of the bounding convergence should be taken as the northern limit.

If, on the other hand, deep features of these waters were considered, other difficulties undoubtedly would arise. Further, the distribution of the organisms that participate in the trophodynamic systems of any of the three areas above complicate the matter even more.

Although the rules marking boundaries of territorial seas and exclusive economic zones can scarcely be applied in this case, perhaps the principles underlying such rules might offer guidelines for decisions. The argument advanced with this suggestion is based on two considerations. First, the basic principle underlying the demarcation of national zones is that areas in which national rights and responsibilities hold are differentiated from those managed under a different regime. Second, demarcation of national areas establishes a basis for a further set of rules directly related to the rights and responsibilities of the involved parties, with regard to exchanges between adjacent national areas and between national areas and nonnational areas. This holds with regard to coastal migratory species and the highly migratory species as well as anadromous and catadromous species.

The demarcation of national areas is made without reference to or effect on the oceanographic status of the broad oceanic system of which those areas are part. Correspondingly, such demarcation carries no necessary implication for the divisibility of the adjacent high seas or otherwise. The international community, however, is bound to find it necessary to make subdivisions of the high seas to serve its own purposes. Already the FAO does this in establishing its statistical areas. Obviously a demarcation ought to be made between the southern seas and the Indian, Pacific, and Atlantic Oceans.

Without prejudice to present or future claims over parts of bounding seas, the name "Southern Ocean" can be given to the southern circumpolar seas extending north to an arbitrarily designated line which will serve the rights and responsibilities of the international community conveniently. This line might well be placed at about 40° south but need not correspond to a single latitude, and its placement should have close regard to the distribution of resources and the management of their exploitation.

Designation of the Southern Ocean in this manner would have no effect on oceanographers' terms "Antarctic waters" and "sub-Antarctic waters" applied to the zones south and north of the Antarctic Convergence or on established names such as "Weddel Sea" and "Ross Sea."

THE RESOURCES OF THE SOUTHERN OCEAN

Resources are considered here in three classes: nonrenewable, renewable, and indestructible, of which the third is not usually taken into account in resource inventories. The greater part of this section is devoted to renewable resources; information on oils and gases is presented in a different paper in this volume.

Nonrenewable Resources

While the greater part of the continental shelf of Antarctica is narrow and much of it permanently overlaid with ice, the general oceanic floor of the Southern Ocean lies deep, at depths of 4,000–5,000 m, its basins broken up by submarine ridges. The vast area of this ocean (on the order of 20×10^6 km² in the zone south of the Antarctic Convergence) undoubtedly includes some proportion of the 1.5×10^{12} MT of ferromanganese nodules estimated to lie on the floor of the Pacific Ocean. Trial drilling in the Ross Sea has already revealed gases indicating the presence of oil.

Renewable Resources

Nutrient Salts and Primary Production
As remarked by Holdgate, "upwelling in the Antarctic brings nutrients into the surface waters in considerable quantity"[10] which are then at levels that "rarely fall below the maxima of temperate regions and are unlikely to be limiting to phytoplankton growth and development."[11] Figures published by El-Sayed show high values of phosphates, silicates, nitrates, and nitrites in areas investigated in the Atlantic and Pacific sectors of Antarctic waters, with lower values in those same sectors of Sub-Antarctic waters.[12] Apparently the greatest proportion of these abundant nutrients is brought in by the warm Deep Water, part of the load being carried to the surface, although Foster remarks that "there does not seem to be any clear-cut correlation between nutrient distribution and possible upwelling regions."[13] The other part of the load passes into bottom water. The abundance and distribution of nutrients in the Southern Ocean result from the major features and processes referred to above and further demonstrate the interrelation of the Southern Ocean and the world ocean.

Antarctic waters have long had a reputation for being very rich, but recent

---

10. Holdgate, cited by Knox, p. 75.
11. Knox, p. 74.
12. Sayed Z. El-Sayed, "On the Productivity of the Southern Ocean," in Holdgate, *Antarctic Ecology*.
13. Foster, n. 4 above.

discoveries "have qualified earlier emphasis on the extreme high productivity of the Antarctic Ocean, and demonstrated great regional variability in biomass, production and productivity."[14] Significant factors determining seasonal variation in production are light and ice. Total darkness prevails for half the year and continuous daylight for the other half. Synchronously the ice-covered area varies from $2.6-3.2 \times 10^6$ km^2 in summer to 26×10^6 km^2 in winter.[15] In contrast, the temperature regime is extremely constant: Antarctic surfaces have an average temperature, just south of the convergence, of 3°–5° C in summer and 1°–2° C in winter and near the continent vary only about $-1.0°$ to $-1.9°$ C. Regional variation in productivity is very great, with considerable differences between Antarctic and Sub-Antarctic waters and between oceanic and neritic regions. El-Sayed has discussed the effect of the Antarctic Convergence on the distribution and concentration of nutrient salts, showing that "the phytoplankton standing crop and primary productivity are much higher south of the Convergence than north of it," yet "although the concentration of the nutrient elements north of the Convergence is lower than in the Antarctic waters proper, even the lowest levels of concentrations are higher, in general, than are the winter maxima of temperate regions." He goes on to say that "the proverbial richness of the Antarctic waters is factual only with regard to coastal and inshore regions, and not with regard to the oceanic regions."[16]

From such results El-Sayed calculated an average value of primary production for the inshore and offshore regions of Antarctic waters which was about six times the average gross production over all the oceans estimated by Steeman Nielsen.[17] He went on to calculate the annual production of Antarctic waters and arrived at an average of 0.33×10^{10} MT carbon per year, which represented about 20 percent of the average gross production of all oceans and meant that, on the average, Antarctic waters per unit area are four to five times more productive than the rest of the oceans. Gulland took El-Sayed's figures and those of other workers and with a simple conversion factor calculated a total production of $4-20 \times 10^9$ MT liveweight.[18] However, Gulland calculated liveweight from carbon fixation with a ratio of 1:10, whereas it is generally accepted that the calculation is by way of 1g C:2.5g dry weight:25g wet weight. If he had taken this other ratio, he would have obtained an estimate of total production at $1-5 \times 10^{10}$ MT liveweight, still only, respectively, 5 percent and 24 percent of corresponding figures resulting from El-Sayed's production per year. These figures relate to waters south of the Antarctic Convergence. Obviously the production of the Southern Ocean, as I propose to define it, must be

14. Holdgate, *Antarctic Ecology*, p. 117.
15. Knox, pp. 72–73.
16. El-Sayed, p. 133.
17. Ibid; and E. Steeman-Nielsen, "The Use of Radioactive Carbon (C^{14}) for Measuring Organic Production in the Sea," *Journal du Conseil* 18 (1952): 117–40.
18. J. A. Gulland, "The Development of the Resources of the Antarctic Seas," in Holdgate, *Antarctic Ecology*, p. 219.

substantially greater, even if productivity is less north of the Convergence than south.[19]

Special mention must be made of the epontic algal flora which have been closely studied by Bunt and others and to the crustacean and fish fauna associated with it. This community plays a special role in the interaction between the land and sea systems.

Zooplankton
Although Holdgate remarked that "taken as a whole, despite the high standing crop in many areas, the Antarctic oceans do not show an outstandingly high

19. Appreciation of figures such as these presents difficulty not only for lay readers. Apart from unavoidable imprecision and inaccuracies, two sets of figures, obtained by two workers and employing different techniques, may differ because the techniques lead to different results even on one material. The figures are from samples and generally are averages; projection from them to obtain an estimate of some total value for an extended period, say, from an hourly rate to annual production, or for a whole area, say, from a rate per square meter to some tens of millions of square kilometers, is subject to the representativeness of the sample figures for the whole period or area. These are sampling problems; other difficulties arise from the nature of the measurement reported. Plankton can be measured as to volume, wet weight (drained of free water), dry weight, or carbon content (after calcining). Primary production can be measured inter alia as oxygen production and as carbon uptake; the relation between such figures and the produced biomass depends upon the physiological state of the phytoplankton and upon environmental conditions. Factors have been calculated for conversion of figures of one kind of measurement into those of another kind. As an aid to the various kinds of calculation, the following table gives some representative conversions among carbon, dry weight, and wet weight for different unit areas.

CONVERSION FACTORS: m^2, km^2, 10^6 km^2

mg	$1\ mg/m^2$	$1\ kg/km^2$	$10^3\ MT/10^6\ km^2$
g	$1\ g/m^2$	$1\ MT/km^2$	$10^6\ MT/10^6\ km^2$

Converting Chlorophyll a Value ($1\ mg/m^2$)

To carbon value:
Minimum	...............$35\ mg/m^2$	$35\ kg/km^2$	$35 \times 10^3\ MT/10^6\ km^2$
Average	................$50\ mg/m^2$	$50\ kg/km^2$	$50 \times 10^3\ MT/10^6\ km^2$
High	..................$100\ mg/m^2$	$100\ kg/km^2$	$100 \times 10^3\ MT/10^6\ km^2$

To dry weight:*
Minimum	.............$87.5\ mg/m^2$	$87.5\ kg/km^2$	$87.5\ MT/10^6\ km^2$
Average	............$125.0\ mg/m^2$	$125.0\ kg/km^2$	$125.0\ MT/10^6\ km^2$
High	................$250.0\ mg/m^2$	$250.0\ kg/km^2$	$250.0\ MT/10^6\ km^2$

To wet weight:†
Minimum	................$875\ g/m^2$	$875\ kg/km^2$	$875 \times 10^3\ MT/10^6\ km^2$
Average	..............$1.25\ g/m^2$	$1.25\ kg/km^2$	$1.25 \times 10^6\ MT/10^6\ km^2$
High	...................$2.50\ g/m^2$	$2.50\ kg/km^2$	$2.5 \times 10^6\ MT/10^6\ km^2$

rate of primary production,"[20] and his estimates are held to be "in general of the right order,"[21] the usual estimate is that "zooplankton abundance is high."[22] Holdgate's estimates, shown in his food-chain diagram (fig. 3), nevertheless point to various problems. His figure of 105 mg/m³ includes 5 mg of carnivores and therefore signifies a ratio of 3.2:1 phytoplankton to herbivorous zooplankton, whereas the ratio is generally thought to be 10:1 (see Gulland);[23] this discrepancy reflects both the uncertainties with regard to the ratio between phytoplankton and herbivores and the inadequacy of data on zooplankton, remarked by virtually all researchers. Again, the figure of 105 mg/m³ from Holdgate's text seems to hold for at least a 50-m column and for Antarctic waters (not the Southern Ocean) which are held by various authors (e.g., Hempel)[24] to have an area of 2×10^7 km²; multiplication of these figures gives an estimate of 1.05×10^8 MT average zooplankton biomass, of which 0.50×10^8 MT are of *Euphausia superba*. Other workers have put zooplankton biomass in Antarctic waters at a figure at least one order of magnitude greater, and the figure of 0.5×10^8 MT for *E. superba* is probably the lowest of the values suggested by other workers.

The foregoing detail has been given in order to emphasize the difficulties encountered in assessing the resources of the Southern Ocean. Research workers' estimates differ by one and even two orders of magnitude. Klumov, for example, calculated the total stocks of Antarctic Euphausiids at 5×10^9 MT.[25] Even if all workers had full access to all relevant data, it could not be expected that they would arrive at one estimate of any characteristic of the zooplankton

Converting Carbon Value			
To dry weight value:			
1 mg ....2.5 mg/m²		2.5 kg/km²	2.5 MT/10⁶ km²
1 g ....2.5 g/m²		2.5 MT/km²	2.5×10⁶ MT/10⁶ km²
To wet weight value:			
1 mg ....25 mg/m²		25 kg/km²	25 MT/10⁶ km²
1 g ....25 g/m²		25 MT/km²	25×10⁶ MT/10⁶ km²

NOTE.—The table is to be read as follows: e.g., 1 mg chlorophyll per square meter corresponds to 35 mg carbon per square meter.
*Calculated from dry weight:2.5 carbon.
†Calculated from wet weight:10 dry weight.

20. Holdgate, "The Antarctic Ecosystem," p. 366.
21. Knox, p. 90.
22. G. Hempel, "The Antarctic," in *The Fish Resources of the Ocean* (London: Fishing News [Books], 1971), p. 163.
23. Gulland, p. 220.
24. Hempel, p.163.
25. S. K. Klumov, "Nutrition and Helminthological Fauna of Whalebone Whales (Mysticceti) in the Main Fisheries Areas of the World Ocean," cited by Knox, p. 80.

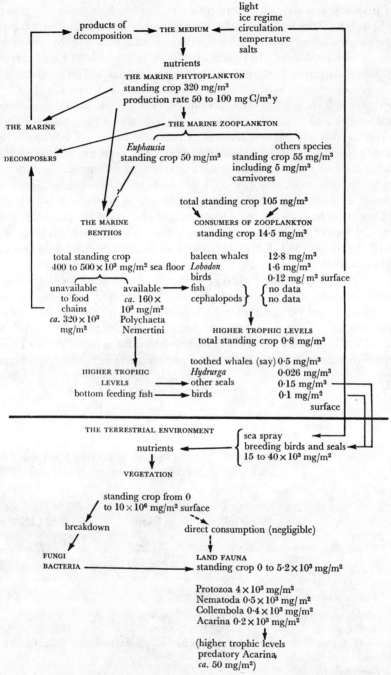

Fig. 3.—Food-chain diagram for the more important interrelationships in the Antarctic seas. (From M. W. Holdgate, "The Antarctic Ecosystem," *Philosophical Transactions of the Royal Society of London* 252B [1967]: 380.)

of the Antarctic, say, its average biomass. The characteristics of the data from sampling admit of various interpretations with regard to how representative are the quantity and composition of the zooplankton over time (with diurnal, seasonal, and annual variation at any one place) and space. Considering the immensity of the area and the difficult conditions prevailing there, sampling programs can never be adequate to account for variability in time and space. Moreover, efficiency and selectivity of sampling equipment vary in imprecisely measured ways. It therefore remains a matter of individual judgment to weight the separate data and arrive at acceptable overall estimates.

These difficulties are especially apparent in measurement of krill, for which reliable estimates of average biomass, annual production, and allowable yield are urgently needed. Krill are only part of the planktonic community and indeed are more accurately considered macroplankton or micronekton. The term is Norwegian and signifies the crustacean component of plankton. As such, it refers to a number of species; in the Antarctic it refers almost exclusively to *E. superba*. Nevertheless, some 90 species of crustacea have been identified in samples of krill. It seems to be generally accepted that krill constitute about half the total zooplankton of Antarctic waters, although there is evidence that the proportion varies considerably according to place and time. Application of the generally accepted proportion to zooplankton estimates will of course vary with those estimates.

Estimates of krill biomass have been made directly (by sampling and the use of acoustic equipment) and indirectly with reference either to phytoplankton or to predators which feed on krill; the indirect estimates generally are of production rather than of average stock biomass. Allen suggests that the ratio of production to average stock biomass is between 1.8 and 2.1 instead of the 1:1 assumed by Gulland.[26] The many estimates of production range from 50×10^6 MT (or even less) based on feeding by whales, through various values in hundreds of million tonnes, to a figure of 9×10^9 MT which would result from applying Allen's ratio to values given by Russian workers for average stock biomass. Narrowing this range cannot be effected by a review of the literature, and it is unlikely that such a review would lead to nomination of any one estimate as most accurate; what is required is a cooperative enterprise by the research workers concerned to standardize the methods of calculation and establish common ground in evaluation of available data.

The value of krill as a resource depends not only on the magnitude of their production but also on their distribution and behavior. Krill apparently inhabit all waters south of the Antarctic Convergence, but their density varies considerably within this zone. However, it is difficult to construct a coherent picture of distribution even of only the postlarval forms, and there is much debate about the distribution of eggs and early larvae. The greatest concentrations of postlarval forms have been observed in the western Atlantic sector, in the East Wind

---

26. K. R. Allen, "Relation between Production and Biomass," *Journal of the Fisheries Research Board of Canada* 28, no. 10 (1971): 1573–81.

Drift, Scotia Sea, Weddel Drift, and South Georgia areas,[27] but large concentrations have been observed elsewhere by British, Japanese, and Russian scientists. Furthermore, there is debate about whether the various concentrations are of self-maintaining populations or even of distinct races.

The swarming behavior of the krill, although well known since first observation of these organisms, is still only partially described. The swarms vary in shape and area—and hence volume and biomass; the density of swarms also varies greatly. Swarming intensity varies in a diurnal pattern (greatest at night) but also with light intensity and season.

Cephalopods

Two species of Ommastrephids, *Martiala hyadesi* and *Notodarus sloani*, were noted by Voss for "the Southern Ocean south of the subtropical convergence," and he remarks that the two most numerous squids are oceanic, small and of no commercial value.[28] Although Voss considered that no estimate of the potential of cephalopods of this region was possible, his calculation of the consumption of squid by the world population of sperm whales was adopted by Everson for a calculation that "at least 13.5×10^6 t per year of squid production is eaten by sperm whale" from the Antarctic. From this figure and information that squid feed on krill, he calculates, with a conversion ratio of 10:1, that squid might consume more than 1.0×10^8 MT of krill and thus be the greatest consumer of krill at present.[29]

Fishes

Some 60 species of pelagic fishes and 90 benthic species have been recorded for Antarctic and Sub-Antarctic waters, but only a minority are of economic importance. A large proportion of the list comprises species endemic to the area, but two important species, *Micromesistius australis* and *Meruccius hubbsii*, have been shown to be migrants from the Patagonian region into Antarctic waters—specifically into the Scotia Sea—to feed on krill.

Although important biological information has been obtained concerning many species, especially of the two just mentioned and of *Notothenia* spp. and *Dissostichus* spp., relatively little is known of the abundance of any of the species. Gulland's figure of 1.0×10^8 MT of production of first-stage carnivores might be taken as an indicator of the magnitude of the stocks were it not for the errors of calculation noted above;[30] nevertheless, it might be of the right order of

27. I. Everson, "Review Paper: Krill" (paper presented at the SCAR/SCOR meeting, Woods Hole, Massachusetts, August 17–21, 1976; see n. 36 below).
28. G. L. Voss, *Cephalopod Resources of the World*. FAO Fisheries Circular no. 149 (Rome: Food & Agriculture Organization, 1973), pp. 59–61.
29. I. Everson, "Review Paper: Squids" (paper presented at the SCAR/SCOR meeting, Woods Hole, Massachusetts, August 17–21, 1976; see n. 36 below).
30. Gulland, p. 220.

accuracy. Some indications of abundance may be found in reports of total catches and catch rates, but the usefulness of these data is diminished by uncertainties as to the areas in which catches were taken. Catches in excess of 5.0×10^5 MT "unspecified Demersal Percomorphs" were reported to have been taken from Antarctic and adjacent waters in 1970, but the quantities were considerably less in later years, only 1.2×10^5 MT in 1973. The usefulness of these data would be greater if they were accompanied by data on effort expended, since with such data it would be possible to decide whether the differences of catch from year to year reflected anything more than changes in effort. The stocks of the two species of the Patagonian region that migrate into Antarctic waters have been estimated to allow a sustainable yield in excess of 1.3 (perhaps 2.6) $\times 10^6$ MT, but the trophodynamic significance of these species for the Antarctic system cannot be assessed until a measure is obtained of the proportion migrants comprise of the total stocks.

Marine Mammals

The variety and abundance of marine mammal life of the southern seas are well known across the world and attract a large proportion of the attention currently aroused by the efforts of environmentalists to "save the whale." Because of the richness of these resources, the Antarctic has been the scene of considerable exploitative activity which has been the object of a great deal of international debate and negotiation.

Zenkovich summarized some of the critical data in a paper on whales and plankton in Antarctic waters (see table 1). From these figures he calculated that the original stocks had a total weight of 23.25×10^6 MT.[31] However, in a paper considered by the ACMRR (FAO Advisory Committee on Marine Resources Research) Working Party on Marine Mammals, Gambell showed that the total biomass of southern hemisphere whales which enter the Southern Ocean and

TABLE 1.—WHALE CATCH

	Catch 1904–5 to 1065–66	Original Stock
Blue	331,142	100,000
Fin	671,092	200,000
Humpback	145,424	50,000
Sei	87,284	75,000*

*No. at time of writing.

31. B. A. Zenkovich, "Whales and Plankton in Antarctic Waters," in Holdgate, *Antarctic Ecology*.

feed on krill declined steadily, as a result of whaling, from about 45×10^6 MT in 1920 to about 8×10^6 MT in 1970.[32]

Antarctic whaling has been a major concern of the International Whaling Commission, whose reports, together with the statistical compilations of the Norwegian Bureau of Whale Statistics, constitute the most complete record available of the whaling industry as well as of the development, in the commission's Scientific Committee and its Committee of Three, of views on the structure and dynamics of whale populations. The scientific views and the commission's response to them have been challenged from time to time; they came under scrutiny by the ACMRR Working Party on Marine Mammals and the Scientific Consultation on the Conservation and Management of Marine Mammals and Their Environment.[33] Currently, there is very strong agitation, in many parts of the world, for a moratorium on whaling.

In addition to the great whales, there are populations of small cetaceans and seals. Much of the available information in the latter group was assembled for the ACMRR Working Party on Marine Mammals by an ad hoc group. Elephant and fur seals were hunted intensively during the eighteenth and nineteenth centuries, and the slaughter, thought to have been of hundreds of thousands in some years, severely reduced populations (to extinction at some sites), but there has been substantial recovery of these stocks during the past 50 years.

The Benthos

Holdgate states that, in contrast to the open ocean in which the total biomass is about 440 mg/m^3 averaged over the top 50 m, benthic biomass in shallow waters is a great deal more, at least in the most productive areas. He quotes Belyaev and Ushakov as indicating the following values at different depths off Sabrina Coast:[34]

200–300 m	183–1,383 g/m^2
100–500 m	400–500 g/m^2
2,000 m	2.8 g/m^2
3,500 m	1.4 g/m^2

32. This information was cited by S. J. Holt in a communication of limited circulation; I have not seen the ACMRR paper; the information was also given at the SCAR/SCOR meeting (n. 36, below), and I have seen it in a draft of the full report of that meeting.

33. Food and Agriculture Organization, "Report of the Advisory Committee on Marine Resources Research Working Party on Marine Mammals," FAO Fisheries Reports, no. 194 (Rome: Food and Agriculture Organization, 1977); and Scientific Consultation on the Conservation and Management of Marine Mammals and Their Environment, a conference held in Bergen, Norway, August 31–September 9, 1976.

34. Holdgate, "The Antarctic Ecosystem"; G. M. Belyaev, *Some Patterns in the Quantitative Distribution of the Bottom Fauna in the Antarctic,* Soviet Antarctic Expedition Information Bulletin, no. 1, 1958 (Amsterdam: Elsevier, 1964), pp. 119–21, cited in Holdgate.

After referring to results with respect to the Ross Sea and off McMurdo, he concludes that, "even if the standing crop over the area as a whole is only 10% of the average off the Sabrina Coast, it will aproximate in biomass to the standing crop in the entire water column above it." However, he also refers to estimates "that as much as 60% of the standing crop of benthos may not be directly available as food for other organisms,"[35] although there is some difference of opinion on this point.

Stocks of large crustacea inhabit shelf areas of Sub-Antarctic islands but are not large.

Ecosystemic Aspects

No mention is made of birds in the preceding paragraphs; this is because birds do not after all dwell in the sea like the other groups, and because they are not economically exploited. Nevertheless, they are an important element of the Antarctic biota, as can be seen from the report of the SCAR/SCOR meeting at Woods Hole, Massachusetts, August 17–21, 1976.[36]

Birds are a principal agent in the interaction between the Antarctic land and sea systems. By their excretion, they and seals and wind-deposited spray, as the major mechanism of interaction, bring nutrients from sea to land. While this nutrient supply supports the land vegetation, some part of it must return to the sea through the flow and melt of ice and through the epontic algae and associated fauna.[37] Similar reciprocity holds between the Southern Ocean and waters to the north of it, nutrient transfer into and out of the Southern Ocean being effected by physical transport and (mainly out of it) by biotic processes.

Complex relationships prevail within the Southern Ocean itself. In generalized form, these are represented in figures 4 and 5. Krill of course play a major role in these systems. Although phytoplankton are the major agent of utilization of the abundant supply of nutrients, it is through krill, as food for so many wide-ranging species, that much material is taken out of the system. Such relationships, of course, are not peculiar to this ocean; but their significance is greater because of their universality. Institutional arrangements for management of exploitation of resources and for general care of these waters ought to take full account of these relationships.

---

35. Holdgate, "The Antarctic Ecosystem."
36. SCOR and SCAR combined with the International Association of Biological Organizations (IABO) and the ACMRR to form SCOR Working Group 54 on Living Resources of the Southern Ocean. The working group met in Woods Hole and was combined with a conference on the Living Resources of the Southern Ocean, attended by 59 scientists. The full report of the Woods Hole meeting is included in this volume as part of the Appendix covering "Documents and Proceedings" and should be read as an adjunct to this chapter.
37. Holdgate, *Antarctic Ecology*.

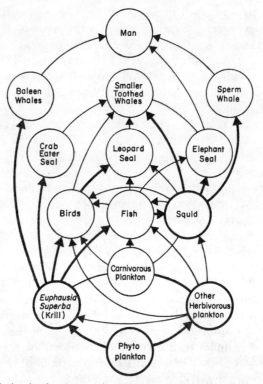

Fig. 4.—Food chains in the Antarctic. Heavy arrows indicate the probable main diet of the groups to which they point. (With permission from G. A. Knox, "Antarctic Marine Ecosystems," in *Antarctic Ecology*, ed. M. W. Holdgate [London: Academic Press, 1970], p. 87. Copyright by Academic Press Inc. [London] Ltd.)

Indestructible Resources

By this term I refer to water and energy. Consideration has been given to the practicability of moving icebergs from the Antarctic to the coasts of arid parts of the world. Salinity gradients have been shown to be a source of energy which should be considered in the present search for alternatives to fossil fuels;[38] if this should become a reality, the Antarctic would seem to offer a major opportunity for its application.

REALIZATION OF POTENTIALS

Very considerable attention is now being given to krill as well as the fish resources of the Southern Ocean, while debate goes on as to whether whaling should be continued. Much optimism is manifested about the benefits to be

38. Papers on salinity gradients as a source of energy were presented at the Joint Oceanographic Assembly held in Edinburgh, September 1976 (see Appendix B, pp. 697–701).

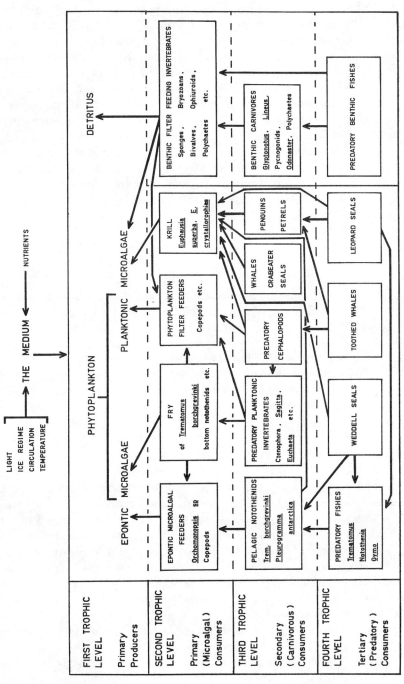

FIG. 5.—Food-chain relationships in the pack-ice zone

gained from exploiting krill and fish, while caution is urged with regard to expectations as well as technological and economic problems of utilizing these resources. Without doubt a great deal of work will have to be done toward developing equipment and methods of catching, handling, and processing krill; much of this will be to improve already established techniques. Common opinion is that the necessary advances can and will be made.

Many problems of a different kind will attend the establishment of arrangements for distribution of the products; still others will surround the organization of enterprises and the setting up of institutional arrangements for administration of whatever industry may be established and for attention to matters such as protection against pollution. Yet while successful realization of potentials will be in proportion to the degree to which solutions are found to these problems, the formulation of problems and the manner of attacking them are themselves difficult matters which involve even the identification of the potentials.

Potentials

The "potential" of a renewable resource is a particular and arbitrarily set value out of a wide range of values which the yield capacity of the resource can take. The yield capacity of a resource varies in response to changes in its environment, especially those changes introduced by man's exploitative activities not only of the particular resource but also of other, related resources and by his pollution of the habitat. Thus no living resource has a single, unique, and intrinsic potential; potentials for growth and reproduction are very labile, and how they are realized depends greatly upon man's strategy with regard to the resource itself, to associated resources, and to the habitat. The average stock biomass (standing crop), production, productivity, and sustainable yield of a resource are different quantities, all related—but each in its own way—to population structure and the potentials for growth and reproduction; and the entire complex depends upon food supply and environmental conditions.

Much work has yet to be done to identify and evaluate these relations—it has not yet been done for any stock of fish. What is clear, however, is that man's interventions have profound, far-reaching effects with consequences which oblige him to modify his strategy. In some cases the adjustments of his strategies are of monotonic effect, headed straight for elimination of a resource. This result is perhaps inevitable whenever a strategy is related to only a part of a natural system and directed to meeting the criteria of only some part of man's own social system—for example, in exploiting one or a few species of a multispecies complex to meet interest-rate criteria in the maximization of profits.

The foregoing seems to point to an impasse: while the design of a strategy should be based on expectations of potentials, potentials cannot be estimated until strategy is stated. Obviously a plan for exploitation of resources should be developed by a backing-and-filling procedure which would progressively establish the features of the strategy while adjusting and refining the estimates of potentials. Such a procedure is especially necessary in the case of fishery resources, because, given the state of the theory of natural populations and the special characteristics and circumstances of fish populations, the estimate of potentials depends heavily, as experience has shown, on evidence obtained from actual exploitation.

These arguments hold with exceptional force for the resources of the Southern Ocean, where some effects of more than 50 yr of hunting whales are recognized and approximately measured, while most effects are not even known, let alone measured. Changes in size and structure of whale populations are known, perhaps with no great accuracy; changes in growth rate and age at first maturity have also been measured; but the consequences of these changes remain unknown. The reduction in numbers of whales is thought to have left a large proportion of annual krill production unconsumed and to have been followed by an increase in the populations of other organisms which feed on krill; but this argument is not generally accepted, because it is not known whether and if so to what degree the other populations have increased or, if they have, to what extent such increase was a consequence of reduced krill predation by whales. It is also argued that the reduced whale populations, of lower average age and faster growing rate, may still be consuming an absolute quantity close to that consumed in their unexploited state.

In light of the foregoing, it will be understood that I can give no estimates here of the potentials of the living resources of the Southern Ocean and that acceptable estimates will not be made until a common strategy is decided upon and put to test under a reliable monitoring system, accompanied by appropriate research. In principle, the initial strategy should be aimed at certain broad objectives and should be designed as an experiment following the lines suggested in proposals put to the Bergen meeting.[39] With regard to objectives, perhaps the major issue is whether whale stocks are to be left free to recover their numbers before exploitation and whether there is to be any hunting of them now or in the future.

In part the necessary decisions as to strategy will be determined by reference to matters outside the Southern Ocean, although the line of argument is essentially the same. The argument above relates to the effects of different exploitation strategies on different elements of a set of resources—to take krill or whales or seals or to take none. Decisions must also be made about the use of other resources in fishing operations—materials and labor for equipment, fuels, and manpower. Under some circumstances, the costs in this sense of

39. See n. 33 above.

taking krill could be held to be prohibitive, and in that case krill would remain a nonresource. It is probable that many species of organisms of minor abundance in the Antarctic will not become effective resources. These considerations will have to be taken into account regarding stocks of fishes, crustacea, and other groups of organisms in the Southern Ocean.

Although no estimates of resource potentials are offered in this essay, some indications of the magnitude of stocks of various groups of organisms which are held to be resources are given in the pages preceding this one.

Figures such as these can be converted into estimates of prospective rewards only by taking into consideration matters, outside the Southern Ocean, which will influence the decisions as to whether the benefit of exploitation will outweigh its cost.

RESEARCH

Probably the prime requirement is the conduct of research on the ecosystems of the Southern Ocean; the equipment and techniques of fishing, handling, and processing; the utilization of the catches; and associated economic matters.

A general review of what is known of the living resources of the Southern Ocean was made at the August 1976 SCAR/SCOR meeting. The meeting's summary report pointed to deficiencies in knowledge of all groups of organisms and to almost all aspects of each group. As a result, the meeting proposed the mounting of "an integrated and well coordinated Biological Investigation of Marine Antarctic Systems and Stocks (BIOMASS)." The objectives proposed for the program can be found on pages 769–770 of this volume. The Woods Hole meeting made detailed proposals for implementation of the program of investigation and for international coordination of and cooperation in its work; for these purposes it made specific recommendations which can be found on pages 771–772.

In addition to the international scientific planning group, the BIOMASS program would have, according to the proposals of the Woods Hole meeting, an international coordination group set up by the IOC (Intergovernmental Oceanographic Commission), with membership from SCAR, SCOR, and IABO under the ICSU umbrella, the Antarctic Treaty member states with membership in SCAR, UNESCO/IOC, and member states of the IOC, FAO, and IWC (International Whaling Commission). The meeting also contemplated formation of ad hoc working groups of the planning group, designation of national organizations for BIOMASS, and nomination of national coordinators.

The IOC already has an International Coordination Group for the Southern Ocean, and several of the commission's global programs such as GIPME (Global Investigation of Pollution in the Marine Environment) and IGOSS

(Integrated Global Ocean Station System) have relevance to the Southern Ocean.

The program of investigations for which this complex network of planning and coordination would be established would have components for physical, chemical, and biological environments; krill; marine mammals and birds; fishes; cephalopods; benthic invertebrates; seaweeds; and remote sensing.

The work would be carried out by national institutions and would involve coordinated multiship operations, supporting ship-based programs and shore-based operations, and arrangements for data reporting and handling. In addition to the proposals for international coordination and cooperation, proposals were formulated for the provision of scientific advice.

For the purposes of the present essay, the main effect of these proposals is a clear demonstration of the scope of required studies of the resources of the Southern Ocean. It is obvious that much more has to be discovered about the living resources of the Southern Ocean and their habitat before reliable estimates of potentials can be matched to a strategy acceptably oriented to the goals of the international community.

Meanwhile, research is going on in technology. At the Woods Hole meeting the FAO presented a review of the problems of harvesting and utilizing Antarctic krill. An examination was made of the problems of search and detection, design of fishing gear, navigation and handling of gear in fishing operations, bringing catch aboard, buffer storage and sorting as phases of in-board handling, and various aspects of processing, especially that of product development. It was pointed out that development and testing of product, processes, and plants are likely to be costly and time consuming, more so than the work needed in connection with search, detection, and catching. However, even greater difficulties are foreseen in connection with effecting full-scale technical development.

> Generally speaking, the harvesting of marine fish resources is most effectively and economically accomplished by the use of the smallest and simplest vessels practicable, and there is a general trend away from long-distance fishing fleets and toward the adoption of more appropriate technology. In the case of the Antarctic krill, however, the distances, the severe environment, and the nature and behaviour of the animal itself combine to make it necessary to use large and powerful ships and high technology for its successful harvesting and utilization. Operation of such vessels requires the exercise of some rather special and unusual human skills, abilities, and experience, that are as yet largely uncommunicable by formal processes of education or in text-books; those who possess them have their own special attitudes and motivations and respond best to certain kinds of incentive. That is to say, those developed nations that are already capable and experienced in high-seas fishing in polar waters may

490 *Regional Developments*

well have to undertake, for some years to come, the actual catching and processing of the Antarctic krill, even if the operation is largely for the direct or indirect benefit of the peoples of the Third World. Hopefully, there will be participation in various ways and to an increasing extent by other advanced and developing countries as the fishery develops.[40]

Among the problems to be met in eventual industry development are those relating to the seasonality of the fishery; transfer (probably at sea) of processed catches; fuel and stores; achievement of an effective compromise among product, processing system, and vessel size, including the difficult matter of rapid reduction of bulk of the catch; and conduct of consumer-acceptance trials. Finally, attention has been drawn to a set of problems which, although not peculiar to the Antarctic, will be especially acute there; these include providing logistic, medical, communications, navigational, and rescue services; the establishment of management regimes; and various matters relating to the world food situation and the new international economic order.

This study summarizes matters that will have to be dealt with if a krill fishery is to be developed. Through its references to research, pilot-scale operations, and full-scale development, the study foresees adjustment of plans in the light of evidence gained from the work itself. In particular, while recognizing the contributions to research on the dynamics of the exploited populations that may be expected from trial fishing operations, it advocates granting highest priority in the allocation of funds, ship time, and accommodation to food technologists, process engineers, and food scientists. This view conforms to the opinion that development of a krill fishery will depend chiefly on whether krill can be converted into a product acceptable to a very large market, presumably in the developing countries.

Development of fisheries for other resources of the Southern Ocean will encounter many of the same problems, but specifically those of distance, isolation, and operating conditions special to this ocean. The development of a krill fishery would probably encourage the development of other fisheries along with it.

POLICY AND INSTITUTIONS

The sense of the preceding section is that, if a krill fishery is to be developed in the Southern Ocean, a great deal of information must be obtained by research in the natural sciences and technology and that, even so, the information obtained might indicate that establishment of such a fishery would be impracticable or uneconomic. Nevertheless, this is not the end of the matter or even

---

40. "The Problems of Harvesting and Utilization of Antarctic Krill," mimeographed (Rome: Fisheries Industries Division, Department of Fisheries, Food and Agriculture Organization, 1976); submitted to the meeting referred to in n. 36 above.

the beginning, since both research and applications of its results are matters which await the resolution of institutional problems.

These problems concern the distribution of powers and responsibilities with regard to (1) organization and conduct of research; (2) regime of the high seas, especially the management of exploitation; (3) formulation of resource-use policy; (4) environmental protection; and (5) distribution of benefits.

POLITICAL CONSIDERATIONS

Although a definition of the Southern Ocean can be made in the manner suggested above, without prejudice to claims which have been made to territory in the Antarctic, the converse is not true; that is, the status of the waters of the Southern Ocean would not remain unaffected by these claims.

All of the Antarctic is subject to some claim, except the sector 150° west (fig. 6), and if these claims were recognized and rights under them were exercised, quite a considerable sea area would become national. Although the Antarctic Treaty declares that nothing in the treaty "shall prejudice or in any way affect the rights, or the exercise of the rights of any State under international law with regard to the high seas" within the area south of 60° south latitude, the article from which this is quoted includes all ice shelves in the area to which the treaty applies and to which the quoted exclusion does not apply. This already excludes from the high seas a substantial area, since the ice shelf lies over and in the sea.

Claims on the Antarctic would appear to be in abeyance and intended to remain so for at least 30 years from the date of the treaty's entry into force. Article IV declares that the treaty is not to be taken as affecting the claims or attitudes to claims of the contracting parties, and although article I reserves Antarctica for "peaceful purposes only" and article V expressly prohibits nuclear explosions and disposal of radioactive waste, most articles deal with the conduct of research, and the treaty has been understood to reserve Antarctica as a refuge for scientific research. This view was reinforced to some extent by a recommendation adopted by the treaty powers in a meeting of their Consultative Committee in Oslo in 1975, urging "states and persons to refrain from actions of commercial exploration and exploitation" while the participating powers seek "timely agreed solutions to the problems raised by the possible presence of valuable mineral resources in the Antarctic Treaty Area." Meanwhile geologists of the United States and West Germany have been searching for uranium in the area, especially west of McMurdo Sound. In addition, a convention has been prepared for the conservation of Antarctic seals and awaits ratification; by January 1977 it had been ratified by five nations and needed two more signatories. This convention would authorize the harvesting of certain species in designated areas under a quota system. Even without these developments, it would not require a large measure of cynicism to conclude that the Antarctic Treaty, for all its admirable features is little more than a

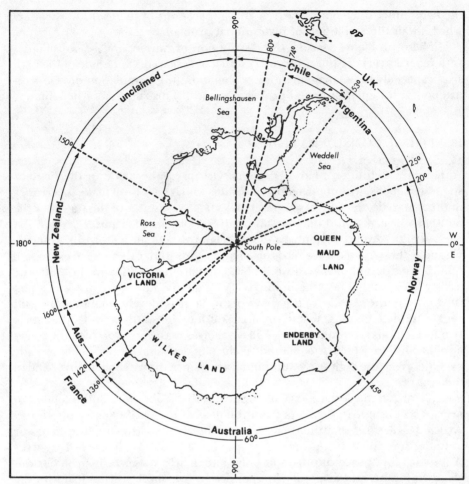

FIG. 6.—The Southern Ocean. (Based on a figure from the *New York Times*, January 17, 1977.)

holding operation pending the accumulation of information and exploration of the grounds on which suitable agreements could be reached.

Claims on Antarctic territory are extremely tenuous and the more doubtful because of the likely magnitude of rewards from resource exploitation there and still more because the Great Powers, which do not recognize any claim, have yet to declare their intentions. It would be extremely optimistic to believe that the international community, through the United Nations, could substantially influence the outcome to protect the area in perpetuity in ways that would fairly benefit the entire world.

APPENDIX

THE ANTARCTIC TREATY[1]

The Governments of Argentina, Australia, Belgium, Chile, the French Republic, Japan, New Zealand, Norway, the Union of South Africa, the Union of Soviet Socialist Republics, the United Kingdom of Great Britain and Northern Ireland, and the United States of America,
   Recognising that it is in the interest of all mankind that Antarctica shall continue forever to be used exclusively for peaceful purposes and shall not become the scene or object of international discord;
   Acknowledging the substantial contributions to scientific knowledge resulting from international cooperation in scientific investigation in Antarctica;
   Convinced that the establishment of a firm foundation for the continuation and development of such cooperation on the basis of freedom of scientific investigation in Antarctica as applied during the International Geophysical Year accords with the interests of science and the progress of all mankind;
   Convinced also that a treaty ensuring the use of Antarctica for peaceful purposes only and the continuance of international harmony in Antarctica will further the purposes and principles embodied in the Charter of the United Nations;
   Have agreed as follows:

Article I

1. Antarctica shall be used for peaceful purposes only. There shall be prohibited, *inter alia,* any measures of a military nature, such as the establishment of military bases and fortifications, the carrying out of military maneuvers, as well as the testing of any type of weapons.
   2. The present Treaty shall not prevent the use of military personnel or equipment for scientific research or for any other peaceful purpose.

Article II

Freedom of scientific investigation in Antarctica and cooperation toward that end, as applied during the International Geophysical Year, shall continue, subject to the provisions of the present Treaty.

   1. Antarctic Treaty between the United States of America and Other Governments, Signed at Washington, December 1, 1959. Conference on Antarctica, Washington, D.C., 1959. Treaties and Other International Acts Series, no. 4780 (Washington, D.C.: Government Printing Office, 1961).

Article III

1. In order to promote international cooperation in scientific investigation in Antarctica, as provided for in Article II of the present Treaty, the Contracting Parties agree that, to the greatest extent feasible and practicable:

   (a) Information regarding plans for scientific programs in Antarctica shall be exchanged to permit maximum economy and efficiency of operations;
   (b) Scientific personnel shall be exchanged in Antarctica between expeditions and stations;
   (c) Scientific observations and results from Antarctica shall be exchanged and made freely available.

2. In implementing this Article, every encouragement shall be given to the establishment of cooperative working relations with those Specialised Agencies of the United Nations and other international organisations having a scientific or technical interest in Antarctica.

Article IV

1. Nothing contained in the present Treaty shall be interpreted as:

   (a) A renunciation by any Contracting Party of previously asserted rights of or claims to territorial sovereignty in Antarctica;
   (b) A renunciation or diminution by any Contracting Party of any basis of claim to territorial sovereignty in Antarctica which it may have whether as a result of its activities or those of its nationals in Antarctica, or otherwise;
   (c) Prejudicing the position of any Contracting Party as regards its recognition of any other State's right of or claim or basis of claim to territorial sovereignty in Antarctica.

2. No acts or activities taking place while the present Treaty is in force shall constitute a basis for asserting, supporting or denying a claim to territorial sovereignty in Antarctica or create any rights of sovereignty in Antarctica. No new claim, or enlargement of an existing claim, to territorial sovereignty in Antarctica shall be asserted while the present Treaty is in force.

Article V

1. Any nuclear explosions in Antarctica and the disposal there of radioactive waste material shall be prohibited.

2. In the event of the conclusion of international agreements concerning the use of nuclear energy, including nuclear explosions and the disposal of radioactive waste material, to which all of the Contracting Parties whose representatives are entitled to participate in the meetings provided for under Article IX are parties, the rules established under such agreements shall apply in Antarctica.

Article VI

The provisions of the present Treaty shall apply to the area south of 60° South Latitude, including all ice shelves, but nothing in the present Treaty shall prejudice or in any way affect the rights, or the exercise of the rights, of any State under international law with regard to the high seas within that area.

Article VII

1. In order to promote the objectives and ensure the observance of the provisions of the present Treaty, each Contracting Party whose representatives are entitled to participate in the meetings referred to in Article IX of the Treaty shall have the right to designate observers to carry out any inspection provided for by the present Article. Observers shall be nationals of the Contracting Parties which designate them. The names of observers shall be communicated to every other Contracting Party having the right to designate observers, and like notice shall be given of the termination of their appointment.
2. Each observer designated in accordance with the provisions of paragraph 1 of this Article shall have complete freedom of access at any time to any or all areas of Antarctica.
3. All areas of Antarctica, including all stations, installations and equipment within those areas, and all ships and aircraft at points of discharging or embarking cargoes or personnel in Antarctica, shall be open at all times to inspection by any observers designated in accordance with paragraph 1 of this Article.
4. Aerial observation may be carried out at any time over any or all areas of Antarctica by any of the Contracting Parties having the right to designate observers.
5. Each Contracting Party shall, at the time when the present Treaty enters into force for it, inform the other Contracting Parties, and thereafter shall give them notice in advance, of

   (a) All expeditions to and within Antarctica, on the part of its ships or nationals, and all expeditions to Antarctica organised in or proceeding from its territory;
   (b) All stations in Antarctica occupied by its nationals; and

(c) Any Military personnel or equipment intended to be introduced by it into Antarctica subject to the conditions prescribed in paragraph 2 of Article I of the present Treaty.

Article VIII

1. In order to facilitate the exercise of their functions under the present Treaty, and without prejudice to the respective positions of the Contracting Parties relating to jurisdiction over all other persons in Antarctica, observers designated under paragraph 1 of Article VII and scientific personnel exchanged under subparagraph 1 (b) of Article III of the Treaty, and members of the staffs accompanying any such persons, shall be subject only to the jurisdiction of the Contracting Party of which they are nationals in respect of all acts or omissions occurring while they are in Antarctica for the purpose of exercising their functions.

2. Without prejudice to the provisions of paragraph 1 of this Article, and pending the adoption of measures in the pursuance of subparagraph 1 (e) of Article IX, the Contracting Parties concerned in any case of dispute with regard to the exercise of jurisdiction in Antarctica shall immediately consult together with a view to reaching a mutually acceptable solution.

Article IX

1. Representatives of the Contracting Parties named in the preamble to the present Treaty shall meet at the City of Canberra within two months after the date of entry into force of the Treaty, and thereafter at suitable intervals and places, for the purpose of exchanging information, consulting together on matters of common interest pertaining to Antarctica, and formulating and considering, and recommending to their Governments, measures in furtherance of the principles and objectives of the Treaty, including measures regarding:

(a) Use of Antarctica for peaceful purposes only;
(b) Facilitation of scientific research in Antarctica;
(c) Facilitation of international scientific cooperation in Antarctica;
(d) Facilitation of the exercise of the rights of inspection provided for in Article VII of the Treaty;
(e) Questions relating to the exercise of jurisdiction in Antarctica;
(f) Preservation and conservation of living resources in Antarctica.

2. Each Contracting Party which has become a party to the present Treaty by accession under Article XIII shall be entitled to appoint representatives to participate in the meetings referred to in paragraph 1 of the present Article,

during such time as that Contracting party demonstrates its interest in Antarctica by conducting substantial scientific research activity there, such as the establishment of a scientific station or the dispatch of a scientific expedition.

3. Reports from the observers referred to in Article VII of the present Treaty shall be transmitted to the representatives of the Contracting Parties participating in the meetings referred to in paragraph 1 of the present Article.

4. The measures referred to in paragraph 1 of this Article shall become effective when approved by all the Contracting Parties whose representatives were entitled to participate in the meetings held to consider those measures.

5. Any or all of the rights established in the present Treaty may be exercised as from the date of entry into force of the Treaty whether or not any measures facilitating the exercise of such rights have been proposed, considered or approved as provided in this Article.

Article X

Each of the Contracting Parties undertakes to exert appropriate efforts, consistent with the Charter of the United Nations, to the end that no one engages in any activity in Antarctica contrary to the principles or purposes of the present Treaty.

Article XI

1. If any dispute arises between two or more of the Contracting Parties concerning the interpretation or application of the present Treaty, those Contracting Parties shall consult among themselves with a view to having the dispute resolved by negotiation, inquiry, mediation, conciliation, arbitration, judicial settlement or other peaceful means of their own choice.

2. Any dispute of this character not so resolved shall, with the consent, in each case, of all parties to the dispute, be referred to the International Court of Justice for settlement; but failure to reach agreement on reference to the International Court shall not absolve parties to the dispute from the responsibility of continuing to seek to resolve it by any of the various peaceful means referred to in paragraph 1 of this Article.

Article XII

1. *(a)* The present Treaty may be modified or amended at any time by unanimous agreement of the Contracting Parties whose representatives are entitled to participate in the meetings provided for under Article IX. Any such modification or amendment shall enter into force when the depositary Government has received notice from all such Contracting Parties that they have ratified it.

(b) Such modification or amendment shall thereafter enter into force as to any other Contracting Party when notice of ratification by it has been received by the depositary Government. Any such Contracting Party from which no notice of ratification is received within a period of two years from the date of entry into force of the modification or amendment in accordance with the provisions of subparagraph 1 *(a)* of this Article shall be deemed to have withdrawn from the present Treaty on the date of the expiration of such period.

2. *(a)* If after the expiration of thirty years from the date of entry into force of the present Treaty, any of the Contracting Parties whose representatives are entitled to participate in the meetings provided for under Article IX so requests by a communication addressed to the depositary Government, a Conference of all the Contracting Parties shall be held as soon as practicable to review the operation of the Treaty.

(b) Any modification or amendment to the present Treaty which is approved at such a Conference by a majority of the Contracting Parties there represented, including a majority of those whose representatives are entitled to participate in the meetings provided for under Article IX, shall be communicated by the depositary Government to all the Contracting Parties immediately after the termination of the Conference and shall enter into force in accordance with the provisions of paragraph 1 of the present Article.

(c) If any such modification or amendment has not entered into force in accordance with the provisions of subparagraph 1 *(a)* of this Article within a period of two years after the date of its communication to all the Contracting Parties, any Contracting Party may at any time after the expiration of that period give notice to the depositary Government of its withdrawal from the present Treaty; and such withdrawal shall take effect two years after the receipt of the notice by the depositary Government.

Article XIII

1. The present Treaty shall be subject to ratification by the signatory States. It shall be open for accession by any State which is a Member of the United Nations, or by any other State which may be invited to accede to the Treaty with the consent of all the Contracting Parties whose representatives are entitled to participate in the meetings provided for under Article IX of the Treaty.

2. Ratification of or accession to the present Treaty shall be effected by each State in accordance with its constitutional processes.

3. Instruments of ratification and instruments of accession shall be deposited with the Government of the United States of America, hereby designated as the depositary Government.

4. The depositary Government shall inform all signatory and acceding States of the date of each deposit of an instrument of ratification or accession,

and the date of entry into force of the Treaty and of any modification or amendment thereto.

5. Upon the deposit of instruments of ratification by all the signatory States, the present Treaty shall enter into force for those States and for States which have deposited instruments of accession. Thereafter the Treaty shall enter into force for any acceding State upon the deposit of its instrument of accession.

6. The present Treaty shall be registered by the depositary Government pursuant to Article 102 of the Charter of the United Nations.

Article XIV

The present Treaty, done in the English, French, Russian, and Spanish languages, each version being equally authentic, shall be deposited in the archives of the Government of the United States of America, which shall transmit duly certified copies thereof to the Governments of the signatory and acceding States.

In witness whereof, the undersigned Plenipotentiaries, duly authorised, have signed the present Treaty.

Done at Washington this first day of December, one thousand nine hundred and fifty-nine.

(Here follow the signatures.)

Regional Developments

Legal Implications of Petroleum Resources of the Antarctic Continental Shelf

F. M. Auburn
University of Auckland

Daunting obstacles face any Antarctic petroleum industry. Remoteness from the world's centers of commerce is one. From McMurdo to Wellington, New Zealand is a distance of 2,200 nautical miles, from the Antarctic Peninsula to Buenos Aires is 1,800 miles, and from McMurdo to Los Angeles is 8,000 miles. Apart from rare medical emergency flights there is no transport to the Ross or Weddell Sea coasts during the 6-month-long winter when the pack-ice belt reaches a width of 900 miles. Minimum sea-ice cover extends over 1,900,000 square miles in March, with a maximum of nearly 8 million square miles in September. Three icebreakers are required to open a shipping channel to McMurdo, the largest U.S. base, in summer. This access only remains usable for 10 weeks.

At least 95 percent of the land area of the continent is covered by an ice sheet with a mean thickness of at least 6,600 feet and a maximum of approximately 14,000 feet. Regular steady winds of 70 miles per hour are encountered. Apart from the Antarctic Peninsula, the mean temperature of the warmest month on the coast is $0°$ C, the coldest being from $-20°$ C to $-30°$ C. Inland temperatures are even lower, dropping to a recorded minimum of $-88.3°$ C.

Despite these hazards, a number of recent developments have focused interest on the most likely economic resource—petroleum on the continental shelf. Concern for offshore oil in Antarctica may be dated from approximately 1969. In July of that year an abbreviated version of a review of Antarctic resources was published.[1] Within a few weeks two companies had approached the New Zealand government, which commenced informal discussions with several other of the 12 consultative parties to the Antarctic Treaty. According to the report itself, offshore drilling for petroleum was impractical because icebergs would destroy rigs. The continental shelf is narrow, and much of it is too deep for drilling platforms. Flexible pipes and drilling apparatus might permit sinking wells beneath the ice shelves. Under such conditions exploitation of Antarctic oil fields appeared impractical without new and probably

1. N. Potter, "Economic Potentials of the Antarctic," *Antarctic Journal* 4 (1969): 61–68.

© 1978 by The University of Chicago. 0-226-06602-9/77/1978-1021$01.44

expensive technologies. Because of the existence of large accessible oil deposits elsewhere, the outlay for Antarctic drilling might never prove worthwhile.[2]

No great urgency was attached to the matter by the consultative parties. At the Wellington treaty meeting in 1972 it was recommended that the topic "Antarctic Resources—Effects of Mineral Exploration" be carefully studied and included on the agenda of the next meeting.[3] Within a few months the situation had changed completely. Three of four holes drilled in the Ross Sea floor by the research vessel *Glomar Challenger* for the Deep Sea Drilling Project showed traces of gaseous hydrocarbons. Scientific evaluation of the holes emphasized that there was no present way of assessing their significance and it would be extremely premature to attach any economic significance to Ross Sea hydrocarbons on this basis.[4] No meaningful estimate of the Ross Sea's potential can be derived from these holes, nor can any of the geophysical surveys be considered as resource exploration.[5]

But the political effect of the work of the *Glomar Challenger* went far beyond its scientific impact. In May 1973 a symposium of experts from the consultative party nations took place under the auspices of the Nansen Foundation in Norway. Although termed an "informal discussion," this meeting was for all practical purposes a reaction to the hydrocarbon results in the Ross Sea. Its significance was underlined by the extensive coverage of legal issues.[6] Continued interest in offshore oil was guaranteed by the sharp rise in prices after the Yom Kippur War in 1973.

At the same time, the issue was considered of sufficient importance for the Antarctic Policy Group of the U.S. National Security Council to prepare a secret study of Antarctic resources. An estimate of 45 billion barrels of petroleum and 115 trillion cubic feet of natural gas for the Ross, Weddell, and Bellingshausen Seas in a confidential U.S. Geological Survey document. Although not intended for release, the figures became public.[7] With considerable justification, the Office of Energy Resources of the Geological Survey pointed out that the figures were expected to serve only as orders of magnitude. From the available data, or the absence thereof, one might also add that other estimates, such as one suggesting 20 billion barrels of oil and 80 trillion cubic feet of gas, could be cited. Any figures based mainly on geological

2. N. Potter, *Natural Resource Potentials of the Antarctic* (Burlington, Vt.: Lane, 1969), pp. 28–29.

3. Recommendation VII-6.

4. D. E. Hayes and L. A. Frakes, "General Synthesis," in *Initial Reports of the Deep Sea Drilling Project* (La Jolla, Calif.: Scripps Institute of Oceanography, 1975), 28:940.

5. D. E. Hayes, "Antarctic Marine Geology and Geophysics," in *Framework for Assessing Environmental Impacts of Possible Antarctic Mineral Development*, ed. Institute of Polar Studies (Columbus: Ohio State University, 1977), pt. 2, app. B7-B10.

6. F. Sollie, "Antarctic Resources," in U.S. Congress, Senate, *U.S. Antarctic Policy: Hearing before the Subcommittee on Oceans and International Environment of the Committee on Foreign Relations*, 94th Cong., 1st sess., 1975, p. 68.

7. J. Spivak, "Frozen Assets?" *Wall Street Journal* (February 21, 1974).

formations in other continents can only be a guess. Furthermore, the existence of a resource has no relation to the question of whether it can be commercially exploited. This point is particularly pertinent to the areas of the Weddell and Bellingshausen Seas, which are permanently inaccessible to ships. On the other hand, the participation of the Department of the Interior and the Federal Energy Office indicated a practical interest in the utilization of Antarctic minerals.

No estimate of oil and gas resources appeared in the Geological Survey's review of the continent's minerals, which was published in 1974. Utilizing geological comparisons from neighboring continents, the Ross Sea continental shelf was seen as most promising, followed by the Weddell Sea, with the Bellingshausen Sea being regarded as holding only natural gas.[8]

At the Oslo consultative meeting in 1975, minerals were dealt with at considerable length. It was resolved to convene a special preparatory meeting in Paris in 1976 on "Antarctic Resources—the Question of Mineral Exploration and Exploitation." Environmental implications of mineral resource activities in the treaty area were to be studied. The Scientific Committee on Antarctic Research (SCAR) was invited to assess the possible environmental impact of mineral exploration and exploitation on the basis of available information.[9] As part of the final report, but not in the form of a recommendation which could become legally binding, states and persons were urged to refrain from actions of commercial exploration and exploitation while the consultative parties were seeking timely agreed solutions.[10]

The Scientific Committee on Antarctic Research presented an interim report to the Paris meeting, but no results of the discussions were made public. Perhaps this was an indication of the polarization of views already apparent at the Nansen Foundation meeting in 1973. Apparently the SCAR suggestions for considerable augmentation of the earth-science research programs, baseline studies, and investigation of key species[11] were the subject of general agreement, if not of immediate action.

Seven nations have made claims to extensive areas of Antarctica. In the South American sector, the British Antarctic Territory extends over the Antarctic Peninsula and the Weddell Sea coast south of 60° south, between 20° and 80° west. Argentina's claim (from 25° to 74° west) and that of Chile (from 53° to 90° west) cannot be reconciled with one another or with the British assertions. All three include the Antarctic Peninsula, which is the most accessible area of this portion of the continent. Nearly half the total number of

8. N. A. Wright and P. L. Williams, *Mineral Resources of Antarctica* (Reston, Va.: Geological Survey, 1974), pp. 16–17.

9. Recommendation VIII-14.

10. "Final Report of the Eighth Antarctic Treaty Consultative Meeting," mimeographed (Oslo, 1975), p. 5.

11. G. Knox, "Antarctic Resources: Implications for the Antarctic Treaty and New Zealand," *New Zealand International Review* (July–August 1976), pp. 18–22.

Antarctic bases are to be found on the peninsula and surrounding islands. In part this may be explained by the less strenuous conditions of the so-called Antarctic banana belt;[12] but the major reason for this clustering of stations is international rivalry, exemplified by Operation Tabarin in 1943, when several huts were erected in the sector to support British claims to sovereignty.

Norway's Dronning Maud Land extends from 20° west to 45° east. Australia's claim covers the area between 45° east and 160° east, being divided into two portions by the French Adélie Land (from 136° to 142° east). Between 160° east and 150° west the Ross Dependency is administered by New Zealand. Marie Byrd and Ellsworth Lands (from 150° to 90° west), sometimes referred to as the "American" sector, have not been the object of formal claims. Presumably those nations which had purported to appropriate territory in the Antarctic wished to leave this sector for the United States. It is hardly surprising that no such action was taken. This region is the least approachable portion of the continent.

Had the United States proclaimed sovereignty over the sector it would, for all practical purposes, have recognized the Antarctic territories of other nations. This would have left the most attractive areas, the Ross Sea coast and the Antarctic Peninsula, under the control of those states, even though U.S. legal arguments with regard to both sectors are of considerable weight when compared with those of the nations which have formally lodged claims. American policy has been to refuse to recognize the Antarctic claims of other states and to reserve its own rights to make claims arising from the substantial efforts of U.S. expeditions and bases. A similar attitude has been taken by the Soviet Union since 1948.

Shortly before and immediately after the Second World War, U.S. expeditions were sent with the aim of laying the foundation for formal claims. Had the interested parties been confined to the seven countries already asserting sovereignty, there might have been the possibility of a territorial settlement. But once the Soviet Union intervened such a solution became difficult, if not impossible. It is correct that no country has done more than the United States to explore the Antarctic, and also that America displayed a lack of vision in not making formal claims equal to that of the critics of Seward's purchase of Alaska.[13] Yet for the practical politician this option apparently expired in 1948.

International law does not provide any easy solution to territorial disputes in the Antarctic. Traditionally, discovery of unknown lands and a formal taking of possession did not suffice for the acquisition of sovereignty. Actual settlement was required. But a number of twentieth-century cases have diluted the effective occupation test. Symbolic annexation gave title over a small,

12. D. Lewis, *Ice Bird* (London: Collins, 1975), p. 142.
13. E. W. Cole, "Claims of Sovereignty over the Antarctic" (thesis, Judge Advocate General's School, Charlottesville, Va., 1958), p. 2.

remote, and uninhabited island in the *Clipperton Island* case.[14] Elsewhere it has been stated that sovereignty cannot be exercised at every moment in every part of the territory.[15] Of particular relevance to Antarctica was the dispute between Norway and Denmark over the *Legal Status of Eastern Greenland*.[16] Considerable stress was placed on the intention and will to exercise sovereignty and the manifestation of state activity. On this test, polar claimants could invoke the passing of laws, issuing of permits, and other activities not necessarily involving actual settlement.

Leaving aside for the moment the effect of the Antarctic Treaty, the most persuasive arguments would, today, be put forward on behalf of the United States and the Soviet Union, whose Antarctic programs far exceed those of other countries. It has been contended that nations whose possessions reach up to the Arctic should have a right to all the territory between their eastern and western extremities on the one hand and the North Pole on the other. Transposed to the Antarctic, the sector theory would avoid the problem of the absence of actual settlement apart from a few small and isolated scientific stations. It would provide an argument for those countries which came first but whose activities are now on a much smaller scale than those of the United States and the Soviet Union. Even if the high seas are excluded, there does not appear to be any legal justification for this approach.[17]

By the early 1950s, conflict seemed inevitable in Antarctica, particularly in the South American sector, where protest notes were exchanged, warships appeared, huts were destroyed, and shots were fired. This was avoided by the cooperation before and during the International Geophysical Year (1957–58). Numerous stations were established during this period, including Soviet bases in the Australian Antarctic Territory and U.S. stations at McMurdo (Ross Dependency) and at the South Pole. Presumably one reason for establishing the latter was to give the United States a foothold in each of the national claims.

As the Soviet Union clearly intended to maintain its Antarctic program, the United States could not withdraw from the major bases established for the International Geophysical Year (IGY). Russian presence was seen as the main stimulus for more activity.[18] Uncritical praise of the Antarctic Treaty of 1959 as a major step forward in international relations was unduly optimistic. Objective analysis of the physical setting and the political outcome of the International Geophysical Year supports the view that the treaty was a modest, limited, and relatively costless attempt at international control.[19] It was essentially a formal

---

14. *American Journal of International Law* 26 (1932): 390–94.
15. J. B. Scott, in *Hague Court Reports*, 2d ser. (1932), p. 94.
16. *Legal Status of Eastern Greenland* (1933) P.C.I.J., ser. A/B, no. 53, p. 22.
17. O. Svarlien, "The Sector Principle in Law and Practice," *Polar Record* 10 (1960–61): 248–63.
18. R. D. Hayton, "The Antarctic Settlement of 1959," *American Journal of International Law* 54 (April 1960): 348–71.
19. H. J. Taubenfeld, "A Treaty for Antarctica," *International Conciliation* 531 (January 1961): 245–322.

version of the status quo reached during the IGY.[20] Evidence may be found in the provision for freedom of scientific research (Art. II) and the freezing of claims (Art. IV). Confronted with conflicting territorial demands in the South American sector and the prospect of the introduction of Cold War politics in Antarctica, the treaty was devised as a means of accommodating the requirements of the parties at the time.

For these reasons the treaty did not purport to solve the main outstanding issues, but attempted to shelve them. Hence the opaque wording of Article IV on the effect of the treaty on claims. Several nations, fearing even the implication of internationalization of the continent, insisted that there should be no permanent body to carry out the aims of the agreement.[21] As the preamble indicated, the treaty was primarily concerned with furthering international cooperation in scientific investigation. Even for this end it is short, phrased in unspecific terms, and does not provide complete solutions for those problems which could have been envisaged in 1959, such as jurisdiction over criminal offenses. Certainly there is nothing in the treaty to provide a legal framework for mining, or indeed for any type of commercial activity.

Antarctic policy is decided upon by consultative meetings of representatives of the 12 original contracting parties to the treaty at approximately 2-year intervals. Under the rules of procedure formulated at the first consultative meeting in 1961, its recommendations must accepted by all representatives present. Even then, recommendations only become effective when approved by all 12 original signatories.[22] Therefore, any measures to be taken under the treaty are subject to the contrary vote of any one of these nations, even if that country's representative has endorsed the proposal. In the present context, this means that if New Zealand and the United States were to agree to encourage oil exploration in the Ross Sea within the framework of the Antarctic Treaty, then Chile, France, South Africa, or the Soviet Union could each put a stop to the project—hardly a state of affairs calculated to encourage investment of large capital sums in a most hazardous environment. In practice the consultative meetings have moved very cautiously, finding particular difficulty in dealing with controversial topics.[23] General confidence in the work of the meetings is not encouraged by the secrecy of the actual proceedings. As both the United States and the Soviet Union participate, there is no question of national security. Presumably the purpose is to avoid appraisal and possible criticism by the general public.

Despite the calling of preparatory meetings, it is clear that the multifarious issues of international cooperation involved in the Antarctic require some type

---

20. G. Battaglini, *La condizione dell'Antartide nel diritto internazionale* (Padua: Cedam, 1971), p. 301.
21. R. E. Guyer, "The Antarctic System," *Hague Recueil* 139 (1973): 149–226.
22. Article IX(4), Antarctic Treaty.
23. T. Hanevold, "The Antarctic Treaty Consultative Meetings—Form and Procedure," *Cooperation and Conflict* (1971), pp. 183–99.

of permanent organization. Research activity on the continent is coordinated by SCAR, which was established during the IGY. The Scientific Committee on Antarctic Research is not an intergovernmental body. Its functions are to a large degree performed by a number of permanent working groups and groups of specialists. In general, all 12 consultative parties' nationals are to be found on each working group. For all practical purposes it may be concluded that SCAR has, as one of its primary functions, the carrying out of the recommendations of the treaty consultative meetings, despite the absence of any formal constitutional linkage with the treaty.

It has already been pointed out that the treaty's purpose was to continue international cooperation in scientific research but not to solve the basic issues of Antarctic international law. In the short run such a narrow approach may have been seen as a prerequisite to securing agreement, but in the long run the difficult problems which were either swept under the carpet or not envisaged render the treaty system very vulnerable to disruption. In the present context it is only possible to isolate a few of the more salient points.

It is far from clear that the treaty covers offshore resources. As its provisions apply to the area south of 60° south, including all ice shelves, but without prejudice to rights to the high seas within that area,[24] the claimant states would regard rights over the continental shelf as appurtenant to their national territory. Following the logic of the official U.S. position, the State Department denies that any nation exercises such rights over the Antarctic continental shelf.[25] A similar approach may be expected from the Soviet Union. With regard to the continental shelves of Marie Byrd and Ellsworth Lands, it would appear that, as they are not appurtenant to the territory of any country, they are unclaimed and that all parties to the treaty would concur on this point.

It would therefore seem that, on the interpretation of the great powers, the entire continental shelf of Antarctica is beyond national jurisdiction. If this is so, then it should come under the control of the projected International Seabed Authority discussed by the Third United Nations Conference on the Law of the Sea. One can safely assume that the U.S. government does not wish this to occur.[26]

There may also be strong sentiment among the other 11 consultative parties to ensure that offshore minerals are regulated under the treaty. If any other body dealt with the matter, the power of the 12, even if represented, would be considerably diluted if not totally offset. At the consultative meetings each consultative party has in effect a double veto, giving it a degree of control which would certainly not be granted by an International Seabed Authority,

---

24. Article VI, Antarctic Treaty.
25. Sollie (n. 6 above), p. 19.
26. J. Rose, "Antarctic Condominium: Building a New Legal Order for Commercial Interests," *Marine Technology Society Journal* 10 (January 1976): 19–27.

particularly to one of the smaller countries. South Africa is obviously in a much better negotiating position in the Antarctic Treaty system than it could hope for in any United Nations organization. As long as it is a question of reaching agreement among 12 countries, even those as dissimilar as the original contracting parties, one can envisage the possibility of a regime for Antarctic minerals. It is more difficult to see a consensus among more than 140 nations, many of whom have no direct Antarctic interest or experience and some of whom might well regard their consent to an Antarctic mineral regime as a trade-off for concessions in totally different matters. For the 12 to agree to an International Seabed Authority controlling offshore minerals in Antarctica would be contrary to their past attitude to UN interest in the continent.

With considerable justification, the consultative parties have been called a club of 12,[27] an exclusive group.[28] These countries participate in the decision-making process without any necessity for qualification.[29] Nations subsequently acceding to the treaty have to demonstrate their interest in Antarctica by conducting substantial scientific research activity there, such as the establishment of a scientific station, or the dispatch of a scientific expedition.[30] If the main criterion is taken to be the maintenance of bases, then two of the 12, Belgium and Norway, would not have qualified for consultative status had they acceded to the treaty. On the other hand, some of the acceding countries, such as the German Democratic Republic and Poland, have carried out a considerable volume of Antarctic research. Governmental and private work has also been undertaken by nontreaty nations, one example being Italy.

As long as the consultative meetings were primarily concerned with scientific research, their exclusive nature did not raise serious difficulties. But when a possible economic resource of substantial value is to be dealt with, it may be assumed that nonconsultative parties to the treaty and countries which have not adhered to it will express their interest. Informal moves in this direction have already taken place. As the prospect of exploration becomes clearer, this concern will increase. It will be pointed out that the treaty has been in force for 16 years and no serious public discord has been voiced. Past experience is no reliable guide. For the treaty system the real test will come when an issue arises which is of sufficient worth to arouse international rivalries. Offshore oil and gas is just this type of problem. Recent examples are the disputes over the Senkaku Islands and the continental shelf of the Aegean Sea.

It has been pointed out that the treaty does not make specific provision for commercial activities, but the consultative parties will probably wish to have

27. R-J. Dupuy, "Le Traité sur l'Antarctique," *Annuaire Français de Droit International* (1960), pp. 111–32.
28. G. Skagestad, "The Frozen Frontier: Models for International Cooperation," *Cooperation and Conflict* 10 (1975): 167–87.
29. Article IX(1), Antarctic Treaty.
30. Article IX(2), Antarctic Treaty.

continental shelf resources dealt with under the treaty regime. There have been various views on the question of whether the treaty and recommendations of the consultative meetings would actually affect exploration and exploitation. Several countries have argued that commercial mineral exploration cannot be initiated until new international measures are drafted. Otherwise scientific research would be disrupted and threats to the ecosystem created.[31] It is difficult to find clear legal authority in the treaty for this approach. Another view would demand the prior consent of all consultative parties.[32] Here reliance is placed on the preamble, which states that the continent shall not become the subject of international discord. Furthermore, Article X provides that each party undertake to exert appropriate efforts, consistent with the Charter of the United Nations, to the end that no one engages in any activity in Antarctica contrary to the principles or purposes of the treaty. Once again the legal reasoning is hardly persuasive. Put in simple terms, the argument says that if a country peacefully seeks Antarctic minerals and a consultative party does not agree with the project, then there is international discord. As regards third-party countries, there is no general acceptance of the proposition that they can be bound by a treaty to which they have not acceded. Even if they were so bound, there is no prima facie ground for the view that mineral exploration and exploitation must be contrary to the principles or purposes of the Antarctic Treaty. Detailed analysis of the treaty and recommendations suggests that they can be complied with.

In a lawyer's eyes the approach of the United States, that commercial activities are permitted provided the treaty is complied with,[33] would seem preferable to the two previous views. If commercial interest in Antarctic petroleum results in pressure to permit exploration and the consultative parties cannot agree on a suitable framework, then the question will become of considerable importance. Precedent for such pressure may be found in the development of the ocean mining industry.

Basic to any consideration of a mineral regime is the present status of claims. At the 1959 conference which drafted the treaty at least one country, New Zealand, indicated that it would be prepared to relinquish national rights in favor of an international regime.[34] At that time, and subsequently, the hope has been expressed that Article IV of the treaty freezing claims for the duration would lead to a de facto internationalization. Enough time has passed since the treaty came into force on June 23, 1961, to allow a firmly based opinion on any possible trend in this direction.

Insofar as the evidence is publicly available, it would seem that those nations which were firmly committed to maintaining national rights in

---

31. Sollie (n. 6 above), p. 79.
32. Ibid.
33. Rose (n. 26 above), p. 24.
34. Department of State, *The Conference on Antarctica* (Washington, D.C.: Department of State, 1960), p. 10.

Antarctica in 1959 have not changed their opinion. If there has been any shift it has been toward an increased emphasis on national claims, even bringing suggestions in the press of several Latin American countries that new claims might be made.

Some 50 requests for exploration licences have been filed with the United Kingdom for the Falkland (Malvinas) Islands continental shelf. Ever since the British took the islands in 1833, Argentina has demanded their return, finding very considerable support after 1960 in the United Nations. Until recently it seemed that the British stake in this colony did not warrant the continued friction with Argentina. Between 1971 and 1974 a series of agreements between the two countries improved communications between the islands and Argentina. In February 1976 the Argentinian destroyer *Almirante Storni* fired warning shots across the bows of the British research ship *Shackleton* which was, in the Argentinian opinion, carrying out geological surveying with a view to the exploitation of hydrocarbons on the Argentinian continental shelf of the Malvinas Islands.[35] Whether the Argentinian or the British interpretation is correct is a difficult question of international law.[36] Argentinian reaction to such research is an indication of that country's attitude to its Antarctic claim, which is directly related to the Falkland Islands dispute.

Brazil acceded to the treaty in 1975. In 1972 a private scientific expedition from that country to the Antarctic was under consideration. Public opinion in Chile feared that Brazil would lodge a claim. A few months later Argentina's president and cabinet traveled to one of that country's Antarctic bases, which was declared the temporary capital of the nation as an affirmation of sovereignty.[37] Chile's president visited Antarctica in January 1977,[38] presumably for the same purpose.

As the result of a Norwegian expedition in 1964, the government of that country initiated contacts with other treaty nations to regulate the possible revival of commercial exploitation of seals in the offshore Antarctic pack ice. On any estimate, this would be a small operation by international fishing standards. From 1964 to 1970 successive consultative meetings discussed guidelines and then a draft convention on the topic. Ultimately the convention was drafted at a conference in 1972 outside the framework of the Antarctic Treaty. One explanation given was that the conservation of seals in the sea does not fall within the scope of the treaty.[39] Apparently some countries insisted on a narrow interpretation of the territorial ambit of the treaty to prevent incursions

---

35. U.N. Doc. A/31/55 (February 24, 1976).
36. "The Shackleton Incident Could Profit International Law," *Nature* 259 (February 1976): 435.
37. "Base Marambio: Capital accidental de la republica," *Antartida* (May 1974), pp. 19–23.
38. W. Sullivan, "19 Nations to Discuss Antarctica Resources," *New York Times* (January 17, 1977), p. 35.
39. "The Final Report of the Sixth Antarctic Treaty Consultative Meeting, mimeographed (Tokyo, 1970), p. 3.

on their sovereignty. This episode is also of considerable relevance in illustrating the difficulties of reaching agreement on an offshore mineral regime.

In the past, Antarctic environmental protection has left much to be desired. McMurdo station's rubbish dump was severely criticized,[40] although measures have been taken since 1969 to alleviate local pollution.[41] Particular care has been taken in unique areas such as Lake Vanda[42] and the Dry Valleys generally. Even the relatively small scientific programs carried out to date have already had significant effects, such as the reduction of the population of penguin rookeries. Antarctica's environment is unique and finely balanced. Life is manifest in large numbers of individuals of a small range of kinds.[43] Confined ice-free areas serve for breeding during a short summer season. Sea life is prolific but restricted to a few species, resulting in a food cycle peculiarly vulnerable to pollution at the lower levels.

Long-term effects of oil spills in polar areas are unknown. Krill, small shrimplike zooplankton, is the key species, being the main food source for squid, fish, some seals, penguins, other birds, and whales. Commercial fishing by the Soviet Union has already commenced, with Japan, Taiwan, Poland, West Germany, the United Kingdom, and Norway carrying out research. Krill promises to be one of the world's major fish resources. Stock estimates range between 800 million and 5 billion tonnes, with an annual biological production of from 200 million to 300 million tonnes. Krill protein could equal that of all other sources of the world's oceans.[44] Sublethal damage to krill from oil could thus cause very heavy direct economic losses, apart from possible damage to animals higher in the food chain. Fur seals' insulation would be affected, as would birds' feathers. Benthic communities would suffer from heavy hydrocarbons sinking to the ocean floor. Adverse impacts would be extended due to the slow rate of microbial degradation of oil at low temperatures. As there are relatively few ice-free areas, local effects would be particularly severe.[45]

Likelihood of damage from the oil industry in Antarctica is considerably greater than in temperate regions. Even small ice formations will prove most costly to deflect from collisions with drilling rigs. Three U.S. icebreakers once moved a small berg 2½ miles, but it took 12 hours work.[46] Attempts at destruction by bombing and other means have not succeeded to date.[47] Towing

---

40. D. Braxton, *The Abominable Snow-Women* (Rutland, Vt.: Reed, 1969), p. 123.

41. B. C. Parker, ed., *Conservation Problems in Antarctica* (Lawrence, Kans.: Allen, 1972), p. 277.

42. R. B. Thomson, "United States and New Zealand Cooperation in Environmental Protection," *Antarctic Journal* 6 (May–June 1971): 59–61.

43. R. C. Murphy, "The Urgency of Protecting Life on and around the Great Southerly Continent," *Natural History* 76 (1967): 21–31.

44. H. Black, "Antarctica—Storm Clouds Are Gathering," *Current Affairs Bulletin* 1 (June 1976): 4–15.

45. Hayes (n. 5 above), p. xiv.

46. N. Potter, *Natural Resource Potentials of the Antarctic* (n. 2 above), p. 54.

47. G. M. Schultz, *Icebergs and Their Voyages* (New York: Morrow, 1975), p. 67.

experiments have recently been carried out off the Canadian coast. But the only presently proven means of dealing with the approach of a large iceberg is to remove the drilling rig from its path.[48] For scientific sampling, such as that done by the *Glomar Challenger,* or exploratory commercial operations, this action may be feasible. In the case of conventional exploitation platforms this is not possible.[49] As icebergs may scour the sea bottom to a depth of 40 feet or more, undersea storage tanks and pipelines would have to be installed well below this depth, involving new technology[50] and greatly increased costs. Gross damage to the ecosystem could well take centuries to recover. Extensive onshore facilities would be required in the choicest areas, reducing the small ice-free coastal zone available for penguins, seals, and scientific stations and research. Construction of tanks, plant, and living quarters could not be landscaped with trees and plants.

Further hazards may be predicated from the depth of the Antarctic continental shelf, whose seaward edge averages 1,650 feet depth in contrast to the world average of 600 feet.[51] For all practical purposes, there is no transport to the continent in winter. If there were a blowout at the end of the Antarctic summer, it might be impossible to drill a relief well for 7 months. In the Arctic it has been postulated that severe ice conditions, as in 1975, might prevent access in the following summer. For profitable commercial exploitation of hydrocarbons in so extreme and isolated an area, one must assume a substantial field by international standards and a fleet of supertankers operating in the short Antarctic summer. There would therefore be a considerable routine operational discharge of oil, apart from the significant probability of the accidental loss of a fully loaded 250,000-ton tanker.[52]

Perhaps the major environmental task is obtaining basic data. There is a general absence of information on all topics relevant to the exploration and exploitation of Antarctic offshore petroleum, whether it is a question of ice and iceberg dynamics or the long-term effects of hydrocarbons on the ecosystem. Some assistance will be sought from high-latitude Arctic drilling experience, but available guidance from this source is of only limited utility. Canadian exploratory drilling is the only real experience of heavy ice work available. Environmental assessment of the impact of such activities in the Arctic has only been caried out on a limited scale. Recommendation VIII-14 of the 1975 consultative meeting at Oslo urged governments to study the environmental implications of mineral resource activities in the Antarctic Treaty area and other related matters, including joint studies. But until there are specific and well-funded programs, over and above current research, there does not appear to be any likelihood of a comprehensive and satisfactory assessment of the impact of future commercial activities.

48. Sollie (n. 6 above), p. 71.
49. Hayes (n. 5 above), app. V-21.
50. Sollie, pp. 71–72.
51. Wright and Williams (n. 8 above), p. 17.
52. Hayes, p. xvii.

512   Regional Developments

If scientific evaluations of the effects of an oil industry are complex, legal issues are more so. Even the relatively mild requirements of the 1964 Agreed Conservation Measures have only the force of voluntarily accepted guidelines. Several countries have not approved them due to legal difficulties.[53] Still, Antarctica is subject to the most comprehensive protection of any major region.[54] For instance, it is not permissible to bring budgerigars to the continent. Enforcing minor restrictions on scientists and support personnel working in programs whose logistic support is almost entirely dependent on governments is relatively simple. Compelling major oil companies to spend very large additional sums for operations on the other side of the world is another.

It has been suggested that the Antarctic should properly be regarded as an international park.[55] Some support may be found in the preamble to the Agreed Conservation Measures of 1964 by which the treaty area is considered a "Special Conservation Area." At the Second World Conference on National Parks in 1972, the Antarctic Treaty nations were urged to negotiate to establish the continent and the surrounding seas as the first world park under the auspices of the United Nations. This proposal was supported by the South Pacific Conference on National Parks and Reserves in 1975. Predictably, the concept was not put on the agenda of the 1975 consultative meeting, whose views may be gathered from the preamble to Recommendation VIII-13 on the Antarctic Environment which recognized that prime responsibility for Antarctic matters, including protection of the environment, lies with the states active in the area which are parties to the treaty.

Two reasons may be assigned for the refusal to consider the world park proposal. First, it is a form of internationalization which is not acceptable to several treaty countries. Second, repeated attempts to give the United Nations or one of its agencies a direct or indirect voice in Antarctic affairs have been rejected by the treaty nations. Some examples are a proposal put to the Committee on Natural Resources of the Economic and Social Council in 1971 to study Antarctic mineral resources[56] and the suggestion put to the United Nations Environment Program for its involvement in Antarctica.[57]

One legal issue has already been settled. It appears that U.S. government departments have accepted that the National Environmental Protection Act applies to Antarctic minerals. But this important concession can only be viewed as an initial step. Some other treaty countries do have legislation on the topic, but discussion and litigation, even of municipal law issues, are generally at an

53. Guyer (n. 21 above), p. 195.
54. D. Andersen, "The Conservation of Wildlife under the Antarctic Treaty," *Polar Record* 14 (January 1968): 25–32.
55. R. C. Murphy (n. 43 above), p. 22.
56. U.N. Doc. E/C.7/5 (January 25, 1971).
57. Sollie (n. 6 above), p. 20.

early stage. For instance, in Australia it is not yet clear to what extent the Commonwealth Environmental Protection (Impact of Proposals) Act of 1974 applies to possible proposals in the Antarctic.[58] New Zealand procedures are not legally binding, making it difficult to persuade departments to prepare environmental impact reports in unusual cases or at the stage of preliminary international negotiations.

For a country to enforce its laws against its own citizens in respect of actions in Antarctica may raise some problems.[59] To take action against the nationals of another state would add a political dimension and invite a sovereignty dispute. For instance, in 1961 scientists from the University of Canterbury, New Zealand, found that the small Adelie penguin colony at Cape Royds in McMurdo Sound had been severely reduced and attributed this to the frequent landing of helicopters with visitors within a few yards of the colony, scattering breeding birds in all directions. As it was thought that New Zealand laws could not be enforced against nationals of other countries, suggestions for the protection of the penguins were presented through the Antarctic Division of the Department of Scientific and Industrial Research to U.S. naval authorities in Christchurch.[60] It is unlikely that legal issues involving hundreds of millions of dollars and so valuable an international commodity as petroleum could be solved as easily.

Major emphasis has been placed on the environment in the treaty nations' discussion of minerals. At times this consideration has outweighed interest in the resource itself. Such government attitudes are unusual, particularly since the same governments have taken the opposite points of view in other parts of the world. In the Antarctic this is not suprising. Scientific work on the continent has traditionally been regarded as pure rather than applied research. Science itself was seen as the chief industry.[61] The treaty was based on such an interpretation. Underlying problems of sovereignty and jurisdiction could be avoided so long as politically neutral activities of this nature were carried on. In the name of environmental protection the consultative parties might unite to assert control over minerals, without having to tackle the potentially divisive legal issues such as the extent of the treaty area or whether commercial activities can be subsumed under the treaty at all.

It may be concluded that the 12 consultative parties have formed an exclusive club. There is no logical reason for outsiders to accept any mineral regime set up under the Antarctic Treaty. It would be unwise to extrapolate

---

58. Cf. Stephen J. in *Murphyores Inc. Pty. Ltd.* v. *Commonwealth* (1976) 9 Australian Law Reports 199 at 208.

59. F. M. Auburn, "International Law and Sea-Ice Jurisdiction in the Arctic Ocean," *International and Comparative Law Quarterly* 22 (July 1973): 552–57.

60. B. Stonehouse, "Animal Conservation in Antarctica," *New Zealand Science Review* 23, no. 1 (1965): 3–7.

61. N. Potter, "The Antarctic: Any Economic Future?" *Science and Public Affairs* 26 (December 1970): 94–99.

from the past success of the treaty in establishing a regime for scientific research to any future regulation of commercial enterprise. Mineral exploitation will inevitably entail severe local pollution damage which can only be mitigated by expensive precautions. Enforcement measures will be of doubtful legal value in view of the unresolved territorial disputes.

No precedent exists for the joint control by 12 states of large-scale commercial enterprises extracting a resource in great demand. If, as appears likely, the prime exploration target will be the Ross Sea, treaty countries which have no territorial claims or activities in the sector will have a veto over possible proposals of the claimant (New Zealand) and the chief sponsor of research (United States). As both countries are making major efforts to find energy sources under their own control, there will be internal pressure for them to act outside the treaty, whether alone or in concert. In view of the serious legal doubts as to whether the treaty covers the continental shelf, such activities would have both moral and judicial support. Should the treaty nations be unable to agree upon a mineral regime, there will be no further national support for unilateral action.

Currently the treaty system, including the recommendations of consultative meetings, provides no guidance for commercial enterprise. Whether a comprehensive regime for exploration and exploitation of offshore petroleum can be agreed upon by the consultative parties is far from clear. Regulation requires licensing. Licenses must be issued by a competent authority, usually a government department. It will be difficult to avoid the unresolved sovereignty disputes, particularly since the resource concerned is in such demand. Elsewhere it has been pointed out that the lack of resolution of the fundamental legal issues renders the Antarctic regime unstable. For instance, a serious criminal offense on the continent could bring the territorial differences to the surface, as illustrated by a U.S. case of alleged killing on Fletcher's Ice Island in the Arctic Ocean.[62] Commercial activities would seriously increase the likelihood of such an incident.

No one can predict when Antarctic minerals will be exploited. If forecasts of the exhaustion of current petroleum reserves within several decades are correct, then commercial exploration in Antarctica could well take place within 15–20 years. Whatever may be the future, it is clear that the putative offshore resources are already of political importance to the consultative parties.

ADDITIONAL REFERENCES

Auburn, F. M. *The Ross Dependency.* The Hague: Nijhoff, 1972.
Da Costa, J.-F. *Souveraineté sur l'Antarctique.* Paris: Libraire Générale de Droit et de Jurisprudence, 1958.

62. C. O. Holmquist, "The T-3 Incident," *U.S. Naval Institute Proceedings* 98 (September 1972): 45–53.

Dollot, R. "Le Droit international des espaces polaires." *Hague Recueil* 75 (1949): 118–200.
Hambro, E. "Some Notes on the Future of the Antarctic Treaty Collaboration." *American Journal of International Law* 68 (April 1974): 217–26.
Hatherton, T., ed. *Antarctica*. London: Methuen, 1965.
Holdgate, M. W., ed. *Antarctic Ecology*. London: Academic Press, 1970.
Mouton, M. "The International Regime of the Polar Regions." *Hague Recueil* 107 (1962): 164–286.
Pharand, D. *The Law of the Sea of the Arctic with Special Reference to Canada*. Ottawa: University of Ottawa Press, 1973.
Quam, L. O., ed. *Research in the Antarctic*. Washington, D.C.: American Association for the Advancement of Science, 1971.
Smedal, G. *Acquisition of Sovereignty over Polar Areas*. Oslo: Dybwad, 1931.
Smith, P. M., ed. "Antarctica since the IGY." *Science and Public Affairs* 26 (December 1970): 1–104.
Waldock, C. H. M. "Disputed Sovereignty in the Falkland Islands Dependencies." *British Yearbook of International Law* (1948), pp. 311–53.

Appendix A

Reports from Organizations

The reports and abridged reports contained in this Appendix come from only some of the many organizations which deal with ocean-related matters. They are collected here in a central reference section to provide the reader with basic coverage of the important programs and directions being pursued by major international organizations. Future editions of the *Ocean Yearbook* will include reports from other organizations as well as subsequent accounts from the bodies reported on in this edition.

THE EDITORS

Reports from Organizations

Engineering Committee on Oceanic Resources (ECOR): An International Nongovernmental Society

The concept of an international organization of engineers interested in marine affairs originated during studies under the joint auspices of the National Academy of Sciences and the National Academy of Engineering of the United States. The studies led to publication of a report, "An Oceanic Quest," in May 1969. The report dealt with proposals for the International Decade of Ocean Exploration. The lack of an adequate mechanism for providing engineering review and evaluation of international programs and for involving representatives of industry was noted. As a result of the studies the National Academy of Engineering proposed establishment of ECOR as a nongovernmental international society of professional engineers with interest in marine affairs.

This proposal was brought to the attention of the Intergovernmental Oceanographic Commission (IOC). During an IOC meeting in Paris in September 1969 a working group reported favorably on the proposal, and the IOC adopted a resolution encouraging formation of ECOR. In November 1969, the WFEO (World Federation of Engineering Organizations) authorized its executive committee to review the possibility of a formal relationship between WFEO and ECOR.

The statutes of ECOR were ratified by the founding members (the national committees of the Federal Republic of Germany, France, Japan, the Netherlands, Portugal, the United Kingdom, and the United States and the international societies IADC and ISSC) in time for the Founding Council Meeting in Bordeaux, March 1971. At this meeting the council elected officers, established bylaws, arranged for incorporation in the Netherlands, and scheduled the first general assembly to be held in London in March 1972.

During a meeting in Paris in October 1971, the IOC recognized ECOR as an advisory body to IOC.

In accordance with its statutes, ECOR is an international, nongovernmental, professional engineering body whose purpose is to provide an international focus for professional engineering interests in marine affairs, with particular emphasis on:

1. Establishing and maintaining international, professional engineering communication in marine affairs.

2. Providing advice, from an engineering viewpoint, on policy, program, and organizational matters to international and intergovernmental organizations concerned with marine affairs, or providing such advice directly to individual nations on behalf of these organizations.

3. Assisting the engineering profession in the development of its capability

in the use of the ocean and in the enhancement of the quality of the marine environment, while recognizing that engineering is practiced within legitimate proprietary interests.

The interests of ECOR include all aspects of engineering practice (such as design, management, operation, planning, and research) and all engineering disciplines (such as biological, chemical, civil, electrical, mechanical, mining, ocean naval architecture, and transportation) as they relate to the marine environment. ECOR now comprises 13 national committees (Argentina, Australia, Canada, the Federal Republic of Germany, France, Japan, Mexico, the Netherlands, Norway, Portugal, South Africa, the United Kingdom, and the United States) and five international associations (FIP, IADC, IAHR, IAWPR, and ISSC).

The officers of ECOR are: President, Dr. Kenji Okamura, Mitsubishi Heavy Industries, Tokyo; Vice President, Cdr. M. B. F. Ranken, Aquamarine International, London; Past President, Prof. R. L. Wiegel, University of California, Berkeley, California; Secretary, Capt. J. W. Boller, National Academy of Engineering, Washington, D.C.; Treasurer, Ir. G. A. Heyning, Jr., Royal Institution of Engineers, The Hague; Asst. Secretary, Capt. D. L. Keach, University of Southern California, Los Angeles, California; Asst. Treasurer, Dr. J. G. Th. Linssen, Manager IADC, The Hague.

At the first general assembly of ECOR in London, March 1972, Adhering Bodies reported on the selected main theme "Engineering Design to Prevent Abuse of the Ocean." The combined report was forwarded to IOC and presented to the United Nations Conference on the Human Environment in Stockholm, June 1972. For the second general assembly, held in Tokyo, May 1975, "Engineering Practice for Offshore Structures" was the selected main theme. The theme of the third general assembly, to be held in Washington, D.C. in May 1978, will be "Critical Elements in the Exchange of Ocean Engineering Technology."

In addition to the triennial assemblies, working groups on various aspects of ocean engineering are convened periodically. Periodic conferences are also held on engineering problems of significant concern to a number of the adhering bodies. The next conference, on ocean instrumentation, will be held in Washington, D.C. in early 1978.

Quarterly newsletters on the activities of ECOR are published.

Reports from Organizations

Maritime Activities of the ILO*

UNITED NATIONS

The ILO, which was established in 1919 to advance the cause of social justice by evolving international labour standards, became the first of the specialised agencies of the United Nations in 1946. After 1964, when the United Nations Conference on Trade and Development (UNCTAD) became one of the permanent organs of the United Nations General Assembly, international shipping affairs were one of its primary concerns in relation to its objectives of aiding developing countries to advance their economic growth through the expansion of trade. UNCTAD therefore appointed a Committee on Shipping, which carried out a review of the activities of the United Nations specialised agencies in order to obviate any duplication of effort in these fields. The ILO has continued to be represented at sessions of the Committee on Shipping, with particular reference to its responsibilities for ensuring fair and equitable conditions of employment for workers in the shipping, ports and inland water transport industries of the world.

At the regional level the ILO collaborated with the United Nations Economic and Social Commission for Asia and the Pacific (ESCAP) during 1971 in carrying out a survey of maritime training facilities in that region. Participation in the annual meetings of the Transport and Communications Committee and the Committee on Trade at ESCAP has been effected on a regular basis through the ILO Regional Office in Bangkok. In 1975, in the same region, the ILO was represented at the second meeting of experts in maritime training among member countries of the Co-ordinating Committee of South-East Asian Senior Officials on Transport and Communications (COORDCOM), jointly organised by ESCAP and the South-East Asian Agency for Regional Transport and Communications Development (SEATAC). The meeting adopted certain recommendations on marine engineering and nautical matters relating to a uniform system of training and qualification for marine deck and engineer officers and for minimum officer manning requirements, which are currently under consideration by the governments of the COORDCOM member countries.[1]

In 1972 the ILO also participated in the meeting of United Nations Development Programme (UNDP) Resident Representatives for Africa, held

---

*From the *Report of the Director-General, ILO* (Geneva, 1976).

1. COORDCOM member countries are Indonesia, Democratic Kampuchea, Republic of Korea, Lao Republic, Malaysia, Philippines, Singapore, Thailand and Republic of South Viet-Nam.

at Addis Ababa, at which the problems of African countries in the field of shipping were reviewed. Subsequently an inter-agency meeting was convened in 1974, at which representatives of the United Nations Economic Commission for Africa, IMCO, UNCTAD and the ILO were present. A Joint Maritime Mission was agreed upon, to be financed by UNDP, a preparatory mission on training facilities and advisory services in the field of maritime transport being carried out in 1975 in 17 countries on the west and east coasts of Africa. The preparatory mission originated 15 proposed projects on maritime legislation and administration; seafarer officers' and ratings' training establishments; seamen's registration, employment and welfare; port operations management and labour training; marine pollution control and navigational aids and pilotage.

WORLD HEALTH ORGANISATION

The Fifth Session of the Joint ILO/WHO Committee on the Health of Seafarers was held in Geneva in September 1973, eight years after the Fourth Session in 1965. The agenda for this session included the questions of medical and first-aid training for ship personnel; the preventive care of teeth and mouth and emergency dental facilities for seafarers in port; medical examinations of crew members on tankers carrying chemicals in bulk; existing medical centres for seafarers; and a general survey of problems connected with immersion hypothermia. The WHO sent copies of the report "Medical and First-Aid Training for Ship Personnel" to a number of its correspondents for comment, and subsequently the Joint Committee adopted a resolution drawing the attention of member States to the need for vocational training for specified crew members, in the absence of a doctor or of full-time medical or nursing personnel on board cargo vessels, both in first-aid treatment and in more advanced medical training under the supervision of a physician familiar with seafarers' problems and in the use of the radio-medical services. In regard to dental care, the Joint Committee's resolution recommended that general health education should be initiated during a seafarer's vocational training, and particular attention should be paid to dental and oral hygiene, both in regard to its dietary aspects and that of regular examinations to ensure dental fitness. It was proposed that the WHO should assist in the promotion of audio-visual aids in dental health education in various languages, specially designed for seafarers.

In view of the fact that the special dangers to seafarers on tankers carrying chemicals in bulk had been considered by other organisations, the Joint Committee recommended that special medical pre-employment examinations as to fitness, including periodical examinations, in accordance with the risks to which crew members were exposed, should be carried out. The Joint Committee took note of the valuable guidance provided in the IMCO *Code for construction and equipment of ships carrying dangerous chemicals in bulk,* in the International Chamber of Shipping's *Tanker safety guide (chemicals),* and in the IMCO/WHO/ILO *Medical first-aid guide for use in accidents involving dangerous goods.* In respect of the training of seafarers serving in ships carrying dangerous chemicals in bulk, the Joint Committee also noted the provisions of the ILO/IMCO "Document for guidance—1970". A separate resolution on hypothermia following immersion in cold water, the risk of which may affect seafarers on vessels plying in the frigid zones of the world, emphasised the methods of mitigating the effects of hypothermia before immersion and the immediate emergency treatment necessary to help preserve life. It was recommended that detailed guidance on this subject should be incorporated in the *International medical guide for ships.*

A final resolution on the future work of the Joint Committee recommended that a review be carried out of the contents of the ship's medicine chest in the light of the increased employment, or presence, of women aboard ships and in the light of the increased numbers of chemical carriers in service. In ships where children are regularly carried as members of a seafarer's family, it was recommended that the WHO should consider what additional pediatric medicaments might be necessary in these ships' medicine chests. It was lastly agreed that the WHO should initiate a revised draft of the *International medical guide for ships* in order to bring it up to date, to be considered, together with the subject of the medical recording of seafarers (deferred since 1965) at an early session of the Joint Committee. It has not yet been possible to hold a further session of the Joint Committee, but the revised draft is being completed and submitted to WHO, IMCO and ILO experts and certain nongovernmental international organisations for their comments.

INTER-GOVERNMENTAL MARITIME CONSULTATIVE ORGANISATION

The Joint IMCO/ILO Committee on Training was officially established by a decision of the Governing Body of the International Labour Office at its 172nd Session in 1968 and of the Council of IMCO at its 18th Session in 1967. Committee members are appointed by each organisation; those from IMCO are government marine administration officials and those from the ILO are representatives of shipowners' and seafarers' organisations nominated by the two groups in the Joint Maritime Commission. The Joint Committee is therefore a forum in which training for all grades of seafarers is a matter for joint discussion and agreement between the three parties concerned in the world shipping industry. The Second Session of the Joint Committee was held in May 1970 in Geneva, when a previous "Document for guidance—1968", an international maritime training guide prepared jointly by ILO and IMCO, was revised and brought up to date. The resulting "Document for guidance—

1970" was communicated to all member States of the ILO and IMCO, with a request that it be given wide distribution among shipowners' and seafarers' organisations as well as national training institutions for seafarers.

In 1973 the Third Session was held in London and considered, inter alia, proposals to incorporate in the "Document for guidance—1970" certain recommendations concerning the training and qualifications of officers and crews carrying hazardous or noxious chemicals in bulk, and a recommendation was made to this end. Other matters on the agenda of this meeting related to the preparatory work of the IMCO Subcommittee on Standards of Training and Watchkeeping. The Joint Committee had been provided by IMCO, for information, with two preliminary draft proposals on the principles relating to navigational watchkeeping and proposals for establishing mandatory minimum requirements for the certification of officers in charge of a navigational watch, and for engineers in charge of a watch in a traditionally manned engine-room or the designated duty engineer in periodically unmanned engine-rooms. While IMCO members took the view that such matters had been referred to the Joint Committee for information only, ILO members made clear that in such matters shipowners' and seafarers' organisations should be fully involved and that existing ILO instruments in the field of nautical training should be taken into account in the framing of new standards. In view of the fact that a constitutional issue appeared to exist, each Organisation's members made their position clear by separate statements, which left the matter, however, as a subject for further clarification as to the responsibilities of the Joint Committee.

A formal understanding between the heads of the two organisations was arrived at in May 1974 and subsequently approved by the governing bodies of both organisations, under which proposals originating with either the ILO or IMCO, or both, concerning the formulation of international maritime training, qualification or certification standards will be examined by the Joint Committee in an advisory capacity. Certain steps are also envisaged if a divergence of opinion on any point of substance occurs.

The Fourth Session of the Joint Committee was held in Geneva in 1975, where it was recommended that the "Document for Guidance—1970" should be further elaborated by the addition of a section on engineer officers. New proposals were originated by IMCO regarding mandatory minimum requirements for the certification of a number of grades of masters, deck and engineer officers and junior ratings, all of which, save one which was not acceptable to the seafarers' representatives, were adopted by the Joint Committee. Discussions were held also on manning standards, where the seafarers' and shipowners' representatives were in disagreement, on radio officers, container ships, qualifications for engine-room ratings and other matters.

IMCO intends to convene a Conference on Standards of Crew Training early in 1978, to consider the adoption of one or more international instruments concerning standards of training, certification and watchkeeping for seafarers. In accordance with the terms of the understanding concluded by the executive heads of the ILO and IMCO, any conference of this nature convened by either Organisation shall be convened either jointly or with the direct participation of the other Organisation.

TRIPARTITE SUBCOMMITTEE OF THE JOINT MARITIME COMMISSION ON SEAFARERS' WELFARE

Since the 55th (Maritime) Session in 1970 no meeting of the Subcommittee on Seafarers' Welfare has been convened, the Third Session having been held in Oslo in October 1966.

The Preparatory Technical Maritime Conference held in 1969 had considered a report concerning seafarers' welfare and adopted conclusions for consideration by the Conference at its 55th (Maritime) Session in 1970, proposing the adoption of a Recommendation concerning seafarers' welfare at sea and in port. After further discussion and amendment, these proposals were approved by the General Conference and resulted in the adoption of Recommendation No. 138. Two further resolutions on sports activities for seafarers, and international co-operation in the field of seafarers' welfare, were adopted for proposed action by the Joint Maritime Commission and the International Labour Office respectively. It is envisaged that provision will be made in the draft Medium-Term Plan of the ILO (1976–81) for convening a session of the Tripartite Subcommittee on Seafarers' Welfare during that period to consider, inter alia, these resolutions and other questions in this field. Recent developments in the international and national spheres have been referred to earlier in the section on welfare in Chapter 2 of this Report.

JOINT MARITIME COMMISSION

In November 1972, the 21st Session of the Joint Maritime Commission was held in Geneva, and had before it the agenda approved by the Governing Body at its 185th (February-March 1972) Session as follows:
1. Industrial relations in the shipping industry.
2. Seafarers' holidays with pay.
3. The protection of young seafarers.
4. Flags of convenience.
5. The minimum basic wage of able seamen.
6. Continuity of employment of seafarers.

At its 42nd Session, held in Geneva in March 1972, the Committee of Experts on the Application of Conventions and Recommendations had considered the result of a general survey concerning the Seafarers' Engagement (Foreign Vessels) Recommendation, 1958 (No. 107), and the Social

Conditions and Safety (Seafarers) Recommendation, 1958 (No. 108), which had been carried out following the adoption of a resolution concerning flags of convenience at the 55th (Maritime) Session of the Conference in October 1970. The reports received from governments under this maritime survey were considered at the 57th (1972) Session of the Conference, when the Workers' Group suggested that the Committee on the Application of Conventions and Recommendations should express its concern at the many important maritime nations which had failed to provide reports on the above two Recommendations Nos. 107 and 108. The Employers' members were also in agreement that the information made available by the governments was inadequate. It had therefore been requested of the governments concerned that further information should be made available prior to the 1972 Session of the Joint Maritime Commission, which was thus assisted in its deliberations on item 4 of its agenda by a summary of the reports which were forthcoming from 65 countries on the application of the above two Recommendations. Summaries of replies to questionnaires sent out by the International Labour Office in relation to items 1 to 3 and 6 of the agenda served as the bases of other discussions for the Commission, while in respect of agenda item 5, the Office had prepared a study from official statistics of the consumer price indices in 44 maritime countries for the opening months of 1972, with 1970 as a base year, and of the change in purchasing power of the US dollar in various countries between 1970 and 1972. The Commission adopted a resolution as regards item 5 that the minimum basic wage for able seamen, as set out in Paragraph 2 of Recommendation No. 109, should be raised to pounds sterling 48 or US dollars 115. Resolutions on the remaining five agenda items were also adopted in conjunction with a request to the Governing Body that a Preparatory Technical Maritime Conference should be held in 1974, followed by a Maritime Session of the International Labour Conference in 1975. In the event, as reported below, it was later decided to hold these Conferences in 1975 and 1976 respectively.

In the ILO draft Medium-Term Plan (1976–81) priority work under the Maritime Workers' programme was taken note of by the Governing Body of the ILO at its 192nd Session in 1974. In respect of seafarers the proposed programme consisted of: *(a)* consideration of social problems arising from new technology on board ship with a view ultimately to the possible adoption of new standards on methods of improving the attractiveness of seafaring as a career; the sharing by seafarers of benefits other than wages arising from modernisation; and the alleviation of human stresses and strains of modern shipboard work and life; *(b)* a meeting of the Tripartite Subcommittee of the Joint Maritime Commission on Seafarers' Welfare; *(c)* a meeting of the Joint Maritime Commission to consider the report of the above Subcommittee and (as at each session) to consider the question of the minimum basic wage of able seamen; *(d)* the desirability of holding a regional maritime conference; and *(e)* any other questions that may be referred to at the present Maritime Session of the Conference. It was also proposed that in order to give effect to the recommendations of the Joint Maritime Commission, provision should be made for holding a regional maritime conference and a Preparatory Technical Maritime Conference, and for the holding of one or more sessions of the Joint IMCO/ILO Committee on Training. At its 199th (March 1976) Session the Governing Body, in its review of the ILO draft Medium-Term Plan, also took note of the resolution of the Preparatory Technical Maritime Conference in 1975, which asked the ILO to carry out a survey concerning the extent to which provisions concerning social security and conditions of employment are applied on flag-of-convenience ships. The views and observations of the present Conference on the above matters will be of particular value to the Office at this time.

PREPARATORY TECHNICAL MARITIME CONFERENCE

The Governing Body of the International Labour Office having considered the resolutions of the Joint Maritime Commission's 21 Session, at its 189th (February-March 1973) and 194th (November 1974) Sessions, decided that the Preparatory Technical Maritime Conference should be held in Geneva in October 1975 with the following agenda:
1. Industrial relations in the shipping industry.
2. Revision of the Paid Vacations (Seafarers) Convention (Revised), 1949 (No. 91), in the light of, but not necessarily restricted to, the Holidays with Pay Convention (Revised), 1970 (No. 132).
3. The protection of young seafarers.
4. Continuity of employment of seafarers.
5. Substandard vessels, particularly those registered under flags of convenience.

The Conference was attended by delegates and advisers from 32 member countries. Representatives of some 15 international inter-governmental and non-governmental organisations were also present.

In regard to the first item on its agenda, the Conference adopted a resolution embodying its views that in the light of existing instruments on industrial relations generally it would not at this time be desirable to contemplate the adoption of additional instruments concerning industrial relations in the shipping industry. Approval of this particular resolution, as referred to in the Introduction to this Report, by the Governing Body, as well as those on the remaining items of its agenda put forward by the Preparatory Technical Maritime Conference, effectively reduced to four the number of technical questions on the agenda of this maritime session of the Conference. Summary reports on holidays with pay for seafarers, the protection of young seafarers and continuity of employment of seafarers, each embodying proposals for the adoption of an international instrument, were thereafter prepared by the International Labour Office in accordance with the finally approved agenda. In regard to the item on substandard vessels, it was concluded that further consultations with governments appeared to be necessary. The Office was therefore authorised to submit a questionnaire to governments, from which a final report on this subject has been prepared for submission to delegates to this maritime session of the Conference.

The Preparatory Technical Maritime Conference also adopted a resolution requesting the Governing Body to instruct the Director-General to carry out a survey on the extent to which provisions concerning social security and conditions of employment are applied on board flag-of-convenience ships. At its 198th (November 1975) Session the Governing Body agreed that the Director-General should bring forward this proposal when submitting the future work plan.

REGIONAL SEMINARS ON MARITIME MANPOWER PLANNING AND DEVELOPMENT

With the co-operation of the Danish International Development Agency (DANIDA) and the Government of Jamaica, the ILO organised in November 1972 a regional Caribbean Seminar on Maritime Manpower Planning and Development, which was held at Kingston. The Seminar was attended by 19 participants from 12 countries and territories of the Central American and Caribbean region, and by 11 observers from intergovernmental or non-governmental organisations, and was conducted in both English and Spanish. Lectures by five Jamaican national and nine international experts in maritime training were followed in each session by discussions or reports on contemporary trends for the various categories and grades of seafarers, with particular reference to seafaring manpower assessment and policy; planning and operation of maritime training facilities; standards of instruction and methods for use in training with particular regard to new shipboard technologies; methods of working and the operation of modern vessels. The countries or territories concerned were Antigua, Dominican Republic, El Salvador, France (Martinique), Guatemala, Guyana, Jamaica, Mexico, Netherlands Antilles, Panama, Surinam and Trinidad and Tobago.

In March 1975, at Catia La Mar in Venezuela, a Seminar on Maritime Training for South American countries was conducted by the ILO, again with the assistance of DANIDA and of the Venezuelan Ministry of Communications. Sessions were held at the Venezuela Nautical School. The following 8 South American countries provided some 15 official participants in the seminar: Argentina, Brazil, Colombia, Chile, Ecuador, Peru, Uruguay and Venezuela. One observer from IMCO and 22 from nautical training institutions and other non-governmental organisations were also present. Papers were presented by each country's delegation and some 11 contributions were read by other participants on various aspects of maritime training as conducted in the seafarers' or officers' training institutes in their countries. Two committees were established, which concluded that for South American countries the priorities in training needs were those of pre-sea induction courses and the evolution of a system of further training aimed at providing security and permanency in the seafarer's career. A final conclusion arrived at was that a similar seminar should be held at an early opportunity related to the problems of training and working conditions of fishermen in South American countries.

TECHNICAL CO-OPERATION ACTIVITIES

At its 55th (Maritime) Session in 1970, the International Labour Conference adopted a resolution concerning technical co-operation, which, in the light of the International Seafarers' Code which had evolved in the past 50 years, drew attention to the experience of the ILO as a source of assistance to those countries in the process of developing or contemplating the establishment of national merchant fleets. The maritime member States concerned were accordingly informed of the terms of this resolution and of the 41 offices of the ILO located in various regions of the world where further details relating to technical co-operation programmes might be obtained. During the past six years, a small but wide range of maritime training and related programmes has been completed, under which international expert assistance has been provided, equipment, books and teaching aids supplied, or fellowship studies carried out overseas. These projects have been financed either by multilateral aid from a country's UNDP funds, by multi-bilateral aid from the Danish, Norwegian and Swedish Governments, or under the ILO's own budget. In addition, as co-ordinators of maritime technical co-operation in particular parts of the world, three regional advisers were appointed abroad. A maritime adviser, serving the Spanish-speaking countries of Latin America, and another, serving the English-speaking countries of the Caribbean sub-region, were assigned to Lima and Trinidad during the period from 1972 to 1975, and between 1973 and 1975 a consultant was based for short periods of each year in Thailand, with a roving commission arising from his responsibilities for the Asian region and Pacific Basin.

Countries in North Africa, the Asian region and the Pacific Basin also requested and received ILO technical co-operation in the maritime training field. At the Bangladesh Marine Diesel Training Centre, four ILO experts in teaching techniques, shipbuilding crafts, steam and gas turbines and marine engineering were appointed, each for 24 months, during 1976. At the request of the Government, an ILO maritime adviser carried out in 1973 a survey of the Merchant Marine Academy, Jaldia, Chittagong, with a view to re-activating its courses on a full scale for nautical and engineer pre-sea cadets. In Egypt a specialised instructor for establishing upgrading courses for ships' ratings was posted for 18 months to the Arab Maritime Transport Academy, Alexandria, during 1975-76. In Malaysia, at Penang, the existing seafarers' training centre was assisted by the services of an expert for 30 months during 1970–73. During 1973–74 in the Philippines, a pre-sea ratings' school at Manila was assisted by the appointment of a training expert for 18 months, four fellowships for counterpart teaching staff were provided in Danish and Norwegian marine schools, and items of teaching aids, equipment and books were supplied to the school.

In the Solomon Islands the Honiara Technical Institute was provided in 1970–73 for a period of 46 months with the services of an ILO expert in marine engineering. In Singapore, with the joint co-operation of UNESCO, a radar simulator was installed at the Marine Department of the Polytechnic Institute, and an expert was appointed to establish courses for ships' masters and officers for a period of 24 months during 1972–74. In Western Samoa during 1973, fellowship studies were arranged for seafarers allotted to manning an inter-island ferry; two nautical candidates attended the Derrick Technical Institute in Fiji, and two engineer candidates were sent to the ILO-sponsored Marine School at the Technical Institute, Honiara, Solomon Islands. In Thailand preliminary investigations into the establishment of a merchant marine training centre for all grades of officers and ratings were carried out in 1975. Assistance to an interim pre-sea officer cadets' training scheme was also provided by an ILO maritime expert, later assisted in 1975–76 by two experts assigned by the Norwegian Government.

Other related aspects of technical co-operation were taken up by the regional consultant for Asia and the Pacific; a number of proposals are still pending and awaiting action by the governments concerned. In the Gilbert Islands the development authority at Tarawa was assisted during 1975 by the services of an ILO expert in port and shipping documentation. In Democratic Kampuchea during 1974 the Government requested advice on the setting up of a marine administration, including the registration of seafarers and the training of, and establishment of statutory examinations for, all grades of merchant marine officers and seamen. In Pakistan, during 1973, a report was made to the Government on the re-establishing, jointly on a new site outside Karachi, of the Pakistan Marine Academy and the Seamen's Training Centre. The Government acted on certain of the recommendations made by the regional adviser during 1974–75. The establishment by the Philippines Government of a National Seamen's Board under the auspices of the Labor Department, led to a request for technical advice in 1974. During 1976 an ILO expert in the registration and placement of seamen and in seafarers' employment and training problems, was posted as adviser to the Labor Department for an initial period of 12 months. Three fellowships for staff of the Labor Department were also granted during 1976, to enable them to study established procedures at the Seamen's Recruitment Office in Hong Kong. In the Solomon Islands, at the request of the government in 1973, proposals were made by the regional adviser of the training of indigenous pilots, under the international fellowship scheme, for the ports of Honiara and Gize. An ILO project to set up a marine training school which will involve expert assistance in the fields of seafarers' induction, marine engineering and electronics courses is also under way.

It will be seen from the foregoing that during the period under review the ILO's competence in the fields of seafarers' training, both of officers and ratings, has been of assistance to the governments in a wide range of countries, many of them in the more isolated regions and with the least capabilities of meeting their maritime training needs from their own resources. There is also evidence of a growing demand among developing countries for other forms of maritime technical co-operation in addition to vocational training, namely in respect of seafarers' conditions of employment, collective agreements or individual contracts of service and the regularisation of recruitment, which are particularly applicable to those countries with a surplus of manpower in their shipping industries.

MARITIME WORKERS' EDUCATION

Workers' education for maritime workers, which has been a concern of the ILO for many years, on an equal footing with workers' education for all other groups of workers, was given special attention in the period 1973-75 through the employment of a specialist in maritime workers' education. In co-operation with the International Transport Workers' Federation and maritime trade unions in Norway and Sweden, three projects have been prepared. Two of these have been approved by the Norwegian Government under its bilateral assistance programme (NORAD), and the third has been submitted to the Swedish International Development Agency (SIDA). With NORAD assistance it is proposed that an inter-regional seminar on maritime workers' education be held in Geneva (September-October 1976), attended by 25 maritime trade unionists and workers' education specialists, representative of maritime countries in Africa, the Americas, Asia, Europe and the Middle East. It is envisaged that the seminar will help to determine the needs of maritime workers in the field of workers' education, promote an exchange of methodologies, techniques, experiences and structures, and identify problems and seek solutions. A direct follow-up of this inter-regional seminar will be a regional project on workers' education for maritime workers in Asia. This will be organised in two separate stages, namely a regional seminar tentatively scheduled for March 1977 in a major Asian seaport, and a second stage consisting of two country programmes of, respectively, 18 and 12 months' duration aimed at providing expert, training and equipment assistance towards establishing and implementing workers' education programmes for seafarers (and dockworkers) in the selected countries. The third project, for which assistance from SIDA is currently being sought, consists of a seminar of 30 maritime trade unionists and labour education specialists from Kenya and Tanzania, to be held in Mombasa and to be preceded by a two-month exploratory expert mission to the two countries concerned.

Reports from Organizations

Excerpts from the Annual Report of the Inter-Governmental Maritime Consultative Organization 1976/1977

The activities of the Inter-Governmental Maritime Consultative Organization during the year April 1, 1976, to March 31, 1977, are outlined under the following headings: (1) constitutional developments and activities of major bodies; (2) developments of major programs; (3) coordination with organizations of the United Nations system and relations with other organizations; (4) technical cooperation; (5) administrative and financial questions; (6) conferences, documents, and publications.

* * *

I. CONSTITUTIONAL DEVELOPMENTS AND ACTIVITIES OF MAJOR PROGRAMS

Membership of the Organization.—During the period under review, Bangladesh, Bahrain, the Bahamas, Cape Verde, Gabon, Jamaica, Papua New Guinea, and Surinam accepted the IMCO Convention, bringing the membership of the organization to 101 plus one associate member.

Maritime Safety Committee.—The Maritime Safety Committee held its thirty-fourth and thirty-fifth sessions in May and September 1976. The committee considered the technical work of the organization related to maritime safety, which is described more fully in Section II of this report, and approved a number of recommendations for submission to the tenth assembly.

Marine Environment Protection Committee.—The Marine Environment Protection Committee held its fifth and sixth sessions in May and November 1976; its heavy and wide-ranging program reflects the continuing expansion of IMCO's work in the prevention and control of marine pollution from ships. Under Section II of this report, a fuller description is given.

Facilitation Committee.—At its tenth session, in May 1976, the Facilitation Committee reviewed the substantial volume of work carried out and the activities planned for the future (see Section II, under the heading "Facilitation of Maritime Travel and Transport").

Legal Committee.—During the period under review, the Legal Committee held its twenty-ninth, thirtieth, and thirty-first sessions in July and September 1976. Section II D of this report is devoted to the very considerable body of work with which this committee dealt.

Committee on Technical Co-operation.—The Committee on Technical Co-operation held its twelfth and thirteenth sessions in June and October 1976, following its normal practice of meeting in conjunction with council sessions. This important sector of IMCO's activities is dealt with in Section IV of this report.

International conferences and meetings.—In addition to the meetings of IMCO's regular bodies, the organization convened the following conferences during the period under review: International Conference on the Establishment of an International Maritime Satellite System (third session, September 1–3, 1976); International Conference on Limitation of Liability for Maritime Claims (November 1–19, 1976); and, in conjunction therewith, three short revision conferences to revise the unit of account in the following conventions: (i) International Convention on Civil Liability for Oil Pollution Damage, 1969; (ii) International Convention on the Establishment of an International Fund for Compensation for Oil Pollution Damage, 1971; (iii) Athens Convention Relating to the Carriage of Passengers and Their Luggage by Sea, 1974.

II. DEVELOPMENTS OF MAJOR PROGRAMS

A. Maritime Safety

Routing of ships.—During the period under review, additional routing schemes for various parts of the world have been prepared for adoption by the tenth IMCO assembly; they will be published as a supplement to the volume entitled *Ships' Routeing.* So far, adherence to the provisions of the schemes was on a voluntary basis, but the new Regulations for Preventing Collisions at Sea, 1972 (which came into force on July 15, 1977) institutionalize the routing schemes adopted by the organization and prescribe certain obligations for vessels when navigating in or near them.

Shipborne navigational aids.—The organization has worked out operational standards and requirements for shipborne navigational aids—mainly those which are carried on a mandatory basis—and a publication on this subject has been prepared.

Regional harmonization of buoyage systems.—In association with IMCO, the International Association of Lighthouse Authorities has completed the first part of a study of maritime buoyage systems, which the two organizations have endorsed and circulated to all concerned as "System A: The Combined Cardinal and Lateral System." The study on "System B: Lateral Only System" is progressing. Administrations will have the choice of implementing one or the other, so that eventually buoyage systems would be harmonized and/or unified on a regional basis, which will facilitate recognition and identification by the mariner through the use of standardized methods and types of buoyage.

Matters relating to the 1972 Regulations for Preventing Collisions at Sea.—The 1972 regulations, which came into force on July 15, 1977, include certain provisions concerning the disposition of ships' lights. These have been considered and action recommended to member states so that application of the rules may be uniform.

International coordination in promulgating navigational warnings to shipping.—The completed plan for a world-wide system of promulgating navigational warnings to shipping prepared by IMCO in collaboration with the International Hydrographic Organization has been communicated to member governments for comment. A number of regional systems, based on the general plan, are already in operation, and countries in other areas are in the process of completing their regional plans for similar systems.

Radiocommunications.—During the period under review the preparation of operational standards for shipborne radio equipment was continued. In view of the World Administrative Radio Conference planned by ITU for 1979, the organization has given emphasis to communications matters related to safety.

International Maritime Satellite System.—The third session of the Conference on the Establishment of an International Maritime Satellite System was held in September 1976. It adopted a convention and an operating agreement on the International Maritime Satellite Organization (INMARSAT). The conference also established a Preparatory Committee to consider technical, economic, and organizational matters connected with the functions and operations of the future organization. The committee held its first session in January 1977 and agreed on a program of work which will continue until INMARSAT comes into being.

Lifesaving appliances.—In conjunction with research work which is being carried out primarily by the Nordic countries, chapter III of the International Convention for the Safety of Life at Sea is being revised. This chapter deals with lifesaving appliances, and the revised text includes requirements for lifesaving systems for various types of ships, equipment specifications, and criteria and test procedures for alternative systems and/or appliances.

Search and rescue (SAR).—At its fourth session (October 1976), the Group of Experts on Search and Rescue continued its preparatory work for a Conference on Maritime Search and Rescue planned for early 1978. The group reviewed the drafts of the proposed convention and its technical annex, which sets out the main lines for the organization of an international plan for maritime search and rescue. The group also continued its work on the maritime search and rescue manual, intended to advise administrations on the organization of national SAR services, and the implementation of the convention and its technical annex. This work is being carried out in close cooperation with ICAO. The Merchant Ship SAR Manual (MERSAR) was further improved with the addition of a chapter dealing with evacuation from ships by helicopter and amended provisions concerning radiocommunication facilities for distress purposes and for communicating with assisting ships.

Training of seafarers.—Work has continued on a draft International Convention on Training and Certification of Seafarers. It includes mandatory minimum requirements and qualifications for the certification of masters, officers, and other members of the crew. A number of recommendations are also being prepared. These deal with training in survival techniques and special training for those serving on large vessels and on ships carrying liquefied gases in bulk. The draft convention and the recommendations are intended to be the main working documents for an international conference scheduled for 1978, and the work is being carried out in close cooperation with the ILO and interested nongovernmental organizations.

Procedures for the control of ships.—During the year, the organization considered in detail a number of measures for strengthening the effectiveness of existing arrangements for identifying ships which do not fully comply with the requirements of the conventions related to safety. Following the adoption of Resolution A.321(IX) by the ninth IMCO assembly in November 1975, the Maritime Safety Committee prepared a set of guidelines on control procedures. The guidelines are addressed to the governmental officers authorized by contracting parties to exercise the control functions prescribed by the conventions. Procedures for the control of ships under conventions relating to marine pollution are being considered by the Marine Environment Protection Committee.

Carriage of dangerous goods.—During the period under review, a significant number of amendments and additions to IMCO's International Maritime Dangerous Goods Code (IMDG Code) have been drawn up. In order to harmonize IMCO's work on dangerous goods with that of the United Nations, the UN grouping for packaging purposes will be adopted in future for the IMDG Code and the packaging group will be included in the appropriate schedules.

On the question of fire safety precautions, measures have been drawn up for the carriage of dangerous goods in packaged form. Future work in this sector will cover the goods when carried in containers and portable tanks on board ships which are specifically designed for this purpose and those which are not. The recommendation on safe practice on dangerous goods in ports and harbors, adopted by the eighth IMCO assembly in 1973, has been implemented by a number of administrations. Work has started on emergency procedures for ships carrying dangerous goods, and while consistent treatment of emergencies in all modes of transport is obviously desirable, the special circumstances of the maritime mode have been recognized. On this basis, the pattern of future work has been agreed. In this context, the *Medical First Aid Guide* (jointly worked out by IMCO, WHO, and the ILO) is to be annotated marginally, medical advice being provided for groups of substances by means of a reference number which will be quoted in the schedule for individual substances.

Carriage of bulk cargoes and containers.—Among the amendments which have been made to the Code of Safe Practice for Bulk Cargoes is the inclusion of dangerous substances carried in bulk. The organization has contributed to the meeting of the ad hoc intergovernmental group on container standards, established under the auspices of ECOSOC and UNCTAD. With the forthcoming entry into force of the 1972 Convention for Safe Containers, discussions have taken place to ensure that it is implemented uniformly.

It is important that those who stow cargo in containers destined for multimodal transport which includes a sea leg are fully aware of the stresses imposed by maritime transport; general guidelines, which can also be used for training purposes, are being developed. A study relating to the safe stowage and securing of containerized cargo and other entities on board noncontainer ships has been started with some emphasis laid on roll-on/roll-off ships. When completed, the study may well be the basis for the development of guidelines.

Carriage of dangerous chemicals and liquefied gases in bulk.—During the period under review, the code for existing ships carrying liquefied gases was approved and circulated to member governments for implementation. Work continues on improving the Gas Carrier Code and the Bulk Chemical Code, and a series of amendments to the latter have been approved and circulated.

Ship design and equipment.—During 1976, the Code for Dynamically Supported Craft (air-cushion vehicles and hydrofoil boats) was finalized and passed to the Maritime Safety Committee for approval and submission to the tenth IMCO assembly for adoption. With the increased attention being given to nuclear-powered merchant ships throughout the world, IMCO has commenced consideration of the safety aspects of these vessels with a view to formulating a code for their construction and operation.

There has been a rapid expansion of the exploitation of off-shore oil fields which has led to the design of many new and unconventional types of craft and floating structures. The safety features of these vehicles are currently being investigated, and work is in progress on drafting a safety code for mobile off-shore drilling units and off-shore supply vessels.

Subdivision, stability, and load lines.—During the period under review, further improvements have been made to the 1966 Load Line Convention, and good progress has been made in the consideration of improved stability criteria. As a follow-up to work on requirements for subdivision, stability, and load lines for special-purpose ships, recommendations on subdivision, stability, and load lines for mobile off-shore drilling units and on stability for off-shore supply vessels were finalized.

Tonnage measurement of ships.—With the prospect of the International Convention on Tonnage Measurement of Ships, 1969, entering into force in the near future, consideration has been given to the tonnage parameters used in other conventions for certain types of ships and a draft recommendation on an interim scheme for tonnage measurement of certain ships has been drawn up and was submitted to the tenth IMCO assembly in November 1977. Certain aspects of the application of the tonnage convention in relation to other conventions have also been considered. A resolution on the tonnage measurement of ballast spaces of segregated ballast oil tankers has been prepared for submission to the tenth IMCO assembly.

Safety of fishing vessels.—During 1976, work on the safety of fishing vessels was devoted entirely to preparation of the draft international convention on that subject, which was submitted to the International Conference for the Safety of Fishing Vessels held at Torremolinos from March 7 to April 2, 1977. With the adoption of this convention a gap will be closed, since safety requirements for the construction and equipment of fishing vessels have not so far been regulated by an international instrument.

Fire protection.—During the year under review, the Recommendation on Fire Safety Measures for Passenger Ships carrying not more than 36 passengers was finalized. The recommendation was drafted in convention-style language, so that at the appropriate time it could be adopted as an amendment to the 1974 safety convention to replace Part C of Chapter II-2. Good progress has been made in fire safety measures for the carriage of dangerous goods, and safety provisions for the carriage of these goods in packaged forms as general cargo were finalized, they will eventually form an amendment to the Recommendation on Fire Safety Measures for Cargo Ships (Resolution A.327[IX]).

B. Marine Pollution

The continuing work of IMCO and its Marine Environment Protection Committee (MEPC) in this field is directed mainly toward implementation of the resolutions adopted by the 1973 International Conference on Marine Pollution and resolving problems related to the entry into force of the International Convention for the Prevention of Pollution from Ships, 1973.

Work of the Marine Environment Protection Committee.—The Marine Environment Protection Committee (MEPC) has made considerable progress in implementing the action plan established at its first session in March 1974. In particular, two recommendations were finalized at the committee's sixth session (November 29–December 3, 1976). One recommendation concerned international performance and test specifications for oily water separators and monitoring equipment; it is a revision of earlier specifications and takes into account the new requirements of the 1973 Marine Pollution Convention. When approved by the IMCO assembly, it will be issued as an IMCO publication. The second recommendation concerns international effluent standards and guidelines for performance tests for shipborne sewage treatment plants and will also be published.

Comprehensive manual on oil pollution.—The MEPC has been preparing a comprehensive manual on oil pollution with a view to providing guidelines to assist governments, particularly those of developing countries, in establishing national marine pollution control programs. The manual will consist of five sections: (1) prevention; (2) contingency planning; (3) salvage; (4) practical information on means of dealing with oil spillages; and (5) legal aspects. Section 1 and section 4 were published in 1976 and 1972, respectively.

Guidelines and reception facilities.—Special attention has been given to problems concerning the provision of reception facilities in ports for wastes generated on board ships, including oily mixtures, sewage, garbage, etc.

In order to assist governments to take the necessary steps, the MEPC is developing guidelines on the provision of adequate reception facilities in ports, which consist of the following four parts: part 1, oily wastes; part 2, wastes containing noxious liquid substances; part 3, sewage; part 4, garbage. Part 1 of the guidelines, dealing with oily wastes, was approved by the MEPC at its sixth session and will be published by IMCO in the near future.

Technical cooperation.—As will be seen from the foregoing, the MEPC has been developing information which will assist governments—particularly those of developing countries—in accepting and implementing the 1973 Marine Pollution Convention.

On the basis of replies to an enquiry to governments, IMCO has compiled and is maintaining a roster of appropriate experts and institutions, so that advice can be provided to developing countries on implementing pollution prevention measures. Similarly, lists of experts and institutions who could be called upon in an emergency involving a major incidence or threat of pollution have been compiled.

Other matters.—Work is continuing on other matters of importance, such as the development of improved methods for enforcing the convention requirements and the development of a standard method for identifying the source of discharged oil. Appropriate procedures and arrangements for the discharge of noxious liquid substances are being developed, and the environmental hazards of harmful substances carried by ships are being evaluated with advice from GESAMP. Other questions being studied are the introduction of segregated ballast in existing tankers and the development of guidelines for intervention under the 1969 Intervention Convention and its 1973 protocol. The committee is also developing lists of approved oily water separating equipment and sewage treatment plants.

Symposium on Prevention of Marine Pollution from Ships.—As indicated in the annual report for 1975/76, IMCO, jointly with the government of Mexico, convened the Symposium on Prevention of Marine Pollution from Ships in Acapulco (Mexico) from March 22 to 31, 1976. The symposium was attended by some 270 participants from 69 countries, including wide representation from governments, port authorities, research institutions, maritime organizations, and environmental groups as well as individuals involved in various aspects and problems of marine pollution.

Regional Oil Combating Center for the Mediterranean.—Reference was made in the 1975/76 annual report to the Conference of Plenipotentiaries of Coastal States of the Mediterranean Region on the Protection of the Mediterranean Sea held in Barcelona in February 1976. In implementation of resolution 7 adopted by that conference, IMCO and UNEP have jointly undertaken to establish a Regional Oil Combating Center for the Mediterranean in Malta.

The center was officially opened by the president of the Republic of Malta on December 11, 1976. The principal functions of the center are to help the coastal states of the region to cooperate in combating marine spillages of oil, especially in emergency situations, and to assist the coastal states in developing their own national antipollution capabilities.

Amendments to oil pollution convention.—As a result of continued efforts, the 1969 amendments to the International Convention for the Prevention of Pollution of the Sea by Oil, 1954, have now been accepted by the required two-thirds of the contracting parties to the convention and will come into force on January 20, 1978. Effective enforcement of these amendments should significantly reduce the quantities of oil reaching the sea as a result of tanker operations, such as tank washing, deballasting, etc.

Guidelines for possible intervention under the 1969 intervention convention.— The 1969 International Convention Relating to Intervention on the High Seas in Cases of Oil Pollution Casualties entered into force on May 6, 1976. In order to assist contracting parties in making technical assessments for possible intervention under this instrument and its 1973 protocol (which relates to substances other than oil), the MEPC is developing appropriate technical guidelines.

Ocean dumping convention.—The Convention for the Prevention of Marine Pollution by Dumping of Wastes and Other Matter, 1972, which entered into force on August 30, 1975, has now been accepted by 30 states. The contracting parties to the convention met in a first consultative meeting in September 1976, and during this inaugural session substantial progress was made on organizational and substantive matters, including the adoption of rules of procedure, administrative and financial arrangements, and the formulation of a comprehensive action plan for carrying out future work.

C. Facilitation of Maritime Travel and Transport

During the period under review, IMCO's facilitation committee has begun studies on the standardization of documentation concerning the carriage of dangerous goods by all modes of transport, on the practical and technical aspects of a maritime ports operation system, and on methods of efficiently clearing in-bound ships, with the cooperation of appropriate bodies.

Good progress has been made on the preparation of a manual on the implementation of IMCO Model Forms. This is intended to be a source of information to shipmasters and agents regarding the forms required in various ports of the world for the arrival and departure of ships. Additional measures have been adopted to simplify formalities connected with the arrival, stay, and departure of persons and ships as well as of cargoes and other articles.

D. Legal Work

The Legal Committee held three sessions (the twenty-ninth, thirtieth, and thirty-first sessions) during the period under review.

Review of the 1957 convention relating to the limitation of the liability of the owners of seagoing ships. —The Legal Committee completed work on the preparation of draft articles of a new convention to replace the 1957 convention. On the recommendations of the committee, the council and the assembly approved the holding of a diplomatic conference to consider the draft convention. The conference was held in November 1976; it adopted a new Convention on the Limitation of Liability for Maritime Claims.

On the recommendation of the Legal Committee three short conferences were convened by the organization in conjunction with the November 1976 conference. These conferences adopted protocols to three earlier conventions, namely, the International Convention on Civil Liability for Oil Pollution Damage, 1969; the International Convention on the Establishment of an International Fund for Compensation for Oil Pollution Damage, 1971; and the Athens Convention Relating to the Carriage of Passengers and Their Luggage by Sea, 1974.

The protocols revised the "unit of account" provisions in the respective conventions. In place of the "Poincaré franc," based on the official value of gold, the new unit of account is based on the "Special Drawing Right" of the International Monetary Fund. This change was felt to be necessary because experience had shown that it had become increasingly difficult, if not impossible, to convert the "gold franc" into national currencies on a uniform basis. The protocols, however, permit the use of the gold franc by those contracting parties which are not members of the International Monetary Fund and whose laws do not permit the use of the special drawing right.

Civil liability for pollution damage from substances other than oil. —The Legal Committee has continued to assign top priority to consideration of the question of civil liability for pollution damage from substances other than oil covered by the 1969 Convention on Civil Liability for Oil Pollution Damage.

The Legal Committee has agreed that it would be feasible to extend, by means of a protocol, the regime of the 1969 convention "to oils other than the persistent oils covered by that Convention, and possibly to other substances which have physical or chemical properties approximating those of persistent oils." Consideration of the elaboration of a comprehensive and separate convention to deal with noxious and hazardous substances other than oils will be undertaken at a later stage, as the requisite information on those substances becomes available.

A possible convention on wreck removal and related issues. —

A possible convention on the regime of vessels in foreign ports. —

Study of the legal status of air-cushion vehicles. —

Promotion of early and wider implementation of an international convention relating to stowaways.—

*Future work of the Legal Committee.—*The Legal Committee's work program for the period after the 1976/77 biennium includes: *(a)* legal status of Ocean Data Acquisition Systems (ODAS); *(b)* establishment of international legal machinery to combat marine pollution from ships and other vessels; *(c)* jurisdiction in respect of collisions at sea and other maritime incidents; *(d)* arrest of seagoing ships.

III. COORDINATION WITH ORGANIZATIONS OF THE UNITED NATIONS SYSTEM AND RELATIONS WITH OTHER ORGANIZATIONS

*The United Nations system.—*The Secretariat has played a very active role in the coordination activities carried out by the various components of the UN system. IMCO's efforts have always been aimed at providing the best possible assistance to the developing countries in shipping and related maritime activities. Accordingly, it has maintained close consultation with UNCTAD, ILO, and UNEP in view of the complementary nature of the work of these organizations in the field of shipping and in the protection of the marine environment. The Secretariat of IMCO has followed closely the work of the Third United Nations Conference on the Law of the Sea and was represented during the entire period of the fifth session of the conference.

The organization cooperated with ICAO in the field of search and rescue, with the ILO as regards the training of seafarers, with the ITU in radiocommunications, with UNEP on marine pollution, and with UNCTAD, the Economic Commission for Europe, and ICAO on facilitation matters. IMCO has also made a significant contribution to the preparation of the report of the Secretary-General of the United Nations on the activities of the United Nations system in the field of transport, with was considered by the CPC in May 1977.

Finally, it should be mentioned that IMCO has cooperated with the United Nations Committee on Dangerous Goods and with the Intergovernmental Preparatory Group on a Convention on International Intermodal Transport, as well as the Ad Hoc Intergovernmental Group on Container Standards for International Multimodal Transport, both convened by UNCTAD. Relations with the economic commissions were strengthened and arrangements were established for coordinating IMCO's work in the regions with the Secretariats of the relevant regional commissions.

*Intergovernmental organizations.—*Pursuant to the approval in principle of the ninth IMCO assembly, an agreement of cooperation between IMCO and the Organization of Arab Petroleum Exporting Countries (OAPEC) was formulated during the period under review. It was submitted to the tenth IMCO assembly in November 1977 for approval. Consultations are in progress regarding a similar agreement with the League of Arab States. Cooperation with another intergovernmental body, the International Hydrographic Organization, in formulating the world-wide system of navigational warnings, is mentioned in Section II of this report.

IV. TECHNICAL COOPERATION

The Assembly of IMCO attaches great importance to technical cooperation.

Committee on Technical Co-operation.—The Assembly's attitude was concretized in resolutions adopted at its ninth session in 1975, whereby it decided to amend the IMCO convention in order to enshrine the Committee on Technical Co-operation. For this purpose, as indicated in the annual report for 1975/76, the IMCO Assembly established an ad hoc working group to consider further amendments to the IMCO Convention, concerning inter alia the institutionalization of the Committee on Technical Co-operation. The ad hoc group, which was open to all IMCO members, met at IMCO headquarters, January 31–February 4, 1977, and its report was submitted to the tenth IMCO assembly in November 1977. In the group's view, the promotion of technical cooperation to developing countries is one of the most important roles of IMCO, and the convention should therefore be updated to reflect more accurately the nature and scope of the organization's activities. It was considered that this updating of the convention should include inter alia the institutionalization of the Committee on Technical Co-operation, so that it would be placed on the same footing as the Maritime Safety Committee, the Legal Committee, and the Marine Environment Protection Committee. IMCO is thus the first agency in the United Nations system to provide a constitutional basis for its Committee on Technical Co-operation.

Regional and interregional advisers.—In addition to two regional maritime advisers—one for Africa (English speaking) and the other for Latin America—and the interregional adviser on maritime legislation, the interregional adviser on maritime safety administration took up duties at IMCO headquarters in October 1976. Two other regional maritime advisers (likewise financed by the UNDP) have also been appointed; they are for the Asia and Pacific region and are based in Djakarta, Indonesia (September 1976), and Bangkok, Thailand (November 1976), respectively. The machinery is now in motion to recruit a second regional maritime adviser for Africa (French speaking).

Since taking up their respective posts, the advisers have traveled extensively and advised governments on such matters as maritime safety, navigational aids, prevention and control of marine pollution, maritime legislation, maritime training, etc.

Marine pollution.—A memorandum of understanding concerning cooperation between IMCO and UNEP was signed on November 9, 1976. This agreement will, inter alia, facilitate joint efforts to speed up the entry into force of international conventions and their amendments relating to prevention of marine pollution from ships and by dumping.

Trends.—Despite the financial constraints of the UNDP during the period under review, the Technical Cooperation program of the organization continued to expand, with increasing requests for technical assistance being received from numerous developing countries.

Large-scale projects.—The Arab Maritime Transport Academy is progressing satisfactorily, and limited funds have been made available by the UNDP for its extension for a period of 2 years, from March 1977 to March 1979. These funds will provide, inter alia, for the services of four experts.

The project for expanding and modernizing the training of merchant marine officers and port operators at Rio de Janeiro, Brazil, was completed at the end of 1976. A complementary 3-year project—a new maritime training center—was begun at Belem in northern Brazil in January 1977.

The project on the Shipbuilding Research Design Institute at Varna, Bulgaria, which began in 1971, is progressing very satisfactorily. The research center was officially inaugurated in October 1976 and deliveries of minor equipment associated with the first stage of the project have almost been completed.

An IMCO consultant in maritime training visited a number of CARICOM countries in October–November 1976 for a preliminary study in preparation for the establishment of a Regional Maritime Training Institution for countries of the CARICOM region. The project for a Regional Maritime Training Academy at Accra, Ghana, is being formulated with implementation envisaged for mid-1977. It is intended to cater to the English-speaking countries of West Africa and, at a later stage, possibly to other English-speaking countries on the African continent.

The establishment of a Regional Academy of Sciences and Techniques of the Sea at Abidjan, Ivory Coast, is well under way. It will cater to the French-speaking African countries.

After a preparatory mission to Indonesia in May–June 1976, the IMCO consultant's report on the proposed project for modernizing the existing maritime training academies in that country has been submitted to the government of Indonesia for comment. Good progress has been made on the project for a Shipbuilding Industries Technical Service (SITS) in the Republic of Korea; some of the equipment has been delivered and the installation of the towing tank is nearing completion. Since September 1975, the IMCO coordinator and the construction manager have assisted the government of Malta in the development of a shipbuilding yard and dry dock. An IMCO team of experts, after their initial survey and detailed recommendations on the development of maritime training centers in Iraq, Qatar, and Saudi Arabia for OAPEC, are currently revising their studies.

A preparatory mission of IMCO experts to Lagos, Nigeria, in October–December 1976, under a funds-in-trust agreement, carried out preliminary studies for the establishment of an integrated Nautical Institute in Nigeria. The experts' report has been submitted to the government for comments.

Fellowships.—During the period under review, IMCO's fellowship program has covered the training abroad of approximately 52 fellows from developing countries, including Argentina, Bangladesh, Bulgaria, India, the Republic of Korea, Democratic Yemen, and the Seychelles. The fellows, placed in some 14 countries, have received training in various aspects of maritime affairs. In addition, since January 1977 preliminary placement work has commenced for some 27 nationals from developing countries, including four fellows who have received awards in 1977 under the UNDP/IMCO project in Alexandria (the Arab Maritime Transport Academy), and seven nominees from Malawi. Following the offer by the government of Chile of six fellowships for training in navigation and marine engineering at the Arturo Prat Naval School, IMCO is currently assisting in the preliminary disseminations.

V. ADMINISTRATIVE AND FINANCIAL QUESTIONS

Budget and finance.—Of the total regular budget appropriation of $5,259,800 approved by the ninth assembly for 1976, actual expenditure for that year amounted to $4,138,651, or $1,121,149 less than the original appropriation. For 1977, the estimated expenditure from all sources of funds, based as regards the regular budget on the actual appropriations, is $8,509,600, which represents an increase of 22.2 percent over 1976 actual expenditure, but because of the exchange rate factor savings of a similar order as in 1976 are expected in the regular budget, and the eventual increase is likely to be on the order of 5–6 percent.

Personnel.—The total staff establishment for the present biennium is 206 posts, of which 78 are in the professional and higher categories and 128 in the general service category, as against 175 (65 professional and higher and 110 general service) in 1974–75. The council at its thirty-seventh session approved the Secretary-General's proposal to transfer responsibility for the organization's external relations activities from the office of the Secretary-General and place them under the director of the Legal Division. The Secretariat consists of the office of the Secretary-General, including the internal auditor, together with eight divisions or offices: Marine Safety, Marine Technology, Marine Environment, Legal, External Relations, Administrative, Conference, and Technical Co-operation.

VI. CONFERENCES, DOCUMENTS, AND PUBLICATIONS

Conferences and documents.—Since April 1, 1976, three international conferences have been held: the International Conference on the Establishment of an International Maritime Satellite System (third and final session), September

1–3, 1976; the International Conference on Limitation of Liability for Maritime Claims, 1976, November 1–19, 1976; and the International Conference on Safety of Fishing Vessels, held in Torremolinos, Spain, at the invitation of the Spanish government March 7–April 2, 1977. IMCO convenes an average of two international conferences a year.

In accordance with the decision of the ninth IMCO assembly, Chinese became an official language of the organization in 1976, in addition to the other four languages (English, French, Spanish, and Russian). However, by agreement with the government of the People's Republic of China, this decision will be put into effect gradually.

Publications.—During 1976, the activities of the Publications Department were further expanded and diversified. The turnover has increased from $420,000 in 1975 to $520,000 in 1976, and the number of publications sold has risen from 52,000 to 68,000, an increase of over 30 percent. The International Maritime Dangerous Goods Code is still the largest selling publication. In accordance with the wishes of both member states and private users, the Secretariat has taken the necessary steps to publish the code in loose-leaf form, thereby making the work of updating easier.

Reports from Organizations

Report of the Intergovernmental Oceanographic Commission on Its Activities*

I. INTRODUCTION

1. The Intergovernmental Oceanographic Commission, established within Unesco by its eleventh session of the General Conference, in 1960, continued "to promote scientific investigations with a view to learning more about the nature and resources of the oceans through the concerted action of its members" (Article 1 of its Statutes).

2. In recent years the Commission's activities have been intensified particularly in relation to projects under the International Decade of Ocean Exploration (East Asia, Latin America, Caribbean); at the same time, the Assembly decided to terminate the Co-operative Investigation of the Caribbean and Adjacent Regions (CICAR) and to form an IOC Association for the Caribbean and adjacent regions (IOCARIB), with greater autonomy but remaining under the auspices of the Commission. Charting of the ocean floor has continued with the development of the General Bathymetric Chart of the Oceans (GEBCO) and the International Bathymetric Chart of the Mediterranean (ICBM). Activities relating to pollution research and monitoring and to training, education and mutual assistance continued to receive high priority. The Commission has stated that it is prepared to advise the United Nations Conference on the Law of the Sea on marine scientific and technological matters.

3. Co-operation was sought with other intergovernmental and non-governmental global and regional organizations such as the Committee for Co-ordination of Joint Prospecting for Mineral Resources in Asian Offshore Areas (CCOP) and the Committee for Co-ordination of Joint Prospecting for Mineral Resources in South Pacific Offshore Areas (CCOP/SOPAC) of the Economic and Social Commission for Asia and the Pacific (ESCAP), the Commission for Marine Geology (CMG) of the International Union of Geological Sciences (IUGS), the International Council for the Exploration of the Sea (ICES), the International Commission for Scientific Exploration of the Mediterranean (ICSEM), the Tsunami Committee of the International Union of Geodesy and Geophysics (IUGG), etc.

4. Since the eighteenth session of the General Conference of Unesco, the IOC Assembly met for its ninth session in Paris, from 22 October to 4

*Report submitted to the Nineteenth Session of the General Conference of Unesco, September 30, 1976.

November 1975; in addition, three sessions of its Executive Council (EC-V, Venice, 3–8 March 1975; EC-VI, Paris, 20–21 October 1975; and EC-VII, Bergen, 21–26 June 1976) have been held.

5. The report of the ninth session of the IOC Assembly is available separately (document SC/MD/55)—this report contains the resolutions adopted during the fourth and fifth sessions of the IOC Executive Council; no resolutions were adopted at the sixth session (this was a planning session for the ninth session of the Assembly); the resolutions of the seventh session appear in the Summary Report of that session (document IOC/EC-VII/3).

6. During this same period the following meetings of the main subsidiary bodies of the Commission were held:

Working Committees and International Co-ordination Groups (Category II)

Fourth session of the Working Committee for the Integrated Global Ocean Station System (IGOSS-IV), Paris, 3 February 1975 (document IOC/IGOSS-IV/3), followed by the fourth joint session of the Working Committee for IGOSS together with the WMO Executive Committee Panel on Meterological Aspects of Ocean Affairs (IOC:IGOSS/WMO:MAOA-IV), Paris, 4–12 February 1975 (document IOC:IGOSS/WMO:MAOA-IV/3);

Tenth session of the Co-operative Study of the Kuroshio and Adjacent Regions (CSK-X), Tokyo, 13–18 March 1975 (document IOC/CSK-X/3);

Adhoc Caribbean regional meeting on Training, Education and Mutual Assistance (TEMA), Mexico City, 10–12 April 1975 (document IOC/TEMA-CARIB-I/3);

Seventh session of the Co-operative Investigations of the Caribbean and Adjacent Regions (CICAR-VII), Mexico City, 14–17 April 1975 (document IOC/CICAR-VII/3);

Eighth session of the IOC Working Committee on International Oceanographic Data Exchange (IODE-VIII), Rome, 12–17 May 1975 (document IOC/IODE-VIII/3);

Third session of the International Co-ordination Group for the Global Investigation of Pollution in the Marine Environment (GIPME-III), Paris, 28 May–4 June 1975 (document IOC/GIPME-III/3);

Ad hoc West African regional meeting on Training, Education and Mutual Assistance (TEMA), Casablanca, 3–7 June 1975 (document IOC/TEMA-AFRI-I/3);

Second session of the joint IOC/FAO(GFCM)/ICSEM International Co-ordination Group for the Co-operative Investigations in the Mediterranean (CIM-II), Dubrovnik, 9–13 June 1975 (document IOC-FAO(GFCM)-ICSEM/CIM-II/3);

Ad hoc East African regional meeting on Training, Education and Mutual

Assistance (TEMA), Manila, 15–19 September 1975 (document IOC/TEMA-ASIA-I/3);

Ad hoc Arab States regional meeting on Training, Education and Mutual Assistance (TEMA), Cairo, 4–8 January 1976 (document IOC/TEMA-ARAB-I/3);

Fifth session of the International Co-ordination Group for the Tsunami Warning System in the Pacific (ITSU), Lima, 23–27 February 1976 (document IOC/ITSU-V/3);

Third session of the IOC/WMO Joint Planning Group for IGOSS, Geneva, 29 March–1 April 1976 (document IOC-WMO/IPLAN-III/3).

7. Further category II meetings scheduled to be held before the nineteenth session of the General Conference are:

19 July Caracas	International Co-ordination Group for the Caribbean and Adjacent Regions—Final (eighth) session
20–24 July Caracas	(CICAR-VIII), followed by IOC Association for the Caribbean and adjacent regions—1st session (IOCARIB-I)
October Hamburg	Working Committee for the Global Investigation of Pollution in the Marine Environment—1st session (WC/GIPME-I).

8. During the same period numerous meetings have been held of Groups of Experts and Task Teams, in particular the following Scientific Workshops:

IOC/Unesco/FAO/WMO Workshop on the Phenomenon known as "El Nino", Guayaquil, Ecuador, 4–12 December 1974 (IOC Workshop Report No. 4);

IDOE International Workshop on the Geology and Geophysics of the Caribbean and its Resources, Kingston, Jamaica, 17–22 February 1975 (IOC Workshop No. 5);

CCOP/SOPAC-IOC IDOE International Workshop on Geology Mineral Resources and Geophysics of the South Pacific, 1–6 September 1975 (IOC Workshop No. 6);

Joint IOC/FAO(IOFC)/Unesco/EAC Scientific Workshop to initiate planning for a co-operative investigation in the North and Central Western Indian Ocean, Nairobi, 25 March–2 April 1976 (IOC Workshop Report No. 7);

Joint IOC/FAO(IPFC)/UNEP International Workshop on Marine Pollution in East Asian Waters, Penang, 7–13 April 1976 (IOC Workshop Report No. 8).

9. The following important meetings are scheduled to be held before the nineteenth session of the General Conference:

12–16 July Caracas	Second (terminal) CICAR Symposium (CICAR-II)
12–16 July New York	IOC Scientific Advisory Board—1st session (SAB-I)
8–13 August Mauritius	IOC/CMG/SCOR Second International Marine Geoscience Workshop (Workshop No. 9).

10. Close liaison has been fostered with member agencies of the Intersecretariat Committee on Scientific Programmes Relating to Oceanography (ICSPRO)—United Nations, FAO, Unesco, WMO and IMCO—and three meetings of the Committee have been held:

ICSPRO-XI, Unesco Headquarters, 10–11 February 1975;
ICSPRO-XII, FAO Headquarters, 8–10 October 1975;
ICSPRO-XIII, WMO Headquarters, 29–30 January 1976.

A further meeting (ICSPRO-XIV) is scheduled to be held in FAO Headquarters, 30 August–1 September 1976.

II. STUDIES IN DEPTH

11. During 1975 the Special Committee of the Unesco Executive Board carried out an In-Depth Study of the Marine Sciences, taking as its basis the activities of the Organization in 1974. The report of the Rapporteur (H.E. Mr. Michel Van Ussel) will be found in document 97 EX/2 Add., Annex III.A.

12. The Executive Board at its 97th session, took note of the report and adopted decision 3.1.1 (a) A 4 topic (iii) Marine Sciences (document 97 EX/Decisions).

13. During 1976, the Special Committee of the Unesco Executive Board carried out an In-Depth Study of
    i) Unesco's contribution towards the establishment of a new international economic, social and cultural order:
      b) measures taken to facilitate access by the developing countries to sciences and technology.

14. The following is an extract from the report of the Rapporteur (Mr. Maheshwar Dayal, Deputy for Mr. Gopalaswami Parthasarathi)—document 99 EX/7, Parts II and III:

"'The World Plan of Action for the Application of Science and Technology to Development' (United Nations, 1971) recognized *inter alia* the importance of marine sciences. The Plan called for increased research on fisheries and non-living marine resources. It also called for improved warning systems on natural marine disasters such as tsunamis and hurricanes. To support these, it requires the establishment of information services, data banks, scientific instrument centres, use of satellites, etc. Most importantly, it requires increasing the scientific capacity of developing countries through development of science faculties in universities, of research and experimental development institutions, and of the scientific and technical services needed to serve these universities and institutions (mapping services, fisheries services, meteorological services, seismological services, data banks, libraries, scientific collections, etc.). The Plan of Action specifically refers to the International Decade of Ocean Exploration (IDOE) as a means of stimulating development—as it has. Unesco and its IOC have contributed to the Second Development Decade in providing a framework and help both for doing the necessary marine science research and for building the required scientific infrastructure and manpower without which such research cannot be done. As examples of other relevant activities, the Tsunami Warning System in the Pacific involves seven developing countries and is currently developing plans for strengthening its tidal observation network. Similarly, the Integrated Global Ocean Station System (IGOSS) is designed to speed data from surface stations and satellites to users all over the world who choose to use the service. The IOC is involved in the development and use of several information systems. The Division of Marine Sciences is most active in developing the marine science infrastructure and manpower. All this work continued in 1975.

"Subsequent to the World Plan of Action (1971), increasing importance has been placed by nations on coastal area development and on the degradation of marine environment (such as through oil pollution), areas in which the Member States have been very active through Unesco and its IOC.

"Science and technology activities especially need funds, and these would help directly in the programmes for facilitating access to science and technology by the developing countries. Every effort must be made to give a major allocation of funds for Unesco's science programmes, and to make other arrangements such as fund-in-trust programmes, which would enable more financial resources to be made available."

III. MEMBERSHIP

15. Interest in the Commission's work is increasing steadily. In the two years since the eighteenth session of the General Conference, nine States—Sudan, Kuwait, Costa Rica, Jordan, Iran, Togo, Ethiopia, Haiti and the United Arab Emirates—have joined the Commission, bringing the total membership to 90 (as at 2 June 1976).

IV. ACTIVITIES

Ocean Science—Global Programmes

Listing and Assessment of IDOE Programmes
16. A compendium of programmes declared as components of the "International Decade of Ocean Exploration (IDOE) 1971–80 has been published in the IOC Technical Series (No. 13). This publication will be updated periodically and a second edition is scheduled for publication during 1977. It is intended to maintain a file on all IDOE programmes so as to provide an answering service for inquiries but also to facilitate updating of the above publication.

17. It is further planned to subject all IDOE programmes declared by Member States to an assessment process to determine their input to, or the need for output from, IOC ocean services activities and also their potential for training, education and mutual assistance (TEMA) in marine science for the developing countries.

General Bathymetric Chart of the Oceans (GEBCO)
18. The first sheet of the fifth edition of the General Bathymetric Chart of the Oceans, sheet 5.05 covering the Mediterranean and Northern Indian Ocean, was published in April 1975 in time for it to be displayed at the third session of the Third United Nations Conference on the Law of the Sea. Two hundred copies were presented free of charge to delegates to the Conference. Eight further sheets are at present in various stages of preparation and are expected to be published in 1976–1977 followed by the remaining sheets in the series.

19. An application has been made to the United Nations Development Programme for Funding for an International Geoscience Unit to be set up in Ottawa, where the series is being scribed and printed.

WMO-ICSU Global Atmospheric Research Programme (GARP)
20. The involvement of the Commission in planning, as well as participating in the WMO-ICSU Global Atmospheric Research Programme (GARP) continued to increase as more attention is given to the results of the GARP Atlantic Tropical Experiment (GATE), planning for the First GARP Global Experiment (FGGE) and on activities related to the understanding of the physical basis of climate.

21. Plans for the First GARP Global Experiment (FGGE) are continuing with active support and participation of the Commission. Discussions in this regard have been directed at several of the Commission's programmes. Most importantly, these include (i) IGOSS as a means of collection and exchange of BATHY and TESAC data, and (ii) IDOE projects such as NORPAX, CLIMAP, MONEX, EL NIÑO as complements to the regional and sub-programmes of the Global Experiment. The Commission is also investigating through its

programme on the Southern Oceans how it might best participate in the drifting buoy programme of the Global Experiment. In addition, as increased emphasis is now being given to the GARP second objective in Climate; requirements for additional ocean observations and study are being formulated. As a result, the Commission is actively seeking how it can best respond to these needs.

22. An ad hoc Task Team has been established to develop a Comprehensive Oceanographic Programme related to GARP; this group is working closely with the Scientific Committee on Ocean Research (SCOR) of ICSU.

The Global Investigation of Pollution in the Marine Environment (GIPME)
23. The third session of the IOC for the Global Investigation of Pollution in the Marine Environment (GIPME) was held in Paris from 28 May to 4 June 1975. A Task Team on Marine Pollution Monitoring was formed to develop GIPME advice to IGOSS, to effect the necessary liaison, and to review periodically the IGOSS Pilot Project on Marine Pollution (Petroleum) Monitoring. An implementation plan for GIPME was developed and it was decided to combine it with the "Proposed Outline for a Comprehensive Plan for the Global Investigation of Pollution in the Marine Environment" and to call the whole "The Comprehensive Plan for the Global Investigation of Pollution in the Marine Environment". This will be published during 1976 in the IOC Technical Series.*

24. The Task Team on Training and Technical Assistance Needs presented a preliminary report pending the assembly of specific information at, or as a follow-up to, each of the regional marine pollution workshops now being organized by the Commission in co-operation with relevant regional bodies (see paragraphs 33, 34 and 46 below). An inter-sessional group prepared detailed instructions on the methodology and procedures to be followed during the IGOSS Pilot Project. An IOC/ICES Working Group on Baseline Study Guidelines also presented a draft report which, after revision and final approval, will be published with the Comprehensive Plan for the Global Investigation of Pollution in the Marine Environment.

25. A report on the *State of Health of the Oceans* will be published during 1976. The manuscript, which was prepared by a consultant, has been reviewed by a number of marine pollution scientists.

26. Upon the initiative of UNEP, a study of the feasibility of a programme for monitoring background levels of selected pollutants in open waters has been undertaken by three consultants appointed by UNEP, WMO and IOC. This project will form part of the Global Environmental Monitoring System (GEMS) of UNEP.

*Editor's Note:"A Comprehensive Plan for the Global Investigation of Pollution in the Marine Environment and Baseline Study Guidelines" has been published as IOC Technical Series No. 14, available through the Secretary IOC.

Ocean Science—Regional Programmes

(a) Caribbean and Adjacent Regions

27. Following a study—CICAR past, present and future. An Evaluation Study of a Regional IOC Programme—(document IOC/INF-238), carried out by a consultant, the Commission decided to disband the International Co-ordination Group for the Co-operative Investigations of the Caribbean and Adjacent Regions (CICAR) and to establish on an experimental basis for a period of six years an IOC Association for the Caribbean and adjacent regions for the purpose of continuing and developing regional co-operation in the marine sciences built up over a period of seven years under the CICAR.

28. The IOC Association for the Caribbean and adjacent regions (IOCARIB) which will have its own small secretariat, will, *inter alia:*

> i) be responsible, under the overall supervision of the Commission, for overseeing all the Commission's activities in its region;
> ii) develop a regional programme of activities in the form of scientific projects which are worth while from the viewpoint of international collaboration in the region, to determine the basic objectives of joint investigations and to agree on the character and pathways for using the results;
> iii) work closely with the Working Committee and International Co-ordination Groups of the Commission in the development of the regional programme of activities;
> iv) co-ordinate scientific projects in the region, subject to overall financial implications approved by the Commission;
> v) develop regional projects, in conjunction with Unesco (or other ICSPRO agencies) on behalf of the Commission, for submission to the United Nations Development Programme (UNDP);
> vi) develop working relationships with other bodies involved in marine scientific research in the region, particularly the regional Commissions and Councils of FAO, and the United Nations Environment Programme (UNEP).

29. The CICAR will be terminated by a symposium on the theme "Progress in Marine Research in the Caribbean and Adjacent Regions", to be held in Caracas, Venezuela, 12–16 July 1976, followed by the final (eighth) session of the International Co-ordination Group for CICAR (19 July) and the first session of IOCARIB (20–24 July).

30. A meeting of an ad hoc Group of Experts on the future programmes and future co-operation in marine sciences in the Caribbean and adjacent regions was convened on behalf of the Commission by the University of Puerto Rico in Mayaguez, 1–4 March 1976. The main task of this meeting was the compilation of 18 selected scientific programmes for the future international

co-operation in marine sciences in the region; these will be placed before the first session of the IOC Association for the Caribbean and adjacent regions (IOCARIB). Five recommendations were also adopted, addressed to the above session.

31. An IDOE International Workshop on Marine Geology and Geophysics of the Caribbean Region and its Resources was held in Kingston, Jamaica, 17–22 February 1975. The Workshop was attended by more than 50 scientists from 16 countries mainly of the Caribbean region. The participants reviewed the present status of marine geological and geophysical research in the region, and then concentrated on the development of a programme of future co-operative work. The programme finally developed includes seven field and two compilation projects. The field projects contain requests for deep sea drilling.

32. The Workshop also recommended that data and information exchange be improved, and that assistance be given in the field of training and education to countries in the region such as the Dominican Republic, Haiti, Jamaica, etc., which need this urgently in order to build up their infrastructures.

33. An International Workshop on Marine Pollution in the Caribbean and Adjacent Regions is being planned in co-operation with FAO and UNEP; it is scheduled for late 1976.

(b) Mediterranean

34. The Joint IOC/FAO(GFCM)/ICSEM International Workshop on Marine Pollution in the Mediterranean, held in Monaco from 9 to 14 September 1974, proposed seven Mediterranean regional pilot projects which were approved by the UNEP Intergovernmental Meeting on the Protection of the Mediterranean (Barcelona, 28 January–4 February 1975). The Commission has assumed responsibility for the Pilot Project on Problems of Coastal Transport of Pollutants and, with WMO, for the Pilot Project on Baseline Studies and Monitoring of Oil and Petroleum Hydrocarbons in Marine Waters, under the general co-ordination of UNEP.

35. An Expert Consultation on the Joint Co-ordinated Project on Pollution in the Mediterranean was held in Msida, Malta, 8–13 September 1975, at which principles for participation from institutions were set out, and operational details for both projects were developed (document IOC/MPPP/3).

36. As part of the Commission's attempts to rationalize and improve the structure of its regional mechanisms, and in view of the development by UNEP of the Mediterranean Action Plan, with which the three sponsoring bodies of the Co-operative Investigations in the Mediterranean (CIM)—IOC, FAO through its General Fisheries Council for the Mediterranean (GFCM) and the International Commission for the Scientific Exploration of the Mediterranean (ICSEM)—are associated, UNEP has been invited to become a fourth co-sponsor of CIM and the UNEP Mediterranean Unit will be co-located in the same building in Monaco, by courtesy of the Monegasque Government.

(c) West Africa

37. A steering group has been formed and planning is well in hand with the Fishery Commission of the Eastern Central Atlantic (CECAF) of FAO and the International Council for the Exploration of the Sea (CES) for a follow-up symposium on the Co-operative Investigations of the Northern Part of the Eastern Central Atlantic (CINECA) which is scheduled to be held in Dakar in early 1977.

(d) Tropical Atlantic

38. The second and final atlas of the Tropical Atlantic—Chemical and Biological Oceanography—will be published during 1976, thus bringing to a close the International Co-operative Investigations of the Tropical Atlantic (ICITA). Work in this region has been continued as part of the GARP Atlantic Tropical Experiment (GATE).

(e) East Africa

39. A seminar and scientific workshop was convened jointly with FAO (Indian Ocean Fisheries Commission), Unesco (Division of Marine Sciences) and the East African Community, in Nairobi in March/April 1976, in order to initiate planning for a co-operative investigation in the region.

40. It was agreed that (i) the output of the investigation should provide the countries of the region with an improved understanding of the oceanography of the area, which they can utilize in their programmes of resource investigation and development; and (ii) full attention should be devoted to the needs of the countries of the region for training and for the development of the infrastructures which will be needed to implement future research.

41. The Workshop had the following terms of reference:

i) to review the existing knowledge of the ocean environment in the area of investigation with special reference to the coastal environment; and to identify gaps in our knowledge which need to be filled to ensure an enhanced rational utilization of the marine resources of the region;
ii) to identify the requirements of present development programmes for oceanographic information in the area;
iii) to suggest scientific programmes which can be carried out:
on a national basis with some mechanism for exchange of information and co-ordination;
as inter-country programmes based on regional co-operation;
as ventures between countries from outside and those inside the region participating in CINCWIO;
iv) to review the existing infrastructures and marine science capabilities of the countries of the region and to determine which programmes can be realistically carried out under present conditions;
v) to suggest additional programmes which can be carried out in the future, if the marine science capabilities of the countries in the region are improved through intensified programmes of training and education which should be an integral part of any CINCWIO programme.

(f) Indian Ocean
42. The International Indian Ocean Expedition (IIOE) geological/geophysical atlas has now been published and is on sale in those bookshops which have been officially designated for the distribution of Soviet literature.*

43. The final manuscript of the IIOE Atlas on Phytoplankton Production and some related factors was approved by SCOR in April 1975. The printing of the atlas has been delayed for technical reasons, but it is now expected to be published in 1976.†

(g) East Asia
44. The tenth session of the International Co-ordination Group for the CSK was held in Tokyo, 13–17 March 1975. Amongst the main topics discussed were: the question of future activities in the CSK area, including marine pollution studies; the future structure for continued collaboration in marine science in the region; and training, education and mutual assistance.

45. The Programme of Research which had been developed by the IDOE Workshop on Metallogenesis, Hydrocarbons and Tectonic Patterns in Eastern Asia (Bangkok, September 1973) was further implemented, and the state of implementations evaluated by the CCOP-IOC Joint Working Group on IDOE Studies on East Asia Tectonics and Resources (SEATAR) (first session, Tokyo, 13–14 August 1975), the CCOP (at its twelfth session, Tokyo, 8–22 August 1975), and the IOC Assembly (at its ninth session, Paris, 22 October–4 November 1975); resolution IX-2 refers. The Commission welcomed the great interest of countries from within and outside the region in the Programme of Research and their high level of involvement in the implementation of its projects. There is a general feeling in the Member States that the basic research matters such as data management, the protection of the marine environment, and training, education and mutual assistance, should receive greater attention.

46. An International Workshop on Marine Pollution in East Asian Waters, convened by the IOC and FAO (Indo-Pacific Fisheries Council), with the support of UNEP, was held in Penang, Malaysia, 7–13 April 1976. The Workshop proposed a number of regional and sub-regional marine pollution research pilot projects.

(h) South East Pacific
47. The recommendations of the Workshop on the Phenomenon known as "El Niño", held in Guayaquil in December 1974, are in course of being implemented.

48. The first session of a Group for Co-ordination of Project ERFEN

---

*EDITORS' NOTE: IIOE *Geological–Geophysical Atlas of the Indian Ocean* is now also available from the Secretary IOC.

†EDITORS' NOTE: The *Phytoplankton Production Atlas of the International Indian Ocean Expedition* (1976) is now available from the Secretary IOC.

(Estudio Regional Fenómeno "El Niño") was held in Lima, Peru, 16–18 April 1975. The session was convened by the Comisión Permanente del Pacifico Sur (CPPS), the secretariat of which body has assumed responsibility for day-to-day co-ordination of the regional activities. During this session the Group prepared (see document IOC/ERFEN-I):

i) Project ERFEN—a Long-Term Plan of Action; and
ii) A Programme of Immediate Action.

The Long-Term Plan of Action has been passed to the governments of the region for approval. In view of the long-term benefits this programme is expected to provide for the economies of the countries in the region, in particular Peru, governmental approval is considered to be very desirable and of high priority. The Programme of Immediate Action will be followed until such approval is received.

49. A second session of the Group is scheduled to be held in Lima, 13–16 July 1976, to be followed later in the year by a meeting on data management and exchange for project ERFEN.

50. The Commission for Marine Meterology (CMM) of WMO will meet in Lima, 6–17 December 1976, and will discuss *inter alia* those recommendations of the Workshop which have meteorological implications. Coincident with the second week of this meeting will be the first session of the IOC ad hoc Intergovernmental Working Group on the Investigations of "El Niño" (established by IOC Assembly resolution IX-1), the mandate of which is to "undertake the responsibility of providing a co-ordination mechanism for oceanographic projects conducted *in support* of ERFEN".

(i) South Pacific
51. In co-operation with the Committee for Co-ordination of Joint Prospecting for Mineral Resources in South Pacific Offshore Areas (CCOP/SOPAC), an IDOE International Workshop on the Geology, Mineral Resources and Geophysics of the South Pacific was successfully convened in Suva, Fiji, in September 1975. A Research Programme for the SOPAC region including 20 Field Projects, two Subject-oriented Projects and four Compilation Programmes was developed. Discussions are being held with a number of countries with a view to obtaining funding and support for these projects.

(j) Southern Oceans (SOC)
52. The second session of the International Co-ordination Group for the Southern Oceans (SOC) met in Buenos Aires, Argentina, 15–19 July 1975. Particular attention was paid to the development of a programme on the study of marine living resources in the Southern Ocean, in collaboration with the Scientific Committee on Oceanic Research (SCOR) and the Scientific Committee on Antarctic Research (SCAR).

Training, Education and Mutual Assistance (TEMA)

Regional TEMA Meetings
53. Following resolution VIII-25D—Assessment of Training Needs—of the eighth session of the IOC Assembly, a series of regional meetings have been convened in 1975 and 1976 for the purpose of assessing the training needs at regional level and also studying ways and means by which such needs could be satisfied.

54. Four such meetings have now been held:

Caribbean—Mexico City, 10–12 April 1975
West Africa—Casablanca, 3–5 June 1975
East Asia—Manila, 15–19 September 1975
Arab States—Cairo, 4–8 January 1976.

55. A fifth meeting will be held for South America, in Montevideo, 15–19 November 1976.

56. The first ad hoc regional TEMA meeting for the Caribbean and adjacent regions was attended by representatives from the following countries in the region: Colombia, Costa Rica, Cuba, the Dominican Republic, Jamaica, Mexico, Trinidad and Tobago, United States of America and Venezuela. Fourteen recommendations were adopted concerning the Significance of Marine Science as a National Effort, Strengthening of Institutions in the Region, Implementation of Recommendations, Marine Science Teaching at University Level, Training Courses, Training of Technicians, National Training Contacts, Instrumentation-Calibration and Repair Facilities, Data Centres and Digital Information, Information Centres, Scientific Literature and Other Information, Sorting Centres, Research Vessels, Pollution Monitoring, and Marine Science Programmes.

57. The second ad hoc regional TEMA meeting for West Africa was attended by delegates from the following countries in the region: Gabon, Morocco, Nigeria and Sierra Leone. Ten recommendations were adopted on Development of TEMA in West Africa, Regional Operation of Oceanographic Research Vessels, Strengthening the Institutions of the Region, International Co-operative Investigations, Participation in Oceanographic Cruises, Collection, Sorting and Analysis of Biological Samples, Training in Fisheries Science, Pollution Monitoring, Selection of National Training Contacts and Marine Research Assistants.

58. The third ad hoc regional TEMA meeting for East Asia was attended by delegates from the following countries in the region: Indonesia, Japan, Malaysia and the Philippines. Eighteen recommendations were adopted on National Priorities and Commitments in Marine Science, Strengthening the Universities of the Region, Regional Post-Graduate Programme for Marine

Sciences in South East Asia, Training of Technicians, Marine Science Administrators, Coastal Zone Management Training, Participation in Oceanographic Cruises, Training Courses, Marine Science Assistants, Visiting Professors and Senior Lecturers, Introduction of Marine Science in the Secondary School System, Use of Video Tapes, Films and other Teaching Aids, National Training Contacts, Upgrading the Reference Collections in South East Asia, Training Requirements in Marine Pollution for South East Asia, WESTPAC and its priorities in Research Programmes, Regional Participation in Marine Field Studies, and Marine Science Activities of the Nations.

59. The fourth ad hoc regional TEMA meeting for the Arab States was attended by representatives from the following countries from the region: Egypt, Iraq, Iran, Kuwait, Libya, Saudi Arabia, Sudan, Tunisia and the People's Democratic Republic of the Yemen. Twelve recommendations were adopted concerning an Inventory of Needs and Resources, Co-operation in Post-Graduate Training and Education within the Region, Training in Coastal Zone Management Practices, Publication of a Magazine, Marine Technicians, Visiting Professors and Marine Science Assistants, Participation in Oceanographic Cruises, National Training Contacts, Pollution Monitoring, National Marine Programmes, Proposals for Strengthening Regional Co-operation, and Scheduling of the Second Session of the Working Committee for TEMA.

60. The above recommendations are now being followed up; approaches are being made to funding agencies and Member States are being solicited for contributions to the IOC Trust Fund to assist with their implementation. In addition a Voluntary Assistance Programme (VAP) is being developed.

Participation in Oceanographic Cruises
61. Shipboard training has been carried out aboard research ships of Germany (Federal Republic of), Union of Soviet Socialist Republics and United States of America. Trainees from Egypt, Fiji, India, Israel, Syria, Tanzania and Thailand were included.

Financial Support for Academic Study, etc.
62. Financial support has been provided for a number of developing country scientists for academic study, attendance at training courses, symposia, etc., as, for example:

> Miss Ruth Calienes, Instituto del Mar, Peru, for attendance at an "Interdisciplinary training course in marine science", Duke University, United States of America;
> Mr. Macias Regulado, UNAM, Mexico, completion of Ph.D. degree at the University of Kiel, Federal Republic of Germany;
> Mr. Rojas Beltrán, Colombia, completion of Ph.D. programme, France, study of shrimp populations in Guadeloupe;
> Mr. Martin Mensah, Fishery Research Unit, Tema, Ghana, post-cruise data analysis, University of Kiel, Federal Republic of Germany;

Mr. Ihenyen, Nigeria, post-cruise data analysis and studies, University of Kiel, Federal Republic of Germany;

Mr. Youcef Lalámi, Mrs. Rashida Lalami, Algeria, Département des Sciences Biologiques. Université d'Alger, toward completion of a Ph.D. degree, Université des Sciences et Techniques du Languedoc, Montpellier, France;

Mr. Jorge Antonio Weinborn, Chile, completion of Ph.D. programme at the Universidad Nacional Autónoma de México (UNAM), Mexico City, Mexico.

Master of Marine Affairs Programme, University of Rhode Island, United States of America
63. For the academic year 1975–1976, fellowships to this Programme were awarded to Ms. Choon Sook Kim (Republic of Korea), Mr. Jesada Jiraporn (Thailand) and Mr. Abdul Gani Ilahude (Indonesia); financial assistance was provided to Dr. T. Rajyalakshmi (India).

For the academic year 1976–1977, two further fellowships will be offered for qualified students sponsored by their governments. The recipients will be selected in open competition.

Annotated Bibliography of Textbooks and Reference Materials in Marine Sciences
64. In response to resolution VIII-25G of the eighth session of the Assembly, a provisional edition of the above publication has been issued as an unnumbered volume in the IOC Technical Series; it is now available upon request. This volume will be updated periodically.

Ocean Services

The Integrated Global Ocean Station System (IGOSS)
65. The IGOSS is a joint IOC/WMO operational service programme for the provision of information on the state of the oceans.

66. The operational programme of IGOSS was started in 1972 as a Pilot Project for the collection and exchange of Bathythermograph data (BATHY Pilot Project). An average of 1,550 reports were exchanged via the Global Telecommunication System (GTS) monthly during 1973, with 1,250 per month in 1974 and 1,400 per month in 1975. In 1975 the IOC Executive Council and the WMO Executive Committee approved the conversion of the Pilot Project into an operational programme.

67. Operational instructions required for participation in IGOSS are contained in the following Manuals and Guides, issued in 1974–1975:

Manual on IGOSS Data Archiving and Exchange (IOC Manuals and Guides No. 1);
Guide to Operational Procedures for Collection and Exchange of

Oceanographic Data (BATHY and TESAC) (IOC Manuals and Guides No. 3);
Guide to Oceanographic and Marine Meteorological Instruments and Observing Practices (IOC Manuals and Guides No. 4).

68. A second operational programme, the IGOSS Pilot Project on Marine Pollution (Petroleum) Monitoring was launched in 1975. This initial IGOSS marine pollution monitoring project is an internationally co-ordinated programme for monitoring petroleum-derived oils. An operational plan, containing a description of procedures for sample collection, preservation and analyses, was prepared and circulated to Member States of IOC and WMO in 1974.

69. This plan was also accepted as a basis for the development of the *Joint IOC/WMO/UNEP Pilot Project on Baseline Studies and Monitoring of Oil and Petroleum Hydrocarbons in the Mediterranean,* which is considered as a substantial input to the IGOSS Pilot Project on Marine Pollution (Petroleum) Monitoring (see paragraph 34 above). A first evaluation of the state of implementation of the Pilot Project was made at the Second Workshop on Marine Pollution (Petroleum) Monitoring in June 1976 in Monaco.

70. The above marine pollution monitoring programmes are being developed with the financial support from UNEP and where applicable as part of the UNEP Mediterranean Action Plan.

71. A draft IGOSS *General Plan and Implementation Programme for 1977– 1982* has been prepared and was submitted to the seventh session of the IOC Executive Council (June 1976) and the XXVIIIth session of the WMO Executive Committee (June 1976) for approval. This plan will serve as a guide for further development of the IGOSS programme and as an expression of the manner in which participating nations wish IGOSS to develop.

72. Studies have been made on the design and development of the IGOSS observing system and the IGOSS Data Processing and Service System, as well as on the development of new Ocean Data Acquisition Systems, Aids and Devices (ODAS), an ocean current observation programme, and the provision of support for experiments in the Global Atmospheric Research Programme (GARP) (see paragraph 20 above).

Oceanic Data Management
73. Work in this field was concentrated on further development and improvement of the international system for exchange of oceanographic data consisting of such basic elements as the World Data Centres (Oceanography), National Oceanographic Data Centres, Designated National Agencies, Regional Oceanographic Data Centres and Responsible National Oceanographic Data Centres. A number of inventory forms and formats were developed and recommended for use by Member States either on permanent or experimental basis.

74. A guide for establishing a National Oceanographic Data Centre was prepared and published as No. 5 in the IOC series "Manuals and Guides".

75. After a number of years of successful work in co-operation with other international organizations: WMO, FAO, IHO, WHO, IAEA, ICES, IMCO and UNEP, a concept and mechanism for a *Marine Environmental Data Information (MEDI) Referral System* has been developed and tested. As a result of these joint efforts, a brochure entitled "Guide to International Marine Environmental Services" was prepared and widely distributed to Member States through participating organizations. In addition, a Pilot MEDI Referral Catalogue has been prepared and reviewed by an inter-agency group and by the Working Committee on IODE.

76. Development of the MEDI Referral System was considered by the ninth session of the IOC Assembly. The Assembly in its resolution IX-30 "decided, subject to the concurrence of other member organizations of the Joint Task Team, i.e. WMO, FAO, ICES, IHO and UNEP, that the necessary co-ordinating functions for the Marine Environmental Data Information Referral System be performed by the IOC Secretariat as the UNEP/IRS contact point in the field of marine environmental data and information". The Assembly also recommended that development of the MEDI Referral System be continued in close co-operation with other international organizations.

77. In order to promote exchange of international scientific information to meet the requirements of the marine scientific community, certain actions have been taken to co-operate with FAO in developing the Aquatic Sciences and Fisheries Information System (ASFIS). In particular, in accordance with resolutions VIII-28, EC-V.13 and IX-31, the joint FAO-IOC Panel of Experts on ASFIS was established to advise on the policy, development and further implementation of an effective international system for scientific and technical information in the fields of marine and freshwater science and technology.

V. PROGRAMME AND BUDGET OF THE CURRENT BIENNIUM, 1975–1976

78. The following is a breakdown of Regular Programme (Unesco) funds, made available to the Commission:

		US $	
I.	Secretariat Services (Assembly, Executive meetings, advisory services to the Commission)	148,950	(22%)
II.	Under the Long-Term and Expanded Programme of Oceanic Exploration and Research (LEPOR):		
	i) Ocean Science (Regional Co-operative Investigations GIPME, IDOE projects, GEBCO)	216,000	(32%)

ii) Ocean Services (IGOSS, Data Management, Tsunami Warning)	142,000	(21%)
iii) Training, Education and Mutual Assistance	107,000	(16%)
III. Publications and dissemination of information	61,000	(9%)
TOTAL (Operational funds)	674,950	(100%)
IV. The *staff costs* approved under the Unesco Regular Programme amounted to	714,450	
	1,389,400	

79. The following is a breakdown of the 1975–1976 total budget (all in United States dollars):

	In cash	In kind; for example, for hosting meetings under ICSPRO agreement	Salaries	Total
From Unesco	674,950	...	714,450	1,389,400
United Nations	...	6,000*	...	6,000*
UNEP	424,200*	...	...	424,200*
FAO	...	8,000*	106,000*	114,000*
WMO	...	...	43,000*	43,000*
IMCO	...	...	49,500*	49,500*
Member States et al. (Trust Fund)	197,430	...	...	197,430*
TOTAL	1,296,580*	14,000*	912,950*	2,223,530*

*Estimates only.

80. It should be noted that the above financial statement is based partly on assumptions and simplifications, and therefore is not an accurate representation of income or expenditure.

Reports from Organizations

The Scientific Committee on Oceanic Research (SCOR)*

Warren S. Wooster
University of Washington

During the last 25 years there has been a rapid escalation of interest in the oceans. In part, this manifested itself in a vigorous growth of national activity, but the very magnitude of the problems and the extent of effort required to study them led to the need for wider cooperation. In recognition of this need the Scientific Committee on Oceanic Research was established by the Executive Board of the International Council of Scientific Unions (ICSU) in July 1957 for the purpose of "furthering international scientific activity in all branches of oceanic research." SCOR first met in Woods Hole in August 1957; since then, there have been 13 general meetings and 20 executive meetings.

Membership of SCOR now consists of 84 marine scientists nominated by scientific institutions and 11 scientists nominated by ICSU and by interested international scientific unions. Many other scientists are associated with SCOR through membership in its various working groups.

Many countries (34) have created national committees for oceanic research, which are representative bodies of marine scientists appointed by national scientific institutions. These committees often serve to strengthen and coordinate marine science nationally as well as providing lines of communication between oceanographers and SCOR.

Most nongovernmental organizations concerned with some aspect of marine science are components of ICSU. Although SCOR is associated with a number of ICSU bodies, four are directly affiliated with SCOR, and their presiding officers are ex officio members of the SCOR Executive Committee; these are the International Association for the Physical Sciences of the Ocean (IAPSO) of the International Union of Geodesy and Geophysics, the International Association of Biological Oceanography (IABO) of the International Union of Biological Sciences, the Commission for Marine Geology (CMG) of the International Union of Geological Science, and the International Association of Meteorology and Atmospheric Physics (IAMAP) of the International Union of Geodesy and Geophysics. These affiliations illustrate the unique interdisciplinary character of SCOR.

*Historical sections were prepared by R.I. Currie and G.E. Hemmen for distribution at the Joint Oceanographic Assembly in September 1976. The sections on noteworthy activities in 1976-77 and on future prospects were prepared by W.S. Wooster at the request of the SCOR Executive Committee.

Of the intergovernmental organizations, SCOR's closest relations are with the United Nations Educational, Scientific, and Cultural Organization (UNESCO) and its Intergovernmental Oceanographic Commission (IOC), for both of which SCOR serves as a scientific advisory body. Another advisory body to IOC is the Advisory Committee on Marine Resources Research (ACMRR) of the Food and Agriculture Organization; SCOR and ACMRR have worked together in considering IOC problems of mutual interest. SCOR's efforts have also been joined on specific matters with those of the World Meteorological Organization and the International Council for the Exploration of the Sea.

The direct financial requirements of SCOR have been relatively small because its work is done by volunteers with the help of its national committees. Expenses are met from annual contributions from these committees and from contracts with UNESCO, IOC, etc.

The first major scientific project of SCOR was the International Indian Ocean Expedition, a multidisciplinary exploration of this large and relatively unknown region. Planning of the expedition began in late 1957, and SCOR played the major part in its organization and coordination until mid-1962 when this responsibility was transferred to the Intergovernmental Oceanographic Commission. The IIOE remains one of the largest and most comprehensive international oceanographic efforts ever attempted; the lessons learned and experience gained during the period have been used in the development of most subsequent cooperative expeditions.

During the past decade, SCOR activities have generally fallen into one of the following categories: scientific meetings, working groups, and advice to UNESCO/IOC.

SCIENTIFIC MEETINGS

Soon after it was established, SCOR devoted its energies to the development of a world scientific meeting, which became the First International Oceanographic Congress (New York, 1959). The Second Congress (Moscow, 1966), although organized by UNESCO and IOC, was also based on a number of SCOR recommendations. The Joint Oceanographic Assembly (Tokyo, 1970) was the third in the series of world meetings, and the fourth was held in Edinburgh, 1976; both were organized by SCOR.

The sixth SCOR General Meeting (Halifax, 1963) included symposia on biogeochemistry, intercalibration, and standardization, and on a general scientific framework for world ocean study. Beginning in 1966, interdisciplinary symposia have become essential parts of the General Meetings, topics including variability in the ocean (Rome, 1966); scientific exploration of the South Pacific (La Jolla, 1968); and remote sensing, ocean monitoring, and the benthic boundary (Tokyo, 1970). The 1972 general meeting in Oban, Scotland was linked to the Second International Congress on the History of Oceanography, held in Edinburgh, and in 1974 the general meeting in Guayaquil, Ecuador was the occasion of an IOC workshop on the phenomenon of "El Niño."

Several more specialized scientific meetings have been organized in cooperation with other international organizations: hydrodynamics of plankton samplers (Sydney, 1966), micropaleontology of marine sediments (Cambridge, 1967) and geology of the east Atlantic continental margin (Cambridge, 1970). In the spring of 1971, a symposium on Indian Ocean biology and the International Indian Ocean Expedition was held in Kiel; in 1974 a symposium on the polar oceans was held in Montreal, and one on marine plankton and sediments was held in Kiel.

WORKING GROUPS

Members of working groups are selected by the Executive Committee. Often the groups are cosponsored by other interested international organizations. During each general meeting, the present status of each group is examined and decisions are reached on reconstitution or disbandment.

Many SCOR working groups have been concerned with problems of oceanographic methodology. Topics have included zooplankton sampling and laboratory methods, determination of photosynthetic pigments and other phytoplankton methods, measurements of photosynthetic radiant energy and estimation of primary production, methods of nutrient analysis, continuous current velocity measurements, and the measurement of deep-sea tides. One group has been concerned with the development of oceanographic tables and standards. Recommendations of these groups are published in the SCOR Proceedings or in one of the UNESCO series, *Technical Papers in Marine Science* and UNESCO *Monographs on Oceanographic Methodology*.

Another category of working groups deals with broader scientific questions such as air-sea interaction, micropaleontology of marine sediments, river inputs to ocean systems, east Atlantic continental margins, and the oceanographic basis of ocean monitoring and prediction systems. A group related to the last has examined problems of continuous monitoring in biological oceanography. Consideration also has been given to data exchange problems, particularly those concerning exchange and inventory of biological data.

There have also been several working groups which examined questions of science policy and ocean affairs. In cooperation with other organizations, a general scientific framework for the comprehensive study of the ocean was examined, consideration was given to implementation of UN resolutions on resources of the sea, and the scientific aspects of international ocean research were explored. All of these groups have given rise to special publications.

ADVICE TO INTERGOVERNMENTAL BODIES

In 1959, UNESCO abolished its own advisory committee and invited SCOR to provide scientific advice in the field of oceanography; in 1960 the IOC followed suit. Part of the cost of this activity has been met by UNESCO and IOC under annual contracts. Advice may be specifically requested by either UNESCO or IOC; on occasion, SCOR voluntarily presents the views of its members or national committees. Technical advice is often developed by working groups which have considered problems of importance to international cooperative expeditions, such as standardization and intercalibration of methods, data exchange, and the establishment of tables and standards.

The advisory activities of SCOR provide an opportunity for scientists to influence the programs and policies of the intergovernmental organizations with great potential for promoting marine science. By this means, scientists can help to ensure that these international programs have a sound technical basis and will serve to increase understanding of the ocean and its resources. Recently SCOR has been particularly concerned over the need to maintain freedom for scientific research on the oceans and has represented this view through ICSU to the UN Law of the Sea Conference.

NOTEWORTHY ACTIVITIES DURING 1976–77

Physical Oceanographic Studies

A number of SCOR working groups are concerned with studies of ocean circulation. WG 34, on internal dynamics of the ocean, has participated in planning of the international POLYMODE Experiment, to be held in the North Atlantic in 1977–78. An earlier set of oceanographic experiments in the equatorial Atlantic, organized by WG 43 as part of the GARP (Global Atmospheric Research Program) Atlantic Tropical Experiment, revealed hitherto unknown transient features of the equatorial circulation. Planning is also under way (by WG 47) for oceanographic programs during the First GARP Global Experiment; these are to take place in tropical regions of the Atlantic, Indian, and Pacific Oceans.

Methods in physical oceanography are also receiving attention. Modern profiling devices to measure conductivity and temperature as functions of pressure (depth) and the methods for correcting and calculating other properties are being reviewed by WG 51, and an intercalibration exercise is being considered. Another group, WG 49, is concerned with mathematical modeling of oceanic processes; five issues of a newsletter on ocean modeling have been distributed in the last year.

Still other groups are examining problems occurring in specific ocean regions that include not only physical, but also chemical, biological, and in some cases geological processes. A new group, WG 56, is being organized to look at the time and space variability of equatorial upwelling processes and to suggest appropriate lines of multidisciplinary inquiry. Another group, WG 57, has been established to look at physical processes, and their interaction with other kinds of processes, in studies of estuaries, coastal areas, and shelf seas.

Oceanography and Climate

It has been increasingly recognized that the key to understanding and eventually forecasting climatic changes lies in the surface layer of the ocean. Thus some of the activities discussed above are relevant to the study of climate. SCOR has recently established a Committee on Oceanography and GARP which is intended to identify, stimulate, and coordinate oceanographic programs linked to GARP and especially to the second GARP objective, achieving better understanding of the physical basis of climate. SCOR has associated with the Joint Organizing Committee of GARP in convening (in May 1977) a study conference on general circulation models of the ocean and their relation to climate. Two working groups are concerned with specific aspects of the problem. WG 55 is examining possible prediction schemes and indices for "El Niño," the large-scale atmospheric and oceanic perturbation off western South America that was accompanied in recent years by collapse of the anchoveta fishery. A new group, WG 58, will assess knowledge of, and required research for, the Arctic Ocean heat budget and the processes that control it.

Living Resources

Great interest has been shown in recent years in developing the harvest of krill and other living resources of the Southern Ocean. SCOR, together with other organizations (including the Scientific Committee on Antarctic Research), convened a conference on this subject in August 1976 and, through its WG 54, is giving further consideration to a proposed Biological Investigation of the Marine Antarctic System and Stocks (BIOMASS).

A new group, WG 60, is being established to review the status of mangrove ecosystem studies and to work with other agencies in appraising man's impact on these important tropical systems. The group will also deal with methodological problems. Other SCOR activities on biological methods include a study of methods for estimating micronekton abundance, particularly krill, squid, and juvenile stages of fish (WG 52), and a new group (WG 59) to suggest mathematical methods in marine ecology and in the treatment of biological data collections.

Pollution

While SCOR has concerned itself with promoting basic rather than applied research, several of its activities are more or less directly related to problems of marine pollution. Most direct is the continued study of pollution of the Baltic (WG 42), carried on in cooperation with the International Council for the Exploration of the Sea. SCOR participation in this effort facilitates the interaction of academic and fishery scientists within the region. Other groups are concerned more generally with processes whereby pollutants are introduced into the marine environment, either via the atmosphere (WG 44 on ocean-atmosphere materials exchange) or via rivers (WG 46 on river inputs to ocean systems).

FUTURE PROSPECTS

Intergovernmental action in the realm of ocean affairs has been increasingly complicated in recent years by political issues related to maritime jurisdiction. At the time of writing, these issues have not yet been settled by the United Nations Conference on the Law of the Sea. It is to be hoped that the relevant intergovernmental bodies can eventually refocus on the scientific and technical aspects of ocean affairs. In the meantime, marine scientists have developed a variety of ways to promote international cooperation in the conduct of their investigations. Often informal arrangements are made among scientists and their institutions. Occasionally, international research projects are carried out within the framework of bilateral intergovernmental arrangements. In many cases, joint action on scientific problems of international interest can be facilitated by an organization such as SCOR. It seems likely that this mix of approaches will continue in the future and that SCOR will continue to have an important role to play.

Reports from Organizations

United Nations, Ocean Economics and Technology Office: Activities in the Field of Marine Science and Its Applications*

I. DEPARTMENT OF ECONOMICS AND SOCIAL AFFAIRS

The Ocean Economics and Technology Office of the Department of Economic and Social Affairs is the only organizational unit in the United Nations with a separate program in the field of marine science and its applications.

Seabed Mineral Development

A report entitled "Economic Implications of Sea-bed Mining in the International Area" (document A/CONF.62/37, 18 February 1975) was completed in 1975 and submitted to the Geneva session of the Third Conference on the Law of the Sea. The preparation of a working paper on "mineral resources of the continental margin beyond a 200 mile economic zone" was near completion by the end of 1975.

Coastal Area Development

A report entitled "Coastal Area Management and Development" (document E/5648), which was prepared by the OETO in close cooperation with the United Nations Headquarters units and UN organizations and agencies concerned, was submitted to the ECOSOC at its fifty-ninth session.†

Work was initiated in 1975 for preparation of a manual on coastal area development that will deal with planning techniques and legislative/institutional arrangements for integrated coastal area development. Also initiated in 1975 were the preparations for a handbook that will provide a comprehensive, world-wide listing of education and training courses pertaining to coastal area management and development.

---

*Report dated January 22, 1976.

†EDITOR'S NOTE: The text of more recent report entitled "Coastal Area Development and Management and Marine and Coastal Technology" (document E/5971) is produced elsewhere in this volume.

Uses of the Sea

A report entitled, "Uses of the Sea" (document E/5650), prepared in close cooperation with the UN Headquarters units and UN organizations and agencies concerned, was submitted to the ECOSOC at its fifty-ninth session.

Marine and Coastal Technology

A report entitled "Description of Some Types of Marine Technology and Possible Methods for Their Transfer" (document A/CONF.62/C.3/L.22) was submitted to the Law of the Sea Conference at its Geneva session.

Planning was begun in 1975 for the development of a triangular program (UN/recipients/suppliers) for the application of marine and coastal technologies. Substantive preparations have at the initial stages been directed to obtaining information about marine and coastal technology suppliers.

Information

Two "fact sheets" on marine affairs, two additional chapters for the marine affairs bibliography, and several articles *(Ocean Management, IMS Newsletter)* were prepared in 1975.

The UN(ESA/OETO) also hosted, and participated as an observer at, the first session of the Joint FAO/IOC Panel of Experts on the Aquatic Sciences and Fisheries Information System, 2–5 September 1975.

The OETO participated in the planning of an ESA information system and has coordinated its efforts in the information field with related activities in the Centre for Natural Resources, Energy and Transport.

Other Reports

The ACC report entitled "Marine Science and Its Applications: Spheres of Competence and Work Programmes of the United Nations Organizations and Agencies" (document E/5676), prepared in close cooperation with the Office for Inter-agency Affairs and Co-ordination and the UN organizations and agencies concerned, by the UN (ESA/OETO), was submitted to the ECOSOC at its fifty-ninth session.

At the request of the ACC Preparatory Committee, the UN (ESA/OETO), in cooperation with the Office for Interagency Affairs and Co-ordination, prepared a draft background paper entitled "Institutional and Co-ordinative Arrangements in the Field of Marine Affairs" for consideration by the ACC Sub-Committee on Marine Science and Its Applications at its sixteenth session.

Technical Cooperation Activities

1. Projects
   Persian Gulf.—As the result of a recommendation made by the Group of Experts on Coastal Area Development which met at UN Headquarters in November 1974, a draft project document for a regional Prefeasibility Study for Coastal Area Development in the Persian Gulf was prepared by the OETO. The project document was circulated to the specialized agencies, various economic commissions, as well as concerned officers within the UN for comments. The project consists of a small, interdisciplinary team of experts nominated by, and in the case of UNEP, supported by, the specialized agencies (UNESCO, IMCO, WHO, FAO, WMO). Recruitment procedures are presently under way and the 2-month study is expected to commence in March 1976.
   United Arab Emirates.—Another project, still in the draft stage, resulting from the Group of Experts Meeting, is a national project for coastal area development in the UAE. An expert mission to the country was undertaken in May 1975 to identify needs and priorities for coastal area planning.
   Caribbean.—A background paper on coastal area development in the Caribbean, prepared by the OETO, was strongly endorsed by the Caribbean Development Co-operation Committee of the Economic Commission for Latin America. A regional project for coastal and marine resources development in the Caribbean is now being formulated. The project will be based on a comprehensive review of existing marine and coastal activities in the region; this review will be prepared by OETO. A complete analysis of the institutional framework will be concurrently prepared by UNEP. These two studies will be submitted to an interagency meeting in 1976.

2. Institutes
In 1975, the project manager for the Marine Affairs Institute in Trinidad and Tobago was recruited and became operational.

3. Training Courses
Preparations were well advanced in 1975 for the implementation in 1976 of a 2-month training course at Rhode Island University on coastal area management and development. Owing, however, to the financial difficulties being experienced by UNDP, the course has been postponed until 1977.
   At the second meeting of ICSPRO Officers Concerned with TEMA (20–21 October 1975), the UN agreed to participate in the organization of a "Joint Inter-Agency Training Programme in the Field of Marine Pollution" for East Africa.

4. Seminars

Interregional Seminar on Development and Management of Resources of Coastal Areas.—Planning for the Interregional Seminar was brought to near completion on 27 May 1975 at UN Headquarters during a meeting with representatives from the German Foundation for International Development, the Permanent Mission of the Federal Republic of Germany to the United Nations, and UN substantive offices.

The Seminar will take place in Berlin (West), Hamburg, Kiel, and Bremen and has been scheduled so that the participants will have the opportunity also to participate in "InterOcean '76" at Düsseldorf, 15–19 June 1976. Seventeen confirmed speakers, including representatives from UN Headquarters, IMCO, FAO, UNESCO, WMO, and WHO will participate. In addition, papers will be presented by representatives from private and public organizations.

Southeast Asia.—Discussions with CCOP representatives were held in Bangkok to plan a coastal area survey and symposium for East Asia (CCOP member countries). The project proposal is in draft form, and it is anticipated that the project will be executed jointly by CCOP and the UN. The host country has not yet been identified, but Indonesia has indicated interest in participating in the convening of such a symposium.

Reports from Organizations

UNESCO: Activities of the Division of Marine Sciences during 1976*

Unesco's marine science activities can be divided into two main categories: those of the Division of Marine Sciences (OCE), and those of the Intergovernmental Oceanographic Commission (IOC), a separate body within the framework of Unesco. This report deals with the Division's activities, especially as they complement and support the IOC. For the Division's 1977–78 Programme and Budget and Unesco's 1977–82 Medium Term Plan in marine science, see document IOC/EC-VIII/8.

The programme of Unesco's Division of Marine Sciences is designed to respond to marine science needs of all Unesco Member States and especially to those of the developing countries. The Division does this through a network involving four indispensable components: the scientists in Member States, the governments, the international scientific community, and Unesco (with other components of the intergovernmental UN system). The Division develops international co-operation between scientists (and their governments) at three levels—globally, regionally, and nationally—with the object of strengthening marine science at all three levels, which is done closely with the Intergovernmental Oceanographic Commission.

The Division's programme is closely complementary to that of the Intergovernmental Oceanographic Commission (the secretariats formed one unit until 1972). Many of the activities are executed directly in association with IOC or in response to specific IOC recommendations. Virtually all of the Division's other activities fall within the general IOC resolutions recommending marine science assistance to Member States. Similarly, the Division works closely with the Scientific Committee on Oceanic Research (SCOR) of the International Council of Scientific Unions (ICSU). Finally, the Division works with certain other United Nations bodies (such as FAO, UN, UNEP, IAEA, WMO, IMCO).

The Division is working with Member States and SCOR to develop scientific programmes that are sound scientifically and are also relevant to a nation's development needs. Such programmes will allow a scientist both to contribute responsibly to his country's development and also to contribute to the advancement of science at the same time. An example is the mangrove programme being developed with Thailand and SCOR in order to provide a

---

*Informative report to the Executive Council of the IOC (Paris, April 4–8, 1977) followed by a Summary by Country of Activities of the Division of Marine Sciences, 1976.

scientific basis for the more applied aspects, such as fisheries investigations and management of the mangrove environment. This mangrove programme already serves as the nucleus for regional co-operation and related projects are being established by nations on other continents. The national efforts are buttressed by international workshops, working groups and research projects, partly within the context of Unesco's Man and Biosphere Programme.

The global programme with IOC and SCOR on plankton research is another long-established and successful example. Here the national efforts are reinforced through SCOR working groups, an advisory panel, support for sorting and study of plankton samples (at centres in Mexico, Singapore and India established in support of IOC regional projects), research projects, study grants and publications such as the new Monograph on Oceanographic Methodology. Other evolving specific topics include coral reefs, coastal lagoons, biogeochemistry of estuarine sediments, and marine ecosystem modelling in the Mediterranean.

The Division's marine science infrastructure and manpower development efforts include consultant advice, equipment, minor research support, travel grants, fellowships, training, publications, etc.

The Division's activities in 1976 were affected by three types of global problems. Firstly, the financial crisis of UNDP required considerable administrative time of the Division in dealing with the repercussions on national projects. Secondly, problems between nations delayed or prevented the execution of several activities. Finally, a shortage of professional staff caused a very heavy workload with commensurate delays—four Professional posts were occupied full-time, but two posts were vacant or frozen most of the year.

PUBLICATIONS

(i) "International Marine Science Newsletter", January, June, September, issues 11–13. Reports on IOC activities, those of ICSPRO agencies and other relevant news.
(ii) "Zooplankton Fixation and Preservation"
Monographs on Oceanographic Methodology 4, 1976, in co-operation with SCOR.
(iii) "Symposium on the Eastern Mediterranean", in *Acta Adriatica,* for printing in 1977.
(iv) "Proceedings of the International Symposium on Tsunami Research", co-publication with the New Zealand Royal Society. 1976. Relevant to IOC tsunami project.
(v) Pacific Sheet of the Geological Atlas of the World, with the Commission for the Geological Map of the World, 1976. An important synthesis of the results of IOC co-operation.

(vi) Unesco Technical Papers in Marine Sciences:
No. 24—"Seventh report of the joint panel on oceanographic tables and standards", 1976, in co-operation with SCOR.
No. 25—"Marine science programme for the Red Sea", 1976.
No. 26—"Marine sciences in the Gulf area", 1976.
No. 27—"Collected reprints of the joint panel on oceanographic tables and standards", 1976.
Four issues of the Technical Papers were reprinted in 1976: Nos. 11, 15, 19 and 22.
(vii) With FAO, support for the publication of the monthly "Marine Science Contents Tables" and further development of the Aquatic Sciences and Fisheries Information System (ASFIS), including a directory of marine scientists, in co-operation with IOC.

MEETINGS OF UNESCO

(i) Unesco/SCOR Symposium on Warm Water Zooplankton, Goa, India, 14–19 October 1976; 78 scientists attended from 13 countries; hosted by the National Institute of Oceanography. This is a natural outgrowth of the International Indian Ocean Expedition.

(ii) Eleventh meeting of the Advisory Panel for International Biological Centres Sponsored by Unesco, Cochin, India, 21–23 October 1976; 15 scientists attended from 9 countries; hosted by NIO. Most of these centres started as sorting centres servicing IOC Regional Investigations.

(iii) Third Unesco Regional Training Course in Marine Environment, Tokai, Japan, 12 July–30 August 1976; 9 participants from 5 countries and lecturer provided; hosted by Tokai University; jointly funded with Japan.

(iv) Unesco Regional Training Course in Sampling Design for Marine Biologists, Cebu City, Philippines, 26 September–16 October 1976; 17 participants from 8 countries; hosted by University of San Carlos. Organized on the basis of recommendations of the IOC ad hoc Regional TEMA meeting.

(v) Interdisciplinary Training Course in Marine Science, Beaufort, USA; 1 April–31 August 1976; 17 participants from 13 countries; hosted by Duke University; funded through extrabudgetary funds of UNDP and the Rockefeller Foundation.

(vi) Second Sub-regional Workshop on Marine Ecosystem Modelling of the Eastern Mediterranean, Dubrovnik, Yugoslavia, 18–22 October 1976; 39 participants from 15 countries.

(vii) Unesco/UNEP Consultation Meeting on Marine Ecosystems Modelling in the Mediterranean, Unesco, Paris, 13–17 December 1976; 13 scientists from 7 countries. Applications to the UNEP Mediterranean Action Plan in which IOC is active.

MEETINGS OF IOC JOINTLY SUPPORTED BY THE DIVISION

(i) Second CICAR Symposium on Progress in Marine Research in the Caribbean and Adjacent Regions, Caracas, Venezuela, 12–16 July 1976; 26 scientists from 9 countries.
(ii) Joint IOC/FAO(IOFC)/Unesco/EAC Seminar and Workshop on Co-operative Investigation in the North and Central Western Indian Ocean, Nairobi, Kenya, 25 March–2 April 1976; 48 participants from 13 countries.
(iii) IOC/SCOR/CMG Second International Geoscience Workshop, Mauritius, 9–13 August 1976; 26 participants from 9 countries.

ASSISTANCE BY UNESCO TO ALLOW INTERNATIONAL PARTICIPATION IN MEETINGS ORGANIZED BY MEMBER STATES

(i) First Uruguayan Workshop on Marine Sciences, Montevideo, Uruguay; 14–30 April 1976; 20 scientists from 4 countries.
(ii) Seminar on South American Pacific Ocean, Cali, Colombia; 1–5 September 1976; 7 scientists supported from 7 countries.
(iii) First Thai Workshop on Mangrove Ecology, Phuket, Thailand; 10–15 January 1976.
(iv) Second Thai Workshop on Mangrove Ecology, Phuket, Thailand; 24–29 October 1976; 21 scientists from 7 countries.
(v) Preparations for 1977 ALECSO/Unesco Training Course on Marine Ecology of the Red Sea, Chardaqa, Egypt.

ACTIVITIES WITH SCOR

(i) Joint Oceanographic Assembly. SCOR/CMG/IABO/IAMAP/IAPSO/ACMRR/ACOMR/ECOR/UKRoyal Society. Edinburgh, Scotland. 13–24 September 1976. Twenty-six scientists were supported from 16 countries by Unesco, IOC, FAO, UNEP and WMO, in descending order of support.

The Joint Oceanographic Assembly was the major event on the Division's calendar in terms of contact with the marine scientists and is the fourth such oceanographic conference: New York, 1962; Moscow, 1966; Tokyo, 1970 and Edinburgh, 1976. Unesco and IOC have co-operated closely with SCOR and other non-governmental organizations (NGOs) in these conferences, which are associated with business meetings of the NGOs as well. For this Assembly, the Division provided major travel support for participants, provided publications, participated in an ICSPRO exhibit, and participated in the meetings.

(ii) Unesco/SCOR co-operative programme on the preparation of new marine research programs relevant to marine scientists in developing countries:

(a) SCOR/Unesco ad hoc Advisory Panel on Mangrove Ecology, Phuket, Thailand, 15–18 January 1976.

(b) Initial preparation of mangrove bibliography by consultant on the recommendation of the above Panel.

(c) SCOR/Unesco/NOAA Seminar on Biochemistry of Estuarine Sediments, Melreux, Belgium, 29 November–3 December 1976 (35 scientists from 14 countries).

(d) 'Inquiry on Existing Coastal Lagoon Research Programmes in the World', organized by the SCOR/Unesco ad hoc Advisory Panel on Coastal Lagoons. Project in progress.

(iii) Continuing support to SCOR in return for advice, especially through several working groups, which also provide general advice on research programmes to marine scientists who, in turn, work in Member States on IOC projects.

(iv) Other activities are noted where appropriate elsewhere in this report.

GENERAL ACTIVITIES

(i) Support has been provided to marine research and educational institutions in six Member States for equipment and research projects.

(ii) Travel grants have been provided to 25 scientists to attend seven NGO scientific meetings (including a SCOR training course) and three IOC expert meetings.

(iii) Travel grants were provided to nine scientists of eight developing countries to organize or carry out marine research projects.

(iv) Support was provided to three visiting lecturers to present courses and assist young marine scientists/students in seven developing countries.

(v) At the request of the Member State in question, six consultants assisted five developing countries (Sudan, Ethiopia, Nigeria, Uruguay, Argentina) to develop plans for marine research and educational centres. In part, these consultant missions are in direct support of requests made by States to IOC and delegated to the Division for execution under the ICSPRO umbrella.

(vi) Unesco is co-operating with the Open University, United Kingdom, in the preparation of an audio-visual educational course in the marine sciences suitable for presentation in developing countries. It should be ready in late 1978. Within the context of TEMA, IOC is assessing the relevant need in its Member States to help Unesco deliver the course to appropriate institutions through extra-budgetary means.

(vii) Unesco held the CASTARAB Conference (Conference of Ministers in Arab States responsible for the Application of Science and Technology to

Development), Rabat, Morocco, 16–25 August 1976, to which contributions were made, among others, relative to marine science in the Mediterranean, Red Sea and Gulf regions.
(viii) A Unesco staff member was provided as Curator of the Mexico Oceanic Sorting Centre (CPOM) at the Universidad Nacional Autonoma de Mexico, as a contribution to CICAR.
(ix) IMCO/FAO/Unesco/WMO/WHO/IAEA/UN Joint Group of Experts on the Scientific Aspects of Marine Pollution, London, 24–30 April 1975 and Rome, 21–27 April 1976. GESAMP forms a back-up for GIPME; Unesco provides five experts and IOC an observer.
(x) Contract with the IAEA Laboratory of Marine Radioactivity in Monaco for intercalibration and standardization of methodology concerning marine pollution monitoring.

FELLOWSHIPS

(i) Six Regular Programme Fellowships of 6–12 months each to Fellows from Argentina, Uruguay, India, Sudan, Turkey and Yugoslavia.
(ii) Three Participation Programme Fellowships of 2–4 months each to Fellows from Tunisia, Ukrainian SSR, Jordan and Poland.
(iii) Three Netherlands/Unesco Fellowships of 6–12 months each, held by Fellows from Philippines, Nigeria and Ghana.
(iv) Three UNDP Country Programme Fellowships of 12–24 months each, held by Fellows from Ecuador, Indonesia and Sierra Leone.

EXTRABUDGETARY DEVELOPMENT PROJECTS

As the result of a reorganization in 1976, such projects are now executed by the Operations Division, in co-operation with the Division of Marine Sciences.
(i) Mexico (UNDP/MEX/74/004). National Plan for the Creation of a Marine Science and Technology Infrastructure (9 experts during 1976).
(ii) Ecuador (Funds-in-Trust). Assistance to Fisheries Institute (one expert).
(iii) Korea (UNDP/ROK/72/027). Korea Ocean Research and Development Institute (one expert and equipment).
(iv) Indonesia (UNDP/INS/72/038). Marine Research in Indonesia (one expert and equipment). Also an associate expert who has since transferred to the Caribbean as the IOC Regional Secretary.
(v) Regional Expert in Marine Sciences for Southeast Asia (IOC Trust Fund), Jakarta.

STAFF MISSIONS

Staff missions are a necessary instrument for the execution of the programme through servicing meetings, developing projects with Member States and providing co-ordination in the marine science programme of the UN organizations. Twenty-two multipurpose missions from Headquarters and seven from the field were carried out which provided services in 39 Member States. Other States were visited by staff of the Unesco Regional Offices. Fifteen Headquarter's missions and 5 Field missions were in support of, in the joint execution of, or for co-ordination with the IOC programme in activities related to either IOC's role within Unesco or its function in co-ordinating marine science.

The following table lists those countries with sea borders and gives the following information:
1. Whether or not the country is a Member State of Unesco.
2. Whether or not the country is a Member State of the IOC.
3. Whether a country is considered to be one of the least developed countries.
4. Those countries with Unesco-executed extra-budgetary (XB) marine science development projects.
5. Experts' and consultants' advisory missions (long or short) to countries funded by extra-budgetary funds.
6. Consultants' advisory and lecture missions funded by the Division's Regular Programme (RP).
7. Countries visited on Staff missions.
8. Country of origin of the experts, consultants and Professional Staff.
9. Fellowships provided through Division's Programme.
10. Host country of study for fellowship recipient.
11. Participants in the training courses.
12. Participants in meetings organized by the Division, or whose travel was supported by the Division.
13. Countries who hosted meetings or training courses organized by the Division.
14. Grants for research, equipment, books and research-related travel.
15. Other Regular Programme activity, such as involvement with publications, curricula development, etc.
16. Indication of involvement with any of the Division's activities during 1976.
17. Indication of involvement with any of the Division's activities during 1975 and 1976.

SUMMARY BY COUNTRY OF ACTIVITIES OF DIVISION OF MARINE SCIENCE, 1976

1976	\multicolumn{2}{c}{COUNTRIES WITH COASTS}							WESTERN EUROPE									EASTERN EUROPE							EASTERN MEDITERRANEAN						
	Iceland	Ireland	United Kingdom	Norway	Sweden	Finland	F.R. Germany	Denmark	Netherlands	Belgium	France	Monaco	Italy	Spain	Portugal	USSR	Ukraine	Poland	D.R. Germany	Romania	Bulgaria	Malta	Yugoslavia	Albania	Greece	Turkey	Cyprus	Syria	Lebanon	Jordan	Egypt	Israel
1. Unesco member	x	x	x	x	x	x	x	x	x	x	x	x	x	x	x	x	x	x	x	x	x	x	x	x	x	x	x	x	x	x	x	x
2. IOC member	x		x	x	x	x	x	x	x	x	x	x	x	x	x	x	x	x	x	x	x	x	x		x	x	x	x	x	x	x	x
3. Least developed country																																
4. Extra-budgetary project																																
5. Extra-budgetary experts																																
6. Reg. Prog. consultant mission			x				x		x	x	x		x															x				
7. Staff missions			5	2			2		1	1						1	1	1					1									
8. Origin experts, consultants, staff	2		1						2		2	3																				
9. Fellowships			4				1		3																							
10. Fellowships—host country							10	3	5		2		10	4		15				4		1	8		6	6		3	1		9	2
11. Participation training courses			3																													
12. Participation meetings			26		2		10	3	5	2	3		10	4		15		1		4		1	8		6	6		3	1		9	2
13. Host of trg. course or meeting			x						x														x						1	1	1	
14. Research-education grant																																
15. Other activity																																
16. Summary for 1976	x		x	x	x	x	x	x	x	x	x	x	x	x	x	x	x	x		x	x	x	x		x	x		x	x	x	x	x
17. Summary for 1975–1976	x	x	x	x	x	x	x	x	x	x	x	x	x	x	x	x	x	x		x	x	x	x		x	x		x	x	x	x	x

UNESCO, Division of Marine Sciences

1976	Libya	Tunisia	Algeria	Morocco	Mauritania	Senegal	Gambia	Guinea-Bissau	Guinea	Sierra Leone	Liberia	Ivory Coast	Ghana	Togo	Benin	Nigeria	Camaroon	Eq. Guinea	Gabon	Congo	Zaire	Angola	Namibia	South Africa	Sudan	Ethiopia	Yemen	D. Yemen	Somalia	Kenya	Tanzania	Mozambique	Madagascar	Mauritius	Seychelles	Iraq	Kuwait	Saudi Arabia	Bahrain	Qatar	Un. Arab. Emir.	Oman	Iran
1. Unesco member	×	×	×	×	×	×	×	×	×	×	×	×	×	×	×	×	×		×	×	×	×	×	×	×	×	×	×	×	×	×	×	×	×	×	×	×	×	×	×	×	×	×
2. IOC member	×	×	×	×	×	×				×		×	×	×		×	×			×					×	×			×	×	×		×	×		×	×	×			×		×
3. Least developed country															×										×	×	×	×	×		×												
4. Extra-budgetary project								×	×																																		
5. Extra-budgetary experts										×																															×		
6. Reg. Prog. consultant mission																2									2	1					1												
7. Staff missions			×																							×				×	×						×	×			×		
8. Origin experts, consultants, staff										1			1			1																											
9. Fellowships																									1																		
10. Fellowships—host country																																											
11. Participation training courses																																											
12. Participation meetings	2	1																											2	14	8		1	2				1					
13. Host of trg. course or meeting																														×	1			×									
14. Research-education grant																2																											
15. Other activity				×																																							
16. Summary for 1976	×	×	×	×						×			×			×									×	×			×				×	×		×	×		×	×	×	×	×
17. Summary for 1975–1976	×	×	×	×	×	×				×		×	×			×									×	×			×				×	×		×	×	×	×	×	×	×	×

SUMMARY BY COUNTRY OF ACTIVITIES OF DIVISION OF MARINE SCIENCE, 1976 (continued)

1976	SOUTH & SOUTHWEST ASIA												EAST ASIA				OCEANIA						
	Pakistan	India	Sri Lanka	Bangladesh	Burma	Thailand	Kampuchea	Vietnam	Malaysia	Singapore	Indonesia	Philippines	China	DPR Korea	Rep. Korea	Japan	Australia	New Zealand	Papua-New Guinea	Fiji	Tonga	W. Samoa	Others
1. Unesco member	x	x	x			x		x	x	x	x	x	x	x	x	x	x	x	x	x	x	x	x
2. IOC member	x	x	x			x		x	x	x	x	x	x		x	x	x	x		x			
3. Least developed country				x															x				
4. Extra-budgetary project								x	x		x		x		x								
5. Extra-budgetary experts											3				1								
6. Reg. Prog. consultant mission																							
7. Staff missions	x	x				x			x		x				x								
8. Origin experts, consultants, staff											2												
9. Fellowships	1																	1					
10. Fellowships—host country																	1						
11. Participation training courses	1	1				4		2	2		6	9			3		4						
12. Participation meetings	1	70	1			6		3	3		1	1				6	3	1	1				
13. Host of trg. course or meeting		2				x					x	x				x							
14. Research-education grant	2	1				2		2	1														
15. Other activity						x		x										x					
16. Summary for 1976	x	x				x		x	x	x	x	x			x	x	x	x	x	x	x	x	x
17. Summary for 1975–1976	x	x	x			x		x	x	x	x	x			x	x	x	x	x	x	x	x	x

UNESCO, Division of Marine Sciences 583

1976	Canada	USA	Mexico	Guatemala	Honduras	El Salvador	Nicaragua	Costa Rica	Panama	Cuba	Jamaica	Haiti	Dominican Rep.	Barbados	Trinidad-Tobago	Br. E. Carib. Group	Grenada	Colombia	Ecuador	Peru	Chile	Venezuela	Surinam	Guyana	Brazil	Uruguay	Argentina	TOTAL ELEMENTS	TOTAL COUNTRIES
1. Unesco member	x	x	x	x	x	x	x	x	x	x	x	x	x	x	x	x	x	x	x	x	x	x	x	x	x	x	x		
2. IOC member	x	x	x	x					x	x	x	x	x		x			x	x	x	x	x			x	x	x		
3. Least developed country												x																	
4. Extra-budgetary project			x																x									8	8
5. Extra-budgetary experts			9																1									14	4
6. Reg. Prog. consultant mission			x			x									x					1	2	1			1		2	17	13
7. Staff missions		x																2	1	x	x	x			1		x	38	38
8. Origin experts, consultants, staff		8																			2					1		32	14
9. Fellowships																												13	12
10. Fellowships—host country	3																										1	19	9
11. Participation training courses			18	1		1			1		3			1	1			2	1	1	3	1			2		2	41	16
12. Participation meetings	x 38	x																2	2	4	x	1+			2		x	326	50
13. Host of trg. course or meeting		x																				x						12	11
14. Research-education grant																		1	x	x	x						1	23	11
15. Other activity	x	x	x						x	x	x				x			x	x	x	x	x			x	x	x	22	22
16. Summary for 1976	x	x	x						x	x	x				x			x	x	x	x	x			x	x	x	80	80
17. Summary for 1975–1976	x	x	x						x	x	x				x			x	x	x	x	x			x	x	x	88	88

Reports from Organizations

UN, Environmental Programme: Activities for the Protection and Development of the Mediterranean Region*

1.1 The Mediterranean region was selected by UNEP as a "concentration area" where UNEP has attempted to fulfill its catalytic role to assist states in this region in an ambitious and consistent manner.

1.2 After extensive preparatory activities involving a number of UN bodies, UNEP convened the Intergovernmental Meeting on the Protection of the Mediterranean (Barcelona, 28 January to 4 February 1975). The meeting was attended by representatives of 16 States bordering on the Mediterranean Sea.[1] At the end of the two-week meeting they approved an Action Plan[2] consisting of four components:

—legal (framework convention and related protocols),
—scientific (research and monitoring),
—integrated planning,
—institutional and financial arrangements.

1.3 All components of the Action Plan are interdependent and provide a framework for comprehensive action to assist both the protection and the continued development of the Mediterranean ecoregion. No component is an end in itself. Each activity is intended to help the Mediterranean Governments to improve the quality of the information base on which national development policies are formulated: to improve the ability of each Government to identify various options and to make rational choices among alternative patterns of development, and appropriate allocations of resources.

1.4 Set forth below in Section I is a general description of the components of the Action Plan. Section II is a more detailed review of the scientific component. Section III briefly outlines future activities.

---

*Report dated October 1976.

1. Algeria, Egypt, France, Greece, Israel, Italy, Lebanon, Libyan Arab Republic, Malta, Monaco, Morocco, Spain, Syrian Arab Republic, Tunisia, Turkey, Yugoslavia.

2. Report of the Intergovernmental Meeting on the Protection of the Mediterranean, UNEP/WG.2/5. Annex.

SECTION I

Environmental Legislation

1.5 During preparatory meetings for the 1972 UN Conference on the Human Environment the value of a regional approach to marine pollution problems was stressed and the representatives of many governments bordering the Mediterranean Sea recognized the value of agreements among interested States to secure international co-operation in order to deal with common environmental problems.[3] Intergovernmental consultations began in 1973 at FAO towards an agreement on guidelines to be taken into account in the negotiation of an international framework convention for the protection of the Mediterranean Sea. On the basis of the agreed guidelines, a tentative text of a framework convention and of two protocols was presented to Governments at the Barcelona Conference in January 1975, and this text, according to the Governments' directives, was further revised by UNEP and FAO, together with legal experts from Mediterranean Governments.[4]

1.6 In accordance with the request of the Mediterranean Governments, the Executive Director of UNEP convened the Conference of Plenipotentiaries of the Coastal States of the Mediterranean Region on the Protection of the Mediterranean Sea in Barcelona, 2–16 February 1976, 16 Governments attended and approved the texts[5] of three legal instruments listed below, after which 12 Governments[6] signed them:

—Convention for the Protection of the Mediterranean Sea against Pollution.
—Protocol for the Prevention of Pollution of the Mediterranean Sea by Dumping from Ships and Aircraft. (This protocol was signed by 11 Governments).
—Protocol concerning Co-operation in Combating Pollution of the Mediterranean Sea by Oil and other Harmful Substances in Cases of Emergency.

3. Report of Intergovernmental Working Group on Marine Pollution, London, June 1971. A/CONF.48/IWGMP.1/5, 21 June 1971.
4. Report of the Intergovernmental Meeting on the Protection of the Mediterranean, UNEP/WG.2/5, Annex, p. 4.
5. Cyprus, Egypt, France, Greece, Israel, Italy, Lebanon, Libyan Arab Republic, Malta, Monaco, Morocco, Spain, Syrian Arab Republic, Tunisia, Turkey, Yugoslavia.
6. Cyprus, Egypt, France, Greece, Israel, Italy, Lebanon, Malta, Monaco, Morocco, Spain, Turkey. Greece did not sign the protocol for the Prevention of Pollution of the Mediterranean Sea by Dumping from Ships and Aircraft. As of 15 September 1976, Tunisia, Yugoslavia, and the European Economic Communities have signed the Convention and Protocols in Madrid.

1.7 On the basis of national processes of ratification now under way, it can be expected that these legal instruments will enter into force in 1978.

1.8 For each particular source of marine pollution a number of specific actions are contemplated, including the preparation of additional protocols to be associated with the Convention. As a first effort to develop other protocols UNEP will convene an Intergovernmental Consultation in Athens, Greece (7–11 February 1977) to discuss principles for a draft protocol for the protection of the Mediterranean Sea from land-based sources of pollution.

Environmental Assessment

1.9 It was as early as 1969 that the General Fisheries Council for the Mediterranean (GFCM) of FAO formed a Working Party on Marine Pollution in the Mediterranean and its Effects on Living Resources and Fisheries (later renamed as the Working Party on Marine Pollution in Relation to the Protection of Living Resources). In co-operation with the International Commission for the Scientific Exploration of the Mediterranean (ICSEM) in 1972 this Working Party produced the first comprehensive review on the state of marine pollution in the Mediterranean.[7]

1.10 The next important step towards action was the UNEP sponsored International Workshop on Marine Pollution in the Mediterranean[8] which was convened in Monaco (9–14 September 1974) by IOC, GFCM and ICSEM. This meeting, attended by 40 scientists from Mediterranean research centres, defined pollution of coastal waters as the main environmental problem in the Mediterranean Sea and attributed it to the general lack of adequate systems for the treatment and disposal of domestic and industrial waste, to the input of pesticides and petroleum hydrocarbons, and to the presence of pathogenic microorganisms. The Workshop reviewed information on current sub-regional programmes as well as existing research and monitoring facilities in the Mediterranean.

1.11 Based on the recommendation of the Monaco Workshop and a subsequent feasibility study of the capabilities of existing national research institutions, conducted by IOC on behalf of UNEP, the 1975 Intergovernmental Meeting in Barcelona approved a Co-ordinated Mediterranean Pollution Monitoring and Research Programme consisting of seven pilot projects and requested UNEP's Executive Director to implement it in close collaboration with the relevant specialized UN bodies (GFCM of FAO, IOC of UNESCO, WHO, WMO and IAEA).

7. GFCM, The State of Marine Pollution in the Mediterranean and Legislative Controls. Stud. Rev. 51.GFCM, 1972.
8. Intergovernmental Oceanographic Commission, Workshop Report no. 3, UNESCO 1975.

1.12 Following governmental approval for the creation of seven networks of co-operating national research centres to carry out the work on the seven pilot projects, a number of technical meetings have been held at which operational documents for the pilot projects have been drawn up, and at present work has started in the cooperating institutions.

1.13 The seven pilot projects of the Co-ordinated Mediterranean Pollution Monitoring and Research Programme deal mainly with the coastal waters of the Mediterranean. An additional pilot project dealing with pollution levels of the open waters of the Mediterranean and the biogeocycle of the most important pollutants, is also under preparation.

1.14 Throughout the period of planning, and in particular during the two-year pilot project phase, a high degree of co-operation will be maintained between UNEP, acting as overall co-ordinator, and the specialized UN bodies (GFCM, IOC, WHO, WMO and IAEA) which have major roles in the implementation of the pilot projects.

1.15 The pilot projects are being executed primarily through activities of existing national institutions. Participation in the projects is open to all institutions in the region, subject to approval from their national authorities. At present 63 research centres from 14 Mediterranean countries have been identified as active participants in the projects. Support has already been provided to some of them so as to enable their full participation in the agreed projects. To date, this support includes a large training programme, distribution of sophisticated analytical instruments, organization of an intercalibration exercise, and provision of common maintenance services. As an aid to participants several technical guidelines are under preparation and a Directory of Mediterranean Marine Research Centres, describing more than 100 institutions has been prepared and issued by UNEP.

1.16 The first results of the two-year pilot projects will be reviewed in 1977, when a decision will be taken on more permanent arrangements to produce on a regular basis evaluated information for the Mediterranean Governments on the state of pollution in the Mediterranean Sea.

1.17 A related project was recently initiated by UNEP on Pollutants from Landbased Sources in the Mediterranean. This project is a concrete example of the linkage between environmental assessment and management as it is intended to produce data which will assist governments in the negotiation of the regional protocol on land-based pollutants. The project, which will be executed in close co-operation with the Governments of the region and a number of specialized UN bodies (including ECE, UNIDO, FAO, UNESCO, WHO, IAEA) has, as its ultimate objective, to provide the Governments of the Mediterranean coastal States with appropriate information on the type and quantity of pollution inputs from major land-based sources and through rivers, and on the present status of waste discharge and water pollution management practices. The project also provides for the preparation of an inventory of land-based sources of pollutants being discharged into the Mediterranean.

1.18 Since 1975 UNESCO and UNEP have been undertaking a project on the role of sedimentation in the pollution of the Mediterranean Sea with special emphasis on the assessment of current knowledge in this field and on the development of guidelines for environmental impact assessment.

1.19 Additional projects to study the input of airborne pollutants into the Mediterranean and to assess the potential fisheries resources in the Mediterranean and the effects of pollutants on this potential are also planned.

Environmental Management

1.20 A large number of present and planned activities around the Mediterranean have impacts on the quality of the environment. Efforts are under way to identify those activities and to learn how to evaluate the severity of their likely impact, and, where indicated, how to apply additional measures to reduce either the risk or the severity of the effect. Many such activities are financed through UNDP, the World Bank and other sources of international aid. UNEP, in close co-operation with such sources of assistance and the States concerned, is studying the means to assess the likely environmental impacts, and to design and apply appropriate safeguards. Many internationally supported projects which have clearly beneficial environmental impacts are already under way, such as the various fisheries projects of FAO, the environmental sanitation activities of WHO, and the assistance in industrial waste treatment provided through UNIDO. Projects of a similar nature are also being carried out by intergovernmental organizations, such as OECD, as well as by non-governmental organizations.

1.21 In connection with the problem of accidental spills of oil or other harmful substances in the Mediterranean, Governments at the Conference of Plenipotentiaries of the Coastal States of the Mediterranean Region on the Protection of the Mediterranean Sea (Barcelona, February 1976) decided to establish a regional oil-spill control centre on Malta. The primary objective of the centre is to help coastal States of the region take co-operative and timely steps to prevent damage to their coastal resources from massive and accidental pollution. The Malta Centre is expected to become operational in October 1976. IMCO has been entrusted with the responsibility as co-operating agency for the establishment and operation of the centre, which at present is being funded by UNEP.

1.22 In co-operation with IUCN steps are also under way to assist Mediterranean States to identify marine parks and wetland areas which deserve greater attention and protection. Information in the form of guidelines for the management of these areas will be generated so that their protection may be secured while their potential for tourism and scientific research is simultaneously realized.

1.23 Through such means as illustrated above, national governmental and non-governmental institutions around the Mediterranean are encouraged to co-operate in attacking common problems as they arise.

1.24 Meanwhile a separate, longer-term planning exercise is being prepared: the "Blue Plan". As the result of an initiative by the Government of France, UNEP and Mediterranean Governments are engaged in consultations on proposals to launch a series of prospective studies by national research institutions that would specify the likely developments until the year 2000 in a variety of sectors, including urbanization, industrialization, agricultural development, transportation (maritime and coastal), offshore exploration and exploitation, and energy production and use. An Intergovernmental Meeting will be convened in Split, Yugoslavia (31 January–4 February 1977) to discuss and decide upon content and proposed phases of the Blue Plan. The first phase would involve the completion of sectoral studies by networks of co-operating national institutions in such a way that the results of each study could be interrelated with the results of the others. A second phase would call for an analysis based on the earlier findings of the likely impacts on the Mediterranean, in whole or in part, of an aggregation of the predicted effects. For example, an analysis of the results of the combined impacts on a part of the Mediterranean of forecast demographic increase, tourism development, intense agricultural activity, etc. Such studies, properly conducted and supported by Governments would aid national planners and decision makers by improving the quality of information on which their actions are based. Preliminary results from the "Blue Plan" should be available in approximately two years time.

Supporting Activities

1.25 A review of present and planned activities reveals numerous examples of the need for initial support to institutions and Governments, especially those of developing countries, to enable them to participate in environmental activities so that in due course they can take on fuller responsibility. In accordance with the wishes of the Mediterranean Governments, national institutions have been helped to participate in regionally co-ordinated activities in all parts of the Mediterranean, despite the economic conditions in many Mediterranean States which do not yet permit adequate national funding. UNEP is also financially supporting the establishment of the Regional Oil-Combating Centre in Malta, the Secretariat serving the Convention and Protocols, and the Unit co-ordinating the research and monitoring programme in the early stages of these activities. This support by UNEP rests on the assumption that the

Governments of the region will themselves gradually cover the operating costs of such activities as UNEP's initial catalytic role is completed.

1.26 Supporting measures for the pollution research and monitoring programme include the provision of technical assistance in the form of education and training, and the provision of research equipment, such as equipment needed to sample and analyze water and marine organisms. Of equal importance in this area is the intercalibration of equipment used for research and monitoring to ensure compatibility of results obtained by various means, as well as the organization of common maintenance services for sophisticated analytical instruments.

1.27 Supporting measures are also required for training in the area of integrated planning, to assist officials of developing countries to include environmental dimensions in their economic development programmes and projects and to secure the proper management of natural resources on a sustainable basis.

SECTION II

Co-ordinated Mediterranean Pollution Monitoring and Research Programme

2.1 The following is a short description of the seven pilot projects which constitute the Co-ordinated Mediterranean Pollution Monitoring and Research Programme.

Joint IOC/WMO/UNEP Pilot Project on Baseline Studies and Monitoring of Oil and Petroleum Hydrocarbons in Marine Waters

2.2 The pollution of the Mediterranean by oil and petroleum hydrocarbons is a serious problem for beaches and other coastal, recreational areas, and as yet too little is known about the present levels of the pollution and about its effects on the Mediterranean ecosystem. So far as the levels of pollution are concerned, it appears appropriate to initiate a regional monitoring programme as a contribution to the pilot project for the monitoring of marine pollution by petroleum of the Integrated Global Ocean Station System (IGOSS) of the IOC and the World Meteorological Organization (WMO). The pilot project will involve initially the visual observation of oil slicks and other floating pollutants by ocean weather ships, research vessels, voluntary observing ships, fishing vessels and their supporting ships, and by observers on suitable offshore platforms and aircraft; tar ball sampling by ocean weather ships, research vessels and other vessels designated by Member States, and by staff at coastal stations, on islands and offshore platforms; survey of tar on beaches by staff of participating institutions of Member States; and sea water sampling by research vessels, ocean weather ships and other vessels suitably staffed and equipped. Although the pilot project calls for the sampling of seawater for the determination of dissolved petroleum hydrocarbons, some problems of sampling and, in particular, of chemical analysis have yet to be resolved.

2.3 The measurement of present levels of petroleum in all its forms in the Mediterranean assumes greater importance in view of the reopening of the Suez Canal to the passage of oil tankers. The value of initiating the pilot project in the Mediterranean region rests mainly on three facts: (i) the observational system has already been developed (IGOSS); (ii) by using a common system of observations the various subregions of the Mediterranean can be readily compared; and (iii) within a common system of observation, the Mediterranean can be truly compared with other areas (e.g., the North Atlantic) with quite different oceanographic regimes, in which the possibilities for dispersion and dilution are greater and the possibilities of evaporation generally lower.

2.4 The operational document for this pilot project was developed at a joint IOC/WMO/UNEP Expert Consultation (Malta, 8–13 September 1975) which was attended by 36 participants from 12 Mediterranean countries.

2.5 At present 11 countries have expressed a wish to participate in the pilot project, nominating 25 national laboratories as participants in the network dealing with the pilot project.

2.6 Work on this pilot project will start this autumn.

Joint FAO(GFCM)/UNEP Pilot Project on Baseline Studies and Monitoring of Metals, Particularly Mercury and Cadmium, in Marine Organisms

2.7 Metals, and particularly heavy metals like mercury, are more or less toxic to man. They can reach man through the food chain, and the source of greatest concern is, therefore, the level of concentration of such metals in fish, shell-fish and other edible marine organisms.

2.8 It is recognised that the Mediterranean is a tectonically rich region and that some metals manifest high natural levels and great variations in their concentration in sea water and sediments. The bluefin tuna, as well as other tuna, is known to accumulate mercury and, although there is no strong evidence that the Mediterranean stock is separate from the Atlantic stock as a whole, Mediterranean tuna apparently have much higher levels than those from the Atlantic.

2.9 The pilot project will deal only with the concentration of selected metals, particularly mercury and cadmium, in marine organisms. In addition to these the measurement of the levels of copper, lead, manganese, selenium and zinc are recommended, particularly when detection methods providing for multi-elemental analysis are used. The striped mullet, the

Mediterranean mussel and the bluefin tuna are selected for the monitoring programme so that representatives of different ecotypes are included. The sampling frequency is seasonal.

2.10 The operational document for this pilot project was formulated at a joint FAO(GFCM)/UNEP Expert Consultation (Rome, 23–27 June 1975) attended by 35 participants from 13 Mediterranean countries.

2.11 Thirteen countries have expressed a wish to take part in the pilot project nominating 36 national laboratories as participants in the network dealing with the pilot project.

2.12 Work on this pilot project started in late autumn 1975.

Joint FAO(GFCM)/UNEP Pilot Project on Baseline Studies and Monitoring of DDT, PCBs and Other Chlorinated Hydrocarbons in Marine Organisms

2.13 Similar arguments to those advanced for the monitoring of metals apply to chlorinated hydrocarbons; they are persistent; they are usually accumulated by organisms; they are usually harmful to man indirectly, through effects on the stocks of marine organisms he exploits. Even less is known about the present concentration of these chemicals than about the concentrations of heavy metals. Since virtually all chlorohydrocarbons are generated by man, natural background levels of these substances are not a problem in baseline studies.

2.14 The pilot project will deal with levels of selected organochlorine compounds which are considered as specially relevant to representative members of the Mediterranean ecosystem. DDT, PCBs, dieldrin and their metabolites are singled out as falling into this category. Whenever possible, other persistent organic compounds will also be identified in analysed samples. The organisms selected as monitoring targets (striped mullet, Mediterranean mussel, pink shrimp) are representative of the different Mediterranean ecotypes, are of great economic importance and are almost ubiquitous for the whole Mediterranean. The sampling frequency is seasonal.

2.15 The operational document for this pilot project was developed by the same Expert Consultation which formulated the proceeding pilot project.

2.16 Currently 12 countries have nominated 25 national laboratories to participate in the pilot project.

2.17 Work on this pilot project started in late autumn 1975.

Joint FAO(GFCM)/UNEP Pilot Project on Research on the Effects of Pollutants on Marine Organisms and their Populations

2.18 The marine environment is characterized by relatively constant physical and chemical conditions. Most marine organisms are therefore not adapted to sudden changes in their environmental conditions, to certain substances not normally present in sea water, or to unusually high concentrations of substances which normally appear only as sea water microconstituents.

2.19 The project will not deal with acute toxicity experiments unless the organisms cannot be kept long enough under culture conditions to allow long-term toxicity tests. Instead, long-term experiments are envisaged with the aim of investigating the sub-lethal effects of potential pollutants, and functional as well as morphological changes.

2.20 The experiments will not be limited to individual organisms but should rather cover populations where subtle changes in the behavioural pattern could serve as early warning signs and lead to the possibility of predicting the moment at which the organisms will be harmed at the population level. The influence transmitted through the trophic chains, particularly in experiments on populations, will not be neglected.

2.21 Due attention will be paid to establish the most sensitive stages in the life cycle of the tested organisms. Physiological and biochemical studies will be conducted in order to provide information on the mechanisms involved in the effects and transport of pollutants.

2.22 Damage to the genetic materials of individuals and their populations will be studied.

2.23 The ultimate aim of all these tests is to develop the necessary background for biological monitoring and to contribute data required for the development of water quality criteria. Naturally, these criteria cannot be based solely on biological tests but the expected results might provide a basis for a better understanding of the potential hazard for the ecosystem, including man, from the increased level of pollutants in the marine environment.

2.24 The operational document for this pilot project was developed at a joint FAO(GFCM)/UNEP Expert Consultation (Rome, 30 June–4 July 1975) attended by 25 participants from 13 Mediterranean countries.

2.25 Until now, 10 countries have expressed a desire to participate in the pilot project nominating 22 national laboratories as participants in the network dealing with the pilot project.

2.26 The work on this pilot project started in late autumn 1975.

Joint FAO(GFCM)/UNEP Pilot Project on Research on the Effects of Pollutants on Marine Communities and Ecosystems

2.27 Theoretically several types of marine communities and ecosystems could be studied in the frame of the proposed pilot project. For practical purposes, the project will deal with natural marine communities and ecosystems under stress in coastal waters, including lagoons and brackish coastal lakes, in areas where ecosystem changes may be anticipated as a consequence of man's activities and with ecosystems in relatively unpolluted areas, such as marine parks, for reference.

2.28 Ecosystems will be particularly investigated in areas which were repetitively studied in the past in order to detect long-term changes.

2.29 To the largest possible extent the ecosystems will be studied as integral units taking into account the dynamic interactions among their various components. Special attention will be paid to the role of those organisms which will be used in the monitoring pilot projects in the transport of pollutants through the trophic levels.

2.30 The parameters and effects to be studied will vary depending on the studied community and ecosystem. The most common ones will be: community structure, functional indices and body burden of pollutants.

2.31 The operational document for this pilot project was formulated at the same Expert Consultation which developed the previous pilot project.

2.32 Until now, 11 countries expressed a wish to participate in the pilot project nominating 21 national laboratories as participants in the network dealing with the pilot project.

2.33 Work on this pilot project started in late autumn 1975.

Joint IOC/UNEP Pilot Project on Problems of Coastal Transport of Pollutants

2.34 The general pattern of sea surface transport in the Mediterranean is cyclonic (counterclockwise) in both the eastern and western basins. Pollutants discharged into coastal waters tend to be transported along the coasts, thus restricting advection from the coasts towards the open sea. At the same time, floating marine litter and tar balls in the open sea will tend to be centrifuged towards the coasts. Water leaves the Mediterranean at depth and enters at the surface through the Straits of Gibraltar. Since most pollutants are most abundant in the upper layers of the sea, the loss by transport through the Straits is relatively small. The average residence time of entering seawater is estimated to be about 80 years, on the basis of the general hydrography of the Mediterranean and of mass transport measurements in the Straits, though the duration probably ranges from a few years to several hundred.

2.35 Although the general nature of the mass transport of seawater in the Mediterranean is reasonably well understood, our knowledge of local circulation patterns is still sparse. The former may serve in studies of the distribution of pollutants entering the sea via the atmosphere, but the latter is much more important in studies of the distribution of pollutants entering the sea via rivers.

2.36 The main objective of this pilot project will be the investigation of water circulation in coastal areas and exchange of water between the coastal and offshore regions, in order to provide the necessary information on the physical processes contributing to the transport of pollutants in the Mediterranean Sea.

2.37 The operational document for this pilot project was developed by the same Expert Consultation which formulated the pilot project on Baseline Studies and Monitoring of Oil and Petroleum Hydrocarbons in Marine Waters.

2.38 Until now, 12 countries expressed a desire to participate in the pilot project nominating 23 national laboratories as participants in the network dealing with the pilot project.

2.39 Work on this pilot project will start this autumn.

Joint WHO/UNEP Pilot Project on Coastal Water Quality Control

2.40 The serious and rapidly growing pollution of the coastal waters of the Mediterranean is having an increasing impact on the social and economic well-being of the countries bordering it. In addition to the millions of inhabitants living along the coastline of the Mediterranean, millions of tourists spend their holidays on the shores of the sea, and there is a considerable potential for exchange of pathogenic agents. The present situation constitutes a significant health hazard in many places; salmonellosis, dysentery, viral hepatitis and poliomyelitis have all been endemic in the Mediterranean area, and during recent years there have been a number of cholera outbreaks. There is a distinct need for better statistics concerning correlation between diseases and water pollution. There is ample evidence that contaminated shellfish are an important concern to public health. It is also certain that contamination of seafood by chemicals and heavy metals has to be taken into consideration, but this aspect is dealt with by other pilot projects within the framework of the Co-ordinated Mediterranean Pollution Monitoring and Research Programme. The risk of infection from swimming and other recreational activities in coastal waters is enhanced in certain areas because of the absence or inadequacy of beach sanitary facilities. Thus the subsequent actual and potential health effects are of prime importance.

2.41 The overall objective of the pilot project is to produce statistically significant data, scientific information and technical principles which are required for the assessment of the present level of coastal pollution as it concerns human health. The most important immediate objectives are to design and implement a programme for the sanitary and health surveillance of coastal recreational areas and of shellfish growing waters in selected coastal areas and to initiate a scientific study on the epidemiological evidence of health effects caused by inadequate sanitary conditions in coastal areas.

2.42 The operational document for this pilot project was prepared at a joint

WHO/UNEP Expert Consultation (Geneva, 15–19 December 1975) which was attended by 35 participants from 15 countries.

2.43 Up to the present, 7 countries have expressed a desire to participate in the pilot project nominating 13 national institutions as participants in the network dealing with the pilot project.

2.44 The work on this pilot project will start this autumn.

SECTION III

Future Developments

3.1 While it is difficult to forecast future developments, particularly in the light of the many political differences among the various States in the Mediterranean region, it is apparent that by 1978 there may be a series of simultaneous developments in various aspects of the Mediterranean Action Plan.

3.2 Thus, if progress continues at the present pace, it can be expected that the Barcelona Convention together with one or more of the related protocols will enter into force in 1978. This will bring about the first meeting of the Parties to the Convention and could well coincide with the completion of the pilot project phase on the research and monitoring activities. The meeting should present an opportunity for intergovernmental consideration and decision on the establishment of an operational phase thereafter. At approximately the same time, initial results should be available from both the environment-development activities and the Blue Plan studies, providing national decision makers with an authoritative statement about short and long-term environmental implications of ongoing development activities throughout the region.

3.3 In this way an opportunity should be created in several years time for the results of collective assessment activities throughout the Mediterranean to be presented in a form which will be useful to the Mediterranean Governments; governments on whose management decisions rest the responsibility for environmentally sound and sustainable development throughout the region. Increasingly responsible roles will have been assumed by national institutions, on whose co-operative endeavours the successful implementation of the programme depends. Though additional international financial and other support may be sought, the ultimate aim is to make the programme self-supporting within the regional context, that is to say, not only to develop institutional capabilities to perform the required tasks but also to support these activities with training, provision of equipment, and other forms of assistance from within the region.

3.4 As the Mediterranean regional activity becomes self-supporting, UNEP will continue to retain a strong interest, due to both its continuing concern with this critical region and to UNEP's global responsibilities, to which the Mediterranean programme is a major contribution. On a continuing basis UNEP will ensure that data and information generated within this region are compatible with those from other regions of the world. Steps have already been taken to initiate comprehensive action plans in other regions: the Persian Gulf, the Caribbean Sea, the Red Sea, the West African coast and East Asian Waters. The Mediterranean approach may be used as a model for a comprehensive programme aiming at the protection and the development of these regional seas. However, it is recognized that the approach used in the Mediterranean region cannot be copied mechanically in all areas and that each region must develop its own Action Plan based on variations in the state of knowledge, the information and human resources available, and other regional characteristics.

Reports from Organizations

UN Institute for Training and Research (UNITAR): Activities Related to Ocean Problems

The United Nations Institute for Training and Research (UNITAR) is an autonomous institution within the United Nations system, established for the purpose of carrying out independent and objective research aimed at enhancing the capability of the United Nations to achieve its major objectives. The institute's activities include specialized training for diplomats, international officials, and officials of Member States to promote the effectiveness of their participation in United Nations activities. Apart from specific area studies and training programs, the institute is also engaged in high-level discussions and exploration of "issues for the future" likely to affect the capabilities of the United Nations system in general.

As part of this program, the institute has undertaken a major research project and published several monographs on peaceful settlement of international disputes. The following studies prepared at the institute during the period 1975–77 are related to ongoing developments in the Law of the Sea:

"Ocean Resources: Peaceful Settlement of Disputes—an Analysis of Part IV of the Single Negotiating Text": Dr. K. Venkata Raman, Fellow UNITAR, 1976, draft, 164 pp.

"Sharing the World's Resources": Prof. Oscar Schachter, formerly Director of Studies, UNITAR (Columbia University Press, 1977).

"The Protection of the Human Environment: Procedures and Principles for Preventing and Resolving International Conflicts": Dr. Aida L. Levin, UNITAR, 1977, draft, 187 pp.

A preliminary draft of Dr. Raman's study on the president's proposals concerning dispute settlement mechanisms for the Law of the Sea Convention (Part IV of the Revised Single Negotiating Text) was made available to the key delegations actively involved in the negotiations during the fifth session of the Third United Nations Law of the Sea Conference. The study was discussed during a high-level panel meeting convened by UNITAR on September 10, 1976. The critical appraisal of the proposals under consideration and the specific suggestions advanced to improve the text were found useful by the diplomats and international officials who participated in the discussion. The revised version of Part IV of the Single Negotiating Text issued by the president of the conference during November 1976 reflected many of the changes discussed. The study is now being brought up to date, incorporating the developments that have taken place since the fifth session and enlarging the scope of the subject matter to cover the whole range of issues concerning conflict management related to the oceans.

During the fifth session of the conference in New York, UNITAR in conjunction with the International Ocean Institute organized a colloquium for diplomats and senior officials involved in the conference at which the various proposals under consideration in the three main committees were discussed.

Earlier, a briefing seminar on Law of the Sea issues was organized by UNITAR's Department of Training in Geneva. The seminar took place on March 6 and 7, 1975 immediately before the opening of the third session of the conference. It was designed particularly for members of the Permanent Missions who had not previously been involved in Law of the Sea matter but who were likely to work on them in some capacity because the conference was taking place at Geneva. A description of the main events leading up the conference, a discussion of the terminology in use, and an outline of the structure of the conference itself preceded an assessment of the second session at Caracas by participants in each of the three main committees. This was followed, on the second day, by an introduction to some of the technical developments that had been taking place in off-shore oil and gas exploration, deep seabed mining, and so on.

"The Future of the Oceans and World Order" by Gerard J. Mangone, Director of the Center for the Study of Marine Policy at the University of Delaware, was presented as a background paper at an International Conference on the Future organized by UNITAR in Moscow in June 1974 in cooperation with the Institute of World Economy and International Relations of the Academy of Sciences of the USSR and the USSR State Committee for Science and Technology. The paper appears in *The United Nations and the Future* (UNITAR Conference Report no. 6, Moscow, 1976).

A special issue of *UNITAR News* (vol. 6, no. 1) devoted to "UN and the Sea," published in 1974, contained among other things a map of the surface distribution of ferromanganese deposits on the ocean floor.

A monograph on "Marine Pollution Problems and Remedies," by Oscar Schachter and Daniel Serwer, was published by UNITAR in 1970 (UNITAR Research Report no. 4, New York, 1970) and subsequently in the January 1971 issue of the *American Journal of International Law*. Among other things the study reviews existing international controls and future prospects for pollution control.

A report on the planning and development aspects of fishery and mineral resources and related problems was presented to the "Pacem in Maribus" Convention in 1970 following an expert meeting on the subject convened by UNITAR in association with the Center for the Study of Democratic Institutions, Santa Barbara, California.

Reports from Organizations

WHO: Recent Activities in the Field of Marine Pollution and Related Subjects*

1. Previous work carried out by WHO in the field of marine pollution and related subjects has appeared in two publications released in 1976. One is concerned with an environmental sanitation plan for the Mediterranean seaboard which consists of a revised version of the text originally presented by Professor Brisou to the Interparliamentary Conference of Coastal States on the Control of Pollution in the Mediterranean Sea and which was done in collaboration with the United Nations Environment Programme. It deals with the main sources of pollution, the fate of pollutants with special regard to micro-organisms in the marine environment and their effects on man's health. The health situation around the Mediterranean is briefly reviewed and an outline of a plan to protect environmental health in the coastal region is given. The other one is a guide to shellfish hygiene which describes the shellfish industry and gives some details of the diseases transmitted by shellfish as well as the marine environmental factors which affect their quality, and recommends measures for ensuring the production and marketing of wholesome shellfish.

2. WHO continued to act as an executing agency for a number of UNDP supported projects, which included feasibility studies for designs for sewerage schemes in coastal towns and the sanitary disposal of waste water in the marine environment.

Algeria	Sector studies for water supply and sewerage
Argentina	Sector studies for water supply and sewerage
Bahrain	Sewage disposal scheme (in preparation)
Bangladesh	Sector studies for water supply and sewerage
Bermuda	Sector studies for water supply and sewerage
Bolivia	Sector studies for water supply and sewerage
Brazil	Sector studies for water supply and sewerage
Cameroon	Sector studies for water supply and sewerage
Chile	Sector studies for water supply and sewerage
Costa Rica	Sector studies for water supply and sewerage
Democratic Yemen	Sector studies for water supply and sewerage
Egypt	Sector studies for water supply and sewerage
El Salvador	Sector studies for water supply and sewerage
Ethiopia	Sector studies for water supply and sewerage
Gabon	Master plan for water supply and sewerage Accra/Tema (completed)

*Report dated December 1976.

Greece	Environmental Pollution Control (in operation)
Guinea	Water supply and sewerage, Conakry (in preparation)
Guyana	Development of potable water supply, sanitary sewerage and storm drainage (in operation)
Honduras	Sector studies for water supply and sewerage
India	Survey of water supply resources for Greater Calcutta (completed)
India (Madhya Pradesh)	Sector studies for water supply and sewerage
India (Uttar Pradesh)	Sector studies for water supply and sewerage
Indonesia	Sector studies for water supply and sewerage
Iran	Sector studies for water supply and sewerage
Ivory Coast	Water supply and sewerage for Abidjan (in operation)
Kenya (Sida)	National programming community water supply, sewerage water pollution control (in operation)
Korea, Rep. of	Sector study for water supply and sewerage
Lebanon	National wastes management plan (in operation)
Malaysia	Sector studies for water supply and sewerage
Maldive Islands	Water supply, drainage and sewerage (in preparation)
Malta	Wastes disposal and water supply (completed)
Mexico	Sector studies for water supply and sewerage
Nepal	Sector studies for water supply and sewerage
Nicaragua	Sector studies for water supply and sewerage
Oman	Sector studies for water supply and sewerage
Pakistan	Sector studies for water supply and sewerage
PDR Yemen	Pre-investment survey of sewerage requirements for the Aden Peninsular (in operation)
Philippines	Sewerage system for the Manila Metropolitan area
Portugal	Sector studies for water supply and sewerage
Senegal	Establishment of a Master Plan for water supply and sewerage for Dakar and surrounding areas (in operation)
Somali	Sector studies for water supply and sewerage
Sri Lanka	Public water supply, drainage and sewerage for southwest coastal area (completed)
Sudan	Sector studies for water supply and sewerage
Surinam	Wastes disposal and water supply (in operation)
Tanzania	Sector studies for water supply and sewerage
Thailand	Sector studies for water supply and sewerage
Turkey	Sector studies for water supply and sewerage
Turkey (Istanbul Region)	Master Plan for water supply and sewerage (completed)
Yemen	Sector studies for water supply and sewerage

Zaire	Sector studies for water supply and sewerage
Zambia	Sector studies for water supply and sewerage

3. WHO collaboration with UNEP and other UN agencies in the "Plan of Action" for the Mediterranean:

(a) *"Blue Plan."* WHO has been invited to actively participate and prepare documentation for the Intergovernmental Meeting of Mediterranean Coastal States on the "Blue Plan" as well as at the meeting of governmental experts for drafting protocol (as part of the Barcelona Convention) to limit the discharge of pollutants from land-based sources in the Mediterranean.

(b) *Coastal Water Quality Control in the Mediterranean.* The objectives of this project are:
   (i) The surveillance of the quality of coastal water with respect to its physical, chemical and microbiological characteristics and the assessment of risk to human health.
   (ii) Monitoring and study of the bio-accumulation of various micro-organisms including pathogenic ones, inedible marine-organisms, etc.
   (iii) Patterns of inspection of the sanitary quality of beaches and coastlines in coastal areas most frequented by tourists.
   (iv) Elaboration of principles for attaining, through various measures, the desired quality of coastal areas.

(c) *Guidelines for the Drafting of Protocol for the Prevention of Pollution from Land-based Sources.* WHO is preparing the topical documentation which will be submitted to a meeting of governmental experts in 1977 on this topic. The documentation includes:
   (i) Draft guidelines for the protocol.
   (ii) A technical paper suggesting lists of substances to be dealt with in various ways with scientific justification.
   (iii) A compendium of principal international instruments relevant to pollution from land-based sources.
   (iv) A comparative survey of national legislation of Mediterranean countries, accompanied by a review of each country's national legislation on the prevention and control of pollution affecting coastal waters.

(d) *Inventory and Assessment of Pollutants from Land-based Sources in the Mediterranean.* WHO is playing a leading role in the project in which several other United Nations agencies are collaborating. The main purpose of this project is to assess and evaluate the kind, amount and if any, the degree of treatment of wastes discharged into the Mediterranean from land-based sources.

4. In addition, WHO has participated in other meetings sponsored by UNEP for their regional seas and coastal area development programmes and more particularly at the meeting which took place in Nairobi on the legal aspect of international conventions for regional seas; a meeting of the coastal states of the Gulf area in order to draft a plan of action, a meeting in Bahrain at a proposed legal agreement to protect the Gulf from pollution, and a first meeting of experts to discuss a preliminary plan of action for the Caribbean Sea and adjacent areas.

5. WHO continues to support the International Reference Centre for Wastes Disposal, which is located at the Federal Institute for Water Resources and Water Pollution Control, Dubendorf, Switzerland. This centre is affiliated to 44 collaborating institutions throughout the world and major attention is directed to improvement of all aspects of wastes treatment and disposal.

6. WHO continues to collaborate with other international organizations concerned with coastal pollution control and coastal area development especially with the United Nations in this and other related fields.

7. Training of personnel in water pollution assessment, evaluation and control techniques continues to be one of the most important activities of WHO. In 1976 the fifth interregional WHO training course on coastal pollution control was held under the auspices of DANIDA and a monograph containing the text of the lectures has been published.

8. WHO endeavors to disseminate technical information on all subjects related to health and marine activities.

Appendix B

Selected Documents and Proceedings

The eight documents and proceedings included in this separate reference section along with the "Treaty on the Prohibition of the Emplacement of Nuclear Weapons and Other Weapons of Mass Destruction on the Sea-Bed and the Ocean Floor and in the Subsoil Thereof" and the "Antarctic Treaty" (appended to the essays of Jozef Goldblat and G. L. Kesteven, respectively) represent only a sampling of the important international developments on ocean-related issues. Space permitting, future editions of the *Ocean Yearbook* will include a selection of international agreements, legislation, and proceedings of international conferences, which the Editors consider of general interest.

THE EDITORS

Selected Documents and Proceedings

United Nations Third Conference on the Law of the Sea: Informal Composite Negotiating Text*

TABLE OF CONTENTS

PREAMBLE 612
PART I. USE OF TERMS 612
Article 1. Use of terms
Part II. TERRITORIAL SEA AND CONTIGUOUS ZONE 613
                Section 1. General
Article 2.   Juridical status of the territorial sea, of the air space over the territorial sea and of its bed and subsoil
                Section 2. Limits of the territorial sea
Article 3.   Breadth of the territorial sea
Article 4.   Outer limit of the territorial sea
Article 5.   Normal baseline
Article 6.   Reefs
Article 7.   Straight baselines
Article 8.   Internal waters
Article 9.   Mouths of rivers
Article 10.   Bays
Article 11.   Ports
Article 12.   Roadsteads
Article 13.   Low-tide elevations
Article 14.   Combination of methods for determining baselines
Article 15.   Delimitation of the territorial sea between States with opposite or adjacent coasts
Article 16.   Charts and lists of geographical co-ordinates
                Section 3. Innocent passage in the territorial sea
                Subsection A. Rules applicable to all ships
Article 17.   Right of innocent passage
Article 18.   Meaning of passage
Article 19.   Meaning of innocent passage
Article 20.   Submarines and other underwater vehicles
Article 21.   Laws and regulations of the coastal State relating to innocent passage
Article 22.   Sea lanes and traffic separation schemes in the territorial sea
Article 23.   Foreign nuclear-powered ships and ships carrying nuclear or other inherently dangerous or noxious substances
Article 24.   Duties of the coastal State
Article 25.   Rights of protection of the coastal State
Article 26.   Charges which may be levied upon foreign ships
                Subsection B. Rules applicable to merchant ships and government ships operated for commercial purposes
Article 27.   Criminal jurisdiction on board a foreign ship
Article 28.   Civil jurisdiction in relation to foreign ships
                Subsection C. Rules applicable to warships and other government ships operated for non-commercial purposes
Article 29.   Definition of warships
Article 30.   Non-observance by warships of the laws and regulations of the coastal State
Article 31.   Responsibility of the flag State for damage caused by a warship or other government ship operated for non-commercial purposes
Article 32.   Immunities of warships and other government ships operated for non-commercial purposes
                Section 4. Contiguous Zone
Article 33.   Contiguous Zone

*Sixth session, New York, May 23–July 15, 1977 (A/CONF.62/WP.10 and ADD.1)

PART III. STRAITS USED FOR INTERNATIONAL NAVIGATION 617
                Section 1. General
Article 34.   Juridical status of waters forming straits used for international navigation
Article 35.   Scope of this Part
Article 36.   High seas routes or routes through exclusive economic zones through straits used for international navigation
                Section 2.   Transit passage
Article 37.   Scope of this section
Article 38.   Right of transit passage
Article 39.   Duties of ships and aircraft during their passage
Article 40.   Research and survey activities
Article 41.   Sea lanes and traffic separation schemes in straits used for international navigation
Article 42.   Laws and regulations of States bordering straits relating to transit passage
Article 43.   Navigation and safety aids and other improvements and the prevention and control of pollution
Article 44.   Duties of States bordering straits
                Section 3.   Innocent passage
Article 45.   Innocent passage
PART IV. ARCHIPELAGIC STATES 619
Article 46.   Use of terms
Article 47.   Archipelagic baselines
Article 48.   Measurement of the breadth of the territorial sea, the contiguous zone, the exclusive economic zone and the continental shelf
Article 49.   Juridical status of archipelagic waters, of the airspace over archipelagic waters, and of their bed and subsoil
Article 50.   Delimitation of internal waters
Article 51.   Existing agreements, traditional fishing rights and existing submarine cables
Article 52.   Right of innocent passage
Article 53.   Right of archipelagic sea lanes passage
Article 54.   Duties of ships and aircraft during their passage, research and survey activities duties of the archipelagic State and laws and regulations of the archipelagic State relating to archipelagic sea lanes passage
PART V. EXCLUSIVE ECONOMIC ZONE 621
Article 55.   Specific legal régime of the exclusive economic zone
Article 56.   Rights, jurisdiction and duties of the coastal State in the exclusive economic zone
Article 57.   Breadth of the exclusive economic zone
Article 58.   Rights and duties of other States in the exclusive economic zone
Article 59.   Basis for the resolution of conflicts regarding the attribution of rights and jurisdiction in the exclusive economic zone
Article 60.   Artificial islands, installations and structures in the exclusive economic zone
Article 61.   Conservation of the living resources
Article 62.   Utilization of the living resources
Article 63.   Stocks occurring within the exclusive economic zones of two or more coastal States or both within the exclusive economic zone and in an area beyond and adjacent to it
Article 64.   Highly migratory species
Article 65.   Marine mammals
Article 66.   Anadromous stocks
Article 67.   Catadromous species
Article 68.   Sedentary species
Article 69.   Right of land-locked States
Article 70.   Right of certain developing coastal States in a subregion or region
Article 71.   Non-applicability of articles 69 and 70
Article 72.   Restrictions on transfer of rights
Article 73.   Enforcement of laws and regulations of the coastal State
Article 74.   Delimitation of the exclusive economic zone between adjacent or opposite States
Article 75.   Charts and lists of geographical co-ordinates
PART VI. CONTINENTAL SHELF 625
Article 76.   Definition of the continental shelf
Article 77.   Rights of the coastal State over the continental shelf
Article 78.   Superjacent waters and air space
Article 79.   Submarine cables and pipelines on the continental shelf
Article 80.   Artificial islands, installations and structures on the continental shelf
Article 81.   Drilling on the continental shelf
Article 82.   Payments and contributions with respect to the exploitation of the continental shelf beyond 200 miles
Article 83.   Delimitation of the continental shelf between adjacent or opposite States

Article 84. Charts and lists of geographical co-ordinates
Article 85. Tunnelling
PART VII. HIGH SEAS 626
      Section 1. General
Article 86. Application of the provisions of this Part
Article 87. Freedom of the high seas
Article 88. Reservation of the high seas for peaceful purposes
Article 89. Invalidity of claims of sovereignty over the high seas
Article 90. Right of navigation
Article 91. Nationality of ships
Article 92. Status of ships
Article 93. Ships flying the flag of the United Nations, its specialized agencies and the International Atomic Energy Agency
Article 94. Duties of the flag States
Article 95. Immunity of warships on the high seas
Article 96. Immunity of ships used only on government non-commercial service
Article 97. Penal jurisdiction in matters of collision
Article 98. Duty to render assistance
Article 99. Prohibition of the transport of slaves
Article 100. Duty to co-operate in the repression of piracy
Article 101. Definition of pirate
Article 102. Piracy by a warship, government ship or government aircraft whose crew has mutinied
Article 103. Definition of a pirate ship or aircraft
Article 104. Retention or loss of the nationality of a pirate ship or aircraft
Article 105. Seizure of a pirate ship or aircraft
Article 106. Liability for seizure without adequate grounds
Article 107. Ships and aircraft which are entitled to seize on account of piracy
Article 108. Illicit traffic in narcotic drugs or psychotropic substances
Article 109. Unauthorized broadcasting from the high seas
Article 110. Right of visit
Article 111. Right of hot pursuit
Article 112. Right to lay submarine cables and pipelines
Article 113. Breaking or injury of a submarine cable or pipeline
Article 114. Breaking or injury by owners of a submarine cable or pipeline of another submarine cable or pipeline
Article 115. Indemnity for loss incurred in avoiding injury to a submarine cable or pipeline
      Section 2. Management and conservation of the living resources of the high seas
Article 116. Right to fish on the high seas
Article 117. Duty of States to adopt with respect to their nationals measures for the conservation of the living resources of the high seas
Article 118. Co-operation of States in the management and conservation of living resources
Article 119. Conservation of the living resources of the high seas
Article 120. Marine mammals
PART VIII. REGIME OF ISLANDS 631
Article 121. Régime of islands
PART IX. ENCLOSED OR SEMI-ENCLOSED SEAS 631
Article 122. Definition
Article 123. Co-operation of States bordering enclosed or semi-enclosed seas
PART X. RIGHT OF ACCESS OF LAND-LOCKED STATES TO AND FROM THE SEA AND FREEDOM OF TRANSIT 631
Article 124. Use of terms
Article 125. Right of access to and from the sea and freedom of transit
Article 126. Exclusion of application of the most-favoured-nation clause
Article 127. Customs duties, taxes and other charges
Article 128. Free zones and other customs facilities
Article 129. Co-operation in the construction and improvement of means of transport
Article 130. Measures to avoid or eliminate delay or other difficulties of a technical nature in traffic in transit
Article 131. Equal treatment in maritime ports
Article 132. Grant of greater transit facilities
PART XI. THE AREA 632
      Section 1. General
Article 133. Use of terms
Article 134. Scope of this Part

Article 135. Legal status of the superjacent waters and air space
Section 2. Principles governing the Area
Article 136. Common heritage of mankind
Article 137. Legal status of the Area and its resources
Article 138. General conduct of States in relation to the Area
Article 139. Responsibility to ensure compliance and liability for damage
Article 140. Benefit of mankind
Article 141. Use of the Area exclusively for peaceful purposes
Article 142. Rights and legitimate interests of coastal States
Section 3. Conduct of activities in the Area
Article 143. Marine scientific research
Article 144. Transfer of technology
Article 145. Protection of the marine environment
Article 146. Protection of human life
Article 147. Accommodation of activities in the Area and in the marine environment
Article 148. Participation of developing countries in activities in the Area
Article 149. Archaeological and historical objects
Section 4. Development of resources of the Area
Article 150. Policies relating to activities in the Area
Article 151. Functions of the Authority
Article 152. Periodic review
Article 153. The Review Conference
Section 5. The Authority
Subsection 1. General
Article 154. Establishment of the Authority
Article 155. Nature and fundamental principles of the Authority
Article 156. Organs of the Authority
Subsection 2. The Assembly
Article 157. Composition, procedure and voting
Article 158. Powers and functions
Subsection 3. The Council
Article 159. Composition, procedure and voting
Article 160. Powers and functions
Article 161. Organs of the Council
Article 162. The Economic Planning Commission
Article 163. The Technical Commission
Article 164. The Rules and Regulations Commission
Subsection 4. The Secretariat
Article 165. The Secretary-General
Article 166. The staff of the Authority
Article 167. International character of the Secretariat
Article 168. Consultation and co-operation with non-governmental organizations
Subsection 5. The Enterprise
Article 169. The Enterprise
Subsection 6. Finance
Article 170. General fund
Article 171. Annual budget of the Authority
Article 172. Expenses of the Authority
Article 173. Special fund
Article 174. Borrowing powers of the Authority
Article 175. Annual audit
Subsection 7. Legal status, privileges and immunities
Article 176. Legal status
Article 177. Privileges and immunities
Article 178. Immunity from legal process
Article 179. Immunity from search and any form of seizure
Article 180. Property and assets free from restrictions, regulations, control and moratoria
Article 181. Immunities of certain persons connected with the Authority
Article 182. Immunities of persons appearing in proceedings before the Tribunal
Article 183. Inviolability of archives
Article 184. Exemption from taxation and customs duties
Subsection 8. Suspension of rights of members
Article 185. Suspension of voting rights
Article 186. Suspension of privileges and the rights of membership
Section 6. Settlement of disputes
Article 187. Jurisdiction of the Sea-bed Disputes Chamber of the Law of the Sea Tribunal

Article 188. Submission of disputes to arbitration
Article 189. Disputes involving States Parties or their nationals
Article 190. Advisory opinions
Article 191. Scope of jurisdiction with regard to certain decisions adopted by the Assembly or by the Council
Article 192. Rights of States Parties when their nationals are parties to a dispute

PART XII. PROTECTION AND PRESERVATION OF THE MARINE ENVIRONMENT 646

Section 1. General provisions

Article 193. General obligation
Article 194. Sovereign right of States to exploit their natural resources
Article 195. Measures to prevent, reduce and control pollution of the marine environment
Article 196. Duty not to transfer damage or hazards or transform the type of pollution
Article 197. Use of technologies or introduction of alien or new species

Section 2. Global and regional co-operation

Article 198. Co-operation on a global or regional basis
Article 199. Notification of imminent or actual damage
Article 200. Contingency plans against pollution
Article 201. Promotion of studies, research programmes and exchange of information and data
Article 202. Scientific criteria and regulations

Section 3. Technical assistance

Article 203. Scientific and technical assistance to developing States
Article 204. Preferential treatment for developing States

Section 4. Monitoring and environmental assessment

Article 205. Monitoring of the risks or effects of pollution
Article 206. Publication of reports
Article 207. Assessment of potential effects of activities

Section 5. International rules and national legislation to prevent, reduce and control pollution of the marine environment

Article 208. Pollution from land-based sources
Article 209. Pollution from sea-bed activities
Article 210. Pollution from activities in the Area
Article 211. Dumping
Article 212. Pollution from vessels
Article 213. Pollution from or through the atmosphere

Section 6. Enforcement

Article 214. Enforcement with respect to land-based sources of pollution
Article 215. Enforcement with respect to pollution from sea bed activities
Article 216. Enforcement with respect to pollution from activities in the Area
Article 217. Enforcement with respect to dumping
Article 218. Enforcement by flag States
Article 219. Enforcement by port States
Article 220. Measures relating to seaworthiness of vessels to avoid pollution
Article 221. Enforcement by coastal States
Article 222. Measures relating to maritime casualties to avoid pollution
Article 223. Enforcement with respect to pollution from or through the atmosphere

Section 7. Safeguards

Article 224. Measures to facilitate proceedings
Article 225. Exercise of powers of enforcement
Article 226. Duty to avoid adverse consequences in the exercise of the powers of enforcement
Article 227. Investigation of foreign vessels
Article 228. Non-discrimination of foreign vessels
Article 229. Suspension and restrictions on institution of proceedings
Article 230. Institution of civil proceedings
Article 231. Monetary penalties and the observance of recognized rights of the accused
Article 232. Notification to flag States and other States concerned
Article 233. Liability of States arising from enforcement measures
Article 234. Safeguards with respect to straits used for international navigation

Section 8. Ice-covered areas

Article 235. Ice-covered areas

Section 9. Responsibility and liability

Article 236. Responsibility and liability

Section 10. Sovereign immunity

Article 237. Sovereign immunity

Section 11. Obligations under other Conventions on the protection and preservation of the marine environment

Article 238. Obligations under other Conventions on the protection and preservation of the marine environment

Informal Composite Negotiating Text   *611*

PART XIII. MARINE SCIENTIFIC RESEARCH 655
                Section 1. General provisions
Article 239. Right to conduct marine scientific research
Article 240. Promotion of marine scientific research
Article 241. General principles for the conduct of marine scientific research
Article 242. Marine scientific research activities not constituting the legal basis for any claim
                Section 2. Global and regional co-operation
Article 243. Promotion of international co-operation
Article 244. Creation of favourable conditions
Article 245. Publication and dissemination of information and knowledge
                Section 3. Conduct and promotion of marine scientific research
Article 246. Marine scientific research in the territorial sea
Article 247. Marine scientific research in the exclusive economic zone and on the continental shelf
Article 248. Research project under the auspices of, or undertaken by, international organizations
Article 249. Duty to provide information to the coastal State
Article 250. Duty to comply with certain conditions
Article 251. Communications concerning research project
Article 252. General criteria and guidelines
Article 253. Implied consent
Article 254. Cessation of research activities
Article 255. Rights of neighbouring land-locked and geographically disadvantaged States
Article 256. Measures to facilitate marine scientific research and assist research vessels
Article 257. Marine scientific research in the Area
Article 258. Marine scientific research in the water column beyond the exculsive economic zone
                Section 4. Legal status of scientific research installations and equipment in the marine environment
Article 259. Deployment and use
Article 260. Legal status
Article 261. Safety zones
Article 262. Non-interference with shipping routes
Article 263. Identification markings and warning signals
                Section 5. Responsibility and liability
Article 264. Responsibility and liability
                Section 6. Settlement of disputes
Article 265. Settlement of disputes
Article 266. Interim measures

PART XIV. DEVELOPMENT AND TRANSFER OF MARINE TECHNOLOGY 658
                Section 1. General provisions
Article 267. Promotion of the development and transfer of marine science and marine technology
Article 268. Protection of legitimate interests
Article 269. Basic objectives
Article 270. Measures to achieve the basic objectives
                Section 2. International co-operation
Article 271. Ways and means of international co-operation
Article 272. Guidelines, criteria and standards
Article 273. Co-ordination of international programmes technology
Article 274. Co-operation with international organizations and the Authority in the transfer of technology to developing States
Article 275. Objectives of the Authority with respect to the transfer of technology
                Section 3. Regional marine scientific and technological centres
Article 276. Establishment of regional centres
Article 277. Functions of regional centres
                Section 4. Co-operation among international organizations
Article 278. Co-operation among international organizations

PART XV. SETTLEMENT OF DISPUTES 660
                Section 1
Article 279. Obligation to settle disputes by peaceful means
Article 280. Settlement of disputes by means chosen by the parties
Article 281. Obligation to exchange views
Article 282. Obligations under general, regional or special agreements
Article 283. Procedure when dispute is not settled by means chosen by the parties
Article 284. Conciliation
Article 285. Application of this section to disputes submitted pursuant to Part XI
Article 286. Application of section 1 and proceedings under this section

Article 287. Choice of procedure
Article 288. Competence
Article 289. Expert advice and assistance
Article 290. Provisional measures
Article 291. Access
Article 292. Prompt release of vessels
Article 293. Applicable law
Article 294. Exhaustion of local remedies
Article 295. Finality and binding force of decisions
Article 296. Limitations on applicability of this section
Article 297. Optional exceptions
PART XVI. FINAL CLAUSES 664
Article 298. Ratification
Article 299. Accession
Article 300. Entry into force
Article 301. Status of annexes
Article 302. Authentic texts
Article 303. Testimonium clause, place and date
              Transitional provisions

ANNEXES
Annex I.   Highly migratory species 664
Annex II.   Basic conditions of exploration and exploitation 665
Annex III. Statute of the Enterprise 673
Annex IV.   Conciliation 676
Annex V.   Statute of the Law of the Sea Tribunal 677
Annex VI.   Arbitration 681
Annex VII. Special arbitration procedure 683

PREAMBLE

"The States Parties to the present Convention,

Considering that the General Assembly of the United Nations, by its resolution 2749 (XXV) of 17 December 1970, adopted the Declaration of Principles Governing the Sea-Bed and the Ocean Floor, and the Subsoil Thereof, beyond the Limits of National Jurisdiction,

Believing that the codification and progressive development of the law of the sea achieved in the present Convention will contribute to the maintenance of international peace and security, in accordance with the purposes and principles of the United Nations as set forth in the Charter,

Having regard to the Declaration on Principles of International Law concerning Friendly Relations and Co-operation among States in accordance with the Charter of the United Nations,

Affirming that the rules of customary international law continue to govern matters not expressly regulated by the provisions of the present Convention,

Have agreed as follows:

PART I
Article 1
Use of terms

1. For the purposes of the present Convention:
   (1) "Area" means the sea-bed and ocean floor and subsoil thereof beyond the limits of national jurisdiction.
   (2) "Authority" means the International Sea-Bed Authority.
   (3) "Activities in the Area" means all activities of exploration for, and exploitation of, the resources of the Area.
   (4) "Pollution of the marine environment" means the introduction by man, directly or indirectly of substances or energy into the marine environment (including estuaries) which results or is likely to result in such deleterious effects as harm to living resources and marine life, hazards to human health, hindrance to marine activities, including fishing and other legitimate uses of the sea, impairment of quality for use of sea water and reduction of amenities.
   (5) (a) "Dumping" means:
       (i) Any deliberate disposal including incineration of wastes or other matter from vessels, aircraft, platforms or other man-made structures at sea;
       (ii) Any deliberate disposal of vessels, aircraft, platforms or other man-made structures at sea.
       (b) "Dumping" does not include:
       (i) The disposal of wastes or other matter incidental to, or derived from the normal operations of vessels, aircraft, platforms or other man-made structures at sea and their equipment, other than wastes or other matter transported by or to vessels,

aircraft, platforms or other man-made structures at sea, operating for the purpose of disposal of such matter or derived from the treatment of such wastes or other matter on such vessels, aircraft, platforms or structures;
(ii) Placement of matter for a purpose other than the mere disposal thereof, provided that such placement is not contrary to the aims of the present Convention.
(c) The disposal of wastes or other matter directly arising from or related to the exploration, exploitation and associated off-shore processing of sea-bed mineral resources will not be covered by the provisions of the present Convention.

PART II
TERRITORIAL SEA AND CONTIGUOUS ZONE
SECTION 1. GENERAL
Article 2
Juridical status of the territorial sea, of the air space over the territorial sea and of its bed and subsoil

1. The sovereignty of a coastal State extends beyond its land territory and internal waters, and in the case of an archipelagic State, its archipelagic waters, over an adjacent belt of sea described as the territorial sea.
2. This sovereignty extends to the air space over the territorial sea as well as to its bed and subsoil.
3. The sovereignty over the territorial sea is exercised subject to the present Convention and to other rules of international law.

SECTION 2. LIMITS OF THE TERRITORIAL SEA
Article 3
Breadth of the territorial sea

Every State has the right to establish the breadth of its territorial sea up to a limit not exceeding 12 nautical miles, measured from baselines determined in accordance with the present Convention.

Article 4
Outer limit of the territorial sea

The outer limit of the territorial sea is the line every point of which is at a distance from the nearest point of the baseline equal to the breadth of the territorial sea.

Article 5
Normal baseline

Except where otherwise provided in the present Convention, the normal baseline for measuring the breadth of the territorial sea is the low-water line along the coast as marked on large-scale charts officially recognized by the coastal State.

Article 6
Reefs

In the case of islands situated on atolls or of islands having fringing reefs, the baseline for measuring the breadth of the territorial sea is the seaward low-water line of the reef, as shown by the appropriate symbol on official charts.

Article 7
Straight baselines

1. In localities where the coastline is deeply indented and cut into, or if there is a fringe of islands along the coast in its immediate vicinity, the method of straight baselines joining appropriate points may be employed in drawing the baseline from which the breadth of the territorial sea is measured.
2. Where because of the presence of a delta and other natural conditions the coastline is highly unstable, the appropriate points may be selected along the furthest seaward extent of the low-water line and, notwithstanding subsequent regression of the low-water line, such baselines shall remain effective until changed by the coastal State in accordance with the present Convention.
3. The drawing of such baselines must not depart to any appreciable extent from the general direction of the coast, and the sea areas lying within the lines must be sufficiently closely linked to the land domain to be subject to the régime of internal waters.
4. Straight baselines shall not be drawn to and from low-tide elevations, unless lighthouses or similar installations which are permanently above sea level have been built on them or except in instances where the drawing of baselines to and from such elevations has received general international recognition.
5. Where the method of straight baselines is applicable under paragraph 1 account may be taken, in determining particular baselines, of economic interests peculiar to the region concerned, the reality and the importance of which are clearly evidenced by a long usage.
6. The system of straight baselines may not be applied by a State in such a manner as to cut off from the high seas or the exclusive economic zone the territorial sea of another State.

Article 8
Internal waters

1. Except as provided in Part IV, waters on the landward side of the baseline of the territorial sea form part of the internal waters of the State.
2. Where the establishment of a straight baseline in accordance with article 7 has the effect of enclosing as internal waters areas which had not previously been considered as such, a right of innocent passage as provided in the present Convention shall exist in those waters.

Article 9
Mouths of rivers

If a river flows directly into the sea, the baseline shall be a straight line across the mouth of the river between points on the low-tide line of its banks.

Article 10
Bays

1. This article relates only to bays the coasts of which belong to a single State.
2. For the purposes of the present Convention, a bay is a well-marked indentation whose penetration is in such proportion to the width of its mouth as to contain land-locked waters and constitute more than a mere curvature of the coast. An indentation shall not, however, be regarded as a bay unless its area is as large as, or larger than, that of the semi-circle whose diameter is a line drawn across the mouth of that indentation.
3. For the purpose of measurement, the area of an indentation is that lying between the low-water mark around the shore of the indentation and a line joining the low-water mark of its natural entrance points. Where, because of the presence of islands, an indentation has more than one mouth, the semi-circle shall be drawn on a line as long as the sum total of the lengths of the lines across the different mouths. Islands within an indentation shall be included as if they were part of the water area of the indentation.
4. If the distance between the low-water marks of the natural entrance points of a bay does not exceed 24 miles a closing line may be drawn between these two low-water marks, and the waters enclosed thereby shall be considered as internal waters.
5. Where the distance between the low-water marks of the natural entrance points of a bay exceeds 24 miles a straight baseline of 24 miles shall be drawn within the bay in such a manner as to enclose the maximum area of water that is possible with a line of that length.
6. The foregoing provisions do not apply to so-called "historic" bays, or in any case where the system of straight baselines provided for in article 7 is applied.

Article 11
Ports

For the purpose of delimiting the territorial sea, the outermost permanent harbour works which form an integral part of the harbour system are regarded as forming part of the coast. Off-shore installations and artificial islands shall not be considered as permanent harbour works.

Article 12
Roadsteads

Roadsteads which are normally used for the loading, unloading, and anchoring of ships, and which would otherwise be situated wholly or partly outside the outer limit of the territorial sea, are included in the territorial sea.

Article 13
Low-tide elevations

1. A low-tide elevation is a naturally formed area of land which is surrounded by and above water at low tide but submerged at high tide. Where a low-tide elevation is situated wholly or partly at a distance not exceeding the breadth of the territorial sea from the mainland or an island, the low-water line on that elevation may be used as the baseline for measuring the breadth of the territorial sea.
2. Where a low-tide elevation is wholly situated at a distance exceeding the breadth of the territorial sea from the mainland or an island, it has no territorial sea of its own.

Article 14
Combination of methods for determining baselines

The coastal State may determine baselines in turn by any of the methods provided for in the foregoing articles to suit different conditions.

Article 15
Delimitation of the territorial sea between States with opposite or adjacent coasts

Where the coasts of two states are opposite or adjacent to each other, neither of the two States is entitled, failing agreement between them to the contrary, to extend its territorial sea beyond the median line every point of which is equidistant from the nearest points on the baselines from which the breadth of the territorial seas of each of the two States is measured. This article does not apply, however, where it is necessary by reason of historic title or other special circumstances to delimit the territorial seas of the two States in a way which is at variance with this provision.

Article 16
Charts and lists of geographical co-ordinates

1. The baselines for measuring the breadth of the territorial sea determined in accordance with articles 7, 9 and 10, or the limits derived therefrom, and the lines of delimitation drawn in accordance with articles 12 and 15, shall be shown on charts of a scale or scales adequate for determining them. Alternatively, a list of geographical co-ordinates of points, specifying the geodetic datum, may be substituted.
2. The coastal State shall give due publicity to such charts or lists of geographical co-ordinates and shall deposit a copy of each such chart or list with the Secretary-General of the United Nations.

SECTION 3. INNOCENT PASSAGE IN THE TERRITORIAL SEA
Subsection A. Rules applicable to all ships
Article 17
Right of innocent passage

Subject to the present Convention, ships of all States, whether coastal or land-locked, enjoy the right of innocent passage through the territorial sea.

Article 18
Meaning of passage

1. Passage means navigation through the territorial sea for the purposes of:
   (a) Traversing that sea without entering internal waters or calling at a roadstead or port facility outside internal waters; or
   (b) Proceeding to or from internal waters or a call at such a roadstead or port facility.
2. Passage shall be continuous and expeditious. However, passage includes stopping and anchoring, but only in so far as the same are incidental to ordinary navigation or are rendered necessary by *force majeure* or distress or for the purpose of rendering assistance to persons, ships or aircraft in danger or distress.

Article 19
Meaning of innocent passage

1. Passage is innocent so long as it is not prejudicial to the peace, good order or security of the coastal State. Such passage shall take place in conformity with the present Convention and with other rules of international law.
2. Passage of a foreign ship shall be considered to be prejudicial to the peace, good order or security of the coastal State, if in the territorial sea it engages in any of the following activities:
   (a) Any threat or use of force against the sovereignty, territorial integrity or political independence of the coastal State, or in any other manner in violation of the principles of international law embodied in the Charter of the United Nations;
   (b) Any exercise or practice with weapons of any kind;
   (c) Any act aimed at collecting information to the prejudice of the defence or security of the coastal State;
   (d) Any act of propaganda aimed at affecting the defence or security of the coastal State;
   (e) The launching, landing or taking on board of any aircraft;
   (f) The launching, landing or taking on board of any military device;
   (g) The embarking or disembarking of any commodity, currency or person contrary to the customs, fiscal, immigration or sanitary regulations of the coastal State;
   (h) Any act of wilful and serious pollution, contrary to the present Convention;
   (i) Any fishing activities;
   (j) The carrying out of research or survey activities;
   (k) Any act aimed at interfering with any systems of communication or any other facilities or installations of the coastal State;
   (l) Any other activity not having a direct bearing on passage.

Article 20
Submarines and other underwater vehicles

In the territorial sea, submarines and other underwater vehicles are required to navigate on the surface and to show their flag.

Article 21
Laws and regulations of the coastal State relating to innocent passage

1. The coastal State may make laws and regulations, in conformity with the provisions of the present Convention and other rules of international law, relating to innocent passage through the territorial sea, in respect of all or any of the following:
   (a) The safety of navigation and the regulation of marine traffic;
   (b) The protection of navigational aids and facilities and other facilities or installations;
   (c) The protection of cables and pipelines;
   (d) The conservation of the living resources of the sea;
   (e) The prevention of infringement of the fisheries regulations of the coastal State;
   (f) The preservation of the environment of the coastal State and the prevention, reduction and control of pollution thereof;
   (g) Marine scientific research and hydrographic surveys;
   (h) The prevention of infringement of the customs, fiscal, immigration, or sanitary regulations of the coastal State.
2. Such laws and regulations shall not apply to the design, construction, manning or equipment of foreign ships unless they are giving effect to generally accepted international rules or standards.
3. The coastal State shall give due publicity to all such laws and regulations.
4. Foreign ships exercising the right of innocent passage through the territorial sea shall comply with all such laws and regulations and all generally accepted international regulations relating to the prevention of collisions at sea.

Article 22
Sea lanes and traffic separation schemes in the territorial sea

1. The coastal State may, where necessary having regard to the safety of navigation, require foreign ships exercising the right of innocent passage through its territorial sea to use such sea lanes and traffic separation schemes as it may designate or prescribe for the regulation of the passage of ships.
2. In particular, tankers, nuclear-powered ships and ships carrying nuclear or other inherently dangerous or noxious substances or materials may be required to confine their passage to such sea lanes.
3. In the designation of sea lanes and the prescription of traffic separation schemes under this article the coastal State shall take into account:
   (a) The recommendations of competent international organizations;
   (b) Any channels customarily used for international navigation;
   (c) The special characteristics of particular ships and channels; and
   (d) The density of traffic.

4. The coastal State shall clearly indicate such sea lanes and traffic separation schemes on charts to which due publicity shall be given.

Article 23
Foreign nuclear-powered ships and ships carrying nuclear or other inherently dangerous or noxious substances

Foreign nuclear-powered ships and ships carrying nuclear or other inherently dangerous or noxious substances shall, when exercising the right of innocent passage through the territorial sea, carry documents and observe special precautionary measures established for such ships by international agreements.

Article 24
Duties of the coastal State

1. The coastal State shall not hamper the innocent passage of foreign ships through the territorial sea except in accordance with the present Convention. In particular, in the application of the present Convention or of any laws or regulations made under the present Convention, the coastal State shall not:
   (a) Impose requirements on foreign ships which have the practical effect of denying or impairing the right of innocent passage; or
   (b) Discriminate in form or in fact against the ships of any State or against ships carrying cargoes to, from or on behalf of any State.
2. The coastal State shall give appropriate publicity to any dangers to navigation, of which it has knowledge, within its territorial sea.

Article 25
Rights of protection of the coastal State

1. The coastal State may take the necessary steps in its territorial sea to prevent passage which is not innocent.
2. In the case of ships proceeding to internal waters or a call at a port facility outside internal waters, the coastal State also has the right to take the necessary steps to prevent any breach of the conditions to which admission of those ships to internal waters or such a call is subject.
3. The coastal State may, without discrimination amongst foreign ships, suspend temporarily in specified areas of its territorial sea the innocent passage of foreign ships if such suspension is essential for the protection of its security. Such suspension shall take effect only after having been duly published.

Article 26
Charges which may be levied upon foreign ships

1. No charge may be levied upon foreign ships by reason only of their passage through the territorial sea.
2. Charges may be levied upon a foreign ship passing through the territorial sea as payment only for specific services rendered to the ship. These charges shall be levied without discrimination.

Subsection B. Rules applicable to merchant ships and government ships operated for commercial purposes

Article 27
Criminal jurisdiction on board a foreign ship

1. The criminal jurisdiction of the coastal State should not be exercised on board a foreign ship passing through the territorial sea to arrest any person or to conduct any investigation in connection with any crime committed on board the ship during its passage, save only in the following cases:
   (a) If the consequences of the crime extend to the coastal State;
   (b) If the crime is of a kind to disturb the peace of the country or the good order of the territorial sea;
   (c) If the assistance of the local authorities has been requested by the captain of the ship or by the diplomatic agent or consular officer of the flag State; or
   (d) If such measures are necessary for the suppression of illicit traffic in narcotic drugs or psychotropic substances.
2. The above provisions do not affect the right of the coastal State to take any steps authorized by its laws for the purpose of an arrest or investigation on board a foreign ship passing through the territorial sea after leaving internal waters.
3. In the cases provided for in paragraphs 1 and 2, the coastal State shall, if the captain so requests, advise the diplomatic agent or consular officer of the flag State before taking any steps, and shall facilitate contact between such agent or officer and the ship's crew. In cases of emergency this notification may be communicated while the measures are being taken.
4. In considering whether or how an arrest should be made, the local authorities shall pay due regard to the interests of navigation.
5. Except as provided in Part XII or with respect to violations of laws and regulations enacted in accordance with Part IV, the coastal State may not take any steps on board a foreign ship passing through the territorial sea to arrest any person or to conduct any investigation in connexion with any crime committed before the ship entered the territorial sea, if the ship, proceeding from a foreign port, is only passing through the territorial sea without entering internal waters.

Article 28
Civil jurisdiction in relation to foreign ships

1. The coastal State should not stop or divert a foreign ship passing through the territorial sea for the purpose of exercising civil jurisdiction in relation to a person on board the ship.
2. The coastal State may not levy execution against or arrest the ship for the purpose of any civil proceedings, save only in respect of obligations or liabilities assumed or incurred by the ship itself in the course or for the purpose of its voyage through the waters of the coastal State.

3. Paragraph 2 is without prejudice to the right of the coastal State, in accordance with its laws, to levy execution against or to arrest, for the purpose of any civil proceedings, a foreign ship lying in the territorial sea or passing through the territorial sea after leaving internal waters.

Subsection C. Rules applicable to warships and other government ships operated for non-commercial purposes

Article 29
Definition of warships

For the purposes of the present Convention, "warship" means a ship belonging to the armed forces of a State bearing the external marks distinguishing such ships of its nationality, under the command of an officer duly commissioned by the Government of the State and whose name appears in the appropriate service list or its equivalent, and manned by a crew which is under regular armed forces discipline.

Article 30
Non-observance by warships of the laws and regulations of the coastal State

If any warship does not comply with the laws and regulations of the coastal State concerning passage through the territorial sea and disregards any request for compliance which is made to it, the coastal State may require it to leave the territorial sea immediately.

Article 31
Responsibility of the flag State for damage caused by a warship or other government ship used for non-commercial purposes

The flag State shall bear international responsibility for any loss or damage to the coastal State resulting from the non-compliance by a warship or other government ship operated for non-commercial purposes with the laws and regulations of the coastal State concerning passage through the territorial sea or with the provisions of the present Convention or other rules of international law.

Article 32
Immunities of warships and other government ships operated for non-commercial purposes

With such exceptions as are contained in subsection A and in articles 30 and 31, nothing in the present Convention affects the immunities of warships and other government ships operated for non-commercial purposes.

SECTION 4. CONTIGUOUS ZONE
Article 33
Contiguous zone

1. In a zone contiguous to its territorial sea, described as the contiguous zone, the coastal State may exercise the control necessary to:
   (a) Prevent infringement of its customs, fiscal, immigration or sanitary regulations within its territory or territorial sea;
   (b) Punish infringement of the above regulations committed within its territory or territorial sea.
2. The contiguous zone may not extend beyond 24 nautical miles from the baselines from which the breadth of the territorial sea is measured.

PART III
STRAITS USED FOR INTERNATIONAL NAVIGATION
SECTION 1. GENERAL
Article 34
Juridical status of waters forming straits used for international navigation

1. The régime of passage through straits used for international navigation established in this Part shall not in other respects affect the status of the waters forming such straits or the exercise by the States bordering the straits of their sovereignty or jurisdiction over such waters and their air space, bed and subsoil.
2. The sovereignty or jurisdiction of the States bordering the straits is exercised subject to this Part and to other rules of international law.

Article 35
Scope of this Part

Nothing in this Part shall affect:
   (a) Any areas of internal waters within a strait, except where the establishment of a straight baseline in accordance with article 7 has the effect of enclosing as internal waters areas which had not previously been considered as such;
   (b) The status of the waters beyond the territorial seas of States bordering straits as exclusive economic zones or high seas; or
   (c) The legal régime in straits in which passage is regulated in whole or in part by long-standing international conventions in force specifically relating to such straits.

Article 36
High seas routes or routes through exclusive economic zones through straits used for international navigation

This Part does not apply to a strait used for international navigation if a high seas route or a route through an exclusive economic zone of similar convenience with respect to navigational and hydrographical characteristics exists through the strait.

SECTION 2. TRANSIT PASSAGE
Article 37
Scope of this section

This section applies to straits which are used for international navigation between one area of the high seas or an exclusive economic zone and another area of the high seas or an exclusive economic zone.

Article 38
Right of transit passage

1. In straits referred to in article 37, all ships

and aircraft enjoy the right of transit passage, which shall not be impeded, except that, if the strait is formed by an island of a State bordering the strait and its mainland, transit passage shall not apply if a high seas route or a route in an exclusive economic zone of similar convenience with respect to navigational and hydrographical characteristics exists seaward of the island.
2. Transit passage is the exercise in accordance with this Part of the freedom of navigation and overflight solely for the purpose of continuous and expeditious transit of the strait between one area of the high seas or an exclusive economic zone and another area of the high seas or an exclusive economic zone. However, the requirement of continuous and expeditious transit does not preclude passage through the strait for the purpose of entering, leaving or returning from a State bordering the strait, subject to the conditions of entry to that State.
3. Any activity which is not an exercise of the right of transit passage through a strait remains subject to the other applicable provisions of the present Convention.

Article 39
Duties of ships and aircraft during their passage
1. Ships and aircraft, while exercising the right of transit passage, shall:
   (a) Proceed without delay through or over the strait;
   (b) Refrain from any threat or use of force against the sovereignty, territorial integrity or political independence of States bordering straits, or in any other manner in violation of the principles of international law embodied in the Charter of the United Nations;
   (c) Refrain from any activities other than those incident to their normal modes of continuous and expeditious transit unless rendered necessary by *force majeure* or by distress;
   (d) Comply with other relevant provisions of this Part.
2. Ships in transit shall:
   (a) Comply with generally accepted international regulations, procedures and practices for safety at sea, including the International Regulations for Preventing Collisions at Sea;
   (b) Comply with generally accepted international regulations, procedures and practices for the prevention, reduction and control of pollution from ships.
3. Aircraft in transit shall:
   (a) Observe the Rules of the Air established by the International Civil Aviation Organization as they apply to civil aircraft; State aircraft will normally comply with such safety measures and will at all times operate with due regard for the safety of navigation;
   (b) At all times monitor the radio frequency assigned by the appropriate internationally designated air traffic control authority or the appropriate international distress radio frequency.

Article 40
Research and survey activities
During their passage through straits, foreign ships, including marine research and hydrographic survey ships, may not carry out any research or survey activities without the prior authorization of the States bordering straits.

Article 41
Sea lanes and traffic separation schemes in straits used for international navigation
1. In conformity with this Part, States bordering straits may designate sea lanes and prescribe traffic separation schemes for navigation in straits where necessary to promote the safe passage of ships.
2. Such States may, when circumstances require, and after giving due publicity thereto, substitute other sea lanes or traffic separation schemes for any sea lanes or traffic separation schemes previously designated or prescribed by them.
3. Such sea lanes and traffic separation schemes shall conform to generally accepted international regulations.
4. Before designating or substituting sea lanes or prescribing or substituting traffic separation schemes, States bordering straits shall refer proposals to the competent international organization with a view to their adoption. The organization may adopt only such sea lanes and traffic separation schemes as may be agreed with the States bordering the straits, after which the States may designate, prescribe or substitute them.
5. In respect of a strait where sea lanes or traffic separation schemes are proposed through the waters of two or more States bordering the strait, the States concerned shall co-operate in formulating proposals in consultation with the organization.
6. States bordering straits shall clearly indicate all sea lanes and traffic separation schemes designated or prescribed by them on charts to which due publicity shall be given.
7. Ships in transit shall respect applicable sea lanes and traffic separation schemes established in accordance with this article.

Article 42
Laws and regulations of States bordering straits relating to transit passage
1. Subject to the provisions of this section, States bordering straits may make laws and regulations relating to transit passage through straits, in respect of all or any of the following:
   (a) The safety of navigation and the regulation of marine traffic, as provided in article 41;
   (b) The prevention, reduction and control of pollution, by giving effect to applicable international regulations regarding the discharge of oil, oily wastes and other noxious substances in the strait;

(c) With respect to fishing vessels, the prevention of fishing, including the stowage of fishing gear;
(d) The taking on board or putting overboard of any commodity, currency or person in contravention of the customs, fiscal, immigration or sanitary regulations of States bordering straits.
2. Such laws and regulations shall not discriminate in form or in fact amongst foreign ships or in their application have the practical effect of denying, hampering or impairing the right of transit passage as defined in this section.
3. States bordering straits shall give due publicity to all such laws and regulations.
4. Foreign ships exercising the right of transit passage shall comply with such laws and regulations.
5. The flag State of a ship or aircraft entitled to sovereign immunity which acts in a manner contrary to such laws and regulations or other provisions of this Part shall bear international responsibility for any loss or damage which results to States bordering straits.

Article 43
Navigation and safety aids and other improvements and the prevention, reduction and control of pollution

User States and States bordering a strait should by agreement co-operate:
(a) In the establishment and maintenance in a strait of necessary navigation and safety aids or other improvements in aid of international navigation; and
(b) For the prevention, reduction and control of pollution from ships.

Article 44
Duties of States bordering straits

States bordering straits shall not hamper transit passage and shall give appropriate publicity to any danger to navigation or overflight within or over the strait of which it has knowledge. There shall be no suspension of transit passage.

SECTION 3. INNOCENT PASSAGE
Article 45
Innocent passage

1. The régime of innocent passage, in accordance with section 3 of Part II, shall apply in straits used for international navigation:
(a) Excluded under paragraph 1 of article 38 from the application of the régime of transit passage; or
(b) Between one area of the high seas or an exclusive economic zone and the territorial sea of a foreign State.
2. There shall be no suspension of innocent passage through such straits.

PART IV
ARCHIPELAGIC STATES
Article 46
Use of terms

For the purposes of the present Convention:
(a) "Archipelagic State" means a State constituted wholly by one or more archipelagos and may include other islands;
(b) "Archipelago" means a group of islands, including parts of islands, interconnecting waters and other natural features which are so closely interrelated that such islands, waters and other natural features form an intrinsic geographical, economical and political entity, or which historically have been regarded as such.

Article 47
Archipelagic baselines

1. An archipelagic State may draw straight archipelagic baselines joining the outermost points of the outermost islands and drying reefs of the archipelago provided that within such baselines are included the main islands and an area in which the ratio of the area of the water to the area of the land, including atolls, is between one to one and nine to one.
2. The length of such baselines shall not exceed 100 nautical miles, except that up to three per cent of the total number of baselines enclosing any archipelago may exceed that length, up to a maximum length of 125 nautical miles.
3. The drawing of such baselines shall not depart to any appreciable extent from the general configuration of the archipelago.
4. Such baselines shall not be drawn to and from low-tide elevations, unless lighthouses or similar installations which are permanently above sea level have been built on them or where a low-tide elevation is situated wholly or partly at a distance not exceeding the breadth of the territorial sea from the nearest island.
5. The system of such baselines shall not be applied by an archipelagic State in such a manner as to cut off from the high seas or the exclusive economic zone the territorial sea of another State.
6. The archipelagic State shall clearly indicate such baselines on charts of a scale or scales adequate for determining them. The archipelagic State shall give due publicity to such charts and shall deposit a copy of each such chart with the Secretary-General of the United Nations.
7. If a certain part of the archipelagic water of an archipelagic State lies between two parts of an immediately adjacent neighboring State, existing rights and all other legitimate interests which the latter State has traditionally exercised in such waters and all rights stipulated under agreement between those States shall continue and be respected.
8. For the purposes of computing the ratio of water to land under paragraph 1, land areas

may include waters lying within the fringing reefs of islands and atolls, including that part of a steep-sided oceanic plateau which is enclosed or nearly enclosed by a chain of limestone islands and drying reefs lying on the perimeter of the plateau.

Article 48
Measurement of the breadth of the territorial sea, the contiguous zone, the exclusive economic zone and the continental shelf

The breadth of the territorial sea, the contiguous zone, the exclusive economic zone and the continental shelf shall be measured from the baselines drawn in accordance with article 47.

Article 49
Juridical status of archipelagic waters, of the air space over archipelagic waters and of their bed and subsoil

1. The sovereignty of an archipelagic State extends to the waters enclosed by the baselines, described as archipelagic waters, regardless of their depth or distance from the coast.
2. This sovereignty extends to the air space over the archipelagic waters, the bed and subsoil thereof, and the resources contained therein.
3. This sovereignty is exercised subject to this Part.
4. The régime of archipelagic sea lanes passage established in this Part shall not in other respects affect the status of the archipelagic waters, including the sea lanes, or the exercise by the archipelagic State of its sovereignty over such waters and their air space, bed and subsoil, and the resources contained therein.

Article 50
Delimitation of internal waters

Within its archipelagic waters, the archipelagic State may draw closing lines for the delimitation of internal waters, in accordance with articles 9, 10 and 11.

Article 51
Existing agreements, traditional fishing rights and existing submarine cables

1. Without prejudice to article 49, archipelagic States shall respect existing agreements with other States and shall recognize traditional fishing and other legitimate activities of the immediately adjacent neighboring States in certain areas falling within archipelagic waters. The terms and conditions of the exercise of such rights and activities, including the nature, the extent and the areas to which they apply, shall, at the request of any of the States concerned, be regulated by bilateral agreements between them. Such rights shall not be transferred to or shared with third States or their nationals.
2. Archipelagic States shall respect existing submarine cables laid by other States and passing through their waters without without making a landfall. Archipelagic States shall permit the maintenance and replacement of such cables upon receiving due notice of the location of such cables and the intention to repair or replace them.

Article 52
Right of innocent passage

1. Subject to article 53 and without prejudice to article 50, ships of all States enjoy the right of innocent passage through archipelagic waters, in accordance with section 3 of Part II.
2. The archipelagic State may, without discrimination in form or in fact amongst foreign ships, suspend temporarily in specified areas of its archipelagic waters the innocent passage of foreign ships if such suspension is essential for the protection of its security. Such suspension shall take effect only after having been duly published.

Article 53
Right of archipelagic sea lanes passage

1. An archipelagic State may designate sea lanes and air routes thereabove, suitable for the safe, continuous and expeditious passage of foreign ships and aircraft through or over its archipelagic waters and the adjacent territorial sea.
2. All ships and aircraft enjoy the right of archipelagic sea lanes passage in such sea lanes and air routes.
3. Archipelagic sea lanes passage is the exercise in accordance with the present Convention of the rights of navigation and overflight in the normal mode solely for the purpose of continuous, expeditious and unobstructed transit between one part of the high seas or an exclusive economic zone and another part of the high seas or an exclusive economic zone.
4. Such sea lanes and air routes shall traverse the archipelagic waters and the adjacent territorial sea and shall include all normal passage routes used as routes for international navigation or overflight through the archipelagic waters and, within such routes, so far as ships are concerned, all normal navigational channels, provided that duplication of routes of similar convenience between the same entry and exit points shall not be necessary.
5. Sea lanes shall be defined by a series of continuous axis lines from the entry points of passage routes to the exit points. Ships and aircraft in archipelagic sea lanes passage shall not deviate more than 25 nautical miles to either side of such axis lines during passage, provided that ships and aircraft shall not navigate closer to the coasts than 10 per cent of the distance between the nearest points on islands bordering the sea lane.
6. An archipelagic State which designates sea lanes under this article may also prescribe traffic separation schemes for the safe passage of ships through narrow channels in such sea lanes.
7. An archipelagic State may, when circumstances require, after giving due publicity thereto, substitute other sea lanes or traffic separation schemes for any sea lanes or traffic separation schemes previously designated or prescribed by it.

8. Such sea lanes and traffic separation schemes shall conform to generally accepted international regulations.

9. In designating or substituting sea lanes or prescribing or substituting traffic separation schemes, an archipelagic State shall refer proposals to the competent international organization with a view to their adoption. Thē organization may adopt only such sea lanes and traffic separation schemes as may be agreed with the archipelagic State, after which the archipelagic State may designate, prescribe or substitute them.

10. The archipelagic State shall clearly indicate the axis of the sea lanes and the traffic separation schemes designated or prescribed by it on charts to which due publicity shall be given.

11. Ships in transit shall respect applicable sea lanes and traffic separation schemes established in accordance with this article.

12. If an archipelagic State does not designate sea lanes or air routes, the right of archipelagic sea lanes passage may be exercised through the routes normally used for international navigation.

Article 54
Duties of ships and aircraft during their passage, research and survey activities, duties of the archipelagic State and laws and regulations of the archipelagic State relating to archipelagic sea lanes passage

Articles 39, 40, 42 and 44 apply *mutatis mutandis* to archipelagic sea lanes passage.

PART V
EXCLUSIVE ECONOMIC ZONE
Article 55
Specific legal régime of the exclusive economic zone

The exclusive economic zone is an area beyond and adjacent to the territorial sea, subject to the specific legal régime established in this Part, under which the rights and jurisdictions of the coastal State and the rights and freedoms of other States are governed by the relevant provisions of the present Convention.

Article 56
Rights, jurisdiction and duties of the coastal State in the exclusive economic zone

1. In the exclusive economic zone, the coastal State has:
   (a) sovereign rights for the purpose of exploring and exploiting, conserving and managing the natural resources, whether living or non-living, of the sea-bed and subsoil and the superjacent waters, and with regard to other activities for the economic exploitation and exploration of the zone, such as the production of energy from the water, currents and winds;
   (b) jurisdiction as provided for in the relevant provisions of the present Convention with regard to:
      (i) the establishment and use of artificial islands, installations and structures;
      (ii) marine scientific research;
      (iii) the preservation of the marine environment;
   (c) other rights and duties provided for in the present Convention.

2. In exercising its rights and performing its duties under the present Convention in the exclusive economic zone, the coastal State shall have due regard to the rights and duties of other States and shall act in a manner compatible with the provisions of the present Convention.

3. The rights set out in this article with respect to the sea-bed and subsoil shall be exercised in accordance with Part VI.

Article 57
Breadth of the exclusive economic zone

The exclusive economic zone shall not extend beyond 200 nautical miles from the baselines from which the breadth of the territorial sea is measured.

Article 58
Rights and duties of other States in the exclusive economic zone

1. In the exclusive economic zone, all States, whether coastal or land-locked, enjoy, subject to the relevant provisions of the present Convention, the freedoms referred to in article 87 of navigation and overflight and of the laying of submarine cables and pipelines, and other internationally lawful uses of the sea related to these freedoms such as those associated with the operation of ships, aircraft and submarine cables and pipelines, and compatible with the other provisions of the present Convention.

2. Articles 88 to 115 and other pertinent rules of international law apply to the exclusive economic zone in so far as they are not incompatible with this Part.

3. In exercising their rights and performing their duties under the present Convention in the exclusive economic zone, States shall have due regard to the rights and duties of the coastal State and shall comply with the laws and regulations established by the coastal State in accordance with the provisions of this Convention and other rules of international law in so far as they are not incompatible with this Part.

Article 59
Basis for the resolution of conflicts regarding the attribution of rights and jurisdiction in the exclusive economic zone

In cases where the present Convention does not attribute rights or jurisdiction to the coastal State or to other States within the exclusive economic zone, and a conflict arises between the interests of the coastal State and any other State or States, the conflict should be resolved on the basis of equity and in the light of all the relevant circumstances, taking into account the respective importance of the interests involved to the parties as well as to the international community as a whole.

Article 60
Artificial islands, installations and structures in the exclusive economic zone

1. In the exclusive economic zone, the coastal State shall have the exclusive right to construct

and to authorize and regulate the construction, operation and use of:
  (a) Artificial islands;
  (b) Installations and structures for the purposes provided for in article 56 and other economic purposes;
  (c) Installations and structures which may interfere with the exercise of the rights of the coastal State in the zone.
2. The coastal State shall have exclusive jurisdiction over such artificial islands, installations and structures, including jurisdiction with regard to customs, fiscal, health, safety and immigration regulations.
3. Due notice must be given of the construction of such artificial islands, installations or structures, and permanent means for giving warning of their presence must be maintained. Any installations or structures which are abandoned or disused must be entirely removed.
4. The coastal State may, where necessary, establish reasonable safety zones around such artificial islands, installations and structures in which it may take appropriate measures to ensure the safety both of navigation and of the artificial islands, installations and structures.
5. The breadth of the safety zones shall be determined by the coastal State, taking into account applicable international standards. Such zones shall be designed to ensure that they are reasonably related to the nature and function of the artificial islands, installations or structures, and shall not exceed a distance of 500 metres around them, measured from each point of their outer edge, except as authorized by generally accepted international standards or as recommended by the appropriate international organizations.
6. All ships must respect these safety zones and shall comply with generally accepted international standards regarding navigation in the vicinity of artificial islands, installations, structures and safety zones. Due notice shall be given of the extent of safety zones.
7. Artificial islands, installations and structures and the safety zones around them may not be established where interference may be caused to the use of recognized sea lanes essential to international navigation.
8. Artificial islands, installations and structures have no territorial sea of their own and their presence does not affect the delimitation of the territorial sea, the exclusive economic zone or the continental shelf.

Article 61
Conservation of living resources

1. The coastal State shall determine the allowable catch of the living resources in its exclusive economic zone.
2. The coastal State, taking into account the best scientific evidence available to it, shall ensure through proper conservation and management measures that the maintenance of the living resources in the exclusive economic zone is not endangered by over-exploitation. As appropriate, the coastal State and relevant subregional, regional and global organizations shall co-operate to this end.
3. Such measures shall also be designed to maintain or restore populations of harvested species at levels which can produce the maximum sustainable yield, as qualified by relevant environmental and economic factors, including the economic needs of coastal fishing communities and the special requirements of developing countries, and taking into account fishing patterns, the interdependence of stocks and any generally recommended subregional, regional or global minimum standards.
4. In establishing such measures the coastal State shall take into consideration the effects on species associated with or dependent upon harvested species with a view to maintaining or restoring populations of such associated or dependent species above levels at which their reproduction may become seriously threatened.
5. Available scientific information, catch and fishing effort statistics, and other data relevant to the conservation of fish stocks shall be contributed and exchanged on a regular basis through subregional, regional and global organizations where appropriate and with participation by all States concerned, including States whose nationals are allowed to fish in the exclusive economic zone.

Article 62
Utilization of the living resources

1. The coastal State shall promote the objective of optimum utilization of the living resources in the exclusive economic zone without prejudice to article 61.
2. The coastal State shall determine its capacity to harvest the living resources of the exclusive economic zone. Where the coastal State does not have the capacity to harvest the entire allowable catch, it shall, through agreements or other arrangements and pursuant to the terms, conditions and regulations referred to in paragraph 4, give other States access to the surplus of the allowable catch.
3. In giving access to other States to its exclusive economic zone under this article, the coastal State shall take into account all relevant factors, including, *inter alia,* the significance of the living resources of the area to the economy of the coastal State concerned and its other national interests, the provisions of articles 69 and 70, the requirements of developing countries in the subregion or region in harvesting part of the surplus and the need to minimize economic dislocation in States whose nationals have habitually fished in the zone or which have made substantial efforts in research and identification of stocks.
4. Nationals of other States fishing in the exclusive economic zone shall comply with the conservation measures and with the other terms and conditions established in the regulations of the coastal State. These regulations

shall be consistent with the present Convention and may relate, *inter alia*, to the following:

(a) Licencing of fishermen, fishing vessels and equipment, including payment of fees and other forms of remuneration, which in the case of developing coastal States, may consist of adequate compensation in the field of financing, equipment and technology relating to the fishing industry;

(b) Determining the species which may be caught, and fixing quotas of catch, whether in relation to particular stocks or groups of stocks or catch per vessel over a period of time or to the catch by nationals of any State during a specified period;

(c) Regulating seasons and areas of fishing, the types, sizes and amount of gear, and the numbers, sizes and types of fishing vessels that may be used;

(d) Fixing the age and size of fish and other species that may be caught;

(e) Specifying information required of fishing vessels, including catch and effort statistics and vessel position reports;

(f) Requiring, under the authorization and control of the coastal State, the conduct of specified fisheries research programmes and regulating the conduct of such research, including the sampling of catches, disposition of samples and reporting of associated scientific data;

(g) The placing of observers or trainees on board such vessels by the coastal State;

(h) The landing of all or any part of the catch by such vessels in the ports of the coastal State;

(i) Terms and conditions relating to joint ventures or other co-operative arrangements;

(j) Requirements for training personnel and transfer of fisheries technology, including enhancement of the coastal State's capability of undertaking fisheries research;

(k) Enforcement procedures.

5. Coastal States shall give due notice of conservation and management regulations.

Article 63
Stocks occurring within the exclusive economic zones of two or more coastal States or both within the exclusive economic zone and in an area beyond and adjacent to it

1. Where the same stock or stocks of associated species occur within the exclusive economic zones of two or more coastal States, these States shall seek either directly or through appropriate subregional or regional organizations to agree upon the measures necessary to co-ordinate and ensure the conservation and development of such stocks without prejudice to the other provisions of this Part.

2. Where the same stock or stocks of associated species occur both within the exclusive economic zone and in an area beyond and adjacent to the zone, the coastal State and the States fishing for such stocks in the adjacent area shall seek either directly or through appropriate subregional or regional organizations to agree upon the measures necessary for the conservation of these stocks in the adjacent area.

Article 64
Highly migratory species

1. The coastal State and other States whose nationals fish in the region for the highly migratory species listed in annex I shall co-operate directly or through appropriate international organizations with a view to ensuring conservation and promoting the objective of optimum utilization of such species throughout the region, both within and beyond the exclusive economic zone. In regions where no appropriate international organization exists, the coastal State and other States whose nationals harvest these species in the region shall co-operate to establish such an organization and participate in its work.

2. The provisions of paragraph 1 apply in addition to the other provisions of this Part.

Article 65
Marine mammals

Nothing in the present Convention restricts the right of a coastal State or international organization, as appropriate, to prohibit, regulate and limit the exploitation of marine mammals. States shall co-operate either directly or through appropriate international organizations with a view to the protection and management of marine mammals.

Article 66
Anadromous stocks

1. States in whose rivers anadromous stocks originate shall have the primary interest in and responsibility for such stocks.

2. The State of origin of anadromous stocks shall ensure their conservation by the establishment of appropriate regulatory measures for fishing in all waters landwards of the outer limits of its exclusive economic zone and for fishing provided for in subparagraph (b) of paragraph 3. The State of origin may, after consultation with other States fishing these stocks, establish total allowable catches for stocks originating in its rivers.

3. (a) Fisheries for anadromous stocks shall be conducted only in the waters landwards of the outer limits of exclusive economic zones, except in cases where this provision would result in economic dislocation for a State other than the State of origin.

(b) The State of origin shall co-operate in minimizing economic dislocation in such other States fishing these stocks, taking into account the normal catch and the mode of operations of such States, and all the areas in which such fishing has occurred.

(c) States referred to in subparagraph (b), participating by agreement with the State of origin in measures to renew anadromous stocks, particularly by expenditures for that purpose, shall be given special consideration by the State of origin in the harvesting of stocks originating in its rivers.

(d) Enforcement of regulations regarding anadromous stocks beyond the exclusive economic zone shall be by agreement between the State of origin and the other States concerned.

4. In cases where anadromous stocks migrate into or through the waters landwards of the outer limits of the exclusive economic zone of a State other than the State of origin, such State shall co-operate with the State of origin with regard to the conservation and management of such stocks.

5. The State of origin of anadromous stocks and other States fishing these stocks shall make arrangements for the implementation of the provisions of this article, where appropriate, through regional organizations.

Article 67
Catadromous species

1. A coastal State in whose waters catadromous species spend the greater part of their life cycle shall have responsibility for the management of these species and shall ensure the ingress and egress of migrating fish.

2. Harvesting of catadromous species shall be conducted only in waters landwards of the outer limits of exclusive economic zones. When conducted in exclusive economic zones, harvesting shall be subject to this article and the other provisions of the present Convention concerning fishing in these zones.

3. In cases where catadromous fish migrate through the exclusive economic zone of another State or States, whether as juvenile or maturing fish, the management, including harvesting, of such fish shall be regulated by agreement between the State mentioned in paragraph 1 and the State or States concerned. Such agreement shall ensure the rational management of the species and take into account the responsibilities of the State mentioned in paragraph 1 for the maintenance of these species.

Article 68
Sedentary species

This Part does not apply to sedentary species as defined in paragraph 4 of article 77.

Article 69
Right of land-locked States

1. Land-locked States shall have the right to participate in the exploitation of the living resources of the exclusive economic zones of adjoining coastal States on an equitable basis, taking into account the relevant economic and geographical circumstances of all the States concerned. The terms and conditions of such participation shall be determined by the States concerned through bilateral, subregional or regional agreements. Developed land-locked States shall, however, be entitled to exercise their rights only within the exclusive economic zones of adjoining developed coastal States.

2. This article is subject to the provisions of articles 61 and 62.

3. Paragraph 1 is without prejudice to arrangements agreed upon in regions where the coastal States may grant to land-locked States of the same region equal or preferential rights for the exploitation of the living resources in the exclusive economic zones.

Article 70
Right of certain developing coastal States in a subregion or region

1. Developing coastal States which are situated in a subregion or region whose geographical peculiarities make such States particularly dependent for the satisfaction of the nutritional needs of their populations upon the exploitation of the living resources in the exclusive economic zones of their neighboring States and developing coastal States which can claim no exclusive economic zones of their own shall have the right to participate, on an equitable basis, in the exploitation of living resources in the exclusive economic zones of other States in a subregion or region.

2. The terms and conditions of such participation shall be determined by the States concerned through bilateral, subregional or regional agreements, taking into account the relevant economic and geographical circumstances of all the States concerned, including the need to avoid effects detrimental to the fishing communities or to the fishing industries of the States in whose zones the right of participation is exercised.

3. This article is subject to the provisions of articles 61 and 62.

Article 71
Non-applicability of articles 69 and 70

The provisions of articles 69 and 70 shall not apply in the case of a coastal State whose economy is overwhelmingly dependent on the exploitation of the living resources of its exclusive economic zone.

Article 72
Restrictions on transfer of rights

1. Rights provided under articles 69 and 70 to exploit living resources shall not be directly or indirectly transferred to third States or their nationals by lease or licence, by establishing joint collaboration ventures or in any other manner which has the effect of such transfer unless otherwise agreed upon by the States concerned.

2. The foregoing provision does not preclude the States concerned from obtaining technical or financial assistance from third States or international organizations in order to facilitate the exercise of the rights pursuant to articles 60 and 70, provided that it does not have the effects referred to in paragraph 1.

Article 73
Enforcement of laws and regulations of the coastal State

1. The coastal State may, in the exercise of its

sovereign rights to explore, exploit, conserve and manage the living resources in the exclusive economic zone, take such measures, including boarding, inspection, arrest and judicial proceedings, as may be necessary to ensure compliance with the laws and regulations enacted by it in conformity with the present Convention.
2. Arrested vessels and their crews shall be promptly released upon the posting of reasonable bond or other security.
3. Coastal State penalties for violations of fisheries regulations in the exclusive economic zone may not include imprisonment, in the absence of agreement to the contrary by the States concerned, or any other form of corporal punishment.
4. In cases of arrest or detention of foreign vessels the coastal State shall promptly notify, through appropriate channels, the flag State of the action taken and of any penalties subsequently imposed.

Article 74
Delimitation of the exclusive economic zone between adjacent or opposite States

1. The delimitation of the exclusive economic zone between adjacent or opposite States shall be effected by agreement in accordance with equitable principles, employing, where appropriate, the median or equidistance line, and taking account of all the relevant circumstances.
2. If no agreement can be reached within a reasonable period of time, the States concerned shall resort to the procedures provided for in Part XV.
3. Pending agreement or settlement, the States concerned shall make provisional arrangements, taking into account the provisions of paragraph 1.
4. For the purposes of the present Convention, "median or equidistance line" means the line every point of which is equidistant from the nearest points of the baselines from which the breadth of the territorial sea of each State is measured.
5. Where there is an agreement in force between the States concerned, questions relating to the delimitation of the exclusive economic zone shall be determined in accordance with the provisions of that agreement.

Article 75
Charts and lists of geographical co-ordinates

1. Subject to this Part, the outer limit lines of the exclusive economic zone and the lines of delimitation drawn in accordance with article 74 shall be shown on charts of a scale or scales adequate for determining them. Where appropriate, lists of geographical co-ordinates of points, specifying the geodetic datum, may be substituted for such outer limit lines or lines of delimitation.
2. The coastal State shall give due publicity to such charts or lists of geographical co-ordinates and shall deposit a copy of each such chart or list with the Secretary-General of the United Nations.

PART VI
CONTINENTAL SHELF

Article 76
Definition of the continental shelf

The continental shelf of a coastal state comprises the sea-bed and subsoil of the submarine areas that extend beyond its territorial sea throughout the natural prolongation of its land territory to the outer edge of the continental margin, or to a distance of 200 nautical miles from the baselines from which the breadth of the territorial sea is measured where the outer edge of the continental margin does not extend up to that distance.

Article 77
Rights of the coastal State over the continental shelf

1. The coastal State exercises over the continental shelf sovereign rights for the purpose of exploring it and exploiting its natural resources.
2. The rights referred to in paragraph 1 are exclusive in the sense that if the coastal State does not explore the continental shelf or exploit its natural resources, no one may undertake these activities without the express consent of the coastal State.
3. The rights of the coastal State over the continental shelf do not depend on the occupation, effective or notional, or on any express proclamation.
4. The natural resources referred to in this Part consist of the mineral and other non-living resources of the sea-bed and subsoil together with living organisms belonging to sedentary species, that is to say, organisms which, at the harvestable stage, either are immobile on or under the sea-bed or are unable to move except in constant physical contact with the sea-bed or the subsoil.

Article 78
Superjacent waters and air space

The rights of the coastal State over the continental shelf do not affect the legal status of the superjacent waters or of the air space above those waters.

Article 79
Submarine cables and pipelines on the continental shelf

1. All States are entitled to lay submarine cables and pipelines on the continental shelf, in accordance with the provisions of this article.
2. Subject to its right to take reasonable measures for the exploration of the continental shelf, the exploitation of its natural resources and the prevention, reduction and control of pollution from pipelines, the coastal State may not impede the laying or maintenance of such cables or pipelines.
3. The delineation of the course for the laying of such pipelines on the continental shelf is subject to the consent of the coastal State.
4. Nothing in this Part affects the right of the

coastal State to establish conditions for cables or pipelines entering its territory or territorial sea, or its jurisdiction over cables and pipelines constructed or used in connexion with the exploration of its continental shelf or exploitation of its resources or the operations of artificial islands, installations and structures under its jurisdiction.

5. When laying submarine cables or pipelines, States shall pay due regard to cables or pipelines already in position. In particular, possibilities of repairing existing cables or pipelines shall not be prejudiced.

Article 80
Artificial islands, installations and structures on the continental shelf

Article 60 applies *mutatis mutandis* to artificial islands, installations and structures on the continental shelf.

Article 81
Drilling on the continental shelf

The coastal State shall have the exclusive right to authorize and regulate drilling on the continental shelf for all purposes.

Article 82
Payments and contributions with respect to the exploitation of the continental shelf beyond 200 miles

1. The coastal State shall make payments or contributions in kind in respect of the exploitation of the non-living resources of the continental shelf beyond 200 nautical miles from the baselines from which the breadth of the territorial sea is measured.
2. The payments and contributions shall be made annually with respect to all production at a site after the first five years of production at that site. For the sixth year, the rate of payment or contribution shall be one per cent of the value or volume of production at the site. The rate shall increase by one per cent for each subsequent year until the tenth year and shall remain at five per cent thereafter. Production does not include resources used in connexion with exploitation.
3. A developing country which is a net importer of a mineral resource produced from its continental shelf is exempt from making such payments or contributions in respect of that mineral resource.
4. The payments or contributions shall be made through the Authority, which shall distribute them to States Parties to the present Convention on the basis of equitable sharing criteria, taking into account the interests and needs of developing countries, particularly the least developed and the land-locked amongst them.

Article 83
Delimitation of the continental shelf between adjacent or opposite States

1. The delimitation of the continental shelf between adjacent or opposite States shall be effected by agreement in accordance with equitable principles, employing, where appropriate, the median or equidistance line, and taking account of all the relevant circumstances.
2. If no agreement can be reached within a reasonable period of time, the States concerned shall resort to the procedures provided for in Part XV.
3. Pending agreement or settlement, the States concerned shall make provisional arrangements, taking into account the provisions of paragraph 1.
4. Where there is an agreement in force between the States concerned, questions relating to the delimitation of the continental shelf shall be determined in accordance with the provisions of that agreement.

Article 84
Charts and lists of geographical co-ordinates

1. Subject to this Part the outer limit lines of the continental shelf and the lines of delimitation drawn in accordance with article 82 shall be shown on charts of a scale or scales adequate for determining them. Where appropriate, lists of geographical co-ordinates of points, specifying the geodetic datum, may be substituted for such outer limit lines or lines of delimitation.
2. The coastal State shall give due publicity to such charts or lists of geographical co-ordinates and shall deposit a copy of each such chart or list with the Secretary-General of the United Nations.

Article 85
Tunnelling

This Part does not prejudice the right of the coastal State to exploit the subsoil by means of tunnelling, irrespective of the depth of water above the subsoil.

PART VII
HIGH SEAS
SECTION 1. GENERAL
Article 86
Application of the provisions of this Part

The provisions of this Part apply to all parts of the sea that are not included in the exclusive economic zone, in the territorial sea or in the internal waters of a State, or in the archipelagic waters of an archipelagic State. This article does not entail any abridgement of the freedoms enjoyed by all States in the exclusive economic zone in accordance with article 58.

Article 87
Freedom of the high seas

1. The high seas are open to all States, whether coastal or land-locked. Freedom of the high seas is exercised under the conditions laid down by the present Convention and by other rules of international law. It comprises, *inter alia*, both for coastal and land-locked States:
   (a) Freedom of navigation;
   (b) Freedom of overflight;
   (c) Freedom to lay submarine cables and pipelines, subject to Part VI;
   (d) Freedom to construct artificial islands

and other installations permitted under international law, subject to Part VI;
(e) Freedom of fishing, subject to the conditions laid down in section 2;
(f) Freedom of scientific research, subject to Parts VI and XIII.
2. These freedoms shall be exercised by all States, with due consideration for the interests of other States in their exercise of the freedom of the high seas, and also with due consideration for the rights under the present Convention with respect to activities in the Area.

Article 88
Reservation of the high seas for peaceful purposes
The high seas shall be reserved for peaceful purposes.

Article 89
Invalidity of claims of sovereignty over the high seas
No State may validly purport to subject any part of the high seas to its sovereignty.

Article 90
Right of navigation
Every State, whether coastal or landlocked, has the right to sail ships under its flag on the high seas.

Article 91
Nationality of ships
1. Each State shall fix the conditions for the grant of its nationality to ships, for the registration of ships in its territory, and for the right to fly its flag. Ships have the nationality of the State whose flag they are entitled to fly. There must exist a genuine link between the State and the ship.
2. Each State shall issue to ships to which it has granted the right to fly its flag documents to that effect.

Article 92
Status of ships
1. Ships shall sail under the flag of one State only and, save in exceptional cases expressly provided for in international treaties or in the present Convention, shall be subject to its exclusive jurisdiction on the high seas. A ship may not change its flag during a voyage or while in a port of call, save in the case of a real transfer of ownership or change of registry.
2. A ship which sails under the flags of two or more States, using them according to convenience, may not claim any of the nationalities in question with respect to any other State, and may be assimilated to a ship without nationality.

Article 93
Ships flying the flag of the United Nations, its specialized agencies and the International Atomic Energy Agency
The preceding articles do not prejudice the question of ships employed on the official service of the United Nations, its specialized agencies or the International Atomic Energy Agency, flying the flag of the Organization.

Article 94
Duties of the flag States
1. Every State shall effectively exercise its jurisdiction and control in administrative, technical and social matters over ships flying its flag.
2. In particular every State shall:
(a) Maintain a register of shipping containing the names and particulars of ships flying its flag, except those which are excluded from generally accepted international regulations on account of their small size; and
(b) Assume jurisdiction under its internal law over each ship flying its flag and its master, officers and crew in respect of administrative, technical and social matters concerning the ship.
3. Every State shall take such measures for ships flying its flag as are necessary to ensure safety at sea with regard, *inter alia*, to:
(a) The construction, equipment and seaworthiness of ships;
(b) The manning of ships, labour conditions and the training of crews, taking into account the applicable international instruments;
(c) The use of signals, the maintenance of communications and the prevention of collisions.
4. Such measures shall include those necessary to ensure:
(a) That each ship, before registration and thereafter at appropriate intervals, is surveyed by a qualified surveyor of ships, and has on board such charts, nautical publications and navigational equipment and instruments as are appropriate for the safe navigation of the ship;
(b) That each ship is in the charge of a master and officers who possess appropriate qualifications, in particular in seamanship, navigation, communications, and marine engineering, and that the crew is appropriate in qualification and numbers for the type, size, machinery and equipment of the ship;
(c) That the master, officers and, to the extent appropriate, the crew are fully conversant with and required to observe the applicable international regulations concerning the safety of life at sea, the prevention of collisions, the prevention, reduction and control of marine pollution, and the maintenance of communications by radio.
5. In taking the measures called for in paragraphs 3 and 4 each State is required to conform to generally accepted international regulations, procedures and practices and to take any steps which may be necessary to secure their observance.
6. A State which has clear grounds to believe that proper jurisdiction and control with respect to a ship have not been exercised may report the facts to the flag State. Upon receiving such a report, the flag State shall investigate the matter and, if appropriate, take any action necessary to remedy the situation.
7. Each State shall cause an inquiry to be held by or before a suitably qualified person or

persons into every marine casualty or incident of navigation on the high seas involving a ship flying its flag and causing loss of life or serious injury to nationals of another State or serious damage to shipping or installations of another State or to the marine environment. The flag State and the other State shall co-operate in the conduct of any inquiry held by that other State into any such marine casualty or incident of navigation.

Article 95
Immunity of warships on the high seas

Warships on the high seas have complete immunity from the jurisdiction of any State other than the flag State.

Article 96
Immunity of ships used only on government non-commercial service

Ships owned or operated by a State and used only on government non-commercial service shall, on the high seas, have complete immunity from the jurisdiction of any State other than the flag State.

Article 97
Penal jurisdiction in matters of collision

1. In the event of a collision or any other incident of navigation concerning a ship on the high seas, involving the penal or disciplinary responsibility of the master or of any other person in the service of the ship, no penal or disciplinary proceedings may be instituted against such person except before the judicial or administrative authorities either of the flag State or of the State of which such person is a national.
2. In disciplinary matters, the State which has issued a master's certificate or a certificate of competence or licence shall alone be competent, after due legal process, to pronounce the withdrawal of such certificates, even if the holder is not a national of the State which issued them.
3. No arrest or detention of the ship, even as a measure of investigation, shall be ordered by any authorities other than those of the flag State.

Article 98
Duty to render assistance

1. Every State shall require the master of a ship sailing under its flag, in so far as he can do so without serious danger to the ship, the crew or the passengers:
    (a) To render assistance to any person found at sea in danger of being lost;
    (b) To proceed with all possible speed to the rescue of persons in distress, if informed of their need of assistance, in so far as such action may reasonably be expected of him;
    (c) After a collision, to render assistance to the other ship, its crew and its passengers and, where possible, to inform the other ship of the name of his own ship, its port of registry and the nearest port at which it will call.
2. Every coastal State shall promote the establishment, operation and maintenance of an adequate and effective search and rescue service regarding safety on and over the sea and, where circumstances so require, by way of mutual regional arrangements co-operate with neighbouring States for this purpose.

Article 99
Prohibition of the transport of slaves

Every State shall adopt effective measures to prevent and punish the transport of slaves in ships authorized to fly its flag and to prevent the unlawful use of its flag for that purpose. Any slave taking refuge on board any ship, whatever its flag, shall, *ipso facto*, be free.

Article 100
Duty to co-operate in the repression of piracy

All States shall co-operate to the fullest possible extent in the repression of piracy on the high seas or in any other place outside the jurisdiction of any State.

Article 101
Definition of piracy

Piracy consists of any of the following acts:
(a) Any illegal acts of violence, detention or any act of depredation, committed for private ends by the crew or the passengers of a private ship or a private aircraft, and directed:
 (i) On the high seas, against another ship or aircraft, or against persons or property on board such ship or aircraft;
 (ii) Against a ship, aircraft, persons or property in a place outside the jurisdiction of any State;
(b) Any act of voluntary participation in the operation of a ship or of an aircraft with knowledge of facts making it a pirate ship or aircraft;
(c) Any act of inciting or of intentionally facilitating an act described in subparagraphs (a) and (b).

Article 102
Piracy by a warship, government ship or government aircraft whose crew has mutinied

The acts of piracy, as defined in article 101, committed by a warship, government ship or government aircraft whose crew has mutinied and taken control of the ship or aircraft are assimilated to acts committed by a private ship.

Article 103
Definition of a pirate ship or aircraft

A ship or aircraft is considered a pirate ship or aircraft if it is intended by the persons in dominant control to be used for the purpose of committing one of the acts referred to in article 101. The same applies if the ship or aircraft has been used to commit any such act, so long as it remains under the control of the persons guilty of that act.

Article 104
Retention or loss of the nationality of a pirate ship or aircraft

A ship or aircraft may retain its nationality although it has become a pirate ship or aircraft. The retention or loss of nationality is determined by the law of the State from which such nationality was derived.

Article 105
Seizure of a pirate ship or aircraft

On the high seas, or in any other place outside the jurisdiction of any State, every State may seize a pirate ship or aircraft, or a ship taken by piracy and under the control of pirates, and arrest the persons and seize the property on board. The courts of the State which carried out the seizure may decide upon the penalties to be imposed, and may also determine the action to be taken with regard to the ships, aircraft or property, subject to the rights of third parties acting in good faith.

Article 106
Liability for seizure without adequate grounds

Where the seizure of a ship or aircraft on suspicion of piracy has been effected without adequate grounds, the State making the seizure shall be liable to the State the nationality of which is possessed by the ship or aircraft, for any loss or damage caused by the seizure.

Article 107
Ships and aircraft which are entitled to seize on account of piracy

A seizure on account of piracy may only be carried out by warships or military aircraft, or other ships or aircraft clearly marked and identifiable as being on government service and authorized to that effect.

Article 108
Illicit traffic in narcotic drugs or psychotropic substances

1. All States shall co-operate in the suppression of illicit traffic in narcotic drugs and psychotropic substances by ships on the high seas contrary to international conventions.
2. Any State which has reasonable grounds for believing that a vessel flying its flag is engaged in illicit traffic in narcotic drugs or psychotropic substances may request the co-operation of other States to suppress such traffic.

Article 109
Unauthorized broadcasting from the high seas

1. All States shall co-operate in the suppression of unauthorized broadcasting from the high seas.
2. Any person engaged in unauthorized broadcasting from the high seas may be prosecuted before the court of the flag State of the vessel, the place of registry of the installation, the State of which the person is a national, any place where the transmissions can be received or any State where authorized radio communications is suffering interference.
3. On the high seas, a State having jurisdiction in accordance with paragraph 2 may, in conformity with article 110, arrest any person or ship engaged in unauthorized broadcasting and seize the broadcasting apparatus.
4. For the purposes of the present Convention, "unauthorized broadcasting" means the transmission of sound radio or television broadcasts from a ship or installation on the high seas intended for reception by the general public contrary to international regulations, but excluding the transmission of distress calls.

Article 110
Right of visit

1. Except where acts of interference derive from powers conferred by treaty, a warship which encounters on the high seas a foreign ship, other than a ship entitled to complete immunity in accordance with articles 95 and 96, is not justified in boarding her unless there is reasonable ground for suspecting:
   (a) That the ship is engaged in piracy;
   (b) That the ship is engaged in the slave trade;
   (c) That the ship is engaged in unauthorized broadcasting and the warship has jurisdiction under article 109;
   (d) That the ship is without nationality; or
   (e) That, though flying a foreign flag or refusing to show its flag, the ship is, in reality, of the same nationality as the warship.
2. In the cases provided for in paragraph 1, the warship may proceed to verify the ship's right to fly its flag. To this end, it may send a boat, under the command of an officer, to the suspected ship. If suspicion remains after the documents have been checked, it may proceed to a further examination on board the ship, which must be carried out with all possible consideration.
3. If the suspicions prove to be unfounded, and provided that the ship boarded has not committed any act justifying them, it shall be compensated for any loss or damage that may have been sustained.
4. These provisions shall apply *mutatis mutandis* to military aircraft.
5. These provisions shall also apply to any other duly authorized ships or aircraft clearly marked and identifiable as being on government service.

Article 111
Right of hot pursuit

1. The hot pursuit of a foreign ship may be undertaken when the competent authorities of the coastal State have good reason to believe that the ship has violated the laws and regulations of that State. Such pursuit must be commenced when the foreign ship or one of its boats is within the internal waters, the territorial sea or the contiguous zone of the pursuing State, and may only be continued outside the territorial sea or the contiguous zone if the pursuit has not been interrupted. It is not necessary, that at the time when the foreign ship within the territorial sea or the contiguous zone receives the order to stop, the ship giving the order should likewise be within the territorial sea or the contiguous zone. If the foreign ship is within a contiguous zone, as defined in article 33, the pursuit may only be undertaken if there has been a violation of the rights for the protection of which the zone was established.
2. The right of hot pursuit shall apply *mutatis mutandis* to violations in the exclusive economic zone or on the continental shelf, including

safety zones around continental shelf installations, of the laws and regulations of the coastal State applicable in accordance with the present Convention to the exclusive economic zone or the continental shelf, including such safety zones.

3. The right of hot pursuit ceases as soon as the ship pursued enters the territorial sea of its own country or of a third State.

4. Hot pursuit is not deemed to have begun unless the pursuing ship has satisfied itself by such practicable means as may be available that the ship pursued or one of its boats or other craft working as a team and using the ship pursued as a mother ship are within the limits of the territorial sea, or as the case may be, within the contiguous zone or the exclusive economic zone or above the continental shelf. The pursuit may only be commenced after a visual or auditory signal to stop has been given at a distance which enables it to be seen or heard by the foreign ship.

5. The right of hot pursuit may be exercised only by warships or military aircraft, or other ships or aircraft clearly marked and identifiable as being on government service and specially authorized to that effect.

6. Where hot pursuit is effected by an aircraft:
(a) The provisions of paragraphs 1 to 4 shall apply *mutatis mutandis*;
(b) The aircraft giving the order to stop must itself actively pursue the ship until a ship or aircraft of the coastal State, summoned by the aircraft, arrives to take over the pursuit, unless the aircraft is itself able to arrest the ship. It does not suffice to justify an arrest outside the territorial sea that the ship was merely sighted by the aircraft as an offender or suspected offender, if it was not both ordered to stop and pursued by the aircraft itself or other aircraft or ships which continue the pursuit without interruption.

7. The release of a ship arrested within the jurisdiction of a State and escorted to a port of that State for the purposes of an inquiry before the competent authorities may not be claimed solely on the ground that the ship, in the course of its voyage, was escorted across a portion of the exclusive economic zone or the high seas, if the circumstances rendered this necessary.

8. Where a ship has been stopped or arrested outside the territorial sea in circumstances which do not justify the exercise of the right of hot pursuit, it shall be compensated for any loss or damage that may have been thereby sustained.

Article 112
Right to lay submarine cables and pipelines

1. All States shall be entitled to lay submarine cables and pipelines on the bed of the high seas beyond the continental shelf.
2. Paragraph 5 of article 79 applies to such cables and pipelines.

Article 113
Breaking or injury of a submarine cable or pipeline

Every State shall take the necessary legislative measures to provide that the breaking or injury by a ship flying its flag or by a person subject to its jurisdiction of a submarine cable beneath the high seas done wilfully or through culpable negligence, in such a manner as to be liable to interrupt or obstruct telegraphic or telephonic communications, and similarly the breaking or injury of a submarine pipeline or high-voltage power cable shall be a punishable offence. This provision shall apply also to conduct calculated or likely to result in such breaking or injury. However, it shall not apply to any break or injury caused by persons who acted merely with the legitimte object of saving their lives or their ships, after having taken all necessary precautions to avoid such break or injury.

Article 114
Breaking or injury by owners of a submarine cable or pipeline of another submarine cable or pipeline

Every State shall take the necessary legislative measures to provide that, if persons subject to its jurisdiction who are the owners of a cable or pipeline beneath the high seas, in laying or repairing that cable or pipeline, cause a break in or injury to another cable or pipeline, they shall bear the cost of the repairs.

Article 115
Indemnity for loss incurred in avoiding injury to a submarine cable or pipeline

Every State shall take the necessary legislative measures to ensure that the owners of ships who can prove that they have sacrificed an anchor, a net or any other fishing gear, in order to avoid injuring a submarine cable or pipeline, shall be indemnified by the owner of the cable or pipeline, provided that the owner of the ship has taken all reasonable precautionary measures beforehand.

SECTION 2. MANAGEMENT AND CONSERVATION OF THE LIVING RESOURCES OF THE HIGH SEAS

Article 116
Right to fish on the high seas

All States have the right for their nationals to engage in fishing on the high seas subject to:
(a) Their treaty obligations;
(b) The rights and duties as well as the interests of coastal States provided for, *inter alia*, in paragraph 2 of article 63 and articles 64 to 67; and
(c) The provisions of this section.

Article 117
Duty of States to adopt with respect to their nationals measures for the conservation of the living resources of the high seas

All States have the duty to adopt, or to co-operate with other States in adopting, such measures for their respective nationals as may be necessary for the conservation of the living resources of the high seas.

Article 118
Co-operation of States in the management and conservation of living resources

States shall co-operate with each other in

the management and conservation of living resources in the areas of the high seas. States whose nationals exploit identical resources, or different resources in the same area, shall enter into negotiations with a view to adopting the means necessary for the conservation of the living resources concerned. They shall, as appropriate, co-operate to establish subregional or regional fisheries organizations to this end.

Article 119
Conservation of the living resources of the high seas
1. In determining the allowable catch and establishing other conservation measures for the living resources in the high seas, States shall:
 (a) Adopt measures which are designed, on the best scientific evidence available to the States concerned, to maintain or restore populations of harvested species at levels which can produce the maximum sustainable yield, as qualified by relevant environmental and economic factors, including the special requirements of developing countries, and taking into account fishing patterns, the interdependence of stocks and any generally recommended subregional, regional or global minimum standards;
 (b) Take into consideration the effects on species associated with or dependent upon harvested species with a view to maintaining or restoring populations of such associated or dependent species above levels at which their reproduction may become seriously threatened.
2. Available scientific information, catch and fishing effort statistics, and other data relevant to the conservation of fish stocks shall be contributed and exchanged on a regular basis through subregional, regional and global organizations where appropriate and with participation by all States concerned.
3. States concerned shall ensure that conservation measures and their implementation do not discriminate in form or in fact against the fishermen of any State.

Article 120
Marine mammals
Article 65 also applies to the conservation and management of marine mammals in the high seas.

PART VIII
REGIME OF ISLANDS
Article 121
Régime of islands
1. An island is a naturally formed area of land, surrounded by water, which is above water at high tide.
2. Except as provided for in paragraph 3, the territorial sea, the contiguous zone, the exclusive economic zone and the continental shelf of an island are determined in accordance with the provisions of the present Convention applicable to other land territory.
3. Rocks which cannot sustain human habitation or economic life of their own shall have no exclusive economic zone or continental shelf.

PART IX
ENCLOSED OR SEMI-ENCLOSED SEAS
Article 122
Definition
For the purposes of this Part, "enclosed or semi-enclosed sea" means a gulf, basin, or sea surrounded by two or more States and connected to the open seas by a narrow outlet or consisting entirely or primarily of the territorial seas and exclusive economic zones of two or more coastal States.

Article 123
Co-operation of States bordering enclosed or semi-enclosed seas
States bordering enclosed or semi-enclosed seas should co-operate with each other in the exercise of their rights and duties under the present Convention. To this end they shall endeavour, directly or through an appropriate regional organization:
 (a) To co-ordinate the management, conservation, exploration and exploitation of the living resources of the sea;
 (b) To co-ordinate the implementation of their rights and duties with respect to the preservation of the marine environment;
 (c) To co-ordinate their scientific research policies and undertake where appropriate joint programmes of scientific research in the area;
 (d) To invite, as appropriate, other interested States or international organizations to co-operate with them in furtherance of the provisions of this article.

PART X
RIGHT OF ACCESS OF LAND-LOCKED STATES TO AND FROM THE SEA AND FREEDOM OF TRANSIT
Article 124
Use of terms
1. For the purposes of the present Convention:
 (a) "Land-locked State" means a State which has no seacoast;
 (b) "Transit State" means a State, with or without a seacoast, situated between a land-locked State and the sea through whose territory "traffic in transit" passes;
 (c) "Traffic in transit" means transit of persons, baggage, goods and means of transport across the territory of one or more transit States, when the passage across such territory, with or without trans-shipment, warehousing, breaking bulk or change in the mode of transport, is only a portion of a complete journey which begins or terminates within the territory of the land-locked State;
 (d) "Means of transport" means
  (i) Railway rolling stock, sea, lake and river craft and road vehicles;
  (ii) Where local conditions so require, porters and pack animals.
2. Land-locked States and transit States may, by agreement between them, include as means of transport pipelines and gas lines and means of transport other than those included in paragraph 1.

Article 125
Right of access to and from the sea and freedom of transit

1. Land-locked States shall have the right of access to and from the sea for the purpose of exercising the rights provided for in the present Convention including those relating to the freedom of the high seas and the common heritage of mankind. To this end, land-locked States shall enjoy freedom of transit through the territories of transit States by all means of transport.
2. The terms and modalities for exercising freedom of transit shall be agreed between the land-locked States and the transit States concerned through bilateral, subregional or regional agreements.
3. Transit States, in the exercise of their full sovereignty over their territory, shall have the right to take all necessary measures to ensure that the rights and facilities, provided for in this Part for land-locked States shall in no way infringe their legitimate interests.

Article 126
Exclusion of application of the most-favoured-nation clause

Provisions of the present Convention, as well as special agreements relating to the exercise of the right of access to and from the sea, establishing rights and facilities on account of the special geographical position of land-locked States, are excluded from the application of the most-favoured-nation clause.

Article 127
Customs duties, taxes and other charges

1. Traffic in transit shall not be subject to any customs duties, taxes or other charges except charges levied for specific services rendered in connexion with such traffic.
2. Means of transport in transit and other facilities provided for and used by land-locked States shall not be subject to taxes or charges higher than those levied for the use of means of transport of the transit State.

Article 128
Free zones and other customs facilities

For the convenience of traffic in transit, free zones or other customs facilities may be provided at the ports of entry and exit in the transit States, by agreement between those States and the land-locked States.

Article 129
Co-operation in the construction and improvement of means of transport

Where there are no means of transport in the transit States to give effect to the freedom of transit or where the existing means, including the port installations and equipment, are inadequate in any respect, transit States and the land-locked States concerned may co-operate in constructing or improving them.

Article 130
Measures to avoid or eliminate delay or other difficulties of a technical nature in traffic in transit

1. Transit States shall take all appropriate measures to avoid delays or other difficulties of a technical nature in traffic in transit.
2. Should such delays or difficulties occur, the competent authorities of the transit States and of land-locked States shall co-operate towards their expeditious elimination.

Article 131
Equal treatment in maritime ports

Ships flying the flag of land-locked States shall enjoy treatment equal to that accorded to other foreign ships in maritime ports.

Article 132
Grant of greater transit facilities

The present Convention does not entail in any way the withdrawal of transit facilities which are greater than those provided for in the present Convention and which are agreed between States Parties to the present Convention or granted by a State Party. The present Convention also does not preclude such grant of greater facilities in the future.

PART XI
THE AREA
SECTION 1. GENERAL

Article 133
Use of terms

For the purpose of this Part of the present Convention:

(a) "Activities in the Area" means all activities of exploration for, and exploitation of, the resources of the Area.

(b) "Resources" means mineral resources *in situ*. When recovered from the Area, such resources shall, for the purposes of this Part of the present Convention, be regarded as minerals.

(c) Minerals shall include the following categories:
   (i) Liquid or gaseous substances such as petroleum, gas, condensate, helium, nitrogen, carbon dioxide, water, steam, hot water, and also sulphur and salts extracted in liquid form in solution;
   (ii) Useful minerals occurring on the surface of the sea-bed or at depths of less than three metres beneath the surface and also concretions of phosphorites and other minerals;
   (iii) Solid minerals in the ocean floor at depths of more than three metres from the surface;
   (iv) Ore-bearing silt and brine.

Article 134
Scope of this Part

1. This Part of the present Convention shall apply to the "Area".
2. States Parties shall notify the Authority established pursuant to article 154, of the limits referred to in paragraph 1 of this article and determined by co-ordinates of latitude and longitude and shall indicate the same on appropriate large-scale charts officially recognized by that State.
3. The Authority shall register and publish

such notification in accordance with rules adopted by it for the purpose.
4. Nothing in this article shall affect the validity of any agreement between States with respect to the establishment of limits between opposite or adjacent States.
5. Activities in the Area shall be governed by the provisions of this Part of the present Convention.

Article 135
Legal status of the superjacent waters and air space
Neither the provisions of this Part of the present Convention nor any rights granted or exercised pursuant thereto shall affect the legal status of the waters superjacent to the Area or that of the air space above those waters.

SECTION 2. PRINCIPLES GOVERNING THE AREA
Article 136
Common heritage of mankind
The Area and its resources are the common heritage of mankind.

Article 137
Legal status of the Area and its resources
1. No State shall claim or exercise sovereignty or sovereign rights over any part of the Area or its resources, nor shall any State or person, natural or juridical, appropriate any part thereof. No such claim or exercise of sovereignty or sovereign rights, nor such appropriation shall be recognized.
2. All rights in the resources of the Area are vested in mankind as a whole, on whose behalf the Authority shall act. These resources are not subject to alienation. The minerals derived from the Area, however, may only be alienated in accordance with this Part of the present Convention and the rules and regulations adopted thereunder.
3. No State or person, natural or juridical, shall claim, acquire or exercise rights with respect to the minerals of the Area except in accordance with the provisions of this Part of the present Convention. Otherwise, no such claim, acquisition or exercise of such rights shall be recognized.

Article 138
General conduct of States in relation to the Area
The general conduct of States in relation to the Area shall be in accordance with the provisions of this Part of the present Convention, and other pertinent rules of international law, including the Charter of the United Nations, in the interests of maintaining peace and security and promoting international co-operation and mutual understanding.

Article 139
Responsibility to ensure compliance and liability for damage
1. States Parties shall have the responsibility to ensure that activities in the Area, whether undertaken by States Parties, or state enterprises, or persons natural or juridical which possess the nationality of States Parties or are effectively controlled by them or their nationals, shall be carried out in conformity with the provisions of this Part of the present Convention. The same responsibility applies to international organizations for activities in the Area undertaken by such organizations. Without prejudice to applicable principles of international law and paragraph 16 of annex II, damage caused by the failure of a State Party to carry out its responsibilities under this Part of the present Convention shall entail liability. A State Party shall not however be liable for damage caused by any failure to comply by a person whom it has sponsored under article 151, paragraph 2 (ii) if the State Party has taken all necessary and appropriate measures to secure effective compliance under article 151, paragraph 4.
2. A group of States Parties or a group of international organizations, acting together, shall be jointly and severally responsible under these articles.
3. States Parties shall take appropriate measures to ensure that the responsibility provided for in paragraph 1 of this article shall apply *mutatis mutandis* to international organizations.

Article 140
Benefit of mankind
Activities in the Area shall be carried out for the benefit of mankind as a whole, irrespective of the geographical location of States, whether coastal or land-locked, and taking into particular consideration the interests and needs of the developing countries as specifically provided for in this Part of the present Convention.

Article 141
Use of the Area exclusively for peaceful purposes
The Area shall be open to use exclusively for peaceful purposes by all States, whether coastal or land-locked, without discrimination and without prejudice to the other provisions of this Part of the present Convention.

Article 142
Rights and legitimate interests of coastal States
1. Activities in the Area, with respect to resource deposits in the Area which lie across limits of national jurisdiction, shall be conducted with due regard to the rights and legitimate interests of any coastal State across whose jurisdiction such resources lie.
2. Consultations, including a system of prior notification, shall be maintained with the State concerned, with a view to avoiding infringement of such rights and interests. In cases where activities in the Area may result in the exploitation of resources lying within national jurisdiction, the prior consent of the coastal State concerned shall be required.
3. Neither the provisions of this Part of the present Convention nor any rights granted or exercised pursuant thereto shall · affect the rights of coastal States to take such measures consistent with the relevant provisions of Part XII of the present Convention as may be necessary to prevent, mitigate or eliminate grave and imminent danger to their coastlines

or related interests from pollution or threat thereof or from other hazardous occurrences resulting from or caused by any activities in the Area.

SECTION 3. CONDUCT OF ACTIVITIES IN THE AREA

Article 143
Marine scientific research

1. Marine scientific research in the Area shall be carried out exclusively for peaceful purposes and for the benefit of mankind as a whole, in accordance with Part XII of the present Convention.
2. States Parties shall promote international co-operation in marine scientific research in the Area exclusively for peaceful purposes by:
   (a) Participation in international programmes and encouraging co-operation in marine scientific research by personnel of different countries and of the Authority;
   (b) Ensuring that programmes are developed through the Authority or other international bodies as appropriate for the benefit of developing countries and technologically less developed countries with a view to
   (i) Strengthening their research capabilities;
   (ii) Training their personnel and the personnel of the Authority in the techniques and applications of research;
   (iii) Fostering the employment of their qualified personnel in activities of research in the Area;
   (c) Effective dissemination of the results of research and analysis when available, through the Authority or other international channels when appropriate.

Article 144
Transfer of technology

The authority and States Parties shall co-operate in promoting the transfer of technology and scientific knowledge relating to activities in the Area so that the Enterprise and all States benefit therefrom. In particular they shall initiate and promote:
   (a) Programmes for the transfer of technology to the Enterprise and to developing countries with regard to activities in the Area, including, *inter alia,* facilitating the access of the Enterprise and of developing countries to the relevant technology, under fair and reasonable terms and conditions;
   (b) Measures directed towards the advancement of the technology of the Enterprise and the domestic technology of developing countries, particularly through the opening of opportunities to personnel from the Enterprise and from developing countries for training in marine science and technology and their full participation in activities in the Area.

Article 145
Protection of the marine environment

With respect to activities in the Area, necessary measures shall be taken in order to ensure effective protection for the marine environment from harmful effects which may arise from such activities in accordance with Part XII of the present Convention:
   (a) The prevention of pollution and contamination, and other hazards to the marine environment, including the coastline, and of interference with the ecological balance of the marine environment, particular attention being paid to the need for protection from the consequences of such activities as drilling, dredging, excavation, disposal of waste, construction and operation or maintenance of installations, pipelines and other devices related to such activities;
   (b) The protection and conservation of the natural resources of the Area and the prevention of damage to the flora and fauna of the marine environment.

Article 146
Protection of human life

With respect to activities in the Area, necessary measures shall be taken in order to ensure effective protection of human life. To that end the Authority shall adopt appropriate rules, regulations and procedures to supplement existing international law as reflected in specific treaties which may be applicable.

Article 147
Accommodation of activities in the Area and in the marine environment

1. Activities in the Area shall be carried out with reasonable regard for other activities in the marine environment.
2. Stationary and mobile installations relating to the conduct of activities in the Area shall be subject to the following conditions:
   (i) Such installations shall be erected, emplaced and removed solely in accordance with the provisions of this Part of the present Convention and subject to rules and regulations adopted by the Authority. The erection, emplacement and removal of such installations shall be the subject of timely notification through Notices to Mariners or other generally recognized means of notification;
   (ii) Such installations shall not be located in the Area where they may obstruct passage through sea lanes of vital importance for international shipping or in areas of intense fishing activity;
   (iii) Safety zones shall be established around such installations with appropriate markings to ensure the safety both of the installations themselves and of shipping. The configuration and location of such safety zones shall not be such as to form a belt impeding the lawful access of shipping to particular maritime zones or navigation along international sea lanes;
   (iv) Such installations shall be used exclusively for peaceful purposes;
   (v) Such installations shall not possess the status of islands. They shall have no territorial sea, nor shall their presence

affect the determination of territorial or jurisdictional limits of any kind.

3. Other activities in the marine environment shall be conducted with reasonable regard for activities in the Area.

Article 148
Participation of developing countries in activities in the Area

The effective participation of developing countries in the activities in the Area shall be promoted as specifically provided for in this Part of the present Convention, having due regard to their special needs and interests, and in particular, the special needs of the landlocked and geographically disadvantaged States among them in overcoming obstacles arising from their disadvantaged location, including access to and from the Area.

Article 149
Archaeological and historical objects

All objects of an archaeological and historical nature found in the Area shall be preserved or disposed of for the benefit of the international community as a whole, particular regard being paid to the preferential rights of the State or country of origin, or the State of cultural origin, or the State of historical and archaeological origin.

SECTION 4. DEVELOPMENT OF RESOURCES OF THE AREA
Article 150
Policies relating to activities in the Area

1. Activities in the Area shall be carried out in accordance with the provisions of this Part of the present Convention in such a manner as to foster healthy development of the world economy and balanced growth of international trade, and to promote international co-operation for the over-all development of all countries, especially the developing countries and specifically so as to ensure:

(a) orderly and safe development and rational management of the resources of the Area, as well as the efficient conduct of activities in the Area in accordance with sound principles of conservation and the avoidance of waste;

(b) the expanding of opportunities for participation in such activities consistent with articles 144 and 148;

(c) transfer of revenues and technology to the Authority;

(d) just, stable and remunerative prices for raw materials originating in the Area which are also produced outside the Area, and increasing availability of those minerals so as to promote equilibrium between supply and demand;

(e) security of supplies to consumers of raw materials originating in the Area which are also produced outside the Area;

(f) the enhancing of opportunities for all States Parties, irrespective of their social and economic systems or geographical location, to participate in the development of the resources of the Area and preventing monopolization of the exploration and exploitation of the resources of the Area; and

(g) the protection of developing countries from any adverse effects on their economies or on their earnings resulting from a reduction in the price of an affected mineral, or in the volume of that mineral exported, to the extent that such reductions are caused by activities in the Area, through the following:

A. The Authority, acting through existing forums or such new arrangements or agreements as may be appropriate and in which all interested parties participate, shall take measures necessary to achieve growth, efficiency and stability of markets for those classes of commodities produced from the Area, at prices remunerative to producers and fair to consumers. All parties shall co-operate to this end. The Authority shall have the right to participate in any commodity conference dealing with the categories of minerals produced in the Area. The Authority shall have the right to become a party to any such arrangement or agreement resulting from such conferences as are referred to above. The participation by the Authority in any organs established under the arrangements or agreements referred to above shall be in respect of the production in the Area and in accordance with the rules of procedure established for such organs.

B. (i) The Authority shall limit in an interim period specified below, total production of minerals from nodules in the Area so as not to exceed for the first seven years of that period the projected cumulative growth segment of the world nickel demand. After the first seven years of the interim period total production of minerals from nodules in the Area shall on a yearly basis not exceed 60 per cent of the cumulative growth segment of the world nickel demand, as projected from the beginning of the interim period, provided however that this shall not affect such production under contracts already awarded, as is permitted under the production limit referred to above for the first seven years of the interim period. The cumulative growth segment for the purpose of this Part of the present Convention shall be computed in accordance with subparagraph (iii) below. The interim period referred to shall begin on 1 January 1980 and shall terminate on the day when such new arrangements or agreements as referred to in subparagraph (A) above, in which all affected parties participate, enter into force. The Authority shall resume the power to limit the production of minerals from nodules in the Area if the said arrangements or agreements should lapse or become ineffective for any reason whatsoever.

(ii) The Authority shall carry out the decisions taken by such organs as referred to in subparagraph (A) above and apply the interim production limit provided for in subparagraph (i) above, in a manner which assures a uniform and non-discriminatory implementation in respect of all production in the Area of the minerals concerned. In doing so, the Authority shall act in a manner consistent with the terms of existing contracts and approved plans of work of the Enterprise.

(iii) The rate of increase in world nickel demand projected for the interim period referred to in subparagraph (i) above shall, for the first five years of the interim period, be the annual constant percentage rate of increase in world demand during the 20-year period to 1 January 1980. The calculation of such a rate of increase in world nickel demand shall be made by application of the least squares method using definitive data from the latest 20-year period prior to that date, and for which such data are available. Thereafter this rate of increase shall be adjusted every five years on the basis of a recalculation applying the aforesaid method and using definitive data from the latest 10-year period prior to the commencement of any such five-year period, and for which such data are available.

(iv) The cumulative growth segment of the world nickel demand referred to in subparagraph (i) above shall be computed by applying the rate of increase determined pursuant to subparagraph (iii) to a base amount calculated by projecting world nickel demand for the year immediately preceding 1 January 1980 by applying the aforesaid rate of increase to the average of world nickel demand during the latest five-year period prior to the aforesaid date, and for which definitive data are available. Thereafter the base amount shall be adjusted every five years on the basis of the most recent definitive data available for the five-year period immediately preceding any such five-year period applying the method specified in this subparagraph.

C. The Authority may regulate production of minerals from the Area, other than minerals from nodules, under such conditions and applying such methods as may be appropriate.

D. Following recommendations from the Council on the basis of advice from the Economic Planning Commission, the Assembly shall establish a system of compensation for developing countries which suffer adverse effects on their export earnings or economies resulting from a reduction in the price of an affected mineral or the volume of that mineral exported, to the extent that such reduction is caused by activities in the Area.

2. (a) The Authority shall avoid discrimination in the exercise of its powers and functions, including the granting of opportunities for activities in the Area.

(b) Special considerations for developing countries, including particular consideration for the land-locked and geographically disadvantaged among them, specifically provided for in this Part of the present Convention, shall not be deemed to be discrimination.

(c) All rights granted shall be fully safeguarded in accordance with the provisions of this Convention.

Article 151
Functions of the Authority

1. Activities in the Area shall be carried out by the Authority on behalf of mankind as a whole in accordance with the provisions of this article as well as other relevant provisions of this Part of the present Convention and its annexes, and the rules, regulations and procedures of the Authority adopted under articles 158, paragraph 2 (xvi) and 160, paragraph 2 (xiv).

2. Activities in the Area shall be carried out on the Authority's behalf as prescribed in paragraph (c) below:

(i) by the Enterprise, and
(ii) in association with the Authority by States Parties or State Entities, or persons natural or juridical which possess the nationality of States Parties or are effectively controlled by them or their nationals, when sponsored by such States, or any group of the foregoing which, through contractual or other arrangements, undertake, in accordance with this Part of the present Convention, to contribute the technological capability, financial and other resources necessary to enable the Authority to fulfil its functions pursuant to paragraph 1 of this article.

3. Activities in the Area shall be carried out in

accordance with a formal written plan of work drawn up in accordance with annex II and approved by the Council after review by the Technical Commission. In the case of activities in the Area carried out on behalf of the Authority by the entities specified in paragraph 2 (ii) of this article such a plan of work shall in accordance with annex II, paragraph 3 (b), be in the form of a contract. Such contracts may provide for joint arrangements in accordance with annex II, paragraph 5 (i) and (j) (iii).

4. The Authority shall exercise such control over activities in the Area as is necessary for the purpose of securing compliance with the relevant provisions of this Part of the present Convention, including its annexes, and the rules, regulations and procedures of the Authority adopted under articles 158, paragraph 2 (xv) and 160, paragraph 2 (xiv) and the plans of work approved in accordance with paragraph 3 of this article. States Parties shall assist the Authority by taking all measures necessary to ensure such compliance.

5. The Authority shall have the right to take at any time any measures provided for under this Part of the present Convention to ensure compliance with its terms, and the performance of the control and regulatory functions assigned to it thereunder or under any contract. The Authority shall have the right to inspect all facilities in the Area used in connection with any activities in the Area.

6. A contract under paragraph 3 of this article shall provide for security of tenure. Accordingly, it shall not be cancelled, revised, suspended or terminated except in accordance with paragraphs 12 and 13 of annex II.

7. The Authority shall carry out marine scientific research concerning the Area and its resources, and may enter into contracts for that purpose. The Authority shall promote and encourage the conduct of marine scientific research in the Area, harmonize and coordinate such research, and arrange the effective dissemination of the results thereof.

8. The Authority shall take measures in accordance with this Convention:

(a) to acquire technology and scientific knowledge relating to activities in the Area; and

(b) to promote and encourage the transfer of such technology and scientific knowledge so that all States benefit therefrom.

9. The Authority shall establish a system for the equitable sharing of benefits derived from the Area, taking into special consideration the interests and needs of the developing countries and peoples, particularly the land-locked and geographically disadvantaged among them, and countries which have not attained full independence or other self-governing status.

Article 152
Periodic review

Every five years from the entry into force of the Convention, the Assembly shall undertake a general and systematic review of the manner in which the international régime of the Area established in the Convention has operated in practice. In the light of the said review the Assembly may adopt, or recommend other organs to adopt, measures in accordance with the provisions and procedures of this Part of the present Convention and its annexes which will lead to the improvement of the operation of the régime.

Article 153
The Review Conference

1. Twenty years from the entry into force of the present Convention, the Assembly shall convene a Conference for the review of those provisions of this Part of the present Convention and the annexes thereto which govern the system of exploration and exploitation of the resources of the Area. The Conference shall consider in detail, in the light of the experience acquired during that period, whether the provisions of this Part of the Convention governing the system of exploration and exploitation of the resources of the Area have achieved their aims in all respects, in particular whether they have contributed to a just distribution of the resources of the Area, whether they have not resulted in an excessive concentration of these resources in the hands of a small number of States, whether the economic principles set forth in article 150 have been complied with and whether the régime has benefited the developing countries.

2. In particular, the Conference shall consider whether, during the 20-year period, a balance has been maintained between the areas reserved for the Authority and developing countries, and the contract areas exploited by States, State entities, natural or juridical persons in association with the Authority.

3. If the Conference decides to amend the provisions of this Part of the present Convention governing the system of exploration and exploitation of the resources of the Area, it shall in any event ensure that the principles of the common heritage of mankind, the international régime designed to ensure its equitable exploitation for the benefit of all countries, especially the developing countries, and an Authority to conduct, organize and control activities in the Area are maintained. It shall also ensure the maintenance of the principles laid down in this Part of the present Convention with regard to the exclusion of claims or exercise of sovereignty over any part of the Area, the general conduct of States in relation to the Area, the prevention of monopolization of activities in the Area, the use of the Area exclusively for peaceful purposes, economic aspects of activities in the Area, scientific research, transfer of technology, protection of the marine environment, and of human life, rights of coastal States, the legal status of the superjacent waters and air space and accommodation as between the various forms of activities in the Area and in the marine environment.

4. With respect to the amendment referred to in paragraph 3 above, the Conference shall determine the system of voting, and entry into force procedures, provided that the majorities required under its voting system shall be the same as the majorities required for decisions by voting of the Third United Nations Conference on the Law of the Sea, under rule 39 of the rules of procedure for that Conference. The Conference shall make every effort to reach agreement on substantive matters by way of consensus and there shall be no voting on such matters until all efforts at consensus have been exhausted.

5. Amendments adopted by the Conference under the provisions of this article shall not affect rights acquired under existing contracts. In adopting rules, regulations and procedures on duration of activities under paragraph 11 (b) (2) of annex II to this Part of the present Convention, the Authority shall however take into account the possibility of the Convention being amended, provided that in all cases a reasonable time for return on capital shall be given.

6. If the Conference fails to amend or to reach agreement within five years on the provisions of this Part of the present Convention governing the system of exploration and exploitation of the resources of the Area, activities in the Area shall be carried out by the Authority through the Enterprise and through joint ventures negotiated with the State and entities referred to in paragraph 2 (ii) of article 151, on terms and conditions to be agreed upon between the parties thereto, provided however that the Authority shall exercise effective control over such activities.

SECTION 5. THE AUTHORITY
Subsection 1. General
Article 154
Establishment of the Authority

1. There is hereby established the International Sea-Bed Authority which shall function in accordance with the provisions of this Part of the present Convention.
2. All States Parties are *ipso facto* members of the Authority.
3. The seat of the Authority shall be at Jamaica.[1]
4. The Authority may establish such regional centres or offices as it deems necessary for the performance of its functions.

Article 155
Nature and fundamental principles of the Authority

1. The Authority is the organization through which States Parties shall organize and control activities in the Area, particularly with the view towards the administration of the resources of the Area, in accordance with this Part of the present Convention.
2. The Authority is based on the principle of the sovereign equality of all of its members.
3. All members, in order to ensure to all of them the rights and benefits resulting from membership, shall fulfil in good faith the obligations assumed by them in accordance with this Part of the present Convention.

Article 156
Organs of the Authority

1. There are hereby established as the principal organs of the Authority, an Assembly, a Council and a Secretariat.
2. There is hereby established the Enterprise, the organ through which the Authority shall directly carry out activities in the Area.
3. Such subsidiary organs as may be found necessary may be established in accordance with this Part of the present Convention.
4. The principal organs shall each be responsible for exercising those powers and functions which have been conferred on them. In exercising such powers and functions each organ shall act in a manner compatible with the distribution of powers and functions among the various organs of the Authority, as provided for in *this Part* of the present Convention.

Subsection 2. The Assembly
Article 157
Composition, procedure and voting

1. The Assembly shall consist of all the members of the Authority.
2. The Assembly shall meet in regular session every year and in such special sessions as may be determined by the Assembly, or convened by the Secretary-General at the request of the Council or of a majority of the members of the Assembly.
3. Sessions shall take place at the seat of the Authority unless otherwise determined by the Assembly. At such sessions, each member shall have one representative who may be accompanied by alternates and advisers.
4. The Assembly shall elect its President and such other offices as may be required, at the beginning of each regular session. They shall hold office until the new President and other officers are elected at the next regular session.
5. Each member of the Assembly shall have one vote.
6. All decisions on questions of substance shall be taken by a two-thirds majority of the members present and voting, provided that such majority includes at least a majority of the members participating in that session of the Assembly. When the issue arises as to whether the question is one of substance or not, the question shall be treated as one of substance unless otherwise decided by the Assembly by the majority required for questions of substance.

---

1. Malta and Fiji have also proposed that the seat of the Authority be located in their countries.

7. Decisions on questions of procedure, including the decision to convene a special session of the Assembly, shall be made by a majority of the representatives present and voting.

8. When a matter of substance comes up for voting for the first time, the President may, and shall, if requested by at least one fifth of the members of the Assembly, defer the question of taking a vote on such matter for a period not exceeding five calendar days. This rule may be applied only once on the matter, and shall not be applied so as to defer questions beyond the end of the session.

9. A majority of the members of the Assembly shall constitute a quorum.

10. Upon a request in writing to the President sponsored by not less than one quarter of the members of the Authority for an advisory opinion on the conformity with this Convention of a proposed action before the Assembly on any matter, a vote on that action shall be deferred pending reference to the Sea-Bed Disputes Chamber for such an opinion. Voting on that action shall be stayed pending delivery of an advisory opinion by the Chamber. If the advisory opinion is not received by the final week of the session in which it is requested, the Assembly shall decide when it will meet to vote upon the deferred matter.

Article 158
Powers and functions

1. The Assembly is the supreme organ of the Authority, and as such shall have the power to establish the general policies in conformity with the provisions of this Part of the present Convention, to be pursued by the Authority on any questions or matters within the competence of the Authority. The Assembly may discuss any such question or matter, and may decide which organ shall deal with any such question or matter not specifically entrusted by the provisions of the present Convention to a particular organ of the Authority.

2. In addition, the powers and functions of the Assembly shall be:

   (i) Election of the members of the Council in accordance with article 159,

   (ii) Election of the Secretary-General from among the candidates proposed by the Council,

   (iii) Selection of the 11 members of the Sea-Bed Disputes Chamber from among the members of the Law of the Sea Tribunal,

   (iv) Appointment, upon the recommendation of the Council, of the members of the Governing Board of the Enterprise as well as the Director-General of the Enterprise,

   (v) Establishment, as appropriate, of such subsidiary organs as may be found necessary for the performance of its functions in accordance with the provisions of this Part of the Convention. In the composition of such subsidiary organs due account shall be taken of the principle of equitable geographical distribution and of special interests and the need for members qualified and competent in the relevant technical questions dealt with by such organs,

   (vi) Assessment of the contributions of members to the Administrative budget of the Authority in accordance with an agreed general assessment scale until the Authority shall have sufficient income for meeting its administrative expenses,

   (vii) Adoption of the financial regulations of the Authority, including rules on borrowing, upon the recommendation of the Council,

   (viii) Consideration and approval of the budget of the Authority on its submission by the Council,

   (ix) Adoption of its rules of procedure,

   (x) Examination of periodic reports from the Council and from the Enterprise and of special reports requested from the Council and from any other organs of the Authority,

   (xi) Studies and recommendations for the purpose of promoting international co-operation concerning activities in the Area and encouraging the progressive development of international law relating thereto and its codification,

   (xii) Adoption of rules, regulations and procedures for the equitable sharing of financial and other economic benefits derived from activities in the Area, taking into particular consideration the interests and the needs of the developing countries,

   (xiii) Consideration of problems of a general nature in connexion with activities in the Area in particular for developing countries, as well as of such problems for States in connexion with activities in the Area as are due to their geographical location, including land-locked and geographically disadvantaged countries,

   (xiv) Establishment, upon the recommendation of the Council on the basis of advice from the Economic Planning Commission of a system of compensation as provided in article 150, paragraph 1 (g) (D),

   (xv) Suspension of members pursuant to article 186,

   (xvi) Final adoption of the rules, regulations and procedures, and amendments thereto, provisionally adopted by the Council in accordance with the provisions of paragraph 11 of annex II and pursuant to article 160 (2) (xiv).

Subsection 3. The Council
Article 159
Composition, procedure and voting

1. The Council shall consist of 36 members of

the Authority elected by the Assembly, the election to take place in the following order:

(a) four members from among countries which have made the greatest contributions to the exploration for, and the exploitation of, the resources of the Area, as demonstrated by substantial investments or advanced technology in relation to resources of the Area, including at least one State from the Eastern (Socialist) European region.

(b) four members from among countries which are major importers of the categories of minerals to be derived from the Area, including at least one State from the Eastern (Socialist) European region.

(c) four members from among countries which on the basis of production in areas under their jurisdiction are major exporters of the categories of minerals to be derived from the Area, including at least two developing countries.

(d) six members from among developing countries, representing special interests. The special interests to be represented shall include those of States with large populations, States which are land-locked or geographically disadvantaged, States which are major importers of the categories of minerals to be derived from the Area, and least developed countries.

(e) eighteen members elected according to the principle of ensuring an equitable geographical distribution of seats in the Council as a whole, provided that each geographical region shall have at least one member elected under this subparagraph. For this purpose the geographical regions shall be Africa, Asia, Eastern Europe (Socialist), Latin America and Western Europe and others.

2. In electing the members of the Council in accordance with paragraph 1 above, the Assembly shall ensure that land-locked and geographically disadvantaged States are represented to a degree which is reasonably proportionate to their representation in the Assembly.

3. Elections shall take place at regular sessions of the Assembly, and each member of the Council shall be elected for a term of four years. In the first election of members of the Council, however, one half of the members of each category shall be chosen for a period of two years.

4. Members shall be eligible for re-election; but due regard should be paid to the desirability of rotating seats.

5. The Council shall function at the seat of the Authority, and shall meet as often as the business of the Authority may require, but not less than three times a year.

6. Each member of the Council shall have one vote.

7. All decisions on questions of substance shall be taken by a three-fourths majority of the members present and voting, provided that such majority includes a majority of the members participating in that session. When the issue arises as to whether the question is one of substance or not, the question shall be treated as one of substance unless otherwise decided by the Council by the majority required for questions of substance. Decisions on matters of procedure shall be decided by a majority of the members present and voting.

8. A majority of the members of the Council shall constitute a quorum.

9. The Council shall establish a procedure whereby a member of the Authority not represented on the Council may send a representative to attend a meeting of the Council when a request is made by such member, or a matter particularly affecting it is under consideration. Such a representative shall be entitled to participate in the deliberations but not to vote.

Article 160
Powers and functions

1. The Council is the executive organ of the Authority, having the power to establish in conformity with the provisions of this Part of the present Convention and the general policies established by the Assembly, the specific policies to be pursued by the Authority on any questions or matters within the competence of the Authority.

2. In addition the Council shall:
  (i) Supervise and co-ordinate the implementation of the provisions of this Part of the present Convention and invite the attention of the Assembly to cases of non-compliance.
  (ii) Propose to the Assembly a list of candidates for the election of the Secretary-General.
  (iii) Recommend to the Assembly candidates for appointment as members of the Governing Board of the Enterprise as well as the Director-General of the Enterprise.
  (iv) Establish, as appropriate, and with due regard to economy and efficiency, in addition to the Commissions provided for in article 161, paragraph 1, such subsidiary organs as may be found necessary for the performance of its functions in accordance with the provisions of this Part of the present Convention. In the composition of such subsidiary organs, emphasis shall be placed on the need for members qualified and competent in the relevant technical matters dealt with by such organs provided that due account shall be taken of the principle of equitable geographical distribution and of special interests.
  (v) Adopt its rules of procedure.
  (vi) Enter into agreements with the United Nations or other intergovernmental organizations on behalf of the Authority, subject to approval by the Assembly.
  (vii) Examine the reports of the Enterprise and transmit them to the Assembly with its recommendations.

(viii) Present to the Assembly annual reports and such special reports as the Assembly may require.
(ix) Issue directives to the Enterprise and exercise control over its activities in accordance with paragraph 4 of article 151.
(x) Approve on behalf of the Authority, after review by the Technical Commission, formal written plans of work, for the conduct of activities in the Area, drawn up in accordance with paragraph 3 of article 151. In so doing the Council shall act expeditiously. The plan of work shall be deemed to have been approved, unless a decision to disapprove it is taken within 60 days of its submission by the Technical Commission.
(xi) Exercise control over activities in the Area in accordance with paragraph 4 of article 151.
(xii) Adopt on the recommendation of the Economic Planning Commission necessary and appropriate measures in accordance with paragraph 1 (g) of article 150 to protect against adverse economic effects specified therein.
(xiii) Make recommendations to the Assembly on the basis of advice from the Economic Planning Commission for a system of compensation as provided in paragraph 1 (g) (D) of article 150.
(xiv) Adopt and apply provisionally, pending final adoption by the Assembly, rules, regulations and procedures, and any amendments thereto, in accordance with the provisions of paragraph 11 of annex II, taking into account the recommendations of the Rules and Regulations Commission. Such rules, regulations and procedures shall remain in effect on a provisional basis until final adoption by the Assembly or amendment by the Council in the light of any views expressed by the Assembly.
(xv) Review the collection of all payments to be made by or to the Authority in connexion with operations pursuant to this Part of the present Convention and recommend to the Assembly the financial regulations of the Authority, including rules on borrowing.
(xvi) Submit to the Assembly for its approval the budget of the Authority.
(xvii) Make recommendations concerning the policies and measures required to give effect to the principles of this Part of the present Convention.
(xviii) Make recommendations to the Assembly concerning suspension of the privileges and rights of membership for gross and persistent violations of the provisions of this Part of the Convention upon a finding of the Sea-Bed Disputes Chamber.

Article 161
Organs of the Council

1. There are hereby established as organs of the Council:
   (a) An Economic Planning Commission, constituted in accordance with article 162;
   (b) A Technical Commission and Rules and Regulations Commission, each of which shall be composed of 15 members appointed by the Council with due regard to the need for members qualified and competent in the relevant technical matters which may arise in such organs and to the principle of equitable geographical distribution and special interests.
2. The Council shall invite States Parties to submit nominations for appointment to each of the Commissions referred to in paragraph 1 above.
3. Appointment to each Commission shall take place not less than 60 days before the end of a calendar year and the members of a Commission shall hold office from the commencement of the next calendar year following their appointment until the end of the third calendar year thereafter. The first appointments to a Commission, however, shall take place not less than 30 days after the entry into force of the present Convention, and five of those so appointed shall hold office until the end of the calendar year next following the year of their appointment, while five other members shall hold office until the end of the second calendary year following their appointment.
4. In the event of the death, incapacity or resignation of a member of a Commission prior to the expiry of his term of office, the Council shall appoint a member from the same area of interest who shall hold office for the remainder of the previous member's term.
5. Members of a Commission shall be eligible for reappointment for one further term of office.
6. Each Commission shall appoint its Chairman and two Vice-Chairmen, who shall hold office for one year.
7. The Council shall approve, on the recommendation of a Commission, such rules and regulations as may be necessary for the efficient conduct of the functions of the Commission.
8. Decisions of the Commission shall be by a two-thirds majority of members present and voting.
9. Each Commission shall function at the seat of the Authority and shall meet as often as shall be required for the efficient performance of its functions.

Article 162
The Economic Planning Commission

1. The Economic Planning Commission shall be composed of 18 experts appointed by the Council, upon nomination by States Parties. Such experts shall have appropriate qualifications and experience relevant to mining and the management of mineral resource activities, and

international trade and finance. In the composition of the Economic Planning Commission, the Council shall take into account the need for equitable geographical distribution. The Council shall also ensure at all times a fair and equitable balance within the Commission between those experts appointed from countries which export and which import minerals which are also derived from the Area.

2. The Economic Planning Commission shall submit its recommendations to the Council upon an affirmative vote of two thirds of members present and voting. In those cases where a recommendation is not adopted by consensus, any dissenting opinions and analyses shall be forwarded to the Council together with the recommendation.

3. The Economic Planning Commission, in consultation with the competent organs of the United Nations, its specialized agencies and any other intergovernmental organization with responsibilities relating to minerals which are also derived from the Area, shall review the trends of, and factors affecting, supply, demand and prices of raw materials which may be obtained from the Area, bearing in mind the interests of both importing and exporting countries, and in particular the developing countries among them.

4. The Commission shall make such special studies and reports as may be required by the Council from time to time.

5. Any situation likely to lead to such adverse effects as referred to in paragraph 1 (g) of article 150 may be brought to the attention of the Economic Planning Commission by the State Party or State Parties concerned. The Commission shall forthwith investigate this situation and shall make recommendations, in consultations with affected States Parties and with the competent intergovernmental organizations, to the Council in accordance with paragraph 6 of this article.

6. On the basis of studies, reports and reviews referred to above, the Commission shall advise the Council as to the exercise of its powers and functions pursuant to subparagraph (xii) of paragraph 2 of article 160.

7. The Commission shall propose to the Council for submission to the Assembly a system of compensation for developing countries who suffer adverse effects caused by activities in the Area, as provided in paragraph 1 (g) (D) of article 150. After adoption by the Assembly of such system of compensation the Economic Planning Commission shall make such recommendations to the Council as are necessary for the application of the system in concrete cases.

Article 163
The Technical Commission

1. Members of the Technical Commission shall have appropriate qualifications and experience in economics, the management of mineral resources, ocean and marine engineering and mining and mineral processing technology and practices, operation of related marine installations, equipment and devices, ocean and environmental sciences and maritime safety, accounting and actuarial techniques.

2. Subject to such guidelines and directives as the Council may adopt, the Technical Commission shall:

(i) Make recommendations to the Council with regard to the carrying out of the Authority's functions with respect to scientific research and transfer of technology;

(ii) Prepare special studies and reports at the request of the Council;

(iii) Advise the Rules and Regulations Commission on all technical aspects of its work;

(iv) Prepare assessments of the environmental implications of activities in the Area;

(v) Supervise, on a regular basis all operations with respect to activities in the Area, where appropriate in consultation and collaboration with any entity carrying out such activities or State or States concerned;

(vi) Initiate on behalf of the Authority proceedings before the Sea-Bed Disputes Chamber in cases of non-compliance;

(vii) Upon a finding by the Sea-Bed Disputes Chamber in proceedings resulting from subparagraph (vi) above, notify the Council and make recommendations with respect to measures to be taken;

(viii) Inspect and audit all books, records and accounts related to financial obligations to the Authority concerning activities in the Area and collect all payments to the Authority prescribed in annex II;

(ix) Advise the Council, the Economic Planning Commission and the Rules and Regulations Commission on financial aspects of their work;

(x) Direct and supervise a staff of inspectors who shall inspect all activities in the Area to determine whether the provisions of this Part of the present Convention, the rules, regulations and procedures prescribed thereunder, and the terms and conditions of any contract with the Authority are being complied with;

(xi) Issue emergency orders, which may include orders for the suspension of operations, to prevent serious harm to the marine environment arising out of any activity in the Area;

(xii) Disapprove areas for exploitation by contractors or the Enterprise in cases where substantial evidence indicates the risk of irreparable harm to a unique environment;

(xiii) Take into account views on protection of the environment of recognized experts in the field before making recommendations to the Council on the above matters as they relate to the protection of the marine environment;

(xiv) Review formal written plans of work for activities in the Area in accordance with article 151, paragraph 3.
3. States Parties and other parties concerned shall facilitate the exercise by the members of the Commission and its staff of their functions which shall not be delayed or otherwise impeded.
4. The Members of the Commission and its staff shall, upon request by any State Party or other party concerned, be accompanied by a representative of such State Party or other party concerned when carrying out their supervision and inspection functions.

Article 164
The Rules and Regulations Commission
1. Members of the Rules and Regulations Commission shall have appropriate qualifications in legal matters, including those relating to ocean mining and other marine matters.
2. The Rules and Regulations Commission shall:
   (i) Formulate and submit to the Council the rules, regulations and procedures referred to in subparagraph (xiv) of paragraph 2 of article 160, taking into account all relevant factors, including assessments prepared by the Technical Commission of the environmental implications of activities in the Area;
   (ii) Keep such rules, regulations and procedures under review and recommend to the Council from time to time such amendments thereto as it may deem necessary or desirable;
   (iii) Prepare special studies and reports at the request of the Council.

Subsection 4. The Secretariat
Article 165
The Secretary-General
1. The secretariat shall comprise a Secretary-General and such staff as the Authority may require. The Secretary-General shall be appointed by the Assembly upon the recommendation of the Council. He shall be the chief administrative officer of the Authority.
2. The Secretary-General shall act in that capacity in all meetings of the Assembly and of the Council, and of any subsidiary organs established by them, and shall perform such other functions as are entrusted to him by any organ of the Authority.
3. The Secretary-General shall make an annual report to the Assembly on the work of the organization.

Article 166
The staff of the Authority
1. The staff of the Authority shall consist of such qualified scientific and technical and other personnel as may be required to fulfil the administrative functions of the Authority. The Authority shall be guided by the principle that its staff shall be kept to a minimum.
2. The paramount consideration in the recruitment and employment of the staff and in the determination of their conditions of service shall be to secure employees of the highest standards of efficiency, competence and integrity. Subject to this consideration, due regard shall be paid to the importance of recruiting staff on as wide a geographical basis as possible.
3. The staff shall be appointed by the Secretary-General. The terms and conditions on which the staff shall be appointed, remunerated and dismissed shall be in accordance with regulations made by the Council, and to general rules approved by the Assembly on the recommendation of the Council.

Article 167
International character of the secretariat
1. In the performance of their duties, the Secretary-General and the staff shall not seek or receive instructions from any Government or from any other source external to the Authority. They shall refrain from any action which might reflect on their position as international officials of the Authority responsible only to the Authority. They shall have no financial interest whatsoever in any activity relating to exploration and exploitation in the Area. Subject to their responsibilities to the Authority, they shall not disclose any industrial secret or data which is proprietary in accordance with paragraph 8 of annex II or other confidential information coming to their knowledge by reason of their official duties for the Authority. Each State Party undertakes to respect the exclusively international character of the responsibilities of the Secretary-General and the staff and not to seek to influence them in the discharge of their responsibilities.
2. Any violation of the responsibilities set forth in paragraph 1 of this article shall be considered a grave disciplinary offence and shall, in addition, entail personal liability for damages. Any State Party or natural or juridical person sponsored by a State Party may bring an alleged violation of this article before the Sea-Bed Disputes Chamber which may order monetary penalties or the assessment of damages. Upon such order, the Secretary-General shall dismiss the staff member concerned. The elaboration of the provisions of this paragraph shall be included in the staff regulations of the Authority.

Article 168
Consultation and co-operation with non-governmental organizations
1. The Secretary-General shall, on matters within the competence of the Authority, make suitable arrangements, with the approval of the Council, for consultation and co-operation with non-governmental organizations recognized by the Economic and Social Council of the United Nations.
2. Any organization with which the Secretary-General has entered into an arrangement under paragraph 1 may designate representatives to attend as observers meetings of the organs of the Authority in accordance with the rules of procedure of any such organ.

Procedures shall be established for obtaining the views of such organizations in appropriate cases.
3. Written reports submitted by these non-governmental organizations on subjects in which they have special competence and which are related to the work of the Authority may be distributed by the Secretary-General to States Parties.

Subsection 5. The Enterprise
Article 169
The Enterprise

1. The Enterprise shall be the organ of the Authority which shall carry out activities in the Area directly, pursuant to article 151, paragraph 2 (i) and in accordance with the general policies laid down by the Assembly. Activities conducted by the Enterprise shall be subject to the directives and control of the Council.
2. The Enterprise shall, within the framework of the international legal personality of the Authority, have such legal capacity and functions as provided for in the Statute set forth in annex III to this Part of the present Convention, and shall in all respects be governed by the provisions of this Part of the Convention. Appointment of the members of the Governing Board shall be made in accordance with the provisions of the Statute set forth in annex III.
3. The Enterprise shall have its principal place of business in the seat of the Authority.
4. The Enterprise shall in accordance with article 173, paragraph 3 and annex III, paragraph 10, be provided with such funds as it may require to carry out its functions, and shall receive technology as provided in article 144, and other relevant provisions of the present Convention.

Subsection 6. Finance
Article 170
General Fund

1. The Assembly shall establish the General Fund of the Authority.
2. All receipts of the Authority arising from activities in the Area, including any excess of revenues of the Enterprise over its expenses and costs in such proportion as the Council shall determine, shall be paid into the General Fund.

Article 171
Annual budget of the Authority

The Council shall submit for the approval of the Assembly annual budget estimates for the expenses of the Authority. To facilitate the work of the Council in this regard, the Secretary-General shall initially prepare the budget estimates.

Article 172
Expenses of the Authority

1. Expenses of the Authority comprise:
(a) Administrative expenses, which shall include costs of the staff of the Authority, costs of meetings, and expenditure on account of the functioning of the organs of the Authority;
(b) Expenses not included in the foregoing, incurred by the Authority in carrying out the functions entrusted to it under this Part of the Convention.
2. The expenses referred to in paragraph 1 of this article shall be met to an extent to be determined by the Assembly on the recommendation of the Council, out of the General Fund, the balance of such expenses to be met out of contributions by members of the Authority in accordance with a scale of assessment adopted by the Assembly pursuant to the subparagraph (vi) of paragraph 2 of article 158.

Article 173
Special Fund

1. Any excess of revenues of the Authority over its expenses and costs to an extent determined by the Council, all payments received pursuant to article 170 and any voluntary contributions made by members of the Authority shall be credited to a Special Fund.
2. Amounts in the Special Fund shall be apportioned and made available equitably in such manner and in such currencies, and otherwise in accordance with criteria, rules, regulations and procedures adopted by the Assembly pursuant to subparagraph (xii) of paragraph 2 of article 158.
3. In establishing the criteria, rules, regulations and procedures referred to in paragraph 2 of this article the Assembly shall take into account the need to earmark for the Enterprise a part of the funds received by the Authority from contractors in accordance with annex II, paragraph 7, in order to enable the Enterprise to explore and exploit directly the resources of areas reserved for it in accordance with annex II, paragraph 5 (j) (ii) as well as other parts of the Area.

Article 174
Borrowing powers of the Authority

Subject to such limitations as may be approved by the Assembly in the financial regulations adopted by it pursuant to subparagraph (vii) of paragraph 2 of article 158, the Council may exercise borrowing powers on behalf of the Authority without, however, imposing on members of the Authority any liability in respect of loans entered into pursuant to this paragraph, and accept voluntary contributions made to the Authority.

Article 175
Annual audit

The records, books and accounts of the Authority, including its annual financial statements, shall be subject to an annual audit by a recognized independent auditor.

Subsection 7. Legal Status, Privileges and Immunities
Article 176
Legal status

The Authority shall have international legal personality, and such legal capacity as may be necessary for the exercise of its functions and the fulfilment of its purpose.

Article 177
Privileges and immunities

To enable the Authority to fulfil its functions it shall enjoy in the territory of each State Party, the immunities and privileges set forth herein except as provided in annex III of the present Convention with respect to operations of the Enterprise.

Article 178
Immunity from legal process

The Authority, its property and assets, shall enjoy, in the territory of each State Party, immunity from legal proess, except when the Authority waives its immunity.

Article 179
Immunity from search and any form of seizure

The property and assets of the Authority, wheresoever located and by whomsoever held, shall be immune from search, requisition, confiscation, expropriation or any form of seizure by executive or legislative action.

Article 180
Property and assets free from restrictions, regulations, control and moratoria

All property and assets of the Authority shall be free from restrictions, regulations, controls and moratoria of any nature.

Article 181
Immunities of certain persons connected with the Authority

The President and members of the Assembly, the Chairman and members of the Council, members of any organ of the Assembly, or the Council, and members of the Sea-Bed Disputes Chamber and the Secretary-General and staff of the Authority, shall enjoy in the territory of each member State:

(a) Immunity from legal process with respect to acts performed by them in the exercise of their functions, except when this immunity is waived;

(b) Not being local nationals, the same immunities from immigration restrictions, alien registration requirements and national service obligations, the same facilities as regards exchange restrictions and the same treatment in respect of travelling facilities as are accorded by States Parties to the representatives, officials and employees of comparable rank of other States Parties.

Article 182
Immunities of persons appearing in proceedings before the Sea-Bed Disputes Chamber

The provisions of the preceding article shall apply to persons appearing in proceedings before the Sea-Bed Disputes Chamber as parties, agents, counsel, advocates, witnesses or experts; provided, however, that subparagraph (b) thereof shall apply only in connexion with their travel to and from, and their stay at, the place where the proceedings are held.

Article 183
Inviolability of archives

1. The archives of the Authority shall be inviolable, wherever they may be.
2. All proprietary data, industrial secrets or similar information and all personnel records shall not be placed in archives open to public inspection.
3. With regard to its official communications, the Authority shall be accorded by each State Party treatment no less favourable than that accorded to other international organizations.

Article 184
Exemption from taxation and customs duties

1. The Authority, its assets, property and income, and its operations and transactions authorized by the present Convention, shall be exempt from all taxation and customs duties. The Authority shall also be exempt from liability for the collection or payment of any taxes or customs duties.
2. Except in the case of local nationals, no tax shall be levied on or in respect of expense allowances paid by the Authority to the President or members of the Assembly, or in respect of salaries, expense allowances or other emoluments paid by the Authority to the Chairman and members of the Council, members of the Sea-Bed Disputes Chamber, members of any organ of the Assembly or of the Council and the Secretary-General and staff of the Authority.

Subsection 8. Suspension of rights of members
Article 185
Suspension of voting rights

A member which is in arrears in the payment of its financial contributions to the Authority shall have no vote in the Authority if the amount of its arrears equals or exceeds the amount of the contribution due from it for the preceding two years. The Assembly may permit such a member to vote if it is satisfied that the failure to pay is due to conditions beyond the control of the State Party.

Article 186
Suspension of privileges and the rights of membership

1. A State Party which has grossly and persistently violated the provisions of this Part of the Convention or of any agreement or contractual arrangement entered into by it pursuant to this Part of the Convention, may be suspended from the exercise of the privileges and the rights of membership by the Assembly upon recommendation by the Council.
2. No action may be taken under this article until the Sea-Bed Disputes Chamber has found that a State Party has grossly and persistently violated the provisions of this Part of the present Convention.

SECTION 6. SETTLEMENT OF DISPUTES
Article 187
Jurisdiction of the Sea-Bed Disputes Chamber of the Law of the Sea Tribunal

1. The Sea-Bed Disputes Chamber of the Law of the Sea Tribunal shall have jurisdiction pursuant to articles 187 to 192 of this Part. The establishment of the Chamber, and the manner in which it shall exercise its jurisdiction shall be governed by the provisions of Part XV of the present Convention.
2. The Chamber shall have jurisdiction with respect to:

   (a) Disputes between a State Party and the Authority concerning an allegation that a decision or measure taken by the Assembly, the Council or any of its organs is in violation of this Part of the Convention, rules, regulations or procedures promulgated in accordance therewith, or that the Assembly, the Council or such organ lacks jurisdiction in respect of such decision or measure, or has misused its power.

   (b) Disputes arising out of any of the grounds stated in subparagraph (a) above between a national of a State Party and the Authority relating to a decision or measure directed specifically to that person, or, in the case of a person referred to in article 151, paragraph 2 (ii) conducting activities in the Area or seeking a contract to do so, a decision or measure which is of direct concern to him though not directed to him.

   (c) Disputes, other than those referred to in subparagraphs (a) and (b) above, between the Authority and a State Party or between the Authority and a national of a State Party relating to the interpretation or application of any contract concerning activities in the Area.

   (d) Disputes between the Authority and a State Party concerning alleged violations by the State Party of the provisions of this Part of the present Convention relating to activities in the Area.

   (e) Disputes arising out of matters referred to in articles 167 and 186.

Article 188
Submission of disputes to arbitration

Where the parties to any dispute referred to in article 187 so agree for any specific dispute, or have so agreed under a contract or a general arbitration clause, such dispute shall be submitted to arbitration in accordance with the provisions of annex VI of the present Convention, or to any other arbitration procedure agreed upon.

Article 189
Disputes involving States Parties or their nationals

1. The Sea-Bed Disputes Chamber of the Law of the Sea Tribunal shall also have jurisdiction with regard to:
   (i) Disputes between States Parties concerning the interpretation or application of this Part of the Convention in respect of activities in the Area;
   (ii) Disputes brought by a State Party against a national of another State Party, or between nationals of different State Parties, regarding the interpretation or application of any contract between them, or in respect of their activities in the Area.
2. Where a respondent party so elects within one month of receiving notice of the referral of a dispute to the Chamber under paragraph 1 above, the dispute shall instead be submitted to arbitration in accordance with the provisions of annex VI of the present Convention.

Article 190
Advisory opinions

The Sea-Bed Disputes Chamber of the Law of the Sea Tribunal shall give advisory opinions when requested to do so by the Assembly, the Council or any of its organs, on any legal question arising within the scope of their activities. Such advisory opinions shall be rendered as a matter of urgency.

Article 191
Scope of jurisdiction wth regard to decisions adopted by the Assembly or Council

In exercising its jurisdiction pursuant to articles 187 and 189 the Sea-Bed Disputes Chamber of the Law of the Sea Tribunal shall not pronounce itself on the question whether any rules, regulations or procedures adopted by the Assembly or by the Council are in conformity with the provisions of the present Convention. Its jurisdiction with regard to such rules, regulations and procedures shall be confined to their application to individual cases. The Sea-Bed Disputes Chamber shall have no jurisdiction with regard to the exercise by the Assembly or by the Council or any of its organs of their discretionary powers under this Part of the present Convention; in no case shall it substitute its discretion for that of the Authority.

Article 192
Rights of States Parties when their nationals are parties to a dispute

When in a dispute referred to in articles 187 and 189, a national of a State Party is a party, the sponsoring State shall be given notice thereof, and shall have a right to intervene in the proceedings.

PART XII
PROTECTION AND PRESERVATION OF THE MARINE ENVIRONMENT
SECTION 1. GENERAL PROVISIONS
Article 193
General obligation

States have the obligation to protect and preserve the marine environment.

Article 194
Sovereign right of States to exploit their natural resources

States have the sovereign right to exploit their natural resources pursuant to their environmental policies and in accordance with their duty to protect and preserve the marine environment.

Article 195
Measures to prevent, reduce and control pollution of the marine environment

1. States shall take all necessary measures consistent with the present Convention to prevent, reduce and control pollution of the marine environment from any source using for this purpose the best practicable means at their disposal and in accordance with their capabilities, individually or jointly as appropriate, and they shall endeavour to harmonize their policies in this connexion.
2. States shall take all necessary measures to ensure that activities under their jurisdiction or control are so conducted that they do not cause damage by pollution to other States and their environment, and that pollution arising from incidents or activities under their jurisdiction or control does not spread beyond the areas where they exercise sovereign rights in accordance with the present Convention.
3. The measures taken pursuant to this Chapter shall deal with all sources of pollution of the marine environment. These measures shall include, *inter alia,* those designed to minimize to the fullest possible extent:
    (a) Release of toxic, harmful and noxious substances, especially those which are persistent:
        (i) from land-based sources;
        (ii) from or through the atmosphere;
        (iii) by dumping.
    (b) Pollution from vessels, in particular for preventing accidents and dealing with emergencies, ensuring the safety of operations at sea, preventing intentional and unintentional discharges, and regulating the design, construction, equipment, operation and manning of vessels;
    (c) Pollution from installations and devices used in the exploration or exploitation of the national resources of the sea-bed and subsoil, in particular for preventing accidents and dealing with emergencies, ensuring the safety of operations at sea, and regulating the design, construction, equipment, operation and manning of such installations or devices;
    (d) Pollution from all other installations and devices operating in the marine environment, in particular for preventing accidents and dealing with emergencies, ensuring the safety of operations at sea, and regulating the design, construction, equipment, operation and manning of such installations or devices.
4. In taking measures to prevent, reduce or control pollution of the marine environment, States shall refrain from unjustifiable interference with activities in pursuance of the rights and duties of other States exercised in conformity with the present Convention.

Article 196
Duty not to transfer damage or hazards or transform one type of pollution into another

In taking measures to prevent, reduce and control pollution of the marine environment, States shall so act as not to transfer, directly or indirectly, damage or hazards from one area to another or transform one type of pollution into another.

Article 197
Use of technologies or introduction of alien or new species

1. States shall take all necessary measures to prevent, reduce and control pollution of the marine environment resulting from the use of technologies under their jurisdiction or control, or the intentional or accidental introduction of species, alien or new, to a particular part of the marine environment, which may cause significant and harmful changes thereto.
2. This article shall not affect the application of the present Convention regarding the prevention, reduction and control of pollution of the marine environment.

SECTION 2. GLOBAL AND REGIONAL CO-OPERATION

Article 198
Co-operation on a global or regional basis

States shall co-operate on a global basis and, as appropriate, on a regional basis, directly or through competent international organizations, global or regional, in formulating and elaborating international rules, standards and recommended practices and procedures consistent with the present Convention, for the protection and preservation of the marine environment, taking into account characteristic regional features.

Article 199
Notification of imminent or actual damage

A State which becomes aware of cases in which the marine environment is in imminent danger of being damaged or has been damaged by pollution shall immediately notify other States it deems likely to be affected by such damage, as well as the competent international organizations, global or regional.

Article 200
Contingency plans against pollution

In the case referred to in article 199, States in the area affected, in accordance with their capabilities, and the competent international organizations, global or regional, shall co-operate, to the extent possible, in eliminating the effects of pollution and preventing or minimizing the damage. Towards that end, States shall jointly promote and develop contingency plans for responding to pollution incidents in the marine environment.

Article 201
Promotion of studies, research programmes and exchange of information and data

States shall co-operate directly or through competent international organizations, global or regional, for the purpose of promoting studies, undertaking programmes of scientific research and encouraging the exchange of information and data acquired about pollution of the marine environment. They shall endeavour to participate actively in regional and international programmes to acquire knowledge for the assessment of the nature and extent of pollution and the pathways and risks of, exposures to and the remedies for pollution.

Article 202
Scientific criteria and regulations
In the light of the information and data acquired pursuant to article 201, States shall co-operate directly or through competent international organizations, global or regional, in establishing appropriate scientific criteria for the formulation and elaboration of rules, standards and recommended practices and procedures for the prevention of pollution of the marine environment.

SECTION 3. TECHNICAL ASSISTANCE
Article 203
Scientific and technical assistance to Developing States
States shall directly or through competent international or regional organizations, global or regional:
(a) Promote programmes of scientific, educational, technical and other assistance to developing States for the protection and preservation of the marine environment and the prevention, reduction and control of marine pollution. Such assistance shall include, *inter alia*:
 (i) Training of their scientific and technical personnel;
 (ii) Facilitating their participation in relevant international programmes;
 (iii) Supplying necessary equipment and facilities;
 (iv) Enhancing the capacity of developing States to manufacture such equipment;
 (v) Developing facilities for and advice on research, monitoring, educational and other programmes;
(b) Provide appropriate assistance, especially to developing States, for the minimization of the effects of major incidents which may cause serious pollution in the marine environment;
(c) Provide appropriate assistance, in particular to developing States, concerning the preparation of environmental assessments.

Article 204
Preferential treatment for Developing States
Developing States shall, for purposes of the prevention of pollution of the marine environment or the minimization of its effects, be granted preference in:
(a) The allocation of appropriate funds and technical assistance facilities of international organizations, and
(b) The utilization of their specialized services.

SECTION 4. MONITORING AND ENVIRONMENTAL ASSESSMENT
Article 205
Monitoring of the risks or effects of pollution
1. States shall, consistent with the rights of other States, endeavour, as far as practicable, individually or collectively through the competent international organizations, global or regional, to observe, measure, evaluate and analyse, by recognized methods, the risks or effects of pollution of the marine environment.
2. In particular, States shall keep under surveillance the effect of any activities which they permit or in which they engage to determine whether these activities are likely to pollute the marine environment.

Article 206
Publication of reports
States shall publish reports of the results obtained relating to risks or effects of pollution of the marine environment, or provide at appropriate intervals such reports to the competent international or regional organizations, which should make them available to all States.

Article 207
Assessment of potential effects of activities
When States have reasonable grounds for expecting that planned activities under their jurisdiction or control may cause substantial pollution of, or significant and harmful changes to, the marine environment, they shall, as far as practicable, assess the potential effects of such activities on the marine environment and shall communicate reports of the results of such assessments in the manner provided in article 206.

SECTION 5. INTERNATIONAL RULES AND NATIONAL LEGISLATION TO PREVENT, REDUCE AND CONTROL POLLUTION OF THE MARINE ENVIRONMENT
Article 208
Pollution from land-based sources
1. States shall establish national laws and regulations to prevent, reduce and control pollution of the marine environment from land-based sources including rivers, estuaries, pipelines and outfall structures, taking into account internationally agreed rules, standards and recommended practices and procedures.
2. States shall also take other measures as may be necessary to prevent, reduce and control pollution of the marine environment from land-based sources.
3. States shall endeavour to harmonize their national policies at the appropriate regional level.
4. States, acting in particular through competent international organizations or diplomatic conference, shall endeavour to establish global and regional rules, standards and recommended practices and procedures to prevent,

reduce and control pollution of the marine environment from land-based sources, taking into account characteristic regional features, the economic capacity of developing States and their need for economic development. Such rules, standards and recommended practices and procedures shall be re-examined from time to time as necessary.

5. Laws, regulations, measures, rules, standards and recommended practices and procedures referred to in paragraphs 1, 2 and 4 respectively shall include those designed to minimize, to the fullest possible extent, the release of toxic, harmful and noxious substances, especially persistent substances, into the marine environment.

Article 209
Pollution from sea-bed activities

1. Coastal States shall establish national laws and regulations to prevent, reduce and control pollution of the marine environment arising from or in connexion with sea-bed activities subject to their jurisdiction and from artificial islands, installations and structures under their jurisdiction, pursuant to articles 60 and 80.
2. States shall also take other measures as may be necessary to prevent, reduce and control such pollution.
3. Such laws, regulations and measures shall be no less effective than international rules, standards and recommended practices and procedures.
4. States shall endeavour to harmonize their national policies at the appropriate regional level.
5. States, acting in particular through competent international organizations or diplomatic conference, shall establish global and regional rules, standards and recommended practices and procedures to prevent, reduce and control pollution of the marine environment arising from or in connexion with sea-bed activities subject to their jurisdiction and from artificial islands, installations and structures under their jurisdiction referred to in paragraph 1. Such rules, standards and recommended practices and procedures shall be re-examined from time to time as necessary.

Article 210
Pollution from activities in the Area

1. International rules, standards and recommended practices and procedures shall be established in accordance with the provisions of Part XI, to prevent, reduce and control pollution of the marine environment from activities relating to the exploration and exploitation of the Area. Such rules, standards and recommended practices and procedures shall be re-examined from time to time as necessary.
2. Subject to other relevant provisions of this Section, States shall establish national laws and regulations to prevent, reduce and control pollution of the marine environment from activities relating to the exploration and exploitation of the Area undertaken by vessels, installations, structures and other devices flying their flag or of their registry. The requirements of such laws and regulations shall be no less effective than the international rules, standards and procedures referred to in paragraph 1 of this article.

Article 211
Dumping

1. States shall establish national laws and regulations to prevent, reduce and control pollution of the marine environment from dumping.
2. States shall also take other measures as may be necessary to prevent, reduce and control such pollution.
3. Such laws, regulations and measures shall ensure that dumping is not carried out without the permission of the competent authorities of States.
4. States, acting in particular through competent international organizations or diplomatic conference, shall endeavour to establish global and regional rules, standards and recommended practices and procedures to prevent, reduce and control pollution of the marine environment by dumping. Such rules, standards and recommended practices and procedures shall be re-examined from time to time as necessary.
5. Dumping, within the territorial sea and the exclusive economic zone or onto the continental shelf shall not be carried out without the express prior approval of the coastal State, which has the right to permit, regulate and control such dumping after due consultation with other States which by reason of their geographical situation may be adversely affected thereby.
6. National laws, regulations and measures shall be no less effective in preventing, reducing and controlling pollution from dumping than global rules and standards.

Article 212
Pollution from vessels

1. States, acting through the competent international organization or general diplomatic conference, shall establish international rules and standards for the prevention, reduction and control of pollution of the marine environment from vessels. Such rules and standards shall, in the same manner, be re-examined from time to time as necessary.
2. States shall establish laws and regulations for the prevention, reduction and control of pollution of the marine environment from vessels flying their flag or vessels of their registry. Such laws and regulations shall at least have the same effect as that of generally accepted international rules and standards established through the competent international organization or general diplomatic conference.
3. Coastal States may, in the exercise of their

sovereignty within their territorial sea, establish national laws and regulations for the prevention, reduction and control of marine pollution from vessels. Such laws and regulations shall, in accordance with section 3 of Part II not hamper innocent passage of foreign vessels.

4. Coastal States, for the purpose of enforcement as provided for in section 6 of this Part of the present Convention, may in respect of their economic zones establish laws and regulations for the prevention, reduction and control of pollution from vessels conforming to and giving effect to generally accepted international rules and standards established through the competent international organization or general diplomatic conference.

5. Where international rules and standards referred to in paragraph 1 are inadequate to meet special circumstances and where coastal States have reasonable grounds for believing that a particular, clearly defined area of their respective exclusive economic zones is an area where, for recognized technical reasons in relation to its oceanographical and ecological conditions, as well as its utilization or the protection of its resources, and the particular character of its traffic, the adoption of special mandatory methods for the prevention of pollution from vessels is required, coastal States, after appropriate consultations through the competent international organization with any other countries concerned, may for that area, direct a communication to the competent international organization, submitting scientific and technical evidence in support, and information on necessary reception facilities. The organization shall, within twelve months after receiving such a communication, determine whether the conditions in that area correspond to the requirements set out above. If the organization so determines, the coastal State may, for that area, establish laws and regulations for the prevention, reduction and control of pollution from vessels, implementing such international rules and standards or navigational practices as are made applicable through the competent international organization for special areas. Coastal States shall publish the limits of any such particular, clearly defined area, and laws and regulations applicable therein shall not become applicable in relation to foreign vessels until fifteen months after the submission of the communication to the competent international organization. Coastal States, when submitting the communication for the establishment of a special area within their respective exclusive economic zones, shall at the same time, notify the competent international organization if it is their intention to establish additional laws and regulations for that special area for the prevention, reduction and control of pollution from vessels. Such additional laws and regulations may relate to discharges or navigational practices but shall not require foreign vessels to observe design, construction, manning or equipment standards other than generally accepted international rules and standards and shall become applicable in relation to foreign vessels 15 months after the submission of the communication to the competent international organization, and provided the organization agrees within twelve months after submission of the communication.

Article 213
Pollution from or through the atmosphere

1. States shall, within air space under their sovereignty or with regard to vessels or aircraft flying their flag or of their registry, establish national laws and regulations to prevent, reduce and control pollution of the marine environment from or through the atmosphere, taking into account internationally agreed rules, standards and recommended practices and procedures.

2. States shall also take other measures as may be necessary to prevent, reduce and control such pollution.

3. States, acting in particular through competent international organizations or diplomatic conference shall endeavour to establish global and regional rules, standards and recommended practices and procedures to prevent, reduce and control pollution of the marine environment from or through the atmosphere.

SECTION 6. ENFORCEMENT
Article 214
Enforcement with respect to land-based sources of pollution

States shall enforce their laws and regulations established in accordance with article 208 of the present Convention and shall adopt the necessary legislative, administrative and other measures to implement applicable international rules and standards established through competent international organizations or diplomatic conference for the protection and preservation of the marine environment from land-based sources of marine pollution.

Article 215
Enforcement with respect to pollution from sea-bed activities

States shall enforce their laws and regulations established in accordance with article 209 of the present Convention and shall adopt the necessary legislative, administrative and other measures to implement applicable international rules and standards established through competent international organizations or diplomatic conference for the protection and preservation of the marine environment from pollution arising from sea-bed activities subject to their jurisdiction and from artificial islands, installations and structures under their jurisdiction, pursuant to articles 60 and 80.

Article 216
Enforcement with respect to pollution from activities in the Area

Enforcement of international rules, standards and recommended practices and procedures established to prevent, reduce and control pollution of the marine environment from activities concerning exploration and exploitation of the Area pursuant to Part XI of the present Convention shall be governed by the provisions of that Part.

Article 217
Enforcement with respect to dumping

1. Laws and regulations adopted in accordance with the present Convention and applicable international rules and standards established through competent international organizations or diplomatic conference for the prevention, reduction and control of pollution of the marine environment from dumping shall be enforced:

   (a) by the coastal State with regard to dumping within its territorial sea or its exclusive economic zone or onto its continental shelf;

   (b) by the flag State with regard to vessels and aircraft registered in its territory or flying its flag;

   (c) by any State with regard to acts of loading of wastes or other matter occurring within its territory or at its off-shore terminals.

2. This article shall not impose on any State an obligation to institute proceedings when such proceedings have already been commenced by another State in accordance with this article.

Article 218
Enforcement by flag States

1. States shall ensure compliance with applicable international rules and standards established through the competent international organization or general diplomatic conference and with their laws and regulations established in accordance with the present Convention for the prevention, reduction and control of pollution of the marine environment, by vessels flying their flag or vessels of their registry and shall adopt the necessary legislative, administrative and other measures for their implementation. Flag States shall provide for the effective enforcement of such rules, standards, laws and regulations, irrespective of where the violation occurred.

2. Flag States shall, in particular, establish appropriate measures in order to ensure that vessels flying their flags or vessels of their registry are prohibited from sailing, until they can proceed to sea in compliance with the requirements of international rules and standards referred to in paragraph 1 for the prevention, reduction and control of pollution from vessels, including the requirements in respect of design, construction, equipment and manning of vessels.

3. States shall ensure that vessels flying their flags or of their registry carry on board certificates required by and issued pursuant to international rules and standards referred to in paragraph 1. Flag States shall ensure that their vessels are periodically inspected in order to verify that such certificates are in conformity with the actual condition of the vessels. These certificates shall be accepted by other States as evidence of the condition of the vessel and regarded as having the same force as certificates issued by them, unless there are clear grounds for believing that the condition of the vessel does not correspond substantially with the particulars of the certificates.

4. If a vessel commits a violation of rules and standards established through the competent international organization or general diplomatic conference, the flag State, without prejudice to articles 219, 221, and 229 shall provide for immediate investigation and where appropriate cause proceedings to be taken in respect of the alleged violation irrespective of where the violation occurred or where the pollution caused by such violation has occurred or has been spotted.

5. Flag States may seek in conducting investigation of the violation the assistance of any other State whose co-operation could be useful in clarifying the circumstances of the case. States shall endeavour to meet the appropriate request of flag States.

6. Flag States shall, at the written request of any State, investigate any violation alleged to have been committed by their vessels. If satisfied that sufficient evidence is available to enable proceedings to be brought in respect of the alleged violation, flag States shall without delay cause such proceedings to be taken in accordance with their laws.

7. Flag States shall promptly inform the requesting State and the competent international organization of the action taken and its outcome. Such information shall be available to all States.

8. Penalties specified under the legislation of flag States for their own vessels shall be adequate in severity to discourage violations wherever the violations occur.

Article 219
Enforcement by port States

1. When a vessel is voluntarily within a port or at an off-shore terminal of a State, that State may undertake investigations and, where warranted by the evidence of the case, cause proceedings to be taken in respect of any discharge from that vessel in violation of applicable international rules and standards established through the competent international organization or general diplomatic conference, outside the internal waters, territorial sea, or exclusive economic zone of that State.

2. No proceedings pursuant to paragraph 1 shall be taken in respect of a discharge violation in the internal waters, the territorial sea or exclusive economic zone of another State unless requested by that State, the flag State, or the State damaged or threatened by a discharge violation, or unless the violation has caused or is likely to cause pollution in the internal waters, territorial sea or exclusive economic zone of the State instituting the proceedings.

3. A State, whenever a vessel is voluntarily within one of its ports, or off-shore terminals, shall, as far as practicable comply with requests from any State for investigation of discharge violations of international rules and standards

referred to in paragraph 1, believed to have occurred in, caused, or threatens damage to the internal waters, territorial sea or exclusive economic zone of the State making such a request, and likewise, shall, as far as practicable, comply with requests from the flag State for investigation of such violations, irrespective of where the violations occurred.

4. The records of the investigation carried out by a port State pursuant to the provisions of this article shall be transferred to the flag State or to the coastal State at their request. Any proceedings initiated by the port State on the basis of such an investigation, subject to the provisions of section 7 of this Part, may be suspended at the request of a coastal State, when the violation has occurred within the internal waters, territorial sea or exclusive economic zone of that State and the evidence and records of the case and any bond posted with the authorities of the port State shall be transferred to the coastal State. Such transfer shall preclude the continuation of proceedings in the port State.

Article 220
Measures relating to seaworthiness of vessels to avoid pollution

Subject to the provisions of section 7 of this Part of the present Convention, States which have ascertained, upon request or on their own initiative, that a vessel within their ports or at their off-shore terminals is in violation of applicable international rules and standards relating to seaworthiness and thereby threatens damage to the marine environment shall, as far as practicable, take administrative measures to prevent the vessel from sailing. Such States may permit the vessel to proceed only to the nearest appropriate repair yard and upon rectification of the causes of the violation, shall permit the vessel to continue immediately.

Article 221
Enforcement by coastal States

1. When a vessel is voluntarily within a port or at an off-shore terminal of a State, that State may, subject to the provisions of section 7 of this Part of the Convention cause proceedings to be taken in respect of any violation of national laws and regulations established in accordance with the present Convention or applicable international rules and standards for the prevention, reduction and control of pollution from vessels when the violation has occurred within the territorial sea or the exclusive economic zone of that State.

2. Where there are clear grounds for believing that a vessel navigating in the territorial sea of a State has, during its passage therein, violated national laws and regulations established in accordance with the present Convention or applicable international rules and standards for the prevention, reduction and control of pollution from vessels, that State, without prejudice to the application of the relevant provisions of section 3 of Part II may undertake physical inspection of the vessel relating to the violation and may, when warranted by the evidence of the case, cause proceedings, including arrest of the vessel, to be taken in accordance with its laws, subject to the provisions of section 7 of this Part of the present Convention.

3. Where there are clear grounds for believing that a vessel navigating in the exclusive economic zone or the territorial sea of a State has, in the exclusive economic zone, violated applicable international rules and standards or national laws and regulations conforming and giving effect to such international rules and standards for the prevention, reduction and control of pollution from vessels, that State may require the vessel to give information regarding the identification of the vessel and its port of registry, its last and next port of call and other relevant information required to establish whether a violation has occurred.

4. Flag States shall take legislative, administrative and other measures so that their vessels comply with requests for information as set forth in paragraph 3.

5. Where there are clear grounds for believing that a vessel navigating in the exclusive economic zone or the territorial sea of a State has, in the exclusive economic zone, violated applicable international rules and standards or national laws and regulations conforming and giving effect to such international rules and standards for the prevention, reduction and control of pollution from vessels and the violation has resulted in a substantial discharge into and, insignificant pollution of, the marine environment, that State may undertake physical inspection of the vessel for matters relating to the violation if the vessel has refused to give information or if the information supplied by the vessel is manifestly at variance with the evident factual situation and if the circumstances of the case justify such inspection.

6. Where there are clear grounds for believing that a vessel navigating in the exclusive economic zone or the territorial sea of a State has, in the exclusive economic zone, committed a flagrant or gross violation of applicable international rules and standards or national laws and regulations conforming and giving effect to such international rules and standards for the prevention, reduction and control of pollution from vessels, resulting in discharge causing major damage or threat of major damage to the coastline or related interests of the coastal State, or to any resources of its territorial sea or exclusive economic zone, that State may, subject to the provisions of Section 7 of this Part of the Convention provided that the evidence so warrants, cause proceedings to be taken in accordance with its laws.

7. Notwithstanding the provisions of paragraph 6, whenever appropriate procedures have been established either through the competent international organization or as otherwise agreed, whereby compliance with requirements for bonding or other appropriate financial security has been assured, the coastal

State if bound by such procedures shall allow the vessel to proceed.
8. The provisions of paragraphs 3, 4, 5, 6 and 7 shall apply correspondingly in respect of national laws and regulations established pursuant to paragraph 5 of article 212.

Article 222
Measures relating to maritime casualties to avoid pollution

1. Nothing in this Chapter shall affect the right of States to take measures, in accordance with international law, beyond the limits of the territorial sea for the protection of coastlines or related interests, including fishing, from grave and imminent danger from pollution or threat of pollution following upon a maritime casualty or acts related to such a casualty.
2. Measures taken in accordance with this article shall be proportionate to the actual or threatened damage.

Article 223
Enforcement with respect to pollution from or through the atmosphere

States shall, within air space under their sovereignty or with regard to vessels or aircraft flying their flag or of their registry, enforce their laws and regulations established in accordance with the provisions of the present Convention and shall adopt the necessary legislative, administrative and other measures to implement applicable international rules and standards established through competent international organizations or diplomatic conference to prevent, reduce and control pollution of the marine environment from and through the atmosphere, in conformity with all relevant international rules and standards concerning the safety of air navigation.

SECTION 7. SAFEGUARDS
Article 224
Measures to facilitate proceedings

In proceedings pursuant to this Chapter, States shall take measures to facilitate the hearing of witnesses and the admission of evidence submitted by authorities of another State, or by the competent international organization and shall facilitate the attendance at such proceedings of official representatives of the competent international organization or of the flag State, or of any State affected by pollution arising out of any violation. The official representatives attending such proceedings shall enjoy such rights and duties as may be provided under national legislation or applicable international law.

Article 225
Exercise of powers of enforcement

The powers of enforcement against foreign vessels under this Part of the present Convention may only be exercised by officials or by warships or military aircraft or other ships or aircraft clearly marked and identifiable as being on government service and authorized to that effect.

Article 226
Duty to avoid adverse consequences in the exercise of the powers of enforcement

In the exercise of their powers of enforcement against foreign vessels under the present Convention, States shall not endanger the safety of navigation or otherwise cause any hazard to a vessel, or bring it to an unsafe port or anchorage, or cause an unreasonable risk to the marine environment.

Article 227
Investigation of foreign vessels

1. States shall not delay a foreign vessel longer than is essential for purposes of investigation provided for in articles 217, 219 and 221 of this Part of the present Convention. If the investigation indicates a violation of applicable laws and regulations or international rules and standards for the preservation of the marine environment release shall be made subject to reasonable procedures such as bonding or other appropriate financial security. Without prejudice to applicable international rules and standards relating to the seaworthiness of ships, the release of a vessel may, whenever it would present an unreasonable threat of damage to the marine environment, be refused or made conditional upon proceeding to the nearest appropriate repair yard.
2. States shall co-operate to develop procedures for the avoidance of unnecessary physical inspection of vessels at sea.

Article 228
Non-discrimination of foreign vessels

In exercising their right and carrying out their duties under this Part of the present Convention, States shall not discriminate in form or in fact against vessels of any other State.

Article 229
Suspension and restrictions on institution of proceedings

1. Proceedings to impose penalties in respect of any violation of applicable laws and regulations or international rules and standards relating to the prevention, reduction and control of pollution from vessels committed by a foreign vessel beyond the territorial sea of the State instituting proceedings shall be suspended upon the taking of proceedings to impose penalties under corresponding charges by the flag State within six months of the first institution of proceedings, unless those proceedings relate to a case of major damage to the coastal State or the flag State in question has repeatedly disregarded its obligations to enforce effectively the applicable international rules and standards in respect of violations committed by its vessels. The flag State shall in due course make available to the first State instituting proceedings a full dossier of the case and the records of the proceedings, whenever the flag

State has requested the suspension of proceedings in accordance with the provisions of this article. When proceedings by the flag State have been brought to a conclusion, the suspended proceedings shall be finally terminated. Upon payment of costs incurred in respect of such proceedings, any bond posted or other financial security provided in connexion with the suspended proceedings shall be released by the coastal State.

2. Proceedings to impose penalties on foreign vessels shall not be instituted after the expiry of a period of three years from the date on which the violation was committed, and shall not be taken by any State in the event of proceedings having been instituted by another State subject to the provisions set out in paragraph 1.

3. The provisions of this article shall be without prejudice to the right of the flag State to adopt any measures, including the taking of proceedings to impose penalties, according to its laws irrespective of prior proceedings by another State.

Article 230
Institution of civil proceedings

Nothing in the present Convention shall affect the institution of civil proceedings in respect of any claim for loss or damage resulting from pollution of the marine environment.

Article 231
Monetary penalties and the observance of recognized rights of the accused

1. Only monetary penalties may be imposed with respect to violations of national laws and regulations, or applicable international rules and standards, for the prevention, reduction and control of pollution from vessels committed by foreign vessels beyond the internal waters.

2. In the conduct of proceedings to impose penalties in respect of such violations committed by a foreign vessel, recognized rights of the accused shall be observed.

Article 232
Notification to flag States and other States concerned

States shall promptly notify the flag State and any other State concerned of any measures taken pursuant to section 6 of this Part of the Convention against foreign vessels, and shall submit to the flag State all official reports concerning such measures. However, with respect to violations committed in the territorial sea, the foregoing obligations of the coastal State shall apply only to such measures as are taken in proceedings. The consular officers or diplomatic agents, and where possible the maritime authority of the flag State, shall be immediately informed of any such measures.

Article 233
Liability of States arising from enforcement measures

States shall be liable for damage or loss attributable to them arising from measures taken pursuant to section 6 of this Part of the Convention, when such measures were unlawful or exceeded those reasonably required in the light of available information. States shall provide for recourse in their courts for actions in respect of such damage or loss.

Article 234
Safeguards with respect to straits used for international navigation

Nothing in sections 5, 6 and 7 of this Part of the Convention shall affect the legal régime of straits used for international navigation. However, if a foreign ship other than those referred to in section 10 of this Part of the present Convention has committed a violation of the laws and regulations referred to in subparagraphs 1 (a) and (b) of article 42 of Part III of the present Convention causing or threatening major damage to the marine environment of the straits, the States bordering the straits may take appropriate enforcement measures and if so shall respect *mutatis mutandis* the provisions of section 7 of this Part of the Convention.

SECTION 8. ICE-COVERED AREAS
Article 235
Ice-covered areas

Coastal States have the right to establish and enforce non-discriminatory laws and regulations for the prevention, reduction and control of marine pollution from vessels in ice-covered areas within the limits of the exclusive economic zone, where particularly severe climatic conditions and the presence of ice covering such areas for most of the year create obstructions or exceptional hazards to navigation, and pollution of the marine environment could cause major harm to or irreversible disturbance of the ecological balance. Such laws and regulations shall have due regard to navigation and the protection of the marine environment based on the best available scientific evidence.

SECTION 9. RESPONSIBILITY AND LIABILITY
Article 236
Responsibility and liability

1. States are responsible for the fulfilment of their international obligations concerning the protection and preservation of the marine environment. They shall be liable in accordance with international law for damage attributable to them resulting from violations of these obligations.

2. States shall ensure that recourse is available in accordance with their legal systems for prompt and adequate compensation or other relief in respect of damage caused by pollution of the marine environment by persons, natural or juridical, under their jurisdiction.

3. States shall co-operate in the development of international law relating to criteria and procedures for the determination of liability, the assessment of damage, the payment of

compensation and the settlement of related disputes.

SECTION 10. SOVEREIGN IMMUNITY
Article 237
Sovereign immunity

The provisions of the present Convention regarding pollution of the marine environment shall not apply to any warship, naval auxiliary, other vessels or aircraft owned or operated by a State and used, for the time being, only on government non-commercial service. However, each State shall ensure by the adoption of appropriate measures not impairing operations or operational capabilities of such vessels or aircraft owned or operated by it, that such vessels or aircraft act in a manner consistent, so far as is reasonable and practicable, with the present Convention.

SECTION 11. OBLIGATIONS UNDER OTHER CONVENTIONS ON THE PROTECTION AND PRESERVATION OF THE MARINE ENVIRONMENT
Article 238
Obligations under other Conventions on the protection and preservation of the marine environment

1. The provisions of this Part of the Convention shall be without prejudice to the specific obligations assumed by States under special conventions and agreements concluded previously which relate to the protection and preservation of the marine environment and to agreements which may be concluded in furtherance of the general principles set forth in the present Convention.
2. Specific obligations assumed by States under special conventions, with respect to the protection and preservation of the marine environment, should be applied in a manner consistent with the general principles and objectives of the present Convention.

PART XIII
MARINE SCIENTIFIC RESEARCH
SECTION 1. GENERAL PROVISIONS
Article 239
Right to conduct marine scientific research

States, irrespective of their geographical location, and competent international organizations have the right to conduct marine scientific research subject to the rights and duties of other States as provided for in the present Convention.

Article 240
Promotion of marine scientific research

States and competent international organizations shall promote and facilitate the development and conduct of marine scientific research in accordance with the present Convention.

Article 241
General principles for the conduct of marine scientific research

In the conduct of marine scientific research the following principles shall apply:
(a) Marine scientific research activities shall be conducted exclusively for peaceful purposes;
(b) Such activities shall be conducted with appropriate scientific methods and means compatible with the present Convention;
(c) Such activities shall not unjustifiably interfere with other legitimate uses of the sea compatible with the present Convention and shall be duly respected in the course of such uses;
(d) Such activities shall comply with all relevant regulations established in conformity with the present Convention including those for the protection and preservation of the marine environment.

Article 242
Marine scientific research activities not constituting the legal basis for any claim

Marine scientific research activities shall not form the legal basis for any claim to any part of the marine environment or its resources.

SECTION 2. GLOBAL AND REGIONAL CO-OPERATION
Article 243
Promotion of international co-operation

States and competent international organizations shall, in accordance with the principle of respect for sovereignty and on the basis of mutual benefit, promote international co-operation in marine scientific research for peaceful purposes.

Article 244
Creation of favourable conditions

States and competent international organizations shall co-operate with one another, through the conclusion of bilateral, regional and multilateral agreements, to create favourable conditions for the conduct of marine scientific research in the marine environment and to integrate the efforts of scientists in studying the essence of and the interrelations between phenomena and processes occurring in the marine environment.

Article 245
Publication and dissemination of information and knowledge

1. States and competent international organizations shall, in accordance with the present Convention, make available information on proposed major programmes and their objectives as well as knowledge resulting from marine scientific research by publication and dissemination through appropriate channels.
2. For this purpose, States shall, both individually and in co-operation with other States and with competent international organizations, actively promote the flow of scientific data and

information and the transfer of knowledge resulting from marine scientific research in particular to developing states, as well as the strengthening of the autonomous marine research capabilities of developing states through, *inter alia*, programmes to provide adequate education and training of their technical and scientific personnel.

SECTION 3. CONDUCT AND PROMOTION OF MARINE SCIENTIFIC RESEARCH

Article 246
Marine scientific research in the territorial sea

Coastal States, in the exercise of their sovereignty, have the exclusive right to regulate, authorize and conduct marine scientific research in their territorial sea. Marine scientific research activities therein shall be conducted only with the express consent of and under the conditions set forth by the coastal State.

Article 247
Marine scientific research in the exclusive economic zone and on the continental shelf

1. Coastal States, in the exercise of their jurisdiction, have the right to regulate, authorize and conduct marine scientific research in their exclusive economic zone and on their continental shelf in accordance with the relevant provisions of the present Convention.
2. Marine scientific research activities in the exclusive economic zone and on the continental shelf shall be conducted with the consent of the coastal State.
3. Coastal States shall, in normal circumstances grant their consent for marine scientific research projects by other states or competent international organizations in their exclusive economic zone or on their continental shelf to be carried out in accordance with the present Convention exclusively for peaceful purposes and in order to increase scientific knowledge of the marine environment for the benefit of all mankind. To this end, coastal States shall establish rules and procedures ensuring that such consent will not be delayed or denied unreasonably.
4. Coastal States may however in their discretion withold their consent to the conduct of a marine scientific research project of another State or competent international organization in the exclusive economic zone or on the continental shelf of the coastal State if that project:

(a) is of direct significance for the exploration and exploitation of natural resources, whether living or non-living;

(b) involves drilling into the continental shelf, the use of explosives or the introduction of harmful substances into the marine environment;

(c) involves the construction, operation or use of artificial islands, installations and structures as referred to in articles 60 and 80;

(d) contains information communicated pursuant to article 254 regarding the nature and objectives of the project which is inaccurate or if the researching state or competent international organization has outstanding obligations to the coastal State from a prior research project.

5. Marine scientific research activities referred to in this article shall not unjustifiably interfere with activities undertaken by coastal States in accordance with their sovereign rights and jurisdiction as provided for in the present Convention.

Article 248
Research project under the auspices of, or undertaken by international organizations

A coastal State which is a member of a regional or global organization or has a bilateral agreement with such an organization, and in whose exclusive economic zone or on whose continental shelf the organization wants to carry out a marine scientific research project, shall be deemed to have authorized the project to be carried out, upon notification to the duly authorized officials of the coastal State by the organization, if that State approved the project when the decision was made by the organization for the undertaking of the project or is willing to participate in it.

Article 249
Duty to provide information to the coastal State

States and competent international organizations which intend to undertake marine scientific research in the exclusive economic zone or on the continental shelf of a coastal State shall, not less than six months in advance of the expected starting date of the research project, provide that State with a full description of:

(a) the nature and objectives of the research project;

(b) the method and means to be used, including name, tonnage, type and class of vessels and a description of scientific equipment;

(c) the precise geographical areas in which the activities are to be conducted;

(d) the expected date of first appearance and final departure of the research vessels, or deployment of the equipment and its removal, as appropriate;

(e) the name of the sponsoring institution, its director, and the person in charge of the research project; and

(f) the extent to which it is considered that the coastal State should be able to participate or to be represented in the research project.

Article 250
Duty to comply with certain conditions

1. States and competent international organizations when undertaking marine scientific research in the exclusive economic zone or on the continental shelf of a coastal State shall comply with the following conditions:

(a) Ensure the rights of the coastal State, if it so desires, to participate or be represented in the research project, especially on board research vessels and other craft or installations, when practicable, without payment of any remuneration to the scientists of the coastal State and without obligation to contribute towards the costs of the research project;

(b) Provide the coastal State, at its request, with preliminary reports, as soon as practicable, and with the final results and conclusions after the completion of the research;

(c) Undertake to provide access for the coastal State, at its request, to all data and samples derived from the research project and likewise to furnish it with data which may be copied and samples which may be divided without detriment to their scientific value;

(d) If requested, assist the coastal State in assessing such data and samples and the results thereof;

(e) Ensure, subject to paragraph 2 of this article, that the research results are made internationally available through appropriate national or international channels, as soon as feasible;

(f) Inform the coastal State immediately of any major change in the research programme;

(g) Unless otherwise agreed remove the scientific installations or equipment once the research is completed.

2. This article is without prejudice to the conditions established by the laws and regulations of the coastal State for the granting of consent where the coastal State, not withstanding the provisions of article 247 nevertheless, grants its consent to the project in question.

Article 251
Communications concerning research project

Communications concerning the research project shall be made through appropriate official channels unless otherwise agreed.

Article 252
General criteria and guidelines

States shall seek to promote through competent international organizations the establishment of general criteria and guidelines to assist States in ascertaining the nature and implications of marine scientific research.

Article 253
Implied consent

States or competent international organizations may proceed with a research project upon the expiry of six months from the date upon which the information required pursuant to article 249 was provided to the coastal State unless within four months of the receipt of the communication containing such information the coastal State has informed the State or organization conducting the research that:

(a) it has withheld its consent under the provisions of article 247; or

(b) the information given by the State or competent international organization in question regarding the nature or objectives of the research project does not conform to the manifestly evident facts; or

(c) it requires supplementary information relevant to the conditions and the information provided for under article 249 and 250; or

(d) outstanding obligations exist with respect to a previous research project carried out by that State or organization, with regard to conditions established in article 250.

Article 254
Cessation of research activities

1. The coastal State shall have the right to require the cessation of any research activities in progress within its exclusive economic zone or on its continental shelf if:

(a) the research project is not being conducted in accordance with the information initially communicated to the coastal State as provided under article 249 regarding the nature, objectives, method, means or geographical areas of the project; or

(b) the State or competent international organization conducting the research project fails to comply with the provisions of article 250 concerning the rights of the coastal State with respect to the project and compliance is not secured within a reasonable period of time.

Article 255
Rights of neighbouring land-locked and geographically disadvantaged States

1. States and competent international organizations conducting marine scientific research in the exclusive economic zone or on the continental shelf of a coastal State shall take into account the interests and rights of neighbouring land-locked and other geographically disadvantaged States, as provided for in the present Convention and shall notify these States of the proposed research project as well as provide, at their request, relevant information and assistance as specified in article 249 and subparagraph (d) and (f) of article 250.

2. Such neighbouring land-locked and other geographically disadvantaged States shall, at their request, be given the opportunity to participate, whenever feasible, in the proposed research project through qualified experts appointed by them.

Article 256
Measures to facilitate marine scientific research and assist research vessels

For the purpose of giving effect to bilateral or regional and other multilateral agreements and in a spirit of international co-operation to promote and facilitate marine scientific research activities conducted in accordance with the present Convention, coastal States shall adopt reasonable and uniformly applied rules, regulations and administrative procedures applicable to States and competent international organizations desiring to carry out research activities in the exclusive economic zone or on the continental shelf and shall, for the same purpose, adopt measures to facilitate access to their harbours and to promote assistance for marine scientific research vessels carrying out such activities, in accordance with the present Convention.

Article 257
Marine scientific research in the Area

States, irrespective of their geographical location, as well as competent international organizations, shall have the right, in conformity with the provisions of Part XI of the present Convention, to conduct marine scientific research in the Area.

Article 258
Marine scientific research in the water column beyond the exclusive economic zone

States, irrespective of their geographical location, as well as competent international organizations, shall have the right, in conformity with the present Convention, to conduct marine scientific research in the water column beyond the limits of the exclusive economic zone.

SECTION 4. LEGAL STATUS OF SCIENTIFIC RESEARCH INSTALLATIONS AND EQUIPMENT IN THE MARINE ENVIRONMENT

Article 259
Deployment and use

The deployment and use of any type of scientific research installations or equipment in any area of the marine environment shall be subject to the same conditions as those for the conduct of marine scientific research in such area, as provided for in the present Convention.

Article 260
Legal status

The installations or equipment referred to in this section shall not have the status of islands or possess their own territorial sea, and their presence shall not affect the delimitation of the territorial sea, exclusive economic zone and continental shelf of the coastal State.

Article 261
Safety zones

Safety zones of a reasonable width not exceeding a distance of 500 metres may be created around scientific research installations in accordance with the relevant provisions of the present Convention. All States shall ensure that such safety zones are respected by their vessels.

Article 262
Non-interference with shipping routes

The deployment and use of any type of scientific research installations or equipment shall not constitute an obstacle to established international shipping routes.

Article 263
Identification markings and warning signals

Installations or equipment referred to in this section shall bear identification markings indicating the State of registry or the international organization to which they belong and shall have adequate internationally agreed warning signals to ensure safety at sea and the safety of air navigation, taking into account the principles established by competent international organizations.

SECTION 5. RESPONSIBILITY AND LIABILITY

Article 264
Responsibility and liability

1. States and competent international organizations shall be responsible for ensuring that marine scientific research, whether undertaken by them or on their behalf, is conducted in accordance with the present Convention.
2. States and competent international organizations shall be responsible and liable for the measures they undertake in contravention of the present Convention in respect of marine scientific research activities conducted by other States, their natural or juridical persons or by competent international organizations, and shall provide compensation for damage resulting from such measures.
3. States and competent international organizations shall be responsible and liable pursuant to the principles set forth in article 236 for damage arising out of marine scientific research undertaken by them or on their behalf.

SECTION 6. SETTLEMENT OF DISPUTE

Article 265
Settlement of disputes

Unless otherwise agreed or settled by the parties concerned, disputes relating to the interpretation or application of the provisions of the present Convention with regard to marine scientific research shall be settled in accordance with section II of Part XV of the present Convention, except that the coastal State shall not be obliged to submit to such settlement any dispute arising out of:

(a) the exercise by the coastal State of a right or discretion in accordance with article 247; or

(b) the decision by the coastal State to terminate a research project in accordance with article 254.

Article 266
Interim measures

Pending settlement of a dispute in accordance with article 265, the State or competent international organization authorized to conduct a research project shall not allow research activities to commence or continue without the express approval of the coastal State concerned.

PART XIV
DEVELOPMENT AND TRANSFER OF MARINE TECHNOLOGY
SECTION 1. GENERAL PROVISIONS

Article 267
Promotion of development and transfer of marine technology

1. States, directly or through appropriate international organizations, shall co-operate within their capabilities to *promote* actively the

development and transfer of marine science and marine technology on fair and reasonable terms and conditions.

2. States shall promote the development of the marine scientific and technological capacity of States which may need and request technical assistance in this field, particularly developing States, including land-locked and geographically disadvantaged States, with regard to the exploration, exploitation, conservation and management of marine resources, the preservation of the marine environment, marine scientific research and other uses of the marine environment compatible with the present Convention, with a view to accelerating the social and economic development of the developing States.

3. States shall endeavour to foster favourable economic and legal conditions for the transfer of marine technology for the benefit of all parties concerned on an equitable basis.

Article 268
Protection of legitimate interests

States, in promoting such co-operation, shall have proper regard for all legitimate interests including, *inter alia*, the rights and duties of holders, suppliers and recipients of marine technology.

Article 269
Basic objectives

States, directly or through competent international organizations, shall promote:

(a) the acquisition, evaluation and dissemination of marine technological knowledge and facilitate access to such information and data;

(b) the development of appropriate marine technology;

(c) the development of the necessary technological infrastructure to facilitate the transfer of marine technology;

(d) the development of human resources through training and education of nationals of developing States and countries and especially of the least developed among them; and

(e) international co-operation at all levels, particularly at the regional, subregional and bilateral levels.

Article 270
Measures to achieve the basic objectives

In order to achieve the above-mentioned objectives, States, directly or through competent international organizations, shall, *inter alia*, endeavour to:

(a) establish programmes of technical co-operation for the effective transfer of all kinds of marine technology to States which may need and request technical assistance in this field, particularly the developing land-locked and other geographically disadvantaged States, as well as other developing States which have not been able to either establish or develop their own technological capacity in marine science and in the exploration and exploitation of the marine resources, and to develop the infrastructure of such technology;

(b) promote favourable conditions for the conclusion of agreements, contracts and other similar arrangements, under equitable and reasonable conditions;

(c) hold conferences, seminars and symposia on scientific and technological subjects, in particular, on policies and methods for the transfer of marine technology;

(d) promote the exchange of scientists, technologists and other experts;

(e) undertake projects, promote joint ventures and other forms of bilateral and multilateral co-operation.

SECTION 2. INTERNATIONAL CO-OPERATION

Article 271
Ways and means of international co-operation

International co-operation for the development and transfer of marine technology shall, where feasible and appropriate, be carried out through existing bilateral, regional or multilateral programmes, and also through expanded and new programmes in order to facilitate marine scientific research and the transfer of marine technology, particularly in new fields and appropriate international funding for ocean research and development.

Article 272
Guidelines, criteria and standards

States, directly or through competent international organizations, shall promote the establishment of generally accepted guidelines, criteria and standards for the transfer of marine technology on a bilateral basis or within the framework of international organizations and other fora, taking into account, in particular, the interests and needs of developing States.

Article 273
Co-ordination of international programmes

In the field of transfer of marine technology, States shall endeavour to ensure that competent international organizations co-ordinate their activities in this field, including any regional or global programmes taking into account the interests and needs of developing States, particularly land-locked and geographically disadvantaged States.

Article 274
Co-operation with international organizations and the Authority in the transfer of technology to developing States

States shall co-operate actively with competent international organizations and the Authority, to encourage and facilitate the transfer to developing States, their nationals and the Enterprise of skills and technology with regard to the exploration of the Area, the exploitation of its resources and other related activities.

Article 275
Objectives of the Authority with respect to the transfer of technology

Subject to all legitimate interests including, *inter alia*, the rights and duties of holders, suppliers and recipients of technology, the

Authority shall, with regard to the exploration of the Area and the exploitation of its resources, ensure:

(a) that on the basis of the principle of equitable geographical distribution, nationals of developing States, whether coastal, land-locked or geographically disadvantaged, shall be taken on for the purposes of training as members of the managerial, research and technical staff constituted for its undertaking;

(b) that the technical documentation on the relevant equipment, machinery, devices and processes be made available to all States, in particular developing States which may need and request technical assistance in this field;

(c) that adequate provision is made by the Authority to facilitate the acquisition by States which may need and request technical assistance in the field of marine technology, in particular developing States and the acquisition by their nationals of the necessary skills and know-how, including professional training;

(d) that States which may need and request technical assistance in this field, in particular developing States, are assisted in the acquisition of necessary equipment, processes, plant and other technical know-how through any financial arrangements provided for in the present Convention.

SECTION 3. REGIONAL MARINE SCIENTIFIC AND TECHNOLOGICAL CENTRES
Article 276
Establishment of regional centres

1. States shall, in co-ordination with the competent international organizations, the Authority and national marine scientific and technological institutions, promote the establishment, especially in developing States, of regional marine scientific and technological research centres in order to stimulate and advance the conduct of marine scientific research by developing States and foster the transfer of technology.
2. All States of the region shall duly co-operate with the regional centres in order to ensure the more effective achievement of their objectives.

Article 277
Functions of regional centres

The functions of such regional centres shall include, *inter alia*:

(a) training and educational programmes at all levels on various aspects of marine scientific and technological research, particularly marine biology, including conservation and management of living resources, oceanography, hydrography, engineering, sea-bed geological exploration, mining and desalination technologies;

(b) management studies;

(c) study programmes related to the protection and preservation of the marine environment, the prevention, reduction and control of pollution;

(d) organization of regional conferences, seminars and symposia;

(e) acquisition and processing of marine scientific and technological data and information;

(f) prompt dissemination of results of marine scientific and technological research in readily available publications;

(g) publicizing national policies with regard to the transfer of technology and systematic comparative study of those policies;

(h) compilation and systematization of information on the marketing of technology and on contracts and other arrangements concerning patents;

(i) technical co-operation with other countries of the region.

SECTION 4. CO-OPERATION AMONG INTERNATIONAL ORGANIZATIONS
Article 278
Co-operation among international organizations

The competent international organizations referred to in Parts XIII and XIV of the Convention shall take all appropriate measures to ensure, either directly or in close co-operation among themselves, the effective discharge of the functions and responsibilities assigned to them under this Part.

PART XV
SETTLEMENT OF DISPUTES
SECTION 1
Article 279
Obligation to settle disputes by peaceful means

The States Parties shall settle any dispute between them relating to the interpretation or application of the present Convention in accordance with paragraph 3 of article 2, and shall seek a solution through the peaceful means indicated in paragraph 1 of article 33, of the Charter of the United Nations.

Article 280
Settlement of disputes by means chosen by the parties

Nothing in this Part shall impair the right of any States Parties to agree at any time to settle a dispute between them relating to the interpretation or application of the present Convention by any peaceful means of their own choice.

Article 281
Obligation to exchange views

1. If a dispute arises between States Parties relating to the interpretation or application of the present Convention, the parties to the dispute shall proceed expeditiously to exchange views regarding settlement of the dispute through negotiations in good faith or other peaceful means.
2. Similarly, the parties shall proceed to an exchange of views whenever a procedure for the settlement of a dispute has been terminated without a settlement of the dispute, or where a settlement has been reached and the circumstances require further consultation regarding the manner of its implementation.

Article 282
Obligations under general, regional or special agreements

If States Parties which are parties to a dispute relating to the interpretation or application of the present Convention have accepted, through a general, regional or special agreement or some other instrument or instruments, an obligation to settle such dispute by resort to a final and binding procedure, such dispute shall, at the request of any party to the dispute, be referred to such procedure. In this case any other procedure provided in this Part shall not apply, unless the parties to the dispute otherwise agree.

Article 283
Procedure when dispute is not settled by means chosen by the parties

1. If States Parties which are parties to a dispute relating to the interpretation or application of the present Convention have agreed to seek a settlement of such dispute by a peaceful means of their own choice, the procedure specified in this Part shall apply only where no settlement has been reached, and the agreement between the parties does not preclude any further procedure.
2. If the parties have also agreed on a time-limit for such a procedure, the provisions of paragraph 1 shall apply only upon the expiration of that time-limit.

Article 284
Conciliation

1. Any State Party which is party to a dispute relating to the interpretation or application of the present Convention may invite the other party or parties to the dispute to submit the dispute to conciliation in accordance with the procedure in annex IV or with some other procedure.
2. If the other party accepts this invitation and the parties agree upon the procedure, any party to the dispute may submit it to the agreed procedure.
3. If the other party does not accept the invitation or the parties do not agree upon the procedure, the conciliation proceedings shall be deemed to be terminated.
4. When a dispute has been submitted to conciliation, such conciliation proceedings may only be terminated in accordance with the provisions of annex IV or other agreed conciliation procedure, as the case may be.

Article 285
Application of this section to disputes submitted pursuant to Part XI

The provisions of this section shall apply to any dispute which pursuant to section 6 of Part XI is to be settled in accordance with procedures provided for in this Part. If an entity other than a State Party is a party to such a dispute, this section shall apply *mutatis mutandis*.

SECTION 2
Article 286
Application of section 1 and proceedings under this section

Subject to the provisions of articles 296 and 297, any dispute relating to the interpretation or application of the present Convention shall, where no settlement has been reached by recourse to the provisions of section 1, be submitted, at the request of any party to the dispute, to the court or tribunal having jurisdiction under the provisions of this section.

Article 287
Choice of procedure

1. A State Party, when signing, ratifying or otherwise expressing its consent to be bound by the present Convention, or at any time thereafter, shall be free to choose, by means of a written declaration, one or more of the following means for the settlement of disputes relating to the interpretation or application of the present Convention:
   (a) The Law of the Sea Tribunal constituted in accordance with annex V;
   (b) The International Court of Justice;
   (c) An arbitral tribunal constituted in accordance with annex VI;
   (d) A special arbitral tribunal constituted in accordance with annex VII for one or more of the categories of disputes specified therein.
2. Any declaration made under paragraph 1 shall not affect or be affected by the obligation of a State Party to accept the jurisdiction of the Sea-Bed Disputes Chamber of the Law of the Sea Tribunal to the extent and in the manner provided for in section 6 of Part XI.
3. A State Party, which is a party to a dispute not covered by a declaration in force, shall be deemed to have accepted arbitration in accordance with annex VI.
4. If the parties to a dispute have accepted the same procedure for the settlement of such dispute, it may be submitted only to that procedure, unless the parties otherwise agree.
5. If the parties to the dispute have not accepted the same procedure for the settlement of such dispute, it may be submitted only to arbitration in accordance with annex VI, unless the parties otherwise agree.
6. Any declaration made under this article shall remain in force until three months after notice of revocation has been deposited with the Secretary-General of the United Nations, who shall transmit copies thereof to the States Parties.
7. When a dispute has been submitted to a court or tribunal having jurisdiction under this article, a new declaration or notice of revocation of a declaration or expiration of a declaration, shall not affect in any way the proceedings so pending, unless the parties otherwise agree.
8. Declarations and notices referred to in this article shall be deposited with the Secretary-General of the United Nations, who shall transmit copies thereof to the States Parties.

Article 288
Competence

1. Any court or tribunal provided for in article 287 shall have jurisdiction in any dispute relating to the interpretation or application of the present Convention which is submitted to it in accordance with the provisions of this Part.
2. Any court or tribunal provided for in article 287 shall have jurisdiction in any dispute relating to the interpretation or application of an international agreement related to the purposes of the present Convention, which is submitted to it in accordance with the provisions of such agreement.
3. The Sea-Bed Disputes Chamber of the Law of the Sea Tribunal constituted in accordance with annex V, or an arbitral tribunal constituted in accordance with annex VI of this Part shall have jurisdiction in any matter provided for in section 6 of Part XI which is submitted to it in accordance with that Part.
4. Any disagreement as to whether a court or tribunal has jurisdiction, shall be settled by the decision of that court or tribunal.

Article 289
Expert advice and assistance

In any dispute involving scientific or technical matters, a court or tribunal exercising jurisdiction under this section may, at the request of a party to the dispute or on its own initiative, and in consultation with the parties, select not less than two scientific or technical experts from the appropriate list prepared in accordance with article 2 of annex VII, to sit with such court or tribunal but without the right to vote.

Article 290
Provisional measures

1. If a dispute has been duly submitted to any court or tribunal which considers *prima facie* that it has jurisdiction under this Part, or section 6 of Part XI, such court or tribunal shall have the power to prescribe any provisional measures which it considers appropriate under the circumstances to preserve the respective rights of the parties to the dispute or to prevent serious harm to the marine environment, pending final adjudication.
2. Any provisional measures under this article may only be prescribed, modified or revoked upon the request of a party to the dispute and after giving the parties an opportunity to be heard. Notice of any provisional measures, or of their modification or revocation, shall be given forthwith by the court or tribunal to the parties to the dispute and to such other States Parties as it considers appropriate.
3. Pending the constitution of an arbitral or special arbitral tribunal to which a dispute has been submitted under this section, any court or tribunal agreed upon by the parties or, failing such agreement within two weeks from the date of the request for provisional measures, the Law of the Sea Tribunal or, when appropriate, its Sea-Bed Disputes Chamber, shall have the power to prescribe provisional measures in conformity with paragraphs 1 and 2, if it considers *prima facie* that the tribunal to which the dispute has been submitted would have jurisdiction and that the urgency of the situation so requires. As soon as it has been constituted, the tribunal to which the dispute has been submitted may affirm, modify or revoke such provisional measures, acting in conformity with paragraphs 1 and 2.
4. As soon as the circumstances justifying the provisional measure have changed or ceased to exist, such provisional measures may be modified or revoked.
5. Any provisional measures prescribed or modified under this article shall be promptly complied with by the parties to the dispute.

Article 291
Access

1. All the dispute settlement procedures specified in this Part shall be open to States Parties.
2. The dispute settlement procedures specified in this Part shall be open to entities other than States Parties as provided for in section 6 of Part XI.

Article 292
Prompt release of vessels

1. Where the authorities of a State Party have detained a vessel flying the flag of another State Party and it is alleged that the coastal State has failed, neglected or refused to comply with the relevant provisions of the present Convention for the prompt release of the vessel or its crew upon the posting of a reasonable bond or other financial security, the question of release from detention may be brought before any court or tribunal agreed upon by the parties. Failing such agreement within 10 days from the time of detention, the question of release may be brought before any court or tribunal accepted by the detaining State under article 287 or before the Law of the Sea Tribunal, unless the parties otherwise agree.
2. An application for such release may only be brought by or on behalf of the flag State of the vessel.
3. The question of release shall be dealt with promptly by such court or tribunal which shall deal only with the question of release, without prejudice to the merits of any case before the appropriate domestic forum against the vessel, its owner or its crew. The authorities of the detaining State shall remain competent to release the vessel or its crew at any time.
4. The decision of such court or tribunal as to the release of the vessel or its crew shall be promptly complied with by the authorities of the detaining State upon the posting of the bond or other financial security determined by the court or tribunal.

Article 293
Applicable law

1. The court or tribunal having jurisdiction under this section shall apply the present Convention and other rules of international law not incompatible with the present Convention.

2. If the parties to a dispute so agree, the court or tribunal having jurisdiction under this section shall make its decision *ex aequo et bono*.

Article 294
Exhaustion of local remedies

Any dispute between States Parties relating to the interpretation or application of the present Convention may be submitted to the procedures provided for in this section only after local remedies have been exhausted as required by international law.

Article 295
Finality and binding force of decisions

1. Any decision rendered or measure prescribed by a court or tribunal having jurisdiction under this section shall be final and shall be complied with by all the parties to the dispute.
2. Any such decision or measure shall have no binding force except between the parties and in respect of that particular dispute.

Article 296
Limitations on applicability of this section

1. Without prejudice to the obligations arising under section 1, disputes relating to the exercise by a coastal State of sovereign rights or jurisdiction provided for in the present Convention shall only be subject to the procedures specified in the present Convention when the following conditions have been complied with:

    (a) that in any dispute to which the provisions of this article apply, the court or tribunal shall not call upon the other party or parties to respond until the party which has submitted the dispute has established *prima facie* that the claim is well founded;

    (b) that such court or tribunal shall not entertain any application which in its opinion constitutes an abuse of legal process or is frivolous or vexatious; and

    (c) that such court or tribunal shall immediately notify the other party to the dispute that the dispute has been submitted and such party shall be entitled, if it so desires, to present objections to the entertainment of the application.

2. Subject to the fulfilment of the conditions specified in paragraph 1, such court or tribunal shall have jurisdiction to deal with the following cases:

    (a) When it is alleged that a coastal State has acted in contravention of the provisions of the present Convention in regard to the freedoms and rights of navigation or overflight or of the laying of submarine cables and pipelines and other internationally lawful uses of the sea specified in article 58; or

    (b) When it is alleged that any State in exercising the aforementioned freedoms, rights or uses has acted in contravention of the provisions of the present Convention or of laws or regulations established by the coastal State in conformity with the present Convention and othe rules of international law not incompatible with the present Convention; or

    (c) When it is alleged that a coastal State has acted in contravention of specified international rules and standards for the protection and preservation of the marine environment which are applicable to the coastal State and which have been established by the present Convention or by a competent international organization or diplomatic conference acting in accordance with the present Convention.

3. No dispute relating to the interpretation or application of the provisions of the present Convention with regard to marine scientific research shall be brought before such court or tribunal unless the conditions specified in paragraph 1 have been fulfilled; provided that:

    (a) when it is alleged that there has been a failure to comply with the provision of articles 247 and 254, in no case shall the exercise of a right or discretion in accordance with article 247, or a decision taken in accordance with article 254, be called in question; and

    (b) the court or tribunal shall not substitute its discretion for that of the coastal State.

4. No dispute relating to the interpretation or application of the provisions of the present Convention with regard to the living resources of the sea shall be brought before such court or tribunal unless the conditions specified in paragraph 1 have been fulfilled; provided that:

    (a) when it is alleged that there has been a failure to discharge obligations arising under articles 61, 62, 69 and 70, in no case shall the exercise of a discretion in accordance with articles 61 and 62 be called in question; and

    (b) the court or tribunal shall not substitute its discretion for that of the coastal State; and

    (c) in no case shall the sovereign rights of a coastal State be called in question.

5. Any dispute excluded by the previous paragraphs may be submitted to the procedures specified in section 2 only by agreement of the parties to such dispute.

Article 297
Optional exceptions

1. Without prejudice to the obligations arising under section 1, a State Party when signing, ratifying or otherwise expressing its consent to be bound by the present Convention, or at any time thereafter, may declare that it does not accept any one or more of the procedures for the settlement of disputes specified in the present Convention with respect to one or more of the following categories of disputes:

    (a) Disputes concerning sea boundary delimitations between adjacent or opposite States, or those involving historic bays or titles, provided that the State making such a declaration shall therein undertake to indicate, and shall, when such dispute arises, indicate that for the settlement of such disputes it accepts a regional or other third party procedure entailing a binding decision, to which all parties to the dispute have access; and provided further that such procedure or decision shall exclude the determination of any claim to sovereignty or other rights with respect to continental or insular land territory;

(b) Disputes concerning military activities, including military activities by government vessels and aircraft engaged in non-commercial service and, subject to the exceptions referred to in article 296, law enforcement activities in the exercise of sovereign rights or jurisdiction provided for in the present Convention;

(c) Disputes in respect of which the Security Council of the United Nations is exercising the functions assigned to it by the Charter of the United Nations, unless the Security Council decides to remove the matter from its agenda or calls upon the parties to settle it by the means provided for in the present Convention.

2. A State Party which has made a declaration under paragraph 1 may at any time withdraw it, or agree to submit a dispute excluded by such declaration to any procedure specified in the present Convention.

3. Any State Party which has made a declaration under paragraph 1 shall not be entitled to submit any dispute falling within the excepted category of disputes to any procedure in the present Convention as against any other State Party, without the consent of that party.

4. If one of the States Parties has made a declaration under subparagraph 1 (a), any other State Party may submit any dispute falling within an excepted category against the declarant party to the procedure specified in such declaration.

5. When a dispute has been submitted to any procedure in accordance with this article, a new declaration, or the withdrawal of a declaration shall not affect in any way the proceedings so pending, unless the parties otherwise agree.

6. Declarations and withdrawals under this article shall be deposited with the Secretary-General of the United Nations, who shall transmit copies thereof to the States Parties.

PART XVI
FINAL CLAUSES
Article 298
Ratification

The present Convention is subject to ratification. The instruments of ratification shall be deposited with the Secretary-General of the United Nations.

Article 299
Accession

The present Convention shall remain open for accession by any State. The instruments of accession shall be deposited with the Secretary-General of the United Nations.

Article 300
Entry into force

1. The present Convention shall enter into force on the ... day following the date of deposit of the ... instrument of ratification or accession.

2. For each State ratifying or acceding to the Convention after the deposit of the ... instrument of ratification or accession, the Convention shall enter into force on the ... day after the deposit by such State of its instrument of ratification or accession.

Article 301
Status of annexes

The annexes form an integral part of the present Convention, and unless expressly provided otherwise, a reference to the present Convention constitutes a reference to its annexes.

Article 302
Authentic texts

The original of the present Convention, of which the Arabic, Chinese, English, French, Russian and Spanish texts are equally authentic, shall be deposited with the Secretary-General of the United Nations, who shall send copies thereof to all States.

Article 303
Testimonium clause, place and date

IN WITNESS WHEREOF the undersigned plenipotentiaries, being duly authorized thereto by their respective Governments, have signed the present Convention.

DONE AT CARACAS, this ... day of ..., one thousand nine hundred and seventy ....

Transitional provision

1. The rights recognized or established by the present Convention to the resources of a territory whose people have not attained either full independence or some other self-governing status recognized by the United Nations, or a territory under foreign occupation or colonial domination, or a United Nations Trust Territory, or a territory administered by the United Nations, shall be vested in the inhabitants of that territory, to be exercised by them for their own benefit and in accordance with their own needs and requirements.

2. Where a dispute over the sovereignty of a territory under foreign occupation or colonial domination exists, in respect of which the United Nations has recommended specific means of solution, rights referred to in paragraph 1 shall not be exercised except with the prior consent of the parties to the dispute until such dispute is settled in accordance with the purposes and principles of the Charter of the United Nations.

3. A metropolitan or foreign power administering, occupying or purporting to administer or occupy a territory may not in any case exercise, profit, or benefit from or in any way infringe the rights referred to in paragraph 1.

4. Reference in this article to a territory includes continental territories and islands.

ANNEX I
Highly migratory species
1. Albacore tuna: *Thunnus alalunga*
2. Bluefin tuna: *Thunnus thynnus*

3. Bigeye tuna: *Thunnus obesus*
4. Skipjack tuna: *Katsuwonus pelamis*
5. Yellowfin tuna: *Thunnus albacares*
6. Blackfin tuna: *Thunnus atlanticus*
7. Little tuna: *Euthynnus alletteratus; Euthynnus affinis*
8. Frigate mackeral: *Auxis thazard; Auxis rochei*
9. Pomfrets: Family Bramidae
10. Marlins: *Tetrapturus angustirostris; Tetrapturus belone; Tetrapturus pfluegeri; Tetrapturus albidus; Tetrapturus audax; Tetrapturus georgei; Makaira mazara ; Makaira indica; Makaira nigricans*
11. Sail-fishes: *Istiophorus platypterus; Istiophorus albicans*
12. Swordfish: *Xiphias gladius*
13. Sauries: *Scomberesox saurus; Cololabis saira; Cololabis adocetus; Scomberesox saurus scombroides*
14. Dolphin: *Coryphaena hippurus; Coryphaena equiselis*
15. Oceanic sharks: *Hexanchus griseus; Cetorhinus maximus;* Family Alopiidae; *Rhincodon typus;* Family Carcharhinidae; Family Sphyrnidae; Family Isurida
16. Cetaceans: Family Physeteridae; Family Balaenopteridae; Family Balaenidae; Family Eschrichtiidae; Family Monodontidae; Family Ziphiidae; Family Delphinidae

ANNEX II
Basic conditions of exploration and exploitation

Title to minerals and processed substances
1. Title to the minerals shall normally be passed upon recovery of the minerals pursuant to a contract of exploration and exploitation. In the case of contracts pursuant to paragraph 3 (b) for stages of operations, title to the minerals or processed substances shall pass in accordance with the contract. This paragraph is without prejudice to the rights of the Authority under paragraph 7.

Prospecting
2. (a) The Authority shall encourage the conduct of prospecting in the Area. Prospecting shall be conducted only after the Authority has received a satisfactory written undertaking that the proposed prospector shall comply with this Part of the Convention and the relevant rules and regulations of the Authority concerning protection of the marine environment, the transfer of data to the Authority, the training of personnel designated by the Authority and accepts verification of compliance by the Authority with all of its rules and regulations in so far as they relate to prospecting. The proposed prospector shall, together with the undertaking, notify the Authority of the broad area or areas in which prospecting is to take place. Prospecting may be carried out by more than one prospector in the same area or areas simultaneously. The Authority may close a particular area for prospecting when the available data indicates the risk of irreparable harm to a unique environment or unjustifiable interference with other uses of the Area.
(b) Prospecting shall not confer any preferential, proprietary or exclusive rights on the prospector with respect to the resources or minerals.

Exploration and exploitation
3. (a) Exploration and exploitation shall only be carried out in areas specified in plans of work referred to in article 151, paragraph 3 and approved by the Authority in accordance with the provisions of this Annex and the relevant rules, regulations and procedures adopted pursuant to paragraph 11.
(b) Contracts shall normally cover all stages of operations. If the applicant for a contract applies for a specific stage or stages, the contract may only comprise such stage or stages. Nothing in this paragraph shall in any way limit the discretion of the Enterprise.
(c) Every contract entered into by the Authority shall:
  (i) Be in strict conformity with this Part of the Convention and the rules and regulations prescribed by the Authority;
  (ii) Ensure control by the Authority at all stages of operations in accordance with article 151, paragraph 4;
  (iii) Confer exclusive rights on the Contractor in the contract area in accordance with the rules and regulations of the Authority.

Qualifications of applicants
4. (a) The Authority shall adopt appropriate administrative procedures and rules and regulations for making an application and for the qualifications of an applicant. Such qualifications shall include financial standing, technological capability and satisfactory performance under any previous contracts with the Authority.
(b) The procedures for assessing the qualifications of States Parties which are applicants shall take into account their character as States.
(c) Every applicant without exception shall:
  (i) Undertake to comply with and to accept as enforceable the obligations created by the provisions of this Part of the Convention, the rules and regulations adopted by the Authority, and the decisions of it organs and the terms of contracts, and to accept control by the Authority in accordance therewith;
  (ii) Undertake to negotiate upon the conclusion of the contract, if the Authority shall so request, an agreement making available to the Enterprise under licence, the technology used or to be used by the applicant, in carrying out activities in the Area on fair and reasonable terms in accordance with paragraph 5 (j) (iv) of this annex;

(iii) Accept control by the Authority in accordance with paragraph 3 (c) (ii);
(iv) Provide the Authority with satisfactory assurances that its obligations covered by the contract entered into by it will be fulfilled in good faith.

Selection of applicants

5. (a) On the first day of the sixth month after the entry into force of this Part of the Convention, and thereafter each fourth month on the first day of that month, the Authority shall take up for consideration applications received for contracts with respect to activities of exploration and exploitation.

(b) When considering an application for a contract with respect to exploration and exploitation the Authority shall first ascertain whether
   (i) the applicant has complied with the procedures established for applications in accordance with paragraph 4 and has given the Authority the commitments and assurances required by that paragraph. In cases of non-compliance with these procedures or of absence of any of the commitments and assurances referred to, the applicant shall be given 20 days to remedy such defects;
   (ii) the applicant possesses the requisite qualifications pursuant to paragraph 4.

(c) Once it is established that the conditions referred to in subparagraph (b) above are met, the Authority shall determine whether more than one application has been received within the preceding time period as provided in subparagraph (a) above in respect of substantially the same area and category of minerals and whether the granting of a contract would be in conformity with the provisions of article 150, paragraph 1 (g) and the relevant decisions of the Authority in implementation thereof. If no competing application has been received, and if the granting of a contract would be in conformity with article 150, paragraph 1 (g), the Authority shall without delay enter into negotiations with the applicant with a view to concluding a contract.

(d) The negotiations referred to in subparagraph (c) above shall, within the framework of the provisions of this Part of the Convention and the rules, regulations and procedures of the Authority adopted under article 158, paragraph 2 (xvi) and 160, paragraph 2 (xiv), deal with:
   (i) operational requirements under regulations adopted pursuant to paragraph 11 of this annex such as duration of activities, size of area, performance requirements and protection of the marine environment;
   (ii) the financial contribution to be made by the applicant under the financial arrangements established in paragraph 7 of this annex, and participation in the project by developing countries, on the basis of the incentives for such participation established in paragraph 7;
   (iii) transfer of technology under programmes and measures pursuant to article 144, and paragraph 4 (c) (ii) of this annex.

(e) In the course of the negotiations referred to in subparagraph (d) above, and prior to the conclusion of a contract, the Authority shall ensure that such contract would be in full conformity with the provisions of this Part of the Convention and the rules, regulations and procedures of the Authority adopted under articles 158, paragraph 2 (xvi) and 160, paragraph 2 (xiv), in particular the provisions, rules, regulations and procedures on the issues enumerated in subparagraph (d) above, and the provisions of article 150, paragraph 1 (g) and the relevant decisions of the Authority in implementation thereof.

(f) The negotiations referred to in subparagraph (d) above shall be conducted as expeditiously as possible. As soon as the issues under negotiation in accordance with subparagraph (d) above have been settled, the Authority shall conclude the corresponding contract with the applicant. In cases of a refusal of contract the Authority shall state the reasons for such refusal.

(g) If the Authority receives within the applicable time period as provided in subparagraph (a) above more than one application in respect of substantially the same part of the Area and category of minerals, or if the applications received within that time period cannot all be accommodated within the production limits established in article 150, paragraph 1 (g), selection from among the applicants shall be made on a comparative basis. In accordance with subparagraphs (c) and (d), the Authority shall enter into negotiations with the applicants in order to make its selection on the basis of a comparative evaluation of their applications and qualification. In so doing the Authority shall also take into account the need to give reasonable priority to applicants who are ready to enter into such joint arrangements with the Enterprise as referred to in subparagraphs (i) and (j) (iii) below. Once the selection is made, the Authority shall enter into negotiations with the selected applicant or applicants on the terms of a contract in accordance with subparagraphs (c) and (d) above.

(h) If the Contractor in accordance with paragraph 3 (b) of this annex has entered into a contract with the Authority for separate stages of operations, he shall have a preference and a priority among applicants for a contract for subsequent stages of operations with regard to the same areas and resources; provided, however, that where the Contractor's performance has not been satisfactory such preference or priority may be withdrawn.

(i) Contracts for the exploration and exploitation of the resources of the Area may provide for joint arrangements between the Contractor and the Authority through the

Enterprise, in the form of joint ventures, production sharing or service contracts, as well as any other form of joint arrangement for the exploration and exploitation of the resources of the Area.

(j) (i) The proposed contract area shall be sufficiently large and of sufficient value to allow the Authority to determine that one half of it shall be reserved solely for the conduct of activities by the Authority through the Enterprise or in association with developing countries. Upon such determination by the Authority the Contractor shall indicate the co-ordinates dividing the area into two halves of equal estimated commercial value and the Authority shall designate the half which is to be reserved. The Contractor may, alternatively, submit two non-contiguous areas of equal estimated commercial value, of which the Authority shall designate one as the reserved area. The designation by the Authority of one half of the area, or of one of two non-contiguous areas, as the case may be, in accordance with the provisions of this subparagraph, shall be made as soon as the Authority has been able to examine the relevant data as may be necessary to decide that both parts are equal in estimated commercial value.

(ii) Areas designated by the Authority as reserved areas in accordance with this subparagraph, may be exploited only through the Enterprise or in association with developing countries. The Enterprise shall be given an opportunity to decide whether it wishes itself to conduct the activities in the designated area. When considering applications from developing countries, or from a group of applicants which include developing countries, for areas designated under this subparagraph, and not selected by the Enterprise, the Authority shall, before entering into a contract, ensure that the developing countries will obtain substantial benefit therefrom.

(iii) In conducting activities in areas reserved in accordance with this subparagraph, the Enterprise may enter into joint arrangements of the kind referred to in subparagraph (i) above with other entities referred to in article 151, paragraph 2 (ii). In such joint arrangements appropriate provision shall be made for participation from developing countries. The nature and extent of such participation shall be determined by the Authority.

(iv) The Authority may require that the Contractor make available to the Enterprise the same technology to be used in the Contractor's operations on fair and reasonable terms and conditions in accordance with paragraph 4 (c) (ii) above. If the Authority requests an agreement pursuant to this subparagraph and the negotiations do not lead to an agreement within a reasonable time, the matter shall be referred to binding arbitration in accordance with the provisions of annex VI of the present Convention. In the event that the Contractor does not accept, or fails to implement the arbitral decision, the Contractor shall be liable in accordance with the provisions of paragraph 12 of this annex.

(v) Nothing in this subparagraph shall be interpreted as preventing the Enterprse from carrying out activities in accordance with the present annex in any part of the Area not subject to contract or joint arrangement.

(k) Contractors entering into such joint arrangements with the Enterprise as referred to in subparagraphs (i) and (j) (iii) above shall receive financial incentives as provided for in the financial arrangements established in paragraph 7 of this annex.

(l) While the inclusion of a quota or anti-monopoly provision appears to be acceptable in principle, its detailed formulation has yet to be fully negotiated.

Activities conducted by the Enterprise

6. Activities in the Area conducted under article 151, paragraph 2 (i) through the Enterprise shall be governed by the provisions of Part XI of the present Convention including the resource policy set forth in article 150 and the relevant decisions of the Authority in implementation thereof, as well as the statutes of the Enterprise and by the rules, regulations and procedures adopted under articles 158, paragraph 2 (xvi) and 160, paragraph 2 (xiv).

Financial terms of contracts[2]

7. (a) In adopting rules, regulations and procedures concerning the financial terms of a

---

2. The text of paragraph 7 is a preliminary draft submitted after consultations with experts and further work needs to be done on this subject

contract between the Authority and the entities referred to in article 151, paragraph 2 (ii) of the present Convention, and in negotiating those terms within the framework of the provisions of Part XI of the present Convention, and of those rules, regulations and procedures, the Authority shall be guided by the following objectives:
  (i) to ensure optimum revenues for the Authority;
  (ii) to attract investments and technology into the exploration and exploitation of the Area;
  (iii) to ensure equality of financial treatment and comparable financial obligations on the part of all States and other entities which obtain contracts;
  (iv) to provide incentives on a uniform and non-discriminatory basis for contractors to undertake joint arrangements with the Enterprse and developing countries or their nationals, and to stimulate the transfer of technology thereto;
  (v) to enable the Enterprise to engage in sea-bed mining effectively from the time of entry into force of this Convention;

(b) A fee shall be levied in respect of the administrative cost of processing an application for a contract and shall be fixed by the Authority at an amount not exceeding . . . per contract application.

(c) (i) The financial contribution of a Contractor shall be made up of an annual fixed charge to mine, a production charge and a share of net proceeds.
  (ii) The authority shall not establish any fees or charges to be applied to the Contractor other than those determined under (i) above, the fee referred to in subparagraph (b) above and the guarantee referred to in paragraph 11 (a) 2 (iii) of this annex.

(d) (i) An annual fixed charge to mine in respect of each year that the Contractor holds rights under the contract to commercial production from the contract area. The charge shall be based on the rate of . . . per annum per contract area. No charge shall be payable for the first three years following the date of the entry into force of the contract and thereafter the charge may be deducted from any production charge under (ii) below paid in the same year.
  (ii) A production charge of . . . per cent of the market value or . . . per cent of the amount of the processed metals extracted from the contract area. For this purpose, the market value shall be the product of the quantity of the recoverable metals produced and the average price for that amount of metal during the relevant account period. Where the Authority determines that an international commodity exchange provides a representative pricing mechanism, the average price on such exchange shall be used in the calculation of the price of each unit of production. In all other cases, the Authority, after consultation with the Contractor, shall determine the average price.
  (iii) A share of net proceeds to be determined by deducting from the proceeds of operations in the Area the costs incurred by the Contractor in respect of those operations and applying a percentage to the balance according to the rate of return on investment to the Contractor as set out in subparagraph D below.
  A. Proceeds of operations shall be assessed in terms of the value of production in the following manner _____ and shall include any proceeds from the disposal of capital assets not deducted from costs under B (1) below or, the market value of those capital assets at the relevant time which are no longer required for operations under the contract and which are not sold.
  B. The costs incurred by the Contractor in respect of those operations shall comprise:
    1. Development costs: that is all expenditures incurred prior to the commencement of commercial production from the contract area which are directly related to the development of the productive capacity of the contract area, including, *inter alia*, costs of machinery, equipment, ships, buildings, land, roads, exploration and feasibility studies and other research and development construction, interest, required leases, licences, and, subsequent to the commencement of commercial production, similar costs required for the replacement of equipment and machinery, maintenance and improvement of productive capacity and improvement of

performance; less proceeds from the disposal of capital assets;

2. Operating costs: that is all expenditures incurred in the operation of the productive capacity of the contract area, including, *inter alia*, expenditures for wages, salaries, employee benefits, supplies, materials, services, transportation, sale of products, interest, charges to mine and production charges paid under litra 1 and 2 of this subparagraph, utilities, purchases, and overhead and administrative costs specifically related to the operations of the contract area and any net operating losses carried forward from prior accounting periods;

Provided that:

(a) Payments in respect of acquisition of assets referred to in 1 and 2 above shall not be allowed as costs to the extent that the acquisition was not the result of an arm's length transaction between the parties concerned;

(b) The costs referred to in 1 and 2 above in respect of interest paid by the Contractor may only be allowed if the debt-equity ratio of the project is reasonable in all the circumstances, and the rates of interest may be no greater than those approved by the Authority as reasonable having regard to existing commercial rates;

(c) The costs referred to in 1 and 2 above shall not be interpreted as including payments in respect of taxes or similar charges levied by States in respect of the operations of the contractor.

C. The net proceeds for a given accounting period shall be determined for each contract area by deducting from the proceeds of operations in that area the development costs and operating costs for that contract area, in accordance with the rules and regulations and in accordance with the following:

1. Operating costs for a given accounting period shall include any loss from the previous accounting period and shall be deductible in the accounting period in which they occur.

2. Development costs shall be deductible in the given accounting period as a depreciation charge on such percentage basis per annum as is agreed in the contract, provided that any such agreement must provide an opportunity for the Contractor to recover initial development costs i.e. development costs as at the commencement of commercial production within ——— years from the start of commercial production.

D. 1. In each year, the share of Net Proceeds to be received by the Authority shall be determined according to the rate of return on the Contractor's investment hereinafter referred to as the Rate of Return. The Rate of Return shall be calculated by dividing the sum of the Contractor's portions of Net Proceeds in all preceding years by the total number of completed years from the date of commencement of commercial production, and expressing this average as a percentage of adjusted Development Costs. Adjusted Development Costs shall be equal to actual development costs less the sum of all amounts deducted by the Contractor as development costs up to the end of the year in question;

2. If for any year, the Rate of Return thus determined is zero or negative, the Contractor's Rate of Return shall be "minimal". If the Rate of Return is greater than zero but less than 10 per cent, the Contractor's Rate of Return status shall be "low". If the Rate of Return is 10 per cent or more, but less than 20 per cent, the status shall be "medium". If the Rate of Return is 20 per cent or more, the status shall be "high".

3. (a) Where the status is "minimal", the Authority shall not be entitled to any payment under this paragraph;

(b) Where the status is

"low", the Authority shall be entitled to _____;
    (c) Where the status is "medium", the Authority shall be entitled to _____;
    (d) Where the status is "high". the Authority shall be entitled to _____;

(iv) A. Where both parties agree, the Authority may elect to receive as its share of the net proceeds, a share of the deemed profits of the Contractor instead of a share of the net proceeds calculated under the method set out in litra (iii) of this subparagraph.

B. For the purpose of A above, the deemed profits of the Contractor in any one year shall be _____ per cent of the imputed value of assessed metal content of nodules mined from the contract area in that year.

C. For the purpose of B above, the imputed value of the metal content shall be _____ per cent of the market value of the processed metal, such market vaue being calculated in accordance with subparagraph (d) (ii) E of this paragraph, and the assessed metal content shall be determined in a manner to be agreed between the parties.

D. The share of the deemed profits of the Contractor in any one year to be paid to the Authority shall be _____ per cent of those deemed profits.

(e) The Authority may, taking into account any recommendations of the Economic Planning Commission and the Technical Commission, adopt rules and regulations that provide for incentives to Contractors that may be applied on a uniform and non-discriminatory basis in cases where such incentives would further the objectives set out in subparagraph (a) above. Such incentives may include reducing or eliminating the fixed charge or the production charge, or both of them, or reducing its percentage of net proceeds, or consenting to accelerated depreciation of development costs.

(f) (i) The amounts referred to in subparagraph (b) and subparagraph (d) (i) shall be in constant 1st-January 1980 U.S. dollars.

(ii) The payments to the Authority under (ii) (iii) and (iv) of subparagraph (d) above may be made either in a currency agreed upon between the Authority and the Contractor, or in the equivalents of processed metals at current market value. The market value shall be ascertained in accordance with (ii) of subparagraph (d) above.

(g) The Authority shall adopt rules and regulations regarding the method of selection of auditors responsible for attesting to the conformity of the Contractor with these related financial terms and the related rules and procedures of the Authority.

Transfer of data

8. The Contractor shall transfer in accordance with the rules and regulations and the terms and conditions of the contract to the Authority at time intervals determined by the Authority all data which are both necessary and relevant to the effective implementation of the powers and functions of the organs of the Authority in respect of the contract area. Transferred data in respect of the contract area, deemed to be proprietary, shall not be disclosed by the Authority, and may only be used for the purposes set forth above in this subparagraph. Data which are necessary for the promulgation of rules and regulations concerning protection of the marine environment and safety shall not be deemed to be proprietary. Except as otherwise agreed between the Authority and the Contractor, the Contractor shall not be obliged to disclose proprietary equipment design data.

Training programmes

9. The Contractor shall draw up practical programmes for the training of personnel of the Authority and developing countries, including the participation of such personnel in all activities covered by the contract.

Exclusive right to explore and exploit in the contract area

10. The Authority shall, pursuant to Part XI of the present Convention and the rules and regulations prescribed by the Authority, accord the Contractor the exclusive right to explore and exploit the contract area with the Authority in respect of a specified category of minerals and shall ensure that no other entity operates in the same contract area for a different category of minerals in a manner which might interfere with the operations of the Contractor. The Authority shall not, during the continuance of a contract, permit any other entity to carry out activities in the same area for the same category of minerals. The Contractor shall have security of tenure in accordance with article 151, paragraph 5.

Rules, regulations and procedures

11. (a) The Authority shall adopt and uniformly apply rules, regulations and procedures for the implementation of Part XI of the present Convention including these basic conditions, on the following matters:

(1) *Administrative procedures relating to prospecting, exploration and exploitation in the area*

(2) *Operations*
 (i) Size of area;
 (ii) Duration of activities;
 (iii) Performance requirements and guarantees;
 (iv) Categories of minerals;
 (v) Renunciation of areas;
 (vi) Progress reports;
 (vii) Submission of data;
 (viii) Inspection and supervision of operations;
 (ix) Passing of title pursuant to paragraph 1;
 (x) Prevention of interference with other uses of the sea and of the marine environment;
 (xi) Transfer of rights by a Contractor;
 (xii) Procedures for transfer of technology to developing countries and for their direct participation;
 (xiii) Mining standards and practices including those relating to operational safety, conservation of the resources and the protection of the marine environment;
 (xiv) Continuity of operations in the event of disputes;
 (xv) Definition of commercial production.

(3) *Financial matters*
 (i) Establishment of uniform and non-discriminatory costing and accounting rules;
 (ii) Apportionment of proceeds of operations;
 (iii) The incentives referred to in paragraph 7.

(4) *Rules, regulations and procedures to implement decisions of the Council taken in pursuance of articles 150 and 162.*

(b) Regulations on the following items shall fully reflect the objective criteria set out below:

(1) *Size of area*
The Authority shall determine the appropriate size of areas for exploration which may be up to twice as large as those for exploitation in order to permit intensive exploration operations. Areas for exploitation shall be calculated to satisfy stated production requirements over the term of the contract taking into account the state of the art of technology then available for ocean mining and the relevant physical characteristics of the area. Areas shall neither be smaller nor larger than are necessary to satisfy this objective. In cases where the Contractor has obtained a contract for exploitation, the area not covered by such contract shall be relinquished to the Authority.

(2) *Duration of activities*
 (i) Prospecting shall be without time-limit;
 (ii) Exploration should be of sufficient duration as to permit a thorough survey of the specific area, the design and construction of mining equipment for the area, the design and construction of small and medium-size processing plants for the purpose of testing mining and processing systems;
 (iii) The duration of exploitation should be related to the economic life of the mining project, taking into consideration such factors as the depletion of the ore, the useful life of mining equipment and processing facilities and commercial viability. Exploitation should be of sufficient duration as to permit commercial extraction of minerals of the area and should include a reasonable time period for construction of commercial scale mining and processing systems, during which period commercial production should not be required. The total duration of exploitation, however, should also be short enough to give the Authority an opportunity to amend the terms and conditions of the contract at the time it considers renewal in accordance with rules and regulations which it has issued subsequent to entering into the contract.

(3) *Performance requirements*
The Authority shall require that during the exploration stage, periodic expenditures be made by the Contractor which are reasonably related to the size of the contract area and the expenditures which would be expected of a *bona fide* Contractor who intended to bring the area into commercial production within the time-limits established by the Authority. Such required expenditures should not be established at a level which would discourage prospective operators with less costly technology than is prevalently in use. The Authority shall establish a maximum time interval after the exploration stage is completed and the exploitation stage begins to achieve commercial production. To determine this interval, the Authority should take into consideration that construction of large-scale mining and processing systems cannot be initiated until after the termination of the exploration stage and the commencement of the exploitation stage. Accordingly, the interval to bring an area into commercial production should take into account the time necessary for this construction after the completion of the exploration stage and reasonable allowance should be made for unavoidable delays in the construction schedule.

Once commercial production is achieved in the exploitation stage, the Authority shall within reasonable limits and taking into consideration all relevant factors require the Contractor to maintain commercial production throughout the period of the contract.

(4) *Categories of minerals*

In determining the category of mineral in respect of which a contract may be entered into, the Authority shall give emphasis *inter alia* to the following characteristics:
  (i) Resources which require the use of similar mining methods; and
  (ii) Resources which can be developed simultaneously without undue interference between Contractors in the same area developing different resources.

Nothing in this paragraph shall deter the Authority from granting a contract for more than one category of mineral in the same contract area to the same applicant.

(5) *Renunciation of areas*

The Contractor shall have the right at any time to renounce without penalty the whole or part of his rights in the contract area.

(6) *Protection of the marine environment*

Rules and regulations shall be drawn up in order to secure effective protection of the marine environment from harmful effects directly resulting from Activities in the Area or from shipboard processing immediately above a minesite of minerals derived from the minesite, taking into account the extent to which such harmful effects may directly result from drilling, dredging, coring and excavation as well as disposal, dumping and discharge into the marine environment of sediment, wastes or other effluents.

(7) *Commercial production*

Commercial production shall be deemed to have begun if an operator engages in activity of sustained large-scale recovery operations which yield a sufficient quantity of materials as to indicate clearly that the principal purpose is large-scale production rather than production intended for information gathering, analysis or equipment or plant-testing.

Penalties

12. (a) A Contractor's rights under the contract concerned may be suspended or terminated only in the following cases:
  (i) If the Contractor has conducted his activities in such a way as to result in gross and persistent or serious, persistent and wilful violations of the fundamental terms of the contract, Part XI of the present Convention and rules and regulations, which were not caused by circumstances beyond his control; or
  (ii) If a Contractor has failed to comply with a final binding decision of the dispute settlement body applicable to him.

(b) The Authority may impose upon the Contractor monetary penalties proportionate to the seriousness of the violation in lieu of suspension or termination or in any case not covered under subparagraph (a) above.

(c) Except in cases of emergency orders as provided for in article 163 paragraph 2 (xi), the Authority may not execute a decision involving monetary penalties, suspension or termination until the Contractor has been accorded a reasonable opportunity to exhaust his judicial remedies in the Sea-Bed Dispute Chamber. The Sea-Bed Dispute Chamber may, however, order execution of a decision regarding monetary penalties or suspension pending final adjudication of the matter.

Revision of Contract

13. (a) When circumstances have arisen, or are likely to arise, which, in the opinion of either party, would render the contract inequitable or make it impracticable or impossible to achieve the objectives set out in the contract or in Part XI of the present Convention, the parties shall enter into negotiation to adjust it to new circumstances in the manner prescribed in the contract.

(b) Any contract entered into in accordance with article 151, paragraph 3 may only be revised if the parties involved have given their consent.

Transfer of Rights

14. The rights and obligations arising out of a contract shall be transferred only with the consent of the Authority, and in accordance with the rules and regulations adopted by it. The Authority shall not withhold consent to the transfer if the proposed transferee is in all respects a qualified applicant, and assumes all of the obligations of the transferor.

Applicable law

15. The law applicable to the contract shall be the provisions of Part XI of the present Convention, the rules and regulations prescribed by the Authority and the terms and conditions of the contract. The rights and obligations of the Authority and of the Contractor shall be valid and enforceable in the territory of each State Party. No State Party may impose conditions on a Contractor that are inconsistent with Part XI of the present Convention. However, the application by a State Party of environmental regulations to sea-bed miners it sponsors or to ships flying its flag, more stringent than those imposed by the Authority pursuant to paragraph 11 (b) (6) of this annex, shall not be deemed inconsistent with Part XI of the present Convention.

Liability

16. Any responsibility or liability for wrongful damage arising out of the conduct of operations by the Contractor shall lie with the Contractor. It shall be a defence in any proceeding against a Contractor that the damage was the result of an act or omission of the Authority. Similarly, any responsibility or liability for wrongful damage arising out of the exercise of the powers and

functions of the Authority shall lie with the Authority. It shall be a defence in any proceeding against the Authority that the damage was a result of an act or omission of the Contractor. Liability in every case shall be for the actual amount of damage.

ANNEX III
Statute of the Enterprise
Purpose

1. (a) The Enterprise shall carry out activities of the Authority in the Area in the performance of its functions in implementation of article 169.

(b) In the performance of its functions and in carrying out its purposes, the Enterprise shall act in accordance with the provisions of Part XI of the present Convention and its annexes, including article 151 and the resource policy set forth in article 150 and the relevant decisions of the Authority in implementation thereof.

Relationship to the Authority

2. (a) Pursuant to article 169 the Enterprise shall be subject to the general policies laid down by the Assembly and the directives and control of the Council.

(b) Nothing in this Convention shall make the Enterprise liable for the acts or obligations of the Authority, or the Authority liable for the acts or obligations of the Enterprise.

Limitation of liability

3. No member of the Authority shall be liable by reason only of its membership for the acts or obligations of the Enterprise.

Structure of the Enterprise

4. The Enterprise shall have a Governing Board, a Director-General and such staff as may be necessary for the performance of its duties.

Governing Board

5. (a) The Governing Board shall be responsible for the conduct of operations of the Enterprise, and for this purpose shall exercise all the powers given to it by this annex.

(b) The Governing Board shall be composed of 15 qualified, competent and experienced members elected by the Assembly. Election of these members shall be based on the principle of equitable geographical representation, taking special interests into account.

(c) Members of the Board shall be elected for a period of four years and shall be eligible for re-election. Due regard should be paid to the desirability of rotating seats.

(d) Each member of the Board shall have one vote. All matters before the Board shall be decided by a majority of the votes cast.

(e) Each member of the Board shall appoint an alternate with full powers to act for him when he is not present.

(f) Members of the Board shall continue in office until their successors are appointed or elected. If the office of a member of the Board becomes vacant more than 90 days before the end of his term, the Board may appoint another member for the remainder of the term. While the office remains vacant, the alternate of the former member of the Board shall exercise his powers, except that of appointing an alternate.

(g) The Governing Board shall function in continuous session at the principal office of the Enterprise, and shall meet as often as the business of the Enterprise may require.

(h) A quorum for any meeting of the Governing Board shall be two thirds of the members of the Board.

(i) Any member of the Authority may send a representative to attend any meeting of the Board when a request made by, or a matter particularly affecting, that member is under consideration.

(j) Subject to directives from the Council on the matter, the Governing Board may appoint such committees as they deem advisable.

Director-General and staff

6. (a) The Assembly shall, upon the recommendation of the Council, elect a Director-General who shall not be a member of the Board or an alternate. The Director-General shall be the legal representative of the Enterprise. He shall participate in the meetings of the Board but shall have no vote. He may participate in meetings of the Assembly, and the Council, when these organs are dealing with matters concerning the Enterprise, but shall have no vote at such meetings. The Director-General shall hold office for a fixed term not exceeding five years and may be reappointed for one further term.

(b) The Director-General shall be chief of the operating staff of the Enterprise and shall conduct, under the direction of the Governing Board, the ordinary business of the Enterprise. Subject to the general control of the Governing Board, he shall be responsible for the organization, appointment and dismissal of the staff.

(c) The Director-General and the staff of the Enterprise, in the discharge of their offices, owe their duty entirely to the Enterprise and to no other authority. Each member of the Enterprise shall respect the international character of this duty and shall refrain from all attempts to influence any of them in the discharge of their duties.

(d) In appointing the staff the Director-General shall, subject to the paramount importance of securing the highest standards of efficiency and of technical competence, pay due regard to the importance of recruiting personnel on as wide a geographical basis as possible, and shall be guided by the principle that the staff should be kept to a minimum.

Location of offices

7. The principal office of the Enterprise shall be at the seat of the Authority. The Enterprise may establish other offices in the territories of any member, with the consent of that member.

Publication of reports and provision of information

8. (a) The Enterprise shall, not later than three months after the end of each financial year, submit to the Council for its approval an annual report containing an audited statement of its accounts and shall transmit to the Council and circulate to members at appropriate intervals a summary statement of its financial position and a profit and loss statement showing the results of its operations.

(b) The Enterprise shall publish its annual report and such other reports as it deems desirable to carry out its purposes.

(c) Copies of all reports, statements and publications made under this article shall be distributed to members.

Allocation of net income

9. (a) Subject to (b) below, all net disposable income generated by the Enterprise, shall be transferred quarterly to the Authority which shall determine the apportionment and distribution of such proceeds to the Enterprise and to States Parties in accordance with article 158, paragraphs 2 (viii), (xii) and (xiv), and article 160, paragraphs 2 (xiii), (xv) and (xvi).

(b) During an initial period, determined by the Council, required for the Enterprise to become self-supporting the Council, on the recommendation of the Governing Board, shall determine annually what part of the net income of the Enterprise, should be transferred to the Authority.

(c) In determining the amount of net disposable income generated by the Enterprise at any material time, the Council, on the recommendation of the Governing Board, shall make due provision for the reserves and surplus of the Enterprise.

Finance

10. (a) The funds and assets of the Enterprise shall comprise:
   (i) Amounts determined from time to time by the Assembly out of the Special Fund referred to in article 173, including funds to cover its administrative expenses in accordance with articles 158, paragraph 2 (vi) and 172, paragraph 2 and including funds earmarked for the Enterprise in accordance with article 173, paragraph 3.
   (ii) Voluntary contribution made by States Parties to the present Convention specifically for the purpose of financing activities of the Enterprise.
   (iii) Amounts borrowed by the Enterprise in accordance with subparagraph (c) below.
   (iv) Amounts received through the participation in contractual relationships with other entities for the conduct of activities in the Area, including joint arrangements in accordance with article 151, paragraph 3.
   (v) Net income of the Enterprise after transfer of revenues to the Authority in accordance with paragraph 7.
   (vi) Other funds made available to the Enterprise including charges to enable it to carry out its functions and to commence operations as soon as possible.

(b) The Governing Board of the Enterprise shall determine when the Enterprise may commence operation.

(c)
   (i) The Enterprise shall have the power to borrow funds, and in that connexion to furnish such collateral or other security therefore as it shall determine; provided, however, that before making a public sale of its obligations in the markets of a member, the Enterprise shall have obtained the approval of that member and of the member in whose currency the obligations are to be denominated. The total amount and sources of borrowings shall be approved by the Council on the recommendation of the Governing Board.
   (ii) States Parties shall make every effort to support applications by the Enterprise for loans in capital markets, including loans from international financial institutions, and to cause appropriate changes where necessary in the constitutive instruments of such institutions.
   (iii) To the extent that the costs of exploration, development and exploitation of the Enterprise's first site cannot be covered by the funds referred to in subparagraph (a) above, States Parties shall guarantee debts incurred by the Enterprise for the financing of such costs. Under such guarantees States Parties shall be liable on a basis adopted by the Assembly which is proportionate to the United Nations scale of assessments. To the extent necessary for the securing of such loans as referred to above, States Parties undertake to advance as refundable paid-in capital up to . . . per cent of the liability which they have incurred in accordance with this subparagraph.

(d) The funds and assets of the Enterprise shall be kept separate and apart from those of the Authority. The provisions of this paragraph shall not prevent the Enterprise from making arrangements with the Authority regarding facilities, personnel, and services and arrangements for reimbursement of administrative expenses paid in the first instance by either organization on behalf of the other.

Operations

11. (a) The Enterprise shall propose to the Council projects for carrying out activities in

the Area in accordance with article 151, paragraph 2 (i). Such proposals shall include a detailed description of the project, an analysis of the estimated costs and benefits, a draft formal written plan of work, and all such other information and data as may be required from time to time for its appraisal by the Technical Commission and approval by the Council.

(b) Upon approval by the Council, the Enterprise shall execute the project on the basis of the formal written plan of work referred to in subparagraph (a) of this paragraph.

(c) Procurement of goods and services:
  (i) To the extent that the Enterprise does not at any time possess the goods and services required for its operations, it may procure and employ them under its direction and management. Procurement of goods and services required by the Enterprise shall be effected by the award of contracts, based on response to invitations in member countries to tender, to bidders offering the best combination of quality, price and most favourable delivery time.
  (ii) If there is more than one bid offering such a combination, the contract shall be awarded in accordance with the following principles:
    (a) Non-discrimination on the basis of political or similar considerations not relevant to the carrying out of operations with due diligence and efficiency;
    (b) Guidelines approved by the Council with regard to the preferences to be accorded to goods and services originating in the developing countries, including the land-locked or otherwise geographically disadvantaged among them.
  (iii) The Governing Board may adopt rules determining the special circumstances in which the requirement of invitations in member countries to bid may in the best interests of the Enterprise be dispensed with.

(d) The Enterprise shall have title to all minerals and processed substances produced by it. They shall be marketed in accordance with rules, regulations and procedures adopted by the Council in accordance with the following criteria:
  (i) The products of the Enterprise shall be made available on a non-discriminatory basis to States Parties;
  (ii) The Enterprise shall sell its products at not less than international market prices.

(e) Without prejudice or any general or special power conferred on the Enterprise under any other provision of this Convention, the Enterprise shall exercise all such powers incidental to its business as shall be necessary or desirable in the furtherance of its purposes.

(f) The Enterprise and its staff shall not interfere in the political affairs of any member; nor shall they be influenced in their decisions by the political character of the member or members concerned. Only economic considerations shall be relevant to their decisions, and these considerations shall be weighed impartially in order to carry out the purposes specified in paragraph 1 of this annex.

Legal status, immunities and privileges
12. (a) To enable the Enterprise to fulfil the functions with which it is entrusted, the status, immunities and privilges set forth herein shall be accorded to the Enterprise in the territories of each member. To give effect to this principle the Enterprise may, where necessary, enter into special agreements for this purpose.

(b) The Enterprise shall have such legal capacity as is necessary for the performance of its functions and the fulfilment of its purposes and, in particular, the capacity:
  (i) To enter into contracts, forms of association, or other arrangements, including agreements with States and international organizations;
  (ii) To acquire, lease, hold and dispose of immovable and movable property;
  (iii) To be a party to legal proceedings in its own name.

(c) Actions may be brought against the Enterprise only in a court of competent jurisdiction in the territories of a member in which the Enterprise has an office, has appointed an agent for the purpose of accepting service or notice of process, has entered into a contract for goods or services, has issued securities, or is otherwise engaged in commercial activity. The property and assets of the Enterprise shall, wheresoever located and by whomsoever held, be immune from all forms of seizure, attachment of execution before the delivery of final judgement against the Enterprise.

(d)
  (i) The property and assets of the Enterprise, wheresoever located and by whomsoever held, shall be immune from confiscation, expropriation, requisition, and any other form of seizure by executive or legislative action.
  (ii) All property and assets of the Enterprise shall be free from discriminatory restrictions, regulations, controls and moratoria of any nature.
  (iii) The Enterprise and its employees shall respect local laws and regulations in any State or territory in which the Enterprise or its employees may do business or otherwise act.
  (iv) States Parties shall assure that the Enterprise enjoys all rights, immunities and privileges afforded by States to entities conducting business within such States. These rights, immunities and privileges shall be afforded the Enterprise on no less favourable a basis than afforded by

States to similarly engaged commercial entities. Where special privileges are provided by States for developing countries or their commercial entities, the Enterprise shall enjoy such privileges on a similarly preferential basis.

(v) States may provide special incentives, rights, privileges and immunities to the Enterprise without the obligation to provide such incentives, rights, privileges, or immunities to other commercial entities.

(e) The Enterprise, its assets, property, and revenues derived from its operations and transactions authorized by this annex, shall be immune from taxation.

(f) Each member shall take such action as is necessary in its own territories for the purpose of making effective in terms of its own law the principles set forth in this annex and shall inform the Enterprise of the detailed action which it has taken.

(g) The Enterprise in its discretion may waive any of the privileges and immunities conferred under this article or in the special agreements referred to in subparagraph (a) above to such extent and upon such conditions as it may determine.

ANNEX IV
Conciliation

Article 1
Institution of proceedings

If the parties to a dispute have agreed, in accordance with article 284 of Part XV, to submit the dispute to the procedure under this annex, any party to such dispute may institute the proceedings by notification addressed to the other party or parties to the dispute.

Article 2
List of conciliators

A list of conciliators shall be drawn up and maintained by the Secretary-General of the United Nations. Every State Party shall be entitled to nominate four conciliators, each of whom shall be a person enjoying the highest reputation for fairness, competence and integrity. The names of the persons so nominated shall constitute the list. If at any time the conciliators nominated by a State Party in the list so constituted shall be less than four, that State Party shall be entitled to make further nominations as necessary. The name of a conciliator shall remain on the list until withdrawn by the party which made the nomination, provided that such conciliator shall continue to serve on any conciliation commission for which that conciliator has been chosen until the completion of the proceedings before that Commission.

Article 3
Constitution of Conciliation Commission

The Conciliation Commission shall, unless the parties otherwise agree, be constituted as follows:

1. Subject to the provisions of paragraph 7, the Conciliation Commission shall consist of five members.

2. The party submitting the dispute to conciliation shall appoint two conciliators to be chosen preferably from the list and who may be its nationals. Such appointments shall be included in the notification under article 1.

3. The other party to the dispute shall appoint two conciliators in the same manner within 21 days of receipt of notification under article 1. If the appointments are not made within the prescribed period, the party which submitted the dispute to conciliation may, within one week of the expiration of the prescribed period, either terminate the proceedings by notification addressed to the other party or request the Secretary-General to make the appointments in accordance with paragraph 5.

4. Within 30 days following the date of the last of their own appointment, the four conciliators shall appoint a fifth conciliator chosen from the list, who shall be chairman. If the appointment is not made within the prescribed period, either party may, within one week of the expiration of the prescribed period, terminate the proceedings by notification addressed to the other party or, where the proceedings are not so terminated, request the Secretary-General to make the appointment in accordance with paragraph 5.

5. Upon the request of a party to the dispute in accordance with paragraphs 3 or 4, the Secretary-General of the United Nations shall make the necessary appointments within 30 days of the receipt of such request. The Secretary-General shall make such appointments from the list referred to in article 2 and in consultation with the parties to the dispute.

6. Any vacancy shall be filled in the manner prescribed for the initial appointment.

7. Parties in the same interest shall appoint two conciliators jointly by agreement. Where there are several parties having separate interests, or where there is disagreement as to whether they are of the same interest, each of them shall appoint one conciliator.

8. In disputes involving more than two parties, the provisions of paragraphs 1 to 6 shall apply to the maximum extent possible.

Article 4
Procedure to be adopted

The Conciliation Commission shall, unless the parties otherwise agree, decide its own procedure. The Commission, with the consent of the parties to the dispute, may invite any State Party to submit to it its views orally or in writing. Recommendations of the Commission and procedural decisions shall be made by a majority vote of its members.

Article 5
Amicable settlement

The Commission may draw the attention of the parties to the dispute to any measures which might facilitate an amicable settlement.

Article 6
Functions of the Commission

The Commission shall hear the parties, examine their claims and objections, and make proposals to the parties with a view to reaching an amicable settlement.

Article 7
Report

1. The Commission shall report within 12 months of its constitution. Its report shall record any agreements reached and, failing agreement, its conclusions on all questions of fact or law relevant to the matter in dispute and such recommendations as the Commission may deem appropriate for an amicable settlement of the dispute. The report shall be deposited with the Secretary-General of the United Nations and shall immediately be transmitted by him to the parties to the dispute.
2. The report of the Commission, including any conclusions or recommendations, shall not be binding upon the parties.

Article 8
Termination

The conciliation procedure shall be deemed terminated when a settlement has been reached, when the parties have accepted or one party has rejected the recommendations of the report by notification addressed to the Secretary-General, or when a period of three months has expired from the date of transmission of the report to the parties.

Article 9
Facilities, fees and expenses

The fees and expenses of the Commission shall be borne by the parties to the dispute.

Article 10
Right of parties to vary procedure

The parties to the dispute may by agreement vary any provision of this annex.

ANNEX V
Statute of the Law of the Sea Tribunal

Article 1
General provisions

1. The Law of the Sea Tribunal shall be constituted and shall function in accordance with the provisions of the present Convention and this Statute.
2. Any reference of a dispute to the Tribunal shall be subject to the provisions of Parts XI and XV.

SECTION 1. ORGANIZATION OF THE TRIBUNAL

Article 2
Composition of Tribunal

1. The Tribunal shall be composed of a body of 21 independent members, elected from among persons enjoying the highest reputation for fairness and integrity and of recognized competence in matters relating to the law of the sea.
2. In the Tribunal as a whole the representation of the principal legal systems of the world and equitable geographical distribution shall be assured.

Article 3
Election of members

1. No two members of the Tribunal may be nationals of the same State, and a person who for the purposes of membership in the Tribunal could be regarded as a national of more than one State shall be deemed to be a national of the one in which he ordinarily exercises civil and political rights.
2. There shall be not less than three members from each geographical group as established by the General Assembly of the United Nations.

Article 4
Procedure for nomination and election

1. Each State Party may nominate not more than two persons having the qualifications prescribed in article 2. The members of the Tribunal shall be elected from a list of persons thus nominated.
2. At least three months before the date of the election, the Secretary-General of the United Nations in the case of the first election and the Registrar of the Tribunal in the case of subsequent elections shall address a written invitation to the States Parties to submit their nominations for members of the Tribunal within two months. He shall prepare a list in alphabetical order of all the persons thus nominated, with an indication of the States Parties which have nominated them, and shall submit it to the States Parties before the seventh day of the last month before the date of each election.
3. The first election shall be held within six months of the date of entry into force of the present Convention.
4. Elections of the members of the Tribunal shall be by secret ballot. They shall be held at a meeting of the States Parties convened by the Secretary-General in the case of the first election and by a procedure agreed to by the States Parties in the case of subsequent elections. At that meeting, for which two thirds of the States Parties shall constitute a quorum, the persons elected to the Tribunal shall be those nominees who obtain the largest number of votes and a two-thirds majority of votes of the States Parties present and voting, provided that such majority shall include at least a majority of the States Parties.

Article 5
Term of office

1. The members of the Tribunal shall be elected for nine years and may be re-elected; provided, however, that of the members elected at the first election, the terms of seven members shall expire at the end of three years and the terms of seven more members shall expire at the end of six years.
2. The members of the Tribunal whose terms are to expire at the end of the above-mentioned initial periods of three and six years shall be

chosen by lots to be drawn by the Secretary-General of the United Nations immediately after the first election has been completed.
3. The members of the Tribunal shall continue to discharge their duties until their places have been filled. Though replaced, they shall finish any proceedings which they may have begun at the time of their replacement.
4. In the case of the resignation of a member of the Tribunal, the resignation shall be addressed to the President of the Tribunal. The place becomes vacant on the receipt of the letter of resignation.

Article 6
Vacancies
1. Vacancies shall be filled by the same method as that laid down for the first election, subject to the following provision: the Registrar shall, within one month of the occurrence of the vacancy, proceed to issue the invitations provided for in article 4, and the date of the election shall be fixed by the President of the Tribunal after consultation with States Parties.
2. A member of the Tribunal elected to replace a member whose term of office has not expired shall hold office for the remainder of the term of his predecessor.

Article 7
Conditions relating to interests of members
1. No member of the Tribunal may exercise any political or administrative function, or associate actively with or be financially interested in any of the operations of any enterprise concerned with the exploration or exploitation of the resources of the sea or the sea-bed or other commercial use of the sea or the sea-bed.
2. No member of the Tribunal may act as agent, counsel, or advocate in any case.
3. Any doubt on these points shall be decided by a majority of the other members of the Tribunal present.

Article 8
Conditions relating to participation of members
1. No member may participate in the decision of any case in which he has previously taken part as agent, counsel, or advocate for one of the parties, or as a member of a national or international court, or in any other capacity.
2. If, for some special reason, a member of the Tribunal considers that he should not take part in the decision of a particular case, he shall so inform the President of the Tribunal.
3. If the President considers that for some special reason one of the members of the Tribunal should not sit in a particular case, he shall give him notice accordingly.
4. Any doubt on this point shall be decided by a majority of the other members of the Tribunal present.

Article 9
Consequences of ceasing to fulfil conditions
If, in the unanimous opinion of the other members of the Tribunal, a member has ceased to fulfil the required conditions, the President of the Tribunal shall declare the seat vacant.

Article 10
Diplomatic privileges and immunities
The members of the Tribunal, when engaged on the business of the Tribunal, shall enjoy diplomatic privileges and immunities.

Article 11
Delcaration by members
Every member of the Tribunal shall, before taking up his duties, make a solemn declaration in open session that he will exercise his powers impartially and conscientiously.

Article 12
President, Vice-President and Registrar
1. The Tribunal shall elect its President and Vice-President for three years; they may be re-elected.
2. The Tribunal shall appoint its Registrar and may provide for the appointment of such other officers as may be necessary.

Article 13
Seat of Tribunal
1. The seat of the Tribunal shall be determined by the States Parties, provided that the Tribunal shall have the right to sit and exercise its functions elsewhere whenever the Tribunal considers it desirable.
2. The President and the Registrar shall reside at the seat of the Tribunal.

Article 14
Quorum
1. All available members shall sit, but a quorum of eleven members shall be required to constitute the Tribunal.
2. Subject to the provisions of article 18, the Tribunal shall determine which members are available to constitute the Tribunal for the consideration of a particular dispute, having regard to the effective functioning of the Sea-Bed Disputes Chamber and the special chambers as provided in articles 15 and 16.
3. All disputes and applications submitted to the Tribunal shall be heard and determined by the Tribunal, unless article 15 applies, or the parties request that it shall be dealt with in accordance with article 16.

Article 15
Establishment of a Sea-Bed Disputes Chamber
A Sea-Bed Disputes Chamber shall be established in accordance with the provisions of section 4 of this annex. Its jurisdiction, powers and functions shall be as provided for in section 6 of Part XI of the present Convention.

Article 16
Special chambers
1. The Tribunal may form such chambers, composed of three or more members, as the Tribunal may deem necessary for dealing with particular categories of disputes.
2. The Tribunal shall form a chamber for dealing with a particular dispute submitted to it

if the parties so request. The composition of such a chamber shall be determined by the Tribunal with the approval of the parties.
3. With a view to the speedy dispatch of business, the Tribunal shall form annually a chamber composed of five members which may hear and determine disputes by summary procedure. Two alternative members shall be selected for the purpose of replacing members who are unable to participate in a particular proceeding.
4. Disputes shall be heard and determined by the chambers provided for in this article if the parties so request.
5. A judgement given by any of the chambers provided for in this article and in article 15 shall be considered as rendered by the Tribunal.

Article 17
Rules of Tribunal
The Tribunal shall frame rules for carrying out its functions. In particular it shall lay down rules of procedure.

Article 18
Nationality of members
1. Members of the nationality of any of the parties to a dispute shall retain their right to participate as members of the Tribunal.
2. If the Tribunal hearing any dispute includes a member of the nationality of one of the parties, any other party to the dispute may choose a person to participate as a member of the Tribunal.
3. If the Tribunal hearing does not include a member of the nationality of the parties, each of these parties may proceed to choose a member as provided in paragraph 2.
4. The provisions of this article shall apply to articles 15 and 16. In such cases, the President, in consultation with the parties, shall request specified members of the Tribunal forming the chamber, as many as necessary, to give place to the members of the Tribunal of the nationality of the parties concerned, and, failing such, or if they are unable to be present, to the members specially chosen by the parties.
5. Should there be several parties in the same interest, they shall, for the purpose of the preceding provisions, be reckoned as one party only. Any doubt on this point shall be settled by the decision of the Tribunal.
6. Members chosen as laid down in paragraphs 2, 3 and 4 shall fulfil the conditions required by article 2, paragraph 2 of article 8 and article 11. They shall participate in the decision on terms of complete equality with their colleagues.

Article 19
Remuneration of members
1. Each member of the Tribunal shall receive an annual allowance and, for each day on which he exercises his functions, a special allowance, provided that in any year the total sum payable to any member as special allowance shall not exceed the amount of the annual allowance.
2. The President shall receive a special annual allowance.
3. The Vice-President shall receive a special allowance for each day on which he acts as President.
4. The members chosen under article 18, other than members of the Tribunal, shall receive compensation for each day on which they exercise their functions.
5. These allowances and compensation shall be fixed from time to time at a meeting of the States Parties, taking into account the workload of the Tribunal. They may not be decreased during the term of office.
6. The salary of the Registrar shall be fixed at a meeting of ths States Parties on the proposal of the Tribunal.
7. Regulations made at the meeting of the States Parties shall fix the conditions under which retirement pensions may be given to members of the Tribunal and to the Registrar and the conditions under which members of the Tribunal and Registrar shall have their travelling expenses refunded.
8. The above salaries, allowances, and compensation shall be free of all taxation.

Article 20
Expenses of Tribunal
1. The expenses of the Tribunal shall be borne by the States Parties and by the Authority on such terms and in such manner as shall be decided at a meeting of the States Parties.
2. When an entity other than a State Party or the Authority is a party to a dispute submitted to it, the Tribunal shall fix the amount which that party is to contribute towards the expenses of the Tribunal.

SECTION 2. COMPETENCE OF THE TRIBUNAL
Article 21
Parties before the Tribunal
1. States Parties may be parties before the Tribunal.
2. Entities other than States Parties may be parties before the Tribunal in any case expressly provided for in Part XI of the present Convention, or in accordance with any other agreement conferring jurisdiction on the Tribunal and accepted by all the parties to the dispute.

Article 22
Access to Tribunal
The Tribunal shall be open to the States Parties. It shall be open to entities other than States Parties in any case provided for in Part XI of the present Convention or in accordance with any other agreement conferring jurisdiction on the Tribunal and accepted by all the parties to any dispute submitted to the Tribunal.

Article 23
Jurisdiction
The jurisdiction of the Tribunal shall comprise all disputes and applications submitted to

it in accordance with the present Convention and all matters specifically provided for in any other agreement which confers jurisdiction on the Tribunal.

Article 24
Reference of disputes subject to other agreements

If all the parties to a treaty or convention already in force and relating to the subject-matter covered by the present Convention so agree, any disputes relating to the interpretation or application of such treaty or convention may, in accordance with such agreement, be submitted to the Tribunal.

Article 25
Applicable law

The Tribunal shall decide all disputes and applications in accordance with article 293 of Part XV.

SECTION 3. PROCEDURE

Article 26
Institution of proceedings

1. Disputes may be submitted to the Tribunal, as the case may be, either by a written application addressed by a party or parties to the dispute, or by the notification of any special agreement between the parties to the dispute, to the Registrar. In either case the subject of the dispute and the parties shall be indicated.
2. The Registrar shall forthwith communicate the application to all concerned.
3. He shall also notify all States Parties.

Article 27
Provisional measures

1. In accordance with article 290 of Part XV, the Tribunal and its Sea-Bed Disputes Chamber shall have the power to prescribe provisional measures.
2. If the Tribunal is not in session, or a sufficient number of members are not available to constitute a quorum, the provisional measures shall be prescribed by the chamber of summary procedure to be established under paragraph 3 of article 16. Notwithstanding paragraph 4 of that article, such provisional measures may be adopted at the request of any party to the dispute. They shall be subject to review and revision by the Tribunal.

Article 28
Hearing

1. The hearing shall be under the control of the President or, if he is not able to preside, of the Vice-President; if neither is able to preside, the senior judge present shall preside.
2. The hearing shall be public, unless the Tribunal shall decide otherwise, or unless the parties demand that the public be not admitted.

Article 29
Conduct of case

The Tribunal shall make orders for the conduct of the case, shall decide the form and time in which each party must present its arguments, and make all arrangements connected with the receiving of evidence.

Article 30
Default of appearance

When one of the parties does not appear before the Tribunal or fails to defend its case, the other party may request the Tribunal to continue the proceedings and make its decision. Absence or default of a party shall not constitute an impediment to the proceedings. Before making its decision, the Tribunal must satisfy itself not only that it has jurisdiction over the dispute, but also that the decision is well founded in fact and law.

Article 31
Majority for decision

1. All questions shall be decided by a majority of the members of the Tribunal who are present.
2. In the event of an equality of votes, the President or the member who acts in his place shall have a casting vote.

Article 32
Judgement

1. The judgement shall state the reasons on which it is based.
2. It shall contain the names of the members of the Tribunal who have taken part in the decision.
3. If the judgement does not represent in whole or in part the unanimous opinion of the members of the Tribunal, any member shall be entitled to deliver a separate opinion.
4. The judgement shall be signed by the President and by the Registrar. It shall be read in open court, due notice having been given to the parties to the dispute.

Article 33
Request to intervene

1. Should a State Party consider that it has an interest of a legal nature which may be affected by the decision in any dispute, it may submit a request to the Tribunal to be permitted to intervene.
2. It shall be for the Tribunal to decide upon this request.
3. If an application to intervene is granted, the decision of the Tribunal in respect of that dispute will be binding upon the applicant in so far as it refers to matters in respect of which that party intervened.

Article 34
Cases of interpretation or application

1. Whenever the interpretation or application of the present Convention is in question, the Registrar shall notify all States Parties forthwith.
2. Whenever, pursuant to article 23 or 24 of this Statute, the interpretation or application of an international agreement is in question, the Registrar shall notify all parties to the agreement.
3. Every party so notified has the right to intervene in the proceedings; but if it uses this right, the construction given by the judgement will be equally binding upon it.

Article 35
Finality and binding force of decisions
1. The decision of the Tribunal is final and shall be complied with by all the parties to the dispute.
2. Such decision shall have no binding force except between the parties and in respect of that particular dispute.
3. In the event of dispute as to the meaning or scope of the decision, the Tribunal shall construe it upon the request of any party.

Article 36
Costs
Unless otherwise decided by the Tribunal, each party shall bear its own costs.

SECTION 4. THE SEA-BED DISPUTES CHAMBER
Article 37
Composition of the Chamber
1. The Sea-Bed Disputes Chamber shall be established in accordance with article 15 and shall be composed of eleven member, selected from among the members of the Tribunal by the Assembly of the Authority, by a majority specified in accordance with paragraph 6 of article 157 of Part XI, for matters of substance.
2. The Assembly shall assure the representation of the principal legal systems of the world and equitable geographical distribution in the Chamber.
3. The members of the Chamber shall be selected every three years and may be selected for a second term.
4. The Chamber shall elect its Chairman from among its members, who shall serve for the period for which the Chamber has been selected.
5. If any proceedings are still pending at the end of any three year period for which the Chamber has been selected, the Chamber shall complete the proceedings in its original composition.
6. Upon the occurrence of a vacancy in the Chamber, the Tribunal shall select a successor from among its members who shall hold office for the remainder of the term of his predecessor, subject to the approval by the Assembly at its next regular session.
7. A quorum of seven members shall be required to constitute the Chamber.

Article 38
Access
The Chamber shall be open to the States Parties, to the Authority and to nationals of States Parties in accordance with the provisions of section 6 of Part XI.

Article 39
Applicable law
In addition to the provisions of article 293 of Part XV, the Chamber shall apply:
  (a) The rules, regulations and procedures adopted by the Assembly or the Council of the Authority in accordance with the present Convention; and
  (b) The terms of any contracts concerning activities in the Area in any matter relating to such contract.

Article 40
Enforcement of decisions of the Chamber
The decisions of the Chamber shall be enforceable in the territories of the States Parties in the same manner as judgements or orders of the highest court of the State Party where the enforcement is sought.

Article 41
Applicability of the procedure of the Tribunal to the Chamber
1. The provisions of this annex which are not incompatible with this section shall apply to the Chamber.
2. In the exercise of its functions relating to advisory opinions, the Chamber shall be guided by the provisions of this annex relating to procedure before the Tribunal to the extent to which it recognizes them to be applicable.

SECTION 5. AMENDMENT
Article 42
Amendment
1. Amendments to the present Statute shall be effected by the same procedure as provided for amendments to the present Convention.
2. The Tribunal shall have power to propose such amendments to the present Statute as it may deem necessary, through written communications to the States Parties, for consideration in conformity with the provisions of paragraph 1.

ANNEX VI
Arbitration
Article 1
Institution of proceedings
Subject to the provisions of Part XV, any party to a dispute may submit the dispute to the arbitration procedure provided for in this annex by notification addressed to the other party or parties to the dispute.

Article 2
List of arbitrators
A list of arbitrators shall be drawn up and maintained by the Secretary-General of the United Nations. Every State Party shall be entitled to nominate four arbitrators, each of whom shall be a person experienced in maritime affairs and enjoying the highest reputation for fairness, competence and integrity. The names of the persons so nominated shall constitute the list. If at any time the arbitrators nominated by a State Party in the list so constituted shall be less than four, that State Party shall be entitled to make further nominations as necessary. The name of an arbitrator shall remain on the list until withdrawn by the party which made the nomination, provided that such arbitrator shall continue to serve until the completion of any case in which that arbitrator has begun to serve.

Article 3
Constitution of arbitral tribunal

For the purpose of proceedings under this annex, the arbitral tribunal shall, unless the parties otherwise agree, be constituted as follows:

1. Subject to the provisions of paragraph 7, the arbitral tribunal shall consist of five members. Each party to the dispute shall appoint one member, who shall be chosen preferably from the list and may be its national. In the case of the party requesting arbitration, such appointment shall be made at the time of the request. The other three members shall be appointed by agreement of the parties and shall be chosen preferably from the list and shall be nationals of third States, unless the parties otherwise agree. The parties to the dispute shall appoint the President of the arbitral tribunal from among these three members.

2. The party requesting arbitration shall, at the time of making the request, submit a statement of its claim and the grounds on which such claim is based.

3. Should the other party to the dispute fail to appoint a member within a period of 30 days from the date of receipt of the request for arbitration, the appointment shall be made in accordance with paragraph 5, at the request of the party which submitted the dispute to arbitration. Such request shall be made within two weeks of the expiry of the aforementioned period of 30 days.

4. If, within a period of 60 days from the date of receipt of the request for arbitration, the parties are unable to reach agreement on the appointment of one or more of the members of the tribunal to be designated jointly, or on the appointment of the President, the remaining appointment or appointments shall be made in accordance with paragraph 5, at the request of a party to the dispute. Such request shall be made within two weeks of the expiry of the aforementioned period of 60 days.

5. Unless the parties agree that any appointment under paragraphs 3 and 4 be made by some person or a third State chosen by the parties, the President of the Law of the Sea Tribunal shall make such appointment. If the President is unable to act under this paragraph, or is a national of one of the parties to the dispute, the appointment shall be made by the next senior member of the Law of the Sea Tribunal who is available and is not a national of one of the parties. The appointments referred to in this paragraph shall be made from the list of arbitrators within a period of 30 days of the receipt of the request and in consultation with the parties. The members so appointed must be of different nationalities and must not be in the service of, ordinarily resident in the territory of, or nationals of, any of the parties to the dispute.

6. Vacancies which may occur as a result of death, resignation or any other cause shall be filled in such manner as provided for original appointments.

7. Parties in the same interest shall appoint one member of the tribunal jointly by agreement. Where there are several parties having separate interests or where there is disagreement as to whether they are of the same interest, each of them shall appoint one member of the tribunal. The number of members of the tribunal appointed separately by the parties shall always be smaller by one than the number of members of the tribunal to be appointed jointly by the parties.

8. In disputes involving more than two parties, the provisions of paragraphs 1 to 6 shall apply to the maximum extent possible.

Article 4
Functions of arbitral tribunal

An arbitral tribunal constituted under article 3 shall function in accordance with the provisions of the present Convention and of this annex.

Article 5
Procedure to be adopted

In the absence of an agreement to the contrary between the parties to the dispute, the arbitral tribunal shall lay down its own procedure assuring to each party a full opportunity to be heard and to present its case.

Article 6
Duties of parties to a dispute

The parties to the dispute shall facilitate the work of the arbitral tribunal and, in particular, in accordance with their law and using all means at their disposal, shall:

(a) Provide the tribunal with all relevant documents, facilities and information; and

(b) Enable the tribunal when necessary to summon and receive the evidence of witnesses or experts and to visit the localities in question.

Article 7
Expenses

Unless the arbitral tribunal determines otherwise because of the particular circumstances of the case, the expenses of the tribunal, including the remuneration of its members, shall be borne by the parties to the dispute in equal shares.

Article 8
Required majority for decisions

Decisions of the arbitral tribunal shall be taken by a majority vote of its members. The absence or abstention of less than half of the members shall not constitute an impediment to the tribunal reaching a decision. In the event of an equality of votes, the President shall have a casting vote.

Article 9
Default of appearance

When one of the parties to the dispute does not appear before the arbitral tribunal or fails to defend its case, the other party may request the tribunal to continue the proceedings and to make its award. Absence or default of a party shall not constitute an impediment to the proceedings. Before making its award, the arbitral tribunal must satisfy itself not only that it has jurisdiction over the dispute but also that the award is well founded in fact and law.

Article 10
Award

The award of the arbitral tribunal shall be confined to the subject-matter of the dispute, and state the reasons on which it is based. It shall contain the names of the members who have participated and the date of the award. Any member of the tribunal may attach a separate or dissenting opinion to the award.

Article 11
Finality of award

1. The award shall be final and without appeal, unless the parties to the dispute have agreed in advance to an appellate procedure. It shall be complied with by all the parties to the dispute.

Article 12
Interpretation or implementation of award

1. Any controversy which may arise between the parties to the dispute as regards the interpretation or manner of implementation of the award may be submitted by either party for decision to the arbitral tribunal which made the award. For this purpose, any vacancy in the tribunal shall be filled in the manner provided for in the original appointments of the members of the tribunal.
2. Any such controversy may be submitted to another court or tribunal under article 287 of Part XV by agreement of all the parties to the dispute.

Article 13
Application to entities other than States Parties

The provisions of this annex shall apply *mutatis mutandis* to any dispute involving entities other than States Parties.

ANNEX VII
Special arbitration procedure
Article 1
Institution of proceedings

Subject to the provisions of Part XV, any party to a dispute concerning the interpretation or application of the articles of the present Convention relating to (1) fisheries, (2) protection and preservation of the marine environment, (3) marine scientific research, and (4) navigation, including vessel source pollution, may submit the dispute to the special arbitration procedure provided for in this annex by notification addressed to the other party or parties to the dispute.

Article 2
List of experts

Separate lists of experts shall be established and maintained in respect of each of the fields of (1) fisheries, (2) protection and preservation of the marine environment, (3) marine scientific research, and (4) navigation, including vessel source pollution. The lists of experts shall be drawn up and maintained, in the field of fisheries by the Food and Agriculture Organization of the United Nations, in the field of protection and preservation of the marine environment by the United Nations Environment Programme, in the field of marine scientific research by the Inter-Governmental Oceanographic Commission, in the field of navigation by the Inter-Governmental Maritime Consultative Organization, or in each case by the appropriate subsidiary body concerned to which such organization, programme or commission has delegated this function. Every State Party shall be entitled to nominate two experts in each field whose competence in the legal, scientific or technical aspects of such field is established and generally recognized, and who enjoy the highest reputation for fairness and integrity. The names of the persons so nominated in each field shall constitute the appropriate list. If at any time the experts nominated by a State Party in any list so constituted shall be less than two, that State Party shall be entitled to make further nominations as necessary. The name of an expert shall remain on the list until withdrawn by the party which made the nomination, provided that such expert shall continue to serve until the completion of any case in which that expert has begun to serve.

Article 3
Constitution of special arbitral tribunal

For the purpose of proceedings under this annex, a special arbitral tribunal shall, unless the parties otherwise agree, be constituted as follows:
1. Subject to the provisions of paragraph 7, the special arbitral tribunal shall consist of five members. Each party to the dispute shall appoint two members, one of whom may be its national, to be chosen preferably from the appropriate list or lists relating to the matters in dispute. The parties to the dispute shall by agreement appoint the President of the special arbitral tribunal who shall be chosen preferably from the appropriate list and shall be a national of a third State, unless the parties otherwise agree.
2. The party requesting special arbitration shall, at the time of making the request, appoint its members and submit a statement of its claim and the grounds on which such claim is based.
3. Should the other party to the dispute fail to appoint its members within a period of 30 days from the date of receipt of the request for special arbitration, the appointments shall be made in accordance with paragraph 5, at the request of the party which submitted the dispute to arbitration. Such request shall be made within two weeks of the expiry of the aforementioned period of 30 days.
4. If, within a period of 30 days from the date of receipt of the request for special arbitration, the parties are unable to reach agreement on the appointment of the President, such appointment shall be made in accordance with paragraph 5, at the request of a party to the dispute. Such request shall be made within two weeks of the expiry of the aforementioned period of 30 days.

5. Unless the parties agree that any appointment under paragraphs 3 and 4 be made by some person of a third State chosen by the parties, the Secretary-General of the United Nations shall make such appointment, in consultation with the parties to the dispute and the appropriate international intergovernmental organization. The appointments referred to in this paragraph shall be made from the appropriate list or lists of experts within a period of 30 days of the receipt of the request. The members so appointed must be of different nationalities and must not be in the service of, ordinarily resident in the territory of, or nationals of, any of the parties to the dispute.

6. Vacancies which may occur as a result of death, resignation or any other cause shall be filled in such manner as provided for original appointments.

7. Parties in the same interest shall appoint two members of the tribunal jointly by agreement. Where there are several parties having separate interests or where there is disagreement as to whether they are of the same interest, each of them shall appoint one member of the tribunal.

8. In disputes involving more than two parties, the provisions of paragraphs 1 to 6 shall apply to the maximum extent possible.

Article 4
General provisions

The provisions of articles 4 to 12 of annex VI shall apply *mutatis mutandis* to the special arbitration procedure under this annex.

Article 5
Fact finding

1. The parties to a dispute may at any time agree to request a special arbitral tribunal constituted in accordance with article 3, to carry out an inquiry and establish the facts giving rise to any dispute concerning the interpretation or the application of the provisions of the present Convention relating to fisheries, protection and preservation of the marine environment, marine scientific research or navigation.

2. Unless the parties otherwise agree, the findings of fact of the special arbitral tribunal acting in accordance with paragraph 1, shall be considered as conclusive as between the parties. If all the parties to the dispute so request, the special arbitral tribunal may formulate recommendations which, without having the force of a decision, shall only constitute the basis for a review, by the parties concerned, of the questions giving rise to the dispute.

3. Subject to paragraph 2, the special arbitral tribunal shall act in accordance with the preceding provisions of this annex, unless the parties otherwise agree.

INFORMAL COMPOSITE NEGOTIATING TEXT
Explanatory memorandum by the President

At the 78th plenary meeting of the Third United Nations Conference on the Law of the Sea held on 30 June 1977, the Conference decided that the President should undertake, jointly with the chairmen of the three Main Committees, the preparation of an informal composite negotiating text which would bring together in one document the draft articles relating to the entire range of subjects and issues covered by parts I, II, III[3] and IV[4] of the revised single negotiating text. It was agreed that for this purpose the President and the Chairmen of the three Main Committees would form a team under the President's leadership and that the Chairman of the Drafting Committee and the Rapporteur-General would be associated with the team as the former should be fully aware of the considerations that determined the contents of the informal composite negotiating text and the latter should, *ex officio*, be kept informed of the manner in which the work of the Conference has proceeded at all stages.

It was understood that while the President would be free to proffer his own suggestions on the proposed provisions of any part of the composite text, in regard to any matter which fell within the exclusive domain of a particular chairman that chairman's judgement as to the precise formulation to be incorporated in the text should prevail. The adoption of this procedure was a recognition of the fact that each chairman was in the best position to determine, having regard to the negotiations that had taken place, the extent to which changes in his revised single negotiating text should be made in order to reflect the progress achieved in the course of negotiations where, in the chairman's opinion, such progress justified changes in the revised single negotiating text and also to decide, even where the negotiations had not resulted in substantial agreement, whether such progress as had been achieved warranted changes which would be conducive to the ultimate attainment of general agreement. It was also understood that so far as issues on which negotiations had not taken place were concerned, there should be no departure from the revised single negotiating text unless it was of a consequential character. This understanding was scrupulously observed in the course of the preparation of the informal composite negotiating text. There is no question, therefore, of joint responsibility being assumed for the provisions of the text by the President and the chairmen of the three Main Committees. The chairman of each Committee bears the full

---

3. *Official Records of the Third United Nations Conference on the Law of the Sea*, vol. 5 (Sales No. E.76.V.8) document A/CONF.62/WP.8/Rev. 1, parts I, II and III.

4. Ibid., vol. VI (Sales No. E.77.V.2) document A/CONF.62/WP.9/Rev. 2.

responsibility for those provisions of the informal composite negotiating text which are the exclusive and special concern of his Committee. This is not an enunciation of a new doctrine of collective irresponsibility.

The Conference also agreed that the composite negotiating text would be informal in character and would have the same status as the informal single negotiating text and the revised single single negotiating text and would, therefore, serve purely as a procedural device and only provide a basis for negotiation without affecting the right of any delegation to suggest revisions in the search for a consensus. It would be relevant to recall here the observation made in my proposals regarding the preparation of this text that it would not have the character and status of the text which was prepared by the International Law Commission and presented to the Geneva Conference of 1958 and would, therefore, not have the status of a basic proposal that would stand unless rejected by the requisite majority.

Special attention was given, in the course of the preparation of the informal composite negotiating text, to the need for co-ordination between the different parts of the revised single negotiating text where there appeared to be contradictions or unnecessary repetition.

The time available for the preparation of the informal composite negotiating text was so limited that the niceties of draftmanship had to be sacrificed in the interests of speedy completion of the text. This would eventually be the responsibility of the Drafting Committee.

The main purpose of this explanatory memorandum is to convey to the Conference the reasons for the changes that have been effected in, and the deviations from, the revised single negotiating text, as well as to draw pointed attention to the principal issues which are regarded as indispensable elements of the package deal that is envisaged and which require further and intensive negotiation. At a later stage I hope to present to the delegations participating in the Conference a document indicating in greater detail those issues and also suggesting for their consideration the order of priority to be assigned to them for treatment in future negotiations.

The structure of the informal composite negotiating text does not retain the order of the four parts of the revised negotiating text, but has been established on the principle that the most logical progression in the proposed new convention on the law of the sea would be from areas of national jurisdiction, such as the territorial sea, through an intermediate area such as the exclusive economic zone, to the area of international jurisdiction. It is hoped that the structure that has been adopted will constitute a definite advance in the elaboration of a comprehensive convention on the law of the sea and will be in conformity with the considerations that led the General Assembly in its resolution 2749 (XXV) of 17 December 1970 on the Declaration of Principles Governing the Sea-Bed and the Ocean Floor and the Sub-Soil thereof Beyond the Limits of National Jurisdiction to treat the problems of ocean space as closely interrelated and which need to be considered as a whole.

The use of the expression "Geographically Disadvantaged States" which appears in various provisions of the text is contingent upon a decision by the Conference regarding the definition of that term.

The Conference will note that in paragraph 3 of article 154 it has been stated that the seat of the Authority will be at Jamaica. This provision appeared in part I of the informal single negotiating text and was incorporated in it by the chairman of the First Committee because of the position taken in Caracas by the Group of 77. It was decided that this provision be retained with an indication that, subsequently, two other countries, namely Malta and Fiji, had offered to accommodate the seat of the Authority. This question has yet to be discussed in the Conference.

In order to ensure the comprehensive character of the informal composite negotiating text, it was thought fit to include a preamble and final clauses although these two subjects have not yet been discussed in the Conference. Every effort has been made to avoid any provisions in the preamble and final clauses that could lead to needless controversy at this stage and it is hoped that they will be accepted in that spirit.

The memorandum will now deal seriatim with the provisions of the revised single negotiating text, parts I to IV, in relation to those of the informal composite negotiating text.

Part I of the revised single negotiating text

At the time of the preparation of the revised single negotiating text, part I, the Chairman of the First Committee noted in his introduction to that text that there were still subjects that had not been given detailed consideration, namely the annex on the Statute of the Enterprise, the annex on the Statute of the Sea-Bed Disputes Settlement System, articles 33 to 40 of the main body of the text dealing with the Tribunal, and a "Special Appendix" to the annex on basic conditions of prospecting, exploration and exploitation which was to deal with financial arrangements. While these areas of the revised single negotiating text, particularly the Statute of the Enterprise and articles 33 to 40 on the settlement of disputes have since received closer attention, there is still much to be done.

It should also be recalled that following the preparation of the revised single negotiating text at the fifth session, the First Committee devoted itself almost entirely to the question of a system of exploitation and succeeded only in reducing the problem to some basic questions which still remained to be solved. It may be claimed, however, that remarkable progress has since been made in overcoming what threatened to be a complete deadlock.

The issues and subjects falling within the First Committee's purview are now covered in part XI of the informal composite negotiating text. The financial terms of contracts which previously formed a Special Appendix are now included in a paragraph in annex II. This annex as a whole refers only to activities carried out under contractual arrangements. The annex on the Statute of the Sea-Bed Disputes Settlement System was deleted in the light of the agreement reached in the First Committee, following extensive discussions in plenary, that the part of the Convention relating to First Committee matters would deal only with the jurisdictional aspects of settling sea-bed disputes and that the institutional and procedural aspects would be covered by the part of the Convention dealing with the general question of settlement of disputes, viz. part XV and the relevant annexes. There is one exception in that there is now a provision in article 158 whereby the Assembly would select the members of the Sea-Bed Disputes Chamber from among the members of the Law of the Sea Tribunal.

As regards the substantive changes made in the provisions of part I of the revised single negotiating text now appearing as part XI and in annexes II and III, it must be recalled that:

1. At the fifth session the First Committee devoted itself almost exclusively to negotiation of the system of exploitation, namely revised single negotiating text article 22 and related paragraphs of revised single negotiating text, annex I;

2. At intersessional consultations held in Geneva the proposed "mini-package" on the system of exploitation was developed in greater detail to include, most important of all, the setting up and financing of the Enterprise and also the question of refining and revising the temporary or interim system envisaged;

3. At the beginning of the sixth session, in order to facilitate the First Committee's work, the Chairman established that the "mini-package" would comprise the resource policy of the Authority (revised single negotiating text, article 9), the system of exploitation (revised single negotiating text, article 22, related paragraphs of the annex, and provision for a periodic and definitive review of the system as proposed in articles 64 and 65, introduced intersessionally), and the setting up and financing of the Enterprise, particularly in the start-up phase.

The Chairman of the First Committee also established that in addition to this "mini-package" the Committee would need to take up the other elements which make up the larger package, viz. institutional questions and settlement of disputes. The Committee, through a Chairman's Negotiating Group, developed further the elements of a possible compromise on the revised single negotiating text, articles 9, 22 to 32, 41, 49, and 33 to 38, and on revised single negotiating text annex I, paragraphs 8 (new) and 8 *bis*, and on the Statute of the Enterprise. It would thus be clear that most of the changes introduced in the present text have occurred in those articles and paragraphs.

Particular attention must, however, be drawn to the changes made by the Chairman of the First Committee and appearing in articles 150, 151, 153, 159 and 160, 169, 187, and 192 and in paragraphs 4, 5, 6 and 7 of annex II of the composite text. Most of the major changes are intended by the Chairman to overcome the fundamental difficulties that still remain as to the approach which should be adopted towards the temporary system of exploration and exploitation. In this regard, article 151, paragraph 2 (ii) is of special importance. In order to meet the concern that the temporary system might not ensure the balance intended between on the one hand the area reserved for the Enterprise and developing countries and on the other hand the contract areas to be exploited by States Parties and other entities in association with the Authority, which is the most important evaluation to be made by the Review Conference as stated in paragraph 2 of article 153, the Chairman has added to paragraph 2 of article 151 the requirement that the contractual or other arrangements made by the Authority with States Parties and other entities are such as will enable the Authority to fulfil its most important function, as set out in paragraph 1 of that article. While specific reference has been made to technology and financial and other resources, the wording is not intended to determine the actual type or form of contract or other arrangement. It would not, for example, automatically impose joint arrangements.

In addition to the references to technology in paragraph 2 (ii) and paragraph 8 of article 151, there are numerous other references to transfer of technology to the Enterprise or to the develoment of the technological capability of the Authority. The Chairman of the First Committee felt that there was, undoubtedly, need to strengthen this most important requirement of the Authority and also the need to mention this aspect in the context of the qualifications of applicants, while fully realizing that, in the present text as a whole, the numerous references may not all be necessary or be appropriately placed and that, in addition to the problem of their co-ordination, the question as a whole and numerous references may not all be necessary or be appropriately place and that, in addition to the problem of their co-ordination, the question as a whole and the implications, legal and financial, of the acquisition of technology by the Authority would need further and more detailed examination.

The question whether the new provision on scientific research in article 151 is sufficient to indicate the role that the Authority may be expected to play in this activity, which is very important to the international community, may require further discussion.

The Chairman of the First Committee feels

that the present text has made a considerable advance on the revised single negotiating text in two other areas, those concerning the Enterprise and the question of review, and that perhaps the single most important factor in the emergence of a new compromise is to provide for the rapid creation of a viable Enterprise. The text now effectively gives a higher status to the role of the Enterprise as an operating entity vis-à-vis States Parties and other entities. The introduction of joint arrangements as a possible option, the establishment of a special fund to cover the first mine site, and provisions concerning its acquisition of technology are intended essentially to facilitate its start-up.

For the first time the implications of a review clause for the duration and effectiveness of a temporary system of exploitation were considered by the First Committee but, while the need for periodic review was readily acknowledged, there was considerable scepticism as to what could be genuinely reviewed and whether it would be possible actually to revise the system. Many consequently rejected a review clause as a determining element for the acceptance of a temporary system of exploitation. As a possible compromise the Chairman of the First Committee has made a new proposal contained in paragraph 6 of article 153 which in his view is intended to allay that scepticism and also to deal with the legal vacuum that would arise should the review Conference fail to reach agreement. As there have been many and varied references to joint arrangements, and in this case to joint ventures, a thorough discussion of such methods of exploitation and their implications would serve a most useful purpose.

The negotiations on First Committee issues produced a series of suggested formulations. Small expert groups worked on two provisions, the specific measures of a resource policy in article 150, and paragraph 7 of annex II on the financial terms of a contract. For the first time the different formulae for the control of sea-bed production were analysed as to their effects in terms of the number of mine sites that would be available over a 20 year period. A small expert group pursued this analysis by developing a common method for making calculations, to overcome the problem caused by contradictory sets of figures, and by reaching an agreement on the method of interpreting the formula suggested. It is the considered opinion of the Chairman of the First Committee that the new formula now contained in article 150 is a considerable improvement on that contained in the revised single negotiating text; although, of course, the quantitative aspects of the specific measures to protect developing countries from adverse effects require further negotiation.

Similarly, a small group of experts worked on the question of the "Special Appendix" on financial matters, first preparing a paper for discussion in the Chairman's Negotiating Group and later developing that paper in the light of several informal meetings held on that subject in the Group. As with several other items this question still requires further technical work as well as negotiation. A foot-note has been included in order to underline the status of work on that issue (foot-note to para. 7 of annex II of the informal composite negotiating text).

The Chairman of the First Committee would like to emphasize that there is still a difference of approach with regard to the two main methods of payment to the Authority as a result of activities in the contract area, a difference of opinion as to whether those methods would be alternatives or not, and also as to which of the two methods is considered preferable. The first method calls for a system of royalties or what may more appropriately be called, in respect of the international sea-bed area, a *fixed* charge on production. This could take various forms. The second method calls for sharing of profits or, to use language analogous to that used in respect of production within the territory of a State, a tax on profits. Both systems have advangages and disadvantages. The royalty system allows for a definite payment to the Authority at an early stage and a foreseeable sum throughout the contract period irrespective of the amount of profits made. From the contractor's point of view, however, the system involves "front-end payments" or payments at a time when he could least afford them. The second system of payments, namely the share of profits, i.e. a net proceeds system, allows a contractor to pay when he can best afford to pay and could give the Authority a share of very high profits. Trends in modern land-based operations favour this system.

It must be kept in mind that the Authority, not being a sovereign State and lacking the usual range of measures for controlling a foreign-based operator, would need effective powers to scrutinize costs and profits of contractors.

The Chairman of the First Committee does not see the new paragraph (para. 7 of annex II of the informal composite negotiating text) as a compromise text but rather as a considerable step forward from the alternative approaches in the special appendix to the revised single negotiating text. The new paragraph is designed to focus the attention of Governments on the issues and to indicate to them what he regards as the elements of an eventual compromise—royalties and profit-sharing, in addition to a mining fee. It is realized that the acceptability of the proposed arrangements will depend on the figures which will ultimately have to be incorporated into the texts. It was considered premature at this stage to include figures as these would require careful examination and negotiation. Although the informal working group had two specific proposals before it, the figures in each case are based on assumptions as yet to be accepted.

The outstanding questions to be negotiated on financial arrangements are:
(i) whether payment to the Authority should be based on processing

activities, as well as the mining operations themselves:
(ii) whether the modern practice in land-based mining of a comparatively small royalty or production charge, and a greater emphasis on a share of profits, is appropriate to mining of the international sea-bed or whether primary emphasis should be placed on the royalty charge;
(iii) whether an agreed basis can be found that will provide the Authority with effective powers to scrutinize costs and profits of contractors;
(iv) with (i) and (iii) in mind the consideration that should ultimately determine the level of payment to the Authority;
(v) whether there is a need in the Convention for financial arrangements for activities other than those under contracts in the non-reserved area.

As noted above, it will be important to discuss whether and to what extent the system of exploitation will cover processing activities as well as mining activities. It may be useful to ascertain whether or not the system could be envisaged as incorporating the processing stage, at least in certain cases, and particularly where developing countries, and perhaps also the small developed countries, could be involved in operations.

While in the view of the Chairman of the First Committee the new text represents a considerable advance on the stages of negotiation reflected by the revised single negotiating text formulations, much work remains to be done on the corresponding provisions of the composite text. Apart from those questions which have earlier been mentioned as requiring further attention, there is the question of the Enterprise. Given the limited applicability of the new annex II, there may be a need to clarify the institutional provisions pertaining to the plans of work to be drawn up by the Enterprise. The work carried out on the financial terms of contracts may need to be complemented by a similar effort encompassing joint arrangements involving the Enterprise. There may still be some aspects of the general financial structure of the Authority affecting the Enterprise which could benefit from a further discussion, taking into account the report by the Secretary-General. That report will also be useful in further discussions on the question of joint arrangements.

While transitional arrangements do not appear in annex II for practical reasons that the Chairman of the First Committee considers obvious, it may still be necessary to consider how the Authority will fulfil its functions in the first several years of its coming into existence.

With respect to the question of a quota system or anti-monopoly clause, the consultations that have already been initiated will be continued and an addendum to the present text will be proposed. Such consultations might cover the relationship between an anti-monopoly provision and non-discrimination.

With respect to the institutional provisions, the Chairman of the First Committee felt that possibly the most important question, so far as the discussions at the sixth session were concerned, was the composition of the Council. In his opinion, the formula now used does much to answer the doubts and queries raised by those who feel that its composition should ensure that the Council can reach the most appropriate decisions and that smaller States, developed and developing, will have the opportunity of becoming members at some point.

Part II of the revised single negotiating text

The legal status of the exclusive economic zone has proved to be one of the most controversial issues facing the Conference. In the course of the final days of the sixth session, a group consisting of some of the delegations most interested in the issue made a concerted effort to seek a solution. Although the texts elaborated by this group are by no means a negotiated solution, it was felt by the Chairman of the Second Committee that they constituted a better basis for further negotiation than the articles in the revised single negotiating text. The Chairman of the Second Committee, therefore, decided that the texts which resulted from the work of the group and which were discussed in that Committee should be included in the informal composite negotiating text. The relevant provisions, which are contained in articles 55, 56, 58 and 86 and 89 of the composite text, when read with related articles, retain the essential features of the specific legal régme of the exclusive economic zone without upsetting the balance implicit in the revised single negotiating text between the rights and duties of the coastal State and those of other States.

The question of the right of land-locked States and certain coastal States to participate in the exploitation of the living resources of the exclusive economic zone was similarly the subject of intensive negotiation during the sixth session. A possible compromise appeared to be within reach but could not be finally negotiated for want of time. In these circumstances, the Group of Land-locked and Geographically Disadvantaged States experessed a preference for the retention of the existing articles in the revised single negotiating text while expressing their readiness to negotiate further on this question. Consequently, although there was a possibility of introducing as article 71 a related provision agreed upon by the interested delegations and of amending article 72 in regard to restrictions on the transfer of rights, articles 58 and 59 of the revised single negotiating text were retained, unchanged, as articles 69 and 70 of the informal composite negotiating text.

The Chairman of the Second Committee was satisfied that there was widespread agreement that the definition of the continental shelf as appearing in article 76 of the composite text constituted one of the essential elements of the "package deal." On this assumption and in accordance with the terms of that article a need has been recognized for a more precise definition of the outer edge of the continental margin. A specific proposal has been supported by a group of delegations claiming to be most directly interested in this matter. However, despite the need for such a definition, and although no alternative definition which is generally acceptable has been submitted, the inclusion of the suggested wording in the composite text was not considered justifiable at this stage.

On the question of payments and contributions with respect to the exploitation of the continental shelf beyond 200 nautical miles, the Chairman of the Second Committee decided that the relevant provision in article 82 of the composite text should reflect the efforts made in the Second Committee to provide more comprehensive indications of the system which would apply to these payments and contributions. The incorporation of certain elements in the relevant article does not in any way imply that a consensus on this issue has been reached in the Second Committee.

As regards the right of access of land-locked States to and from the sea and freedom of transit the Chairman of the Second Committee was of the opinion that the inclusion in the composite text of the results of the extensive negotiations held during the fifth session of the Conference could have a beneficial effect on the process of negotiation.

On the question of the delimitation of the territorial sea, the exclusive economic zone and the continental shelf between adjacent or opposite States, the Chairman decided that the relevant articles as appearing in the revised single negotiating text should be retained as it had not been possible to devise a formula which would narrow the differences between the opposing points of view. The issue would, therefore, remain open to further negotiation.

Certain changes in the articles on archipelagic States have been introduced as the States most interested in the subject had reached agreement on the point.

Part III of the revised single negotiating text

Protection and preservation of the marine environment

As regards the protection and preservation of the marine environment, the Chairman stated that the provisions contained in the revised single negotiating text constituted a generally acceptable package in which a proper balance was maintained. He considered this to be particularly so with regard to the key question of pollution from vessels.

Numerous proposals relating to the question were either withdrawn or resulted in an inconclusive debate concerning their incorporation in the informal composite negotiating text in the light of the carefully structured compromise which was reflected in the pertinent articles of the revised single negotiating text (articles 21, 27, 28, 30, 31 and 33 to 39).

Consequently, only in two instances (article 28, 1 of part III of the revised single negotiating text, where the word "applicable" was introduced to qualify the meaning of "international rules and standards" and in article 30, 7 of the same text, where the language of the text was replaced by a new provision relating to the release of vessels through bonding or other appropriate financial security) did the negotiations result in full support for explicit changes in the language of the revised single negotiating text. It is not without significance that these two changes are now incorporated in the informal composite negotiating text, though in somewhat clearer terms, but leaving intact the structure of the compromise on the question of vessel source pollution. This was also, in effect, the general thrust of the negotiations on the other articles examined at the sixth session relating to the protection and preservation of the marine environment.

It was apparent that by the introduction of technical changes or additions, the relevant provisions would gain in clarity or precision, or would be better correlated with the rest of the revised single negotiating text. In introducing these technical modifications of a technical character, it was firmly intended to preserve unchanged the substance of the "package" as reflected in the revised single negotiating text. Changes of this nature have been made in the following articles of that text: 1, 6, 9, 10, 11, 12, 14, 18, 20 (foot-note), 21 (4), 21 (5), 27 (8), 28 (1), 28 (3), 28 (4), 29, 30 (4), 34, 38 (1), 38 (2), 39, 40 (2), and 41. For the same reasons, a new provision was included dealing with the institution of civil proceedings.

As a result of some of these changes the provisions relating to international rules and national legislation to prevent, reduce and control pollution of the marine environment (articles 17 to 25 of the revised single negotiating text) as well as those dealing with enforcement (articles 23 to 32 of that text) have been brought closer together in the composite text. The Chairman of the Third Committee believes that the informal composite negotiating text is an improvement on the revised single negotiating text in that it expresses in a more coherent form, both conceptually and textually, the complementary nature of the two principal elements of the same process viz. the establishment of the relevant legal principles and rules and their practical implementation. Concerning the delicate question of the applicability of safeguards with respect to straits used for international navigation, an additional provision was added by the Chairman of the Third Committee to the corresponding article of the revised single negotiating text. This new provision was the result of negotiations by a group of

States most directly concerned with the implications of the safeguards provisions for straits.

Marine scientific research

Negotiations on this question were protracted and extensive. The Chairman of the Third Committee was satisfied that there was general agreement that the régime of marine scientific research in the exclusive economic zone or on the continental shelf of the coastal State had to be compatible with the jurisdiction of the coastal State as provided for in the relevant provisions of part II of the revised single negotiating text. On this principle the coastal State must have the right to regulate, authorize and conduct marine scienfific research in its exclusive economic zone and on its continental shelf and it follows that marine scientific research in the exclusive economic zone and on the continental shelf is to be carried out with the consent of the coastal State. This principle is reflected in article 247 of the informal composite negotiating text.

Given the importance of marine scientific research for increasing mankind's knowledge of the marine environment, it is imperative that the consent of the coastal State shall be granted to a research project which is conducted for peaceful purposes. This fundamental principle is embodied in the corpus of the régime itself in paragraph 3 of article 247. The negotiations had made it clear to the Chairman of the Third Committee that a balance should be established between the right and duty of the coastal State to grant consent and the exercise of its jurisdictional power to withhold it whenever the project is of direct significance for the exploration and exploitation of the natural resources of the economic zone or the continental shelf, or involves drilling into the continental shelf or the use of explosives or the introduction of harmful substances into the marine environment or the construction of artificial islands, installations and structures. The new version of this text, as compared with the revised single negotiating text, contains a provision which gives the coastal State two additional reasons for withholding its consent, namely, when the information regarding the nature and objectives of a project is inaccurate or when the State or international organization conducting the research has outstanding obligations from a prior research project. These explicit conditions for withholding consent could be considered as safeguards in favour of the State conducting marine scientific research activities.

The obligation imposed on States to establish rules and procedures to ensure that consent will not be delayed or denied unreasonably constitutes an additional safeguard.

Under article 253 a marine scientific research project can be commenced if the coastal State has failed within a specified period of time to reply to a request for consent to carry out the project. This notion of implied consent is intended to counterbalance the right of a coastal State to regulate or authorize the conduct of a marine scientific research project in its economic zone or on its continental shelf.

To meet the concerns of several delegations which felt that research projects undertaken under the auspices of or by an international organization should be facilitated through a special régime, a new article—248—has been incorporated in the composite text.

Development and transfer of marine technology

In regard to the development and transfer of marine technology, the Chairman of the Third Committee considered the changes introduced in the composite text to be generally acceptable. In paragraph 1 of article 267 it appeared to him to be sufficient to refer only to fair and reasonable terms and conditions in preference to the cumbersome formulation contained in the corresponding article in the revised single negotiating text.

Article 274 of the composite text is designed to accomodate the concern of several delegations that co-operation in the field of transfer of technology to the developing States shall be extended to other competent international organizations as well as to the Enterprise. Some minor changes and adjustments for purposes of clarification were made in article 275.

Part IV of the revised single negotiating text

In regard to the question of settlement of disputes which is covered by Part XV and annexes IV to VII of the informal composite negotiating text, the fundamental principles that have determined the substance of this part of the informal composite negotiating text have been the freedom of choice of court or tribunal; the agreement of the parties to the dispute on the choice of court or tribunal; the securement of finality in the form of a binding and conclusive settlement; and the designation of a specific procedure where the parties to the dispute fail to agree on the court or tribunal together providing a system of compulsory dispute settlement.

The provisions relating to the settlement of disputes are applicable, with some exceptions, to all the substantive parts of the proposed Convention. For that reason, while the provisions must remain general in their application, it has been necessary, in regard to certain aspects of this question, to maintain a close link with relevant provisions of the other parts, particularly in relation to the exceptions from comprehensive application (articles 296 and 297), activities in the Area (articles 15 and 37 to 41 of annex V) and such specific issues as the release of detained vessels (article 292).

The negotiations revealed that there was a wide measure of agreement that the acceptance of the jurisdiction of the Sea-Bed Disputes Chamber for the resolution of conflicts arising from activities in the Area should not entail acceptance of the jurisdiction of the Law of the Sea Tribunal for other disputes. A provision to that effect has been added to article 287 in paragraph 2. The institutional arrangements for the settlement of disputes relating to activities in the Area have been covered in annex V (articles 15 and 37 to 41 and in annex VI).

Provision has also been made in paragraph 3 of this article, whereby in a dispute which is not covered by a declaration of a State Party regarding choice of procedure, arbitration is deemed to have been accepted.

The new formulation of article 296 is intended to provide safeguards against an abuse of power by a coastal State and at the same time to avoid an abuse of legal process by other States. In paragraph 1 of this article provision has been made through procedural devices to avoid the abuse of legal process. Constraints have also been imposed on the challenge of discretionary powers in relation to living resources and marine scientific research.

The compromise provision appearing in article 18 of the revised single negotiating text has been retained in substance in article 297 of the informal composite negotiating text, due to the fact that there was a nearly equal division of views as to the need, or otherwise, for compulsory binding procedures. Certain new elements have been incorporated in relation to disputes concerning delimitation of sea boundaries between adjacent or opposite States. They are, firstly, the exclusion of adjudication of territorial claims, and secondly, that where a party has chosen a procedure not specified in this Convention, the other party to the dispute must have access to such procedure.

Subparagraph (b) of article 18 (1) of the revised single negotiating text has been amended so as to give law enforcement activities similar immunity to military activities. The corresponding provision is in article 297.1 (b) of the informal composite negotiating text. Article 297.1 (c) has incorporated a change relating to disputes in respect of which the Security Council is exercising the functions assigned to it.

Provision relating to the exhaustion of local remedies, which is a well-recognized principle of international law has been reintroduced in article 294 of the informal composite negotiating text.

There was very little discussion on the annexes to part IV of the revised single negotiating text with the exception of annex II of that part, and only incidental changes have been made in regard to these provisions in the informal composite negotiating text.

* * *

In this explanatory memorandum an effort has been made to present the principal features of the informal composite negotiating text in their relationship to one another. It is hoped that it will simplify and facilitate the task of further negotiations.

(Signed) *H. Shirley Amerasinghe*
President

Selected Documents and Proceedings

International Conventions and Other Agreements for Control of Marine Pollution

Conventions, Treaties, Protocols Regulations, and Standards	Pollutant	Responsible Body	Status April 1977
General marine pollution:			
Convention on the Territorial Sea and Contiguous Zone, 1958	Various	UN	Signed and ratified 1964
Convention on the High Seas, 1958	Oil, wastes from exploration and exploitation of the sea-bed and its subsoil, and radioactive wastes	UN	Signed and ratified 1962
Convention on the Continental Shelf, 1958	Any harmful agents	UN	Signed and ratified 1964
Convention on Fishing and Conservation of the Living Resources of the High Seas, 1958	All deleterious substances	UN	Signed 1966 and in force
Convention on the Prohibition of the Development, Production, and Stockpiling of Bacteriological (Biological) and Toxic Weapons and on Their Destruction, 1972	Biologically hazardous and other toxic substances	UN	Signed in 1972 but not yet in force
Protocol Relating to Intervention on the High Seas in Cases of Marine Pollution by Substances Other than Oil	Other substances than oil	IMCO	Adopted in November 1973; opened for signature January 1974; not yet in force

692

International Conventions and Other Agreements 693

Convention on the Protection of the Marine Environment of the Baltic Sea, 1974	All sources	Norwegian government	Adopted in 1974 but not yet in force
Convention on the Prevention of Marine Pollution from Land-based Sources, 1974	Wastes from land-based sources	French government	Signed but not yet in force
Convention on the Protection of the Environment between Denmark, Finland, Norway, and Sweden	Environmentally harmful substances	Swedish government	Signed but not yet in force
Convention for the Protection of the Mediterranean Sea against Pollution, 1976	All sources	UNEP	Signed but not yet in force
Oil pollution: International Convention for the Prevention of Pollution of the Sea by Oil, 1954	Oil	U.K. until establishment of IMCO in 1958	Ratified and in force 1958
Amendments to the International Convention for the Prevention of Pollution of the Sea by Oil, 1962	Oil	IMCO	Ratified and in force 1967
Amendments to the International Convention for the Prevention of Pollution of the Sea by Oil, 1969	Oil	IMCO	Signed but not yet in force (20 January 1978)
Amendments to the International Convention for the Prevention of Pollution of the Sea by Oil, 1971	Oil	IMCO	Signed but not yet in force

694   Selected Documents and Proceedings

INTERNATIONAL CONVENTIONS. *Continued*

International Convention Relating to Intervention on the High Seas in Cases of Oil Pollution Casualties, 1969	Oil	IMCO	Ratified and in force 1975
International Convention on Civil Liability for Oil Pollution Damage, 1969	Oil	IMCO	Ratified and in force
International Convention on the Establishment of an International Fund for Compensation for Oil Pollution Damage, 1971	Oil	IMCO	Signed but not yet in force
International Convention for the Prevention of Pollution from Ships, 1972	Oil and other substances	IMCO	Not yet in force
Agreement concerning Cooperation in Dealing with Pollution of the North Sea by Oil, 1969	Oil	Government of the Federal Republic of Germany	In force 1969
Agreement concerning Cooperation in Measures to Deal with Pollution of the Sea by Oil, 1971	Oil	Danish government	In force 1971
Convention on Civil Liability for Oil Pollution Damage from Offshore Operations, 1975	Oil	U.K. government	Not yet in force
Radioactivity:			
Convention on Third Party Liability in the Field of Nuclear Energy, 1960	Radioactive materials	UN, IAEA	Ratified and in force 1968

Convention Supplementary to the Paris Convention on Third Party Liability in the Field of Nuclear Energy, 1963	Radioactive materials	UN, IAEA	In force 1974
Convention on the Liability of Operators of Nuclear Ships, 1962	Radioactive materials	UN, IAEA, IMCO	Signed but not yet in force
Convention on Civil Liability for Nuclear Damage, 1963	Radioactive materials	UN, IAEA	Signed in 1963 but not yet in force
Treaty Banning Nuclear Weapons Tests in the Atmosphere, in Outer Space, and Underwater, 1963	Radioactive materials	UN, IAEA	Ratified and in force 1964
Treaty on the Prohibition of the Emplacement of Nuclear Weapons and Other Weapons of Mass Destruction on the Seabed and Ocean Floor and on the Subsoil Thereof, 1971	Radioactive materials	UN, IAEA	Signed; prohibits emplacement of such weapons in waters beyond 12 miles from shore; in force 1972
Convention relating to Civil Liability in the Field, of Maritime Carriage of Nuclear Material, 1971	Radioactive materials	UN, IAEA	In force 1975
Regulations for the Safe Transport of Radioactive Materials	Radioactive materials	UN, IAEA	Adopted
Basic Safety Standards for Radiation Protection	Radioactive materials	UN, IAEA	Adopted
Standardization of Radioactive Waste Categories	Radioactive materials	UN, IAEA	Adopted

INTERNATIONAL CONVENTIONS. *Continued*

Ocean dumping:			
Convention for the Prevention of Marine Pollution by Dumping from Ships and Aircraft, 1972	All wastes and other substances dumped at sea	Instruments placed with Norwegian government	Signed; in force 1974; known as the "Oslo Convention"; involves countries of northwestern Europe
Convention for the Prevention of Marine Pollution by Dumping of Wastes and Other Matter, 1972	All wastes and other matter dumped at sea	Instruments placed with government of Great Britain and Northern Ireland, inter alia	Ratified and in force

Source.—UNEP.

Selected Documents and Proceedings

Joint Oceanographic Assembly*

Following the Joint Oceanographic Assembly, held in Edinburgh, 13–24 September 1976, the following summary of scientific highlights was prepared by Sir George Deacon, Chairman of the organizing committee. The proceedings of the Assembly are being printed by Plenum Press, and persons interested in obtaining copies should contact the publisher at the following address: Black Arrow House, Chandos Road, London NW 10 6NR, United Kingdom.

Biological papers presented at the Joint Oceanographic Assembly demonstrated the very wide scope of biological studies which require many different approaches for their effective execution. Despite their diversity, it is to be hoped that it will be found possible to make concerted attacks on joint problems, as is not common in physical oceanography.

Unified studies deemed desirable

One promising advance was seen in upwelling where unified studies are now underway. Biological modelling goes ahead, but it was shown to require detailed understanding of both the biology and the mathematical techniques.

In this area particular attention is being paid to factors likely to favour larval survival and subsequent growth, and notable advances are being made towards linking biological and physical patterns. Favourable, partly closed, water patterns begin to appear more essential than the particular temperature and salinity ranges for which they are also responsible.

In the study of the origin and distribution of plankton species there seems to be evidence that geological history is also significant. The difficulty of catching representative samples of animals higher up the food chain is proving a hindrance. Here, especially, improvements in methods and design of sampling programmes are tasks in which biologists are collaborating with other disciplines.

Visual observations by divers or from submersibles are providing new information about animal associations and the functions of gelatinous structures and mucus. The 'Alvin lunch' experiment showed that microbial degradation, even at fairly shallow ocean depths, is one or two orders of magnitude less than at shelf depths.

One of the interesting observations during the upwelling programme was of an established phytoplankton growth being dispersed by a newly upwelling water. The Edinburgh meetings had a more balanced outlook on pollution problems than other recent conferences.

*Reprinted from Unesco, *International Marine Science Newsletter*, No. 14, March, 1977.

Physical topics

Some of the physical topics were concerned with events on a very long time scale relating to the evolution of the ocean basins, and some on a micro-scale down to a few decimetres or less and even to the molecular limit.

Most of the interest was in intermediate scales, especially on eddies 100 km across, lasting a month or so, with perhaps a factor or two in each dimension. Paleo-oceanography is based largely on the analysis of cores from the deep-drilling ship *Glomar Challenger,* and the microstructure on fall-out from continuous-profile recordings.

The 100 km eddies must play an important part in the dynamics of ocean circulation and, as was suggested, may contain 99 per cent of the kinetic energy of the ocean. However, it is the residual flows, moving much slower than the eddies, that carry large amounts of heat, salt and nutrients great distances across the ocean to maintain its long-term balance.

How sensitive these are to the spatial inhomogeneities on the 100 km scale will become an increasingly profitable question as we learn more about the eddies themselves. Progress can be expected from GEOSECS, though there is probably much to be done before its quantitative aspects, in terms of time and volume transports, are unquestionable.

Climatological effects

A special symposium on the effect of the ocean on climate seemed to give a clear indication that cause and effect is not likely to be established by statistical studies alone. It appears to require mechanistic studies of actual events when they are simple enough to be fully observed.

Numerical and laboratory experiments also show promise, but an adequate solution will obviously require much effort.

The Joint North Sea Wave Project and the Woods Hole Internal Wave Project, using the new 3-point mooring, have provided more effective data for surface and internal wave studies. Papers on the 'skin of the ocean' pointed to another promising, complex, question that requires interdisciplinary study.

Palaeo-oceanography comes of age

Palaeo-geography and oceanography are now of mature interest to geology and geophysics, and progress is being made by special studies of indicative areas. Much of the evidence comes from deep drilling and correlation of the palaeontological stratification with improved physical and chemical measurements.

Special interest is being shown in the Atlantic Ocean, where the depth of sediments and some availability of climatic records provides reasonable opportunities. Estimates of advance and retreat of the frontal region between temperate and Arctic waters have been used.

Another criterion is found in changes of carbon compensation depth, below which high CO_2 content—as in the Antarctic bottom current—helps to dissolve calcium. The relation between geological events and movements of water in the oceans has done quite a lot to extend interest in quantitative models of oceanic and atmospheric circulation.

Chemistry and sedimentology

Chemical papers dealt mainly with processes which influence the concentrations of reactive substances and trace metals in both oceans and sediments. The effects of transport from the land, by the atmosphere and rivers, were considered, but most attention was paid to events near the sea floor. Among the topics discussed were differences between concentrations in near-bottom waters and sediments, gradients in the sediments and interstitial water, metal-enriched surface oxides, biochemical processes, thermal convection and disturbance by marine animals.

Comparison of sediments along active mid-ocean ridges showed high accumulation rates for manganese compared with adjacent regions, though those on the ridge had less copper, nickel and cobalt. Sedimentation and resuspension in estuaries was considered as well as at the ocean floor.

The engineering talks mentioned a possibility that commercial opportunism might hinder the build-up of basic knowledge. On the physical side the engineers seemed to place most of their emphasis on better understanding of waves, ocean circulation and climate. Waves might be used as a direct source of energy, and indirectly, if suitably controlled, as an agent as well as a hindrance in coastal engineering.

In practical biology we should look for improved use of living resources and warning of overfishing. Monitoring of sea floor disturbance by sand and gravel extraction might have to be extended to the deep ocean when mining and processing moved out beyond the shelf. In view of its potential economic significance, fuller understanding of the origin and associations of deep-sea minerals will obviously increase in practical importance.

Practical applications for the future

Papers and discussion in this area showed little evidence that governments and industry intend to experiment with practical applications of some of the scientific possibilities—enriching an area by artificial upwelling for example. These might well be appropriate topics for the next Assembly, to be held in four or five years.

SCIENTIFIC PROGRAMME OF THE JOINT OCEANOGRAPHIC ASSEMBLY

Edinburgh, Scotland, 13 to 24 September 1976

GENERAL SYMPOSIA

G1 *History of the Oceans*—14 September 1976
    Convener: Dr. E. S. W. Simpson (South Africa)
G2 *Ocean Circulation and Marine Life*—15 September 1976
    Convener: Professor G. Hempel (FRG)
G3 *Natural Variations in the Marine Environment*—17 September 1976
    Convener: Professor A. S. Monin (USSR)
G4 *Man and the Sea*—20 September 1976
    Convener: Dr A. Ayala-Castañares (Mexico)
G5 *New Approaches in Oceanography*—22 September 1976
    Convener: Professor J. D. Isaacs (USA)
G6 *Summary Lectures*—23 September 1976
    Dr M. V. Angel (UK)
    Professor J. D. Woods (UK)
    Dr J. G. Sclater (USA)
    Professor T. D. Patten (UK)

SPECIAL SYMPOSIA

S1   *Paleo-Oceanography*—14 September 1976
     Convener: Professor Tj. H. van Andel (USA)
S2   *Geochemistry and Ocean Mixing*—14 September 1976
     Convener: Professor Y. Horibe (Japan)
S3   *Regional Studies of Dynamics and Productivity*—15 September 1976
     Convener: Dr K. N. Fedorov (USSR)
S4   *Depths of the Ocean*—15 September 1976
     Convener: Dr T. Wolff (Denmark)
S5   *Biological Effects of Ocean Variability*—17 September 1976
     Convener: Dr A. R. Longhurst (UK)
S6   *Effect of the Ocean on Climate and Weather*—17 September 1976
     Convener: Professor H. Charnock (UK)
S7   *Oceanography and Fisheries*—20 September 1976
     Convener: Dr G. Saetersdal (Norway)
S8   *Ocean Engineering*—20 September 1976
     Convener; Ir G. A. Heyning (Netherlands)
S9   *Geoscience, Minerals and Petroleum*—20 September 1976
     Convener: Professor C. Morelli (Italy)
S10 *Large Scale Physical Experiments*—22 September 1976
     Convener: Professor H. Stommel (USA)
S11 *Controlled Ecosystem Experiments*—22 September 1976
     Convener: Professor T. R. Parsons (Canada)

ASSOCIATION SYMPOSIA

C1 *Continental Shelf Dynamics:* IAPSO—16 September 1976
    Convener: Dr G. T. Csanady (USA)

C2 *Microbial Processes of the Sea Floor:* IABO—16 September 1976
    Convener: Professor T. M. Fenchel (Denmark)
C3 *Sea Bed Surveys:* CMG—16 September 1976
    Convener: Dr A. J. A. van Overeem (Netherlands)
C4 *Characteristics and Generation of the Mixed Layer and Seasonal Thermocline:* IAPSO/IAMAP—16 September 1976
    Convener: Dr R. T. Pollard (UK)
C5 *Contributions in Biological Oceanography:* IABO—16 September 1976
    Convener: Dr T. Wolff (Denmark)
C6 *Benthic Processes and the Geochemistry of Interstitial Waters of Marine Deposits:* CMG—16 September 1976
    Convener: Dr F. T. Manheim (USA)
C7 *The Skin of the Ocean:* IAPSO/IAMAP—21 September 1976
    Convener: Dr F. MacIntyre (Australia)
C8 *Dynamics of Ecosystems:* IABO—21 September 1976
    Convener: Professor B. O. Jansson (Sweden)
C9 *Criteria for Correlation of Cretaceous and Cenozoic Marine Deptoits:* CMG—21 September 1976
    Convener: Dr W. W. Hay (USA)
C10 *Forecasting of Ocean Currents:* IAPSO—23 September 1976
    Convener: Dr M. R. Crepon (France)
C11 *Contributions in Biological Oceanography:* IABO—23 September 1976
    Convener: Dr T. Wolff (Denmark)
C12 *Contributed Papers in Marine Geoscience:* CMG—23 September 1976
    Convener: Dr T. F. Gaskell (UK)

In addition, poster sessions were organized by IAPSO, IABO and CMG.

Selected Documents and Proceedings

Conference of Plenipotentiaries of the Coastal States of the Mediterranean Region for the Protection of the Mediterranean Sea*

FINAL ACT
OF THE CONFERENCE OF PLENIPOTENTIARIES OF THE COASTAL STATES OF THE MEDITERRANEAN REGION ON THE PROTECTION OF THE MEDITERRANEAN SEA

1. The Conference of Plenipotentiaries of the Coastal States of the Mediterranean Region on the Protection of the Mediterranean Sea was convened by the Executive Director of the United Nations Environment Programme, in co-operation with the Food and Agriculture Organization of the United Nations and other United Nations agencies concerned, in pursuance of a recommendation adopted by the Intergovernmental Meeting on the Protection of the Mediterranean which had been convened by the Executive Director of the Programme in Barcelona from 28 January to 4 February 1975. The recommendation received the approval of the Governing Council of UNEP at its third session.
2. The Conference met at the Palacio de Congresos, Barcelona, at the invitation of the Government of Spain, from 2 to 16 February 1976.
3. The Mediterranean coastal States that were invited to participate in the Conference were: Albania, Algeria, Cyprus, Egypt, France, Greece, Israel, Italy, Lebanon, Libyan Arab Republic, Malta, Monaco, Morocco, Spain, Syrian Arab Republic, Tunisia, Turkey, Yugoslavia.
4. The following Mediterranean coastal States accepted the invitation and participated in the Conference: Cyprus, Egypt, France, Greece, Israel, Italy, Lebanon, Libyan Arab Republic, Malta, Monaco, Morocco, Spain, Syrian Arab Republic, Tunisia, Turkey, Yugoslavia.
5. Observers for the following States attended the Conference's proceedings: Union of the Soviet Socialist Republics, United Kingdom of Great Britain and Northern Ireland, United States of America.
6. Observers for the following United Nations bodies, specialized agencies and intergovernmental bodies also attended the Conference:

United Nations
The United Nations—Office for Inter-Agency Affairs and Co-ordination,
The Secretariat of the Third United Nations Conference on the Law of the Sea,
The Economic Commission for Europe,
The Economic Commission for Western Asia,
The United Nations Development Programme.

*Barcelona, February 2–16, 1976.

Specialized agencies
The Food and Agriculture Organization of the United Nations,
The Inter-Governmental Maritime Consultative Organization,
The World Health Organization.

Intergovernmental bodies
The Commission of the European Communities,
The Organisation for Economic Co-operation and Development,
The League of Arab States,
The International Council for the Scientific Exploration of the Mediterranean.

7. In the course of an inaugural ceremony the Conference heard a welcoming address by Mr. Salvador Sánchez Terán, Governor of Barcelona, on behalf of the Government of Spain. The Conference was formally opened by Dr. Mostafa K. Tolba, the Executive Director of the United Nations Environment Programme, who served as the Secretary-General of the Conference.

8. The Conference elected Mr. Fernando de Ybarra y López Dóriga, Marqués de Arriluce de Ybarra, head of the Spanish delegation, as its President, and Mr. Joseph Najjar, head of the delegation of Lebanon, and Mr. Tome Kuzmanovski, head of the delegation of Yugoslavia, as its Vice-Presidents.

9. The Conference adopted as its agenda the provisional agenda as proposed by the United Nations Environment Programme secretariat (UNEF/CONF.1/7/Rev.1). The agenda as adopted read as follows:

1. Opening of the Conference
2. Election of the President
3. Adoption of the Rules of Procedure
4. Election of two Vice-Presidents and two Chairmen for the two main committees
5. Adoption of the agenda
6. Appointment of the Credentials Committee
7. Appointment of the Drafting Committee
8. Organization of the Work of the Conference
9. Examination of Draft Convention for the Protection of the Marine Environment against Pollution in the Mediterranean
10. Examination of the Draft Protocol for the Prevention of Pollution of the Mediterranean Sea by Dumping from Ships and Aircraft
11. Examination of Draft Protocol on Co-operation in Combating Pollution of the Mediterranean Sea by Oil and other Harmful Substances in cases of Emergency
12. Examination of the Executive Director's report on the establishment of a Regional Oil Combating Centre in the Mediterranean
13. Consideration of the report of the Credentials Committee
14. Adoption of the Convention and Protocols and of the Final Act of the Conference
15. Signature of the Final Act of the Conference
16. Signature of the Convention and Protocols.

10. The Conference adopted as its rules of procedure the draft rules proposed

by the United Nations Environment Programme secretariat (UNEP/CONF.1/6 and Corr.1 and 2).

11. In conformity with the rules of procedure, the Conference established the following Committees:

General Committee
Chairman: The President of the Conference
Members: The Vice-Presidents of the Conference, the Chairmen of the two Main Committees of the Drafting Committee.

Main Committee I
Chairman: Professor Dr. Hamed Sultan (Egypt)
Rapporteur: Mr. Charles Vella (Malta)

Main Committee II
Chairman: Mr. Alberto Sciolla Lagrange (Italy)
Rapporteur: Mr. M'hamed Malliti (Morocco)

The Credentials Committee
Chairman: The President of the Conference
Members: The Vice-Presidents of the Conference, the Chairmen of the two Main Committees and the Chairman of the Drafting Committee.

The Drafting Committee
Chairman: Mr. Marcel F. Surbiguet (France)
Members: Mr. Mohamed Mouldi Marsit (Tunisia)
Mr. José A. de Yturriaga (Spain)
Mr. Demetre Yiannopoulos (Greece)
Mr. Mehmet Dulger (Turkey)

12. The Conference referred to Main Committee I agenda items 9 and 10 and to Main Committee II items 11 and 12, with the request that they consider these items and report the results of their deliberations to the Plenary of the Conference. Main Committee I referred the Annexes of the draft Protocol for the Prevention of Pollution of the Mediterranean Sea by Dumping from Ships and Aircraft to a special working group for consideration.

13. The main documents which served as the basis for the deliberations of the Conference were:

— Draft Convention for the Protection of the Marine Environment against Pollution in the Mediterranean (UNEP/CONF.1/3 and Corr.1)
— Draft Protocol for the Prevention of Pollution of the Mediterranean Sea by Dumping from Ships and Aircraft (UNEP/CONF.1/4 and Corr.1)
— Draft Protocol on Co-operation in Combating Pollution of the Mediterranean Sea by Oil and Other Harmful Substances in cases of Emergency (UNEP/CONF.1/5 and Corr. 1)
— Report by the Executive Director of the United Nations Environment Programme on the proposed establishment of a regional oil-combating centre for the Mediterranean (UNEP/CONF.1/9 and Corr. 1 and 2, and Add.1-4) prepared in co-operation with the Inter-Governmental Maritime Consultative Organization.

14. In addition, the Conference had before it a number of other documents that were made available to it by the Secretariat of UNEP.[1]

15. The Conference approved the recommendation of its Credentials Committee that the credentials of the representatives of the participating States should be recognized as being in order.

16. On the basis of the deliberations of the two Main Committees as embodied in their reports,[2] the Conference, on 13 February 1976, adopted the Convention for the Protection of the Mediterranean Sea against Pollution, the Protocol for the Prevention of Pollution of the Mediterranean Sea by Dumping from Ships and Aircraft, and the Protocol concerning Co-operation in Combating Pollution of the Mediterranean Sea by Oil and Other Harmful Substances in Cases of Emergency. The Convention and the two Protocols which are annexed to this Final Act will be opened by the Government of Spain as Depositary for signature, in Barcelona on 16 February 1976 and thereafter in Madrid from 17 February 1976 to 16 February 1977, by the coastal States of the Mediterranean Sea Area, by the European Economic Community and by similar regional economic groupings at least one member of which is a coastal State of the Mediterranean Sea Area and which exercise competences in fields covered by this Convention, as well as by any protocol affecting them.

17. The Conference also adopted the following resolutions which are appended to this Final Act:

1. Resolution concerning signature, ratification, acceptance and approval of and accession to the legal instruments
2. Resolution concerning interim arrangements
3. Resolution concerning the application of article 11 of the Protocol for the Prevention of Pollution of the Mediterranean Sea by Dumping from Ships and Aircraft
4. Resolution concerning the establishment of a committee of experts on an inter-State guarantee fund for the Mediterranean Sea Area
5. Resolution concerning reporting by ships and aircraft of pollution incidents
6. Resolution concerning future improvement in a maritime communications system
7. Resolution concerning the establishment of a regional oil-combating centre for the Mediterranean
8. Resolution concerning subregional oil-combating centres
9. Resolution concerning an intergovernmental meeting in 1977
10. Tribute to the Government of Spain.

IN WITNESS WHEREOF the representatives of the following coastal States of the Mediterranean Region have signed this Final Act:

---

1. For the list of documents, see the Appendix.
2. UNEP/ CONF. 1/ CRP. 15 and Add. 1 and 2, and UNEP/ CONF. 1/ CRP. 16/ Rev. 1, CRP. 16/ Rev. 1/ Add. 1 and Corr. 1 and CRP. 16/ Rev. 1/ Add. 2.

DONE AT BARCELONA this sixteenth day of February one thousand nine hundred and seventy six in a single copy in the Arabic, English, French and Spanish languages, the four texts being equally authentic. The original texts shall be deposited with the Government of Spain.

RESOLUTIONS ADOPTED BY THE CONFERENCE

1. *Signature, Ratification, Acceptance and Approval of and Accession to the Legal Instruments*

The Conference
Having concluded and adopted on this thirteenth day of February 1976 the Convention for the Protection of the Mediterranean Sea against Pollution, the Protocol for the Prevention of Pollution of the Mediterranean Sea by Dumping from Ships and Aircraft and the Protocol concerning Co-operation in Combating Pollution of the Mediterranean Sea by Oil and other Harmful Substances in Cases of Emergency (hereinafter respectively referred to as "the Convention" and "the Protocols"),
Desirous of ensuring that the Convention and the Protocols shall begin to produce their effects at the earliest possible moment,
Having regard to the clauses in the Convention and in the Protocols which govern the signature, ratification, acceptance or approval of the said instruments and accession thereto and their entry into force,
Having regard furthermore to the provisions in the Convention which relate to the functions of the Depositary,
Having designated the Government of Spain as Depositary of the Convention and of the Protocols,
1. *Invites* the Government of Spain to open the Convention and the Protocols for signature in Barcelona on 16 February 1976 and in Madrid from 17 February 1976 to 16 February 1977, by all those entitled to sign the said instruments by virtue of their provisions governing signature;
2. *Requests* the Government of Spain to perform all the functions pertaining to the Depositary pursuant to the relevant provisions of article 29 of the Convention;
3. *Urges* all parties that are entitled to sign the Convention and the Protocols to do so as soon as practicable and to complete at the earliest opportunity the constitutional procedures needed for the ratification, acceptance or approval of the Convention under their respective statutory or legislative provisions and to transmit the instruments of ratification, acceptance or approval to the Depositary;
4. *Calls* upon all parties entitled to accede to the Convention and the Protocols to do so as soon as possible after the period specified in article 26 of the Convention.

2. *Interim Arrangements*

The Conference,
Acknowledging the role of the Food and Agriculture Organization of the United

Nations and its General Fisheries Council for the Mediterranean in initiating the work on legal instruments for the protection of the marine environment against pollution in the Mediterranean, and the important contribution of FAO and the Inter-Governmental Maritime Consultative Organization and the Government of Spain in the preparation of these instruments,
Acknowledging the work undertaken by the World Health Organization for the preparation of a draft Protocol for the Protection of the Mediterranean Sea against Pollution from Land-based Sources,
Having regard to the recommendation of the Intergovernmental Meeting for the Protection of the Mediterranean, held in Barcelona from 28 January to 4 February 1975, for the convening of working groups of governmental experts to prepare additional protocols,
1. *Notes with appreciation* the announcement by the Executive Director of the United Nations Environment Programme of his willingness to carry out the secretariat functions relating to the Convention for the Protection of the Mediterranean Sea against Pollution, the Protocol for the Prevention of Pollution of the Mediterranean Sea by Dumping from Ships and Aircraft, and the Protocol concerning Co-operation in Combating Pollution of the Mediterranean Sea by Oil and Other Harmful Substances in Cases of Emergency, and to provide the necessary facilities for this purpose;
2. *Calls* on the Executive Director of UNEP, in co-operation with the international organizations concerned, to continue the preparatory work for a draft Protocol for the Protection of the Mediterranean Sea against Pollution from Land-Based Sources;
3. *Also calls* on the Executive Director, pending the entry into force of the Convention and Protocols, to make such interim arrangements as may be required for the achievement of the objectives of this Convention and to continue to convene working groups of government experts to prepare additional protocols, in co-operation with the international organizations concerned, as recommended in the Action Plan approved at the Intergovernmental Meeting held in Barcelona in 1975.

3. *Application of Article 11 of the Protocol for the Prevention of Pollution of the Mediterranean Sea by Dumping from Ships and Aircraft*

The Conference
Having adopted the text of the Protocol for the Prevention of Pollution of the Mediterranean Sea by Dumping from Ships and Aircraft, which provides in article 11 that each Party shall apply the measures required to implement this Protocol to ships and aircraft registered in its territory and to ships and aircraft loading in its territory;
Recognizing the importance of universal implementation and observation of article 11 by all ships and aircraft;
1. *Invites* the Parties to the said Protocol to prevail upon other States to take appropriate steps so that ships flying their flags and aircraft registered in their countries will observe articles 4, 5 and 6 of the Protocol;
2. *Invites* the Inter-Governmental Maritime Consultative Organization to persuade the other States to act in conformity with the said Protocol.

4. *Establishment of a Committee of Experts on an Inter-state Guarantee Fund for the Mediterranean Sea Area*[3]

The Conference
Conscious of the gravity of the threat posed by the various forms of pollution to the environment of the Mediterranean Sea,
Recognizing that the Barcelona Conference and the legal instruments resulting from it are a first step towards safeguarding and protecting that Sea,
Believing the question of liability and compensation, which is the subject of article 12 of this Convention, to be fundamental and to call, therefore, for appropriate measures,
Convinced of the urgent need to protect the coastal States against all damage caused by pollution, whether of accidental or other origin,
Requests the Organization, as defined in article 2, paragraph (b), of the Convention for the Protection of the Mediterranean Sea against Pollution, upon entry into force of the Convention and at the time of the first meeting of the Contracting Parties, to:
(a) propose that a study should be made of the possibility of establishing an Interstate Guarantee Fund for the Mediterranean Sea Area and that the study should be entrusted to a committee of experts from the Contracting Parties to the Convention;
(b) request the said committee of experts to report to the Contracting Parties concerning the implications of establishment of the Fund, in order that, at a later stage, appropriate legal instruments may be prepared.

5. *Reporting by Ships and Aircraft of Pollution Incidents*

The Conference
Having adopted the text of the Protocol concerning Co-operation in Combating Pollution of the Mediterranean Sea by Oil and Other Harmful Substances in Cases of Emergency which provides in article 8 that the masters of ships flying the flags of Parties and pilots of aircraft registered with Parties are required to report on incidents involving harmful substances,
Noting article 8 of the International Convention for the Prevention of Pollution from Ships, 1973, and Protocol I of that Convention, concerning the reports on incidents involving harmful substances,
Recognizing the importance of the provisions of article 8 of the Protocol first above mentioned being applied also by ships flying the flags of States and aircraft registered in States which are not Parties to the Protocol,
1. *Requests* the Parties to the said Protocol to prevail upon other States to take appropriate steps so that ships flying their flags and aircraft registered in their countries will observe the provision of article 8 of the Protocol;
2. *Requests further* the Parties to the Protocol to encourage charterers of their nationality to insert in charter parties a clause to the effect that the ships in question navigating in the Mediterranean Sea Area shall observe the same provision as a ship flying the flag of a Party;

3. One delegation expressed the reservations of its Government to this resolution.

3. *Invites* the Inter-Governmental Maritime Consultative Organization to assist in the implementation of the above-mentioned measures.

6. *Future Improvement in the Maritime Communications System*

The Conference
Considering that an efficient communications system in the maritime field is a factor of the utmost importance for the prevention of pollution by oil and other noxious substances, and for combating such pollution in the Mediterranean Sea Area,
Conscious of the susceptibility of the telecommunications systems to technical improvement,
Recommends to the coastal States of the Mediterranean Sea Area to encourage the adoption and operation of any maritime communications system which, by combining the possibilities of location and communications, should assist in improving the prevention of accidents, in reducing consequently the risk of pollution and in reinforcing the efficacy in combating pollution in the Mediterranean Sea Area.

7. *Establishment of a Regional Oil-Combating Centre for the Mediterranean*

The Conference,
Conscious of the ever-present and growing threat to the Mediterranean environment from massive oil pollution whether caused by accident or through accumulation,
Realising the lack of regional contingency plans for co-ordinated action for the prevention, control and combating of oil spills, especially in cases of emergencies,
Recognizing the need to develop and strengthen the capacities of the coastal States of the Mediterranean and to facilitate co-operation among them in order to deal effectively with cases of massive pollution,
Noting that the Protocol on Co-operation in Combating Pollution of the Mediterranean Sea by Oil and other Harmful Substances in Cases of Emergency makes provisions for a Regional Centre within the Mediterranean for the execution of some of the functions required by that Protocol,
Noting also the consensus reached, at the Consultation of Experts from Mediterranean States convened by the United Nations Environment Programme at Malta in September 1975, on the advisability of the establishment of a Regional Centre and on the objectives and functions of such a Centre,
Having considered the report of the Executive Director of the United Nations Environment Programme, prepared in co-operation with the Intergovernmental Maritime Consultative Organization, on the Establishment of a Regional Oil-Combating Centre for the Mediterranean,[4] in particular the willingness of the United Nations Environment Programme to assist in the early establishment of an oil-combating centre,

4. UNEP/ CONF. 1/9 and Corr. 1 and Corr. 2, and Add. 1-4.

Taking into consideration the comments and observations made by various delegations during the discussion of the above-mentioned report,
1. *Decides* to accept the offer of the Government of the Republic of Malta to host such a Regional Centre;
2. *Requests* the Executive Director of the United Nations Environment Programme, after consultation with the Government of Malta and the Inter-Governmental Maritime Consultative Organization, to assist in the early establishment of a Regional Oil-Combating Centre in Malta having the objectives and functions set out in the annex to this resolution;
3. *Welcomes* the intention of the Executive Director of the United Nations Environment Programme to entrust to the Inter-Governmental Maritime Consultative Organization the functions and responsibility as Co-operating Agency for the establishment and operation of the aforesaid Regional Centre, it being understood that the exercise of functions and responsibilities by IMCO should not lead to an increase in its budget;
4. *As a consequence requests* the Executive Director of the United Nations Environment Programme to submit, in the light of comments made at the Conference of Plenipotentiaries, a report on the establishment of the Regional Oil-Combating Centre to the Governing Council of UNEP at its fourth session and to seek to obtain such further authorization from the Council as he may need to draw on the Fund of UNEP for the purpose of defraying the expenses involved in the establishment and initial operating costs of the centre. This authorization might be requested on the assumption that the operating expenses of the Centre will be gradually defrayed by means of voluntary multilateral or individual contributions from governments of the Mediterranean Region, from international organizations and from non-governmental organizations. The financing of the centre should be reviewed at the meeting of the Contracting Parties to the Convention and the appropriate Protocol, when these instruments have entered into force;
5. *Further requests* the Executive Director of the United Nations Environment Programme to report to the coastal States of the Mediterranean region at the next intergovernmental meeting of these States and thereafter annually on the work and activities of the Centre.

ANNEX

Objectives and Functions of a Regional Oil-combating Centre

I. *Objectives*
   1. To strengthen the capacities of the coastal States in the Mediterranean region and to facilitate co-operation among them in order to combat massive pollution by oil, especially in case of emergencies in which there is grave and imminent danger to the marine environment.
   2. To assist coastal States of the Mediterranean region, which so request, in the development of their own national capabilities to combat oil pollution and to facilitate information exchange, technological co-operation and training.
   3. A later objective, namely the possibility of initiating operations to combat pollution by oil and eventually by other harmful substances at the regional level, can

be considered. This possibility should be submitted for approval by governments after evaluating the results achieved in the fulfilment of the previous two objectives and in the light of financial resources which could be made available for this purpose.

II. *Functions*

A. To collect and disseminate information on:
  (i) inventories of experts and equipment in each coastal State for combating massive accidental spillages of oil;
  (ii) plans, methods and techniques used for combating oil pollution in order to assist as far as necessary countries of the region in the preparation of their national contingency plans;
  (iii) those areas in the Mediterranean which are especially vulnerable to oil pollution and, with reference to these areas, specification of clean-up methods which can be used with minimum environmental damage in such areas.

B. To prepare and keep up to date, in the light of information collected, emergency plans that could be implemented:
  (i) in cases of massive oil pollution where there is an absence of bilateral or multilateral agreements between coastal States;
  (ii) in sectors of the Mediterranean, yet to be identified, where the risk of massive accidental oil pollution is high or where the capability for prompt counter-action in times of emergency does not presently exist.

C. To develop and maintain a Communications/Information system appropriate to the needs of States being served by the Centre.

D. To develop and encourage technological co-operation and training programmes for combating oil pollution.

E. To assist in strengthening the IRS by developing the capacity to serve as a sectoral focal point which could collect data on the sources of information available in connexion with oil pollution with special emphasis on dealing with massive spills of oil and will make that data available.

F. To develop and maintain close working relationships with other Mediterranean regional activity centres and with the "specialized regional organisms" which play a co-ordinating role as set forth in the Barcelona Action Plan,[5] particularly with the scientific institutions within the region.

G. To co-operate in all appropriate activities which are directed towards the prevention and reduction of pollution in the Mediterranean resulting from oil spills.

8. *Subregional Oil-combating Centres*

The Conference,
Taking note of the relevant paragraphs of the report of the Executive Director of the United Nations Environment Programme on the establishment of a regional oil-combating centre for the Mediterranean,[6]
Taking also note of the offers made by certain States to act as host to subregional oil-combating centres,

---

5. UNEP/WG. 2/5, annex.
6. UNEP/ CONF. 1/9 and Corr. 1 and Corr. 2 and Add. 1-4 and UNEP/CONF. 1/ INF. 8.

1. *Requests* the Executive Director to consult with the coastal States of the Mediterranean Region on the required objectives and functions of such subregional centres and their relations to the regional oil-combating centre;
2. *Further requests* the Executive Director to report his findings to the appropriate intergovernmental meeting of the coastal States of the Mediterranean Region.

9. *Intergovernmental Meeting in 1977*

The Conference,
Considering that the Action Plan approved at the Intergovernmental Meeting on the Protection of the Mediterranean held in Barcelona from 28 January to 4 February 1975[7] provided a valuable basis for the deliberations which have led to the conclusion of the Convention for the Protection of the Mediterranean Sea against Pollution, the Protocol for the Prevention of Pollution of the Mediterranean Sea by Dumping from Ships and Aircraft and the Protocol concerning Co-operation in Combating Pollution of the Mediterranean Sea by Oil and Other Harmful Substances in Cases of Emergency, adopted in Barcelona on 13 February 1976,
Considering that the Action Plan approved by the said Intergovernmental Meeting sets forth a number of additional recommendations concerning activities to be carried out over a period of years and that the Executive Director of the United Nations Environment Programme was entrusted with a number of tasks to be performed in co-operation or consultation, as appropriate, with the Governments of the coastal States of the Mediterranean Sea Area and with the international and regional intergovernmental organizations concerned,
Considering it desirable to make provision for a future review of the way in which the Action Plan is being put into effect,
Desirous of strengthening the efforts of Governments, of UNEP and of the international and regional intergovernmental bodies concerned in protecting the environment of the Mediterranean Sea Area and in enhancing the environment while prompting the development of the Area in keeping with sound principles of environmental management,
1. *Recommends* that the Executive Director of UNEP convene in 1977 an inter-governmental meeting at which he would inform Governments of the coastal States of the Mediterranean Sea Area of steps taken and progress achieved pursuant to all recommendations set forth in the said Action Plan for the Protection of the Mediterranean approved at Barcelona on 4 February 1975;
2. *Requests* these Governments to keep the Executive Director informed of steps they are taking which will assist in the accomplishment of the objectives and in the implementation of the recommendations embodied in the Action Plan;
3. *Accepts* with appreciation the invitation of the Government of Monaco to hold the Intergovernmental Meeting of 1977 at Monte Carlo.

7. UNEP/ WG. 2/5, annex.

10. *Tribute to the Government of Spain*

The Conference,
Having met in Barcelona from 2 to 16 February 1976 at the gracious invitation of the Government of Spain,
Convinced that the efforts made by the Government of Spain and by the civic authorities of Barcelona in providing facilities, premises and other resources contributed significantly to the smooth conduct of its proceedings,
Deeply appreciative of the courtesy and hospitality extended by the Government of Spain and the City of Barcelona to the members of the delegations, observers and the secretariat attending the Conference,
Expresses its sincere gratitude to the Government of Spain, to the authorities of Barcelona and, through them, to the Spanish people and to the population of Barcelona for the cordial welcome which they accorded to the Conference and to those associated with its work and for their contribution to the success of the Conference.

APPENDIX

LIST OF DOCUMENTS BEFORE THE CONFERENCE
(Other than those mentioned in the Final Act)

UNEP/CONF. 1/1	Introductory note
UNEP/CONF. 1/2	Working Group on Draft Legal Instruments for the Protection of the Mediterranean, Geneva, 7-11 April 1975. Report of the Meeting
UNEP/CONF.1/8 and Corr.1	Annotated provisional agenda
UNEP/CONF.1/INF.1	List of documents
UNEP/CONF.1/INF.2	Provisional list of participants
UNEP/CONF.1/INF.3	Progress report of the Executive Director on the implementation of the Mediterranean Action Plan
UNEP/CONF.1/INF.4	Note by the Executive Director (Meeting of Experts to advise the Executive Director on the preparations for the Conference of Plenipotentiaries of the coastal States of the Mediterranean Region, Geneva, 7–9 January 1976)
UNEP/CONF.1/INF.5	Note by the Executive Director (Report of the Intergovernmental Consultation of Experts on a Regional Oil-Combating Centre, Malta, 15–19 September 1975)
UNEP/CONF.1/INF.6	Note by the Executive Director (Existing and proposed International Conventions for the Control of Marine Pollution and their Relevance to the Mediterranean, FAO Legal Office, Background Paper No. 8, 1975)
UNEP/CONF.1/INF.7	Note by the Executive Director (Comparative table of texts relating to the Draft Convention for the Protection of the Marine Environment against Pollution in the Mediterranean, FAO Legal Office, Background paper No. 9, 1975)
UNEP/CONF.1/INF.8	Information concerning possible subregional centres.

CONVENTION FOR THE PROTECTION OF THE MEDITERRANEAN SEA AGAINST POLLUTION

THE CONTRACTING PARTIES,

Conscious of the economic, social, health and cultural value of the marine environment of the Mediterranean Sea Area,

Fully aware of their responsibility to preserve this common heritage for the benefit and enjoyment of present and future generations,

Recognizing the threat posed by pollution to the marine environment, its ecological equilibrium, resources and legitimate uses,

Mindful of the special hydrographic and ecological characteristics of the Mediterranean Sea Area and its particular vulnerability to pollution,

Noting that existing international conventions on the subject do not cover, in spite of the progress achieved, all aspects and sources of marine pollution and do not entirely meet the special requirements of the Mediterranean Sea Area,

Realizing fully the need for close co-operation among the States and international organizations concerned in a co-ordinated and comprehensive regional approach for the protection and enhancement of the marine environment in the Mediterranean Sea Area.

HAVE AGREED AS FOLLOWS:

Article 1
Geographical Coverage

1. For the purposes of this Convention, the Mediterranean Sea Area shall mean the maritime waters of the Mediterranean Sea proper, including its gulfs and seas bounded to the west by the meridian passing through Cape Spartel lighthouse, at the entrance of the Straits of Gibraltar, and to the East by the southern limits of the Straits of the Dardanelles between Mehmeteik and Kumkale lighthouses.

2. Except as may be otherwise provided in any protocol to this Convention the Mediterranean Sea Area shall not include internal waters of the Contracting Parties.

Article 2
Definitions

For the purposes of this Convention:

(a) "pollution" means the introduction by man, directly or indirectly, of substances or energy into the marine environment resulting in such deleterious effects as harm to living resources, hazards to human health, hindrance to marine activities including fishing, impairment of quality for use of sea water and reduction of amenities.

(b) "organization" means the body designated as responsible for carrying out secretariat functions pursuant to article 13 of this Convention.

Article 3
General Provisions

1. The Contracting Parties may enter into bilateral or multilateral agreements, including regional or sub-regional agreements, for the protection

of the marine environment of the Mediterranean Sea against pollution, provided that such agreements are consistent with this Convention and conform to international law. Copies of such agreements between Contracting Parties to this Convention shall be communicated to the Organization.
2. Nothing in this Convention shall prejudice the codification and development of the Law of the Sea by the United Nations Conference on the Law of the Sea convened pursuant to resolution 2750 C (XXV) of the General Assembly of the United Nations, nor the present or future claims and legal views of any State concerning the law of the sea and the nature and extent of coastal and flag State jurisdiction.

Article 4
General Undertakings

1. The Contracting Parties shall individually or jointly take all appropriate measures in accordance with the provisions of this Convention and those protocols in force to which they are party, to prevent, abate and combat pollution of the Mediterranean Sea Area and to protect and enhance the marine environment in that Area.
2. The Contracting Parties shall co-operate in the formulation and adoption of protocols, in addition to the protocols opened for signature at the same time as this Convention, prescribing agreed measures, procedures and standards for the implementation of this Convention.
3. The Contracting Parties further pledge themselves to promote, within the international bodies considered to be competent by the Contracting Parties, measures concerning the protection of the marine environment in the Mediterranean Sea Area from all types and sources of pollution.

Article 5
Pollution Caused by Dumping from Ships and Aircraft

The Contracting Parties shall take all appropriate measures to prevent and abate pollution of the Mediterranean Sea Area caused by dumping from ships and aircraft.

Article 6
Pollution from Ships

The Contracting Parties shall take all measures in conformity with international law to prevent, abate and combat pollution of the Mediterranean Sea Area caused by discharges from ships and to ensure the effective implementation in that Area of the rules which are generally recognized at the international level relating to the control of this type of pollution.

Article 7
Pollution Resulting from Exploration and Exploitation of the Continental Shelf and the Seabed and Its Subsoil

The Contracting Parties shall take all appropriate measures to prevent, abate and combat pollution of the Mediterranean Sea Area resulting from exploration and exploitation of the continental shelf and the seabed and its subsoil.

Article 8
Pollution from Land-based Sources

The Contracting Parties shall take all appropriate measures to prevent, abate and combat pollution of the Mediterranean Sea Area caused by discharges from rivers, coastal establishments or out falls, or emanating from any other land-based sources within their territories.

Article 9
Co-operation in Dealing with Pollution Emergencies

1. The Contracting Parties shall co-operate in taking the necessary measures for dealing with pollution emergencies in the Mediterranean Sea Area, whatever the causes of such emergencies, and reducing or eliminating damage resulting therefrom.
2. Any Contracting Party which becomes aware of any pollution emergency in the Mediterranean Sea Area shall, without delay, notify the Organization and, either through the Organization or directly, any Contracting Party likely to be affected by such emergency.

Article 10
Monitoring

1. The Contracting Parties shall endeavour to establish, in close co-operation with the international bodies which they consider competent, complementary or joint programmes including, as appropriate, programmes at the bilateral or multilateral levels, for pollution monitoring in the Mediterranean Sea Area and shall endeavour to establish a pollution monitoring system for that Area.
2. For this purpose, the Contracting Parties shall designate the competent authorities responsible for pollution monitoring within areas under their national jurisdiction and participate as far as practicable in international arrangements for pollution monitoring in areas beyond national jurisdiction.
3. The Contracting Parties undertake to co-operate in the formulation, adoption and implementation of such annexes to this Convention as may be required to prescribe common procedures and standards for pollution monitoring.

Article 11
Scientific and Technological Co-operation

1. The Contracting Parties undertake as far as possible to co-operate directly, or when appropriate through competent regional or other international organizations, in the fields of science and technology, and to exchange data as well as other scientific information for the purpose of this Convention.
2. The Contracting Parties undertake as far as possible to develop and coordinate their national research programmes relating to all types of marine pollution in the Mediterranean Sea Area and to co-operate in the establishment and implementation of regional and other international research programmes for the purposes of this Convention.
3. The Contracting Parties undertake to co-operate in the provision of technical and other possible assistance in fields relating to marine pollution, with priority to be given to the special needs of developing countries in the Mediterranean region.

Article 12
Liability and Compensation

The Contracting Parties undertake to co-operate as soon as possible in the formulation and adoption of appropriate procedures for the determination of liability and compensation for damage resulting from the pollution of the marine environment deriving from violations of the provisions of this Convention and applicable protocols.

Article 13
Institutional Arrangements

The Contracting Parties designate the United Nations Environment Programme as responsible for carrying out the following secretariat functions:

   (i) To convene and prepare the meetings of Contracting Parties and conferences provided for in article 14, 15, and 16;

   (ii) To transmit to the Contracting Parties notifications, reports and other information received in accordance with articles 3, 9, and 20;

   (iii) To consider inquiries by, and information from, the Contracting Parties, and to consult with them on questions relating to this Convention and the protocols and annexes thereto;

   (iv) To perform the functions assigned to it by the protocols to this Convention;

   (v) To perform such other functions as may be assigned to it by the Contracting Parties;

   (vi) To ensure the necessary co-ordination with other international bodies which the Contracting Parties consider competent, and in particular to enter into such administrative arrangements as may be required for the effective discharge of the secretariat functions.

Article 14
Meetings of the Contracting Parties

1. The Contracting Parties shall hold ordinary meetings once every two years, and extraordinary meetings at any other time deemed necessary, upon the request of the Organization or at the request of any Contracting Party, provided that such requests are supported by at least two Contracting Parties.
2. It shall be the function of the meetings of the Contracting Parties to keep under review the implementation of this Convention and the protocols and, in particular:

   (i) To review generally the inventories carried out by Contracting Parties and competent international organizations on the state of marine pollution and its effects in the Mediterranean Sea Area;

   (ii) To consider reports submitted by the Contracting Parties under article 20;

   (iii) To adopt, review and amend as required the annexes to this Convention and to the protocols, in accordance with the procedure established in article 17;

   (iv) To make recommendations regarding the adoption of any additional protocols or any amendments to this Convention or the protocols in accordance with the provisions of articles 15 and 16;

   (v) To establish working groups as required to consider any matters related to this Convention and the protocols and annexes;

(vi) To consider and undertake any additional action that may be required for the achievement of the purposes of this Convention and the protocols.

Article 15
Adoption of Additional Protocols

1. The Contracting Parties, at a diplomatic conference, may adopt additional protocols to this Convention pursuant to paragraph 2 of article 4.
2. A diplomatic conference for the purpose of adopting additional protocols shall be convened by the Organization at the request of two thirds of the Contracting Parties.
3. Pending the entry into force of this Convention the Organization may, after consulting with the signatories to this Convention, convene a diplomatic conference for the purpose of adopting additional protocols.

Article 16
Amendment of the Convention or Protocols

1. Any Contracting Party to this Convention may propose amendments to the Convention. Amendments shall be adopted by a diplomatic conference which shall be convened by the Organization at the request of two thirds of the Contracting Parties.
2. Any Contracting Party to this Convention may propose amendments to any protocol. Such amendments shall be adopted by a diplomatic conference which shall be convened by the Organization at the request of two thirds of the Contracting Parties to the protocol concerned.
3. Amendments to this Convention shall be adopted by a three-fourths majority vote of the Contracting Parties to the Convention which are represented at the diplomatic conference, and shall be submitted by the Depositary for acceptance by all Contracting Parties to the Convention. Amendments to any protocol shall be adopted by a three-fourths majority vote of the Contracting Parties to such protocol which are represented at the diplomatic conference, and shall be submitted by the Depositary for acceptance by all Contracting Parties to such protocol.
4. Acceptance of amendments shall be notified to the Depositary in writing. Amendments adopted in accordance with paragraph 3 of this article shall enter into force between Contracting Parties having accepted such amendments on the thirtieth day following the receipt by the Depositary of notification of their acceptance by at least three fourths of the Contracting Parties to this Convention or to the protocol concerned, as the case may be.
5. After the entry into force of an amendment to this Convention or to a protocol, any new Contracting Party to this Convention or such protocol shall become a Contraction Party to the instrument as amended.

Article 17
Annexes and Amendments to Annexes

1. Annexes to this Convention or to any protocol shall form an integral part of the Convention or such protocol, as the case may be.
2. Except as may be otherwise provided in any protocol, the following procedure shall apply to the adoption and entry into force of any amendments

to annexes to this Convention or to any Protocol, with the exception of amendments to the Annex on arbitration:

(i) Any Contracting Party may propose amendments to the annexes to this Convention or to protocols at the meetings referred to in article 14;

(ii) Such amendments shall be adopted by a three-fourths majority vote of the Contracting Parties to the instrument in question;

(iii) The Depositary shall without delay communicate the amendments so adopted to all Contracting Parties;

(iv) Any Contracting Party that is unable to approve an amendment to the annexes to this Convention or to any protocol shall so notify in writing the Depositary within a period determined by the Contracting Parties concerned when adopting the amendment;

(v) The Depositary shall without delay notify all Contracting Parties of any notification received pursuant to the preceding sub-paragraph;

(vi) On expiry of the period referred to in sub-paragraph (iv) above, the amendment to the annex shall become effective for all Contracting Parties to this Convention or to the protocol concerned which have not submitted a notification in accordance with the provisions of that sub-paragraph.

3. The adoption and entry into force of a new annex to this Convention or to any protocol shall be subject to the same procedure as for the adoption and entry into force provided that, if any amendment to the Convention or the protocol concerned is involved the new annex shall not enter into force until such time as the amendment to the Convention or the protocol concerned enters into force.

4. Amendments to the annex on arbitration shall be considered to be amendments to this Convention and shall be proposed and adopted in accordance with the procedures set out in article 16 above.

Article 18
Rules of Procedure and Financial Rules

1. The Contracting Parties shall adopt rules of procedure for their meetings and conferences envisaged in articles 14, 15, and 16 above.

2. The Contracting Parties shall adopt financial rules, prepared in consultation with the Organization, to determine, in particular, their financial participation.

Article 19
Special Exercise of Voting Right

Within the areas of their competence, the European Economic Community and any regional economic grouping referred to in article 24 of this Convention shall exercise their right to vote with a number of votes equal to the number of their member States which are Contracting Parties to this Convention and to one or more Protocols; the European Economic Community and any grouping as referred to above shall not exercise their right to vote in cases where the member States concerned exercise theirs, and conversely.

Article 20
Reports

The Contracting Parties shall transmit to the Organization reports on the measures adopted in implementation of this Convention and of Protocols to which they are Parties, in such form and at such intervals as the meetings of Contracting Parties may determine.

Article 21
Compliance Control

The Contracting Parties undertake to co-operate in the development of procedures enabling them to control the application of this Convention and the Protocols.

Article 22
Settlement of Disputes

1. In case of a dispute between Contracting Parties as to the interpretation or application of this Convention or the Protocols, they shall seek a settlement of the dispute through negotiation or any other peaceful means of their own choice.
2. If the parties concerned cannot settle their dispute through the means mentioned in the preceding paragraph, the dispute shall upon common agreement be submitted to arbitration under the conditions laid down in Annex A to this Convention.
3. Nevertheless, the Contracting Parties may at any time declare that they recognize as compulsory *ipso facto* and without special agreement, in relation to any other party accepting the same obligation, the application of the arbitration procedure in conformity with the provisions of Annex A. Such declaration shall be notified in writing to the Depositary, who shall communicate it to the other Parties.

Article 23
Relationship between the Convention and Protocols

1. No one may become a Contracting Party to this Convention unless it becomes at the same time a Contracting Party to at least one of the protocols. No one may become a Contracting Party to a protocol unless it is, or becomes at the same time a Contracting Party to this Convention.
2. Any protocol to this Convention shall be binding only on the Contracting parties to the protocol in question.
3. Decisions concerning any protocol pursuant to articles 14, 16 and 17 of this Convention shall be taken only by the Parties to the protocol concerned.

Article 24
Signature

This Convention, the Protocol for the Prevention of Pollution of the Mediterranean Sea by Dumping from Ships and Aircraft and the Protocol concerning Co-operation in Combating Pollution of the Mediterranean Sea by Oil and Other Harmful Substances in Cases of Emergency shall be open for signature in Barcelona on 16 February 1976 and in Madrid from 17 February

1976 to 16 February 1977 by any State invited as a participant in the Conference of Plenipotentiaries of the Coastal States of the Mediterranean Region on the Protection of the Mediterranean Sea, held in Barcelona from 2 to 16 February 1976, and by any State entitled to sign any protocol in accordance with the provisions of such Protocol. They shall also be open until the same date for signature by the European Economic Community and by any similar regional economic grouping at least one member of which is a coastal State of the Mediterranean Sea Area and which exercise competences in fields covered by this Convention, as well as by any protocol affecting them.

Article 25
Ratification, Acceptance or Approval
This Convention and any protocol thereto shall be subject to ratification, acceptance, or approval. Instruments of ratification, acceptance or approval shall be deposited with the Government of Spain, which will assume the functions of Depositary.

Article 26
Accession
1. As from 17 February 1977, the present Convention, the Protocol for the Prevention of Pollution of the Mediterranean Sea by Dumping from Ships and Aircraft, and the Protocol concerning Co-operation in Combating Pollution of the Mediterranean Sea by Oil and other Harmful Substances in Cases of Emergency shall be open for accession by the States, by the European Economic Community and by any grouping as referred to in article 24.
2. After the entry into force of the Convention and of any protocol, any State not referred to in article 24 may accede to this Convention and to any protocol, subject to prior approval by three-fourths of the Contracting Parties to the protocol concerned.
3. Instruments of accession shall be deposited with the Depositary.

Article 27
Entry into Force
1. This Convention shall enter into force on the same date as the protocol first entering into force.
2. The Convention shall also enter into force with regard to the States, the European Economic Community and any regional economic grouping referred to in article 24 if they have complied with the formal requirements for becoming Contracting Parties to any other protocol not yet entered into force.
3. Any protocol to this Convention, except as otherwise provided in such protocol, shall enter into force on the thirtieth day following the date of deposit of at least six instruments of ratification, acceptance, or approval of, or accession to such protocol by the Parties referred to in article 24.
4. Thereafter, this Convention and any protocol shall enter into force with respect to any State, the European Economic Community and any regional economic grouping referred to in article 24 on the thirtieth day following the date of deposit of the instruments of ratification, acceptance, approval or accession.

Article 28
Withdrawal

1. At any time after three years from the date of entry into force of this Convention, any Contracting Party may withdraw from this Convention by giving written notification of withdrawal.
2. Except as may be otherwise provided in any protocol to this Convention, any Contracting Party may, at any time after three years from the date of entry into force of such protocol, withdraw from such protocol by giving written notification of withdrawal.
3. Withdrawal shall take effect 90 days after the date on which notification of withdrawal is received by the Depositary.
4. Any Contracting Party which withdraws from this Convention shall be considered as also having withdrawn from any protocol to which it was a Party.
5. Any Contracting Party which, upon its withdrawal from a protocol, is no longer a Party to any protocol to this Convention, shall be considered as also having withdrawn from this Convention.

Article 29
Responsibilities of the Depositary

1. The Depositary shall inform the Contracting Parties, any other Party referred in article 24, and the Organizations:

   (i) Of the signature of this Convention and of any protocol thereto, and of the deposit of instruments of ratification, acceptance, approval or accession in accordance with articles 24, 25 and 26;

   (ii) Of the date on which the Convention and any protocol will come into force in accordance with the provisions of article 27;

   (iii) Of notifications of withdrawal made in accordance with article 28;

   (iv) Of the amendments adopted with respect to the Convention and to any protocol, their acceptance by the Contracting Parties and the date of entry into force of those amendments in accordance with the provisions of article 16;

   (v) Of the adoption of new annexes and of the amendment of any annex in accordance with article 17;

   (vi) Of declarations recognizing as compulsory the application of the arbitration procedure mentioned in paragraph 3 of article 22.

2. The original of this Convention and of any protocol thereto shall be deposited with the Depositary, the Government of Spain, which shall send certified copies thereof to the Contracting Parties, to the Organization, and to the Secretary-General of the United Nations for registration and publication in accordance with Article 102 of the United Nations Charter.

   IN WITNESS WHEREOF the undersigned, being duly authorized by their respective Governments, have signed this Convention.

   DONE at Barcelona on 16 February 1976 in a single copy in the Arabic, English, French and Spanish languages, the four texts being equally authoritative.

ANNEX A
Arbitration

Article 1
Unless the Parties to the dispute otherwise agree, the arbitration procedure shall be conducted in accordance with the provisions of this Annex.

Article 2
1. At the request addressed by one Contracting Party to another Contracting Party in accordance with the provisions of paragraph 2 or paragraph 3 of article 22 of the Convention, an arbitral tribunal shall be constituted. The request for arbitration shall state the subject matter of the application including, in particular, the articles of the Convention or the Protocols the interpretation or application of which is in dispute.
2. The claimant party shall inform the Organization that it has requested the setting up of an arbitral tribunal, stating the name of the other party to the dispute and articles of the Convention or the Protocols the interpretation or application of which is in its opinion in dispute. The Organization shall forward the information thus received to all Contracting Parties to the Convention.

Article 3
The arbitral tribunal shall consist of three members: each of the parties to the dispute shall appoint an arbitrator: the two arbitrators so appointed shall designate by common agreement the third arbitrator who shall be the chairman of the tribunal. The latter shall not be a national of one of the parties to the dispute, nor have his usual place of residence in the territory of one of these parties, nor be employed by any of them, nor have dealt with the case in any other capacity.

Article 4
1. If the chairman of the arbitral tribunal has not been designated within two months of the appointment of the second arbitrator, the Secretary-General of the United Nations shall, at the request of the most diligent party, designate him within a further two months' period.
2. If one of the parties to the dispute does not appoint an arbitrator within two months of receipt of the request, the other party may inform the Secretary-General of the United Nations who shall designate the chairman of the arbitral tribunal within a further two months' period. Upon designation, the chairman of the arbitral tribunal shall request the party which has not appointed an arbitrator to do so within two months. After such period, he shall inform the Secretary-General of the United Nations, who shall make this appointment within a further two months' period.

Article 5
1. The arbitral tribunal shall decide according to the rules of international law and, in particular, those of this Convention and the protocols concerned.
2. Any arbitral tribunal constituted under the provisions of this Annex shall draw up its own rules of procedure.

Article 6
1. The decisions of the arbitral tribunal, both on procedure and on substance, shall be taken by majority vote of its members.
2. The tribunal may take all appropriate measures in order to establish the facts. It may, at the request of one of the parties, recommend essential interim measures of protection.

3. If two or more arbitral tribunals constituted under the provisions of this Annex are seized of requests with identical or similar subjects, they may inform themselves of the procedures for establishing the facts and take them into account as far as possible.
4. The parties to the dispute shall provide all facilities necessary for the effective conduct of the proceedings.
5. The absence or default of a party to the dispute shall not constitute an impediment to the proceedings.

Article 7

1. The award of the arbitral tribunal shall be accompanied by a statement of reasons. It shall be final and binding upon the parties to the dispute.
2. Any dispute which may arise between the parties concerning the interpretation or execution of the award may be submitted by the most diligent party to the arbitral tribunal which made the award or, if the latter cannot be seized thereof, to another arbitral tribunal constituted for this purpose in the same manner as the first.

Article 8

The European Economic Community and any regional economic grouping referred to in article 24 of the Convention, like any Contracting Party to the Convention, are empowered to appear as complainants or as respondents before the arbitral tribunal.

PROTOCOL FOR THE PREVENTION OF POLLUTION OF THE MEDITERRANEAN SEA BY DUMPING FROM SHIPS AND AIRCRAFT

THE CONTRACTING PARTIES TO THE PRESENT PROTOCOL,

Being Parties to the Convention for the Protection of the Mediterranean Sea against Pollution,

Recognizing the danger posed to the marine environment by pollution caused by the dumping of wastes or other matter from ships and aircraft,

Considering that the coastal States of the Mediterranean Sea have a common interest in protecting the marine environment from this danger,

Bearing in mind the Convention on the Prevention of Marine Pollution by Dumping of Wastes and other Matter, adopted in London in 1972,

HAVE AGREED AS FOLLOWS:

Article 1

The Contracting Parties to this Protocol (hereinafter referred to as "the Parties") shall take all appropriate measures to prevent and abate pollution of the Mediterranean Sea Area caused by dumping from ships and aircraft.

Article 2

The area to which this Protocol applies shall be the Mediterranean Sea Area as defined in Article 1 of the Convention for the Protection of the Mediterranean Sea against Pollution (hereinafter referred to as "the Convention").

Article 3

For the purposes of this Protocol:
1. "Ships and aircraft" means waterborne or airborne craft of any type

whatsoever. This expression includes air-cushioned craft and floating craft whether self-propelled or not, and platforms and other man-made structures at sea and their equipment.

2. "Wastes or other matter" means material and substances of any kind, form or description.

3. "Dumping" means:

(a) Any deliberate disposal at sea of wastes or other matter from ships or aircraft;

(b) Any deliberate disposal at sea of ships or aircraft.

4. "Dumping" does not include:

(a) The disposal at sea of wastes or other matter incidental to, or derived from the normal operations of vessels, or aircraft and their equipment, other than wastes or other matter transported by or to vessels, or aircraft, operating for the purpose of disposal of such matter or derived from the treatment of such wastes or other matter on such vessels or aircraft;

(b) Placement of matter for a purpose other than the mere disposal thereof, provided that such placement is not contrary to the aims of this Protocol.

3. "Organization" means the body referred to in article 13 of the Convention.

Article 4

The dumping into the Mediterranean Sea Area of wastes or other matter listed in Annex I to this Protocol is prohibited.

Article 5

The dumping into the Mediterranean Sea Area of wastes or other matter listed in Annex II to this Protocol requires, in each case, a prior special permit from the competent national authorities.

Article 6

The dumping into the Mediterranean Sea Area of all other wastes or other matter requires a prior general permit from the competent national authorities.

Article 7

The permits referred to in Articles 5 and 6 above shall be issued only after careful consideration of all the factors set forth in Annex III to this Protocol. The Organization shall receive records of such permits.

Article 8

The provisions of Articles 4, 5 and 6 shall not apply in case of *force majeure* due to stress of weather or any other cause when human life or the safety of a ship or aircraft is threatened. Such dumpings shall immediately be reported to the Organization, either through the Organization or directly to any Party or Parties likely to be affected, together with full details of the circumstances and of the nature and quantities of the wastes or other matter dumped.

Article 9

If a Party in a critical situation of an exceptional nature considers that wastes or other matter listed in Annex I to this Protocol cannot be disposed of on land without unacceptable danger or damage, above all for the safety of human life, the party concerned shall forthwith consult the Organization. The Organization, after consulting the Parties to this Protocol, shall recommend methods of storage or the most satisfactory means of destruction or disposal under the prevailing circumstances. The Party shall inform the Organization of the steps adopted in pursuance of these recommendations. The Parties pledge themselves to assist one another in such situations.

Article 10

1. Each Party shall designate one or more competent authorities:
    (a) Issue the special permits provided for in article 5;
    (b) Issue the general permits provided for in article 6;
    (c) Keep records of the nature and quantities of the wastes or other matter permitted to be dumped and the location, date and method of dumping.
2. The competent authorities of each Party shall issue the permits provided for in articles 5 and 6 in respect of the wastes or other matter intended for dumping:
    (a) Loaded in its territory;
    (b) Loaded by a ship or aircraft registered in its territory or flying its flag, when the loading occurs in the territory of a State not Party to this Protocol.

Article 11

1. Each Party shall apply the measures required to implement this Protocol to all:
    (a) Ships and aircraft registered in its territory or flying its flag;
    (b) Ships and aircraft loading in its territory wastes or other matter which are to be dumped;
    (c) Ships and aircraft believed to be engaged in dumping in areas under its jurisdiction in this matter.
2. This Protocol shall not apply to any ships or aircraft owned or operated by a State Party to this Protocol and used for the time being only on Government non-commercial service. However, each Party shall ensure by the adoption of appropriate measures not impairing the operations or operational capabilities of such ships or aircraft owned or operated by it, that such ships and aircraft act in a nammer consistent, so far as is reasonable and practicable, with this Protocol.

Article 12

Each Party undertakes to issue instructions to its maritime inspection ships and aircraft and to other appropriate services to report to its authorities any incidents or conditions in the Mediterranean Sea Area which give rise to suspicions that dumping in contravention of the provisions of this Protocol has occurred or is about to occur. That Party shall, if it considers it appropriate, report accordingly to any other Party concerned.

Article 13

Nothing in this Protocol shall affect the right of each Party to adopt other measures, in accordance with international law, to prevent pollution due to dumping.

Article 14

1. Ordinary meetings of the Parties to this Protocol shall be held in conjunction with ordinary meetings of the Contracting Parties to the Convention held pursuant to Article 14 of the Convention. The Parties to this Protocol may also hold extraordinary meetings in conformity with Article 14 of the Convention.
2. It shall be the function of the meetings of the Parties to this Protocol:
   (a) To keep under review the implementation of this Protocol, and to consider the efficacy of the measures adopted and the need for any other measures, in particular in the form of annexes;
   (b) To study and consider the records of the permits issued in accordance without Articles 5, 6 and 7, and of the dumping which has taken place;
   (c) To review and amend as required any Annex to this Protocol;
   (d) To discharge such other functions as may be appropriate for the implementation of this Protocol.
3. The adoption of amendments to the Annexes to this Protocol pursuant to Article 17 of the Convention shall require a three-fourths majority of the Parties.

Article 15

1. The provisions of the Convention relating to any Protocol shall apply with respect to the present Protocol.
2. The rules of procedure and the financial rules adopted pursuant to Article 18 of the Convention shall apply with respect to this Protocol, unless the Parties to this Protocol agree otherwise.

IN WITNESS WHEREOF the undersigned, being duly authorized by their respective Governments, have signed this Protocol.

DONE at Barcelona on 16 February 1976 in a single copy in the Arabic, English, French and Spanish languages, the four texts being equally authoritative.

ANNEX I

A. The following substances and materials are listed for the purpose of article 4 of the Protocol.
   1. Organohalogen compounds and compounds which may form substances in the marine environment, excluding those which are non-toxic or which are rapidly converted in the sea into substances which are biologically harmless, provided that they do not make edible marine organisms unpalatable.
   2. Organosilicon compounds and compounds which may form such substances in

the marine environment excluding those which are non-toxic or which are rapidly converted in the sea into substances which are biologically harmless, provided that they do not make edible marine organisms unpalatable.
3. Mercury and mercury compounds.
4. Cadmium and cadmium compounds.
5. Persistent plastic and other persistent synthetic materials which may materially interfere with fishing or navigation, reduce amenities, or interfere with other legitimate uses of the sea.
6. Crude oil and hydrocarbons which may be derived from petroleum, and any mixtures containing any of these, taken on board for the purpose of dumping.
7. High- and medium- and low-level radioactive wastes or other high- and medium- and low-level radioactive matter to be defined by the International Atomic Energy Agency.
8. Acid and alkaline compounds of such composition and in such quantity that they may seriously impair the quality of sea water. The composition and quantity to be taken into consideration shall be determined by the Parties in accordance with the procedure laid down in article 14, paragraph 3, of this Protocol.
9. Materials in whatever form (e.g. solids, liquids, semi-liquids, gases, or in a living state) produced for biological and chemical warfare, other than those rapidly rendered harmless by physical, chemical or biological processes in the sea provided that they do not:
    (i) Make edible marine organisms unpalatable; or
    (ii) Endanger human or animal health.
B. This Annex does not apply to wastes or other materials, such as sewage sludge and dredge spoils, containing the substances referred to in paragraphs 1-6 above as trace contaminants. The dumping of such wastes shall be subject to the provisions of Annexes II and III as appropriate.

ANNEX II

The following wastes and other matter the dumping of which requires special care are listed for the purposes of article 5.
1. (i) Arsenic, lead, copper, zinc, beryllium, chromium, nickel, vanadium, selenium, antimony and their compounds;
    (ii) Cyanides and fluorides;
    (iii) Pesticides and their by-products not covered in Annex I;
    (iv) Synthetic organic chemicals, other than those referred to in Annex I, likely to produce harmful effects on marine organisms or to make edible marine organisms unpalatable.
2. (i) Acid and alkaline compounds the composition and quantity of which have not yet been determined in accordance with the procedure referred to in Annex I, paragraph A. 8.
    (ii) Acid and alkaline compounds not covered by Annex I, excluding compounds to be dumped in quantities below thresholds which shall be determined by the Parties in accordance with the procedure laid down in article 14, paragraph 3 of this Protocol.
3. Containers, scrap metal and other bulky wastes liable to sink to the sea bottom which may present a serious obstacle to fishing or navigation.
4. Substances which, though of a non-toxic nature may become harmful owing to the quantities in which they are dumped, or which are liable to reduce amenities seriously or to endanger human life or marine organisms or to interfere with navigation.
5. Radioactive waste or other radioactive matter which will not be included in Annex I.

In the issue of permits for the dumping of this matter, the Parties should take full account of the recommendations of the competent international body in this field, at present the International Atomic Energy Agency.

ANNEX III

The factors to be considered in establishing criteria governing the issue of permits for the dumping of matter at sea taking into account article 7 include:

A. *Characteristics and composition of the matter*
   1. Total amount and average compositions of matter dumped (e.g. per year).
   2. Form (e.g. solid, sludge, liquid or gaseous).
   3. Properties: physical (e.g. solubility and density), chemical and biochemical (e.g. oxygen demand, nutrients) and biological (e.g. presence of viruses, bacteria, yeasts, parasites).
   4. Toxicity.
   5. Persistence: physical, chemical and biological.
   6. Accumulation and biotransformation in biological materials or sediments.
   7. Susceptibility to physical, chemical and biochemical changes and interaction in the aquatic environment with other dissolved organic and inorganic materials.
   8. Probability of production of taints or other changes reducing marketability of resources (fish, shell-fish etc.).

B. *Characteristics of dumping site and method of deposit*
   1. Location (e.g. co-ordinates of the dumping area, depth and distance from the coast), location in relation to other areas (e.g. amenity areas, spawning, nursery and fishing areas and exploitable resources).
   2. Rate of disposal per specific period (e.g. quantity per day, per week, per month).
   3. Methods of packaging and containment, if any.
   4. Initial dilution achieved by proposed method of release, particularly the speed of the ship.
   5. Dispersal characteristics (e.g. effects of currents, tides and wind on horizontal transport and vertical mixing).
   6. Water characteristics (e.g. temperature, pH, salinity, stratification, oxygen indices of pollution—dissolved oxygen (DO), chemical oxygen demand (COD), biochemical oxygen demand (BOD)—nitrogen present in organic and mineral form, including ammonia, suspended matter, other nutrients and productivity).
   7. Bottom characteristics (e.g. topography, geochemical and geological characteristics and biological productivity).
   8. Existence and effects of other dumpings which have been made in the dumping area (e.g. heavy metal background reading and organic carbon content).
   9. When issuing a permit for dumping, the Contracting Parties shall endeavor to determine whether an adequate scientific basis exists for assessing the consequences of such dumping in the area concerned, in accordance with the foregoing provisions and taking into account seasonal variations.

C. *General considerations and conditions*
   1. Possible effects on amenities (e.g. presence of floating or stranded material, turbidity, objectionable odour, discolouration and foaming).
   2. Possible effects on marine life, fish and shellfish culture, fish stocks and fisheries, seaweed harvesting and culture.
   3. Possible effects on other uses of the sea (e.g. impairment of water quality for industrial use, underwater corrosion of structure, interference with ship operations from floating materials, interference with fishing or navigation through deposit of

waste or solid objects on the sea floor and protection of areas of special importance for scientific or conservation purposes).

4. The practical availability of alternative land-based methods of treatment disposal or elimination, or of treatment to render the matter less harmful for sea dumping.

PROTOCOL CONCERNING CO-OPERATION IN COMBATING POLLUTION OF THE MEDITERRANEAN SEA BY OIL AND OTHER HARMFUL SUBSTANCES IN CASES OF EMERGENCY

THE CONTRACTING PARTIES TO THE PRESENT PROTOCOL,

Being Parties to the Convention for the Protection of the Mediterranean Sea against Pollution,

Recognizing that grave pollution of the sea by oil and other harmful substances in the Mediterranean Sea Area involves a danger for the coastal States and the marine eco-system,

Considering that the co-operation of all the coastal States of the Mediterranean is called for to combat this pollution,

Bearing in mind the International Convention for the Prevention of Pollution from Ships, 1973, the International Convention relating to Intervention on the High Seas in Cases of Oil Pollution Casualties, 1969, as well as the Protocol relating to Intervention on the High Seas in Cases of Marine Pollution by Substances Other than Oil, 1973,

Further taking into account the International Convention on Civil Liability for Oil Pollution Damage, 1969,

HAVE AGREED AS FOLLOWS:

Article 1

The Contracting Parties to this Protocol (hereinafter referred to as "the Parties") shall co-operate in taking the necessary measures in cases of grave and imminent danger to the marine environment, the coast or related interests of one or more of the Parties due to the presence of massive quantities of oil or other harmful substances resulting from accidental causes or an accumulation of small discharges which are polluting or threatening to pollute the sea within the area defined in Article 1 of the Convention for the Protection of the Mediterranean Sea against Pollution (hereinafter referred to as "the Convention").

Article 2

For the purpose of this Protocol, the term "related interests" means the interests of a coastal State directly affected or threatened and concerning, among others:

(a) activities in coastal waters, in ports or estuaries, including fishing activities;
(b) the historical and tourist appeal of the area in question, including water sports and recreation;
(c) the health of the coastal population;
(d) the preservation of living resources.

Article 3

The Parties shall endeavour to maintain and promote, either individually or through bilateral or multilateral co-operation, their contingency plans and means for combating pollution of the sea by oil and other harmful substances. These means shall include, in particular, equipment, ships, aircraft and manpower prepared for operations in cases of emergency.

Article 4

The Parties shall develop and apply, either individually or through bilateral or multilateral co-operation, monitoring activities covering the Mediterranean Sea Area in order to have as precise information as possible on the situations referred to in Article 1 of this Protocol.

Article 5

In the case of release or loss overboard of harmful substances in packages, freight containers, portable tanks or road and rail tank wagons, the Parties shall co-operate as far as practicable in the salvage and recovery of such substances so as to reduce the danger of pollution of the marine environment.

Article 6

1. Each party undertakes to disseminate to the other Parties information concerning:
    (a) The competent national organization or authorities responsible for combating pollution of the sea by oil and other harmful substances;
    (b) The competent national authorities responsible for receiving reports of pollution of the sea by oil and other harmful substances and for dealing with matters concerning measures of assistance between Parties;
    (c) New ways in which pollution of the sea by oil and other harmful substances may be avoided, new measures of combating pollution and the development of related research programmes.
2. Parties which have agreed to exchange information directly between themselves shall nevertheless communicate such information to the regional centre. The latter shall communicate this information to the other Parties and, on a basis of reciprocity, to coastal States of the Mediterranean Sea Area which are not Parties to this Protocol.

Article 7

The Parties undertake to co-ordinate the utilization of the means of communication at their disposal in order to ensure, with the necessary speed and reliability, the reception, transmission and dissemination of all reports and urgent information which relate to the occurrences and situations referred to in article 1. The regional centre shall have the necessary means of communication to enable it to participate in this co-ordinated effort and, in particular, to fulfil the functions assigned to it by paragraph 2 of article 10.

Article 8

1. Each Party shall issue instructions to the masters of ships flying its flag and to the pilots of aircraft registered in its territory requiring them to report by the most rapid and adequate channels in the circumstances, and in accordance with Annex I to this Protocol, either to a Party or to the regional centre:
    (a) All accidents causing or likely to cause pollution of the sea by oil or other harmful substances;
    (b) The presence, characteristics and extent of spillages of oil or other harmful substances observed at sea which are likely to present a serious and imminent threat to the marine environment or to the coast or related interests of one or more of the Parties.
2. The information collected in accordance with paragraph 1 shall be communicated to the other Parties likely to be affected by the pollution:
    (a) by the Party which has received the information, either directly or preferably, through the regional centre; or
    (b) by the regional centre.
In case of direct communication between Parties, the regional centre shall be informed of the measures taken by these Parties.

3. In consequence of the application of the provisions of paragraph 2, the Parties are not bound by the obligation laid down in article 9, paragraph 2, of the Convention.

Article 9
1. Any party faced with a situation of the kind defined in article 1 of this Protocol shall:
    (a) Make the necessary assessments of the nature and extent of the casualty or emergency or, as the case may be, of the type and approximate quantity of oil or other harmful substances and the direction and speed of drift of the spillage;
    (b) Take every practicable measure to avoid or reduce the effects of pollution;
    (c) Immediately inform all other Parties, either directly or through the regional centre, of these assessments and of any action which it has taken or which it intends to take to combat the pollution;
    (d) Continue to observe the situation for as long as possible and report thereon in accordance with article 8.
2. Where action is taken to combat pollution originating from a ship, all possible measures shall be taken to safeguard the persons present on board and, to the extent possible, the ship itself. Any Party which takes such action shall inform the Inter-Governmental Maritime Consultative Organization.

Article 10
1. Any Party requiring assistance for combating pollution by oil or other harmful substances polluting or threatening to pollute its coast may call for assistance from other Parties, either directly or through the regional centre referred to in article 6, starting with the Parties which appear likely to be affected by the pollution. This assistance may comprise, in particular, expert advice and the supply to or placing at the disposal of the Party concerned of products, equipment and nautical facilities. Parties so requested shall use their best endeavours to render this assistance.
2. Where the Parties engaged in an operation to combat pollution cannot agree on the organization of the operation, the regional centre may, with their approval, co-ordinate the activity of the facilities put into operation by these Parties.

Article 11
The application of the relevant provisions of articles 6, 7, 8, 9 and 10 of this Protocol relating to the regional centre shall be extended, as appropriate, to sub-regional centres in the event of their establishment, taking into account their objectives and functions and their relationship with the said regional centre.

Article 12
1. Ordinary meetings of the Parties to this Protocol shall be held in conjunction with ordinary meetings of the Contracting Parties to the Convention, held pursuant to Article 14 of the Convention. The Parties to this Protocol may also hold extraordinary meetings as provided in Article 14 of the Convention.
2. It shall be the function of the meetings of the Parties to this Protocol, in particular:
    (a) To keep under review the implementation of this Protocol, and to consider the efficacy of the measures adopted and the need for any other measures, in particular in the form of Annexes;
    (b) To review and amend as required any Annex to this Protocol;
    (c) To discharge such other functions as may be appropriate for implementation of this Protocol.

Article 13
1. The provisions of the Convention relating to any Protocol shall apply with respect to the present Protocol.

2. The rules of procedure and the financial rules adopted pursuant to Article 18 of the Convention shall apply with respect to this Protocol, unless the Parties to this Protocol agree otherwise.

IN WITNESS WHEREOF the undersigned, being duly authorized by their respective Governments, have signed this Protocol.

DONE at Barcelona on 16 February 1976 in a single copy in the Arabic, English, French and Spanish languages, the four texts being equally authoritative.

ANNEX I

Contents of the Report to Be Made Pursuant to Article 8 to This Protocol

1. Each report shall, as far as possible, contain, in general:
    (a) The identification of the source of pollution (identity of the ship, where appropriate);
    (b) The geographic position, time and date of the occurrence of the incident or of the observation;
    (c) The wind and sea conditions prevailing in the area;
    (d) Where the pollution originates from a ship, relevant details respecting the conditions of the ship.
2. Each report shall contain, whenever possible, in particular:
    (a) A clear indication or description of the harmful substances involved, including the correct technical names of such substances (trade names should not be used in place of the correct technical names);
    (b) A statement or estimate of the quantities, concentrations and likely conditions of harmful substances discharged or likely to be discharged into the sea;
    (c) Where relevant, a description of the packaging and identifying marks; and
    (d) The name of the consignor, consignee or manufacturer.
3. Each report shall clearly indicate, whenever possible, whether the harmful substance discharged or likely to be discharged is oil or a noxious liquid, solid or gaseous substance and whether such substance was or is carried in bulk or contained in package form, freight containers, portable tanks, or road and rail tank wagons.
4. Each report shall be supplemented as necessary by any relevant information requested by a recipient of the report or deemed appropriate by the person sending the report.
5. Any of the persons referred to in article 8, paragraph 1, of this Protocol shall:
    (a) Supplement as far as possible the initial report, as necessary, with information concerning further developments; and
    (b) Comply as fully as possible with requests from affected States for additional information.

Selected Documents and Proceedings

Convention on the International Maritime Satellite Organization (INMARSAT)

THE STATES PARTIES TO THIS CONVENTION:

CONSIDERING the principle set forth in Resolution 1721 (XVI) of the General Assembly of the United Nations that communication by means of satellites should be available to the nations of the world as soon as practicable on a global and non-discriminatory basis,

CONSIDERING the relevant provisions of the Treaty on Principles Governing the Activities of States in the Exploration and Use of Outer Space, including the Moon and Other Celestial Bodies, concluded on 27 January 1967, and in particular Article 1, which states that outer space shall be used for the benefit and in the interests of all countries,

TAKING INTO ACCOUNT that a very high proportion of world trade is dependent upon ships,

BEING AWARE that considerable improvements to the maritime distress and safety systems and to the communication link between ships and between ships and their management as well as between crew or passengers on board and persons on shore can be made by using satellites,

DETERMINED, to this end, to make provision for the benefit of ships of all nations through the most advanced suitable space technology available, for the most efficient and economic facilities possible consistent with the most efficient and equitable use of the radio frequency spectrum and of satellite orbits,

RECOGNIZING that a maritime satellite system comprises mobile earth stations and land earth stations, as well as the space segment,

AGREE AS FOLLOWS:

Article 1
Definitions

For the purposes of this Convention:
  (a) "Operating Agreement" means the Operating Agreement on the International Maritime Satellite Organization (INMARSAT), including its Annex.
  (b) "Party" means a State for which this Convention has entered into force.
  (c) "Signatory" means either a Party or an entity designated in accordance with Article 2(3), for which the Operating Agreement has entered into force.
  (d) "Space segment" means the satellites, and the tracking, telemetry,

command, control, monitoring and related facilities and equipment required to support the operation of these satellites.
(e) "INMARSAT space segment" means the space segment owned or leased by INMARSAT.
(f) "Ship" means a vessel of any type operating in the marine environment. It includes *inter alia* hydrofoil boats, air-cushion vehicles, submersibles, floating craft and platforms not permanently moored.
(g) "Property" means anything that can be the subject of a right of ownership, including contractual rights.

Article 2
Establishment of INMARSAT

(1) The International Maritime Satellite Organization (INMARSAT), herein referred to as "the Organization", is hereby established.
(2) The Operating Agreement shall be concluded in conformity with the provisions of this Convention and shall be opened for signature at the same time as this Convention.
(3) Each Party shall sign the Operating Agreement or shall designate a competent entity, public or private, subject to the jurisdiction of that Party, which shall sign the Operating Agreement.
(4) Telecommunications administrations and entities may, subject to applicable domestic law, negotiate and enter directly into appropriate traffic agreements with respect to their use of telecommunications facilities provided pursuant to this Convention and the Operating Agreement, as well as with respect to services to be furnished to the public, facilities, division of revenues and related business arrangements.

Article 3
Purpose

(1) The purpose of the Organization is to make provision for the space segment necessary for improving maritime communications, thereby assisting in improving distress and safety of life at sea communications, efficiency and management of ships, maritime public correspondence services and radiodetermination capabilities.
(2) The Organization shall seek to serve all areas where there is need for maritime communications.
(3) The Organization shall act exclusively for peaceful purposes.

Article 4
Relations between a Party and its Designated Entity

Where a Signatory is an entity designated by a Party:
(a) Relations between the Party and the Signatory shall be governed by applicable domestic law.
(b) The Party shall provide such guidance and instructions as are

appropriate and consistent with its domestic law to ensure that the Signatory fulfils its responsibilities.

(c) The Party shall not be liable for obligations arising under the Operating Agreement. The Party shall, however, ensure that the Signatory, in carrying out its obligations within the Organization, will not act in a manner which violates obligations which the Party has accepted under this Convention or under related international agreements.

(d) If the Signatory withdraws or its membership is terminated the Party shall act in accordance with Article 29(3) or 30(6)

Article 5
Operational and Financial Principles of the Organization

(1) The Organization shall be financed by the contributions of Signatories. Each Signatory shall have a financial interest in the Organization in proportion to its investment share which shall be determined in accordance with the Operating Agreement.

(2) Each Signatory shall contribute to the capital requirements of the Organization and shall receive repayment and compensation for use of capital in accordance with the Operating Agreement.

(3) The Organization shall operate on a sound economic and financial basis having regard to accepted commercial principles.

Article 6
Provision of Space Segment

The Organization may own or lease the space segment.

Article 7
Access to Space Segment

(1) The INMARSAT space segment shall be open for use by ships of all nations on conditions to be determined by the Council. In determining such conditions, the Council shall not discriminate among ships on the basis of nationality.

(2) The Council may, on a case-by-case basis, permit access to the INMARSAT space segment by earth stations located on structures operating in the marine environment other than ships, if and as long as the operation of such earth stations will not significantly affect the provision of service to ships.

(3) Earth stations on land communicating via the INMARSAT space segment shall be located on land territory under the jurisdiction of a Party and shall be wholly owned by Parties or entities subject to their jurisdiction. The Council may authorize otherwise if it finds this to be in the interests of the Organization.

Article 8
Other Space Segments

(1) A Party shall notify the Organization in the event that it or any person

within its jurisdiction intends to make provision for, or initiate the use of, individually or jointly, separate space segment facilities to meet any or all of the purposes of the INMARSAT space segment, to ensure technical compatibility and to avoid significant economic harm to the INMARSAT system.
(2) The Council shall express its views in the form of a recommendation of a non-binding nature with respect to technical compatibility and shall provide its views to the Assembly with respect to economic harm.
(3) The Assembly shall express its views in the form of recommendations of a non-binding nature within a period of nine months from the date of commencing the procedures provided for in this Article. An extraordinary meeting of the Assembly may be convened for this purpose.
(4) The notification pursuant to paragraph (1), including the provision of pertinent technical information, and subsequent consultations with the Organization, shall take into account the relevant provisions of the Radio Regulations of the International Telecommunication Union.
(5) This Article shall not apply to the establishment, acquisition, utilization or continuation of separate space segment facilities for national security purposes, or which were contracted for, established, acquired or utilized prior to the entry into force of this Convention.

Article 9
Structure

The organs of the Organization shall be:
   (a) The Assembly.
   (b) The Council.
   (c) The Directorate headed by a Director General.

Article 10
Assembly—Composition and Meetings

(1) The Assembly shall be composed of all the Parties.
(2) Regular sessions of the Assembly shall be held once every two years. Extraordinary sessions shall be convened upon the request of one-third of the Parties or upon the request of the Council.

Article 11
Assembly—Procedure

(1) Each Party shall have one vote in the Assembly.
(2) Decisions on matters of substance shall be taken by a two-thirds majority, and on procedural matters by a simple majority, of the Parties present and voting. Parties which abstain from voting shall be considered as not voting.
(3) Decisions whether a question is procedural or substantive shall be taken by the Chairman. Such decisions may be overruled by a two-thirds majority of the Parties present and voting.
(4) A quorum for any meeting of the Assembly shall consist of a majority of the Parties.

Article 12
Assembly—Functions

(1) The functions of the Assembly shall be to:
   (a) Consider and review the activities, purposes, general policy and long-term objectives of the Organization and express views and make recommendations thereon to the Council.
   (b) Ensure that the activities of the Organization are consistent with this Convention and with the purposes and principles of the United Nations Charter, as well as with any other treaty by which the Organization becomes bound in accordance with its decision.
   (c) Authorize, on the recommendation of the Council, the establishment of additional space segment facilities the special or primary purpose of which is to provide radiodetermination, distress or safety services. However, the space segment facilities established to provide maritime public correspondence services can be used for telecommunications for distress, safety and radiodetermination purposes without such authorization.
   (d) Decide on other recommendations of the Council and express views on reports of the Council.
   (e) Elect four representatives on the Council in accordance with Article 13(1)(b).
   (f) Decide upon questions concerning formal relationships between the Organization and States, whether Parties or not, and international organizations.
   (g) Decide upon any amendment to this Convention pursuant to Article 34 or to the Operating Agreement pursuant to Article XVIII thereof.
   (h) Consider and decide whether membership be terminated in accordance with Article 30.
   (i) Exercise any other functions conferred upon it in any other Article of this Convention or the Operating Agreement.

(2) In performing its functions the Assembly shall take into account any relevant recommendations of the Council.

Article 13
Council—Composition

(1) The Council shall consist of twenty-two representatives of Signatories as follows:
   (a) Eighteen representatives of those Signatories, or groups of Signatories not otherwise represented, which have agreed to be represented as a group, which have the largest investment shares in the Organization. If a group of Signatories and a single Signatory have equal investment shares, the latter shall have the prior right. If by reason of two or more Signatories having equal investment shares the

number of representatives on the Council would exceed twenty-two, all shall nevertheless, exceptionally, be represented.
   (b) Four representatives of Signatories not otherwise represented on the Council, elected by the Assembly, irrespective of their investment shares, in order to ensure that the principle of just geographical representation is taken into account, with due regard to the interests of the developing countries. Any Signatory elected to represent a geographical area shall represent each Signatory in that geographical area which has agreed to be so represented and which is not otherwise represented on the Council. An election shall be effective as from the first meeting of the Council following that election, and shall remain effective until the next ordinary meeting of the Assembly.
(2) Deficiency in the number of representatives on the Council pending the filling of a vacancy shall not invalidate the composition of the Council.

Article 14
Council—Procedure

(1) The Council shall meet as often as may be necessary for the efficient discharge of its functions, but not less than three times a year.
(2) The Council shall endeavour to take decisions unanimously. If unanimous agreement cannot be reached, decisions shall be taken as follows: Decisions of substantive matters shall be taken by a majority of the representatives on the Council representing at least two-thirds of the total voting participation of all Signatories and groups of Signatories represented on the Council. Decisions on procedural matters shall be taken by a simple majority of the representatives present and voting, each having one vote. Disputes whether a specific matter is procedural or substantive shall be decided by the Chairman of the Council. The decision of the Chairman may be overruled by a two-thirds majority of the representatives present and voting, each having one vote. The Council may adopt a different voting procedure for the election of its officers.
(3) (a) Each representative shall have a voting participation equivalent to the investment share or shares he represents. However, no representative may cast on behalf of one Signatory more than 25 per cent of the total voting participation in the Organization except as provided in sub-paragraph (b)(iv).
   (b) Notwithstanding Article V(9), (10) and (12) of the Operating Agreement:
      (i) If a Signatory represented on the Council is entitled, based on its investment share, to a voting participation in excess of 25 per cent of the total voting participation in the Organization, it may offer to other Signatories any or all of its investment share in excess of 25 per cent.
      (ii) Other Signatories may notify the Organization that they are prepared to accept any or all of such excess investment share. If

the total of the amounts notified to the Organization does not exceed the amount available for distribution, the latter amount shall be distributed by the Council to the notifying Signatories in accordance with the amounts notified. If the total of the amounts notified does exceed the amount available for distribution, the latter amount shall be distributed by the Council as may be agreed among the notifying Signatories, or, failing agreement, in proportion to the amounts notified.

(iii) Any such distribution shall be made by the Council at the time of determinations of investment shares pursuant to Article V of the Operating Agreement. Any distribution shall not increase the investment share of any Signatory above 25 per cent.

(iv) To the extent that the investment share of the Signatory in excess of 25 per cent offered for distribution is not distributed in accordance with the procedure set forth in this paragraph, the voting participation of the representative of the Signatory may exceed 25 per cent.

(c) To the extent that a Signatory decides not to offer its excess investment share to other Signatories, the corresponding voting participation of that Signatory in excess of 25 per cent shall be distributed equally to all other representatives on the Council.

(4) A quorum for any meeting of the Council shall consist of a majority of the representatives on the Council, representing at least two-thirds of the total voting participation of all Signatories and groups of Signatories represented on the Council.

Article 15
Council—Functions

The Council shall have the responsibility, having due regard for the views and recommendations of the Assembly, to make provision for the space segment necessary for carrying out the purposes of the Organization in the most economic, effective and efficient manner consistent with this Convention and the Operating Agreement. To discharge this responsibility, the Council shall have the power to perform all appropriate functions, including:

(a) Determination of maritime satellite telecommunications requirements and adoption of policies, plans, programmes, procedures and measures for the design, development, construction, establishment, acquisition by purchase or lease, operation, maintenance and utilization of the INMARSAT space segment, including the procurement of any necessary launch services to meet such requirements.

(b) Adoption and implementation of management arrangements which shall require the Director General to contract for technical and operational functions whenever this is more advantageous to the Organization.

(c) Adoption of criteria and procedures for approval of earth stations on land, on ships and on structures in the marine environment for access to the INMARSAT space segment and for verification and monitoring of performance of earth stations having access to and utilization of the INMARSAT space segment. For earth stations on ships, the criteria should be in sufficient detail for use by national licensing authorities, at their discretion, for type-approval purposes.

(d) Submission of recommendations to the Assembly in accordance with Article 12(1)(c).

(e) Submission to the Assembly of periodic reports on the activities of the Organization, including financial matters.

(f) Adoption of procurement procedures, regulations and contract terms and approval of procurement contracts consistent with this Convention and the Operating Agreement.

(g) Adoption of financial policies, approval of the financial regulations, annual budget and annual financial statements, periodic determination of charges for use of the INMARSAT space segment, and decisions with respect to all other financial matters, including investment shares and capital ceiling consistent with this convention and the Operating Agreement.

(h) Determination of arrangements for consultation on a continuing basis with bodies recognized by the Council as representing shipowners, maritime personnel and other users of maritime telecommunications.

(i) Designation of an arbitrator where the Organization is a party to an arbitration.

(j) Exercise of any other functions conferred upon it in any other Article of this Convention or the Operating Agreement or any other function appropriate for the achievement of the purposes of the Organization.

Article 16
Directorate

(1) The Director General shall be appointed, from among candidates proposed by Parties or Signatories through Parties, by the Council, subject to confirmation by the Parties. The Depositary shall immediately notify the Parties of the appointment. The appointment is confirmed unless within sixty days of the notification more than one-third of the Parties have informed the Depositary in writing of their objection to the appointment. The Director General may assume his functions after appointment and pending confirmation.

(2) The term of office of the Director General shall be six years. However, the Council may remove the Director General earlier on its own authority. The Council shall report the reasons for the removal to the Assembly.

(3) The Director General shall be the chief executive and legal representative of the Organization and shall be responsible to and under the direction of the Council.
(4) The structure, staff levels and standard terms of employment of officials and employees and of consultants and other advisers to the Directorate shall be approved by the Council.
(5) The Director General shall appoint the members of the Directorate. The appointment of senior officials reporting directly to the Director General shall be approved by the Council.
(6) The paramount consideration in the appointment of the Director General and other personnel of the Directorate shall be the necessity of ensuring the highest standards of integrity, competency and efficiency.

Article 17
Representation at Meetings

All Parties and Signatories which, under this Convention or the Operating Agreement, are entitled to attend and/or participate at meetings of the Organization shall be allowed to attend and/or participate at such meetings as well as any other meeting held under the auspices of the Organization, regardless of where the meeting may take place. The arrangements made with any host country shall be consistent with these obligations.

Article 18
Costs of Meetings

(1) Each Party and Signatory shall meet its own costs of representation at meetings of the Organization.
(2) Expenses of meetings of the Organization shall be regarded as an administrative cost of the Organization. However, no meeting of the Organization shall be held outside its headquarters, unless the prospective host agrees to defray the additional expenditure involved.

Article 19
Establishment of Utilization Charges

(1) The Council shall specify the units of measurement for the various types of utilization of the INMARSAT space segment and shall establish charges for such utilization. These charges shall have the objective of earning sufficient revenues for the Organization to cover its operating, maintenance, and administrative costs, the provision of such operating funds as the Council may determine to be necessary, the amortization of investment made by Signatories, and compensation for use of capital in accordance with the Operating Agreement.
(2) The rates of utilization charge for each type of utilization shall be the same for all Signatories for that type of utilization.

(3) For entities, other than Signatories, which are authorized in accordance with Article 7 to utilize the INMARSAT space segment, the Council may establish rates of utilization charge different from those established for Signatories. The rates for each type of utilization shall be the same for all such entities for that type of utilization.

Article 20
Procurement

(1) The procurement policy of the Council shall be such as to encourage, in the interests of the Organization, world-wide competition in the supply of goods and services. To this end:
- (a) Procurement of goods and services required by the Organization, whether by purchase or lease, shall be effected by the award of contracts, based on responses to open international invitations to tender.
- (b) Contracts shall be awarded to bidders offering the best combination of quality, price and the most favourable delivery time.
- (c) If there are bids offering comparable combinations of quality, price and the most favourable delivery time, the Council shall award the contract so as to give effect to the procurement policy set out above.

(2) In the following cases the requirement of open international tender may be dispensed with under procedures adopted by the Council, provided that in so doing the Council shall encourage in the interests of the Organization world-wide competition in the supply of goods and services:
- (a) The estimated value of the contract does not exceed 50,000 US dollars and the award of the contract would not by reason of the application of the dispensation place a contractor in such a position as to prejudice at some later date the effective exercise by the Council of the procurement policy set out above. To the extent justified by changes in world prices, as reflected by relevant price indices, the Council may revise the financial limit.
- (b) Procurement is required urgently to meet an emergency situation.
- (c) There is only one source of supply to a specification which is necessary to meet the requirements of the Organizations or the sources of supply are so severely restricted in number that it would be neither feasible nor in the best interest of the Organization to incur the expenditure and time involved in open international tender, provides that where there is more than one source they will have an opportunity to bid on an equal basis.
- (d) The requirement is of an administrative nature for which it would be neither practicable nor feasible to undertake open international tender.
- (e) The procurement is for personal services.

Article 21
Inventions and Technical Information

(1) The Organization, in connexion with any work performed by it or on its behalf at its expense, shall acquire in inventions and technical information those rights, but no more than those rights, which are necessary in the common interests of the Organization and of the Signatories in their capacity as such. In the case of work done under contract, any such rights obtained shall be on a non-exclusive basis.

(2) For the purpose of paragraph (1) the Organization, taking into account its principles and objectives and generally accepted industrial practices, shall, in connexion with such work involving a significant element of study, research or development ensure for itself:
   (a) The right to have disclosed to it without payment all inventions and technical information generated by such work.
   (b) The right to disclose and to have disclosed to Parties and Signatories and others within the jurisdiction of any Party such inventions and technical information, and to use and to authorize and to have authorized Parties and Signatories and such others to use such inventions and technical information without payment in connexion with the INMARSAT space segment and any earth station on land or ship station operating in conjunction therewith.

(3) In the case of work done under contract, ownership of the rights in inventions and technical information generated under the contract shall be retained by the contractor.

(4) The Organization shall also ensure for itself the right, on fair and reasonable terms and conditions, to use and to have used inventions and technical information directly utilized in the execution of work performed on its behalf but not included in paragraph (2), to the extent that such use is necessary for the reconstruction or modification of any product actually delivered under a contract financed by the Organization, and to the extent that the person who has performed such work is entitled to grant such right.

(5) The Council may in individual cases approve a deviation from the policies prescribed in paragraphs (2)(b) and (4), where in the course of negotiation it is demonstrated to the Council that failure to deviate would be detrimental to the interests of the Organization.

(6) The Council may also, in individual cases where exceptional circumstances warrant, approve a deviation from the policy prescribed in paragraph (3) where all the following conditions are met:
   (a) It is demonstrated to the Council that failure to deviate would be detrimental to the interests of the Organization.
   (b) The Council determines that the Organization should be able to ensure patent protection in any country.
   (c) Where, and to the extent that, the contractor is unable or unwilling to ensure such patent protection within the time required.

(7) With respect to inventions and technical information in which rights are acquired by the Organization otherwise than pursuant to paragraph (2), the Organization, to the extent that it has the right to do so, shall upon request:
    (a) Disclose or have disclosed such inventions and technical information to any Party or Signatory subject to reimbursement of any payment made by or required of the Organization in respect of the exercise of this right of disclosure.
    (b) Make available to any Party or Signatory the right to disclose or have disclosed to others within the jurisdiction of any Party and to use and to authorize and to have authorized such others to use such inventions and technical information:
        (i) Without payment in connexion with the INMARSAT space segment or any earth station on land or ship operating in conjunction therewith.
        (ii) For any other purpose, on fair and reasonable terms and conditions to be settled between Signatories or others within the jurisdiction of any Party and the Organization or the owner of the inventions and technical information or any other authorized entity or person having a property interest therein, and subject to reimbursement of any payment made by or required of the Organization in respect of the exercise of these rights.

(8) This disclosure and use, and the terms and conditions of disclosure and use, of all inventions and technical information in which the Organization has acquired any rights shall be on a non-discriminatory basis with respect to all Signatories and others within the jurisdiction of Parties.

(9) Nothing in this Article shall preclude the Organization, if desirable, from entering into contracts with persons subject to domestic laws and regulations relating to the disclosure of technical information.

Article 22
Liability

Parties are not, in their capacity as such, liable for the acts and obligations of the Organization, except in relation to non-Parties or natural or juridical persons they might represent in so far as such liability may follow from treaties in force between the Party and the non-Party concerned. However, the foregoing does not preclude a Party which has been required to pay compensation under such a treaty to a non-Party or to a natural or juridical person it might represent from invoking any rights it may have under that treaty against any other Party.

Article 23
Excluded Costs

Taxes on income derived from the Organization by any of the Signatories shall not form part of the costs of the Organization.

Article 24
Audit

The Accounts of the organization shall be audited annually by an independent Auditor appointed by the Council. Any Party or Signatory shall have the right to inspect the accounts of the Organizaion.

Article 25
Legal Personality

The Organization shall have legal personality and shall be responsible for its acts and obligations. For the purpose of its proper functioning, it shall, in particular, have the capacity to contract, to acquire, lease, hold and dispose of movable and immovable property, to be a party to legal proceedings and to conclude agreements with States or international organizations.

Article 26
Privileges and Immunities

(1) Within the scope of activities authorized by this Convention, the Organization and its property shall be exempt in all States Parties to this Convention from all national income and direct national property taxation and from customs duties on communication satellites and components and parts for such satellites to be launched for use in the INMARSAT space segment. Each Party undertakes to use its best endeavours to bring about, in accordance with the applicable domestic procedure, such further exemption from income and direct property taxation and customs duties as is desirable, bearing in mind the particular nature of the Organization.

(2) All Signatories acting in their capacity as such, except the Signatory designated by the Party in whose territory the headquarters is located, shall be exempt from national taxation on income earned from the Organization in the territory of that Party.

(3) (a) As soon as possible after the entry into force of this Convention, the Organization shall conclude, with any Party in whose territory the Organization establishes its headquarters, other offices or installations, an agreement to be negotiated by the Council and approved by the Assembly, relating to the privileges and immunities of the Organization, its Director General, its staff, of experts performing missions for the Organization and of representatives of Parties and Signatories whilst in the territory of the host Government for the purpose of exercising their functions.

   (b) The agreement shall be independent of this Convention and shall terminate by agreement between the host Government and the Organization or if the headquarters of the Organization are moved from the territory of the host Government.

(4) All Parties other than a Party which has concluded an agreement referred to in paragraph (3) shall as soon as possible after the entry into force of this Convention conclude a Protocol on the privileges and immunities of the

Organization, its Director General, its staff, of experts performing missions for the Organization and of representatives of Parties and Signatories whilst in the territory of Parties for the purposes of exercising their functions. The Protocol shall be independent of this Convention and shall prescribe the conditions for its termination.

Article 27
Relationship with Other International Organizations
The Organization shall co-operate with the United Nations and its bodies dealing with the Peaceful Uses of Outer Space and Ocean Area, its Specialized Agencies, as well as other international organizations, on matters of common interest. In particular the Organization shall take into account the relevant Resolutions and Recommendations of the Inter-Governmental Maritime Consultative Organization. The Organization shall observe the relevant provisions of the International Telecommunication Convention and regulations made thereunder, and shall in the design, development, construction and establishment of the INMARSAT space segment and in the procedures established for regulating the operation of the INMARSAT space segment and of earth stations give due consideration to the relevant Resolutions, Recommendations and procedures of the organs of the International Telecommunication Union.

Article 28
Notification to the International Telecommunication Union
Upon request from the Organization, the Party in whose territory the Headquarters of the Organization is located shall co-ordinate the frequencies to be used for the space segment and shall, on behalf of each Party that consents, notify the International Telecommunication Union of the frequencies to be so used and other information, as provided for in the Radio Regulations annexed to the International Telecommunication Convention.

Article 29
Withdrawal
(1) Any Party or Signatory may by written notification to the Depositary withdraw voluntarily from the Organization at any time. Once a decision has been made under applicable domestic law that a Signatory may withdraw, notice of the withdrawal shall be given in writing to the Depositary by the Party of the withdrawal. Withdrawal of a Party, in its capacity as such, shall entail the simultaneous withdrawal of any Signatory designated by the Party or of the Party in its capacity as Signatory, as the case may be.
(2) Upon receipt by the Depositary of a notice to withdraw, the Party giving notice and any Signatory which it has designated, or the Signatory in respect of which notice has been given, as the case may be, shall cease to have any rights of representation and any voting rights in any organ of the Organization and shall

incur no obligation after the date of such receipt. However, a withdrawing Signatory shall remain responsible, unless otherwise decided by the Council pursuant to Article XIII of the Operating Agreement, for contributing its share of the capital contributions necessary to meet contractual commitments specifically authorized by the Organization before the receipt and liabilities arising from acts or omissions before the receipt. Except with respect to such capital contributions and except with respect to Article 31 of this Convention and Article XVI of the Operating Agreement, withdrawal shall become effective and this Convention and/or the Operating Agreement shall cease to be in force for the Party and/or Signatory three months after the date of receipt by the Depositary of the written notification referred to in paragraph (1).

(3) If a Signatory withdraws, the Party which designated it shall, before the effective date of withdrawal and with effect from that date, designate a new Signatory, assume the capacity of a Signatory in accordance with paragraph (4), or withdraw. If the Party has not acted by the effective date, it shall be considered to have withdrawn as from that date. Any new Signatory shall be responsible for all the outstanding capital contributions of the previous Signatory and for the proportionate share of any capital contributions necessary to meet contractual commitments specifically authorized by the Organization, and liabilities arising from acts or omissions, after the date of receipt of the notice.

(4) If for any reason a Party desires to substitute itself for its designated Signatory or to designate a new Signatory, it shall give written notice to the Depositary. Upon assumption by the new Signatory of all the outstanding obligations, as specified in the last sentence of paragraph (3), of the previously designated Signatory and upon signature of the Operating Agreement, that Agreement shall enter into force for the new Signatory and shall cease to be in force for the previous Signatory.

Article 30
Suspension and Termination

(1) Not less than one year after the Directorate has received written notice that a Party appears to have failed to comply with any obligation under this Convention, the Assembly, after considering any representations made by the Party, may decide, if it finds that the failure to comply has in fact occurred and that such failure impairs the effective operation of the Organization, that the membership of the Party is terminated. This Convention shall cease to be in force for the Party as from the date of the decision or at such later date as the Assembly may determine. An extraordinary session of the Assembly may be convened for this purpose. The termination shall entail the simultaneous withdrawal of any Signatory designated by the Party or of the Party in its capacity as Signatory, as the case may be. The Operating Agreement shall cease to be in force for the Signatory on the date on which this Convention ceases to be in force for the Party concerned, except with respect to capital contributions

necessary to meet contractual commitments specifically authorized by the Organization before the termination and liabilities arising from acts or omissions before the termination, and except with respect to Article 31 of this Convention and Article XVI of the Operating Agreement.

(2) If any Signatory, in its capacity as such, fails to comply with any obligation under this Convention or the Operating Agreement, other than obligations under Article III(1) of the Operating Agreement and the failure has not been remedied within three months after the Signatory has been notified in writing of a resolution of the Council taking note of the failure to comply, the Council, after considering any representations made by the Signatory and, if applicable, the Party concerned may suspend the rights of the Signatory. If, after an additional three months and after consideration of any representations made by the Signatory and, if applicable, the Party, the Council finds that the failure to comply has not been remedied, the Assembly may decide on the recommendation of the Council that the membership of the Signatory is terminated. Upon the date of such decision, the termination shall become effective and the Operating Agreement shall cease to be in force for that Signatory.

(3) If any Signatory fails to pay any amount due from it pursuant to Article III(1) of the Operating Agreement within four months after the payment has become due, the rights of the Signatory under this Convention and the Operating Agreement shall be automatically suspended. If within three months after the suspension the Signatory has not paid all sums due or the Party which has designated it has not made a substitution pursuant to Article 29(4), the Council, after considering any representations made by the Signatory or by the Party which has designated it, may decide that the membership of the Signatory is terminated. From the date of such decision, the Operating Agreement shall cease to be in force for the Signatory.

(4) During the period of suspension of the rights of a Signatory pursuant to paragraphs (2) or (3), the Signatory shall continue to have all the obligations of a Signatory under this Convention and the Operating Agreement.

(5) A Signatory shall incur no obligation after termination, except that it shall be responsible for contributing its share of the capital contributions necessary to meet contractual commitments specifically authorized before the termination and liabilities arising from acts or omissions before the termination, and except with respect to Article 31 of this Convention and Article XVI of the Operating Agreement.

(6) If the membership of a Signatory is terminated, the Party which designated it shall, within three months from the date of the termination and with effect from that date, designate a new Signatory, assume the capacity of a Signatory in accordance with Article 29(4), or withdraw. If the Party has not acted by the end of that period, it shall be considered to have withdrawn as from the date of termination, and this Convention shall cease to be in force for the Party as from that date.

(7) Whenever this Convention has ceased to be in force for a Party, settlement

between the Organization and any Signatory designated by that Party or that Party in its capacity as Signatory shall be accomplished as provided in Article XIII of the Operating Agreement.

Article 31
Settlement of Disputes

(1) Disputes arising between Parties, or between Parties and the Organization, relating to rights and obligations under this Convention should be settled by negotiation between the parties concerned. If within one year of the time any party has requested settlement, a settlement has not been reached and if the parties to the dispute have not agreed to submit it to the International Court of Justice or to some other procedure for settling disputes, the dispute may, if the parties to the dispute consent, be submitted to arbitration in accordance with the Annex to this Convention. Any decision of an arbitral tribunal in a dispute between Parties, or between Parties and the Organization, shall not prevent or affect a decision of the Assembly pursuant to Article 30(1), that the Convention shall cease to be in force for a Party.

(2) Unless otherwise mutually agreed, disputes arising between the Organization and one or more Parties under agreements concluded between them, if not settled by negotiation within one year of the time any party has requested settlement, shall, at the request of any party to the dispute, be submitted to arbitration in accordance with the Annex to this Convention.

(3) Disputes arising between one or more Parties and one or more Signatories in their capacity as such, relating to rights and obligations under this Convention or the Operating Agreement may be submitted to arbitration in accordance with the Annex to this Convention if the Party or Parties and the Signatory or Signatories involved agree to such arbitration.

(4) This Article shall continue to apply to a Party or Signatory which ceases to be a Party or Signatory, in respect of disputes relating to rights and obligations arising from its having been a Party or Signatory.

Article 32
Signature and Ratification

(1) This Convention shall remain open for signature in London until entry into force and shall thereafter remain open for accession. All States may become Parties to the Convention by:
    (a) Signature not subject to ratification, acceptance or approval, or
    (b) Signature subject to ratification, acceptance or approval, followed by ratification, acceptance or approval, or
    (c) Accession.

(2) Ratification, acceptance, approval or accession shall be effected by the deposit of the appropriate instrument with the Depositary.

(3) On becoming a Party to this Convention, or at any time thereafter, a State may declare, by written notification to the Depositary, to which Registers of

ships operating under its authority, and to which land earth stations under its jurisdiction, the Convention shall apply.
(4) No State shall become a Party to this Convention until it has signed, or the entity it has designated, has signed the Operating Agreement.
(5) Reservations cannot be made to this Convention or the Operating Agreement.

Article 33
Entry into Force
(1) This Convention shall enter into force sixty days after the date on which States representing 95 per cent of the initial investment shares have become Parties to the Convention.
(2) Notwithstanding paragraph (1), if the Convention has not entered into force within thirty-six months after the date it was opened for signature, it shall not enter into force.
(3) For a State which deposits an instrument of ratification, acceptance, approval or accession after the date on which the Convention has entered into force, the ratification, acceptance, approval or accession shall take effect on the date of deposit.

Article 34
Amendments
(1) Amendments to this Convention may be proposed by any Party. Proposed amendments shall be submitted to the Directorate, which shall inform the other Parties and Signatories. Three months' notice is required before consideration of an amendment by the Council, which shall submit its views to the Assembly within a period of six months from the date of circulation of the amendment. The Assembly shall consider the amendment not earlier than six months thereafter, taking into account any views expressed by the Council. This period may, in any particular case, be reduced by the Assembly by a substantive decision.
(2) If adopted by the Assembly, the amendment shall enter into force one hundred and twenty days after the Depositary has received notices of acceptance from two-thirds of those States which at the time of adoption by the Assembly were Parties and represented at least two-thirds of the total investment shares. Upon entry into force, the amendment shall become binding upon all Parties and Signatories, including those which have not accepted it.

Article 35
Depositary
(1) The Depositary of this Convention shall be the Secretary-General of the Inter-Governmental Maritime Consultative Organization.

(2) The Depositary shall promptly inform all signatory and acceding States and all Signatories of:
- (a) Any signature of the Convention.
- (b) The deposit of any instrument of ratification, acceptance, approval or accession.
- (c) The entry into force of the Convention.
- (d) The adoption of any amendment to the Convention and its entry into force.
- (e) Any notification of withdrawal.
- (f) Any suspension or termination.
- (g) Other notifications and communications relating to the Convention.

(3) Upon entry into force of the Convention the Depositary shall transmit a certified copy to the Secretariat of the United Nations for registration and publication in accordance with Article 102 of the Charter of the United Nations.

IN WITNESS WHEREOF the undersigned, duly authorized by their respective Governments, have signed this Convention.

DONE AT LONDON this third day of September one thousand nine hundred and seventy-six in the English, French, Russian and Spanish languages, all the texts being equally authentic, in a single original which shall be deposited with the Depositary, who shall send a certified copy to the Government of each of the States which were invited to attend the International Conference on the Establishment of an International Maritime Satellite System and to the Government of any other State which signs or accedes to this Convention.

ANNEX
PROCEDURES FOR THE SETTLEMENT OF DISPUTES REFERRED TO IN ARTICLE 31 OF THE CONVENTION AND ARTICLE XVI OF THE OPERATING AGREEMENT

Article 1
Disputes cognizable pursuant to Article 31 of the Convention or Article XVI of the Operating Agreement shall be dealt with by an arbitral tribunal of three members.

Article 2
Any petitioner or group of petitioners wishing to submit a dispute to arbitration shall provide each respondent and the Directorate with a document containing:
- (a) A full description of the dispute, the reasons why each respondent is required to participate in the arbitration, and the measures being requested.
- (b) The reasons why the subject matter of the dispute comes within the competence of a tribunal and why the measures requested can be granted if the tribunal finds in favour of the petitioner.
- (c) An explanation why the petitioner has been unable to achieve a settlement of the dispute by negotiation or other means short of arbitration.
- (d) Evidence of the agreement or consent of the disputants when this is a condition for arbitration.

(e) The name of the person designated by the petitioner to serve as a member of the tribunal.

The Directorate shall promptly distribute a copy of the document to each Party and Signatory.

Article 3

(1) Within sixty days from the date copies of the documents described in Article 2 have been received by all the respondents, they shall collectively designate an individual to serve as a member of the tribunal. Within that period, the respondents may jointly or individually provide each disputant and the Directorate with a document stating their individual or collective responses to the document referred to in Article 2 and including any counterclaims arising out of the subject matter of the dispute.

(2) Within thirty days after the designation of the two members of the tribunal, they shall agree on a third arbitrator. He shall not be of the same nationality as, or resident in the territory of, any disputant, or in its service.

(3) If either side fails to nominate an arbitrator within the period specified or if the third arbitrator is not appointed within the period specified, the President of the International Court of Justice, or, if he is prevented from acting or is of the same nationality as a disputant, the Vice-President, or, if he is prevented from acting or is of the same nationality as a disputant, the senior judge who is not of the same nationality as any disputant, may at the request of either disputant, appoint an arbitrator or arbitrators as the case requires.

(4) The third arbitrator shall act as president of the tribunal.

(5) The tribunal is constituted as soon as the president is selected.

Article 4

(1) If a vacancy occurs in the tribunal for any reason which the president or the remaining members of the tribunal decide is beyond the control of the disputants, or is compatible with the proper conduct of the arbitration proceedings, the vacancy shall be filled in accordance with the following provisions:
 (a) If the vacancy occurs as a result of the withdrawal of a member appointed by a side to the dispute, then that side shall select a replacement within ten days after the vacancy occurs.
 (b) If the vacancy occurs as a result of the withdrawal of the president or of a member appointed pursuant to Article 3(3), a replacement shall be selected in the manner described in paragraph (2) or (3), respectively, of Article 3.

(2) If a vacancy occurs for any other reason, or if a vacancy occurring pursuant to paragraph (1) is not filled, the remainder of the tribunal shall have the power, notwithstanding Article 1, upon request of one side, to continue the proceedings and give the final decision of the tribunal.

Article 5

(1) The tribunal shall decide the date and place of its meetings.

(2) The proceedings shall be held in private and all material presented to the tribunal shall be confidential. However, the Organization and any Party which has designated a Signatory which is a disputant in the proceedings shall have the right to be present and shall have access to the material presented. When the Organization is a disputant in the proceedings, all Parties and all Signatories shall have the right to be present and shall have access to the material presented.

(3) In the event of a dispute over the competence of the tribunal, the tribunal shall deal with that question first.

(4) The proceedings shall be conducted in writing, and each side shall have the right to submit written evidence in support of its allegations of fact and law. However, oral arguments and testimony may be given if the tribunal considers it appropriate.

(5) The proceedings shall commence with the presentation of the case of the petitioner containing its arguments, related facts supported by evidence and the principles of law relied upon. The case of the petitioner shall be followed by the counter-case of the respondent. The petitioner may submit a reply to the counter-case of the respondent and the respondent may submit a rejoinder. Additional pleadings shall be submitted only if the tribunal determines they are necessary.
(6) The tribunal shall hear and determine counter-claims arising directly out of the subject matter of the dispute, if the counter-claims are within its competence as defined in Article 31 of the Convention and Article XVI of the Operating Agreement.
(7) If the disputants reach an agreement during the proceedings, the agreement shall be recorded in the form of a decision of the tribunal given by consent of the disputants.
(8) At any time during the proceedings, the tribunal may terminate the proceedings if it decides the dispute is beyond its competence as defined in Article 31 of the Convention or Article XVI of the Operating Agreement.
(9) The deliberations of the tribunal shall be secret.
(10) The decisions of the tribunal shall be presented in writing and shall be supported by a written opinion. Its rulings and decisions must be supported by at least two members. A member dissenting from the decision may submit a separate written opinion.
(11) The tribunal shall forward its decision to the Directorate, which shall distribute it to all Parties and Signatories.
(12) The tribunal may adopt additional rules of procedure, consistent with those established by this Annex, which are appropriate for the proceedings.

Article 6

If one side fails to present its case, the other side may call upon the tribunal to give a decision on the basis of its presentation. Before giving its decision, the tribunal shall satisfy itself that it has competence and that the case is well-founded in fact and in law.

Article 7

(1) Any Party whose Signatory is a disputant shall have the right to intervene and become an additional disputant. Intervention shall be made by written notification to the tribunal and to the other disputants.
(2) Any other Party, any Signatory or the Organization may apply to the tribunal for permission to intervene and become an additional disputant. The tribunal shall grant permission if it determines that the applicant has a substantial interest in the case.

Article 8

The tribunal may appoint experts to assist it at the request of a disputant or on its own initiative.

Article 9

Each Party, each Signatory and the Organization shall provide all information which the tribunal, at the request of a disputant or on its own initiative, determines to be required for the handling and determination of the dispute.

Article 10

Pending the final decision, the tribunal may indicate any provisional measures which it considers ought to be taken to preserve the respective rights of the disputants.

Article 11

(1) The decision of the tribunal shall be in accordance with international law and be based on:

(a) The Convention and the Operating Agreement.
(b) Generally accepted principles of law.
(2) The decision of the tribunal, including any reached by agreement of the disputant pursuant to Article 5(7), shall be binding on all the disputants, and shall be carried out by them in good faith. If the Organization is a disputant, and the tribunal decides that a decision of any organ of the Organization is null and void as not being authorized by or in compliance with the Convention and the Operating Agreement, the decision of the tribunal shall be binding on all Parties and Signatories.

Selected Documents and Proceedings

Convention concerning Minimum Standards in Merchant Ships

The General Conference of the International Labour Organization,

Having been convened at Geneva by the Governing Body of the International Labour Office and having met in its Sixty-second Session on 13 October 1976, and

Recalling the provisions of the Seafarers' Engagement (Foreign Vessels) Recommendation, 1958, and of the Social Conditions and Safety (Seafarers) Recommendation, 1958, and

Having decided upon the adoption of certain proposals with regard to substandard vessels, particularly those registered under flags of convenience, which is the fifth item on the agenda of the session, and

Having determined that these proposals shall take the form of an international Convention,

adopts this twenty-ninth day of October of the year one thousand nine hundred and seventy-six the following Convention, which may be cited as the Merchant Shipping (Minimum Standards) Convention, 1976:

Article 1

1. Except as otherwise provided in this Article, this Convention applies to every sea-going ship, whether publicly or privately owned, which is engaged in the transport of cargo or passengers for the purpose of trade or is employed for any other commercial purpose.

2. National laws or regulations shall determine when ships are to be regarded as sea-going ships for the purpose of this Convention.

3. This Convention applies to sea-going tugs.

4. This Convention does not apply to—
(a) ships primarily propelled by sail, whether or not they are fitted with auxiliary engines;
(b) ships engaged in fishing or in whaling or in similar pursuits;
(c) small vessels and vessels such as oil rigs and drilling platforms when not engaged in navigation, the decision as to which vessels are covered by this subparagraph to be taken by the competent authority in each country in consultation with the most representative organisations of shipowners and seafarers.

5. Nothing in this Convention shall be deemed to extend the scope of the Conventions referred to in the Appendix to this Convention or of the provisions contained therein.

Article 2

Each Member which ratifies this Convention undertakes—

(a) to have laws or regulations laying down, for ships registered in its territory—
   (i) safety standards, including standards of competency, hours of work and manning, so as to ensure the safety of life on board ship;
   (ii) appropriate social security measures; and
   (iii) shipboard conditions of employment and shipboard living arrangements, in so far as these, in the opinion of the Member, are not covered by collective agreements or laid down by competent courts in a manner equally binding on the shipowners and seafarers concerned;

   and to satisfy itself that the provisions of such laws and regulations are substantially equivalent to the Conventions or Articles of Conventions referred to in the Appendix to this Convention, in so far as the Member is not otherwise bound to give effect to the Conventions in question;

(b) to exercise effective jurisdiction or control over ships which are registered in its territory in respect of—
   (i) safety standards, including standards of competency, hours of work and manning, prescribed by national laws or regulations;
   (ii) social security measures prescribed by national laws or regulations;
   (iii) shipboard conditions of employment and shipboard living arrangements prescribed by national laws or regulations, or laid down by competent courts in a manner equally binding on the shipowners and seafarers concerned;

(c) to satisfy itself that measures for the effective control of other shipboard conditions of employment and living arrangements, where it has no effective jurisdiction, are agreed between shipowners or their organisations and seafarers' organisations constituted in accordance with the substantive provisions of the Freedom of Association and Protection of the Right to Organise Convention, 1948, and the Right to Organise and Collective Bargaining Convention, 1949;

(d) to ensure that—
   (i) adequate procedures—subject to over-all supervision by the competent authority, after tripartite consultation amongst that authority and the representative organisations of shipowners and seafarers where appropriate—exist for the engagement of seafarers on ships registered in its territory and for the investigation of complaints arising in that connection;
   (ii) adequate procedures—subject to over-all supervision by the competent authority, after tripartite consultation amongst that authority and the representative organisations of shipowners and seafarers where appropriate—exist for the investigation of any complaint

made in connection with and, if possible, at the time of the engagement in its territory of seafarers of its own nationality on ships registered in a foreign country, and that such complaint as well as any complaint made in connection with and, if possible, at the time of the engagement in its territory of foreign seafarers on ships registered in a foreign country, is promptly reported by its competent authority of the country in which the ship is registered, with a copy to the Director-General of the International Labour Office;

(e) to ensure that seafarers employed on ships registered in its territory are properly qualified or trained for the duties for which they are engaged, due regard being had to the Vocational Training (Seafarers) Recommendation, 1970;

(f) to verify by inspection or other appropriate means that ships registered in its territory comply with applicable international labour Conventions in force which it has ratified, with the laws and regulations required by subparagraph *(a)* of this Article and, as may be appropriate under national law, with applicable collective agreements;

(g) to hold an official inquiry into any serious marine casualty involving ships registered in its territory, particularly those involving injury and/or loss of life, the final report of such inquiry normally to be made public.

Article 3

Any Member which has ratified this Convention shall, in so far as practicable, advise its nationals on the possible problems of signing on a ship registered in a State which has not ratified the Convention, until it is satisfied that standards equivalent to those fixed by this Convention are being applied. Measures taken by the ratifying State to this effect shall not be in contradiction with the principle of free movement of workers stipulated by the treaties to which the two States concerned may be parties.

Article 4

1. If a Member which has ratified this Convention and in whose port a ship calls in the normal course of its business or for operational reasons receives a complaint or obtains evidence that the ship does not conform to the standards of this Convention, after it has come into force, it may prepare a report addressed to the government of the country in which the ship is registered, with a copy to the Director-General of the International Labour Office, and may take measures necessary to rectify any conditions on board which are clearly hazardous to safety or health.

2. In taking such measures, the Member shall forthwith notify the nearest maritime, consular or diplomatic representative of the flag State and shall, if possible, have such representative present. It shall not unreasonably detain or delay the ship.

3. For the purpose of this Article, "complaint" means information

submitted by a member of the crew, a professional body, an association, a trade union or, generally, any person with an interest in the safety of the ship, including an interest in safety or health hazards to its crew.

Article 5

1. This Convention is open to the ratification of Members which—
(a) are parties to the International Convention for the Safety of Life at Sea, 1960, or the International Convention for the Safety of Life at Sea, 1974, or any Convention subsequently revising these Conventions; and
(b) are parties to the International Convention on Load Lines, 1966, or any Convention subsequently revising that Convention; and
(c) are parties to, or have implemented the provisions of, the Regulations for Preventing Collisions at Sea of 1960, or the Convention on the International Regulations for Preventing Collisions at Sea, 1972, or any Convention subsequently revising these international instruments.

2. This Convention is further open to the ratification of any Member which, on ratification, undertakes to fulfil the requirements to which ratification is made subject by paragraph 1 of this Article and which are not yet satisfied.

3. The formal ratifications of this Convention shall be communicated to the Director-General of the International Labour Office for registration.

Article 6

1. This Convention shall be binding only upon those Members of the International Labour Organisation whose ratifications have been registered with the Director-General.

2. It shall come into force twelve months after the date on which there have been registered ratifications by at least ten Members with a total share in world shipping gross tonnage of 25 per cent.

3. Thereafter, this Convention shall come into force for any Member twelve months after the date on which its ratification has been registered.

Article 7

1. A Member which has ratified this Convention may denounce it after the expiration of ten years from the date on which the Convention first comes into force, by an act communicated to the Director-General of the International Labour Office for registration. Such denunciation shall not take effect until one year after the date on which it is registered.

2. Each Member which has ratified this Convention and which does not, within the year following the expiration of the period of ten years mentioned in the preceding paragraph, exercise the right of denunciation provided for in this Article, will be bound for another period of ten years and, thereafter, may denounce this Convention at the expiration of each period of ten years under the terms provided for in this Article.

Article 8

1. The Director-General of the International Labour Office shall notify all Members of the International Labour Organisation of the registration of all ratifications and denunciations communicated to him by the Members of the Organisation.

2. When the conditions provided for in Article 6, paragraph 2, above have been fulfilled, the Director-General shall draw the attention of the Members of the Organisation to the date upon which the Convention will come into force.

Article 9

The Director-General of the International Labour Office shall communicate to the Secretary-General of the United Nations for registration in accordance with Article 102 of the Charter of the United Nations full particulars of all ratifications and acts of denunciation registered by him in accordance with the provisions of the preceding Articles.

Article 10

At such times as it may consider necessary the Governing Body of the International Labour Office shall present to the General Conference a report on the working of this Convention and shall examine the desirability of placing on the agenda of the Conference the question of its revision in whole or in part.

Article 11

1. Should the Conference adopt a new Convention revising this Convention in whole or in part, then, unless the new Convention otherwise provides—
   (a) the ratification by a Member of the new revising Convention shall *ipso jure* involve the immediate denunciation of this Convention, notwithstanding the provisions of Article 7 above, if and when the new revising Convention shall have come into force;
   (b) as from the date when the new revising Convention comes into force this Convention shall cease to be open to ratification by the Members.

2. This Convention shall in any case remain in force in its actual form and content for those Members which have ratified it but have not ratified the revising Convention.

Article 12

The English and French versions of the text of this Convention are equally authoritative.

APPENDIX

Minimum Age Convention, 1973 (No. 138), or
   Minimum Age (Sea) Convention (Revised), 1936 (No. 58), or
   Minimum Age (Sea) Convention, 1920 (No. 7);

Shipowners' Liability (Sick and Injured Seamen) Convention, 1936 (No. 55), or
  Sickness Insurance (Sea) Convention, 1936 (No. 56), or
  Medical Care and Sickness Benefits Convention, 1969 (No. 130);
Medical Examination (Seafarers) Convention, 1946 (No. 73);
Prevention of Accidents (Seafarers) Convention, 1970 (No. 134) (Articles 4 and 7);
Accommodation of Crews Convention (Revised), 1949 (No. 92);
Food and Catering (Ships' Crews) Convention, 1946 (No. 68) (Article 5);
Officers' Competency Certificates Convention, 1936 (No. 53) (Articles 3 and 4)[1];
Seamen's Articles of Agreement Convention, 1926 (No. 22);
Repatriation of Seamen Convention, 1926 (No. 23);
Freedom of Association and Protection of the Right to Organise Convention, 1948 (No. 87);
Right to Organise and Collective Bargaining Convention, 1949 (No. 98).

1. In cases where the established licensing system or certification structure of a State would be prejudiced by problems arising from strict adherence to the relevant standards of the Officers' Competency Certificates Convention, 1936, the principle of substantial equivalence shall be applied so that there will be no conflict with that State's established arrangements for certification.

Selected Documents and Proceedings

SCAR/SCOR Group on the Living Resources of the Southern Ocean (SCOR Working Group 54)*

Report of a Meeting Held at Woods Hole, USA, 23–24 August 1976

Present: S. Z. El Sayed (chairman); G. Deacon, UK; J. A. Gulland, FAO; G. Hempel, FRG; G. A. Knox, New Zealand; R. M. Laws, UK; T. Nemoto, Japan; G. G. Newman, South Africa; S. Olsen, Norway; D. B. Siniff, USA; A. P. Tomo, Argentina; D. J. Tranter, Australia.

Introduction

The group met in the US National Academy of Sciences Summer Studies Centre in Woods Hole, USA. It was the third meeting of the SCAR Group of Specialists but the first since it became SCAR/SCOR Working Group 54. The meeting was combined with a conference on the Living Resources of the Southern Ocean which was well attended—by 59 scientists—but it was regretted that there was no participation from the USSR.

The chief objective of both the conference and the meeting of the Working Group was to review the present knowledge of the living resources of the Southern Ocean and to develop a proposal for future co-operative studies in this area. Members of the Group of Specialists prepared a draft proposal for an international study of the Living Resources of the Southern Ocean which followed the outline developed at the first meeting of the SCAR Group of Specialists in Cambridge, UK, 6–8 October 1975. A number of background review papers were also delivered at the conference as a basis for discussion. These papers included: 'Physical oceanography of the Southern Ocean: key to understanding its biology' (T. Foster, USA), 'The problems of harvesting and utilization of Antarctic krill' (J. Schärffe, FAO), 'The legal status of the Antarctic' (F. Sollie, Norway), 'Remote sensing of Antarctic living resources' (W. Hovis, USA) and 'Modelling of Antarctic ecosystems' (K. Green, USA).

The growing interest by several countries in marine research in Antarctic waters was reflected by reports and films on Antarctic expeditions in the 1975–76 season carried out by research vessels, commercial factory trawlers and ice breakers of the Federal Republic of Germany, Poland, Japan, France, Argentina and the USA. The results of those expeditions will greatly augment our knowledge of the krill and fish resources of Antarctic waters. The Group of

---

*Reprinted with permission from the *Polar Record*, January 1977.

Specialists regretted the lack of up-to-date information on the recent activities of the USSR. The major part of the conference as well as the subsequent sessions of the Group of Specialists were devoted to the discussions on the further development of the proposal for an international co-operative study of the living resources of the Southern Ocean which led to the BIOMASS Programme.

Finally the group noted that it would probably need a further meeting soon after SCOR, SCAR and IOC had considered its report and suggested programme. The proposed changes in its terms of reference, and the increasing activity in Southern Ocean biological research (including work by countries not presently members of SCAR) are likely to make changes and additions to the group's membership desirable, in order to include sufficient range of expertise, and to ensure that the group is informed of all significant research activities. At the same time, it is desirable to keep the group small, and it will be possible in some cases to bring in the required expertise by including appropriate scientists in specialized sub-groups.

Ecosystem Studies

An understanding of the trophodynamics of the Southern Ocean ecosystem is vital from the viewpoint of wise resource management and conservation and presents a unique opportunity to contribute to an understanding of ecosystem processes in the ocean. One approach to the understanding of the whole ecosystem is to develop suitable models. The development of a crude whole ecosystem model for Antarctic waters might be feasible now with the present data base and some additional biological data, such as field data on biomass of ecologically important elements of the system and experimental data on transfer rates. A model can serve to organize existing information and to point out areas where additional research is needed. At the same time it can be sufficiently flexible to incorporate further information and insight in the course of its future development.

A model has been constructed of the Ross Sea ecosystem (Green, 1975) which describes the interaction of nutrients, light, ice, phytoplankton, zooplankton and larger animals. The development of a general model of the entire system must be accompanied by other activities, such as development of more detailed sub-models of certain elements of the system. To make full use of the modelling process, it is necessary to bring the model builders together with those concerned with field investigations and experimental studies. A process feedback between model development and field studies will serve to enhance the predictive capabilities of the various models as they are updated, validated and improved. It can also serve to identify information needs and provide some direction for future research. In addition to correspondence there seems a need for meetings of small groups to discuss these matters in detail.

Description of the Living Resources

A number of compilations and reviews of the main Antarctic living resources were available to the meeting. These included preliminary drafts of reviews of krill, fish and cephalopods prepared by I. Everson as part of the UNDP (FAO) Southern Ocean project, and a review of Antarctic birds by J. Prévost. Extensive reviews of information on marine mammal stocks have been made by scientists for the International Whaling Commission (IWC) and for the Scientific Consultation on Marine Mammals, Bergen, Norway, September 1976.

With a few exceptions, little is known of the biomass and productivity of the living marine resources other than the marine mammals. Though investigations began shortly after the turn of this century, it was not until recent years that studies of the biomass of such resources as fish, crustaceans and seaweeds were made. The discussions on the various living resources aimed at the identification of those aspects of ecology which are of particular importance for an assessment of the magnitude of the resource and the possible consequences of its exploitation; harvesting should be planned and managed on the basis of a knowledge not only of the population dynamics of the resource itself, but also of its interaction with the other parts of the ecosystem.

Marine Mammals and Birds

The conference and the Group of Specialists concentrated on certain biological aspects which are of particular importance regarding the present role of marine mammals in the Antarctic ecosystem, and on the changes which have taken place since the depletion of the whale stocks and the cessation of sealing. Abundance estimates of seals are fairly reliable and their biology is relatively well known except for rarer species (particularly the Ross Seal). Crabeater Seals are by far the most abundant of all seal species in the world; their diet consists almost exclusively of krill. Most of the other species have a mixed diet consisting of krill, fish and squids.

Table 1 compares the present abundance and krill consumption of whales with those occurring before whaling began. As a consequence of whaling, total biomass of baleen whales and their annual consumption of krill decreased to about one-sixth and one-fifth, respectively, of the original figures. There are indications of increased body growth rates, earlier maturation and increases in pregnancy rates in Blue, Fin and Sei whales as well as in Crabeater Seals.

TABLE 1. BALEEN WHALES IN THE ANTARCTIC

(Rounded figures of initial stocks and their Antarctic food consumption; present figures given in brackets)

	Stock (thousands)		Mean weight (tonnes)		Biomass (millions tonnes)		Krill consumption (millions tonnes)	
Fin	400	(84)	50	(48)	20	(4)	81	(16)
Blue	200	(10)	88	(83)	18	(0.8)	72	(3)
Sei	75	(40)	18	(17)	1.4	(0.7)	6	(3)
Humpback	100	(3)	27	(26)	2.7	(0.01)	11	(0.3)
Minke	200	(200)	7	(7)	1.4	(1.4)	10	(10)
Total	975	(337)			43	(7)	180	(33)

For the first time an attempt has been made to make a global assessment of the bird population of the Southern Ocean. Penguins comprise 99 per cent of the biomass of Antarctic avifauna with Adélie Penguins largely dominating in the pack ice zone; in the sub-Antarctic 83 per cent of the biomass consists of penguins—petrels and albatrosses making up much of the rest. The total biomass of all birds in the Southern Ocean is estimated to be nearly 200 million individuals, but there are considerable differences in accuracy of the censuses of the various populations. It is also estimated that the food consumption of the bird population is about 40 million tonnes of food per year, 54 per cent of which is taken in the sub-Antarctic region. 86 per cent of all the food of the Southern Ocean birds is eaten by penguins. It consists principally of euphausiids; only a few concentrate their diets on fish or squids. Albatrosses apparently feed heavily on pelagic squids. The petrels differ very much in size (0.04–4 kg) and in food, ranging from micro-zooplankton to larger fish and squids. Food consumption is by far the highest in the vicinity of the Antarctic and sub-Antarctic islands and along the continental coast. Penguins may not have profited greatly from the decrease in whales as competitors for krill. It has been suggested that penguin populations are more strongly affected by the availability of food in the winter and nesting sites than the availability of krill in the summer.

In former times, whales were by far the largest group of consumers of Antarctic krill. Presently, whales and sea birds seem to be similar in their yearly consumption of krill, and seals consume about twice as much as either. The consumption figures for birds are less reliable than for seals and whales.

Fish

A dozen species of fish, mainly Nototheniids, are presently exploited or are likely to become attractive for exploitation in the near future. All of them are demersal and live on the narrow shelves and banks of the Antarctic and sub-Antarctic islands and on parts of the continental shelf. The systematic inventory of the Antarctic ichthyofauna seems to be almost complete and some physiological work has been carried out. In certain species of Nototheniids growth rate is not particularly slow and some information indicates that their food often consists mainly of krill. No reliable figures on abundance, stock density and distribution can be provided for any of the Antarctic fish. Amongst pelagic fishes only Mycotophids are relatively frequent in Antarctic waters. They seem to be krill eaters too. Nothing is known about their population dynamics.

Cephalopods

Squids are frequently found in the catches of pelagic trawls taken north of the Antarctic Convergence; they are extremely rare in the samples taken further south. However, this might be due to the fact that most of the southern catches are taken in near-surface layers. There are large numbers of squids recorded in the stomachs of Sperm Whales, seals and birds, particularly in the vicinity of the

Convergence and north of it. Further studies on the stomach contents of these predator species and the use of multiple types of collecting gear, as well as good records on the deposits of squid beaks at the sea bed, are needed to substantiate the belief that cephalopods constitute as large a resource in the Southern Ocean as they do in other parts of the world ocean. Without reliable data on squid abundance, species composition and life history, no estimate is possible on the potential resource, or on the role of squids as consumers of krill. The stocks of octopus in the Antarctic do not appear to be large.

Benthic Invertebrates
So far there is little indication of large resources of decapod crustacea and bivalves south of the Antarctic Convergence. The continental shelf is possibly too deep to sustain large stocks of macrobenthos. Records in the Palmer archipelago and the Scotia Arc are limited. They indicate the presence of scallops and mytilids but give little indication of substantial stocks of shrimps and prawns. Benthic bivalves and crustaceans are far more important around the northernmost islands where rock lobsters, and particularly lithodids and spider crabs, seem to be abundant. Apart from the resource aspect, benthic studies should be encouraged as important contributions to the understanding of the function of the Antarctic ecosystem; for instance, recent observations made at the shore bases showed that krill is consumed by ophiuroids.

Krill
The discussions of krill concentrated on *Euphausia superba,* which is by far the most important species of Antarctic euphausiid. Other species play a major role only at the edge of the Antarctic continent and in the area north of the Antarctic Convergence. In certain areas other planktonic crustacea— amphipods, for example—are abundant and may be grazed upon by large consumers.

In spite of the extensive work on the distribution and life history of krill carried out during and soon after the main whaling period, there are still major gaps in our knowledge. Areas, depth and intensity of spawning are poorly known. It was suggested that major spawning takes place under the ice on the continental shelf and shelf break and/or in the open ocean. The vertical distribution and horizontal transport of early larval stages from egg to first calyptopis has been described by Marr's hypothesis of development ascent (1962). This has to be confirmed and placed on a more quantitative basis in both time and space (in the light of data on water transport at the relevant depths). There is still much dispute on growth rate of krill, whether krill reaches an age of a little more than two years or almost four years, and whether each female spawns only once in her life-time. Experimental work in enclosures and tanks may help to answer these questions. Furthermore, it is not yet known whether *Euphausia superba* consists of one genetically uniform circumpolar stock or, more likely, of a number of more or less self-sustaining units which differ genetically and in population parameters. Present data show that krill of all stages found in the eastern Weddell drift appear to be expatriates. No mechanism has been determined by which these are returned as eggs or adults to the eastward drift zone.

The present size and annual production of the krill population of the Southern Ocean is unknown except within broad limits. Crude guesses have been based on estimates of primary productivity, ecological efficiency, and the role of krill relative to other herbivores in the region. Recent observations indicate that average primary production of the Southern Ocean is lower than assumed and that other herbivores make up a major part of zooplankton at least in certain regions. Estimates of food consumption of the virgin whale stocks are of the order of 200 million tonnes. While these figures are relatively reliable, the share taken by other predators such as small cetaceans, seals, birds, squids and benthic invertebrates remains largely unknown. Furthermore, we do not know enough of the recent changes in the abundance of those predators to produce estimates of an accuracy which goes beyond the statement that krill production has been of the order of several hundreds of million tonnes prior to depletion of the Antarctic baleen whales. It is not known to what extent man might replace whales in the exploitation and regulation of krill production.

Recent observations on geographical distribution of krill in the Antarctic have mainly confirmed the earlier work by Marr, Mackintosh and others. Modern pelagic trawls and acoustic devices indicate that krill concentrations may be found at greater depths than have been previously reported. Although micronekton sampling suggests the presence of single krill outside major swarms, the numerical importance of these 'individualists' is not known. In general, structure, migration, dynamics and continuity of krill swarms is poorly understood.

Seaweeds

The littoral zones of the Antarctic archipelago and the sub-Antarctic islands are the habitat of large populations of macrophytes such as red algae, agarephytes and particularly large brown kelps. For example, average figures of standing stock of 5-10 kg/m^2 have been recorded for large beds of *Macrocystis* and of *Durvillea* off Kerguelen Island.

In summer, large concentrations of smaller algae develop along the Antarctic continent. At the conference, the potential importance of the algae as a resource for industrial, chemical and pharmaceutical use was stressed, together with its ecological importance as a habitat, a source of detritus and a producer of dissolved organic substances. An increase in research on Antarctic and sub-Antarctic seaweed is desirable.

Resource Utilization

Marine Mammals
Whales were the main resource to be harvested between 1930 and 1960, with annual production in the range 1.5 to 2 million tonnes (with a peak of nearly 3

million tonnes in one season). The sustainable harvest, if all stocks were maintained at their most productive level, would be of the same order. Right, Blue and Humpback whales are very scarce, and now receive complete protection. The combination of low predicted rates of increase, the relatively short period of protection in the case of Blue and Humpback whales, and the lack of even moderately precise estimates of absolute or relative abundance have made it impossible to determine whether or not they are recovering as predicted. The other large whales (Fin, Sei, Minke and Sperm) have been depleted to less and varying extents. The management policies of IWC are now in accord with scientific recommendations.

Elephant and Fur seals were heavily exploited in the 18th and 19th centuries. Annual catches are believed to have been in excess of some hundreds of thousand animals in some years. Since the cessation of large scale sealing, these stocks have increased, some extremely rapidly, in the past 50 years. The true Antarctic seals have never been harvested on a significant scale, and though consideration has been given to possible exploitation, it does not seem likely to occur in the near future.

Fish

Large scale harvesting, principally by USSR vessels, has occurred in the last ten years around the sub-Antarctic islands. Peak catches have been 400,000 tonnes in 1970 in the south Atlantic (probably mainly around South Georgia) and 200,000 tonnes in 1971 in the southern Indian Ocean (probably mainly around Kerguelen). While the available data do not allow detailed assessment, the subsequent decline in catches suggests that the stocks have been affected by the fishery. Catches of Southern Poutassou, or Blue Whiting, have increased from 8,000 tonnes in 1970 to 48,000 tonnes in 1973. The exact area of fishing is not known, though the principal grounds appear to be around New Zealand and in the Scotia Sea. No significant fish catches appear to have been taken around the Antarctic continent.

Large Crustacea

Though these are not abundant, their high price has allowed the development of fisheries around several of the sub-Antarctic islands. Total production of rock lobsters has been some 1,500 tonnes annually, which seems to be almost equal to the maximum productivity of the stocks. Management measures have been introduced into most of the fisheries. Interest has been expressed in various species of crab, but no fishery has developed apart from those around South America.

Cephalopods

Though the resource is believed to be large, the group had no evidence to suggest that significant squid harvesting would start in the Antarctic or sub-Antarctic in the near future.

Krill

Extensive studies have been made on the practicability of krill harvesting, and several thousand tonnes have already been taken. The technological difficulties of locating and catching krill have been overcome to the extent that krill could be caught in large quantities (perhaps several hundred tonnes per vessel per day) provided that the fishing operations did not have to be interrupted to handle and process the catch. The development of a process to produce, at reasonable cost, a krill product that would find a wide market still presents formidable technological problems. For the present, therefore, the prospects are of only small-scale krill harvesting.

Conservation

In addition to the harvesting of whales and seals, for which there are arrangements for conservation through existing international conventions (IWC and the Seal Convention), substantial exploitation of fish in the sub-Antarctic has begun and the large-scale harvesting of krill appears to be a reasonable possibility in the not too distant future. Arrangements should be made as soon as possible to ensure the conservation and rational utilization of these important resources. The group recommended that SCAR should draw the attention of the parties concerned to the need for such arrangements, which should include the collection and reporting of information (especially from commercial operations), scientific studies (especially the assessment of the state of stocks), and agreement on specific management measures and their implementation.

Proposed Scientific Programme

Objectives

An integrated and well co-ordinated Biological Investigation of Marine Antarctic Systems and Stocks (BIOMASS) is proposed. In broad terms the objectives of BIOMASS are:

   1. To provide data and information for the conservation and wise management of the living resources of the Southern Ocean.
   2. To improve our understanding of the complex ecosystem on which the resources depend, and to understand the flow of energy through the system. To achieve these objectives it will be necessary to promote an in-depth study of the individual components of the marine ecosystem, as well as to study the entire system as an integrated whole. For reasons of resource management and basic ecological science, attention will be focussed on those particular components that offer actual or potential opportunities for commercial harvest.

   For each of these resources (krill, squids, fish, marine mammals, lobsters, birds and seaweeds) the main objectives will be to assess:

(i) Standing stock and production
(ii) Basic parameters important in the dynamics of the population (for example, growth, mortality, reproduction)
(iii) Trophic relationships (feeding and predation)
(iv) General biological and ecological characteristics, especially those needed to elucidate the preceding points.

The emphasis given to each line of study will differ for each class of resource, and the BIOMASS programme outlines the detailed objectives for the different resources. The programme also aims to develop a general and theoretical understanding of the system as a whole, including the construction of models describing part or the whole of the ecosystem.

Implementation

Between the period of *Discovery* investigations in the 1930's and the recent work on krill, which has been mainly directed towards studying the possibilities of commercial harvesting of krill, biological research in the open waters of the Southern Ocean had been grossly neglected. Greatly increased research efforts are now required, and should involve the close international co-ordination of activities on board research vessels operating in Antarctic waters, and at shore stations.

Specific activities proposed for biological research vessels include observations of key processes, at times and places at present poorly sampled, and wide-ranging surveys using acoustic and other instruments. Most of the other vessels operating in the Southern Ocean (supply vessels, research vessels carrying out physical oceanography or other studies, and commercial and exploratory fishing vessels) can provide platforms for useful observations, especially those that do not interfere with the ship's main objectives. As a minimum, information resulting from exploratory and commercial fishing vessels must include catch and effort statistics in sufficient detail, sample size, and other biological characteristics of the animals caught.

The principal role of shore stations will be to provide information on the near-shore ecosystem, including long time-series of year-round observations at fixed positions. The possibilities of using remote sensing techniques from aircraft or satellites in order to supplement and extend ship-board or shore-based investigations will be closely studied. Analysis of the information collected in the field will be mainly the responsibility of participating institutions; however, arrangements will be made for the central compilation of some types of data.

International Co-ordination and Co-operation

There exists a number of international organizations which have expressed interest in the resources of the Southern Ocean; several of these have biological programmes of one kind or another. The success of BIOMASS will depend on the establishment of effective planning and co-ordination machinery. The group studied the present structure of international co-operation which is outlined in the BIOMASS proposal. Recommendations regarding the international planning and co-ordination of BIOMASS are summarized below. While the scientific planning should be done by a non-governmental SCAR/SCOR Group of Specialists, the co-ordination of BIOMASS should be carried out by an international group for BIOMASS under IOC, whose exact status will need to be determined by IOC.

SUMMARY OF RECOMMENDATIONS OF SCAR/SCOR GROUP ON LIVING RESOURCES OF THE SOUTHERN OCEAN

1. That SCAR and SCOR approve the following amended terms of reference for the Group of Specialists on the Living Resources of the Southern Ocean (SCOR WG 54).
   (a) To encourage and stimulate investigations of the trophodynamics of the Antarctic marine ecosystem and the ecology and population dynamics of organisms at different trophic levels.
   (b) To keep under review the current state of knowledge concerning the Antarctic marine ecosystem from the viewpoint of structure, biomass of organisms, dynamic processes at different trophic levels, and prospects and consequences of exploitation of the marine living resources of the Southern Ocean.
   (c) To advise SCAR and SCOR and through them other international organizations on scientific matters related to the study of the ecosystem and the living resources of the Southern Ocean, and in particular to respond to relevant recommendations of the Antarctic Treaty Consultative Meetings and IOC.
   (d) To act as the international scientific planning group for BIOMASS.
   (e) To recommend standardized methods, techniques and data research for biological investigations in the Southern Ocean.
2. That IOC undertakes the international co-ordination of BIOMASS.
3. That IOC requests countries carrying out research in the Southern Ocean to provide details of proposed cruise tracks and scheduled researches, which would be made available to the Group of Specialists.
4. That SCAR requests the National Agencies operating supply ships to institute a circum-Antarctic programme of underway observations of surface temperature, salinity, chlorophyll and underway collection of records of XBT and biological echo traces.
5. That SCAR collaborates with FAO in drawing the attention of all parties engaged in the exploration and exploitation of living resources of the Southern Ocean to the need for detailed catch and effort statistics to be submitted to FAO.
6. That SCAR informs FAO of its approval of the proposed northward

movement of the boundary lines between statistical areas in the Atlantic and Indian oceans, and that in the interim period before the new regions are formally approved by all interested parties, countries should be requested to distinguish separately, when reporting to FAO, the catches taken (a) in the south Atlantic in the area bounded by 50° to 60°S in 20° to 50°W and 55° to 60°S in 50° to 60°W, and (b) in the Indian Ocean between 40° and 50°S in 30° to 80°E.

7. That SCAR and SCOR should agree as soon as possible on the publication of selected documents submitted as working material for the Woods Hole meetings.

References

GREEN, K. A. 1975. *Simulation of pelagic ecosystem of the Ross Sea, Antarctica: a time varying compartmental model.* PhD dissertation, Texas A & M University, p 187.

MARR, J. W. S. 1962. The natural history and geography of the Antarctic krill *(Euphausia superba* Dana). *Discovery Reports,* No 32, p 33–464.

ACRONYMS

CGMW   Commission for the Geological Map of the World
ICG   Inter-Union Commission on Geodynamics
IOC   Intergovernmental Oceanographic Commission
IUGS   International Union of Geological Sciences
UNDP   United Nations Development Program

Selected Documents and Proceedings

Convention on Limitation of Liability for Maritime Claims, 1976

THE STATES PARTIES TO THIS CONVENTION,

HAVING RECOGNIZED the desirability of determining by agreement certain uniform rules relating to the limitation of liability for maritime claims,

HAVE DECIDED to conclude a Convention for this purpose and have thereto agreed as follows:

CHAPTER 1. THE RIGHT OF LIMITATION
Article 1
Persons entitled to limit liability

1. Shipowners and salvors, as hereinafter defined, may limit their liability in accordance with the rules of this Convention for claims set out in Article 2.
2. The term "shipowner" shall mean the owner, charterer, manager and operator of a seagoing ship.
3. Salvor shall mean any person rendering services in direct connexion with salvage operations. Salvage operations shall also include operations referred to in Article 2, paragraph 1(d), (e) and (f).
4. If any claims set out in Article 2 are made against any person for whose act, neglect or default the shipowner or salvor is responsible, such person shall be entitled to avail himself of the limitation of liability provided for in this Convention.
5. In this Convention the liability of a shipowner shall include liability in an action brought against the vessel herself.
6. An insurer of liability for claims subject to limitation in accordance with the rules of this Convention shall be entitled to the benefits of this Convention to the same extent as the assured himself.
7. The act of invoking limitation of liability shall not constitute an admission of liability.

Article 2
Claims subject to limitation

1. Subject to Articles 3 and 4 the following claims, whatever the basis of liability may be, shall be subject to limitation of liability:
   (a) claims in respect of loss of life or personal injury or loss of or damage to property (including damage to harbour works, basins and waterways and aids to navigation), occurring on board or in direct connexion with the operation of the ship or with salvage operations, and consequential loss resulting therefrom;

(b) claims in respect of loss resulting from delay in the carriage by sea of cargo, passengers or their luggage;
(c) claims in respect of other loss resulting from infringement of rights other than contractual rights, occurring in direct connexion with the operation of the ship or salvage operations;
(d) claims in respect of the raising, removal, destruction or the rendering harmless of a ship which is sunk, wrecked, stranded or abandoned, including anything that is or has been on board such ship;
(e) claims in respect of the removal, destruction or the rendering harmless of the cargo of the ship;
(f) claims of a person other than the person liable in respect of measures taken in order to avert or minimize loss for which the person liable may limit his liability in accordance with this Convention, and further loss caused by such measures.

2. Claims set out in paragraph 1 shall be subject to limitation of liability even if brought by way of recourse or for indemnity under a contract or otherwise. However, claims set out under paragraph 1(d), (e) and (f) shall not be subject to limitation of liability to the extent that they relate to remuneration under a contract with the person liable.

Article 3
Claims excepted from limitation

The rules of this Convention shall not apply to:
(a) claims for salvage or contribution in general average;
(b) claims for oil pollution damage within the meaning of the International Convention on Civil Liability for Oil Pollution Damage, dated 29 November 1969 or of any amendment or Protocol thereto which is in force;
(c) claims subject to any international convention or national legislation governing or prohibiting limitation of liability for nuclear damage;
(d) claims against the shipowner of a nuclear ship for nuclear damage;
(e) claims by servants of the shipowner or salvor whose duties are connected with the ship or the salvage operations, including claims of their heirs, dependants or other persons entitled to make such claims, if under the law governing the contract of service between the shipowner or salvor and such servants the shipowner or salvor is not entitled to limit his liability in respect of such claims, or if he is by such law only permitted to limit his liability to an amount greater than that provided for in Article 6.

Article 4
Conduct barring limitation

A person liable shall not be entitled to limit his liability if it is proved that the loss resulted from his personal act or omission, committed with the intent to cause such loss, or recklessly and with knowledge that such loss would probably result.

Article 5
Counterclaims

Where a person entitled to limitation of liability under the rules of this Convention has a claim against the claimant arising out of the same occurrence, their respective claims shall be set off against each other and the provisions of this Convention shall only apply to the balance, if any.

CHAPTER II. LIMITS OF LIABILITY
Article 6
The general limits

1. The limits of liability for claims other than those mentioned in Article 7, arising on any distinct occasion, shall be calculated as follows:
    (a) in respect of claims for loss of life or personal injury,
        (i) 333,000 Units of Account for a ship with a tonnage not exceeding 500 tons,
        (ii) for a ship with a tonnage in excess thereof, the following amount in addition to that mentioned in (i):
            for each ton from 501 to 3,000 tons, 500 Units of Account;
            for each ton from 3,001 to 30,000 tons, 333 Units of Account;
            for each ton from 30,001 to 70,000 tons, 250 Units of Account; and
            for each ton in excess of 70,000 tons, 167 Units of Account,
    (b) in respect of any other claims,
        (i) 167,000 Units of Account for a ship with a tonnage not exceeding 500 tons,
        (ii) for a ship with a tonnage in excess thereof the following amount in addition to that mentioned in (i):
            for each ton from 501 to 30,000 tons, 167 Units of Account;
            for each ton from 30,001 to 70,000 tons, 125 Units of Account; and
            for each ton in excess of 70,000 tons, 83 Units of Account.
2. Where the amount calculated in accordance with paragraph 1(a) is insufficient to pay the claims mentioned therein in full, the amount calculated in accordance with paragraph 1(b) shall be available for payment of the unpaid balance of claims under paragraph 1(a) and such unpaid balance shall rank rateably with claims mentioned under paragraph 1(b).
3. However, without prejudice to the right of claims for loss of life or personal injury according to paragraph 2, a State Party may provide in its national law that claims in respect of damage to harbour works, basins and waterways and aids to navigation shall have such priority over other claims under paragraph 1(b) as is provided by that law.

4. The limits of liability for any salvor not operating from any ship or for any salvor operating solely on the ship to, or in respect of which he is rendering salvage services, shall be calculated according to a tonnage of 1,500 tons.

5. For the purpose of this Convention the ship's tonnage shall be the gross tonnage calculated in accordance with the tonnage measurement rules contained in Annex I of the International Convention on Tonnage Measurement of Ships, 1969.

Article 7
The limit for passenger claims

1. In respect of claims arising on any distinct occasion for loss of life or personal injury to passengers of a ship, the limit of liability of the shipowner thereof shall be an amount of 46,666 Units of Account multiplied by the number of passengers which the ship is authorized to carry according to the ship's certificate, but not exceeding 25 million Units of Account.

2. For the purpose of this Article "claims for loss of life or personal injury to passengers of a ship" shall mean any such claims brought by or on behalf of any person carried in that ship:
   (a) under a contract of passenger carriage, or
   (b) who, with the consent of the carrier, is accompanying a vehicle or live animals which are covered by a contract for the carriage of goods.

Article 8
Unit of Account

1. The Unit of Account referred to in Articles 6 and 7 is the Special Drawing Right as defined by the International Monetary Fund. The amounts mentioned in Articles 6 and 7 shall be converted into the national currency of the State in which limitation is sought, according to the value of that currency at the date the limitation fund shall have been constituted, payment is made, or security is given which under the law of that State is equivalent to such payment. The value of a national currency in terms of the Special Drawing Right, of a State Party which is a member of the International Monetary Fund, shall be calculated in accordance with the method of valuation applied by the International Monetary Fund in effect at the date in question for its operations and transactions. The value of a national currency in terms of the Special Drawing Right, of a State Party which is not a member of the International Monetary Fund, shall be calculated in a manner determined by that State Party.

2. Nevertheless, those States which are not members of the International Monetary Fund and whose law does not permit the application of the provisions of paragraph 1 may, at the time of signature without reservation as to ratification, acceptance or approval or at the time of ratification, acceptance, approval or accession or at any time thereafter, declare that the limits of liability provided for in this Convention to be applied in their territories shall be fixed as follows:

(a) in respect of Article 6, paragraph 1(a) at an amount of:
  (i) 5 million monetary units for a ship with a tonnage not exceeding 500 tons;
  (ii) for a ship with a tonnage in excess thereof, the following amount in addition to that mentioned in (i):
  for each ton from 501 to 3,000 tons, 7,500 monetary units;
  for each ton from 3,001 to 30,000 tons, 5,000 monetary units;
  for each ton from 30,001 to 70,000 tons, 3,750 monetary units; and
  for each ton in excess of 70,000 tons, 2,500 monetary units; and
(b) in respect of Article 6, paragraph 1(b), at an amount of:
  (i) 2.5 million monetary units for a ship with a tonnage not exceeding 500 tons;
  (ii) for a ship with a tonnage in excess thereof, the following amount in addition to that mentioned in (i):
  for each ton from 501 to 30,000 tons, 2,500 monetary units;
  for each ton from 30,001 to 70,000 tons, 1,850 monetary units; and
  for each ton in excess of 70,000 tons, 1,250 monetary units; and
(c) in respect of Article 7, paragraph 1, at an amount of 700,000 monetary units multiplied by the number of passengers which the ship is authorized to carry according to its certificate, but not exceeding 375 million monetary units.

Paragraphs 2 and 3 of Article 6 apply correspondingly to sub-paragraphs (a) and (b) of this paragraph.

3. The monetary unit referred to in paragraph 2 corresponds to sixty-five and a half milligrammes of gold of millesimal fineness nine hundred. The conversion of the amounts referred to in paragraph 2 into the national currency shall be made according to the law of the State concerned.

4. The calculation mentioned in the last sentence of paragraph 1 and the conversion mentioned in paragraph 3 shall be made in such a manner as to express in the national currency of the State Party as far as possible the same real value for the amounts in Articles 6 and 7 as is expressed there in units of account. States Parties shall communicate to the depositary the manner of calculation pursuant to paragraph 1, or the result of the conversion in paragraph 3, as the case may be, at the time of the signature without reservation as to ratification, acceptance or approval, or when depositing an instrument referred to in Article 16 and whenever there is a change in either.

Article 9
Aggregation of claims

1. The limits of liability determined in accordance with Article 6 shall apply to the aggregate of all claims which arise on any distinct occasion:
  (a) against the person or persons mentioned in paragraph 2 of Article 1

and any person for whose act, neglect or default he or they are responsible; or
(b) against the shipowner of a ship rendering salvage services from that ship and the salvor or salvors operating from such ship and any person for whose act, neglect or default he or they are responsible; or
(c) against the salvor or salvors who are not operating from a ship or who are operating solely on the ship to, or in respect of which, the salvage services are rendered and any person for whose act, neglect or default he or they are responsible.

2. The limits of liability determined in accordance with Article 7 shall apply to the aggregate of all claims subject thereto which may arise on any distinct occasion against the person or persons mentioned in paragraph 2 of Article 1 in respect of the ship referred to in Article 7 and any person for whose act, neglect or default he or they are responsible.

Article 10
Limitation of liability without constitution of a limitation fund

1. Limitation of liability may be invoked notwithstanding that a limitation fund as mentioned in Article 11 has not been constituted. However, a State Party may provide in its national law that, where an action is brought in its Courts to enforce a claim subject to limitation, a person liable may only invoke the right to limit liability if a limitation fund has been constituted in accordance with the provisions of this Convention or is constituted when the right to limit liability is invoked.
2. If limitation of liability is invoked without the constitution of a limitation fund, the provisions of Article 12 shall apply correspondingly.
3. Questions of procedure arising under the rules of this Article shall be decided in accordance with the national law of the State Party in which action is brought.

CHAPTER III. THE LIMITATION FUND
Article 11
Constitution of the fund

1. Any person alleged to be liable may constitute a fund with the Court or other competent authority in any State Party in which legal proceedings are instituted in respect of claims subject to limitation. The fund shall be constituted in the sum of such of the amounts set out in Articles 6 and 7 as are applicable to claims for which that person may be liable, together with interest thereon from the date of the occurrence giving rise to the liability until the date of the constitution of the fund. Any fund thus constituted shall be available only for the payment of claims in respect of which limitation of liability can be invoked.
2. A fund may be constituted, either by depositing the sum, or by producing a guarantee acceptable under the legislation of the State Party where the fund is

constituted and considered to be adequate by the Court or other competent authority.

3. A fund constituted by one of the persons mentioned in paragraph 1(a), (b) or (c) or paragraph 2 of Article 9 or his insurer shall be deemed constituted by all persons mentioned in paragraph 1(a), (b) or (c) or paragraph 2, respectively.

Article 12
Distribution of the fund

1. Subject to the provisions of paragraphs 1, 2 and 3 of Article 6 and of Article 7, the fund shall be distributed among the claimants in proportion to their established claims against the fund.

2. If, before the fund is distributed, the person liable, or his insurer, has settled a claim against the fund such person shall, up to the amount he has paid, acquire by subrogation the rights which the person so compensated would have enjoyed under this Convention.

3. The right of subrogation provided for in paragraph 2 may also be exercised by persons other than those therein mentioned in respect of any amount of compensation which they may have paid, but only to the extent that such subrogation is permitted under the applicable national law.

4. Where the person liable or any other person establishes that he may be compelled to pay, at a later date, in whole or in part any such amount of compensation with regard to which such person would have enjoyed a right of subrogation pursuant to paragraphs 2 and 3 had the compensation been paid before the fund was distributed, the Court or other competent authority of the State where the fund has been constituted may order that a sufficient sum shall be provisionally set aside to enable such person at such later date to enforce his claim against the fund.

Article 13
Bar to other actions

1. Where a limitation fund has been constituted in accordance with Article 11, any person having made a claim against the fund shall be barred from exercising any right in respect of such claim against any other assets of a person by or on behalf of whom the fund has been constituted.

2. After a limitation fund has been constituted in accordance with Article 11, any ship or other property, belonging to a person on behalf of whom the fund has been constituted, which has been arrested or attached within the jurisdiction of a State Party for a claim which may be raised against the fund, or any security given, may be released by order of the Court or other competent authority of such State. However, such release shall always be ordered if the limitation fund has been constituted:
    (a) at the port where the occurrence took place, or, if it took place out of port, at the first port of call thereafter; or

(b) at the port of disembarkation in respect of claims for loss of life or personal injury; or
(c) at the port of discharge in respect of damage to cargo; or
(d) in the State where the arrest is made.

3. The rules of paragraphs 1 and 2 shall apply only if the claimant may bring a claim against the limitation fund before the Court administering that fund and the fund is actually available and freely transferable in respect of that claim.

Article 14
Governing law

Subject to the provisions of this Chapter the rules relating to the constitution and distribution of a limitation fund, and all rules and procedure in connexion therewith, shall be governed by the law of the State Party in which the fund is constituted.

CHAPTER IV. SCOPE OF APPLICATION
Article 15

1. This Convention shall apply whenever any person referred to in Article 1 seeks to limit his liability before the Court of a State Party or seeks to procure the release of a ship or other property or the discharge of any security given within the jurisdiction of any such State. Nevertheless, each State Party may exclude wholly or partially from the application of this Convention any person referred to in Article 1 who at the time when the rules of this Convention are invoked before the Courts of that State does not have his habitual residence in a State Party or does not have his principal place of business in a State Party or any ship in relation to which the right of limitation is invoked or whose release is sought and which does not at the time specified above fly the flag of a State Party.

2. A State Party may regulate by specific provisions of national law the system of limitation of liability to be applied to vessels which are:
   (a) according to the law of that State, ships intended for navigation on inland waterways;
   (b) ships of less than 300 tons.

A State Party which makes use of the option provided for in this paragraph shall inform the depositary of the limits of liability adopted in its national legislation or of the fact that there are none.

3. A State Party may regulate by specific provisions of national law the system of limitation of liability to be applied to claims arising in cases in which interests of persons who are nationals of other States Parties are in no way involved.

4. The Courts of a State Party shall not apply this Convention to ships constructed for, or adapted to, and engaged in, drilling:
   (a) when that State has established under its national legislation a higher limit of liability than that otherwise provided for in Article 6; or
   (b) when that State has become party to an international convention regulating the system of liability in respect of such ships.

In a case to which sub-paragraph (a) applies that State Party shall inform the depositary accordingly.
5. This Convention shall not apply to:
   (a) air-cushion vehicles;
   (b) floating platforms constructed for the purpose of exploring or exploiting the natural resources of the sea-bed or the sub-soil thereof.

CHAPTER V. FINAL CLAUSES
Article 16
Signature, ratification and accession

1. This Convention shall be open for signature by all States at the Headquarters of the Inter-Governmental Maritime Consultative Organization (hereinafter referred to as "the Organization") from 1 February 1977 until 31 December 1977 and shall thereafter remain open for accession.
2. All States may become parties to this Convention by:
   (a) signature without reservation as to ratification, acceptance or approval; or
   (b) signature subject to ratification, acceptance or approval followed by ratification, acceptance or approval; or
   (c) accession.
3. Ratification, acceptance, approval or accession shall be effected by the deposit of a formal instrument to that effect with the Secretary-General of the Organization (hereinafter referred to as "the Secretary-General").

Article 17
Entry into force

1. This Convention shall enter into force on the first day of the month following one year after the date on which twelve States have either signed it without reservation as to ratification, acceptance or approval or have deposited the requisite instruments of ratification, acceptance, approval or accession.
2. For a State which deposits an instrument of ratification, acceptance, approval or accession, or signs without reservation as to ratification, acceptance or approval, in respect of this Convention after the requirements for entry into force have been met but prior to the date of entry into force, the ratification, acceptance, approval or accession or the signature without reservation as to ratification, acceptance or approval, shall take effect on the date of entry into force of the Convention or on the first day of the month following the ninetieth day after the date of the signature or the deposit of the instrument, whichever is the later date.
3. For any State which subsequently becomes a Party to this Convention, the Convention shall enter into force on the first day of the month following the expiration of ninety days after the date when such State deposited its instrument.
4. In respect of the relations between States which ratify, accept, or approve this Convention or accede to it, this Convention shall replace and abrogate the

International Convention relating to the Limitation of the Liability of Owners of Sea-going Ships, done at Brussels on 10 October 1957, and the International Convention for the Unification of certain Rules relating to the Limitation of Liability of the Owners of Sea-going Vessels, signed at Brussels on 25 August 1924.

Article 18
Reservations

1. Any State may, at the time of signature, ratification, acceptance, approval or accession, reserve the right to exclude the application of Article 2, paragraph 1(d) and (e). No other reservations shall be admissible to the substantive provisions of this Convention.
2. Reservations made at the time of signature are subject to confirmation upon ratification, acceptance or approval.
3. Any State which has made a reservation to this Convention may withdraw it at any time by means of a notification addressed to the Secretary-General. Such withdrawal shall take effect on the date the notification is received. If the notification states that the withdrawal of a reservation is to take effect on a date specified therein, and such date is later than the date the notification is received by the Secretary-General, the withdrawal shall take effect on such later date.

Article 19
Denunciation

1. This Convention may be denounced by a State Party at any time after one year from the date on which the Convention entered into force for that Party.
2. Denunciation shall be effected by the deposit of an instrument with the Secretary-General.
3. Denunciation shall take effect on the first day of the month following the expiration of one year after the date of deposit of the instrument, or after such longer period as may be specified in the instrument.

Article 20
Revision and amendment

1. A Conference for the purpose of revising or amending this Convention may be convened by the Organization.
2. The Organization shall convene a Conference of the States Parties to this Convention for revising or amending it at the request of not less than one-third of the Parties.
3. After the date of the entry into force of an amendment to this Convention, any instrument of ratification, acceptance, approval or accession deposited shall be deemed to apply to the Convention as amended, unless a contrary intention is expressed in the instrument.

Article 21
Revision of the limitation amounts and of Unit of Account or monetary unit

1. Notwithstanding the provisions of Article 20, a Conference only for the

purposes of altering the amounts specified in Articles 6 and 7 and in Article 8, paragraph 2, or of substituting either or both of the Units defined in Article 8, paragraphs 1 and 2, by other units shall be convened by the Organization in accordance with paragraphs 2 and 3 of this Article. An alteration of the amounts shall be made only because of a significant change in their real value.
2. The Organization shall convene such a Conference at the request of not less than one-fourth of the States Parties.
3. A decision to alter the amounts or to substitute the Units by other units of account shall be taken by a two-thirds majority of the States Parties present and voting in such Conference.
4. Any State depositing its instruments of ratification, acceptance, approval or accession to the Convention, after entry into force of an amendment, shall apply the Convention as amended.

Article 22
Depositary

1. This Convention shall be deposited with the Secretary-General.
2. The Secretary-General shall:
   (a) transmit certified true copies of this Convention to all States which were invited to attend the Conference on Limitation of Liability for Maritime Claims and to any other States which accede to this Convention;
   (b) inform all States which have signed or acceded to this Convention of:
      (i) each new signature and each deposit of an instrument and any reservation thereto together with the date thereof;
      (ii) the date of entry into force of this Convention or any amendment thereto;
      (iii) any denunciation of this Convention and the date on which it takes effect;
      (iv) any amendment adopted in conformity with Articles 20 or 21;
      (v) any communication called for by any Article of this Convention.
3. Upon entry into force of this Convention, a certified true copy thereof shall be transmitted by the Secretary-General to the Secretariat of the United Nations for registration and publication in accordance with Article 102 of the Charter of the United Nations.

Article 23
Languages

This Convention is established in a single original in the English, French, Russian and Spanish languages, each text being equally authentic.

DONE AT LONDON this nineteenth day of November one thousand nine hundred and seventy-six.

IN WITNESS WHEREOF the undersigned being duly authorized for that purpose have signed this Convention.

Appendix C

Directory of Institutions

This Directory is based upon a register of ocean-related institutions maintained by the Food and Agriculture Organization. Due to limitations of space, this first *Ocean Yearbook* Directory lists only 561 of the more than 1,375 institutions found in the *International Directory of Marine Scientists 1977*, which is now available from the Department of Fisheries, FAO, and from the Secretary IOC.

At least one institution, whatever its size, is shown for each country listed in the register. Where two or more institutions are listed in a given country, only those employing more than four scientists have been included here. However, all international bodies in the register are noted.

When a given institution has two or more levels of ocean-related organizations within it, as in the cases of large universities or government agencies, a partial list of these organizations will be found in parentheses following the parent institution. This means a relevant organization might have been omitted in some listings.

Because diacritical marks were not found in the register, they have been omitted from this Directory.

The methods for its compilation and condensation will be reviewed for subsequent editions of the *Ocean Yearbook*.

Abbreviations used in the English-language entries: Admin. = Administration; Assoc. = Association; Bd. = Board; Bldg. = Building; Br. = Branch; Bur. = Bureau; Comm. = Commission; Coll. = College; Ctr. = Center, Centre; Corp. = Corporation; Dept. = Department; Div. = Division; Found. = Foundation; Inst. = Institute; Lab. = Laboratory; Ltd. = Limited; Min. = Ministry; Obs. = Observatory; Proj. = Project; Rep. = Republic; Res. = Research; Sch. = School; Sta. = Station; Univ. = University.

THE EDITORS

Algeria
CENTRE DE RECHERCHES OCEANOGRAPHIQUES ET DES PECHES, Alger

Antilles (Netherlands)
CARIBBEAN MARINE BIOLOGICAL INST., Piscadera Bay, P.O. Box 2090, Curacao

Argentina
CENTRO DE INVESTIGACION DE BIOLOGIA MARINA, Libertad 1235, Buenos Aires
INSTITUTO ANTARTICO ARGENTINO, Cerrito 1248, Buenos Aires
INSTITUTO DE BIOLOGIA MARINA, Casilla de Correo 175, Playa Grande, Mar del Plata
MUSEO ARGENTINO DE CIENCIAS NATURALES BERNARDINO RIVADAVIA, Avenida Angel Gallardo 470, Buenos Aires
SERVICIO DE HIDROGRAFIA NAVAL, Avenida Montes de Oca 2124, Buenos Aires

Australia
AUSTRALIAN INST. OF MARINE SCIENCE, P.O. Box 1104, Townsville, Qld. 4810
AUSTRALIAN MUSEUM, P.O. Box A285, Sydney South, N.S.W. 2000
AUSTRALIAN NATIONAL UNIV. (Dept. of Geology, Res. Sch. of Biological Sciences, Res. Sch. of Earth Sciences, Res. Sch. of Pacific Studies), Canberra City 2600
BUR. OF MINERAL RESOURCES, GEOLOGY, AND GEOPHYSICS, Box 378, Canberra City 2601
CSIRO (Div. of Fisheries and Oceanography), P.O. Box 21, Cronulla, Sydney, N.S.W. 2230; P.O. Box 20, Cleveland Brisbane, Qld. 4163; P.O. Box 20, North Beach Perth, W.A. 6020
DEPT. OF AGRICULTURE AND FISHERIES (Fisheries Br.), G.P.O. Box 1191, Adelaide, S.A. 5001
DEPT. OF FISHERIES AND WILDLIFE (Western Australian Marine Res. Lab.), P.O. Box 20, North Beach, W.A. 6020
MIN. FOR CONSERVATION (Dept. of Fisheries and Wildlife, Environmental Studies Unit), East Melbourne, Vic. 3002
UNIV. OF MELBOURNE (Dept. of Biochemistry, Dept. of Geography, Dept. of Zoology, Sch. of Botany), Parkville, Vic. 3052
UNIV. OF MONASH (Dept. of Earth Sciences, Dept. of Mechanical Engineering, Dept. of Zoology, Geophysical Fluid Dynamics Lab.), Melbourne, Vic. 3168
UNIV. OF NEW SOUTH WALES (Dept. of Zoology, Sch. of Applied Geology, Sch. of Biological Technology, Sch. of Mathematics), Sydney, N.S.W. 2033
UNIV. OF NORTH QUEENSLAND (Dept. of Botany, Sch. of Biological Sciences), Townsville, Qld. 4810
UNIV. OF QUEENSLAND (Dept. of Botany, Dept. of Microbiology, Dept. of Zoology), Brisbane, Qld. 4067
WESTERN AUSTRALIAN MUSEUM, Francis St., Perth, W.A. 6000

Bermuda
BERMUDA BIOLOGICAL STA. FOR RES., INC., St. George's 1-15
PALISADES GEOPHYSICAL INST. SOFAR STA., St. David's

Brazil
BASE DE PESQUISA CANANEIA, Av. Prof. Wladimir Besnard, 11990 Cananeia, Sao Paulo
CENTRO DE PESQUISA ICTIOLOGICAS DA DIRETORIA DE PESCA E PISCICULTURA DO DNOCS, Av. Duque de Caxias, 1700, 60000 Fortaleza Ceara
COMPANHIA DE PESQUISA E RECURSOS MINERAIS CPRM, Av. Pasteur, 404, 20000 Rio de Janeiro RJ
DIRETORIA DE HIDROGRAFIA E NAVEGACAO, 1. Distrito Naval, Ilha Fiscal, 20000 Rio de Janeiro
FUNDACAO UNIVERSIDADE DO RIO GRANDE, R. Luiz Lorea, 261, Rio Grande 96200 RS
PONTIFICIA UNIVERSIDADE CATOLICA DO RS. MUSEU DE CIENCIAS, Av. Ipiranga, 6681, Porto Alegre 90000 RS
PROJETO REMAC, R. Ramon Franco, 49 Urca RJ
SECRETARIA AGRICULTURA (Instituto de Pesca), Av. Bartolomeu Gusmao, 192, 11100 Santos S.P.
UNIVERSIDADE FEDERAL RIO DE JANEIRO, Bloco A-CCS, Ilha do Fundao, 20000 Rio de Janeiro
UNIVERSIDADE DE SAO PAULO (Instituto Oceanografico), Cidade Universitaria, Butanta, Sao Paulo
UNIVERSIDADE FEDERAL DE PERNAMBUCO (Departamento de Oceanografia), Av. Bernardo Vieira de Melo, 986, Piedade, Recife 50000 Pernambuco
UNIVERSIDADE FEDERAL DO CEARA (Laboratorio de Ciencias do Mar), Av. Abolicao, 3207, Fortaleza, Ceara
UNIVERSIDADE FEDERAL RIO GRANDE DU SUL (Centro de Estudos de Geologia Costeira e Oceanica), Av. Paulo Gama Sn, Porto Alegre 90000 RS

Canada
B. C. RES. (Ctr. for Ocean Engineering), 3650 Wesbrook Mall, Vancouver, British Columbia V6S 2L2
BEDFORD INST. OF OCEANOGRAPHY (Atlantic Geoscience Ctr., Atlantic Oceanographic Lab.), P.O. Box 1006, Dartmouth, Nova Scotia B2Y 4A2

DALHOUSIE UNIV. (Dept. of Biology, Dept. of Geology, Dept. of Oceanography), Halifax, Nova Scotia
DEPT. OF ENERGY, MINES AND RESOURCES (Canada Ctr. for Remote Sensing, Gravity and Geodynamics Div., Polar Continental Shelf Proj., Terrain Sciences Div.), 2464 Sheffield Rd., Ottawa, Ontario K1A; (Regional and Economic Geology Div.), 100 W. Pender St., Vancouver, British Columbia V6B 1R8
DEPT. OF NATIONAL DEFENCE (Defence Res. Bd.), Fleet Mail Office, Victoria, British Columbia V0S 1B0; Ottawa, Ontario K1A 0Z4
DEPT. OF THE ENVIRONMENT (Arctic Biological Sta.), P.O. Box 400, Ste. Anne de Bellevue, Quebec H9X 3L6; (Bedford Inst. of Oceanography), P.O. Box 1006, Dartmouth, Nova Scotia B2Y 4A2; (Biological Sta.), St. Andrews, New Brunswick E0G 2X0; (Biological Sta.), St. John's, Newfoundland A1C 1A1; (Fisheries Res. Br., Marine Sciences and Information Directorate, Ocean and Aquatic Science Affairs Br., Resource Services Directorate, Scientific Information and Publications Br.), Ottawa, Ontario K1A 0H3; (Halifax Lab.), P.O. Box 429, Halifax, Nova Scotia; (Inst. of Ocean Sciences), 1230 Government St., Victoria, British Columbia V8W 1Y4; (Pacific Biological Sta.), P.O. Box 100, Nanaimo, British Columbia V9R 5K6; (Pacific Environment Inst.), 4160 Marine Dr., West Vancouver, _ritish Columbia V7V 1N6; (Vancouver Lab.), 6640 N.W. Marine Dr., Vancouver, British Columbia V6T 1X2; (Ocean and Aquatic Sciences), Burlington, Ontario L7R 4A6
McGILL UNIV. (Marine Sciences Ctr.), P.O. Box 6070, Sta. A, Montreal, Quebec H3C 3G1
MEMORIAL UNIV. OF NEWFOUNDLAND (Ctr. for Cold Ocean Resources Engineering, Faculty of Engineering and Applied Science, Faculty of Science, Marine Sciences Res. Lab.), St. John's, Newfoundland A1C 5S7
MINISTERE DE L'INDUSTRIE ET DU COMMERCE (Direction Generale des Peches Maritimes), 2700, Rue Einstein, Quebec, Quebec G1P 3W8
NATIONAL MUSEUM OF NATURAL SCIENCES (Botany Div., Canadian Aquatic Identification Ctr., Zoology Div.), Ottawa, Ontario K1A 0M8
NATIONAL RES. COUNCIL OF CANADA (Atlantic Regional Lab.), Halifax, Nova Scotia B3H 3Z1

NOVA SCOTIA RES. FOUND. CORP., P.O. Box 790, Dartmouth, Nova Scotia B2Y 3Z7
UNIVERSITE DU QUEBEC (Institut National de la Recherche Scientifique Oceanologie, Section d'Oceanographie), 310, Avenue des Ursulines, Rimouski, Quebec G5L 3A1
UNIVERSITE LAVAL (Department de Biologie, Groupe Interuniversitaire de Recherches Oceanographiques du Quebec), Quebec, Quebec G1K 7P4
UNIV. OF BRITISH COLUMBIA (Dept. of Zoology, Inst. of Animal Resource Ecology, Inst. of Oceanography), Vancouver, British Columbia V6T 1W5
UNIV. OF GUELPH (Coll. of Biological Science), Guelph, Ontario N1G 2W1
UNIV. OF NEW BRUNSWICK (Marine and Estuarine Res. Group), Tucker Park, Saint John, New Brunswick E2L 4L5
UNIV. OF VICTORIA (Dept. of Biology), Victoria, British Columbia V8W 2Y2

Chile

ARMADA DE CHILE (Servicio Meteorologico de la Armada), Correo Naval, El Belloto, Valparaiso
INSTITUTO DE FOMENTO PESQUERO (Departamento de Recursos), Jose Domingo Canas 2277, Casilla 1287, Santiago
INSTITUTO HIDROGRAFICO DE LA ARMADA (Centro Nacional de Datos Oceanograficos de Chile, Departamento de Oceanografia), Casilla 324, Valparaiso
UNIVERSIDAD AUSTRAL DE CHILE (Instituto de Ecologia, Instituto de Zoologia), Casilla 567, Valdivia
UNIVERSIDAD CATOLICA DE CHILE (Departamento de Biologia Ambiental y Poblaciones), Casilla 114-D, Santiago
UNIVERSIDAD CATOLICA DE VALPARAISO (Centro de Investigaciones del Mar, Escuela de Pesquerias y Alimentos), Valparaiso
UNIVERSIDAD DE CHILE (Departamento de Oceanologia), Antofagasta; Santiago
UNIVERSIDAD DE CHILE DE VALPARAISO (Departamento de Oceanologia), Casilla 13-D, Vina del Mar
UNIVERSIDAD DE CONCEPCION (Instituto Central de Biologia), Casilla 1367, Concepcion
UNIVERSIDAD DEL NORTE (Centro de Investigaciones Submarinas), Casilla 117, Coquimbo; (Departamento de Pesquerias), Casilla 1287, Antofagasta

Colombia

ARC '7 DE AGOSTO', Fuerza Naval del Atlantico, Cartagena

ESCUELA NAVAL'ALMIRANTE PADILLA',
  Cartagena
UNIVERSIDAD TECNOLOGICA DEL MAGDALENA,
  A.A. 731, Santa Marta

Costa Rica
UNIVERSIDAD DE COSTA RICA, Ciudad
  Universitatia, San Jose

Cyprus
MIN. OF AGRICULTURE AND NATURAL
  RESOURCES (Dept. of Fisheries), Nicosia

Czechoslovakia
CZECHOSLOVAK ACADEMY OF SCIENCES (Inst.
  of Botany) Brno; (Inst. of Botany,
  Inst. of Parasitology), Prague

Denmark
DANMARKS FISKERI-OG HAVUNDERSOGELSER
  Charlottenlund Slot, 2920 Charlotten-
  lund
GRONLANDS FISKERIUNDERSOGELSER,
  Jaersborg Alle 1-B, 2920 Charlotten-
  lund
UNIVERSITET ZOOLOGISKE MUSEUM,
  Universitetsparken 15 Museum, 2100
  Copenhagen 0
UNIVERSITY OF COPENHAGEN (Marinbiologisk
  Laboratorium), Strandpromenaden, 3000
  Helsingor; (Inst. for Spore Planter,
  Inst. of Petrology, Inst. of Physical
  Oceanography, Inst. of Plant Anatomy
  and Citology), Copenhagen

Dominican Republic
UNIVERSIDAD AUTONOMA DE SANTO DOMINGO,
  Primada de America, Santo Domingo

Finland
INST. OF MARINE RES., Box 166,
  SF-00141 Helsinki 14
UNIV. OF HELSINKI (Dept. of Botany,
  Dept. of Geophysics, Dept. of Lim-
  nology, Dept. of Meteorology, Dept. of
  Radiochemistry, Dept. of Zoology,
  Tvarminne Zoological Sta.), Helsinki
UNIV. OF TURKU (Dept. of Biology, Dept.
  of Geography, Inst. of Archipelago
  Res.), SF-20500 Turku 50

France
CENTRE D'OCEANOGRAPHIE ET STATION MARINE
  D'ENDOUME, Rue de la Batterie des
  Lions, 13007 Marseille
CENTRE NATIONAL POUR L'EXPLOITATION DES
  OCEANS, B.P. 337, 29273 Brest Cedex;
  39 Avenue d'Iena, 75-Paris XVI
CENTRE OCEANOLOGIQUE DU PACIFIQUE,
  B.P. 303, Avenue du Prince Hinoi,
  Papeete, Tahiti
DIRECTION DU S.H.O.M., 3 Avenue
  Octabe Greard, 75007 Paris
INSTITUT SCIENTIFIQUE ET TECHNIQUE DES
  PECHES MARITIMES, B.P. 1049,
  44037 Nantes Cedex
INSTITUT OCEANOGRAPHIQUE, 195, Rue
  Saint Jacques, F-75005 Paris
MUSEUM NATIONAL D'HISTOIRE NATURELLE
  (Centre National d'Histoire pour la
  Protection des Etres Vivants, Labora-
  toire d'Oceanographie Physique,
  Laboratoire de Malacologie, Laboratoire
  des Peches Outre-Mer, Laboratoire de
  Dynamique des Populations Aquatiques),
  Paris
OFFICE DE LA RECHERCHE SCIENTIFIQUE ET
  TECHNIQUE OUTRE-MER, 24, Rue Bayard,
  75008 Paris
SERVICE HYDROGRAPHIQUE ET OCEANO-
  GRAPHIQUE DE LA MARINE, Route de
  Bergot, 29283 Brest
UNIVERSITE DE BORDEAUX (Institut de
  Geologie du Bassin d'Aquitaine,
  Laboratoire de Geophysique Applique a
  l'Oceanographie), 351 Cours de la
  Liberation, 33405 Talence
UNIVERSITE DE BRETAGNE OCCIDENTALE
  (Institut de Droit de l'Economie de
  la Mer, Laboratoire de Physiologie,
  Department de Geographie des Sciences
  de La Mer, Laboratoire d'Oceanographie
  Chimique, Laboratoire d'Oceanologie
  Biologique, Laboratoire de Biologie
  Animale, Laboratoire de Zoologie-
  Aquaculture-Pollutions Marines),
  6, Avenue le Gorgeu, 29283 Brest
  Cedex
UNIVERSITE DE PARIS (Laboratoire
  d'Oceanographie Physique, Laboratoire
  de Biologie Vegetale, Laboratoire de
  Geologie Dynamique), Paris
UNIVERSITE DE PARIS VI (Laboratoire de
  Biologie Vegetale Marine, Laboratoire
  de Cytologie, Laboratoire de Geologie
  Structurale), Paris; (Station Geo-
  dynamique, Station Zoologique), 06230
  Villefranche sur Mer
UNIVERSITE DE PARIS VI (M. ET P. CURIE)
  (Station Biologique de Roscoff),
  29211 Roscoff; (Laboratoire Arago),
  66650 Banyuls sur Mer

German Democratic Republic
AKADEMIE DER WISSENSCHAFTEN DER DDR
  (Institut fur Meereskunde), 253
  Rostock Warnemunde
WILHELM PIECK UNIVERSITAET
  (Sektion Biologie), Wismarsche
  Str. 8, 25 Rostock

German Federal Republic
BIOLOGISCHE ANSTALT HELGOLAND (Zentrale),
  Palmaille 9, 2000 Hamburg 50;

(Litoralstation), 2282 List; (Meeresstation), 2192 Helgoland
BUNDESANSTALT FUER GEOWISSENSCHAFTEN UND ROHSTOFFE, Postfach 54, 3000 Hannover-Buchholz
BUNDESANSTALT FUER GEWAESSERKUNDE, Postfach 309, 54 Koblenz
BUNDESFORSCHUNGSANSTALT FUER FISCHEREI (Informations und Dokumentationsstelle, Institut fuer Biochemie und Technologie, Institut fuer Fangtechnik, Institut fuer Kuesten und Binnenfischerei, Institut fuer Seefischerei, Isotopenlaboratorium), Palmaille 9, 2000 Hamburg 50
DEUTSCHER WETTERDIENST-SEEWETTERANT, Wustland 2, D.2 Hamburg 55
DEUTSCHES HYDROGRAPHISCHES INSTITUT, Bernhard-Noch-Str. 78, 2000 Hamburg 4
FORSCHUNGSSTELLE FUR INSEL-UND KUESTEN-SCHUTZ, An der Muehle 5, 2982 Norderney
INSTITUT FUER MEERESFORSCHUNG, Am Handelshafen 12, 2850 Bremerhaven-G.
INSTITUT FUER MEERESGEOLOGIE UND MEERESBIOLOGIE SENCKENBERG, Schleusenstrasse 39A, 294 Wilhelmshaven
NATURMUSEUM UND FORSCHUNGSINSTITUT, Senckenberg Anlage 25, 6000 Frankfurt am Main 1
TECHNISCHE UNIVERSITAET (Institut fuer Geologie), 8000 Muenchen 2
TECHNISCHE UNIVERSITAET HANNOVER (Franzius-Institut fuer Wasserbau und Kuesteningenieurwesen), Welfengarien 1, 3 Hannover
UNIVERSITAET GOETTINGEN (Geochemisches Institut, Geologisch-Palaeontologisches Institut, Sedimentpetrographisches Institut, Zoologisches Institut und Museum), 34 Goettingen
UNIVERSITAET HAMBURG (Chemische Institut, Geologisch-Palaeontologisches Institut, Institut fuer Allgemeine Botanik, Institut fuer die Physik des Erd- koerpers, Institut fuer Geophysik, Institut fuer Hydrobiologie und Fischereiwissenschaft, Institut fuer Meereskunde, Institut fuer Radiometeorologie und Meteorologie, Institut fuer Schiffbau, Meteorologisches Institut, Zoologisches Institut und Zoologisches Museum), Hamburg
UNIVERSITAET HEIDELBERG (Mineralogisches Institut, Physikalisches Institut), 69 Heidelberg
UNIVERSITAET KIEL (Geologisch-Palaeontologisches Institut, Institut fuer Angewandte Physik, Institut fuer Geophysik, Institut fuer Meereskunde, Institut fuer Reine und Angewandte Kernphysik, Mineralogisch-Petrographisches Institut und Museum, Zoologisches Institut und Museum), Kiel

Ghana
FISHERY RESEARCH UNIT, P.O. Box 62, Tema, Community 2

Honduras
DIRECCION GENERAL DE RECURSOS NATURALES RENOVABLES (Departamento de Pesca), 8A. Av. Entre 11 y 12 Calles, Comayaguela, D.C.

Hong Kong
CANTON ROAD GOVERNMENT OFFICES (Dept. of Agriculture and Fisheries), 12-14 Fl., Canton Rd., Kowloon

Iceland
HAFRANNSOKNASTOFNUNIN, Skulagata 4, P.O. Box 390, 101 Reykjavik

India
BHABHA ATOMIC RES. CTR. (Health Physics Div.), P.O. Barc (Trombay), Bombay 400085
CALCUTTA PORT TRUST (Hydraulic Study Dept.), 20, Garden Reach Rd., Calcutta 43
CENTRAL INST. OF FISHERIES EDUCATION, P.B. N.7392, Kakori Camp, J.P. Rd., Versova, Bombay 58
CENTRAL INST. OF FISHERIES TECHNOLOGY, Kakinada 2; Wellington Islands, Cochin 682003
CENTRAL MARINE FISHERIES RES., Waltair, Vishakhapatnam 530003 (A.P.); 2d Fl., Botawalla Chambers, Sir P.M. Rd., Bombay 1
CENTRAL MARINE FISHERIES RES. INST., P.O. Box N.1912, Cochin 682018; 9, Commander-in-Chief Rd., Egmore, Madras 600008; 93, North Beach Rd., Tuticorin 628001
GEOLOGICAL SURVEY OF INDIA (Offshore Mineral Exploration and Marine Geology Div.), 8th Fl., "B" Block, 4, Chowringee Lane, Calcutta 700016
GOVERNMENT OF INDIA (Exploratory Fisheries Proj.), Botawalla Chambers, Sir P.M. Rd., Bombay 1; (Central Inst. of Fisheries Operatives), Dewan's Rd., Ernakulam, Cochin 16; (Dept. of Agriculture), Krishi Bhavan, New Delhi 110 001
GOVERNMENT OF MAHARASHTRA (Dept. of Fisheries), Netaji Subhash Rd., Bombay 400002
MIN. OF AGRICULTURE AND IRRIGATION (Integrated Fisheries Proj.), P.B. No. 1801, Cochin 16

NATIONAL INST. OF OCEANOGRAPHY,
P.O. NIO, Dona Paula, Goa; Pullepady Cross Rd., P.B. 1913, Ernakulam, Cochin 18; 1st Fl., Sea Shell Bldg., 7 Bungalows, Versova, Bombay 61
NAVAL BASE (Naval Physical and Oceanographic Lab.), Cochin 682004
NAVAL HYDROGRAPHIC OFFICE, 107-A Rajpur Rd., P.B. No. 75, Dehra Dun 248001
PELAGIC FISHERY PROJ., P.B. 1719, Ernakulam, Cochin 682016
UNIV. OF AGRICULTURAL SCIENCE, P.B. No. 2477, Bebbal, Bangalore 560024
UNIV. OF ANDHRA (Dept. of Zoology, Dept. of Geology and Ctr. for Assistance Programme in Marine Geology Sponsored by the U.G.C.), Waltair, Vishakhapatnam 530003
UNIV. OF COCHIN (Dept. of Marine Sciences), Foreshore Rd., Cochin 682016
UNIV. OF MADRAS (Ctr. of Advanced Studies in Botany, Dept. of Zoology), Madras 600005
ZOOLOGICAL SURVEY OF INDIA (School of Ichthyology, Southern Regional Ctr.), Madras

Indonesia

BANDUNG INST. OF TECHNOLOGY (Dept. of Biology), Jalan Ganesha 10, Bandung, West Java
MIN. OF AGRICULTURE (Inst. of Fishery Technology, Marine Fisheries Res. Inst.), Jakarta
NATIONAL INST. OF OCEANOLOGY (Indonesian Inst. of Sciences), P.O. Box 580 Dak, Jakarta
NAVAL HYDRO-OCEANOGRAPHIC SERVICE, Jalan Gunung Sahari 87, Jakarta Pusat

Israel

HEBREW UNIV. OF JERUSALEM (Dept. of Atmospheric Sciences, Dept. of Geology, Dept. of Marine Microbiology, Dept. of Zoology), Jerusalem
ISRAEL OCEANOGRAPHIC AND LIMNOLOGICAL RES. LTD., P.O. Box 1793, Haifa
TEL-AVIV UNIV. (Dept. of Environmental Sciences, Dept. of Geography, Dept. of Zoology), Ramat-Aviv
UNIV. OF HAIFA, Mount Carmel, Haifa; Rehovot
WEIZMANN INST. OF SCIENCE (Isotope Dept.), Rehovot

Japan

EHIME UNIV. (Coll. of Agriculture, Faculty of Engineering), Matsuyama-Shi, Ehime 790
FAR SEAS FISHERIES (Res. Lab.), 1000 Orido, Shimizu, Shizuoka 424
FISHERIES UNIV. OF SHIMONOSEKI, Yoshimi, Shimonoseki-Shi, Yamaguchi 759-65
GEOLOGICAL SURVEY OF JAPAN (Dept. of Marine Geology), Hisamoto, Takatsu-Ku, Kawasaki-Shi, Kanagawa 213
HIROSHIMA UNIV. (Faculty of Fisheries), Fukuyama, Hiroshima 720
HOKKAIDO REGIONAL FISHERIES RES. LAB., Hamanaka 238, Yoichi, Hokkaido 046
HOKKAIDO UNIV. (Faculty of Engineering, Faculty of Science, Inst. of Geology, Inst. of Low Temperature Science), Sapporo 060; (Faculty of Fisheries), Hakodate, Hokkaido 040
INST. OF PHYSICAL AND CHEMICAL RES., Wako-Shi, Saitama 351
JAPAN MARINE SCIENCE AND TECHNOLOGY CTR., Natsushima-Cho, Yokosuka-Shi 237
JAPAN METEOROLOGICAL AGENCY, Otemachi, Chiyoda-Ku, Tokyo 100
JAPAN SEA REGIONAL FISHERIES RES. LAB., Hamaura, Nishifunami-Cho, Niigata 951
JAPAN WEATHER ASSOC., Kanda, Nishiki-Cho, Chiyoda-Ku, Tokyo 101
KAGOSHIMA UNIV. (Coll. of Liberal Arts, Faculty of Engineering, Faculty of Fisheries, Faculty of Science), Kagoshima-Shi 890
KANAZAWA UNIV. (Faculty of Science, Inst. of Geology), Kanazawa 920
KINKI UNIV. (Dept. of Fisheries, Dept. of Chemistry), Kowakae, Higashi, Osaka 577
KOCHI UNIV. (Dept. of Education, Dept. of Science and Literature, Faculty of Agriculture), Kochi
KYOTO UNIV. (Dept. of Chemistry, Disaster Prevention Res. Inst., Dept. of Fisheries, Geophysical Inst., Inst. for Chemical Res., Res. Inst. for Food Science), Kyoto; (Seto Marine Biological Sta.), Shirahama-Cho, Wakayama 649-22
KYUSHU UNIV. (Dept. of Fisheries, Lab. of Earth Science, Fishery Res. Lab., Dept. of Geology, Res. Inst. for Applied Mechanics), Fukuoka 812
MAIZURU MARINE OBSERVATORY, Shimofukui Onobe, Maizuru-Shi, Kyoto 624
METEOROLOGICAL RES. INST., Koenji-Kita 4-35-8, Suginami-Ku, Tokyo 166
MIE UNIV. (Faculty of Agriculture, Faculty of Fisheries), Tsu-Shi, Mie 514
MIN. OF TRANSPORT (Maritime Safety Agency), 5-Chome Tsukiji, Chuo-Ku, Tokyo 104
NAGASAKI UNIV. (Faculty of Fisheries), Bunkyo-Machi 1-14, Nagasaki 852
NAGOYA UNIV. (Dept. of Fisheries, Faculty of Engineering, Geophysical

Inst., Inst. of Earth Science, Water Res. Lab.), Nagoya 464
NANSEI REGIONAL FISHERIES RES. LAB., Sanbashi-Dori 6, Kochi 780; Maruishi, Ohno-Cho, Saeki-Gun, Hiroshima 739-04
NATIONAL INST. OF RADIOLOGICAL SCIENCES, Anagawa, Chiba 280
NATIONAL PEARL OYSTER (Res. Lab.), Kashikojima, Mie-Ken
NATIONAL SCIENCE MUSEUM, Ueno Park, Tokyo 110
NIHON UNIV. (Inst. of Fisheries), Setagaya-Ku, Tokyo 1514
OSAKA CITY UNIV., Osaka-Shi
SEIKAI REGIONAL FISHERIES RES. LAB., Higashi-Yamato-Cho 2, Shimonoseki-Shi, Yamaguchi 750; 49 Kokubu-Cho, Nagasaki 850
TOHOKU REGIONAL FISHERIES RES. LAB., 25 Simomekurakubo Samemachi, Hachinohe-Shi Aomori 031; Niihama-Cho 3 Shiogama, Miyagi 985
TOHOKU UNIV. (Faculty of Agriculture, Inst. of Geology and Palaeontology, Faculty of Science), Sendai-Shi, Miyagi 980
TOKAI REGIONAL FISHERIES RES. LAB., Kachidoki 5, Chuo-Ku, Tokyo 104; Arazaki, Nagai-Cho, Yokosuka, Kanagawa 238-03
TOKAI UNIV. (Coll. of Marine Science and Technology), Orido, Shimizu-Shi, Shizuoka 424
TOKYO KYOIKU UNIV. (Botanical Inst., Geological and Mineral Inst.), 3-29-1 Otsuka, Bunkyo-Ku, Tokyo 112
TOKYO METROPOLITAN UNIV., Setagaya-Ku, Tokyo 158
TOKYO UNIV. OF FISHERIES, 4-5-7 Konan, Minato-Ku, Tokyo 108
UNIV. OF THE RYUKYUS (Coll. of Education, Coll. of Science and Engineering, Sesoko Marine Science Lab.), 3-1 Tonokura-Cho, Naha, Okinawa 903
UNIV. OF TOKYO (Dept. of Fisheries, Dept. of Geography, Dept. of Urban Engineering, Earthquake Res. Inst., Faculty of Agriculture, Faculty of Engineering, Geophysical Inst., Geological Inst., Univ. Museum), Bunkyo-Ku, Tokyo 113; (Ocean Res. Inst.), Nakano-Ku, Tokyo 164

Kenya
FISHERIES DEPT., P.O. Box 40241, Nairobi

Malawi
FISHERIES RES. UNIT, P.O. Box 27, Monkey Bay

Malaysia
FISHERIES RES. INST., Calthrop Rd., Glugor, Penang
UNIV. OF AGRICULTURE MALAYSIA (Div. of Fisheries and Marine Sciences), Serdang, Selangor
UNIV. OF SAINS MALAYSIA (School of Biological Sciences), Minden, Penang
UNIV. OF SINGAPORE (Zoology Dept.), Bukit Timah Rd., Singapore 10

Malta
UNIV. OF MALTA (Dept. of Biology, Dept. of Physics, Dept. of Physiology and Biochemistry, International Ocean Inst.), Msida

Mauritius
MIN. OF FISHERIES (Fisheries Div.), Port Louis

Mexico
CENTRO DE INVESTIGACION CIENTIFICA Y EDUCACION SUPERIOR DE ENSENADA, Av. Gastelum N. 898, Ensenada, B.C.
DIRECCION GENERAL DE IRRIGACION Y CONTROL DE RIOS (Direccion de Aguacultura), Av. Juarez 100, Mexico 1, D.F.
DIRECCION GENERAL DE OCEANOGRAFIA Y SENALAMENTO MARITIMO (Secretaria de Marina), Medellin 10, Mexico 7, D.F.
DIRECCION GENERAL DE USOS DEL AGUA Y PREVENCION DE LA CONTAMINACION (Secretaria de Recursos Hidraulicos), Av. Reforma 107, Mexico 4, D.F.
ESCUELA NACIONAL DE CIENCIAS BIOLOGICAS (Instituto Politecnico Nacional), Prolongacion de Carpio Y Plan de Ayala, Mexico 17, D.F.
ESCUELA SUPERIOR DE CIENCIAS MARITIMAS Y TECNOLOGIA DE ALIMENTOS (Instituto Tecnologico y de Estudios Superiores de Monterrey), Bahia de Bacochibampo, APDO. Postal 484, Guaymas, Son.
INSTITUTO NACIONAL DE PESCA (Subsecretaria de Pesca), Chiapas 121, Mexico 7, D.F.; Calle 20, Cont. Boulevard Sanchez Taboada 605, Guaymas, Son.; APDO. Postal 1306, Ensenada, B.C.; APDO. Postal 140, Campeche, Camp., Mexico 20, D.F.; APDO. Postal 396, Mazatlan, Sin.; Hernandez y Hernandez 1193, Veracruz, Ver.
UNIVERSIDAD AUTONOMA DE BAJA CALIFORNIA (Unidad de Ciencias Marinas), APDO. Postal N.453, KM. 103, Carretera Tijuana-Ensenada, Ensenada, B.C.

UNIVERSIDAD DE SONORA (Centro de Investigaciones Cientificas y Tecnologicas de la Universidad de Sonora), Bermosillo, Son.
UNIVERSIDAD NACIONAL AUTONOMA DE MEXICO (Centro de Ciencias del Mar y Limnologia, Centro de Preclasificacion Oceanoca de Mexico, Instituto de Biologia, Instituto de Geofisica), Ciudad Universitaria, Mexico 20, D.F.

Monaco
CENTRE SCIENTIFIQUE DE MONACO, 16 Boulevard de Suisse, Monaco

Netherlands
DELTA INST. FOR HYDROBIOLOGY, Vierstraat 28, Yerseke
HYDRAULIC LAB., Rotterdamseweg 185, P.O. Box 177, Delft
NETHERLANDS INST. FOR FISHERY INVESTIGATIONS, Haringkade 1, Postbus 68, Ijmuiden-1620
NETHERLANDS INST. FOR SEA RESEARCH, P.O. Box 59, Texel
ROYAL NETHERLANDS METEOROLOGICAL INST., De Bilt
ROYAL NETHERLANDS NAVY (Hydrographic Dept.), Badhuisweg 171, Den Haag
STATE MUSEUM OF NATURAL HISTORY, Raamsteeg 2, Leiden
UNIV. OF AMSTERDAM (Zoological Museum), Plantage Middenlaan 53, Amsterdam

New Caledonia
O.R.S.T.O.M., B.P.A. 5, Noumea

New Zealand
AUCKLAND UNIV. (Dept. of Botany, Dept. of Geology, Dept. of Physics, Dept. of Zoology), Private Bag, Auckland
CANTERBURY UNIV. (Dept. of Geography, Dept. of Physics and Engineering, Dept. of Zoology), Private Bag, Christchurch
D.S.I.R. (N.Z. Geological Survey), P.O. Box 30-368, Lower Hutt
D.S.I.R. (N.A. Oceanographic Inst.), P.O. Box 12-346, Wellington North
DEFENCE SCIENTIFIC ESTABLISHMENT, Devonport, Auckland
MIN. OF AGRICULTURE AND FISHERIES (Fisheries Management Div., Fisheries Research Div.), Wellington
NATIONAL MUSEUM OF NEW ZEALAND, Buckle St., Wellington
OTAGO UNIV. (Dept. of Geology, Dept. of Zoology), P.O. Box 56, Dunedin
VICTORIA UNIV. OF WELLINGTON (Dept. of Physics, Dept. of Zoology, Island Bay Marine Lab.), Private Bag, Wellington

Norway
DET NORSKE METEOROLOGISK INSTITUTT, Niels Henr, Abels Veg 40, Oslo 3
FISKERIDIREKTORATET (Vitaminlaboratoriet), 5000 Bergen
FISKERIDIREKTORATETS HAVFORSKNINGSINSTITUTT, P.O. Box 2906, 5011 Bergen Nordnes
INSTITUTT FOR KONTINENTALSOKKELUNDERSOKELSER, Postboks 1883, Hakon Magnussonsgt. 1B, 7000 Trondheim
NORGES GEOGRAFISKE OPPMALING, St. Olavs Gage 32, Oslo 1
NORSK INSTITUTT FOR VANNFORSKNING, BREKKEVN. 22-24, P.O. Box 260, Blindern Oslo 3
UNIVERSITETET I BERGEN (Biologisk Stasjon), 5065 Blomsterdalen
UNIVERSITETET I BERGEN (Geofysisk Institutt A, Geologisk Institutt B, Inst. for Generell Mikrobiologi, Jordskjelvstasjonen, Norges Fiskeri-Hogskole, Zoologisk Laboratorium, Zoologisk Museum), 5000 Bergen
UNIVERSITETET I OSLO (Botanisk Laboratorium, Institutt for Geofysikk A, Institutt for Marin Biologi og Limnologi, Zoologisk Institutt), Blindern, Oslo 3
UNIVERSITETET I TROMSO (Institutt for Biologi og Geologi, Marinbiologisk Stasjon, Norges Fiskerihogskole), 9000 Tromso
UNIVERSITETET I TRONDHEIM (Institutt for Akustikk, Institutt for Marin Biokjemi, Institutt for Reguleringsteknikk, Institutt for Teknisk Biokjemi, Trondheim Biologiske Stasjon, Zoologisk Avdeling), 7000 Trondheim
VASSDRAGS OG HAVNELABORATORIET, Sintef, Nth, 7000 Trondheim

Pakistan
KARACHI UNIV. (Inst. of Marine Biology), University Rd., Karachi

Papua New Guinea
DEPT. OF AGRICULTURE, STOCK AND FISHERIES, Konedobu, Port Moresby

Peru
MINISTERIO DE PESQUERIA, Lima

Philippines
BUR. OF COAST AND GEODETIC SURVEY, Barraca St., San Nicolas, Manila
BUR. OF FISHERIES AND AQUATIC RESOURCES, Manila
NATIONAL MUSEUM, Liwasang Rizal, Manila

PHILIPPINE ATMOSPHERIC (Geophysics and Astronomical Services Admin.), Quezon Blvd. Extension, Quezon City
UNIV. OF THE PHILIPPINES (Coll. of Arts and Sciences, Coll. of Fisheries), Diliman, Quezon City

Poland
ADADEMIA MEDYCZNA, Sklodowskiej Curie 3A, 80-210 Gdansk
INSTYTUT METEOROLOGII I GOSPODARKI WODNEJ ODDZIAL MORSKI, Waszyngtona 42, P.O. Box 276, 81-963 Gdynia
KOMITET BADAN MORZA PAN (Polish National Scientific Committee on Oceanic Res.), Okrezna 11, 81-822 Sopot
MORSKI INSTYTUT RYBACKI, Al. Zjednoczenia 1, P.O. Box 184, 81-345 Gdynia
POLSKIEJ AKADEMII NAUK (Instytut Budownictwa Wodnego Zaklad Hydrauliki Morskiej), Cysterow 11, P.O. Box 18, 80-953 Gdansk-Oliwa
POLSKIEJ AKADEMII NAUK (Zaklad Oceanologii), Powstancow Warszawy 55, P.O. Box 60, 81-967 Sopot
UNIWERSYTET GDANSKI (Instytut Oceanografii), Al. Czolgistow 46, 81-378 Gdynia

Republic of Chad
O.R.S.T.O.M. (Centre de N'Djamena), Boite Postale N.65.

Republic of Congo
O.R.S.T.O.M., B.P. 1286, Pointe Noire

Republic of Ivory Coast
CENTRE DE RECHERCHES OCEANOGRAPHIQUES D'ABIDJAN, 29, Rue des Pecheurs, B.P. V18, Abidjan

Republic of Korea
FISHERIES RES. AND DEVELOPMENT AGENCY, Busan 600-01
PUSAN FISHERIES COLLEGE, Pusan
SEOUL NATIONAL UNIV. (Dept. of Biology. Dept. of Botany, Dept. of Geology, Dept. of Oceanography, Dept. of Physics, Dept. of Zoology), Seoul 151

Republic of Madagascar
UNIVERSITE DE MADAGASCAR, Station Marine de Tulear, B.P. 141, Tulear

Republic of Senegal
CENTRE DE RECHERCHES OCEANOGRAPHIQUES, B.P. 2241, Dakar
O.R.S.T.O.M., Dakar

Republic of South Africa
COUNCIL FOR SCIENTIFIC AND INDUSTRIAL RES. (National Committee for Environmental Sciences, National Electrical Engineering Res. Inst., Natioanl Inst. for Water Res., National Physical Res. Lab., National Res. Inst. for Mathematical Sciences, South African National Committee for Oceanographic Res.), P.O. Box 395, Pretoria 0001; (National Inst. for Water Res.), P.O. Box 17001, Congella 4013; (National Res. Inst. for Oceanology), P.O. Box 320, Stellenbosch 7600
DEPT. OF INDUSTRIES (Sea Fisheries Br.), P.O. Box 251, Cape Town 8000
DEPT. OF PLANNING AND THE ENVIRONMENT, Private Bag X213, Pretoria 0001
FISHERIES DEVELOPMENT CORP. OF S. AFRICA LTD., P.O. Box 539, Cape Town 8000
OCEANOGRAPHIC RES. INST., P.O. Box 736, Durban 4000
PORT ELIZABETH MUSEUM AND OCEANARIUM, P.O. Box 13147, Humewood 6013
RHODES UNIV. (Dept. of Botany, Dept. of Chemistry, Dept. of Zoology, J. L. B. Smith Inst. of Ichthyology), P.O. Box 94, Grahamstown 6140
SOUTH AFRICAN MUSEUM, P.O. Box 61, Cape Town 8000
UNIV. OF CAPE TOWN (Dept. of Environmental Studies, Dept. of Biochemistry, Dept. of Geochemistry, Dept. of Geology, Dept. of Oceanography, Dept. of Physics, Dept. of Psychology, Dept. of Zoology, Fishing Industry Res. Inst.), Private Bag, Rondebosch 7700; (Dept. of Mines), P.O. Box 10, Cape Town 8000
UNIV. OF DURBAN-WESTVILLE (Dept. of Botany, Dept. of Zoology), Private Bag X54001, Durban 4000
UNIV. OF PORT ELIZABETH (Dept. of Zoology), P.O. Box 1600, Port Elizabeth 6000
UNIV. OF THE WITWATERSRAND (Bernard Price Inst. of Geophysical Res., Dept. of Botany and Microbiology, Dept. of Mechanical Engineering, Dept. of Medicine, Dept. of Zoology), Jan Smuts Ave., Johannesburg 2001

Romania
INSTITUTUL ROMAN DE CERCETARI MARINE, Bulevardul Lenin 300, Constanta

Saudi Arabia
KING ABDUL AZIZ UNIV. (Dept. of Oceanography), P.O. Box 1540 Jeddah

Sierra Leone
UNIV. OF SIERRA LEONE (Inst. of Marine Biology and Oceanography), Freetown

Spain
INSTITUTO ESPANOL DE OCEANOGRAFIA (Administracion General, Departamento Centro Espanol de Datos Oceanograficos, Departamento de Biologia Aplicada, Departamento de Biologia Marina, Departamento de Biotecnologia Pesquera, Departamento de Contaminacion, Departamento de Fisica, Departamento de Geologia, Departamento de Quimica), Alcala, 27-4, Madrid (14)
INSTITUTO HIDROGRAFICO DE LA MARINA, San Fernando, Cadiz
LABORATORIO DE BALEARES, Muelle de Pelaires, S/N Palma de Mallorca, Apartado 291, Palma de Mallorca
LABORATORIO DE CANARIAS, Avda. Jose Antonio, 3, Santa Cruz de Tenerife.
LABORATORIO DE MALAGA, Paseo de la Farola, 27, Malaga
LABORATORIO DE VIGO, Orillamar, 47, Vigo
LABORATORIO DEL MAR MENOR, Magallanes, San Pedro del Pinatar (Murcia)
LABORATORIO DEL NOROESTE, Muelle de las Animas, Apartado 130, La Coruna

Sweden
ASKO LAB., P.O. Box 58, S-150 13 Trosa
BOTANICAL INSTITUTION (Dept. of Marine Botany, Dept. of Marine Microbiology), Carl Skottsbergs Gata 22, S-413 19 Goteborg
CHALMERS UNIV. OF TECHNOLOGY (Dept. of Hydraulics, Dept. of Ship Hydromechanics), S-402 20 Goteborg
NATIONAL BD. OF FISHERIES (Inst. of Marine Res.), S-453 00 Lysekil
NATIONAL BD. OF FISHERIES (Inst. of Marine Res.), Box 4031 S-400 40 Gotenborg 4
SWEDISH METEOROLOGICAL AND HYDROLOGICAL INST., Fack, S-601 01 Norrkoping
UNIV. OF GOTHENBURG (Dept. of Analytical Chemistry, Dept. of Oceanography, Dept. of Structural and Ecological Zoology, Geological Institution), Gotenborg
UNIV. OF LUND (Dept. of Zoology, Dept. of Marine Botany), S-22362 Lund
UNIV. OF STOCKHOLM (Dept. of Geology, Dept. of Zoology), S-113 86 Stockholm VA
UNIV. OF UPPSALA (Dept. of Zoology, Inst. of Physiological Botany, Institution of Plant Biology, Inst. of Zoophysiology), S-751 22 Uppsala

Switzerland
GEOLOGISCHES INSTITUT, ETH, Sonneggstr. 5, 8006 Zurich
UNIVERSITAT BERN (Geologisches Institut, Physikalisches Institut), 3012 Bern
ZOOLOGISCHES INSTITUT, Rheinsprung 9, 4000 Basel

Taiwan
NATIONAL TAIWAN UNIV. (Chinese National Committee on Oceanic Res., Dept. of Geology, Dept. of Zoology, Inst. of Fishery Biology, Inst. of Oceanography), Taipei

Thailand
CHULALONGKORN UNIV. (Dept. of Marine Science, Biology Dept.), Bangkok
KASETSART UNIV. (Dept. of Marine Science, Dept. of Fishery Biology, Dept. of Fishery Products), Bangkok
MIN. OF AGRICULTURE AND COOPERATIVES (Marine Fisheries Div.), Bangkok
PHUKET MARINE BIOLOGICAL CTR., P.O. Box 200, Phuket
ROYAL THAI NAVY (Hydrographic Dept.), Dhonburi

Turkey
BOGAZICI UNIVERSITESI, P.K. 2 Bebek, Istanbul
ISTANBUL TEKNIK UNIVERSITESI (Cevre Bilimleri ve Teknolojisi Kursusu), Istanbul
ISTANBUL UNIVERSITESI (Hidrobiyoloji Arastirma Enstitusu Direktorlugu), Istanbul
SEYIR (Hidrografi ve Osinografi Dairesi Baskanligi Cubuklu), Istanbul

United States of America
ADELPHI UNIV. (Dept. of Biology, Earth Science Dept., Inst. of Marine Science), Garden City, N.Y. 11530
ALABAMA STATE DEPT. OF CONSERVATION AND NATURAL RESOURCES (Marine Resources Div., Claude Peteet Mariculture Ctr.), Dauphin Island, Ala. 36528
AMOCO PRODUCTION CO. (Res. Ctr.), Chicago, Ill. 60680
BOSTON UNIV. (Biological Sciences Ctr., Marine Biological Lab.), Boston, Mass. 02215
CALIFORNIA DEPT. OF FISH AND GAME (Fish and Game Lab., Marine Culture Lab., Marine Resources Lab., State Fisheries Lab.), Eureka, Calif. 95501
CALIFORNIA POLYTECHNIC STATE UNIV. (School of Science and Mathematics), San Luis Obispo, Calif. 93407
CALIFORNIA STATE UNIV. (Dept. of Biological Science, Dept. of Earth Science), Fullerton, Calif. 92634;

Appendix C: Directory of Institutions

(Sch. of Natural Resources, Dept. of Oceanography), Arcata, Calif. 95521; (Dept. of Biology, Dept. of Civil Engineering, Dept. of Geological Sciences, Dept. of Microbiology), Long Beach, Calif. 90840
CITY COLL. OF THE CITY UNIV. OF NEW YORK (Dept. of Biology, Dept. of Earth and Planetary Sciences, Inst. of Marine and Atmospheric Sciences), New York, N.Y. 10031
COLUMIBA UNIV. (Lamont-Doherty Geological Obs.), New York, N.Y. 10027
CORNELL UNIV. (Lengmuir Lab., Dept. of Geological Sciences, Sch. of Civil and Environmental Engineering), Ithaca, N.Y. 14853
DEFENSE MAPPING AGENCY (Hydrographic Ctr.), Washington, D.C. 20390
DEPT. OF THE ARMY (Coastal Engineering Res. Ctr.), Fort Belvoir, Va. 22060
DUKE UNIV. MARINE LAB., Beaufort, N.C. 28516
ENVIRONMENTAL PROTECTION AGENCY (Corvallis Environmental Res. Lab., Environmental Res. Lab.), Johns Island, S.C. 29455
FLORIDA ATLANTIC UNIV. (Dept. of Ocean Engineering), Boca Raton, Fla. 33432
FLORIDA INST. OF TECHNOLOGY (Dept. of Oceanography and Ocean Engineering, Sch. of Marine and Environmental Technology), Melbourne, Fla. 32901
FLORIDA STATE UNIV. (Dept. of Biological Science, Dept. of Oceanography), Tallahassee, Fla. 32306
GULF COAST RES. LAB., Ocean Springs, Miss. 39564
HARVARD UNIV. (Biological Lab., Div. of Engineering and Applied Physics, Museum of Comparative Zoology), Cambridge, Mass. 02138
INTER-AMERICAN TROPICAL TUNA COMM., La Jolla, Calif. 92093
INTERNATIONAL PACIFIC HALIBUT COMM. Seattle, Wash. 98105
JOHNS HOPKINS UNIV. (Chesapeake Bay Inst., Dept. of Earth and Planetary Sciences), Baltimore, Md. 21218
LEHIGH UNIV. (Ctr. for Marine and Environmental Studies, Dept. of Chemistry, Inst. for Pathobiology, Marine Geotechnical Lab.), Bethlehem, Pa. 18015
LOCKHEED MARINE BIOLOGY LAB., Carlsbad, Calif. 92008
LOUISIANA STATE UNIV. (Ctr. for Wetland Resources, Coastal Studies Inst., Dept. of Marine Sciences, Dept. of Geology, Louisiana Sea Grant Legal and Socio-Economic Program, Sch. of Forestry and Wildlife Management), Baton Rouge, La. 70803
MASSACHUSETTS INST. OF TECHNOLOGY (Dept. of Ocean Engineering, Dept. of Civil Engineering, R. M. Parsons Lab. for Water Res., Dept. of Earth and Planetary Science, Dept. of Meteorology), Cambridge, Mass. 02139
MISSISSIPPI STATE UNIV., State College, Miss. 39762
MOSS LANDING MARINE LAB., Moss Landing, Calif. 95039
MOTE MARINE LABORATORY, Sarasota, Fla. 33581
NATIONAL SCIENCE FOUND. (Idoe Office, Oceanography Section), Washington, D.C. 20550
NAVAL COASTAL SYSTEMS LAB., Panama City, Fla. 32401
NAVAL CONSTRUCTION BATTALION CTR. (Civil Engineering Lab., Navy Environmental Support Office), Port Hueneme, Calif. 93043
NAVAL OCEAN RES. AND DEVELOPMENT ACTIVITY (Naval Oceanographic Lab.), Bay St. Louis, Miss. 39520
NAVAL OCEANOGRAPHIC OFFICE, Washington, D.C. 20373
NAVAL POSTGRADUATE SCH. (Dept. of Oceanography), Monterey, Calif. 93940
NAVAL RES. LAB., Washington, D.C. 20375
NAVAL SHIP RES. AND DEVELOPMENT CTR., Annapolis, Md. 21402
NAVAL UNDERSEA CTR., San Diego, Calif. 92132
NAVAL UNDERWATER SYSTEMS CTR. (New London Lab., Newport Lab.), New London, Conn. 06320
NEW YORK OCEAN SCIENCE LAB., Montauk, N.Y. 11954
NEW YORK STATE COLL. OF AGRICULTURE AND LIFE SCIENCES AT CORNELL (Dept. of Food Science, Dept. of Natural Resources), Ithaca, N.Y. 14853
NOAA (Office of Coastal Environment, Office of Environmental Monitoring and Prediciton, Office of Marine Resources), Rockville, Md. 20852; (Office of Coastal Zone Management, Office of the Administrator, Sea Grant Program), Washington, D.C. 20235
NOAA ENVIRONMENTAL DATA SERVICE (Ctr. for Experiment Design and Data Analysis, National Climatic Ctr., National Geophysical and Solar-Terrestrial Data Ctr., National Oceanographic Data Ctr.), Washington, D.C. 20235
NOAA ENVIRONMENTAL RES. LAB. (Atlantic Oceanographic and Meteorological Lab., Geophysical Fluid Dynamics Lab., Joint Tsunami Res. Effort, Marine

Ecosystem Analysis, Pacific Marine Environmental Lab.), Boulder, Colo. 80302
NOAA NATIONAL MARINE FISHERIES SERVICE (Southeast Fisheries Ctr.), Miami, Fla. 33149
NOAA NATIONAL ENVIRONMENTAL SATELLITE SERVICE, Suitland, Md. 20233
NOAA NATIONAL MARINE FISHERIES SERVICE (Atlantic Estuarine Fisheries Ctr., Gulf Coastal Fisheries Ctr., Marmap Field Office, Middle Atlantic Coastal Fisheries Ctr., Northeast Fisheries Ctr., Northeast Utilization Res. Ctr., Northwest Fisheries Ctr., Office of the Director, Pacific Environmental Group, Southeast Fisheries Ctr., Southeast Utilization Res. Ctr., Southwest Fisheries Ctr.), Pascagoula, Miss. 39567
NOAA NATIONAL OCEAN SURVEY (Data Bouy Office, Field Offices), Rockville, Md. 20852
NORTH CAROLINA STATE UNIV. (Dept. of Civil Engineering, Dept. of Geosciences, Dept. of Zoology), Raleigh, N.C. 27607
NOVA UNIV. (Oceanographic Lab.), Dania, Fla. 33004
OCEANOGRAPHIC INST. OF WASHINGTON, Seattle, Wash. 98109
OFFICE OF NAVAL RES., Arlington, Va. 22217
OLD DOMINION UNIV. (Inst. of Oceanography), Norfolk, Va. 23508
OREGON STATE UNIV. (Marine Science Ctr., Sch. of Engineering, Sch. of Oceanography, Seafoods Lab.), Newport, Oreg. 97365
PENNSYLVANIA STATE UNIV. (Applied Res. Lab., Dept. of Biology, Dept. of Civil Engineering, Dept. of Geology, Dept. of Meteorology), Sharon, Pa. 16146
QUEENS COLL. OF THE CITY UNIV. OF NEW YORK (Biology Dept., Chemistry Dept., Dept. of Earth and Environmental Sciences, Dept. of Health and Physical Education), Flushing, N.Y. 11367
RUTGERS UNIV. (Chemistry Dept., Dept. of Biochemistry and Microbiology, Dept. of Biology, Dept. of Botany, Dept. of Chemical and Biochemical Engineering, Dept. of Entomology and Economic Ecology, Dept. of Environmental Resources, Dept. of Environmental Science, Dept. of Geology, Dept. of Meteorology and Physical Oceanography, Dept. of Physiology, Marine Sciences Ctr., Newark Coll., Zoology Dept.), New Brunswick, N.J. 08903

SAN DIEGO STATE UNIV. (Ctr. for Marine Studies, Dept. of Biology, Dept. of Geological Sciences, Dept. of Microbiology, Dept. of Physical Sciences), San Diego, Calif. 92182
SAN FRANCISCO STATE UNIV. (Dept. of Biology), San Francisco, Calif. 94132
SKIDAWAY INST. OF OCEANOGRAPHY, Savannah, Ga. 31406
SMITHSONIAN INSTITUTION (Chesapeake Bay Center for Environmental Studies, National Museum of Natural History, Office of International and Environmental Programs), Edgewater, Md. 21037
SMITHSONIAN TROPICAL RES. INST., Balboa, Canal Zone 92661
SOUTHAMPTON COLL. OF LONG ISLAND UNIV. (Div. of Natural Sciences), Southampton, N.Y. 11968
SOUTHEASTERN MASSACHUSETTS UNIV. (Biology Dept., Dept. of Electrical Engineering), North Dartmouth, Mass. 02747
STANFORD UNIV. (Hopkins Marine Sta., Sch. of Earth Sciences), Stanford, Calif. 94305
STATE UNIV. OF NEW YORK AT STONY BROOK (Marine Sciences Res. Ctr.), Stony Brook, N.Y. 11794
TEXAS A&M UNIV. (Dept. of Oceanography, Moody Coll. of Marine Sciences and Maritime Resources), College Station, Tex. 77843
U.S. COAST GUARD OCEANOGRAPHIC UNIT, Washington, D.C. 20590
U.S. GEOLOGICAL SURVEY (Fisher Island Sta., Ice Dynamics Proj., Inst. of Oceanography, National Ctr.), Woods Hole, Mass. 02543
U.S. NAVY (Office of the Chief of Naval Operations, Office of the Oceanographer of the Navy), Washington, D.C. 20350
UNIV. OF ALASKA (Alaska Sea Grant Program, Inst. of Marine Science), College, Alaska 99701
UNIV. OF CALIFORNIA (Dept. of Mechanical Engineering, Dept. of Civil Engineering, Dept. of Materials Science, Dept. of Naval Architecture, Dept. of Zoology, Dept. of Chemical Engineering), Berkeley, Calif. 94720; (Bodega Marine Lab.), Bodega Bay, Calif. 94923; (Div. of Environmental Studies, Inst. of Marine Resources), Davis, Calif. 95616; (Dept. of Biology, Dept. of Geology, Medical Microbiology and Immunology), Los Angeles, Calif. 90024; (Scripps Institution of Oceanography), La

Jolla, Calif. 92093; (Dept. of Biological Sciences, Marine Science Inst.), Santa Barbara, Calif. 93106; (Coastal Marine Lab.), Santa Cruz, Calif. 95064
UNIV. OF CHICAGO (Dept. of Geography, Dept. of the Geophysical Sciences), Chicago, Ill. 60637
UNIV. OF CONNECTICUT (Marine Res. Lab., Marine Sciences Inst.), Noank, Conn. 06340
UNIV. OF DELAWARE (Coll. of Marine Studies, Lewes Marine Studies Ctr., Dept. of Civil Engineering), Newark, Del. 19711
UNIV. OF FLORIDA (Coastal and Oceanographic Engineering Lab., Coll. of Medicine, Whitney Marine Lab.), Gainesville, Fla. 32611
UNIV. OF GEORGIA (Marine Inst., Marine Resources Extension Ctr.), Athens, Ga. 30602
UNIV. OF GUAM (Marine Lab.), Agana, Guam 96910
UNIV. OF HAWAII (Dept. of Meteorology, Dept. of Ocean Engineering, Dept. of Oceanography, Environmental Ctr., Hawaii Inst. of Geophysics, Hawaii Inst. of Marine Biology), Honolulu, Hawaii 96822
UNIV. OF HOUSTON (Bates Coll. of Law, Chemistry Dept., Coll. of Pharmacy, Cullen Coll. of Engineering, Dept. of Biology, Dept. of Geography, Dept. of Geology, Physics Dept.), Houston, Tex. 77004
UNIV. OF MAINE (Dept. of Botany and Plant Pathology, Dept. of Oceanography, Dept. of Zoology), Orono, Maine 04473
UNIV. OF MARYLAND (Ctr. for Environmental and Estuarine Studies, Dept. of Botany, Mechanical Engineering Dept.), Cambridge, Md. 21613
UNIV. OF MASSACHUSETTS (Marine Sciences Program), Amherst, Mass. 01002
UNIV. OF MIAMI, Rosentiel School of Marine and Atmospheric Science (Div. of Atmospheric Science, Div. of Biology and Living Resources, Div. of Chemical Oceanography, Div. of Fisheries and Applied Estuarine Ecology, Div. of Marine Geology and Geophysics, Div. of Ocean Engineering, Div. of Physical Oceanography), Miami, Fla. 33149
UNIV. OF MICHIGAN (Dept. of Atmospheric and Oceanic Science), Ann Arbor, Mich. 48104
UNIV. OF NEW HAMPSHIRE, Durham, N.H. 03824
UNIV. OF NORTH CAROLINA (Curriculum in Marine Sciences, Dept. of Environmental Sciences and Engineering, Dept. of Geology, Dept. of Zoology, Inst. of Marine Sciences), Chapel Hill, N.C. 27514
UNIV. OF NORTH CAROLINA AT WILMINGTON (Dept. of Biology, Dept. of Earth Sciences, Inst. of Marine Biomedical Res., Program in Marine Sciences and Environmental Studies), Wilmington, N.C. 28401
UNIV. OF PUERTO RICO (Dept. of Marine Sciences), Mayaguez, P.R. 00708
UNIV. OF RHODE ISLAND (Dept. of Ocean Engineering, Graduate Sch. of Oceanography), Kingston, R.I. 02881
UNIV. OF SOUTH CAROLINA (Baruch Inst. for Marine Biology and Coastal Res.), Beaufort, S.C. 29902
UNIV. OF SOUTH FLORIDA (Dept. of Marine Science), Tampa, Fla. 33620
UNIV. OF SOUTHERN CALIFORNIA (Allan Hancock Found., Dept. of Biological Sciences, Dept. of Geological Sciences, Inst. for Marine and Coastal Studies, Santa Catalina Marine Biological Lab.), Los Angeles, Calif. 90007
UNIV. OF TEXAS (Marine Science Inst.), Austin, Tex. 78712
UNIV. OF WASHINGTON (Applied Physics Lab., Coll. of Fisheries, Dept. of Botany, Dept. of Oceanography, Dept. of Zoology, Inst. for Marine Studies), Seattle, Wash. 98195
UNIV. OF WEST FLORIDA (Faculty of Biology), Pensacola, Fla. 32504
UNIV. OF WISCONSIN (Ctr. for Great Lakes Studies, Geophysical and Polar Res. Ctr., Marine Studies Ctr.), Green Bay, Wis. 54302
UNIV. OF FLORIDA, St. Augustine, Fla. 32084
VIRGINIA INST. OF MARINE SCIENCE (Eastern Shore Lab.), Gloucester Point, Va. 23062
WOODS HOLE OCEANOGRAPHIC INSTITUTION, Woods Hole, Mass. 02543
YALE UNIV. (Dept. of Biology, Dept. of Geology and Geophysics, Peabody Museum of Natural History), New Haven, Conn. 06520

USSR

ALEXANDER KOVALEVSKY INST. OF BIOLOGY OF SOUTH SEAS, 2 Nakhimov Prospect, Sevastopol 335000
ALL-UNION RES. INST. OF MARINE FISHERIES AND OCEANOGRAPHY, 17 Krasnoselskaya, Moscow B-140

ARCTIC AND ANTARCTIC RES. INST.,
34 Fontanka, Leningrad D-104,
192104
ATLANTIC RES. INST. OF MARINE FISHERIES
AND OCEANOGRAPHY, 5 Dmitri Donskoi
St., Kaliningrad
AZOV SEA INST. OF FISHERIES,
21/2 Beregovaya St., Rostov-on-Don,
344077
BIELORUSSIAN STATE UNIV., Minsk
HYDROMETEOROLOGICAL CTR., 9/13
Bolshevistskaya, Moscow D-376
INST. OF MARINE BIOLOGY, 159 Stoletia
Vladivostoka Prospect, Vladivostok
690022
KNIPOVITCH POLAR RES. INST. OF MARINE
FISHERIES AND OCEANOGRAPHY, 6 Knipo-
vitch St., Murmansk 183038
LENINGRAD HYDROMETEOROLOGICAL INST.,
98 Malo-Okhtinsky Prospect,
Leningrad K-196, 195156
LENINGRAD STATE UNIV. (Dept. of
Oceanology), 3 Smolny St., Leningrad
193124
MOSCOW STATE UNIV. (Biological Faculty,
Dept. of Earth's Physics, Dept. of
Geomorphology, Dept. of Hydro-
biology, Dept. of Invertebrate
Zoology, Dept. of Oceanology, Dept.
of Physics of Sea and Land Waters,
Geophysical Faculty), Moscow
117234
MURMANSK MARINE BIOLOGICAL INST.,
Dalnie Zalentzy, Murmansk Region
NORTH MARINE GEOLOGICAL SERVICE,
120 Moyka, Leningrad
PACIFIC RES. INST. OF MARINE FISHERIES
AND OCEANOGRAPHY, 20 Lenin St.,
Vladivostok 690000
RES. INST. OF MARINE FISHERIES AND
OCEANOGRAPHY OF THE AZOV AND BLACK
SEAS, 2 Karl Libknecht St., Kerch
SAKHALIN COMPLEX RES. INST., Novo-
Alexandrovsk 694050, Sakhalin Region
SEVERTSOV INST. OF EVOLUTIONARY ANIMAL
MORPHOLOGY AND ECOLOGY, 33 Lenin
Prospect, Moscow 117071
SOVIET OCEANOGRAPHIC COMM., 44 Vavilov
St., Block 2, Moscow B-333, 117333
STATE OCEANOGRAPHIC INST. (Hydro-
meteorological Service), 6 Kropotkin-
sky Pereulok, Moscow G-34
UKRANIAN ACADEMY OF SCIENCES (Marine
Hydrophysical Inst.), 28 Lenin St.,
Sevastopol 335000
USSR ACADEMY OF SCIENCES (Inst. of
Acoustics, Inst. of Geography, Inst.
of Geology, Inst. of Microbiology,
Inst. of Oceanology, Inst. of Physics
of Atmosphere, Inst. of State and
Law, Inst. of Water Studies, Marine
Expedition Dept.), Moscow; (Inst. of
Zoology), 1 Universitetskaya Nabere-
zhnaya, Leningrad V-164, 199164
VERNADSKY INST. OF GEOCHEMISTRY AND
ANALYTICAL CHEMISTRY, 47-A Voro-
bievskoe Shosse, Moscow B-334, 117334

United Kingdom

BRITISH MUSEUM (NATURAL HISTORY) (Dept.
of Botany, Dept. of Mineralogy, Dept.
of Palaeontology, Dept. of Zoology),
Cromwell Rd., London SW7 5BD
CITY OF LONDON POLYTECHNIC (Dept. of
Biology, Dept. of Geology, Dept.
of Mathematics), London
CLYDE RIVER PURIFICATION BD.,
Rivers House, Murray Rd., East
Kilbride, Glasgow G75 0LA
DEPT. OF AGRICULTURE AND FISHERIES FOR
SCOTLAND (Marine Lab.), P.O. Box 101,
Victoria Rd., Aberdeen AB9 8DB
HERIOT-WATT UNIV. (Dept. of Brewing
and Biological Sciences, Dept. of
Civil Engineering, Dept. of Offshore
Engineering, Inst. of Offshore En-
gineering), Edinburgh EH14 4AS
HYDRAULICS RES. STA., Wallingford,
Oxfordshire
IMPERIAL COLL. OF SCIENCE AND TECH-
NOLOGY (Dept. of Geology, Dept. of
Mathematics), London
MARINE BIOLOGICAL ASSOC. OF THE U.K.,
Lab., Citadel Hill, Plymouth, Devon
PL1 2PB
MARINE TECHNOLOGY SUPPORT UNIT,
Aere Harwell, Bldg. 424, Oxford-
shire OX11 0RA
MIN. OF AGRICULTURE, FISHERIES AND
FOOD (Torry Res. Sta.), P.O. Box 31,
Aberdeen AB9 8DG; (Fish Diseases
Lab.), Nothe, Weymouth, Dorset
DT4 8UB. (Fisheries Experiment
Station), Benarth Road, Conwy,
Gwynedd LL32 8UB, (Fisheries La-
boratory), Burnham-on-Crouch,
Essex CMO 8HA,(Fisheries Labora-
tory, Fisheries Radiobiological
Laboratory), Lowestoft, Suffolk
MIN. OF DEFENCE (Hydrographic Dept.),
London SW1A 2EU; Taunton, Somerset
TA1 2DN
NATURAL ENVIRONMENT RES. COUNCIL (Inst.
of Geological Sciences), London;
(British Antarctic Survey, Culture
Ctr. of Algae and Protozoa, Inst. for
Marine Environmental Res.), Cambridge;
(Inst. for Marine Environmental
Res.), Prospect Pl., Plymouth, Devon
PL1 3AX; (Inst. of Oceanographic
Sciences), Brook Rd., Wormley,
Godalming, Surrey GU8 5UB
OPEN UNIV., Walton Hall, Milton Keynes
MK7 6AA

ORIELTON FIELD CTR., Pembroke, Pembrokeshire
POLYTECHNIC OF PLYMOUTH (Sch. of Maritime Studies), Plymouth, Devon PL4 8AA
ROYAL SCOTTISH MUSEUM, Chambers St., Edinburgh EH1 1JF
SCOTTISH MARINE BIOLOGICAL ASSOC. (Donstaffnage Marine Res. Lab.), P.O. Box 3, Oban, Argyll PA34 4AD
UNIV. COLL. OF LONDON (Dept. of Civil and Municipal Engineering, Dept. of Geology), Gower St., London WC1E 6BT
UNIV. COLL. OF NORTH WALES (Marine Science Lab.), Menai Bridge, Anglesey
UNIV. COLL. OF SWANSEA (Dept. of Botany and Microbiology, Dept. of Geology, Dept. of Oceanography, Dept. of Zoology), Singleton Pk., Swansea SA2 8PP, Wales
UNIV. COLL. OF WALES (Dept. of Geography, Dept. of Geology), Penglais, Aberystwyth
UNIV. OF ABERDEEN (Dept. of Zoology, Dept. of Geology), Aberdeen
UNIV. OF BIRMINGHAM (Dept. of Electronic and Electrical Engineering, Dept. of Geological Sciences, Dept. of Zoology and Comparative Physiology), P.O. Box 363, Birmingham B15 2TT
UNIV. OF BRISTOL (Dept. of Geology, Dept. of Zoology, Res. Unit for Comparative Animal Respiration), Bristol
UNIV. OF CAMBRIDGE (Dept. of Applied Mathematics and Theoretical Physics, Dept. of Geodesy and Geophysics, Dept. of Zoology, Scott Polar Res. Inst.), Cambridge
UNIV. OF DUNDEE (Tay Estuary Res. Ctr.), Old Ferry Pier, Newport on Tay, Fife DD6 8EX
UNIV. OF EAST ANGLIA (Sch. of Biological Sciences, Sch. of Environmental Sciences, Sch. of Mathematics and Physics), Norwich
UNIV. OF EDINBURGH (Dept. of Geophysics, Dept. of Mechanical Engineering, Grant Inst. of Geology), Edinburgh
UNIV. OF ESSEX (Dept. of Mathematics, Dept. of Fluid Mechanics), Wivenhoe Park, Culchester, Essex
UNIV. OF GLASGOW (Dept. of Botany, Dept. of Cell Biology, Dept. of Naval Architecture and Ocean Engineering, Dept. of Zoology), Glasgow
UNIV. OF LEEDS (Wellcome Marine Lab.), Robin Hood's Bay, North Yorkshire Y022 4SL
UNIV. OF LIVERPOOL (Dept. of Botany, Dept. of Civil Engineering, Dept. of Oceanography), Liverpool L69 3BX; (Marine Biological Lab.), Port Erin, Isle of Man
UNIV. OF LONDON (Chelsea Coll., King's Coll., Queen Mary Coll.), London
UNIV. OF NEWCASTLE UPON TYNE (Dove Marine Lab.), Cullercoats
UNIV. OF READING (Dept. of Geophysics, Dept. of Mathematics, Dept. of Zoology), Whiteknights, Reading
UNIV. OF SOUTHAMPTON (Dept. of Biology, Dept. of Oceanography), Southampton
UNIV. OF ST. ANDREWS (Gatty Marine Lab.), St. Andrews KY16 8LB, Scotland
UNIV. OF STIRLING (Dept. of Biology), Stirling FK9 4LA, Scotland
UNIV. OF STRATHCLYDE (Dept. of Applied Microbiology Dept. of Civil Engineering), Glasgow; (Dept. of Biology), Dalandhui House, Garelochhead, Dunbartonshire
WATER RES. CTR. (Stevenage Lab.), Elder Way, Stevenage, Hertfordshire

Venezuela
ESTACION DE INVESTIGACIONES MARINAS, Apartado 144, Porlamar
UNIVERSIDAD DE ORIENTE (Instituto Oceanografico), Apartado Postal 94, Cumana

Yugoslavia
INST. FOR OCEANOGRAPHY AND FISHERIES, Rt. Marjana, P.O. Box 114, Split
INSTITUTE "RUDJER BOSKOVIC" ZAGREB AND ROVINJ, Bijenicka Cesta 54, Zagreb
LAB. FOR SEA BIOLOGY, Kotor

International Institutions
EAST AFRICAN MARINE FISHERIES RES. ORGANIZATION, P.O. Box 668, Zanzibar, Tanzania
FAO, c/o UNDP, P.O. Box 46, Zomba, Malawi; (Fisheries Training Proj.), P.O. Box 72, Mangochi
FAO/IAEA DIV., Karntnerring 11, 1010 Vienna, Austria
FAO-UN (Fisheries Dept.), Via delle Terme di Caracalla, 00100 Rome, Italy
FAO REGIONAL OFFICE FOR AFRICA, P.O. Box 1628, Accra, Ghana
FAO REGIONAL OFFICE FOR ASIA AND THE FAR EAST, Bangkok 2, Thailand
FAO FISHERIES DEVELOPMENT, P.O. Box 329, Maiduguri, Nigeria

FAO FISHERY RES. PROJ., c/o UNDP, B.P. 1490, Bujunbura, Burundi

IAEA-UN, P.O. Box 590, 1011 Vienna, Austria; (Musee Oceanographique, International Lab. of Marine Radio-activity), Monaco

INSTITUTO TECNOLOGICO Y DE ESTUDIOS SUPERIORES DE MONTERREY (Escuela de Ciencias Maritimas y Tecnologia de Alimentos), Guaymas, Sonora, Mexico

INTER-AMERICAN TROPICAL TUNA COMMISSION, c/o Scripps Institution of Oceanography, La Jolla, Calif. 92037

INTERNATIONAL COMMISSION FOR THE CONSERVATION OF ATLANTIC TUNAS, General Mola 17, Madrid 1

INTERNATIONAL COMMISSION FOR THE NORTHWEST ATLANTIC FISHERIES, 800 Windmill Rd., P.O. Box 638, Dartmouth, N.S. B2Y 3Y9, Canada

INTERNATIONAL COUNCIL FOR THE EXPLORATION OF THE SEA, Charlottenlund Slot, DK-2920 Charlottenlund, Denmark

INTERNATIONAL HYDROGRAPHIC ORGANIZATION. Monaco

INTERNATIONAL PACIFIC HALIBUT COMM., P.O. Box 5009, University Sta., Seattle, Washington 98105

INTERNATIONAL PACIFIC SALMON FISHERIES COMM., P.O. Box 30, New Westminster. British Columbia V3L 4X9, Canada

INTERNATIONAL WHALING COMM., Great Westminster House, Horseferry Rd., London SW1P 2AE

IOC UNESCO, 2 Place de Fontenoy, 75007 Paris, France

IOC ASSOCIATION FOR THE CARRIBEAN AND ADJACENT REGIONS (Regional Secretary), Trinidad Al Tobago

KOREA RES. INST. OF SHIP AND OCEAN, P.O. Box 131, Dong Dae Mun, Seoul, Korea

SCIENTIFIC COMMITTEE ON OCEANIC RES., 6 Carlton House Terrace, London SW1Y 5AG

UNAM (Centro de Ciencias del Mar y Limnologia). Apartado Postal 70-305, Mexico 20, D.F., Mexico; (Centro de Preclasificacion Oceanica de Mexico), Apartado Postal 70-233, Mexico 20, D.F. Mexico

UNESCO (Div. of Marine Sciences), 7 Place de Fontenoy, 75007 Paris, France; (Regional Office for South Asia), 40B Lodi Estate, New Delhi 3, India

UNITED NATIONS DEVELOPMENT PROGRAMME, B.P. 1041, Yamoussoukro, Abidjan, Ivory Coast; Apartado Postal 6719, Mexico 10, D.F., Mexico

WORLD BANK, 1818 H. St., N.W., Washington, D.C. 20433

Appendix D

Tables, Living Resources

Only five tables on fisheries are included within this Appendix to the first *Ocean Yearbook* because of the large number of tables incorporated within Sidney Holt's essay on "Marine Fisheries," which examines trends in production and consumption over a period of years. In future editions of the *Yearbook*, most of these tables will be transferred to the tabular appendix and others will be added to complement them.

THE EDITORS

TABLE 1D.—WORLD NOMINAL MARINE CATCH,* BY CONTINENT
(1,000 Metric Tons)

	1970	1971	1972	1973	1974	1975	1976	Annual Trend (%)†
Africa	3,131	3,040	3,397	3,426.5	3,386.0	3,000.8	2,872.9	-.9
America, N.	4,750	4,788	4,616	4,688.7	4,552.5	4,650.7	5,170.2	.7
America, S.	14,629	13,018	6,594	4,309.2	6,640.6	5,763.1	7,152.9	-9.7
Asia	19,453	20,836	21,492	22,902.0	23,624.0	23,543.4	24,171.0	4.0
Europe	11,815	11,909	12,121	12,425.0	12,491.2	12,343.4	13,216.8	1.7
Oceania	194	228	221	274.3	288.2	240.0	278.3	6.2
USSR	6,399	6,402	6,887	7,769.2	8,462.7	8,991.6	9,363.4	10.1
World total:‡	61,137	60,953	56,161	56,714.9	60,247.1	59,480.8	63,117.8	.4

SOURCE.—FAO.
*Nominal marine catch is the total nominal catch minus the nominal inland waters catch.
†The slope of a linear regression divided by the intercept, here expressed as a percentage.
‡The world total is derived from reports on major marine fishing areas (see table 2D); the total exceeds the sum of the figures by continent.

Appendix D: Tables, Living Resources   803

TABLE 2D.--WORLD NOMINAL MARINE CATCH, BY MAJOR FISHING AREA

(1,000 Metric Tons)

	1970	1971	1972	1973	1974	1975	1976	Annual Trend (%)*
Arctic	800	N.A.	N.A.	N.A.	N.A.	N.A.	N.A.	N.A.
Atlantic, N.W.	4,234.3	4,362.8	4,327.3	4,463.3	4,020.3	3,769.6	3,461.3	-2.9
Atlantic, N.E.	10,698.0	10,474.9	10,698.6	11,297.3	11,816.6	12,138.5	13,329.2	4.5
Atlantic, W.C.	1,419.3	1,629.6	1,488.1	1,399.9	1,503.3	1,585.9	1,566.3	.9
Atlantic, E.C.	2,768.2	2,969.0	3,111.4	3,302.8	3,480.4	3,490.7	3,557.3	5.0
Mediterranean and Black	1,153.3	1,113.9	1,165.4	1,155.4	1,366.0	1,283.7	1,275.6	3.0
Atlantic, S.W.	1,100.0	770.0	805.0	951.7	1,001.3	1,000.6	1,206.4	4.2
Atlantic, S.E.	2,517.1	2,481.9	3,012.8	3,171.0	2,859.3	2,583.6	2,872.1	1.5
Atlantic, Antarctic	N.A.	N.A.	N.A.	N.A.	N.A.	N.A.	+	N.A.
Atlantic	23,891.1	23,802.1	24,608.6	25,741.4	26,047.1	25,852.7	27,268.3	2.4
Indian, W.	1,724.1	2,040.4	1,808.9	1,961.6	2,215.1	2,106.4	2,111.1	3.5
Indian, E.	812.5	821.4	821.0	882.4	1,043.0	1,097.3	1,175.8	9.7
Indian, Antarctic	N.A.	N.A.	N.A.	N.A.	.6	1.1	2.3	N.A.
Indian	2,536.6	2,861.8	2,629.9	2,844.0	3,258.8	3,204.8	3,289.1	5.2
Pacific, N.W.	15,008.7	14,285.0	14,531.5	16,526.4	16,675.1	16,999.9	17,231.6	5.7
Pacific, N.E.	2,651.8	2,307.5	2,774.8	1,901.8	2,316.2	2,240.4	2,408.8	-1.8

TABLE 2D. (continued)

(1,000 Metric Tons)

	1970	1971	1972	1973	1974	1975	1976	Annual Trend (%)*
Pacific, N.W.	13,008.7	14,285.0	14,531.5	16,526.4	16,675.1	16,999.0	17,231.6	5.7
Pacific, N.E.	2,651.8	2,307.5	2,774.8	1,901.8	2,316.2	2,240.4	2,408.8	-1.8
Pacific, W.C.	4,216.2	4,531.3	4,770.2	5,058.5	5,155.3	5,143.5	5,430.0	4.5
Pacific, E.C.	907.6	901.9	982.5	1,252.6	1,098.7	1,321.9	1,463.3	12.4
Pacific, S.W.	165.3	223.7	275.2	315.7	363.5	305.6	380.1	19.8
Pacific, S.E.	13,760.1	12,040.1	5,588.6	3,074.5	5,332.5	4,412.0	5,646.5	-11.1
Pacific	34,709.7	34,289.5	28,922.8	28,129.5	30,941.2	30,423.3	32,560.4	-1.3
Total	61,137.4	60,953.4	56,161.3	56,714.9	60,247.1	59,480.8	63,117.8	.4

SOURCE.--FAO.
NOTE.--W.C. = Western Central; E.C. = Eastern Central; N.A. = not available.
*The slope of a linear regression divided by the intercept, expressed as a percentage here.
+Value less than 100 metric tons.

TABLE 3D.—WORLD NOMINAL FISH CATCH, DISPOSITION*
(Million Metric Tons)

	1970	1971	1972	1973	1974	1975
Human consumption	43.5	45.4	45.8	48.2	49.5	48.7
Fresh	19.5	20.1	19.8	20.5	21.3	20.7
Frozen	9.7	10.7	11.2	12.5	12.9	12.7
Cured	8.1	8.0	8.0	8.1	8.1	8.1
Canned	6.2	6.6	6.8	7.1	7.2	7.2
Other purposes	26.5	25.5	20.4	18.6	21.0	21.0
Reduced	25.5	24.5	19.4	17.6	20.0	20.0
Miscellaneous	1.0	1.0	1.0	1.0	1.0	1.0
World catch	70.0	70.9	66.2	66.8	70.5	69.7
World marine catch	61.1	61.0	56.2	56.7	60.2	59.5

SOURCE.—FAO.
*The figures for disposition include fresh water catches. The disposition figures for only the marine catches are not available currently. The figures for world marine catch are included for comparison.

TABLE 4D.--WORLD NOMINAL MARINE CATCH, BY COUNTRY*
(1,000 Metric Tons)

	1970	1971	1972	1973	1974	1975	1976
Anglo-America:							
Canada	1,345.4	1,249.7	1,126.7	1,112.0	989.6	986.2	1,092.6
Greenland	39.8	38.4	41.7	44.5	50.9	47.5	44.7
U.S.A.	2,695.7	2,742.0	2,621.9	2,635.0	2,659.4	2,671.1	2,927.6
Other+	.9	1.0	1.0	.7	.7	.5	.5
Total	4,081.8	4,031.1	3,791.3	3,792.2	3,700.6	3,705.3	4,065.4
Latin America:							
Argentina	209.4	223.3	231.9	294.5	286.2	214.2+	271.9
Brazil	432.8	493.9	523.8	619.4+	672.3+	734.2+	834.3+
Chile	1,209.4	1,506.0	817.5	691.0	1,158.2	929.5	1,264.2
Colombia	21.3	18.4	27.9	32.2	25.2	24.5	23.7
Costa Rica	7.0	7.2	10.8	10.7	13.5	13.9+	12.6
Cuba	105.3	125.4	138.6	149.0	162.8+	171.9+	201.8+
Ecuador	91.4	106.7	108.2	153.9	174.4	223.4	223.4
Guyana	17.4	15.5	17.6+	19.0	23.6	20.1	20.1
Jamaica	8.5	9.3	9.5+	9.6	10.1	10.1	10.1
Mexico	379.3	417.1	448.0+	464.4+	427.9	481.5	554.3
Nicaragua	8.5	8.1	10.0+	11.3+	13.6	14.9	13.4
Panama	53.2+	79.1+	69.0	108.7	88.5	111.4	171.6
Peru	12,532.9+	10,526.1+	4,722.2	2,323.1	4,139.3	3,440.7	4,337.8
Puerto Rico	46.0	57.6	85.4	80.0	76.0	80.7	80.5
Uruguay	13.2	14.4	20.6	17.5	15.7	26.0	33.4
Venezuela	122.6	133.2	1.7	153.8	140.9	145.8	139.1
Other+	63.0	62.1	61.2	67.6	64.3	65.7	65.3
Total	15,321.2	13,803.4	7,447.9	5,205.7	7,492.5	6,708.5	8,257.5
Western Europe:							
Belgium	53.4	60.2	59.0	52.7	46.4	49.0	44.4
Denmark	1,217.1	1,388.0	1,427.8	1,450.5	1,822.1	1,750.6	1,896.4
Faeroe Is.	207.8	207.1	208.0	246.5	247.0	285.6	342.0+
Finland	64.0	71.7	69.3	82.2	87.7	86.7	93.5+
France	782.5	757.7	796.7	822.9	807.5	805.8	805.9
Germany, F.R.	597.9	492.6	403.7	463.2	510.7	426.7	439.4

Appendix D: Tables, Living Resources   807

Greece[+]	91.5	101.1	88.5	88.0	78.6	64.2	64.2
Iceland	733.3	684.4	726.0	901.3	944.3	994.3	985.6
Ireland	78.9	74.0	92.0	90.5	89.5	84.9	94.3
Italy	383.2	386.5	412.5	382.0	413.5	397.9	399.0
Netherlands	298.8	318.1	345.8	340.8	322.5	346.1	281.8
Norway	2,985.7	3,074.9	3,185.6	2,987.4	2,644.9	2,550.4	3,435.3
Portugal	500.2	464.7[+]	451.8	482.0	435.7[+]	374.9[+]	339.1[+]
Spain	1,522.2	1,485.8[+]	1,522.9	1,564.0	1,496.0	1,508.9	1,465.6[+]
Sweden	284.2	227.6	216.1	218.4	203.2	204.9	198.2
U.K.	1,113.8	1,123.9	1,100.7	1,151.2	1,103.3	995.6	1,063.3
Other[+]	2.4	2.8	3.6	3.6	3.8	3.9	3.9
Total	10,916.9	10,921.1	11,110.0	11,327.2	11,256.7	10,930.4	11,951.9
Socialist Eastern Europe:							
Bulgaria	84.5	95.9	101.0	95.0	107.5	150.3[+]	159.2[+]
Germany, D.R.	308.2	324.0	321.4	351.5	349.9	361.4[+]	266.1[+]
Poland	451.3	499.2	522.7	557.4	657.1	777.4	726.3
Romania	24.8	36.6	50.7	60.9	87.5	89.8	76.9
USSR	6,398.8	6,401.6	6,886.9	7,769.2	8,462.7	8,991.6	9,363.4
Yugoslavia	26.7	30.9	31.2	31.0	30.7	32.3	34.8
Other[+]	4.0	4.0	4.0	4.0	4.0	4.0	4.0
Total	7,298.3	7,392.2	7,917.9	8,869.0	9,699.4	10,406.8	10,630.7
Near East:							
Algeria	25.7	23.8	28.3	31.2	35.7	37.7	35.1[+]
Egypt	27.2	34.4	38.8	27.8	27.5	25.9[+]	25.9[+]
Iran[+]	18.0[+]	18.0	16.5[+]	16.9	16.9	16.9	16.9
Morocco	250.0[+]	227.4[+]	247.3[+]	399.5	287.9[+]	212.1[+]	281.0[+]
Oman	180.0[+]	180.0[+]	180.0[+]	180.0[+]	180.0[+]	198.8	198.0
Saudi Arabia	21.7	22.3	23.8	26.4	23.6[+]	23.0[+]	23.3[+]
Tunisia	24.3	27.4	28.0	31.0	42.4	42.4[+]	42.4[+]
Turkey	171.1	151.6	176.0	152.9	243.5	184.8	138.2[+]
United Arab Emirates	40.0[+]	43.0	43.0	43.0	68.0	68.0	68.0[+]
Yemen, Dem.	117.5	119.0	125.1	125.1[+]	127.3	127.3	127.3[+]
Other[+]	40.0	45.0	41.5	47.3	50.7	47.3	47.5
Total	915.5	891.9	948.3	1,081.1	1,103.5	984.2	1,003.6

TABLE 4D (continued)

	1970	1971	1972	1973	1974	1975	1976
Sub-Saharan Africa:							
Angola	368.2	316.3	599.1	472.0+	393.3+	153.6+	153.6+
Cameroon	20.8+	24.5+	21.6+	21.6+	21.6	21.6	21.6+
Congo	9.9+	11.0	18.3+	15.5+	14.8+	14.7+	18.4+
Ethiopia	16.3	18.8	25.8	25.8+	25.8+	25.8+	25.8+
Ghana	141.5+	192.7	249.2	182.4	182.2	212.6	195.8+
Ivory Coast	66.5+	70.6	75.4	59.5	69.3	62.5	72.9
Liberia+	10.7	11.8	12.5	12.5	12.6	12.6	12.6+
Madagascar+	13.1	13.8	14.2	21.2	25.4	14.5	13.5
Mauritania	50.2	62.9	32.4	29.4	21.2	21.2	21.2
Mozambique	7.6	10.4	10.4	13.3	15.7	12.5	10.6
Namibia+	711.2	588.9	527.3	709.7	840.4	760.8	574.1
Nigeria+	217.0	231.7	149.4	155.8	158.5	160.1	165.7
Senegal	169.2	221.0	278.6	303.8	347.0	352.9	350.9+
Sierra Leone	29.6+	29.6	50.1+	65.7+	66.7	66.7+	66.7+
Somalia	30.0+	30.0+	30.0+	30.0+	32.6+	32.6+	32.6+
South Africa	699.3	625.7	663.8	710.0	648.4+	636.0	637.9+
Tanzania	18.6	21.	28.9	23.0	28.3	30.2+	30.2+
Togo	6.4	7.6	7.6+	7.9	8.2+	11.4+	11.4
Zaire	14.7	12.7	13.0+	12.2	13.4+	13.4	7.9
Other+	53.5	50.5	59.6	58.8	59.7	58.2	56.6
Total	2,654.3	2,552.4	2,867.2	2,930.1	2,985.1	2,673.9	2,480.0
South Asia:							
Bangladesh+	90.0	90.0	90.0	90.0	90.0	90.0	90.0
India	1,085.6	1,161.4	971.5	1,210.4	1,472.0	1,478.0	1,525.0
Maldives	34.5	58.9	32.2	33.7	37.5	27.9	32.3
Pakistan	150.6	148.4	185.0	209.1	163.3	167.8	177.2
Sri Lanka	89.8	77.0	93.5	93.7	103.0	115.8	123.4
Total	1,450.5	1,535.7	1,372.2	1,636.9	1,865.8	1,879.5	1,947.9
East Asia:							
China+	2,102.0	2,312.0	2,312.0	2,312.0	2,312.0	2,312.0	2,312.0
Hong Kong	133.3	125.0	132.0	124.8	136.0	147.0	152.7

Japan	9,198.7	9,798.5	10,106.6	10,569.1	10,625.2	10,324.9	10,419.2
Korea, D.P.R.+	800.0	800.0	800.0	800.0	800.0	800.0	800.0
Korea, Rep.	842.1	1,071.5	1,339.8	1,682.2	2,022.1	2,124.6	2,390.0
Macau	9.6	10.2	10.1	10.1+	10.1+	10.1+	10.1+
Taiwan‡	540.5	572.4	613.0	651.0	583.4	652.4	675.1
Total	13,626.2	14,689.6	15,313.5	16,149.2	16,488.8	16,371.0	16,759.1
Southeast Asia:							
Burma	311.4	319.8	329.1	338.1	307.6	355.1	367.2
Indonesia	807.2	820.4	837.3	886.4	948.6+	988.4+	1,043.0+
Kampuchea	20.2	22.0	21.0	10.8	10.8+	10.8+	10.8
Malaysia	338.5	363.9	354.3	440.4	522.2	471.5	514.6
Philippines	916.8	953.2	1,030.8	1,151.4	1,195.1	1,259.8	1,319.8
Singapore	17.3	14.3	14.8	17.9	18.4	17.0	15.7
Thailand	1,347.3	1,482.2	1,550.7	1,540.2	1,355.5	1,392.3	1,464.4
Vietnam+	668.3	731.4	810.9	837.2	837.2	837.2	837.2
Other+	1.5	1.5	1.5	1.5	1.5	1.5	1.6
Total	4,428.5	4,708.7	4,950.4	5,223.9	5,197.0	5,333.6	5,574.3
Australasia:							
Australia	100.8	110.3	123.0	129.0	136.0	107.1	112.5
New Zealand	58.3	64.5	56.1	64.7	67.8	62.0	69.0
Papua New Guinea	18.4	33.0	29.4	47.7	49.9	34.6	50.9
Other	19.7	24.4	27.8	32.9	34.6	36.5	46.1
Total	197.2	232.2	236.3	274.3	288.3	240.2	278.5
Other NEI**	247.0	195.1	206.3	225.3	169.4	247.4	168.9
World total	61,137.4	60,953.4	56,161.3	56,714.9	60,247.1	59,480.8	63,117.8

SOURCE.--FAO, unless otherwise indicated.

*Nominal marine catch equals total nominal catch minus nominal catch from inland waters. Countries which reported catches less than 10,000 metric tons in 1975 are included under "Other" for all years.

+Based on FAO estimates.

‡Estimates based on Republic of China, Executive Yuan, Economic Planning Council, Taiwan Statistical Data Book 1977, p. 70.

**NEI = not elsewhere included.

TABLE 5D.—TRADE OF FISHERY COMMODITIES, BY MAJOR IMPORTING AND EXPORTING COUNTRIES*
(in Thousands of Metric Tons)

	1970	1971	1972	1973	1974	1975
Imports:						
U.S.A.	1,053.1	1,050.9	1,364.0	1,109.2	1,042.9	943.3
Germany, F.R.	963.0	977.6	978.0	760.3	821.9	846.8
Japan	252.6	352.7	432.8	592.2	587.5	642.8
U.K.	731.2	676.6	745.9	653.5	548.3	621.4
France	353.7	361.5	428.0	368.5	359.6	397.5
Italy	319.3	321.9	345.9	290.7	272.5	317.9
Netherlands	340.6	337.2	367.3	257.7	246.0	284.2
Spain	173.2	198.8	254.9	183.0	237.3	176.0
Poland	144.4	130.5	193.5	159.5	214.1	217.2
Denmark	176.4	164.5	161.8	188.4	209.9	173.8
Belgium	217.8	223.7	221.6	172.7	160.5	162.9
Sweden	190.9	199.0	201.2	184.7	152.1	161.8
Singapore	111.8	120.1	122.8	132.0	123.5	137.7
Yugoslavia	143.2	92.6	76.3	96.7	120.3+	79.7
Czechoslovakia	166.4	179.3	167.2	106.7	119.3+	113.6
Switzerland	101.6	103.5	114.0	106.6	110.8	119.7
Portugal	84.0	95.9	114.8	92.3	98.8	...
Norway	61.9	78.9	97.0	47.3	95.1	34.5
Ivory Coast	13.7	20.9	30.6	51.5	90.3	96.5
Ghana	46.3	51.8	76.6	111.0	86.8	...
Hong Kong	61.1	73.0	73.9	81.6	78.9	83.4
Malaysia, W.	71.8	59.9	67.1	66.6	74.4	...
Finland	60.9	65.3	68.3	74.5	73.7	64.6
Australia	70.1	83.9	123.8	61.8	70.8	...
Subtotal	5,909.0	6,020.0	6,827.3	5,949.0	5,895.3	5,675.3
Other	1,472.0	1,629.0	1,390.7	1,126.0	1,350.7	...
Total	7,381.0	7,649.0	8,218.0	7,075.0	7,246.0	7,417.0
Exports:						
Peru	2,122.8	2,049.3	1,936.0	391.9	733.7	917.2
Japan	534.3	586.2	650.6	674.2	706.2	593.4

Appendix D: Tables, Living Resources 811

Norway	663.9	690.3	798.4	770.4	577.6	707.1
Denmark	414.9	446.3	487.2	506.8	572.9	610.8
USSR	316.4	327.8	298.3	301.6	411.8	551.0
Iceland	304.9	269.7	277.3	323.1	300.2	368.7
Canada	371.8	361.0	338.5	354.9	299.5	301.6
South Africa	218.3	127.1	203.2	211.6	265.1+	...
U.S.A.	141.6	193.5	174.2	253.2	211.8	196.3
Netherlands	228.4	254.1	302.8	238.4	217.1	235.0
Spain	177.6	179.5	222.4	188.2	195.7	178.4
U.K.	133.4	137.8	136.3	187.1	186.7	159.4
Germany, F.R.	118.1	127.0	140.0	132.8	182.1+	151.8
Angola	98.0	89.9	189.0	162.5	147.1+	...
Korea, Rep.	64.3	63.7	101.4	176.2	144.9	390.6
Chile	125.8	231.3	103.2	29.6	127.0	...
Faeroe Is.	97.3	96.0	101.4	104.4	109.4	...
Morocco	81.4	85.6	78.2	135.8	102.3	87.8
France	56.8	70.6	77.3	119.5	100.6	88.1
Italy	31.0	46.5	66.3	71.9	97.5	87.1
Malaysia, W.	102.9	104.3	111.5	115.1	96.0	87.4
Thailand	44.1	54.5	81.3	105.2	87.6	...
Poland	58.2	50.9	72.4	80.7	86.6	87.7
Sweden	164.3	123.8	108.2	108.3	84.9	82.2
Argentina	10.7	12.2	25.4	51.8	75.7	...
Subtotal	6,682.2	6,778.9	7,080.8	5,795.8	6,130.0	5,881.6
Other	645.8	839.1	1,000.2	1,092.8	952.0	...
Total	7,328.0	7,618.0	8,081.0	6,888.0	7,082.0	7,685.0

SOURCE.—FAO.

*The term "Fishery Commodities" follows FAO usage and includes the seven principal fishery commodity groups. Countries are ranked by 1974 figures. Countries which reported exports or imports of less than 70,000 metric tons in 1974 are included under "Other."

+Editor's estimate (using a linear extrapolation from reported data or, in the case of Angola, using an average of previous years).

Appendix E

Tables, Non-living Resources

This Appendix consists of only four tables; other tables appear in the chapters by Symonds on oil and gas, and by Charlier on other oceanic resources. Both chapters present an overall view going back several years and inclusion of relevant statistical matter in them is appropriate for this, the first edition of the *Ocean Yearbook*. In later editions, several of those tables will be updated and moved to this Appendix, and the relevant chapters will become more focused on the events of the preceding year.

THE EDITORS

TABLE 1E.—WORLD PRODUCTION OF CRUDE OIL, TOTAL AND OFFSHORE
(Barrels, in Thousands)

Year	World Production	Offshore Production	Offshore as % of World
1970	16,718,708	2,749,290	16.4
1971	17,662,793	3,077,520	17.4
1972	18,600,745	3,317,107	17.8
1973	20,367,981	3,782,733	18.6
1974	20,537,727	3,522,313	17.2
1975*	19,473,903	3,295,585	16.9
1976	20,881,650	3,445,472	16.5

SOURCES.—Basic Petroleum Data Book (Washington, D.C.: American Petroleum Institute, 1976). Data for 1976 from Offshore (June 20, 1977).

*Preliminary.

TABLE 2E.—OFFSHORE CRUDE OIL PRODUCTION, BY REGION AND COUNTRY
(Barrels per Day, in Thousands)

Area and Country	1970	1971	1972	1973	1974	1975	1976
World total	7,532.0	8,232.30	8,858.68	10,067.28	9,293.46	8,264.36	9,414.44
Anglo-America:							
U.S.A.	1,577	1,692	1,664.58	1,697.46	1,427.54	909.59	1,064.00*
Latin America:							
Brazil	8.00	8.20	8.20	17.21	20.37	18.95	35.42
Mexico	35.00	37.00	38.00	20.80	11.90	45.00	45.39
Peru	...	22.00	23.80	31.35	34.42	28.86	31.75
Trinidad/Tobago	76	123	155	110.97	135.10	174.04	180.07
Venezuela+	2,460*	2,490	2,500	2,700	2,071.23	1,737.10	1,677.22*
Subtotal	2,579.0	2,680.2	2,725.0	2,880.3	2,273.0	2,004.0	1,969.8
Western Europe:							
Denmark	...	...	...	2.68	1.91	3.30	8.00
Italy	12	10.90	10.90	12.74	10.23	10.41	10.10
Norway	...	22	32.30	32.30	35.62	189.57	242.61
Spain	...	...	...	20	34.08	32.88	33.33
U.K.	...	...	...	...	...	83.00	446.00
Subtotal	12.0	32.9	43.2	67.7	81.8	319.2	740.0
Socialist Eastern Europe:							
USSR	258	250	236	236	231	228*	220*
Near East:							
Abu Dhabi	269	342	345.20	454.43	512.97	462.71	560.00
Dubai	70	126.50	129.50	221.49	132.89	249.32	308.34
Egypt	257	121	157	130.75	147.15	165.00	231.12
Iran	322	444	467.80	452.41	455.19	481.19	426.54
Neutral Zone	...	380	409.80	394.04	333.90*	315.07	247.10
Qatar	172	...	...	...	...	...	...
Saudi Arabia+	1,251	1,210	1,490.70	1,990.40	2,024.59	1,385.81	1,694.80*

TABLE 2E (continued)

Area and Country	1970	1971	1972	1973	1974	1975	1976
Sharjah	...	...	...	...	...	38.36	37.00
Tunisia	0	0	0	0	24.85	43.00	37.00
Subtotal	2,341.0	2,623.5	3,000.0	3,643.5	3,631.5	3,140.5	3,541.9
Sub-Saharan Africa:							
Angola	96	131	137.40	144.23	140.44	143.20	33.61*
Congo	...	.50	7.90	34.80	45.97	37.35	38.03
Gabon	29	27.20	43.60	57.20	59.45	179.88	165.92
Nigeria	275	361	409	518.44	648.92	431.33	525.05
Zaire	...	...	...	...	...	...	19.96
Subtotal	400.0	519.7	597.9	754.7	894.8	791.7	782.6
South Asia:							
India	...	...	...	...	...	...	15.00
East Asia:							
Japan	3	2	1.19	1.00	1.91	.86	3.00
Southeast Asia:							
Brunei/Malaysia	146	138	218	264.16	287.10	...	...
Brunei	...	...	...	...	...	141.22	170.17
Malaysia	...	...	...	...	...	84.49	151.40
Indonesia	...	32	69.90	174.22	247.36	246.45	425.98
Subtotal	146.0	170.0	287.9	438.4	534.5	472.2	747.6
Australasia:							
Australia	216	262	303	348.20	217.28	412.52	348.00

SOURCES.—Offshore (June 20, 1975; June 20, 1976; June 20, 1977). (Data vary somewhat from American Petroleum Institute figures.)

*Estimated.

†Many fields in these countries lie both onshore and offshore. Since production is reported as a single total, it is impossible to precisely separate the exact amount flowing from offshore.

TABLE 3E.--WORLD PRODUCTION OF NATURAL GAS, TOTAL, AND OFFSHORE
(Cubic Feet, in Millions)

Year	World Production	Offshore Production	Offshore as % of World
1970	38,093,961	5,261,020*	13.8
1971	40,810,659	6,148,461*	15.1
1972	43,463,410	6,824,375*	15.7
1973	56,992,337	7,697,045	13.5
1974	47,253,322	8,088,727	17.1
1975	47,029,914	9,532,076	20.3
1976	48,071,000	10,861,000	22.6

SOURCES.--Basic Petroleum Data Book (Washington, D.C.: American Petroleum Institute, 1976). Data for 1976 estimated from Offshore, and from U.N. Statistical Yearbook, 1976.

*Reflects Free World offshore production since production figures of Communist nations are not available.

TABLE 4E.—OFFSHORE NATURAL GAS PRODUCTION, BY REGION AND COUNTRY
(Cubic Feet per Day, in Millions)

Area and Country	1970	1971	1972	1973	1974	1975	1976
World total	14,188.77	16,615.84	13,048.28	14,920.51	16,395.04	17,141.57	29,756.82
Anglo-America:							
U.S.A.	8,591.78	10,046.58	10,294.29	10,891.33	11,588.36	11,664.27	11,864.46
Latin America:							
Brazil	...	...	...	...	21.50	25.00	26.00
Peru	64.66	62.75	61.17	60.79	69.85	77.00	15.00*
Trinidad/Tobago	10.96	10.69	10.69	16.79	18.00	123.00	311.20
Venezuela	...	...	...	1,200	1,000	...	...
Subtotal	75.6	73.4	71.9	1,277.6	1,109.4	225.0	352.2
Western Europe:							
Netherlands	...	...	...	...	...	186.00	298.40*
Norway	...	...	...	...	16.50	16.50	560.00*
U.K.	1,086.30	1,794.25	2,560.36	3,000	3,600	3,600	13,912.10
Subtotal	1,086.30	1,794.25	2,560.36	3,000	3,616.5	3,802.5	14,770.5
Socialist Eastern Europe:							
USSR	...	...	...	670.00	725.00	774.00	897.00
Near East:							
Abu Dhabi	...	...	...	...	...	...	510.00
Egypt	...	...	...	...	...	120.00	...

Appendix E: Tables, Nonliving Resources   819

Iran	3,500*	3,578.9	4,500*	3,360	3,185*	...
Neutral Zone	193.97	201.10	204.13	181.58	176.5*	...
Saudi Arabia+	494.80	587.12	690.17	721.74	711.40	3,825.39
Qatar	107.12	130.60	121.71	124.36	124.90*	...
Subtotal	4,295.9	4,497.7	5,516.0	4,387.7	4,197.8	3,945.4
Sub-Saharan Africa:						
Angola	78.63	105.75	109.67	558.10	547.00*	...
Ghana	...	...	...	...	3.70	3.70
Nigeria	...	...	...	...	...	314.00
Subtotal	78.6	105.8	109.7	558.1	550.7	317.7
South Asia:						
India	...	...	...	...	...	...
Southeast Asia:						
Indonesia	0	0	0	108.00	155.05	158.80
Australasia:						
Australia	60.55	98.10	121.77	173.60	217.28	203.00

	200.00
	710.0
	...
	500.00*
	500.0
	24.00
	345.66
	293.00

SOURCES.—Offshore (June 20, 1975; June 20, 1976; June 20, 1977).
*Estimated.
+Almost all the natural gas produced in Saudi Arabia is associated with oil production, and none of the gas is marketed. The jump in figures is due to reporting changes by the government.

Appendix F

Tables, Transportation and Communications

The tables in this section of the tabular appendix deal only with shipping. In future editions of the *Ocean Yearbook* tables on satellite communication, underwater pipelines, and cables will be included in addition to updated versions of the current tables.

THE EDITORS

TABLE 1F.—WORLD SHIPPING TONNAGE, BY TYPE OF VESSEL

(Million grt)

	1970	1971	1972	1973	1974	1975	1976
Oil tankers	86.1	96.1	105.1	115.4	129.5	150.1	168.2
Liquefied gas carriers*	1.4	1.6	1.9	2.3	2.4	3.0	3.4
Chemical carriers	.5	.6	.6	.7	.7	1.0	1.3
Miscellaneous tankers	N.A.	N.A.	.1	.1	.1	.1	N.A.
Bulk oil carriers+	8.3	10.7	15.1	19.5	22.0	23.7	25.0
Ore and bulk carriers	38.3	43.1	48.4	53.1	57.4	61.8	66.7
General cargo‡	72.4	71.9	70.6	69.5	68.7	70.4	73.6
Miscellaneous cargo ships	N.A.	N.A.	.5	.6	.4	.4	N.A.
Container ships (fully cellular)	1.9	2.8	4.3	5.9	6.3	6.2	6.7
Barge-carrying vessels	N.A.	N.A.	.5	.6	.7	.8	N.A.
Vehicle carriers	N.A.	N.A.	.5	.4	.5	.4	N.A.
Fishing factories, carriers, and trawlers	7.8	9.0	9.6	10.3	10.7	11.3	N.A.
Passenger liners	3.0	3.0	3.1	3.3	2.9	2.8	N.A.
Ferries and other passenger vessels	N.A.	N.A.	3.8	4.2	4.4	4.6	N.A.
All other vessels§	7.8	8.3	4.3	4.5	4.8	5.3	N.A.

SOURCE.—Lloyd's Register of Shipping: Statistical Tables.
NOTE.—grt = gross registered tons; N.A. = not available.
*I.e., ships capable of transporting liquid natural gas or liquid petroleum gas or other similar hydrocarbon and chemical products which are all carried at pressures greater than atmosphere or at subambient temperature or a combination of both.
+Including ore/oil carriers.
‡Including passenger cargo.
§Including livestock carriers, supply ships and tenders, tugs, cable ships, dredgers, ice-breakers, research ships, and others.

TABLE 2F.--ESTIMATED AVERAGE SIZE OF SELECTED TYPES OF VESSELS: EXISTING WORLD FLEETS AND VESSELS ON ORDER

	1970	1971	1972	1973	1974	1975
	Existing World Fleets (Mid-Year, in grt)					
Oil tankers (100 grt and above)	14,110	15,280	16,270	17,460	19,085	21,363
Ore/bulk carriers* (6,000 grt and above)	18,450	19,490	20,830	21,990	22,755	23,052
Container ships (100 grt and above)	11,420	12,040	13,810	14,970	15,270	14,859
Liquefied gas carriers (grt)	4,690	4,960	5,370	6,090	6,052	7,123
All other ships (100 grt and above)	2,110	2,040	1,980	1,920	1,870	1,908
	Vessels on Order (at Year End, in dwt)					
Tankers (10,000 dwt and above)	166,700	176,960	163,720	171,365	163,261	149,486
Ore/bulk carriers (10,000 dwt and above)	69,010	63,430	65,020	55,772	52,947	47,677
Container ships (300+ container capacity)	19,610	19,330	18,580	17,090	18,783	17,801
Liquefied gas carriers (12,000 dwt and above)	19,210	20,160	26,350	51,620	53,123	56,185
All other ships (1,000 grt and above)	9,780	9,930	9,670	9,561	10,554	11,774

SOURCE.--UNCTAD.
NOTE.--grt = gross registered tons; dwt = dead weight tons.
*Including bulk/oil carriers.

TABLE 3F.—WORLD SHIPPING TONNAGE, BY GROUPS OF COUNTRIES
(Mid-Year Figures)

Flags of Registration in Groups of Countries	Tonnage and Shares (in Parentheses)*							
	In grt (Million)				In dwt (Million)		
	1965	1970	1974	1975	1970	1974	1975
1. World total	146.8 (100.0)	217.9 (100.0)	306.1 (100.0)	336.9 (100.0)	326.1 (100.0)	486.9 (100.0)	546.3 (100.0)
2. Developed market-economy countries (excluding Southern Europe)†	90.6 (61.7)	124.2 (57.0)	155.6 (50.8)	165.5 (49.1)	186.4 (57.2)	246.7 (50.7)	266.4
3. Open registry countries‡	22.1 (15.1)	40.9 (18.8)	74.5 (24.3)	88.4 (26.2)	70.3 (21.6)	133.5 (27.4)	161.9
4. Southern Europe (excluding Cyprus)	11.8 (8.0)	17.6 (8.1)	30.8 (10.1)	32.1 (9.6)	25.6 (7.8)	49.2 (10.1)	51.8
5. Total lines 2-4	124.5 (84.8)	182.7 (83.9)	260.9 (85.2)	286.0 (84.9)	282.3 (86.6)	429.4 (88.2)	480.1 (87.9)
6. Socialist countries of Eastern Europe and Asia	10.9 (7.4)	19.5 (8.9)	25.3 (8.3)	28.3§ (8.4)	21.7 (6.6)	28.9 (6.0)	33.1§ (6.1)

7. Developing countries	10.7 (7.3)	14.5 (6.7)	18.5 (6.0)	21.2 (6.3)	20.4 (6.3)	26.5 (5.4)	30.9 (5.6)
Total (excluding open registry countries):							
In Africa	.6	.8	1.5	1.8	1.1	2.0	2.5
In Asia	5.5	8.0	9.0	11.8	11.7	14.7	17.7
In Latin America and the Caribbean	4.6	5.7	7.0	7.5	7.6	9.7	10.6
In Oceania	...	...	.1	.1	0.0	.1	.1
8. Other, unallocated	.7 (.5)	1.2 (.5)	1.4 (.5)	1.4 (.4)	1.7 (.5)	2.1 (.4)	2.2 (.4)

SOURCE.--Lloyd's Register of Shipping: Statistical Tables.
NOTE.--grt = gross registered tons; dwt = dead weight tons.
*Includes vessels of 100 grt and above.
†Excluding U.S. reserve fleet, U.S. Great Lakes fleet, and Canadian Great Lakes fleet.
‡Cyprus, Liberia, Oman, Panama, Singapore, and Somalia.
§The tonnages for the Socialist countries of Eastern Europe, including the USSR, were 25.4 million grt and 38.8 million dwt, respectively, while those of the Socialist countries of Asia were 2.9 million grt and 4.3 million dwt, respectively.

TABLE 4F.—WORLD MERCHANT FLEETS, BY REGION AND COUNTRY
(in grt as of July 1)

	1971	1972	1973	1974	1975	1976	1977
Anglo America:							
Canada*	831,118	847,402	9 0,719	933,388	988,726	2,638,692	2,822,948
U.S.A.+	9,565,669	9,023,873	10,676,606	10,767,679	10,931,002	14,908,445	15,299,681
Sub-total	10,396,787	9,871,275	11,607,325	11,701,067	11,919,728	17,545,137	18,122,629
Latin America:							
Argentina	1,311,847	1,401,075	1,452,552	1,408,129	1,447,165	1,469,754	1,677,169
Bahamas	357,845	205,862	179,494	153,202	189,890	147,817	106,317
Barbados	1,384	1,676	2,958	3,897	3,897	3,897	4,448
Belize	620	620	620	620	620	620	620
Brazil	1,730,877	1,884,537	2,103,319	2,428,972	2,691,408	3,096,293	3,329,951
Cayman Is.	26,643	26,172	44,419	39,717	49,320	78,251	123,787
Chile	387,810	382,013	383,886	364,364	386,322	409,756	405,971
Colombia	208,337	231,994	223,881	211,083	208,407	211,691	247,240
Costa Rica	3,107	4,359	9,062	5,603	5,102	6,257	6,811
Cuba	384,885	398,030	416,305	409,064	476,279	603,750	667,518
Dominican Rep.	8,881	8,881	9,381	11,963	9,920	8,469	569
Ecuador	45,451	56,807	75,975	128,473	142,356	180,623	197,244
El Salvador	4,259	1,506	443	291	1,957	2,128	1,987
Falkland Is.	9,848	9,494	9,494	7,931	7,931	6,937	6,937
Grenada	534	343	226	226	226	226	226
Guatemala	3,629	3,629	8,222	8,222	9,584	8,197	11,854
Guyana	13,647	13,735	15,035	15,869	16,828	19,105	16,274
Honduras	69,683	74,030	67,274	69,561	67,923	71,042	104,903
Jamaica	12,899	13,819	12,899	6,740	6,740	6,892	7,075
Mexico	400,665	416,832	453,024	514,544	574,857	593,875	673,964
Monserrat	711	711	711	949	949	1,130	1,248
Nicaragua	10,877	21,845	21,845	33,240	32,700	26,415	34,588
Panama	6,262,264	7,793,598	9,568,954	11,003,227	13,667,123	15,631,180	19,458,419
Paraguay	21,884	21,884	21,930	21,930	21,930	21,930	21,930
Peru	420,656	446,374	448,325	513,875	518,316	525,137	555,419
St. Kitts-Nevis-Anguilla	396	396	652	256	405	405	256
St. Lucia	517	517	904	904	904	904	928
St. Vincent	664	1,477	2,247	4,808	5,507	5,663	8,428
Surinam	N.A.	N.A.	N.A.	N.A.	N.A.	4,890	7,277

Appendix F: Tables, Transportation and Communications 827

Trinidad	21,263	17,988	15,659	15,574	13,864	13,603	17,192
Turks and Caicos Is.	1,575	1,572	1,572	1,572	1,572	2,405	2,405
Uruguay	162,774	142,828	142,664	130,147	130,998	151,255	192,792
Venezuela	411,696	411,242	478,643	480,230	515,661	543,446	639,396
Virgin Is. (U.K.)	713	713	876	1,127	2,420	2,409	4,057
Subtotal	12,299,341	13,996,559	16,173,451	17,996,310	21,210,181	23,856,352	28,543,100
Western Europe:							
Austria	11,387	30,788	97,769	97,067	75,396	82,982	53,284
Belgium	1,183,081	1,191,555	1,161,609	1,217,707	1,358,425	1,499,431	1,595,489
Denmark	3,520,021	4,019,927	4,106,525	4,460,219	4,478,112	5,143,022	5,331,165
Finland	1,470,825	1,630,473	1,545,626	1,507,582	2,001,618	2,115,322	2,262,095
France	7,001,476	7,419,596	8,288,773	8,834,519	10,745,999	11,278,016	11,613,859
Gibraltar	27,413	21,375	20,855	28,293	28,850	21,526	10,549
Germany, F.R.	8,678,584	8,515,669	7,914,679	7,980,453	8,516,567	9,264,671	9,592,314
Greece	13,065,930	15,328,860	19,295,143	21,759,449	22,527,156	25,034,585	29,517,059
Iceland	125,912	130,561	142,777	148,695	154,381	162,268	166,702
Ireland	174,459	182,319	229,349	208,700	210,389	201,965	211,872
Italy	8,138,521	8,187,323	8,867,205	9,322,015	10,136,989	11,077,549	11,111,182
Malta	34,500	14,641	11,022	38,011	45,950	39,140	100,420
Monaco	17,541	33,203	28,062	27,292	14,588	3,998	...
Netherlands	5,269,145	4,972,244	5,029,443	5,500,932	5,679,413	5,919,892	5,290,360
Norway	21,720,202	23,507,108	23,621,096	24,852,917	26,153,682	27,943,834	27,801,471
Portugal	925,793	1,027,070	1,271,815	1,243,128	1,209,701	1,173,710	1,281,439
Spain	3,934,129	4,300,055	4,833,048	4,949,146	5,433,354	6,027,763	7,186,081
Sweden	4,978,278	5,632,336	5,669,340	6,226,659	7,486,196	7,971,246	7,429,394
Switzerland	199,591	211,728	202,764	199,732	193,657	212,526	252,746
U.K.	27,334,695	28,624,875	30,159,543	31,566,298	33,157,422	32,923,308	31,646,351
Subtotal	107,811,483	114,981,706	122,496,443	130,168,814	139,607,845	148,096,754	152,453,832
Socialist Eastern Europe:							
Albania	56,523	57,001	57,068	57,368	57,368	57,368	55,870
Bulgaria	703,878	741,986	756,749	864,939	937,458	933,361	964,156
Czechoslovakia	82,731	103,049	86,510	116,148	116,148	148,689	148,689
Germany, D.R.	1,016,205	1,198,365	1,219,037	1,223,859	1,389,000	1,437,054	1,486,838
Hungary	33,061	33,811	53,580	49,150	47,943	54,926	63,016
Poland	1,760,397	2,012,659	2,072,531	2,292,318	2,817,129	3,263,206	3,447,517
Romania	363,996	455,622	474,497	610,982	777,309	994,184	1,218,171

TABLE 4F (continued)

	1971	1972	1973	1974	1975	1976	1977
USSR	16,194,326	16,733,674	17,396,900	18,175,918	19,235,973	20,667,892	21,438,291
Yugoslavia	1,543,149	1,587,585	1,667,183	1,778,423	1,873,482	1,943,750	2,284,526
Subtotal	21,754,266	22,923,752	23,784,055	25,169,105	27,251,810	29,500,430	31,107,074
Sub-Saharan Africa:							
Benin	N.A.	206	474	474	656	656	912
Cameroon	1,399	2,334	2,895	3,199	3,199	19,045	78,180
Congo	4,649	1,070	1,210	1,534	1,846	2,453	4,172
Ethiopia	46,307	45,903	48,093	25,034	24,953	24,953	23,989
Gabon	1,307	1,519	12,428	33,159	106,738	98,285	98,645
Gambia	1,135	1,135	1,651	1,337	1,337	1,337	1,608
Ghana	165,748	166,183	165,565	173,018	180,351	183,089	182,696
Guinea	12,468	15,538	15,538	15,538	15,054	15,280	12,597
Guinea-Bissau	N.A.	N.A.	N.A.	N.A.	N.A.	N.A.	219
Ivory Coast	42,156	82,316	88,749	121,276	119,215	114,191	115,717
Kenya	22,658	21,857	21,722	21,829	17,331	15,469	15,192
Liberia	38,552,240	44,443,652	49,904,744	55,321,641	65,820,414	73,477,326	79,982,968
Madagascar	21,424	52,162	63,919	53,409	44,273	49,738	39,850
Mauritania	1,959	1,681	1,681	1,681	1,681	1,113	1,113
Mauritius	26,177	26,088	15,564	33,281	33,105	35,146	37,288
Mozambique	N.A.	N.A.	N.A.	N.A.	149	13,825	27,618
Nigeria	95,938	99,226	110,015	121,301	142,050	181,565	335,540
Senegal	13,685	16,280	17,032	20,499	23,261	26,621	28,044
Seychelles	306	306	306	1,901	1,901	1,901	59,140
Sierra Leone	N.A.	1,795	3,047	5,045	17,209	17,209	7,298
Somalia	592,664	873,209	1,612,656	1,916,273	1,813,313	1,792,900	158,166
South Africa	538,493	511,190	490,751	535,322	565,575	477,011	476,324
Sudan	23,560	35,502	38,278	45,943	45,578	45,578	43,375
Tanzania	18,218	18,718	28,371	28,371	33,449	34,934	35,613
Uganda	5,510	5,510	5,510	5,510	5,510	5,510	5,510
Zaire	39,317	40,221	38,966	38,966	85,232	107,278	109,785
Zambia	5,513	5,513	5,513	5,513	5,513	5,513	5,513
Subtotal	40,232,831	46,469,114	52,694,678	58,531,054	69,108,793	76,747,926	81,887,072
Near East:							
Algeria	94,838	132,756	162,832	239,815	246,432	463,094	1,055,962

Appendix F: Tables, Transportation and Communications   829

Bahrain	10,126	10,126		5,140	3,670	25,096	6,409
Cyprus	1,498,114	2,014,675	2,935,775	3,394,880	3,221,070	3,114,263	2,787,908
Egypt	241,429	242,745	268,747	248,591	301,383	376,066	407,818
Iran	131,667	180,659	192,386	291,928	479,718	683,329	1,002,061
Iraq	46,435	121,399	228,274	229,603	310,594	748,774	1,135,245
Israel	645,585	698,068	645,391	611,300	451,323	481,594	404,651
Jordan	N.A.	200	6,187	200	200	200	696
Kuwait	646,548	656,403	676,879	681,692	990,857	1,106,816	1,831,194
Lebanon	127,235	116,571	119,468	120,180	167,490	213,572	227,009
Libya	4,692	5,932	36,878	160,180	241,725	458,805	673,969
Morocco	55,585	46,907	56,125	52,564	79,863	136,596	270,295
Oman	N.A.	2,013	2,249	2,249	3,159	3,374	6,137
Qatar	803	803	803	928	1,389	75,747	84,710
Saudi Arabia	45,492	50,369	58,530	61,275	180,246	588,745	1,018,713
Syria	1,020	1,659	2,057	2,643	7,531	10,192	20,679
Tunisia	27,933	28,268	28,408	28,561	40,827	62,941	100,128
Turkey	713,767	743,071	756,807	971,682	994,668	1,079,347	1,288,282
United Arab Emirates	N.A.	12,296	10,498	28,445	50,638	143,109	152,100
Yemen, A.R.	2,844	2,844	2,844	1,260	1,260	1,260	1,436
Yemen, Dem.	1,417	1,417	1,680	2,180	5,860	6,654	6,390
Subtotal	4,295,530	5,069,181	6,195,863	7,135,296	7,779,893	9,779,574	12,481,792
South Asia:							
Bangladesh	N.A.	28,888	60,601	115,612	133,016	146,818	244,314
India	2,478,031	2,649,677	2,886,595	3,484,751	3,869,187	5,093,984	5,482,176
Maldive Is.	48,707	62,230	76,963	78,663	95,154	121,462	110,681
Pakistan	581,753	532,637	503,429	494,065	479,358	483,433	475,600
Sri Lanka	10,039	13,017	43,754	54,099	80,862	91,031	92,581
Subtotal	3,118,530	3,286,449	3,571,342	4,227,190	4,657,577	5,936,728	6,405,352
East Asia:							
China	1,022,256	1,181,179	1,478,992	1,870,567	2,828,290	3,88,726	4,245,446
Hong Kong	572,243	457,924	342,529	265,945	418,512	423,218	609,679
Japan	30,509,280	34,929,214	36,785,094	38,707,659	39,739,598	41,663,188	40,035,853
Korea, D.P.R.	50,556	50,55	60,347	60,347	81,782	89,482	89,482
Korea, Rep.	940,009	1,057,408	1,103,925	1,224,679	1,623,532	1,796,106	2,494,724
Taiwan	1,321,758	1,494,900	1,467,300	1,416,833	1,450,000	1,483,981	1,558,713
	34,416,102	39,17,181	41,238,187	43,551,030	46,141,714	49,044,701	49,033,897

TABLE 4F (continued)

	1971	1972	1973	1974	1975	1976	1977
Southeast Asia:							
Brunei	N.A.	N.A.	N.A.	N.A.	283	899	899
Burma	54,617	54,877	54,877	54,877	54,548	68,867	67,502
Indonesia	618,805	618,589	668,964	762,278	859,378	1,046,198	1,163,173
Kampuchea	4,230	1,880	2,090	2,090	1,208	1,208	3,558
Malaysia	85,743	149,30	226,350	337,511	358,795	442,0	563,666
Philippines	945,508	924,564	947,210	766,748	879,043	1,018,065	1,146,529
Singapore	81,777	870,513	2,004,269	2,878,327	3,891,902	5,481,720	6,791,398
Thailand	86,222	108,271	182,043	176,315	182,554	194,993	260,664
Vietnam, D.P.R.	5,002	5,002	,002	9,151	12,011		
Vietnam, Rep.	32,333	31,979	37,980	43,202	57,615	107,456	128,525
Subtotal	2,414,237	2,764,979	4,128,785	5,030,499	6,297,337	8,362,086	10,125,914
Australasia:							
Australia	1,105,236	1,184,010	1,160,205	1,168,367	1,205,248	1,247,172	1,374,197
Fiji	6,380	4,839	7,151	7,041	7,674	10,604	10,879
Gilbert and Ellice Is.	2,193	2,193	2,193	1,518	1,518	1,333	1,333
Nauru	23,761	23,761	46,504	58,265	48,271	48,353	48,353
New Hebrides	3,883	3,678	4,369	4,916	4,916	5,023	12,189
New Zealand	181,046	181,901	156,503	163,399	162,520	164,192	199,462
Papua New Guinea	24,779	26,186	27,827	17,598	14,550	15,329	16,217
Solomon Is.	629	629	629	629	629	1,008	1,746
Tonga	2,502	2,502	2,502	9,081	9,644	13,722	14,180
Subtotal	1,350,409	1,429,699	1,497,883	1,430,814	1,454,970	1,506,734	1,678,556

SOURCE.--Lloyd's Register of Shipping: Statistical Tables.
NOTE.--grt = gross registered tons; N.A. = not available.
*Excluding Great Lakes.
†Estimated active sea-going fleet.

TABLE 5F.--WORLD SHIPBUILDING: MERCHANT VESSELS UNDER CONSTRUCTION, BY COUNTRY

(1,000 grt)

	1970	1971	1972	1973	1974	1975	1976
Belgium	199	228	219	278	231	282	300
British Commonwealth:							
Australia	128	185	181	124	107	136	100
Canada	18	182	277	194	206	245	194
Other*	216	480	590	456	521	735	607
Denmark	511	523	459	448	663	533	254
Finland	204	202	184	297	339	522	599
France	1,222	1,307	1,441	1,450	1,705	1,859	1,267
Germany, F.R.	2,118	1,671	1,504	1,738	2,052	1,775	1,477
Italy	2,081	1,835	1,610	1,470	1,414	1,810	1,030
Japan	6,471	7,137	8,379	9,944	11,164	12,497	8,052
Korea, Rep.	25	26	32	306	513	996	639
Netherlands	656	846	835	978	1,170	759	378
Norway	567	452	734	889	811	760	572
Poland	468	552	541	522	529	710	859
Portugal	20	16	30	33	19	388	275
Spain	1,248	1,433	1,378	1,798	2,301	2,165	2,362
Sweden	1,446	1,675	1,736	2,376	2,253	2,166	2,498
U.K.	1,649	1,627	1,783	2,069	1,934	1,880	1,901
U.S.A.	861	1,050	1,322	1,436	1,718	2,108	2,512
Yugoslavia	602	645	571	983	896	887	962

TABLE 5F (continued)

	1970	1971	1972	1973	1974	1975	1976
World:+				(1,000 grt)			
Steam	9,262	9,541	11,295	13,962	14,610	16,641	9,914
Motor	12,248	13,082	13,056	14,796	17,094	18,000	18,307
World total‡	21,510	22,623	24,351	28,758	31,704	36,641	28,220

SOURCE.--United Nations.
NOTE.--Includes vessels of 100 gross tons and over; but excludes nonpropelled vessels and wooden vessels; grt = gross registered tons.
*British Commonwealth countries other than the U.K.
+Excludes U.S.S.R. and China.
‡Total figures shown for each period represent the sum of the published figures, and, thus, they do not include estimates for periods where figures are not available.

TABLE 6F.--VESSELS LOST, 1976

Flag State	Number of Vessels Lost	Flag State	Number of Vessels Lost
Brazil	2	Libya	2
Canada	3	Maldive Is.	1
Cayman Is.	2	Mexico	3
Chile	1	Netherlands	1
Colombia	1	Norway	4
Cuba	1	Pakistan	1
Cyprus	20	Panama	52
Denmark	1	Philippines	4
Egypt	1	Romania	1
Fiji	1	Saudi Arabia	2
Germany, D.R.	2	Senegal	1
Germany, F.R.	4	Sharjah	1
Greece	25	Singapore	2
India	1	Somalia	1
Indonesia	4	Spain	7
Iran	2	Taiwan	1
Italy	3	Thailand	1
Japan	12	Turkey	1
Korea, Rep.	7	U.K.	1
Kuwait	1	U.S.A.	3
Lebanon	1	USSR	3
Liberia	20		
		Total	208

SOURCE.--Lloyd's Intelligence Department.

Appendix G

Tables, Military Activities

The tables in this section of the tabular appendix fall into two groups.

The first consists of tables 1G–8G which were compiled by Ronald Huisken of SIPRI and are supplementary to the tables included in his chapter on naval forces. In later editions of the *Ocean Yearbook,* a number of the tables in that chapter will be moved to the tabular appendix and updated on a continuing basis.

The second group consists of tables 9G–17G. These were compiled by Frank Barnaby and Andrzej Karkoszka, also of SIPRI. They represent a comprehensive summary of antisubmarine-warfare detection and weapons systems. This cluster is of particular value when perused in association with the text materials in the *Yearbook* on antisubmarine warfare prepared by Mr. Barnaby and members of the SIPRI staff.

THE EDITORS

TABLE 1G.--WORLD STOCK OF FIGHTING SHIPS, ESTIMATED VALUES

(US$ Millions)

	1950	1955	1960	1965	1970	1976
World total	35,092	48,748	48,215	74,131	98,251	129,129
Developed	33,299	46,313	44,857	69,504	92,057	118,194
U.S.A.	20,755	25,717	23,859	37,779	44,285	43,915
Other NATO	7,976	11,049	8,348	9,985	15,071	18,353
Total NATO	28,731	36,766	32,207	47,764	59,356	62,268
USSR	3,168	7,369	9,801	17,890	27,783	50,033
Other WTO	92	150	458	735	1,127	772
Total WTO	3,278	7,519	10,259	18,623	28,910	50,805
Other Europe	779	1,121	1,314	1,608	1,645	2,072
Other developed	511	907	1,076	1,509	2,146	3,049
Total Third World	1,793	2,435	3,358	4,627	6,194	10,935
Middle East	124	171	269	375	666	1,310
South Asia	147	215	339	506	611	994
Far East	358	505	718	1,157	1,508	2,053
Sub-Saharan Africa	...	...	9	63	99	307
North Africa	...	...	4	25	175	220
Central America	207	208	205	318	331	470
South America	867	1,175	1,333	1,460	1,452	2,280
China	89	162	481	722	1,352	3,301

SOURCE.--SIPRI.

NOTE.--Figures may not add up to totals due to rounding; WTO = Warsaw Treaty Organization.

TABLE 2G.—WORLD STOCK OF AIRCRAFT CARRIERS, BY COUNTRY

	1950	1955	1960	1965	1970	1976
World total:						
Attack	44	43	30	29	25	21
ASW/amph.	...	10	16	26	24	17
Other	75	73	26	17	6	1
U.S.A.:						
Attack	27	19	14	16	15	15
ASW/amph.	...	10	14	22	18	12
Other	75	73	26	17	6	1
U.K.:						
Attack	12	16	8	5	4	1
ASW/amph.	...	...	1	2	2	2
France:						
Attack	2	4	3	3	2	2
ASW/amph.	...	...	...	...	1	1
Australia:						
Attack	1	2	2	1	1	1
ASW/amph.	...	...	...	1	1	...
Canada:						
Attack	1	1	1	1	...	...
ASW/amph.	...	...	...	...	...	...
Netherlands:						
Attack	1	1	1	1	...	...
ASW/amph.	...	...	...	...	...	...
Spain:						
Attack	...	...	...	...	...	...
ASW/amph.	...	...	...	...	1	1
Argentina:						
Attack	...	...	1	1	2	1
ASW/amph.	...	...	...	...	...	...
Brazil:						
Attack	...	...	...	...	...	...
ASW/amph.	...	...	1	1	1	1
India:						
Attack	...	...	...	1	1	1
ASW/amph.	...	...	...	...	...	...

SOURCE.—SIPRI.
NOTE.—ASW = antisubmarine warfare; amph. = amphibious assault; Other = escort and utility.

TABLE 3G.—WORLD STOCK OF STRATEGIC SUBMARINES,* BY GROUPS OF COUNTRIES

	1950	1955	1960	1965	1970	1976
World total:						
Nucl.	...	...	7	42	70	104
Conv.	...	...	10	32	26	24
U.S.A.:						
Nucl.	...	...	3	33	41	41
Conv.	...	...	...	...	...	...
Other NATO:						
Nucl.	...	...	...	...	5	8
Conv.	...	...	...	...	1	1
Total NATO:						
Nucl.	...	...	3	33	46	49
Conv.	...	...	...	...	1	1
USSR:						
Nucl.	...	...	4	9	24	55
Conv.	...	...	10	32	25	23
Other WTO:						
Nucl.	...	...	...	...	...	...
Conv.	...	...	...	...	...	...
Total WTO:						
Nucl.	...	...	4	9	24	55
Conv.	...	...	10	32	25	23

SOURCE.—SIPRI.
NOTE.—Nucl. = nuclear powered; Conv. = conventionally powered; WTO = Warsaw Treaty Organization.
*Equipped with medium- or long-range ballistic missiles.

TABLE 4G.—WORLD STOCK OF PATROL SUBMARINES,* BY GROUPS OF COUNTRIES

	1950	1955	1960	1965	1970	1976
World total:						
Nucl.	...	1	15	48	108	157
Conv.	355	532	535	576	535	498
Developed:						
Nucl.	...	1	15	48	108	157
Conv.	351	527	502	516	459	365
U.S.A.:						
Nucl.	...	1	11	22	46	68
Conv.	194	190	158	139	52	11
Other NATO:						
Nucl.	...	...	...	...	4	9
Conv.	105	109	89	78	75	74
Total NATO:						
Nucl.	...	1	11	22	50	77
Conv.	299	299	247	217	127	85
USSR:						
Nucl.	...	...	4	26	58	80
Conv.	46	215	238	274	283	215
Other WTO:						
Nucl.	...	...	...	...	...	...
Conv.	...	...	...	...	7	8
Total WTO:						
Nucl.	...	...	4	26	58	80
Conv.	46	215	238	274	290	223
Other Europe:						
Nucl.	...	...	...	...	...	...
Conv.	3	9	12	18	27	34
Other developed:						
Nucl.	...	...	...	...	...	...
Conv.	3	4	5	7	15	23
Total Third World+	4	5	33	60	76	133
Middle East	...	...	10	12	16	16
South Asia	...	...	...	1	8	11
Far East	...	...	2	12	14	18
Sub-Saharan Africa	...	...	...	...	...	...
North Africa	...	...	...	...	...	...
Central America	...	...	...	...	...	...
South America	4	5	9	13	11	28
China	...	...	12	22	27	60

SOURCE.—SIPRI.
NOTE.—Nucl. = nuclear powered; Conv. = conventionally powered; WTO = Warsaw Treaty Organization.
*Post-World War II submarines displacing 700 tons or more.
+All conventionally powered.

TABLE 5G.—WORLD STOCK OF COASTAL SUBMARINES,* BY GROUPS OF COUNTRIES

	1950	1955	1960	1965	1970	1976
World total	313	317	179	93	72	73
Developed	299	295	162	85	64	73
Third World	14	22	17	8	8	...
U.S.A.	...	...	...	...	...	...
Other NATO	...	3	20	33	34	53
Total NATO	...	3	20	33	34	53
USSR	273	269	127	40	22	17
Other WTO	...	...	...	...	...	...
Total WTO	273	269	127	40	22	17
Other Europe	26	23	15	12	8	3
Other developed	...	...	...	...	...	...

SOURCE.—SIPRI.
NOTE.—WTO = Warsaw Treaty Organization.
*Submarines displacing less than 700 tons; all conventionally powered.

TABLE 6G.—WORLD STOCK OF MAJOR SURFACE WARSHIPS,* BY GROUPS OF COUNTRIES

	1950	1955	1960	1965	1970	1976
World total:						
Miss.	...	2	18	111	191	383
Conv.	1,783	2,042	1,789	1,650	1,414	823
Developed:						
Miss.	...	2	18	111	189	347
Conv.	1,600	1,811	1,520	1,374	1,128	555
U.S.A.:						
Miss.	...	2	15	58	77	122
Conv.	817	835	704	654	478	84
Other NATO:						
Miss.	...	...	...	19	57	113
Conv.	520	582	402	330	294	185
Total NATO:						
Miss.	...	2	15	77	134	235
Conv.	1,337	1,417	1,106	984	772	269
USSR:						
Miss.	...	...	1	23	39	87
Conv.	150	256	260	228	206	187
Other WTO:						
Miss.	...	...	...	...	1	1
Conv.	5	7	14	14	9	4

TABLE 6G (continued)

	1950	1955	1960	1965	1970	1976
Total WTO:						
Miss.	...	...	1	23	40	88
Conv.	155	263	274	242	215	191
Other Europe:						
Miss.	...	...	2	6	6	12
Conv.	67	70	75	69	71	34
Other developed:						
Miss.	...	...	...	5	9	12
Conv.	41	61	63	73	64	61
Total Third World:						
Miss.	...	...	...	...	2	36
Conv.	183	231	269	276	286	268
Middle East:						
Miss.	...	...	...	...	1	7
Conv.	18	20	15	18	21	14
South Asia:						
Miss.	...	...	...	...	...	4
Conv.	16	18	26	27	32	34
Far East:						
Miss.	...	...	...	...	1	5
Conv.	50	53	71	76	109	92
Sub-Saharan Africa:						
Miss.	...	...	...	...	...	...
Conv.	...	...	...	1	1	2
North Africa:						
Miss.	...	...	...	...	...	1
Conv.	...	...	...	2	2	1
Central America:						
Miss.	...	...	...	...	...	...
Conv.	30	30	29	29	22	34
South America:						
Miss.	...	...	...	...	...	9
Conv.	65	94	105	91	78	75
China:						
Miss.	...	...	...	...	...	13
Conv.	4	16	23	22	21	13

SOURCE.--SIPRI.
NOTE.--Miss. = missile armed; Conv. = conventionally armed; WTO = Warsaw Treaty Organization.
*Cruisers, destroyers, frigates, and escorts.

TABLE 7G.—WORLD STOCK OF PATROL BOATS, TORPEDO BOATS, AND GUNBOATS,*
BY GROUPS OF COUNTRIES

	1950	1955	1960	1965	1970	1976
World total:						
Miss.	...	...	5	141	282	533
Conv.	987	1,380	1,849	2,092	2,457	2,729
Developed:						
Miss.	...	...	5	112	189	248
Conv.	822	1,142	1,422	1,340	1,290	1,082
U.S.A.:						
Miss.	...	...	...	...	2	5
Conv.	147	120	35	18	35	24
Other NATO:						
Miss.	...	...	...	...	1	74
Conv.	190	267	230	233	241	198
Total NATO:						
Miss.	...	...	...	...	3	79
Conv.	337	387	265	251	276	222
USSR:						
Miss.	...	...	5	110	150	120
Conv.	395	516	769	653	600	424
Other WTO:						
Miss.	...	...	...	2	28	32
Conv.	16	54	141	191	164	188
Total WTO:						
Miss.	...	...	5	112	178	152
Conv.	411	570	910	844	764	612
Other Europe:						
Miss.	...	...	...	...	8	16
Conv.	60	117	180	194	196	198
Other developed:						
Miss.	...	...	...	...	...	...
Conv.	14	68	67	51	54	50
Total Third World:						
Miss.	...	...	...	29	93	285
Conv.	156	238	427	752	1,141	1,647
Middle East:						
Miss.	...	...	...	3	32	58
Conv.	11	17	77	86	140	133

TABLE 7G (continued)

	1950	1955	1960	1965	1970	1976
South Asia:						
Miss.	...	...	...	...	...	8
Conv.	...	...	1	9	7	34
Far East:						
Miss.	...	...	...	12	12	52
Conv.	55	120	149	260	403	423
Sub-Saharan Africa:						
Miss.	...	...	...	...	...	5
Conv.	...	...	5	20	54	106
North Africa:						
Miss.	...	...	...	...	16	13
Conv.	...	...	2	2	45	48
Central America:						
Miss.	...	...	...	12	18	23
Conv.	16	16	18	49	72	106
South America:						
Miss.	...	...	...	...	...	6
Conv.	42	32	26	47	38	87
China:						
Miss.	...	...	...	2	15	120
Conv.	...	...	150	279	408	710

SOURCE.--SIPRI.
NOTE.--Miss. = missile armed; Conv. = conventionally armed; WTO = Warsaw Treaty Organization.
*Excluding riverine craft.

TABLE 8G.—WARSHIP CONSTRUCTION UNDER WAY OR FIRMLY PLANNED, 1976, BY TYPE AND COUNTRY

	Class and Type	Displacement (Tons)	Numbers Completed	Numbers Under Constr.	Numbers Planned Total	Country of Construction
			Aircraft Carriers			
NATO: U.S.A.	"Nimitz," nuclear	91,400	1	2	4	U.S.A.
	"Tarawa" amph.	39,300	2	3	5	U.S.A.
	VSTOL support ship	22,000 or 33,000	...	...	(12)	U.S.A.
U.K.	Through deck cruiser	19,500	...	1	3	U.K.
France	Helicopter carrier, nuclear	18,000	...	...	1	France
WTO: USSR	"Kuril" ASW cruiser	40,000	1	2	N.A.	USSR
			Submarines			
NATO: U.S.A.	Trident, strategic, nuclear, Miss.	(10,000)	...	1	(10)	U.S.A.
	"Los Angeles," nuclear	5,500	2	10	39	U.S.A.
U.K.	"Swiftsure," nuclear	3,500	3	3	6	U.K.
	N.A., nuclear	N.A.	...	1	...	U.K.

France	"Redoutable," strategic, nuclear, Miss.	7,500	4	1	6	France
	Submarine, nuclear	2,385	...	1	(6)	France
	"Agosta," diesel	1,200	2	2	4	France
Norway	Type 210, diesel	750	...	...	(15)	FR Germany
FR Germany	Type 210, diesel	750	...	...	(6)	FR Germany
Turkey	Type 209, diesel	990	2	1	5	FR Germany/ Turkey
Italy	"Sauro," diesel	1,456	...	2	2	Italy
Greece	Type 209, diesel	990	4	2	7	FR Germany
WTO: USSR	"Delta-II," strategic, nuclear, Miss.	(16,000)	(2)	N.A.	N.A.	USSR
	"Delta-I," strategic, nuclear, Miss.	9,000	(10)	N.A.	N.A.	USSR
	"Papa," nuclear	N.A.	(1)	N.A.	N.A.	USSR
	"Charlie," nuclear, Miss.	4,300	14	N.A.	N.A.	USSR
	"Victor," nuclear	3,600	19	N.A.	N.A.	USSR
	"Tango," diesel	1,900	3	N.A.	N.A.	USSR
Other Europe: Sweden	"A-14," diesel	980	...	3	3	Sweden
Spain	"Agosta," diesel	1,450	...	2	2	Spain
Yugoslavia	N.A., diesel	964	...	2	N.A.	Yugoslavia

TABLE 8G (continued)

Class and Type	Displacement (Tons)	Numbers Completed	Numbers Under Constr.	Planned Total	Country of Construction
		Submarines			
Other developed:					
Australia					
"Oberon," diesel	1,610	4	2	6	U.K.
Japan					
"Uzushio," diesel	1,850	6	1	8	Japan
South Africa					
"Agosta," diesel	1,450	...	2	2	France
Middle East:					
Israel					
Type 206, diesel	420	(1)	2	3	U.K.
South America:					
Brazil					
"Oberon," diesel	1,610	1	2	3	U.K.
Ecuador					
Type 209, diesel	990	...	2	2	FR Germany
Peru					
Type 209, diesel	990	2	N.A.	4	FR Germany
Venezuela					
Type 209, diesel	990	1	(1)	2	FR Germany
Far East:					
North Korea					
"Romeo," diesel	1,100	2	N.A.	N.A.	North Korea
China					
"Han," nuclear	N.A.	...	1	...	China
"Ming," diesel	(1,500)	1	(1)	N.A.	China
"Romeo," diesel	1,100	36	(6)	N.A.	China
		Major Surface Ships			
NATO:					
U.S.A.					
"Virginia," nuclear, Miss.	10,000	1	2	4	U.S.A.
"Spruance," Miss.	7,800	8	22	30	U.S.A.
"Perry," Miss.	3,500	...	1	50	U.S.A.

Appendix G: Tables, Military Activities   847

Country	Type	Displacement				Supplier
U.K.	Type 42, Miss.	3,500	3	3	8	U.K.
	Type 22, Miss.	3,800	...	1	N.A.	U.K.
	Type 21, Miss.	2,500	5	3	8	U.K.
France	C-70, Miss.	3,800	...	3	24	France
	A-69, Miss.	950	3	8	14	France
Netherlands	"Kortenaer," Miss.	3,500	...	4	13	Netherlands
Belgium	E-71, Miss.	1,940	...	4	4	Belgium
FR Germany	Frigate, Miss.	(3,800)	...	...	12	FR Germany
Italy	"Lupo," Miss.	2,210	...	3	4	Italy
Denmark	Corvette, Miss.	(1,000)	...	...	3	Denmark
WTO:						
USSR	"Kara," Miss.	8,200	4	(1)	N.A.	USSR
	"Kresta II," Miss.	6,000	9	(1)	N.A.	USSR
	"Krivak," Miss.	4,800	11	(2)	N.A.	USSR
	"Nanuchka," Miss.	800	14	N.A.	N.A.	USSR
	"Grisha," Miss.	750	18	N.A.	N.A.	USSR
	"Grisha"	750	3	N.A.	N.A.	USSR
Other Europe:						
Spain	F-90, Miss.	(3,000)	...	...	5	Spain
	Frigate, Miss.	(1,200)	1	N.A.	(10)	Spain
Sweden	Corvette, Miss.	(700)	...	...	3	Sweden
Other developed:						
Australia	"Perry," Miss.	3,500	...	...	2	U.S.A.

TABLE 8G (continued)

	Class and Type	Displacement (Tons)	Numbers			Country of Construction
			Completed	Under Constr.	Planned Total	
			Major Surface Ships			
Japan	"Haruna," Miss.	4,700	2	1	4	Japan
	Frigate, Miss.	1,470	10	1	12	Japan
South Africa	A-69, Miss.	950	...	2	(2)	France
	Frigate, Miss.	N.A.	...	...	6	South Africa
Middle East:						
Iran	"Spruance," Miss.	7,800	...	N.A.	4	U.S.A.
South Asia:						
India	"Leander," Miss.	2,450	4	N.A.	6	India
Far East:						
Indonesia	Corvette, Miss.	N.A.	N.A.	N.A.	(3)	Netherlands
North Korea	...	1,200	2	1	N.A.	North Korea
South America:						
Argentina	Type 21, Miss.	2,500	...	...	6	U.K./Argentina
Brazil	"Niteroi," Miss.	3,200	1	5	6	U.K./Brazil
Peru	"Lupo," Miss.	2,210	...	N.A.	4	Italy/Peru
Venezuela	"Lupo," Miss.	2,210	...	(1)	6	Italy
China	"Luta," Miss.	3,250	4	(3)	N.A.	China
	"Hainan"	500	15	N.A.	N.A.	China

Appendix G: Tables, Military Activities

Missile-armed Patrol Boats

NATO:					
U.S.A.	"Pegasus"	220	1	1	U.S.A.
FR Germany	"Pegasus"	220	...	(24)	FR Germany
	Type 143	295	7	(12)	FR Germany
				10	
France	"Trident"	115	...	4	France
Italy	"Sparviero"	62	1	7	Italy
Turkey	N.A.	410	1	N.A.	FR Germany/
					Italy
Denmark	"Willemoes"	220	4	3	Denmark
Norway	"Hauk"	120	...	6	Norway
				10	
				14	
Other Europe:					
Spain	"La Combattante"	180	...	(3)	France
Sweden	"Jägaren"	140	1	N.A.	Norway
Yugoslavia	"Spica"	240	...	N.A.	Yugoslavia
Other developed:					
South Africa	"Reshef"	415	...	4	Israel/South Africa
Middle East:					
Israel	"Reshef"	415	6	N.A.	Israel
Iran	"Kaman"	249	2	6	France
Far East:					
Indonesia	"Paek Ku"	250	...	N.A.	South Korea
				12	
				17	
				10	
				6	
				12	
				12	
				4	

TABLE 8G (continued)

Class and Type	Displacement (Tons)	Numbers			Country of Construction
		Completed	Under Constr.	Planned Total	
Missile-armed Patrol Boats					
Malaysia					
"Spica-M"	(250)	...	N.A.	6	Sweden
South Korea					
"Paek Ku"	250	3	4	7	South Korea
Thailand					
"Type 148"	230	2	1	(4)	Singapore
South America:					
Argentina					
Type 148	230	...	2	2	Argentina
Ecuador					
TNC 45	250	N.A.	N.A.	N.A.	FR Germany
Africa:					
Libya					
N.A.	550	...	(2)	4	Italy
Morocco					
PR 72	540	...	N.A.	10	France
China					
"Hai Dau"	260	1	N.A.	N.A.	China
"Hola"	165	...	...	...	China
"Hoku"	70	...	N.A.	N.A.	China

SOURCE.—SIPRI.
NOTE.—amph. = amphibious assault; VSTOL = vertical or short takeoff and landing; ASW = anti-submarine warfare; N.A. = not available; Miss. = missile armed; numerals in parentheses indicate uncertain data.

TABLE 9G.—U.S. AND SOVIET STRATEGIC MISSILE-LAUNCHING SUBMARINES AND MISSILES

	1967	1968	1969	1970	1971	1972	1973	1974	1975	1976	1977
				Strategic Submarines							
U.S.A.:											
With Polaris A-2	13	13	13	8	8	8	8	6	3	...	...
With Polaris A-3	28	28	28	32	26	21	13	13	13	13	11
With Poseidon C-3	...	...	...	1	7	12	20	22	25	28	30
USSR:											
"Hotel" class	9	9	9	8	8	8	8	8	8	8	8
"Yankee" class	...	(2)	(8)	(14)	(21)	(27)	(33)	34	34	34	34
"Delta I" class	...	...	...	...	...	...	(1)	(8)	(11)	11	11
"Delta II" class	...	...	...	...	...	...	...	...	...	(2)	(7)
Total:											
U.S.A.	41	41	41	41	41	41	41	41	41	41	41
USSR	9	11	17	22	29	35	42	50	53	55	60
				Submarine-launched Ballistic Missiles							
U.S.A.:											
Polaris A-2	208	208	208	128	128	128	128	96	48	...	...
Polaris A-3	448	448	448	512	416	336	208	208	208	208	176
Poseidon C-3	...	...	...	16	112	192	320	352	400	448	480
USSR:											
"SS-N-5"	27	27	27	24	24	24	24	24	24	24	24
"SS-N-6 mod. 1"⎱	...	32	128	224	336	432	528	544	544	544	544
"SS-N-6 mod. 2"⎰											
"SS-N-8"	...	...	...	...	...	...	12	96	132	164	244
Total:											
U.S.A.	656	656	656	656	656	656	656	656	656	656	656
USSR	27	59	155	248	360	456	564	664	700	732	812

SOURCE.—SIPRI.
NOTE.—Numerals in parentheses indicate uncertain data.

TABLE 10G.—U.S. AND SOVIET SUBMARINES, SUBMARINE-LAUNCHED BALLISTIC MISSILES, INTERCONTINENTAL BALLISTIC MISSILES, AND STRATEGIC BOMBERS, 1976

	Introduced	Range (nm)	Payload	CEP (nm)
Strategic Submarines				
U.S.A.:				
With Polaris A-3	1964	N.A.	16 x A-3	...
With Poseidon C-3	1970	N.A.	16 x C-3	...
USSR:				
"Hotel" class	1960	N.A.	3 x "SS-N-5"	...
"Yankee" class	1968	N.A.	16 x "SS-N-6"	...
"Delta I" class	1973	N.A.	12 x "SS-N-8"	...
"Delta II" class	1976	N.A.	16 x "SS-N-8"	...
Submarine-launched Ballistic Missiles				
U.S.A.:				
Polaris A-3	1964	2,500	3 x 200 kt (MRV)	.5-.7
Poseidon C-3	1970	2,500	10 x 40 kt (MIRV)	.3
USSR:				
"SS-N-5"	1963	700	1 x 1 mt	...
"SS-N-6 mod. 1"	1968	1,300	1 x 1 mt	1.5
"SS-N-6 mod. 2"	1974	1,600	3 x 200 kt (MRV)	1.5
"SS-N-8"	1973	4,200	1 x 1 mt	.8
Intercontinental Ballistic Missiles				
U.S.A.:				
Titan II	1962	6,300	1 x 10 mt	.5
Minuteman II	1966	6,950	1 x 2 mt	.3
Minuteman III	1970	7,020	3 x 170 kt (MIRV)	.13
USSR:				
"SS-7 Saddler"	1962	6,000	1 x 5 mt	2
"SS-8 Sasin"	1963	6,000	1 x 5 mt	1.5
"SS-9 Scarp"	1965	6,515	1 x 20 mt	.7
"SS-11 mod. 1"	1966	5,650	1 x 1 mt	1
"SS-13 Savage"	1968	4,350	1 x 1 mt	.7-1
"SS-11 mod. 3"	1973	5,650	3 x 200 kt (MRV)	1
"SS-18 mod. 1"	1976	5,500	1 x 10 mt	.3
"SS-19"	1976	5,500	6 x 1 mt (MIRV)	.3
"SS-17"	1977	...	4 x 1 mt (MIRV)	.3
"SS-18 mod. 2"	1977	...	8 x 1 mt (MIRV)	.3
Strategic Bombers				
U.S.A.:				
B-52C/D/E/F	1956	10,000	27,210 kg	...
B-52G/H	1959	10,860	34,015 kg	...
FB-111	1970	3,300	16,780 kg	...
USSR:				
Mya-4 "Bison"	1955	5,255	9,070 kg	...
Tu-20 "Bear"	1956	6,775	18,140 kg	...
Tu-N.A. "Backfire"	1975	(3,000)	(20,000 kg)	...

SOURCE.—SIPRI.
NOTE.—CEP = circular error probability; N.A. = not available.

TABLE 11G.—USSR ANTISUBMARINE-EQUIPPED SURFACE SHIPS, 1976

Class	Full Load Displacement (Tons)	Max. Speed (Knots)	AS Equipment		
			Sonar	Weapons	Aircraft
Fixed-Wing Aircraft Carriers					
Kiev	45,000	30	N.A.	1 2-bar AS rocket or missile L	(Possible cap.: 35 VTOL aircraft + 35 hel.)
Helicopter Carriers (Minimum Capacity: 4 Helicopters)					
Moskva	18,000	30	VDS	1 2-bar AS rocket or missile L 2 12-bar AS RL	20 Ka-25 ("Hormone") AS hel. (cap.: 30 hel.)

SOURCE.—SIPRI.
NOTE.—AS = antisubmarine; N.A. = not available; VTOL = vertical takeoff and landing.

TABLE 12G.--U.S. ANTISUBMARINE FIXED-WING AIRCRAFT, 1976

Manufacturer, Name, Military Designation	Crew	Engine	Speed: Patrol/ Max. (Knots)	Range: Mission/ Max. (nm)	Max. Endurance (Hours)	AS Equipment Sonobuoy	AS Equipment Other Sensors	Weapons	Notes
Land-based Fixed-Wing Aircraft									
Lockheed Orion P-3: Orion P-3C	12	T	206/475	1,346- 2,070/ 4,500	11	AN/SSQ-41/ 41A;AN/SSQ- 53 DICASS; AN/SSQ-47/ 47B; AN/ SSQ-50 CASS or SSQ-62 DICASS (87 sono- buoys)	AN/ASQ-10A or ASQ-81 MAD; AN/ASA-64 SAD; AN/ASA-65 com- pensator; AN/ AXR-13 LLTV	Bombs Mines DC (incl. MK 101 nuclear DC) Mk 43, 44, or 46 torpedo (3,290-kg internal load + 6 x 900- kg external load)	AS aircraft. Operations led by com- plex A-NEW avionics equipment. Replacing Orion P-3A/ 3B
Lockheed Nep- tune P-2	7	P+J	150/350	N.A./ 3,500	...	AN/SSQ-41/ 41A; AN/ SSQ-47/47B	AN/ASQ-8 or ASQ-10 MAD; AN/ASA-60 com- pensator	Bombs DC Torpedoes (5,000-kg load)	MR aircraft

Appendix G: Tables, Military Activities 855

Carrier-based Fixed-Wing Aircraft

Aircraft							
Grumman Tracker S-2 E/F/G	4	P	130/280	N.A./1,200	9	AN/SSQ-41/41A; AN/SSQ-53 DIFAR;AN/SSQ-47/47B (32 sonobuoys)	AN/ASQ-10 MAD; AN/ASA-60 compensator	Internal: 2 torpedoes or 4 240-kg DC or 1 Mk 101 nuclear DC Wing-mounted: torpedoes or rockets or 150-kg DC	AS aircraft S-2A in service 1953. S-2E, last production version, delivered 1963-67. S-2F refitted 2A/2B. S-2G refitted 2E with new AS equipment added 1972. Data and numbers 1971-73 apply to S-2G

SOURCE.--SIPRI.
NOTE.--AS = antisubmarine; T = turoprop; MAD = magnetic anomaly detector; SAD = submarine anomaly detector; N.A. = not available; MR = maritime reconnaissance; P = piston; J = jet; DC = depth charge.

TABLE 13G.—ANTISUBMARINE-EQUIPPED HELICOPTERS, 1976.

Manufacturer, Name, Military Designation	Crew	Max. T-O Weight (kg)	Max. Speed (Knots)	Range: Mission/ Max. (nm)	Max. Endurance (Hours)	AS Equipment Sonar, Sonobuoys	AS Equipment Other Sensors	Weapons	Notes
U.K.: Westland WG-13 Sea Lynx HAS. Mk 2	2	3,878	179	176/440	N.A.	N.A.	N.A.	2 Mk 44 torpedoes	Shipborne AS search and attack hel.
U.S.A.: Kaman Sea Sprite SH-2D/LAMPS SH-2D/LAMPS Mk 1	3	5,670	146	N.A./387	N.A.	AN/ASQ-13-A AN/SSQ-41/41A AN/SSQ-47/47B (15 sonobuoys)	AN/ASQ-81 MAD AN/ASA-65 compensator	2 Mk 46 torpedo	Shipborne hel. for use primarily in AS cruise missile defense

							and utility roles. Refitted HH-2D (originally introduced into service as UH-2A/2B in 1962-66)
DASH-Drone AS helicopter (Gyrodyne QH-50D)							
USSR: Kamov Ka-25 ("Hormone")	2	7,300	119	N.A./351	3	Dipping Sonar, MAD, AS torpedoes	Land-based and shipborne AS search and attack hel.

SOURCE.—SIPRI.
NOTE.—AS = antisubmarine; N.A. = not available; MAD = magnetic anomaly detector.

TABLE 14G.—U.S. MOBILE ANTISUBMARINE. WARFARE SENSORS, 1976

Designation or Program	Description	Major Components Acoustic Sensors	Deployment
Submarine sonar: AN/BQQ-5	Integrated sonar system; tied in with Subroc weapon system	AN/BQS-13 DNA and AN/BQS-14 sonar transmitter	U.S.A.: "Los Angeles" class SSNs; may be retrofitted to other SSNs
SSBN Unique Sonar System	Integrated sonar developed for U.S. strategic nuclear submarines	AN/BQR-15 towed array sonar (passive); AN/BQR-19 sonar receiver (passive); AN/BQR-2/BQS-4 improvement/DIMUS, improved version of currently installed integrated sonar	U.S.A.: Polaris/Poseidon SSBNs
Surface ship sonar: AN/SQQ-23 PAIR	Integrated sonar; retrofit improving existing AN/SQS-23 through microcircuitry, digital techniques, and modular packaging	N.A.	U.S.A.: To be deployed on surface ships equipped with AN/SQS-23 except "Gearing" class (FRAM I); Australia, FR Germany: "Charles F. Adams" class destroyers
Helicopter-borne dipping sonar: AN/AQS-13/13A	Active sonar. Longer-range, fully transistorized successor to	N.A.	U.S.A. and other: SH-3A/3D Sea King USA SH-2D LAMPS AS hel.

Appendix G: Tables, Military Activities 859

Sonobuoys (airborne):			
AN/SSQ-53 DIFAR	AN/AQS-10. Model 13A with reduced weight and volume Latest passive sonobuoy system; for use aboard fixed-wing AS aircraft equipped with DIFAR processor; can be employed in directional mode	N.A.	U.S.A.: P-3A/3B/3C Orion, S-2G Tracker, and S-3A Viking aircraft
AN/SSQ-47/47B "Julie"	Active echo-ranging miniature sonobuoy used for localization of target	N.A.	U.S.A.: All AS fixed-wing aircraft and hels. (P-2, P-3A/3B/3C, S-2G, S-3A, SH-2D, SH-3H)

Nonacoustic Sensors (Airborne)

MAD/SAD equipment:			
AN/ASQ-81	Latest generation U.S. MAD system. Max. range about 1,000 m	C-6983/ASQ (v) control detecting set; C-6984/ASQ-81 (v) control, reeling machine; AM-4535/ASQ-81 (v) amplifier, power supply; MT-3618/ASQ-81 (v) base, shock mount; DT-323/ASQ-81 (v) detector, magnetic; TB-623/ASQ-81 (v) towed body, magnetic detecting reeling machine, magnetic detector launching	U.S.A.: P-3C Orion AS aircraft and SH-3H Sea King AS hel.; Spain: Hughes 500 AS hel. May also be deployed on U.S. S-3A Viking and SH-2D LAMPS

TABLE 14G (continued)

Designation or Program	Description	Major Components		Deployment
		Nonacoustic Sensors (Airborne)		

Designation or Program	Description	Major Components	Deployment
Other nonacoustic sensors:			
AN/AAR-37 FLIR	Airborne equipment providing TV-like image of heat radiated from submarines or their antennae or periscopes, breaking water, or from warm water heated by passing of a submarine	N.A.	U.S.A.: P-3A/3B Orion AS aircraft

SOURCE.--SIPRI.
NOTE.--DNA = digital multibeam, narrow-band processing, accelerated active search rate sonar; DIMUS = digital multibeam steering; PAIR = performance and integration retrofit; N.A. = not available; DIFAR = directional low-frequency analyzer and ranging; MAD = magnetic anomaly detector; SAD = submarine anomaly detector; FLIR = forward-looking infrared; AS = antisubmarine.

TABLE 15G.—FIXED ACOUSTIC ARRAYS AND SONOBUOYS

Designation or Program	Description	Major Components	Deployment
International (NATO):			
AFAR	Sonars installed on 3 towers at depths of 300–600 m, off island of Santa Maria (southernmost of Azores group), for surveillance of submarines in Mediterranean	French-built towers, U.S. electronic equipment	...
U.S.A.:			
MSS	Air-dropped, command-activated sonobuoys designed to moor to ocean bottom, locate targets by triangulation method, and transmit data to overflying AS aircraft on activation	N.A.	...
SOSUS	Major fixed underwater passive acoustic submarine detection and classification system, comprising a series of hydrophones and special sonars linked by cable to shore-based installations that process received signals. Possible max. range 1,000 nm	Barrier, Bronco: SOSUS units believed to be located on territory of U.S. allies Caesar: first SOSUS system, initial unit located off U.S. East Coast; incl. AN/FQQ-6 and AN/FQQ-9(v) sonars and AN/UQA-5 spectrum analyser.	(Total of 21 Barrier, Bronco, Caesar, and Colossus stations operational in 1971)

862 *Appendix G: Tables, Military Activities*

TABLE 15G (continued)

Designation or Program	Description	Major Components	Deployment
SOSUS		Expanded and updated 4 times since original installment	
		Colossus: advanced version of Caesar, installed 1964-65 and believed to be located off U.S. West Coast; uses AN/FQQ-10(v) sonar	
Sea Spider	Submarine detection systems, incorporating 3-m-dia. hydrophone buoy to be anchored at 5,500-m depth by 3 8,000-m cables, and to be powered by nuclear battery	N.A.	(Was to have been deployed several hundred miles north of Hawaii, but first attempt to moor system (October 1969) was unsuccessful. A report that it was later installed has appeared but not been confirmed)

Appendix G: Tables, Military Activities 863

SAS	Potential new system consisting of a tripod tower, to rest in ocean at depth of 4,900 m, so large that each leg of tripod would be about 10 km from the next, which would bear acoustic arrays	N.A.
SAS	(transducers and hydrophones) of such power that 1 tower per ocean would suffice for complete "insonification"	
USSR: N.A.	Tamir fixed sonar installations deployed off coast of USSR during WW II reported to have been expanded and improved	N.A.

SOURCE.—SIPRI.
NOTE.—AFAR = Azores fixed acoustic range; MSS = moored surveillance system; N.A. = not available; SOSUS = sonar surveillance system; SAS = suspended array system.

TABLE 16G.—STRATEGIC NUCLEAR SUBMARINES, BY COUNTRY AND CLASS

PART A

Class	Submerged Displacement (Tons)	Max. Submerged Speed (Knots)	Diving Depth (m)	Antiland/Antisurface Ship Missile Armament	AS Equipment Sonar	AS Equipment Weapons	Notes
France: Le Redoutable SNLE	9,000	25	300	16 MSBS M-1 BM, range 1,200 nm	N.A.	4 TT	1,700-nm-range MSBS M-2 to be deployed starting 1974
U.K.: Resolution SSBN	8,400	25	N.A.	16 Polaris A-3 BM, (3 MRV warhead) range 2,500 nm	N.A.	6 21-inch TT	...
U.S.A.: Trident	15,000	20-30	N.A.	24 Trident 1 BM (14-17 MIRV warhead), range 4,000 nm	N.A.	N.A.	Reports of max. diving depth up to 3,000 m
Lafayette/ Benjamin Franklin SSBN-616/ 640	8,250	20-30	600-900	16 Poseidon C-3 BM (10-14 MIRV warhead), range 2,500 nm; or	...	4 21-inch TT	Converted from Polaris A-3

Appendix G: Tables, Military Activities 865

Name	Displacement	Speed	Missiles	Sonar	Torpedoes	Notes
Ethan Allen SSBN-608	7,880	20–30	16 Polaris A-3 BM, (3 MRV warhead) range 2,500 nm	...	...	To be converted to Poseidon
	600–900	20–30	16 Polaris A-2 BM, range 1,500 nm	AN/BQR-15 AN/BQR-19/21 AN/BQR-2 BQS-4 DIMUS	4 21-inch TT	To be converted to Polaris A-3
George Washington SSBN-598	6,688	20–30	16 Polaris A-3 BM, (3 MRV warhead), range 2,500 nm	...	6 21-inch TT	...
USSR: "Delta" or "D"	9,000	25	12 "SS-N-8" BM, range 4,200 nm	N.A.	N.A.	...
"Yankee" or "Y"	8,300–9,000	25	16 "SS-N-6" BM, range 1,300 nm	N.A.	8 21-inch TT	...
"Hotel II" or "H II"	4,100	20–25	3 "Serb SS-N-5" BM, range 650 nm	N.A.	4 16-inch AS TT	Longer-range missiles refitted between 1963 and 1967

TABLE 16G (continued)

PART B

Class	First Comm./ Compl.	Number in Active Service as of December 31								
		1969	1970	1971	1972	1973	1974	1975	1976	1977
France:										
Le Redoutable SNLE	1971	...	...	1	1	2	3	4	4	4
U.K.:										
Resolution SSBN	1967	4	4	4	4	4	4	4	4	4
U.S.A.:										
Trident	1978	...	...	...	...	...	...	...	...	...
Lafayette/Benjamin Franklin SSBN-616/640	(1970) 1963	... 31	4 27	10 21	16 15	20 11	22 9	25 6	28 3	31 ...
Ethan Allen SSBN-608	1961	5	5	5	5	5	5	5	5	5
George Washington SSBN-598	1959	5	5	5	5	5	5	5	5	5
USSR:										
"Delta" or "D"	1972	...	...	...	1	4	8	11	13	18
"Yankee" or "Y"	1967	10	17	25	31	33	33	33	34	34
"Hotel II" or "H II"		9-10	9-10	9-10	9-10	8	8	8	8	8

SOURCE.--SIPRI.
NOTE.--AS = antisubmarine; BM = ballistic missile; N.A. = not available; TT = torpedo tubes; DIMUS = digital multibeam steering; year in parentheses uncertain.

TABLE 17G.—NUCLEAR-POWERED ATTACK SUBMARINES, BY COUNTRY AND CLASS

PART A

Class	Submerged Displacement (Tons)	Max. Submerged Speed (Knots)	Diving Depth (m)	Antiland/Antisurface Ship Missile Armament	AS Equipment Sonar	Weapons	Notes
U.K.:							
Valiant/Churchill	4,500	28	N.A.	...	N.A.	6 21-inch AS TT	...
Swiftsure	4,500	30	N.A.	...	N.A.	6 21-inch AS TT	...
Dreadnought	4,000	30	N.A.	...	N.A.	6 21-inch Mk 23 torpedo	Prototype U.K. nuclear-powered submarine
U.S.A.:							
Los Angeles SSN-688	6,900	40	N.A.	...	AN/BQQ-5 AN/BQS-15	4 21-inch TT, Subroc and Mk 48 torpedo	...
Glenard P. Lipscomb SSN-685	5,000+	25	N.A.	...	N.A.	4 21-inch TT, Subroc and Mk 48 torpedo	"Quiet submarine" with new propulsion and machinery
Narwhal SSN-671	5,350	30	N.A.	...	AN/BQQ-2 AN/BQQ-14	4 21-inch TT, Subroc and AS torpedo	Test vessel for water-cooled reactor

TABLE 17G (continued)

PART A

Class	Submerged Displacement (Tons)	Max. Submerged Speed (Knots)	Diving Depth (m)	Antiland/Antisurface Ship Missile Armament	AS Equipment Sonar	AS Equipment Weapons	Notes
Skate SSN-578	2,861	20+	N.A.	...	N.A.	N.A.	...
Seawolf SSN-575	4,200	20+	N.A.	...	N.A.	N.A.	Used mainly for experiment since 1969
Nautilus SSN-571	4,040	20+	N.A.	...	N.A.	N.A.	Prototype nuclear submarine
USSR: "Papa" or "P"	N.A.	N.A.	N.A.	Cruise missiles	N.A.	N.A.	...
"Charlie" or "C"	5,100	30	N.A.	8 "...SS-N-7" cr. miss. range 26 nm	N.A.	N.A.	...
"Echo II" or "E II"	5,600	20-25	N.A.	8 "Shaddock SS-N-3" cr. miss., range 100-200 nm	N.A.	4 16-inch AS TT	...
"Echo I" or "E I"	4,500-5,000	20	N.A.	(6 "Shaddock SS-N-3" cr. miss., range 100-200 nm)	N.A.	4 16-inch AS TT	Cruise missiles being removed
"Victor" or "V"	4,200	30	N.A.	...	N.A.	8 21-inch AS TT	...

Appendix G: Tables, Military Activities   869

Sturgeon SSN-637	4,640	30	N.A.	AN/BQQ-1 Ret. III	...	4 21-inch TT, Subroc and Mk 48 torpedo	...
Permit SSN-594 (former "Thresher" class)	4,300	30	N.A.	AN/BQQ-1 Ret. III	...	4 21-inch TT, Subroc and AS torpedo	...
Skipjack SSN-585	3,513	30+	N.A.	(Original sonar has been modified)	...	AS torpedo	...
Tullibee SSN-597	2,640	15+	N.A.	AN/BQQ-1 Ret. III	...	AS torpedo	...
Halibut SSN-587	5,000	15+	N.A.	N.A.	...	N.A.	Built to carry "Regulus" cruise missile, later removed
"November" or "N"	4,000	25-30	230+	N.A.	...	4 16-inch AS TT	...

TABLE 17G (continued)

PART B

Class	First Comm./Compl.	Number in Active Service as of December 31								
		1969	1970	1971	1972	1973	1974	1975	1976	1977
U.K.:										
Valiant/Churchill	1966	2	3	5	5	5	5	5	5	5
Swiftsure	1973	...	...	...	...	1	2	3	3	4-6
Dreadnought	1963	1	1	1	1	1	1	1	1	1
U.S.A.:										
Los Angeles SSN-688	1974	...	...	...	...	...	1	2	4	10
Glenard P. Lipscomb SSN-685	1974	...	...	...	...	...	1	1	1	1
Narwhal SSN-671	1969	1	1	1	1	1	1	1	1	1
Sturgeon SSN-637	1967	17	20	27	31	34	35	37	37	37
Permit SSN-594 (former "Thresher" class)	1962	13	13	13	13	13	13	13	13	13
Skipjack SSN-585	1959	5	5	5	5	5	5	5	5	5

Appendix G: Tables, Military Activities   871

	Year								
Tullibee SSN-597	1960	1	1	1	1	1	1	1	1
Halibut SSN-587	1960	1	1	1	1	1	1	1	1
Skate SSN-578	1957	4	4	4	4	4	4	4	4
Seawolf SSN-575	1957	1	1	1	1	1	1	1	1
Nautilus SSN-571	1954	1	1	1	1	1	1	1	1
USSR:									
"Papa" or "P"	N.A.	...	...	0–1	1–2	1	1	1	1
"Charlie" or "C"	1969	2	5	8	11	11	11	11	14
"Echo II" or "E II"	1963	25–27	25–27	25–27	25–27	25–27	27	27	27
"Echo I" or "E I"	1960	3–5	3–5	3–5	3–5	3–5	4	4	4
"Victor" or "V"	1962	8	11	14	14	12	12	14	19
"November" or "N"	1958	13	12	12	12	13	13	13	13

SOURCE.--SIPRI.

NOTE.--This table comprises all kinds of nuclear-powered submarines; thus, both countership and hunter-killer (antisubmarine) submarines are included. AS = antisubmarine; N.A. = not available; TT = torpedo tubes.

Contributors

F. M. Auburn is a senior lecturer in law at the University of Auckland, New Zealand. He is author of a major work on Antarctica and of numerous articles on Antarctica, the Law of the Sea, constitutional law, conflict of laws, and computers and the law for journals in the United States, England, Canada, West Germany, France, Australia, New Zealand, Denmark, Argentina, Japan, and Ghana.

Frank Barnaby has been the director of SIPRI (Stockholm International Peace Research Institute) since October 1971. Before that he was the executive secretary of the Pugwash Conferences on Science and World Affairs and a research physicist at University College, London. He is the author and editor of several books on disarmament issues.

Elisabeth Mann Borgese is chairman of the Planning Council of the International Ocean Institute, Malta; an associate of the Center for the Study of Democratic Institutions, Santa Barbara; and advisor to the Delegation of Austria at the Third United Nations Conference on the Law of the Sea. She has written numerous books, monographs, and essays on international ocean affairs and marine resources management, including *The Ocean Regime* (1968), *The Drama of the Oceans* (1976), *The New International Economic Order and the Law of the Sea* (with Arvid Pardo, 1976), and *Seafarm* (1978).

Thomas Busha has been with the Inter-Governmental Maritime Consultative Organization (IMCO) since 1961, the first American to join the Secretariat after it was established in London in 1959. Mr. Busha is deputy director of the Legal Division and has taken part in all the diplomatic conferences convened by IMCO between 1962 and 1977, as well as attending sessions of the Third United Nations Conference on the Law of the Sea and the Seabed Committee which preceded it. He is a member of the International Ocean Institute Planning Council.

R. H. Charlier is professor of oceanography at the Flemish Free University of Brussels, and holds a professorship at Northeastern Illinois University as well. He has specialized in problems relating to the tapping of ocean-based energy, to ocean mining, and to environmental studies. He has participated in several international ocean conferences. His latest book is *Ocean Resources* (1977).

James Dawson is a Canadian whose grandfather was responsible for marine surveys on the east and west coasts of Canada. Since 1927 he has lived in England. He joined Lloyd's of London in 1939 as a marine insurance broker. In 1968 Mr. Dawson began specializing in the insurance of divers, oceanographic ships, submersibles, and related marine activities. He is a member of the Council of the Society for Underwater Technology, the Royal Institute of Navigation, the Marine Technology Society, and the United States Naval Institute.

Contributors

David A. Deese is currently a research fellow at Harvard University's Program for Science and International Affairs and an International Relations Fellow at the Rockefeller Foundation. He received his B.A. in government from Dartmouth College in 1970. From 1970 to 1974 he served as a U.S. naval officer. He holds M.A., M.A.L.D., and Ph.D. degrees in international relations from the Fletcher School of Law and Diplomacy. During the period 1976–1977 he was a Marine Policy and Ocean Management Fellow at the Woods Hole Oceanographic Institution.

Robert A. Frosch is administrator at the National Aeronautics and Space Administration in Washington, D.C. He received his B.A., M.A., and Ph.D. from Columbia University. He was scientist, and later director, of the Hudson Laboratories at Columbia University from 1951 to 1963. He has also served as director of nuclear test detection at the Advanced Research Projects Agency, Defense Department, 1963–1965; deputy director, ARPA, 1965–1966; assistant secretary of the navy for research and development, 1966–73; assistant executive director, United Nations Environment Programme, 1973–75; associate director for applied oceanography, Woods Hole Oceanographic Institution, 1975–77; and, since 1977, as administrator at NASA. He received the Arthur S. Flemming Award in 1966.

Norton Ginsburg is professor of geography at the University of Chicago where he completed a doctoral dissertation in 1949 on Japan's prewar trade and shipping. His interest in the oceans was further stimulated during his tenure as academic dean at the Center for the Study of Democratic Institutions (1971–74) where he became involved in the Pacem in Maribus movement. In 1972 he directed a PIM conference on the Mediterranean Sea and region. He is the author of several articles on ocean-related topics and has written and edited a number of papers and books on other subjects, for example, *An Atlas of Economic Development* (1961). He also has served as chairman of the Committee on Environment of the U.S. National Commission for UNESCO and as consultant on environmental problems to SCOPE and UNESCO.

Jozef Goldblat is a senior member of the research staff at the Stockholm International Peace Research Institute (SIPRI). He has university degrees in international relations, law, and economics. Since the early sixties he has been involved, in different capacities, including service for the United Nations, in disarmament negotiations in Geneva and New York. He also served on international control commissions in Korea and Vietnam. He has written numerous reports, articles, a book, and brochures on truce supervision, the arms race, and disarmament problems. He has been published in Finland, France, Norway, Poland, Sweden, and the U.S.A.

Charles D. Hollister is a marine geologist and associate scientist at the Woods Hole Oceanographic Institution. Since 1973 Dr. Hollister has been a key person in the formation and continued progress of the National Seabed Working Group and chairman of the Site Selection Criteria Task Group. He also maintains an active role in oceanographic research on the Benthic Boundary Layer and has published extensively on this and related topics.

Sidney Holt is adviser on marine affairs in the Office of the Assistant Director-General (Fisheries) of the Food and Agricultural Organization of the United Nations, Rome. He specializes in international and scientific aspects of fisheries research and management, and has recently been engaged in conducting a joint FAO–UN Environment Programme project on whales and other marine mammals. Dr. Holt has served as secretary of UNESCO's Intergovernmental Oceanographic Commission and, as the UN Mediterranean adviser in Malta, helped to establish the International Ocean Institute.

Ronald Huisken received his Bachelor of Economics from the University of Western Australia and his Master of Social Sciences from the University of Stockholm, Sweden. In 1969 he joined the Stockholm International Peace Research Institute (SIPRI) and then spent two years with the University of Malaya. Mr. Huisken returned to SIPRI in 1972. In 1976 he took up a visiting fellowship with the Strategic and Defence Studies Centre, Australian National University. He is currently engaged in doctoral research concerning long-range cruise missiles.

G. L. Kesteven is Australian. On graduation from Sydney University in 1937 he obtained employment as a fisheries biologist, and he has been occupied with fisheries ever since. He did original research in Australia, held a senior post in Australian fisheries administration during World War II, and was a director of fisheries research there from 1960 to 1967. He has spent many years with international organizations, first with UNRRA in the Far East, then with FAO in Southeast Asia and at Headquarters in Rome, and since 1967 in Latin America. He has published many papers on fisheries biology.

A. L. Kolodkin is a Doctor of Law and a prominent authority in the field of the International Law of the Sea. He has published more than seventy articles and books, including *World Oceans: Legal Status and the Main Problems* (Moscow, 1973). He is a member of the Planning Council of the International Ocean Institute. Dr. Kolodkin was a participant in the conference on the establishment of the INMARSAT, and he participates in the activities of the Preparatory Committee of INMARSAT.

Y. M. Kolosov is a Doctor of Law, the author of more than a hundred publications dealing with various aspects of international law, including such monographic studies as "Struggle for Peaceful Outer Space," "Responsibility in International Law," and "Mass Information and International Law" (1974), in which the author investigates existing legal international instruments governing the use of mass media for dissemination of information beyond national boundaries.

Rudolf Kreuzer, born in 1910 in Bavaria, studied biology, food technology, food chemistry, and bacteriology. He always has taken a particular interest in fish as well as in art and cultural history. He began as a fisheries officer at Lake Constance, was a member of the Institut für Fischverarbeitung in Hamburg, and in 1958 joined the FAO, Rome. There he was chief of the Fishery Products and Marketing Branch, worked on the development of fish utilization in numerous countries, organized five international technical conferences, and edited five books on various aspects of fish utilization. Now retired, he has directed his attention to studies on the role fish have had in various cultures during human history.

Nikki Meith-Avcin is a research ecologist and instructor at the Marine Biological Station in Portoroz, Yugoslavia (of Ljubljana University's Institute of Biology), and has worked for UNEP on a variety of projects. She has published papers on organochlorine levels in deep sea fish, the sublethal effects of DDT on barnacles, and general community ecology. Her current research involves effects of sewage on infaunal recruitment in a coastal lagoon community.

Peter R. Odell is professor of economic geography at Erasmus University, Rotterdam, where his main research interests are on the oil and gas industry. He has been much involved in public policy debates on the evolution of Western European energy policies. He has published extensively in his area of interest, including *Oil and World Power* (1976) and *The West European Energy Economy: The Case for Self-Sufficiency 1980–2000* (1976).

Arvid Pardo is a Maltese national, born in Rome in 1914, and a graduate in law from the University of Rome. Pardo joined the United Nations Trusteeship Department in 1945. In 1961 he joined the UN Development Program and served as deputy resident representative in Nigeria and Ecuador. In 1964 he became permanent representative of Malta to the United Nations, and in 1967 was named ambassador to the United States, Russia, and Canada contemporaneously. In 1967 he proposed that the seabed beyond national jurisdiction receive legal status as a common heritage of mankind. Since 1971 Pardo has been a visiting fellow at the Center for the Study of Democratic Institutions, coordinator of marine programs at the Woodrow Wilson International Centre for Scholars, and, since 1975, professor of political science and international law and fellow of the Institute for Marine and Coastal Studies at the University of Southern California in Los Angeles. His publications include *The Common Heritage, The New International Economic Order and the Law of the Sea* (with Elisabeth Mann Borgese), and numerous articles, particularly on ocean politics and policy.

T. V. R. Pillay is head of the Aquaculture Development and Coordination Programme of the Food and Agriculture Organization of the United Nations. After about twenty years of service in different fishery institutions in India, Dr.

Pillay joined the FAO in 1962, where he pioneered a variety of international activities in the field of aquaculture development. He is a world authority on aquaculture and the author of a number of scientific papers and books on the subject.

R. A. Ramsay is chief of the shipping section of the Shipping Division of the UN Conference on Trade and Development (UNCTAD) in Geneva. He has worked on shipping policy questions in both developed and developing countries.

The Lord Ritchie-Calder of Balmashannar, C.B.E., D. Univ., D.Sc., M.A., is a member of the Planning Council of the International Ocean Institute. He is a science writer who became professor of international relations at the University of Edinburgh, was raised to the peerage, and became a senior fellow of the Center for the Study of Democratic Institutions. As a colleague of Elisabeth Mann Borgese there he helped initiate the Pacem in Maribus movement. He was responsible for the book *Pollution of the Mediterranean* and has written some three dozen other volumes on science and society as well as several score articles in the *Center Magazine* and other leading periodicals.

Edward Symonds is a consultant in energy economics and finance with thirty years' experience in government and industry. Before setting up his own business, he served as deputy assistant secretary responsible for energy policy in the U.S. Treasury, and as vice-president, Energy Economics, of Citibank, New York. He writes a monthly review of ocean and onshore financial developments in the internationally read *Petroleum Economist,* published in London.

Peter S. Thacher was programme director for the Stockholm Conference in 1972 which led to the creation of UNEP in 1973. Since then he has served as director of UNEP's Geneva office and, as such, has been responsible for developing and carrying out the Mediterranean Action Plan. As of September 1, 1977, he moved to UNEP's Nairobi headquarters as deputy executive director of the programme, serving directly under UNEP's executive director, Dr. Mostafa K. Tolba.

M. E. Volosov is a candidate of law, engaging in studies of the influence of the technology revolution on maritime law. He is the author of a number of articles and the book *Legal Status of the Continental Shelf* (with T. D. Rosina, 1974). Mr. Volosov has taken part in preparing institutional documents of the International Maritime Satellite Communication System/INMARSAT.

Arthur H. Westing was a senior research fellow at SIPRI, the Stockholm International Peace Research Institute. He is currently professor of ecology and dean of natural science at Hampshire College, Amherst, Massachusetts. The author has had a long-standing interest in the ecological impact of warfare.

Index

Abu Dhabi, 116, 119
Action Plan for the Human Environment (Stockholm), 317-18
Aegean Sea, 121, 125
Afghanistan, 406, 408
Africa: fisheries and, 67-69; mineral resources and, 183, 186, 189, 199, 204-5; naval forces of, 418, 426; navigation and, 218-19; offshore hydrocarbons of, 123, 130
Alaska, Gulf of, 117
Albania, 35, 322
Algeria, 322
Amazon delta, 121
Anchoveta collapse, 38, 40-41, 278, 290
Ancient fisheries. See Fisheries, Sumerian
Angola, 35, 116, 119
Antarctica: legal issues in, 512-14; mineral resources of, 468, 491; pollution of, 502, 510, 512-14; territorial claims to, 468, 491-92, 502-5, 508-9
Antarctic Ocean. See Southern Ocean
Antarctic Treaty, 505-6; area of, 469, 495; dispute settlement in, 497; mineral exploration and, 505-8; scientific research and, 493, 505, 507, 513; Seabed Treaty and, 386, 391, 396, 399; territorial claims and, 491-92, 505, 514; text of, 493-99
Antisubmarine warfare (ASW), 276-385, 417, 420n, 428-29, 439-40; acoustic detection, 378, 381-83, 432; electromagnetic detection, 378, 381; magnetic anomaly detection, 378, 383; weapons systems, 378, 383-85

Aquaculture, 2, 48n, 84-101; ancient, 107; aquarange farming as, 94; of eels, 93; employment in, 88, 100; environmental quality and, 89, 93-94; future of, 99-101; legal status of, 89, 96; nonindigenous species and, 95; organization of, 95-97; of oysters, 92-93, 95; production of, 84-89; regional cooperation in, 98-99; shrimp farming as, 91-92, 95; technological advances in, 90-95; use of wastes in, 93-94
Arab League Educational, Cultural, and Scientific Organization (ALESCO), 335-37
Arctic Ocean, 461
Argentina, 125, 209, 427; Antarctic and, 492-93, 502, 509; energy resources and, 164-65; Seabed Treaty and, 389, 399, 408; sea limits and, 35, 406
Argo Merchant, wreck of, 224, 227
Arms control in the ocean. See Seabed Treaty
Atlantic Ocean, 258, 384; fisheries in, 43, 45, 51, 63-64; mineral resources of, 191, 204; offshore hydrocarbons in, 117, 123; Panama Canal and, 462-63; pollution of, 298, 303, 306, 453-55
Australia, 86, 258, 328, 406, 408; Antarctic and, 492-93, 513; energy resources and, 164-65; mineral resources and, 161-62, 183, 189, 195, 198-99, 205, 209; offshore hydrocarbons of, 116, 119-21, 123, 130; naval forces of, 427, 432

879

Austria, 86, 406, 408

Bahamas, 35, 162, 197, 209
Baltic Sea, pollution of, 303, 306, 337-38, 441, 455
Bangladesh, 35, 86
Barbados, 162
Barents Sea, 125
Beaufort Sea, 122, 177
Belgium, 86, 162, 349, 406, 408; Antarctic and, 493, 507
Bengal, Bay of, 327
Benin, 35, 408
Bering Sea, 117
Bermuda, 297
Biological Investigation of Marine Antarctic Systems and Stocks (BIOMASS), 488
Biscay, Bay of, 123
Black Sea, 45, 51, 300
Blue Plan (for the Mediterranean), 324-25
Bolivia, 406, 408
Botswana, 408
Brazil, 35, 86, 204, 258, 509; offshore hydrocarbons of, 116, 119-21; Seabed Treaty and, 389, 397, 399, 408
Brunei, 116, 119
Bulgaria, 259, 406, 408
Burma, 35, 86, 408
Burundi, 408

Cameroon, 35, 410
Canada, 35, 86, 220, 258, 349, 406, 427; energy resources and, 164-65, 167; mineral resources and, 183, 190, 195, 199, 205, 207-9; offshore hydrocarbons of, 123, 126, 130, 511; pollution and, 303, 309, 442; Seabed Treaty and, 389, 396-97, 399-400, 408
Cape Verde, 35
Caribbean Action Plan, 329
Caribbean Sea, 123; coastal area development of, 361-62, 369; pollution of, 298, 318, 319n, 329-331
Caspian Sea, 122
Celtic Sea, 123, 125
Central African Empire, 86, 93, 408
Chile, 35, 40, 86-87; Antarctic and, 492-93, 502, 509; mineral resources and, 161-62, 209
China, 122, 207, 209, 387, 398, 450-51; aquaculture in, 86-87, 90, 93, 95; naval forces of, 418, 425-26, 439, 452
Circular error probable (CEP), 377
Club of Rome, 467
Coastal area development, 5, 350-75; coastal classification for, 358-59; conflicting uses and, 355; defining the area of, 357-58; international cooperation and, 372-73; manual on, 5, 352-60; multiple jurisdictions and, 356, 360; pollution and, 355; a systems engineering approach to, 357-59; technology transfer and, 365-73. See also United Nations system and coastal area development
Columbia, 406, 408
Committee for Training, Education, and Mutual Assistance of the IOC (TEMA), 364-65, 368, 372-73
"Common heritage of mankind," 9, 29, 237; Antarctic and, 468; extent of, 277; living resources and, 82; offshore hydrocarbons and, 124, Southern Ocean and, 467-68
Comoro Islands, 35
Comprehensive Plan for the Global Investigation of Pollution in the Marine Environment (GIPME), 289, 488
Congo, 35, 116, 119
Continental shelf, convention on, 276-77
Cook Inlet, 117
Costa Rica, 35, 406, 408
Cuba, 35, 86, 162, 259, 406
Curie (Ci), 341
Cyprus, 86, 322-23, 408
Czechoslovakia, 86, 93, 259, 406, 408

Denmark, 86, 258, 274, 408; mineral resources and,

197-98; offshore hydrocarbons of, 116, 119, 125, 144; sea limits and, 35, 406
Desalination, 161-62, 208, 332
Developing countries: aquaculture in, 96-97; coastal area development and, 325, 354, 356, 358-59, 363-75; fisheries and, 42, 53, 61, 66-67, 69-74, 76; krill and, 490; marine sciences and, 271, 291; naval forces of, 416, 418, 425-26, 430-31, 433; navigation and, 238; shipping and, 213-16; UNCLOS and, 10, 26
Dominican Republic, 35, 195, 406, 408
Dubai, 116, 119

East Asia, pollution in, 318, 325-29
Ecology: coastal disruption and, 461; effect of canals on, 462-63; military abuses of, 442-48, 461-63; regional developments and, 6; underwater explosions and, 442-51. See also Pollution; Radioactive waste management
Economic Commission for Africa (ECA), 363
Economic Commission for Latin America (ECLA), 329, 361
Ecuador, 35, 86, 390-91
Egypt, 86, 322-23, 433, 463; offshore hydrocarbons and, 116, 119-21, 125
El Niño, 278
El Salvador, 35, 86, 391
Energy resources, marine, 2, 162-81; currents, 175-76; deuterium, 180, 208; geothermal, 178-79; hydrogen (electrolysis), 180-81; salinity gradients, 179-80, 484; solar, 168-70; thermal, 170-75; tidal, 163-68; waves, 176-78; wind, 170. See also Offshore hydrocarbons
English Channel, 300; pollution of, 219, 231-32

Equatorial Guinea, 408
Estuaries and mangrove swamps, ecology of, 437, 461
Ethiopia, 408
European Economic Commission (EEC), 38, 323-24
Exclusive economic zone (EEZ), 13n, 195, 401; fisheries and, 76-80; naval forces and, 4, 413-14, 434; offshore hydrocarbons and, 124-26; scientific research and, 277, 290

Fiji, 197, 406
Finland, 86, 199, 209, 406, 408
Fisheries, distribution of benefits from, 67-80; EEZ and, 76-80; trade and, 72-78
Fisheries, management of: coastal area development and, 353; improvement of, 62-67; incidental catches and, 65-66; maximum sustainable yield and, 64, 78-82; processing in, 66-67; species interaction and, 63; unconventional resources and, 62-63. See also Krill.
Fisheries potential: estimating, 486-88; geographical distribution of, 50-52; in the Southern Ocean, 474-83
Fisheries production, 40-49, 438; decline in, 2, 40; energy inputs to, 61-62; geographical distribution of, 2, 43, 45-48; species composition of, 43-44, 48-49
Fisheries, Sumerian, 102-113; monetary value of, 105-7; organization of, 106-7; species in, 102, 106; status of fishermen in, 104
Fisheries, values of, 52-62; disposition of catch and, 55-58; as employment, 52-53; as food supply, 58-61; monetary, 53-58
Fishing jurisdictions, 35-37

Fishing resources convention, 64
Flags of convenience, 214, 237n
Food and Agriculture Organization (FAO), 5, 7, 287, 367n; aquaculture and, 88, 97-100; fisheries and, 38, 43, 50, 52-53, 62, 71; pollution and, 320-21, 324-25, 329, 339; Southern Ocean and, 468-69, 488-89
France, 35, 125, 226, 258, 283, 348; Antarctic and, 492-93, 503; aquaculture in, 86, 89, 91; energy resources and, 163-65; mineral resources and, 190, 197, 204, 209; naval forces of, 423-24, 427-29, 433, 439, 452; pollution and, 302, 322-23, 450-51, 461; Seabed Treaty and, 387, 390-91, 398
Freedom of scientific research, 4-5, 290-91. See also Marine sciences

Gabon, 35, 116, 119-20
Gambia, 35, 408
General Bathymetric Chart of the Oceans (GEBCO), 288
Germany, Democratic Republic of, 86, 259, 406, 408, 507
Germany, Federal Republic of, 35, 143, 190, 258, 274, 349; Antarctic and, 491, 510; aquaculture in, 86, 90; naval forces of, 424; pollution and, 302; Seabed Treaty and, 408; shipping and, 452
Ghana, 35, 86, 120, 406, 408
Greece, 86, 223, 258, 322-23, 408; mineral resources and, 162, 209; offshore hydrocarbons of, 121, 125
Greenland, 198, 219, 314
Group of Experts on Coastal Area Development, 360-62, 368
Guatemala, 35, 195, 406, 408
Guinea, 35, 189, 408
Guinea-Bissau, 35, 408
Guinea, Gulf of: coastal area development of, 356, 362-63, 369; pollution of, 318, 333-35
Guyana, 35

Haiti, 35, 406
Half-life of radionuclides, 344
Honduras, 408
Hong Kong, 86, 197, 328
Hungary, 86, 93, 259, 406, 408

Icebergs, fresh water from, 161, 484
Iceland, 35, 160, 176, 303, 406, 408; mineral resources and, 183, 197
India, 35, 122, 164, 258, 328; aquaculture in, 86, 90, 93; mineral resources and, 183, 189, 199; naval forces of, 426-27; Seabed Treaty and, 399, 409
Indian Ocean, 114, 191, 218, 258, 298; fisheries in, 39, 46, 51
Indicative World Plan for Agricultural Development (IWP), 50-52
Indonesia, 221, 233, 328, 426; aquaculture in, 86, 91-93; mineral resources and, 183, 195-96, 199, 208; offshore hydrocarbons of, 116, 119-23
Industrialized countries: aquaculture in, 95-97; coastal area development and, 354, 370-71; fisheries and, 42, 53, 61, 66-67, 70-74, 76; krill and, 489-90; marine sciences and, 271; naval forces of, 416, 431; shipping and, 213-14; UNCLOS and, 30
Informal Composite Negotiating Text, 10-34; archipelagic waters in, 12n; arms control and, 24; coastal state control in, 11-15, 19, 25-27, 32, 34; common heritage regime in, 15, 28-29; continental shelf in, 13n; dispute settlement in, 21-23, 31-33;

economic uses and, 24-31; EEZ in, 12n; flag-states in, 15; geographically disadvantaged states and, 25; high seas regime in, 14-15, 27-28; innocent passage in, 13n; international straits in, 13n; islands and, 13; navigation and, 238; 1958 Geneva Convention and, 12; pollution and, 18-19; scientific research in, 19-20, 27n; seabed area in, 29-31; Seabed Authority in, 15-18, 30-31; shortcomings of, 24-33; unauthorized broadcasting and, 15n

Integrated Global Ocean Station System (IGOSS), 287-88, 488-89

Inter-Governmental Maritime Consultative Organization (IMCO), 3, 5-7, 287; INMARSAT and, 3, 245-48, 250-51, 254; navigation and safety and, 222-234, 236, 238

Intergovernmental Oceanographic Commission (IOC), 4-5, 488; coastal area management and, 362, 364; marine sciences and, 286-89; pollution and, 320-21, 325, 329, 335, 339; TEMA 364-65, 386, 372-73

International Association of Lighthouse Authorities (IALA), 230-31

International Atomic Energy Agency (IAEA), 5, 320, 324, 349, 367n

International Commission for Scientific Exploration of the Mediterranean (ICSEM), 320

International Council for the Exploration of the Sea, 5

International Decade of Ocean Exploration, 287

International Geophysical Year (IGY), 275-76, 504-5

International Hydrographic Bureau, 288

International Indian Ocean Expedition (IIOE), 287

International Maritime Satellite Communication System (INMARSAT), 3, 240-70; Assembly of, 249, 254, 257, 260, 264; capitalization of, 255-56; competitive systems and, 263-64; Conference on the Establishment of, 249, 251-53, 261; Council of, 249, 254-64; Directorate of, 249, 254, 257, 260, 264, 268; dispute settlement in, 253, 266-68; Group of Experts on Maritime Satellites and, 247, 249-51, 269; implementation of, 253-54; increased need for, 242-43; land stations of, 249-50, 254-55, 262-63; liability under, 252-53, 264-66; participation in, 251-56, 264-65; preliminary developments of, 245-50; sanctions under, 264-65; space segment of, 249, 254-55, 258, 260-62; structure of, 249, 257-58, 270; technical standards for, 254, 262; weighted voting in, 253, 258-60

International Ocean Institute, 6, 319n

International Telecommunications Union (ITU), 367n; INMARSAT and, 245-46, 248, 250-51

International Union for the Conservation of Nature (IUCN), 331, 337, 352n

International Whaling Commission (IWC), 38, 40, 488

Iran, 35, 406, 409, 426; offshore hydrocarbons of, 116, 119, 121

Iraq, 409

Ireland, 35, 86, 125, 207, 406, 409

Israel, 125, 322-23, 406; aquaculture in, 86, 90; mineral resources and, 197, 207, 209; naval forces of, 426, 433

Italy, 86, 119, 258, 406, 507; mineral resources and, 204, 207, 209; naval forces of, 424, 433; pollution and, 302, 322-23; Seabed Treaty and, 399-400, 409

Ivory Coast, 86, 170-72, 409

Jamaica, 406, 409
Japan, 35, 53, 349, 406, 409, 425, 438; Antarctic and, 493, 510; aquaculture in, 85-87, 89-90, 92-94; energy resources and, 175, 177; INMARSAT and, 258, 266; mineral resources and, 162, 183, 186, 189-90, 196-97, 199, 207, 209; offshore hydrocarbons of, 116, 119, 123; pollution and, 302, 307, 328; shipping and, 223, 452
Japan, Sea of, 190, 328
Joint Group of Experts on the Scientific Aspects of Marine Pollution (GESAMP), 352n, 368
Joint Oceanographic Assembly, 4
Jordan, 409

Kampuchea, 406, 408
Kenya, 86, 400, 406
Korea, People's Democratic Republic of, 35
Korea, Republic of, 165, 328, 410, 426; aquaculture in, 86-87, 91
Krill, 39, 51, 62-63, 477; production of, 479, 487-88, 510; scientific research on, 479, 484, 489; technology for, 468, 486, 489-90
Kuwait, 161, 258
Kuwait Regional Convention, 333
Kyoto Declaration on Aquaculture, 100-101

Laos, 409
Law of the Sea Institute, 6
Lebanon, 197, 322-23, 409
Lesotho, 406, 409
Liberia, 35, 223, 237n, 406, 409
Libya, 125, 322-23
Liner conferences convention, 214
Liquid natural gas (LNG), shipment of, 133
Luxembourg, 409

Madagascar, 35, 86, 406, 409
Magellan, Strait of, 219-20
Malacca, Strait of: navigation of, 221, 232-33; pollution of, 318, 327-28
Malawi, 406
Malaysia, 328, 406, 409; aquaculture in, 86-87, 93; mineral resources and, 183, 186, 196; navigation and, 221, 233; offshore hydrocarbons of, 116, 119, 125
Maldives, 35
Mali, 409
Malta, 9n, 35, 125, 406, 409; pollution and, 322-24
Manganese nodules. See Mineral resources, marine, polymetallic nodules
Marine and Coastal Technology Information Service (MACTIS), 367, 369-71, 373
Marine and Coastal Technology Programme (MACTECH), 365-75
Marine Environment Data Information Referral System (MEDI), 5
Marine sciences, 271-92; drugs from the sea, 278-79; early development of, 271-72; fish distribution and, 279-80; freedom of research for, 290-91; habitats (undersea), 282-84; international cooperation in, 286-88, 290-91, 488-89; living resources and, 277-78; manned exploration and, 281-84; ocean/atmosphere interface and, 273, 285-86; as a planetary science, 273-74; plate tectonics and, 274-75; pollution and, 288-89, 320-22; remote sensing and, 279-81, 371; research vessels for, 282, 284-85, 290-91; UNCLOS and, 290
Mauritania, 35
Mauritius, 35, 406, 409
Maximum sustainable yield (MSY), 64, 78-82
Mediterranean Action Plan, 318-325

Mediterranean Conference (Barcelona), 6, 322-24
Mediterranean Sea, 462; fisheries in, 45, 51; mineral resources of, 202, 205; offshore hydrocarbons in, 123, 125; pollution of, 297-98, 306, 309, 318-25, 454
Methanol plants, offshore, 133-34
Metula, wreck of, 219-20
Mexico, 35, 61, 86, 406; mineral resources and, 162, 207; offshore hydrocarbons of, 119, 123, 126; Seabed Treaty and, 389, 400
Mexico, Gulf of, 362, 379; energy resources of, 178-79, 201-3; mineral resources of, 162, 198; offshore hydrocarbons in, 117, 121, 123, 126, 129, 131-32, 135; pollution of, 298, 304, 329-31
Middle East, 205; naval forces of, 418, 426, 434; offshore hydrocarbons in, 123, 125, 135
Mineral resources, marine, 182-210; classification of, 188; common heritage and, 3; copper, zinc, and barite, 187, 203; dissolved substances, 204-8, 276, 437-38; economics of mining, 182-86, 194-95, 198, 200-201; EEZ and, 195; iron ore, 187, 196; muds, 187, 203-4; nickel, 187, 199-201; phosphorite, 186-189; polymetallic nodules, 30-31, 185, 187, 190-95, 199-202, 272, 275-77, 474; production of, 184, 202; sands and gravel, 187, 197-99; sulfur, 187, 202-4; tin, 187, 196-97; undersea mines and, 186, 209. See also Energy resources, marine; Offshore hydrocarbons
Monaco, 283, 288, 322-23
Mongolia, 409

Morocco, 35, 121, 189, 322-23, 409
Mozambique, 36
Multinational companies. See Offshore hydrocarbons, international companies and

Namibia, 209
National Aquaculture Information System (NAIS), 97
NATO, 412, 417-19, 423-25, 431
Naval forces, 4, 412-35, 438-40, 464; aircraft carriers, 421-23, 427-29; ASW aircraft, 378, 384, 428; attack submarines, 378-79, 384, 417, 430; distribution of, 417-19; measuring, 414-16; patrol boats, 421-22, 433-34; as political instruments, 413-14, 421-22, 432, 435; SLBMs, 376-77, 420-21; strategic submarines, 376-77, 412, 417, 420, 429, 439; structures of, 419-22; submarines, 421-22, 429-31, 452-53, 466; surface ships, 421-22, 431-33, 452, 466; trends in, 417-18. See also Antisubmarine warfare
Navigation and safety, 217-39; buoyage for, 231-32; charts and, 220-21, 226; efficacy of treaties on, 229-30; episodic waves and, 218-19; fire and, 224-26; human error and, 219-21; INMARSAT and, 234, 240-42, 244, 246, 254, 259; man-made obstacles and, 218; the oil industry and, 222; routeing for, 4, 226, 232-36; sea courts and, 238; UNCLOS and, 236; very large crude carriers (VLCC) and, 220-21, 223, 236-37
Nepal, 86, 93, 406, 409
Netherlands, 86, 168, 197, 258, 349, 406, 409; naval forces of, 426-27; offshore hydrocarbons of, 116, 120-21, 142-43
Neutral Zone, 116, 119

New International Economic Order, 10, 26-27, 30, 490
New Zealand, 406, 409; Antarctic and, 492-93, 500, 503, 508, 513-14; aquaculture in, 86-87; mineral resources and, 189, 199; offshore hydrocarbons of, 116, 500
Nicaragua, 36, 409
Niger, 409
Nigeria, 36, 86, 400, 407; offshore hydrocarbons of, 116, 119-21
North Sea: boundaries in, 114, 125; compared to Persian Gulf, 140-41; drilling activity in, 123; fisheries in, 38, 63; gas production of, 121; hydrocarbon reserves of, 122, 140-57; hydrocarbons in, 129-30, 135, 139-59; international oil companies and, 139-40, 146, 148-49, 151, 155; leasing policies in, 139n; mineral resources of, 205; platform designs for, 131, 154-55; pollution of, 300, 338-39, 459
Norway, 36, 165, 223, 258, 409, 433; Antarctic and, 492-93, 503, 507, 509-10; aquaculture in, 86, 90; mineral resources and, 206-7, 209; offshore hydrocarbons of, 116, 119-20, 122, 125, 129, 139, 145-46, 152
Nuclear weapons testing, 449-51, 464-66

OECD, 360
Offshore hydrocarbons, 2, 474; critical field sizes for, 115, 117, 149, 151, 153; developments in 1976, 121-23; drilling activity and, 122-23; drilling costs for, 128-29, 131; estimating reserves of, 149-55; future of, 135-36; international companies and, 132-33, 139-40, 146, 155; jurisdictional problems and, 114, 123-26; leasing policies and, 128, 130, 135-37, 139n; North Sea experience with, 129-30, 139-59; obstacles to development of, 114-15, 117-18, 123-26, 135-36; pollution threat from, 131-32, 459; production of, 2, 118-121; resources of, 115-17, 140-48, 163; Southern Ocean and, 500-501, 510-11; U.S. experience with, 126-30
Oil spills from ships, 220-21, 226-27, 459. See also Pollution, ship caused
Oman, 36
"Orbital space," concept of, 269

Pacem in Maribus, 6-7, 319n
Pacific Ocean, 258, 384, 462-63; fisheries in, 38, 43, 45, 51, 63-64; mineral resources of, 191-92, 204, 276; offshore hydrocarbons in, 117, 123, 126; pollution of, 298, 306, 339, 453-54
Pakistan, 36, 390, 407, 426
Panama, 36, 224, 407, 409, 462-63
Paraguay, 86, 409
Persian Gulf, 426; coastal area development of, 360-61, 369; offshore hydrocarbons in, 121, 125, 133, 140-41; pollution of, 318, 331-33
Peru, 36, 189; fisheries and, 38, 40-41, 67, 290; offshore hydrocarbons and, 116, 119-20; Seabed Treaty and, 390-91
Petroleum. See Offshore hydrocarbons
Philippines, 36, 53, 328; aquaculture in, 85-87, 90-91; mineral resources and, 195, 199; offshore hydrocarbons of, 122, 125
Plate tectonics, 345-46
Po Hai, 122
Poland, 409; Antarctic and, 507, 510; aquaculture in, 85-86, 93; INMARSAT and, 258-59

Pollution, 288-89, 293-332,
  440-42, 447-61, 463-66;
  from chemical-warfare
  agents, 455-58; coastal
  area development and, 355-
  56, 437; from DDT, 293-94,
  301-4, 321, 327-28, 337;
  definition of, 294; from
  fertilizers, 314, 327-28,
  440; from halogenated
  hydrocarbons, 301-5, 321,
  327-29, 331; from HCB,
  304; from heavy metals,
  294, 305-9, 320, 325, 327-
  29, 334, 337; from hydrocarbons, 294-301, 319-20,
  323, 327-28, 332-35, 338,
  440, 458-61, 510; international cooperation and,
  317-39; of military origin, 441-42, 447-61, 463-
  66; from Mirex, 304-5;
  from munitions, 446-48;
  ocean mining and, 185,
  440; from offshore oil
  production, 131-32, 295-
  96, 327, 339, 459; by
  pathogenic organisms
  (sewage), 307, 309-13,
  319, 321, 324-32, 334,
  440; from PCBs, 294, 301-
  4, 321, 331, 337; from
  radionuclides, 314-15,
  327-28, 448-57, 463-66;
  ship caused, 221, 226,
  295-97, 319, 323, 327,
  332, 334-35, 339, 437,
  440, 452, 458-59, 466;
  from solid wastes, 316-
  17, 327-28, 334, 356;
  thermal, 327-28
Portugal, 36, 407, 409
Puerto Rico, 86, 204

Qatar, 409

Radioactive waste management,
  340-49; Antarctic and,
  491; breeder reactors and,
  343-44; by dispersion in
  sea water, 346-47, 453;
  international problems of,
  348-49; nonproliferation
  and, 343; production requiring, 341-42, 345, 449-
  52; reprocessing and, 343-
  44, 348; by seabed disposal, 346-49; storage and,
  341-42, 348-49
Radioactivity, natural level
  in the ocean of, 454-55
Red Sea, 174, 462; mineral resources of, 205-8, 285;
  pollution of, 318, 335-37
Red tides, pollution and, 326
Romania, 162, 407, 410; aquaculture in, 86-87, 93
Ross Sea, 474, 501-3, 514
Rwanda, 410

Safety at sea, convention for,
  223-26. See also Navigation and safety
Santa Barbara Channel, 117,
  131, 300, 460
Sargasso Sea, pollution of,
  293, 297-98, 302-3
Saudi Arabia, 410; offshore
  hydrocarbons of, 116, 119-
  20
Scientific Committee on Antarctic Research (SCAR),
  468, 488, 502, 506
Scientific Committee on Ocean
  Research (SCOR), 4, 288,
  468, 488
Seabed Treaty, 4, 386-441;
  "allied option" in, 393-
  94; area of, 388-90, 393-
  95, 399-400; coastal
  states and, 389, 391, 393,
  395, 398-402; conventional
  forces and, 392, 398, 400-
  401; countries ratifying,
  408-11; implementation of,
  499-500; negotiation of,
  387-91; prohibitions of,
  391-93; text of, 402-5;
  UNCLOS and, 400; verification in, 388-90, 392, 395-
  99
Senegal, 36, 86, 169, 407, 410
Seychelles, 36
Sharjah, 119
Shipping, 3, 211-16; bulk cargoes and, 211, 216; coastal area development and,
  353, 371; charter parties
  and, 212-13, 216; containerization of, 212, 236;
  contracts of affreightment
  and, 212-13, 216; growth
  of fleet for, 222-23, 241-
  42; industrial carriage

and, 212-13; INMARSAT and, 244-45, 247, 254; liner cargoes and, 211-13; liner conferences and, 213-16; liner services in, 212-14, 216; rate structures in, 213-16; vessel sizes in, 463-64. See also Navigation and safety
Sierra Leone, 36, 407, 410
Singapore, 86, 328, 410; navigation and, 221, 233
Somalia, 36
South Africa, 36, 407, 410; Antarctic and, 493, 507; mineral resources and, 198, 209
South China Sea, 114, 125, 328
Southeast Asia, 130, 135, 363
Southern Ocean, 467-515; Antarctic claims and, 491-92, 506, 509, 514; birds and, 483, 489; cephalopods in, 480; delineation of, 468-73; fisheries catches in, 63, 481; fisheries potential of, 474-83, 486-88, 489; fishing technology and, 467, 489-90; food-chain of, 478, 484-85; hydrocarbon potential of, 501; hydrographic features of, 469-72; ice cover of, 475, 491; International Seabed Authority and, 506-7; mineral resources of, 468, 474, 484, 500; pollution of, 293, 303, 510-11; scientific research on, 467, 479, 488-91; seals in, 482-83, 489, 491, 509. See also Antarctica; Krill; Whales
South Vietnam, 86, 461
Spain, 86, 258, 407, 425; mineral resources and, 189, 209; offshore hydrocarbons of, 116, 119, 121; pollution and, 302-3, 322-23
Sri Lanka, 36, 86, 183, 199, 407
Sudan, 410
Suez, Gulf of, 121, 125, 462
Sumaria, 102-111
Supertanker ports, 134
Supertankers. See Navigation and safety, very large crude carriers; Shipping
Swaziland, 407, 410
Sweden, 197, 233, 258, 410, 425, 442
Switzerland, 86, 349, 407, 410
Syria, 322

Taiwan, 125, 407, 409, 510; aquaculture in, 86-87, 91
Tanzania, 36, 86, 410
Territorial seas, 35-37; convention on, 388, 390, 393-94, 400, 406-7
Thailand, 61, 86, 118, 328, 407; mineral resources and, 183, 196-97
Thailand, Gulf of, 122, 327-28
Thalassothermal energy plants, 169-74
Third World. See Developing countries
Togo, 36, 410
Tonga, 36, 407
Torrey Canyon, wreck of, 220, 224, 226-27, 459
Trade and fisheries, 72-78
Trinidad/Tobago, 407; offshore hydrocarbons of, 116, 119-21
Trinity House, 230-31
Tropical Agricultural Research and Training Centre (CATIE), 331
Tunisia, 87, 407, 410; offshore hydrocarbons of, 116, 119, 121; pollution and, 322-23
Turkey, 125, 209, 322-23, 410, 429n

Uganda, 86, 407
UNESCO, 4, 287; coastal area development and, 357, 367n, 368; pollution and, 320, 324, 335, 337
United Kingdom, 36, 223, 407; Antarctic and, 492-93, 502-3, 509-10; aquaculture in, 86, 91, 96; energy resources and, 163, 165, 177-78; INMARSAT and, 256-57, 263, 265; marine sciences and, 274, 283; mineral resources and, 183, 197, 205, 207, 209; naval

forces of, 423-24, 427-30, 439, 452; navigation and, 222, 224, 226, 228, 230-31; nuclear explosions by, 450-51; offshore hydrocarbons of, 116, 119-22, 125, 129, 139, 143-45, 157-58; pollution and, 301-3, 450-51, 453, 455; radioactive waste management and, 341, 349; Seabed Treaty and, 387, 398-99, 410
United Nations Conference on the Law of the Sea (UNCLOS), 9-34; aquaculture and, 84, 88; arms control and, 400; continental shelf and, 277; disputes at, 31-33; fisheries and, 39, 64, 78-82; mineral resources and, 3, 185; offshore hydrocarbons and, 114, 124-25; Southern Ocean and, 506
United Nations Conference on Trade and Development (UNCTAD), 214-15, 367n, 368
United Nations Department of Economic and Social Affairs (ESA), 332; coastal area development and, 361-64; 367, 370
United Nations Development Programme (UNDP), 88, 98, 238, 468-69; coastal area development and, 361, 363, 368, 372-73
United Nations Environmental Programme (UNEP), 5-7; coastal area development and, 361-63, 367n, 368; pollution and, 294, 317-18, 320-21, 324-25, 329, 331-33, 335, 337, 339
United Nations Industrial Development Organization (UNIDO), 324, 367n, 370
United Nations system and coastal area development, 5-6, 350-75; coordination of, 351-52; MACTECH and, 365-75; MACTIS and, 367, 369-71, 373; program activities for, 360-65; Referral Service of, 369-70

United States, 36, 134, 276-77, 407; Antarctic and, 491, 493, 501, 503-4, 506, 508, 512-14; aquaculture in, 86, 89-91, 93-95; ASW and, 376-79, 381-85; energy resources and, 164-65, 167, 172, 175-76, 179; fisheries and, 65, 438; INMARSAT and, 250, 256-57, 259, 261, 269; marine sciences and, 274, 282-85, 287; mineral resources and, 161-62, 183, 185-86, 189-91, 193-95, 197-209, 438, 501; naval forces of, 376-77, 412-14, 418-35, 439, 452; navigation and, 224-25, 227-28, 233; nuclear explosions by, 450-51; offshore hydrocarbons of, 116, 119-21, 123, 126-30, 132-33, 135-37; pollution and, 301-4, 306, 308-9, 441, 445, 448-51, 453-54, 456-57; radioactive waste management and, 340-43, 345, 348-49, 453; Seabed Treaty and, 388-97, 399, 410; shipping and, 214, 223, 452
Upwelling, 278, 290, 436-37, 474
Uruguay, 36, 407, 410
USSR, 36, 86, 349, 407; Antarctic and, 493, 503-4, 506, 510; ASW and, 376-79, 384; energy resources and, 163, 165, 168; fisheries and, 67-69, 438; INMARSAT and, 243, 247-49, 256-59, 263, 265, 269; marine sciences and, 283-84; mineral resources and, 196-97, 199, 207, 209; naval forces of, 376-77, 412-14, 417-24, 426, 428-31, 433-35, 452; nuclear explosions by, 450-51; offshore hydrocarbons and, 119-20, 123, 123, 129; pollution and, 449-51, 454; Seabed Treaty and, 387-91, 393-97, 410, 439; shipping and, 223, 452

Venezuela, 86, 119, 121, 407

Vietnam, 36, 125, 328, 410. See also South Vietnam

Warsaw Treaty Organization (WTO), 418-19, 423-26
West African coast. See Guinea, Gulf of
Western Europe, 145, 259-61
Whales, 40, 303; in the Southern Ocean, 481-82, 484, 489
World Meteorological Organization (WMO), 5, 286-87, 367n; pollution and, 320-21, 324

World Wildlife Fund, 331

Yemen, Arab Republic of, 410
Yemen, People's Democratic Republic of, 408
Yugoslavia, 407, 425; aquaculture in, 86-87; pollution and, 322-23; Seabed Treaty and, 399, 410

Zaire, 86
Zambia, 86, 410